# Elementary
# Linear Algebra

## Second Edition

W. Keith Nicholson
University of Calgary

McGraw-Hill
Ryerson

Toronto   Montréal   New York   Burr Ridge   Bangkok   Beijing   Bogotá   Caracas   Dubuque
Kuala Lumpur   Lisbon   London   Madison   Madrid   Mexico City   Milan   New Delhi
San Francisco   Santiago   St. Louis   Seoul   Singapore   Sydney   Taipei

The McGraw-Hill Companies

# McGraw-Hill Ryerson

**Elementary Linear Algebra,**
**Second Edition**

ISBN: 0-07-091142-8

1 2 3 4 5 6 7 8 9 10 TCP 0 9 8 7 6 5 4

Printed and bound in Canada

Care has been taken to trace ownership of copyright material contained in this text; however, the publisher will welcome any information that enables them to rectify any reference or credit for subsequent editions.

Vice President, Editorial and Media Technology: Patrick Ferrier
Senior Sponsoring Editor: Catherine Koop
Developmental Editors: Su Mei Ku and Darren Hick
Marketing Manager: David Groth
Supervising Editor: Anne Macdonald
Copy Editor: Edie Franks
Production Coordinator: Jennifer Wilkie
Cover Design: Dianna Little
Cover Image Credit: © Hideki Kuwajima/Photonica
Composition: Pronk&Associates
Printer: Transcontinental Printing Group

National Library of Canada Cataloguing in Publication

Nicholson, W. Keith
    Elementary linear algebra / W. Keith Nicholson. -- 2nd ed.

Includes index.
ISBN 0-07-091142-8

    1. Algebras, Linear.  I. Title.

QA184.N52 2003          512'.5          C2002-904210-0

To Kathleen, Jason, and Mark

# Contents

## Chapter Dependencies

This diagram suggests how the material in each chapter depends on earlier chapters. The dependence of Chapter 4 on Chapter 3 is primarily for motivation and examples. A dotted line in the diagram indicates a minor dependency.

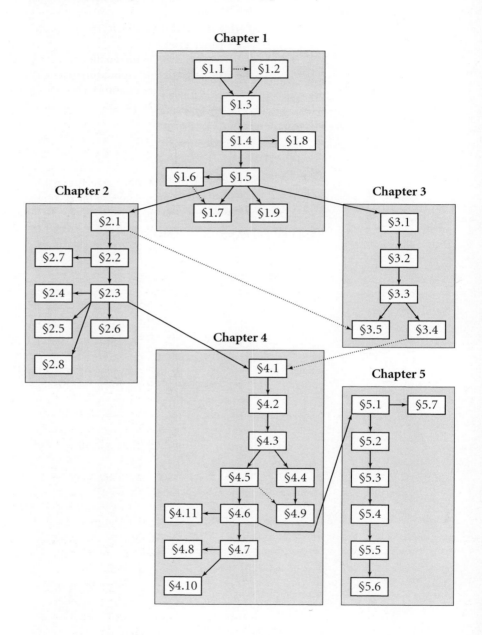

Visit the text website at www.mcgrawhill.ca/college/nicholson.

# Preface

This book springs from three sources: The needs of the users of linear algebra as a service course, the trend away from the abstract view of the subject toward a more matrix approach, and the desire to incorporate the optional use of technology to help with instruction and computation. Let me explain.

In the late 1990s discussions with Engineering and Science faculty members revealed a desire for early introduction of diagonalization and for more emphasis on linear transformations. Consequently, eigenvalues and diagonalization are introduced in Chapter 2 (using only determinants and matrix algebra), motivated by examples of dynamical systems that the students can relate to. Linear transformations of the plane are discussed in Chapter 3, providing a solid introduction to the treatment in Chapter 4 of $\mathbb{R}^n$ and giving geometrical interpretations of multiplication, determinants, and inverses of $2 \times 2$ matrices.

The desire for a more matrix-oriented linear algebra course expressed by the engineers and scientists echos a more general trend in the subject away from the abstract approach. The second half of the book (Chapters 4 and 5) continues this evolution. Every teacher of linear algebra knows that the students "hit the wall" when the notion of an abstract vector space is introduced. One reason for this is that they are coping simultaneously with *two* new ideas: the concept of an abstract structure, and mastering difficult notions like spanning, independence, and linear transformations. This double jeopardy is difficult to deal with for students, even the most talented ones. Consequently, in Chapter 4 topics like independence are discussed first in the context of $\mathbb{R}^n$, so the students can deal with them without having to cope with the abstract baggage of an $n$-dimensional space. Then the general setup is introduced in Chapter 5, with $\mathbb{R}^n$ as the principal motivating example. This has been tested in the classroom, and it works very well pedagogically.

The third theme is to involve technology in a way that is more than merely a "book-in-the-computer," and is more accessible than a computational package. This has been achieved in the creation of LILA (Lyryx—Interactive Linear Algebra). This is an (optional) Internet-based computer tutorial that provides audio-enhanced lessons covering the entire text, interactive explorations allowing the student to "play" with the central concepts or perform the more difficult computations, and self-assessment in the form of labs with feedback to help the student discover where and why an error was made. The system has been class tested, and is available (at a nominal price) to users of this book.[1]

On the other hand, this book serves as a stand-alone text for the traditional linear algebra syllabus (see the Table of Contents). It can be used for a two-semester course for beginners with a working knowledge of high school algebra, or for a single-semester course for more advanced students. For students in many application areas, Chapters 1 through 4 may be all the linear algebra they need. While the text is designed with Science and

---

[1] More information about LILA can be found on pages xiv–xv.

Engineering students in mind, it will also serve the needs of students in Management and the Social Sciences. Calculus is not a prerequisite; the few places where it is needed are clearly marked and can be omitted with no loss of continuity.

Wherever possible, concepts are introduced with real-world examples that are meaningful to the students, and theorems are stated simply with short proofs often preceded by an example. (Longer or more difficult proofs are deferred to the end of the section and can be omitted.) Nearly 350 solved examples are included to aid understanding, introduce techniques, illustrate algorithms, and motivate theorems. These examples are reinforced by the exercises (about 800 in all) at the end of each section, many with answers or solutions at the end of the book.

## Features

- **Early eigenvalues**. Eigenvalues and diagonalization are introduced in Chapter 2 using only determinants and matrix inverses, and not requiring notions like dimension and independence. This allows a number of applications (for example, dynamical systems) that are seen as relevant by the students.

- **Early introduction of linear transformations**. In response to a request from scientists and engineers, rotations, reflections, shears, etc. in $\mathbb{R}^2$ are described using matrix multiplication in Section 3.4. The idea of a linear transformation is introduced in this context, and its relation to matrix multiplication is made explicit. This leads to geometrical interpretations of matrix multiplication, inverses, and determinants, and it provides a basis for the study of linear transformations in $\mathbb{R}^n$ in Chapter 4.

- **Matrix approach**. In keeping with a trend over the past few years, more emphasis is placed on matrix computations. For example, the following topics are discussed: Positive definite matrices, LU-factorization, QR-factorization, Triangular form, Complex matrices, and Singular value decomposition. This has the effect of reducing the overall level of abstraction in the book.

- **Vector space concepts introduced first in $\mathbb{R}^n$**. Independence, spanning, subspaces, and dimension are introduced first in $\mathbb{R}^n$ (in Chapter 4) before the student has to cope with abstract vector spaces. The general notions then come easily in the abstract context (Chapter 5), thus allowing the student to focus on the new conceptual framework.

- **Applications**. These occur both as separate subsections or as examples in the text. Examples: Markov chains, systems of differential equations, directed graphs, linear dynamical systems, computer graphics, statistical principal component analysis, and constrained optimization.

- **Motivating examples**. Wherever possible, concepts are introduced with examples that are not only understandable to the students but are seen as relevant to the real world. A good illustration is the motivation (in Section 2.3) of diagonalization by looking at the possible extinction of a species of birds.

- **Solved examples**. Nearly 350 solved examples are given to help the student to understand the techniques of the subject and to motivate theorems and proofs.

- **Exercises**. Almost 800 exercises are included, which are keyed to the examples. They start with routine, computational problems and progress to more theoretical exercises. Answers or solutions are provided at the end of the book to all even-numbered exercises, or even parts of exercises (including odd-numbered exercises) when the parts are independent problems.

- **Difficult proofs**. These are deferred to the end of the section.

- **Approximation**. Best possible solutions to inconsistent systems are discussed, leading to the normal equations and the method of least squares.

- **Complex matrices**. Optional treatment including Hermitian, unitary, and normal matrices; the Spectral Theorem; and Schur's theorem.

- **Singular value decomposition**. Optional treatment including the fundamental subspaces and the polar decomposition.

- **Appendices**. Brief reviews are included of trigonometry, induction, and polynomials.

## New in the Second Edition

Along with improving the design and the exposition of the textbook, a number of changes have been made, as follows:

- The answers section at the end of the book has been rewritten and now contains solutions or answers to all even-numbered exercises, including the even parts of odd-numbered exercises (where the parts are separate problems).

- The index has had a major overhaul and now contains cross-references.

- A discussion of the LU-factorization of a symmetric matrix has been added in Section 1.7, yielding the so-called LDLT factorization for positive definite matrices.

- A new Section 1.9 has been added in which an informal introduction to finite fields is given, inverses of matrices are discussed, and an application to cryptography is included.

- The cross product is now introduced in Section 3.3 with just enough detail for the application to lines and planes. Section 3.5 on the Cross Product has been made optional.

- The subsection on Computer Graphics in Section 3.4 has been rewritten and more material has been included.

- Section 4.7 on Orthogonal Diagonalization has been rewritten so as to obtain the Principal Axis theorem earlier, and put the main theorems on diagonalization at the end.

- The notion of the inverse of a linear operator on $\mathbb{R}^n$ has been introduced into Section 4.9 to make Chapter 4 more self-contained.

- The general definition of a change matrix is now given in Section 5.5 using the matrix of an operator. The main properties of change matrices are then simple consequences of facts in Section 5.4 for matrices of operators.

- A list of applications and a list of symbols have been added.

- 48 new exercises and 6 new examples have been included.

- Two new applications have been added: one on cryptography and another on balancing chemical equations.

## Supplements

- **Text website**: A website containing information about the text and supplements can be found at www.mcgrawhill.ca/college/nicholson.

- **Instructor solution manual**. Includes an answer or solution to every exercise in the text.

- **Internet tutorial and assessment**. A student tutorial is available over the Internet at a reduced cost to any student using this book. It includes several features to aid student understanding (see description below), and the labs can be used by instructors for secure testing.

- **i-Learning Sales Specialist**. Your *Integrated Learning Sales Specialist* is a McGraw-Hill Ryerson representative who has the experience, product knowledge, training, and support to help you assess and integrate any of the above products, technology, and services into your course for optimum teaching and learning performance. Whether it's how to use our test bank software, helping your students improve their grades, or how to put your entire course online, your i-Learning Sales Specialist is there to help. Contact your local i-Learning Sales Specialist today to learn how to maximize all McGraw-Hill Ryerson resources!

## LILA: Lyryx Interactive Linear Algebra

**LYRYX LEARNING INC**
Online Learning and Assessment
lyryx.com

www.mcgrawhill.ca/college/nicholson

LILA is a self-paced, interactive tutorial and assessment system available on the Internet at a greatly reduced cost to users of this book. It provides a college-level introduction to the concepts of linear algebra. Our objective has been to create appropriate tools to support students and instructors in all aspects of the course.

**Features for Students**

- **Lessons**. These are animated, audio-enhanced tutorials, that reinforce concepts in much the way a patient human tutor would do it. You will find the lessons useful for study or review, and they are available any time of the day or night, whenever you are in the mood to study.

- **Explorations**. These provide hands-on student interaction with the concepts of the course to enhance understanding. We believe that exploring the concepts, and so developing your understanding through concrete manipulation, is an essential part of learning and *doing* mathematics. They have an easy-to-use, cut-and-paste capability that also allows you to do all the calculations in the Labs. Moreover, they utilize the same standard notation used in the text, rather than the artificial programming format of calculators and computing packages, and so helps overcome another barrier to using computing tools for learning mathematics. In particular, matrix objects *look* like a matrix and *behave* as you would expect (when entering data directly, for example).

- **Labs**. These allow you to gauge your understanding of the material by working problems, with feedback explaining how and why a submitted answer was incorrect, and guidance for improvement utilizing the Lessons and Explorations. The lab problems are randomly generated but similar each time, and can be taken as often as desired up to a due date set by the instructor.

- **Message Board**. This is a feature that allows you to communicate with other students and exchange views about the course.

**Features for Instructors**
- **Secure Testing**. The lab grades are held securely on our server, and instructors can access them at any time, along with the labs themselves.

- **Economical**. Using LILA for your course has little or no cost to your college or university. We have found that most students have their own computer, so very little expensive infrastructure is required.

- **Inexpensive for your students**. LILA can be bundled with this textbook as a package. Please ask your i-Learning Sales Specialist for details.

## Chapter Summaries

**Chapter 1. Linear Equations and Matrices**. After a brief introduction to matrix addition, scalar multiplication and transposition, the Gaussian algorithm for solving systems of linear equations is given. Then matrix multiplication is defined and applied to linear systems, and the main features of matrix inverses are presented. General block multiplication is mentioned, but the emphasis is on three basic cases that are used throughout the book. Elementary matrices are included (but may be omitted with no loss of continuity), and optional sections on LU-factorization, Markov chains, and finite fields are presented.

**Chapter 2. Determinants and Diagonalization**. Determinants are defined inductively, the cofactor expansion is stated (motivated by the $2 \times 2$ case), and the usual rules deduced. Hence the student can immediately begin computing determinants using row and column operations. A major innovation is that diagonalization is introduced at this level using only determinants and matrix arithmetic with no need for vector space concepts like basis and dimension. This leads to linear dynamical systems which provide many motivating examples that are meaningful for the students. Because of the possibility of nonreal eigenvalues, this leads naturally into a discussion of complex numbers. The chapter closes with optional applications to linear recurrences and polynomial interpolation.

**Chapter 3. Vector Geometry**. Motivated by examples, vectors in $\mathbb{R}^2$ and $\mathbb{R}^3$ are introduced as matrices and the matrix operations are interpreted geometrically. Then angles, orthogonality, and projections are brought in via the dot product and, after the cross product is defined, are used to describe lines and planes. Next, matrix transformations of $\mathbb{R}^2$ are viewed as linear transformations, leading to geometric interpretations of matrix multiplication, determinants, and inverses. Finally, an optional section on the cross product is included.

**Chapter 4. The Vector Space** $\mathbb{R}^n$. The basic concepts of subspace, spanning, independence, basis, and dimension are all introduced for subspaces of $\mathbb{R}^n$. This allows the student to learn these new ideas in a familiar context, without the extra complication of having to deal with abstract spaces. After discussing rank, the dot product is introduced, the expansion theorem is proved, and orthogonal bases are introduced (including the Gram-Schmidt algorithm). Then orthogonal complements are defined, the projection theorem is proved, and the approximation theorem is given (leading to best possible solutions of inconsistent systems and to least squares). Next, orthogonal diagonalization is developed and the principal axis theorem for symmetric matrices is proved. This leads to quadratic forms, positive definite matrices, and the Cholesky decomposition (and to applications to constrained optimization and statistical principal component analysis). Then linear transformations are discussed for $\mathbb{R}^n$, the effect of a basis change is noted and related to similarity, and the isometries of $\mathbb{R}^3$ are determined. The chapter concludes with optional sections on complex matrices and the singular value decomposition.

**Chapter 5. Vector Spaces.** General vector spaces are introduced, with spaces of matrices, polynomials, and functions as the main new examples, and their basic properties are deduced (with $\mathbb{R}^n$ as the motivating example). Concepts like independence and basis are now routine extensions of their $\mathbb{R}^n$ counterparts, and the student can focus on becoming accustomed to working in the abstract setting. General linear transformations (and their matrices) are introduced, one-to-one and onto mappings are discussed, and isomorphisms are characterized. Then the effect of a basis change on operators is described with its relation to similarity. This leads to a discussion of invariant subspaces and, as an (optional) illustration, to the Cayley-Hamilton theorem. The chapter (and the book) ends with a discussion of general inner products, with applications to numerical integration, Fourier approximation, and orthogonal polynomials.

## List of Applications

1. Balancing chemical reactions (Example 4 §1.3), page 25.
2. Counting paths in directed graphs (Subsection 1.4.5), page 34.
3. Markov Chains (Section 1.8), page 62.
4. Finite Fields and Cryptography (Subsection 1.9), page 68.
5. Population growth (Example 1 §2.3). This motivates the whole section, page 89.
6. Predator-prey model (Subsection 2.4.3), page 105.
7. Parking lot design (Example 3 §2.6)—uses linear recurrences, page 122.
8. Loading a beam (Example 1 §2.7)—uses polynomial interpolation, page 124.
9. Computer graphics (Subsection 3.4.7), page 171.
10. Steel production (Example 7 §4.6)—uses best approximation to the solution of an inconsistent system, page 229.
11. Voter preferences (Example 9 §4.8)—constrained optimization, page 252.
12. Statistical principal component analysis (Subsection 4.8.4), page 254.
13. Numerical integration (Example 9 §5.7), page 357.
14. Statistical trend analysis (Example 12 §5.7), page 360.

## Additional Applications

Two additional applications can be found on the textbook's website (www.mcgrawhill.ca/college/nicholson):

1. Leontief business application.
2. Electric network application.

## List of Symbols and their Page References

# Acknowledgements

It is a pleasure to recognize the contributions of many other people to this book. Thanks are especially due to Claude Laflamme for his enthusiastic support and invaluable comments, and for his drive in creating the LYRYX tutorial. I also want to thank the group in Calgary who brought the LYRYX project to life, Claire Sauvé for her help with the solution manual, and Joanne Longworth for her computer skills.

As well, I would like to acknowledge the contribution of my peers from across the country and internationally. Their reviews helped guide the revision of this edition. Many thanks to:

**Canada**:

Peter W. Aitchison, *University of Manitoba*
Murray Bremner, *University of Saskatchewan*
Deirdre Haskell, *McMaster University*
Sasho Kalaidzievski, *University of Manitoba*
Stanley O. Kochman, *York University*
Robert Milson, *Dalhousie University*
Harold Proppe, *Concordia University*
F.C.Y. Tang, *University of Waterloo*

Anthony Bonato, *Sir Wilfrid Laurier University*
John Fournier, *University of British Columbia*
Barry Jessup, *University of Ottawa*
David Kaminski, *University of Lethbridge*
James Lewis, *University of Alberta*
John Poland, *Carleton University*
Steven Ruuth, *Simon Fraser University*

**United States**:

Larry Grove, *University of Arizona*
Len Miller, *Mississippi State University*
Paul M. Weichsel, *University of Illinois*

Vera Hu-Hyneman, *Pennsylvania State University*
John Schneider, *Hastings College*

**Ireland**:

Martin Stynes, *National University of Ireland*

**Saudi Arabia**:

Fawzi A. Al-Thukair, *King Saud University*

**Taiwan**:

I-Kong Fong, *National Taiwan University*
Chiou-Ting Hsu, *National Tsing Hua University*

Thanks are also due to the Editorial and Production staff at McGraw-Hill Ryerson. First, to Cathy Koop for her enthusiasm and determination at every stage of the project, and to Pat Ferrier, Darren Hick, Jason Stanley, Margaret Henderson, Jeff MacLean, David Groth, and Bill Todd. For the second edition, I also want to thank Anne Macdonald, Su Mei Ku, Edie Franks, and Sharene Bungay for their invaluable assistance.

Finally, this book would not exist without the constant understanding and cooperation of my wife Kathleen.

# CHAPTER 1

# Linear Equations and Matrices

## 1.1 MATRICES

In mathematics and its applications, numbers frequently appear naturally in ordered rectangular arrays. Here are a few examples.

**Example 1** The coordinates of a point in the plane are usually written as an ordered pair $(x, y)$. Here *ordered* means that the order of the coordinates $x$ and $y$ is important. For example, $(2, 3)$ and $(3, 2)$ represent *different* points. Similarly the coordinates of a point in space are written as an ordered triple $(x, y, z)$.

**Example 2** The system of linear equations

$$3x - 5y = 2$$
$$4x + 7y = 1$$

is completely described by the $2 \times 3$ array of numbers

$$\begin{bmatrix} 3 & -5 & 2 \\ 4 & 7 & 1 \end{bmatrix}.$$

In fact, this is how the system of equations is represented in a computer. We will return to this topic in Section 1.2.

**Example 3** Statistical data are often displayed in arrays called tables. For example, the number of male, female, undergraduate, and graduate students at a small college is given in the following table:

|  | undergraduate | graduate |
|---|---|---|
| women | 3013 | 256 |
| men | 3155 | 511 |

Such a table gives a clear, graphic description of the data, and has other uses as we shall see.

**Example 4** Directed graphical networks like the one shown at the right on page 2 occur frequently in the scheduling of large projects. They are described by the adjacency table where we place 1 (or 0) in row $v_i$ and column $v_j$ to indicate an arrow (or none) from vertex $v_i$ to vertex $v_j$. For example, the vertices could represent cities, and the arrows available flights. By eliminating the arrowheads and allowing more than one edge between two vertices, this gives a model for chemical bonding.

|       | $v_1$ | $v_2$ | $v_3$ | $v_4$ |
|-------|-------|-------|-------|-------|
| $v_1$ | 0     | 1     | 1     | 1     |
| $v_2$ | 0     | 0     | 1     | 1     |
| $v_3$ | 0     | 1     | 0     | 0     |
| $v_4$ | 0     | 0     | 0     | 0     |

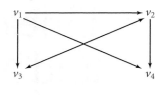

<hr>

### 1.1.1 Matrices

Motivated by these examples, a rectangular array of numbers is called a **matrix**[1], and the numbers themselves are called the **entries**[2] of the matrix. Matrices are usually denoted by upper case letters $A$, $B$, $C$, etc.

**Example 5**   The following are matrices:

$$A = \begin{bmatrix} 1 & 0 & -3 \\ 5 & 3 & 8 \end{bmatrix} \quad B = \begin{bmatrix} 2 & -5 \end{bmatrix} \quad C = \begin{bmatrix} 2 \\ 0 \\ -3 \end{bmatrix} \quad D = \begin{bmatrix} 3 & 1 & -1 \\ 2 & 0 & 5 \\ -2 & -2 & 1 \end{bmatrix}$$

Matrices come in various (rectangular) shapes depending on the number of **rows** and **columns**. For example, the matrix $A$ in Example 5 has 2 rows and 3 columns. In general a matrix with $m$ rows and $n$ columns is called an **$m \times n$** matrix, or is said to have **size $m \times n$**. Thus the matrices $A$, $B$, $C$, and $D$ in Example 5 have size $2 \times 3$, $1 \times 2$, $3 \times 1$, and $3 \times 3$ respectively. Not surprisingly, a matrix of size $1 \times n$ is called a **row matrix** or an **$n$-row**, and one of size $m \times 1$ is called a **column matrix** or an **$m$-column**. A matrix with equal numbers of rows and columns is called a **square matrix**. Thus each square matrix has size $n \times n$ for some $n$.

Each entry of a matrix is located by the row and column in which it lies. The rows of the matrix are numbered from the top down, and the columns are numbered from left to right. Then the entry in row $i$ and column $j$ is called the **$(i, j)$-entry** of the matrix. For example:

The $(1, 2)$-entry of $\begin{bmatrix} 1 & 0 & -3 \\ 5 & 3 & 8 \end{bmatrix}$ is 0, while the $(2, 3)$-entry is 8.

A special notation has been devised for the entries of a matrix $A$. If $A$ is of size $m \times n$, and if the $(i, j)$-entry of $A$ is denoted by $a_{ij}$, then $A$ is displayed as follows:

$$A = \begin{bmatrix} a_{11} & a_{12} & a_{13} & \cdots & a_{1n} \\ a_{21} & a_{22} & a_{23} & \cdots & a_{2n} \\ \vdots & \vdots & \vdots & \ddots & \vdots \\ a_{m1} & a_{m2} & a_{m3} & \cdots & a_{mn} \end{bmatrix}.$$

This is usually written simply as $A = [a_{ij}]$, and $a_{ij}$ is the entry in row $i$ and column $j$. For example, using this notation the general $2 \times 3$ matrix is written

$$A = \begin{bmatrix} a_{11} & a_{12} & a_{13} \\ a_{21} & a_{22} & a_{23} \end{bmatrix}.$$

<hr>

[1] While the term *matrix* was first used in 1848 by James Joseph Sylvester (1814–1897), it was Arthur Cayley (1821–1895) who first considered matrices as single entities in a paper in 1858 entitled "A memoir on the theory of matrices".

[2] Definitions of new terms appear in bold face throughout the text.

Note that, as for ordered pairs, we regard two matrices $A$ and $B$ as **equal** (written $A = B$) if they have the same number of rows and columns and if corresponding entries are equal. More precisely:

$$A = B \quad \text{if and only if} \quad \begin{cases} 1. & \text{$A$ and $B$ have the same size} \\ 2. & \text{Corresponding entries are equal} \end{cases}$$

If $A = [a_{ij}]$ and $B = [b_{ij}]$ are both $m \times n$, this takes the form

$$[a_{ij}] = [b_{ij}] \quad \text{if and only if} \quad a_{ij} = b_{ij} \text{ for all } i \text{ and } j.$$

**Example 6**   Given $A = \begin{bmatrix} 1 & -3 & 0 \\ 2 & 5 & 0 \end{bmatrix}$, $B = \begin{bmatrix} 1 & -3 \\ 2 & 5 \end{bmatrix}$, $C = \begin{bmatrix} x & y \\ z & w \end{bmatrix}$, discuss the possibility that $A = B$, $A = C$, and $B = C$.

**SOLUTION**   $A = B$ is impossible because $A$ and $B$ are of different sizes: $A$ is $2 \times 3$ while $B$ is $2 \times 2$. Thus $A$ and $B$ are *not* equal (written $A \neq B$). Similarly $A \neq C$. However, $B = C$ is possible provided that corresponding entries are equal: $\begin{bmatrix} 1 & -3 \\ 2 & 5 \end{bmatrix} = \begin{bmatrix} x & y \\ z & w \end{bmatrix}$ can happen only if $x = 1$, $y = -3$, $z = 2$, and $w = 5$.

## 1.1.2   Matrix Addition

Matrices can be added in much the same way as numbers. The following example shows how this arises in practice.

**Example 7**   The yearly government tax revenue from Corporation 1 (in millions) is \$175m in corporation tax, \$35m in income tax, and \$17m in sales tax. This data is summarized in the revenue matrix

$$T_1 = [175 \quad 35 \quad 17].$$

Similarly, Corporation 2 has revenue matrix

$$T_2 = [190 \quad 41 \quad 22].$$

The combined revenue matrix for both corporations is thus

$$T = [175 + 190 \quad 35 + 41 \quad 17 + 22] = [365 \quad 76 \quad 39].$$

This new matrix is obtained by adding corresponding entries in $T_1$ and $T_2$.

In general, if $A$ and $B$ are two matrices of the same size, their **sum** $A + B$ is defined to be the matrix of the same size formed by adding corresponding entries. If $A = [a_{ij}]$ and $B = [b_{ij}]$ this takes the form

$$[a_{ij}] + [b_{ij}] = [a_{ij} + b_{ij}].$$

Similarly, the **difference** $A - B$ of $A$ and $B$ is defined by subtracting corresponding entries:

$$[a_{ij}] - [b_{ij}] = [a_{ij} - b_{ij}].$$

As in arithmetic, $A - B$ is said to be the result of **subtracting** $B$ from $A$. Note that addition and subtraction are not defined for matrices of different sizes. In fact, we adopt the following convention:

*When we write $A + B$ or $A - B$, it is assumed that $A$ and $B$ are the same size.*

We will use this convention without comment throughout the book.

**Example 8**    If $A = \begin{bmatrix} 2 & -1 & 7 \\ -3 & 5 & 0 \end{bmatrix}$ and $B = \begin{bmatrix} 4 & 6 & -2 \\ 8 & 1 & 9 \end{bmatrix}$ then

$$A + B = \begin{bmatrix} 2+4 & -1+6 & 7-2 \\ -3+8 & 5+1 & 0+9 \end{bmatrix} = \begin{bmatrix} 6 & 5 & 5 \\ 5 & 6 & 9 \end{bmatrix}$$

$$A - B = \begin{bmatrix} 2-4 & -1-6 & 7-(-2) \\ -3-8 & 5-1 & 0-9 \end{bmatrix} = \begin{bmatrix} -2 & -7 & 9 \\ -11 & 4 & -9 \end{bmatrix}$$

Many general properties of numerical addition also hold for matrix addition. If $A$, $B$, and $C$ are matrices of the same size, then:

$$A + B = B + A \qquad \textbf{(Commutative Law)}$$
$$A + (B + C) = (A + B) + C \qquad \textbf{(Associative Law)}$$

In fact, if $A = [a_{ij}]$, and $B = [b_{ij}]$, then the $(i, j)$-entries of $A + B$ and $B + A$ are $a_{ij} + b_{ij}$ and $b_{ij} + a_{ij}$ respectively. Since these are equal for all $i$ and $j$, we obtain

$$A + B = [a_{ij} + b_{ij}] = [b_{ij} + a_{ij}] = B + A.$$

This proves the commutative law, and the associative law is similarly verified.

The $m \times n$ matrix in which every entry is zero is called the **zero matrix** of that size, and is denoted by $0$ (or $0_{mn}$ if the size must be noted)[3]. It is clear that

$$0 + A = A \quad \text{for every } m \times n \text{ matrix } A.$$

The **negative** of the $m \times n$ matrix $A$ (denoted $-A$) is the $m \times n$ matrix obtained by negating every entry of $A$. If $A = [a_{ij}]$ this can be written

$$-[a_{ij}] = [-a_{ij}].$$

In other words, the $(i, j)$-entry of $-A$ is $-a_{ij}$. Since matrix addition is carried out component-wise, we have

$$A + (-A) = 0 \quad \text{for every } m \times n \text{ matrix } A$$

where $0$ denotes the $m \times n$ zero matrix. The following theorem collects these four basic properties of matrix addition for reference later.

**THEOREM 1**    If $A$, $B$, and $C$ denote arbitrary $m \times n$ matrices, then

(1) $A + B = B + A$.
(2) $A + (B + C) = (A + B) + C$.
(3) $0 + A = A$ where $0$ is the $m \times n$ zero matrix.
(4) $A + (-A) = 0$.

**Example 9**    If $A = \begin{bmatrix} 3 & 1 \\ 6 & 4 \end{bmatrix}$ and $B = \begin{bmatrix} 1 & -2 \\ 0 & 4 \end{bmatrix}$, find a matrix $X$ such that $X + B = A$.

**SOLUTION**    To solve a numerical equation $x + b = a$, one simply subtracts $b$ from both sides to obtain $x = a - b$. Similarly, the matrix equation $X + B = A$ leads to

$$X = A - B = \begin{bmatrix} 3-1 & 1-(-2) \\ 6-0 & 4-4 \end{bmatrix} = \begin{bmatrix} 2 & 3 \\ 6 & 0 \end{bmatrix}.$$

The reader can check that this matrix $X$ does indeed satisfy the original equation $X + B = A$.

---

[3] The same symbol, 0, is commonly used for a zero matrix and for the number zero. This causes very little confusion in practice since the meaning is nearly always clear from the context.

The properties in Theorem 1 have several useful consequences. For example, property (2) asserts that the sum $A + (B + C) = (A + B) + C$ is the same no matter how it is formed, and so it is written simply as $A + B + C$. Similarly, the sum $A + B + C + D$ is independent of how it is formed. For example, it equals both $(A + B) + (C + D)$ and $A + [B + (C + D)]$. Furthermore, property (1) in Theorem 1 ensures that, for example, $B + D + A + C = A + B + C + D$. In other words, the *order* in which these four matrices are added does not matter. Similar remarks apply to sums of five or more matrices.

---

### 1.1.3   Scalar Multiplication

**Example 10**   Suppose a corporation has a tax revenue matrix $T = [350 \quad 65 \quad 40]$ by which (as in Example 7) we mean that the government tax revenue from the corporation is \$350m in corporate tax, \$65m in income tax, and \$40m in sales tax. If the government decided to increase all forms of tax by half, the new tax revenue matrix would be

$$T_1 = \left[ \tfrac{3}{2} \cdot 350 \quad \tfrac{3}{2} \cdot 65 \quad \tfrac{3}{2} \cdot 40 \right] = [525 \quad 97.5 \quad 60].$$

This new matrix is computed from $T$ by multiplying each entry by $\tfrac{3}{2}$.

In general, if $A$ is a matrix and $c$ is a number, the **scalar product**[4] $cA$ is the matrix formed from $A$ by multiplying each entry of $A$ by the number $c$. If $A = [a_{ij}]$, this takes the form

$$cA = c[a_{ij}],$$

that is, the $(i, j)$-entry of $cA$ is $c\,a_{ij}$. Note that the matrix $cA$ is the same size as $A$.

**Example 11**   If $A = \begin{bmatrix} 1 & 0 & -3 \\ 2 & 7 & 3 \end{bmatrix}$ and $B = \begin{bmatrix} 5 & -6 & 1 \\ 8 & 8 & 9 \end{bmatrix}$, we have

$$3A = \begin{bmatrix} 3 & 0 & -9 \\ 6 & 21 & 9 \end{bmatrix}, \quad \tfrac{1}{2}B = \begin{bmatrix} \tfrac{5}{2} & -3 & \tfrac{1}{2} \\ 4 & 4 & \tfrac{9}{2} \end{bmatrix}, \text{ and}$$

$$5A - 2B = \begin{bmatrix} 5 & 0 & -15 \\ 10 & 35 & 15 \end{bmatrix} - \begin{bmatrix} 10 & -12 & 2 \\ 16 & 16 & 18 \end{bmatrix} = \begin{bmatrix} -5 & 12 & -17 \\ -6 & 19 & -3 \end{bmatrix}.$$

If $A$ is any matrix, the following facts are clear:

$$1A = A \quad \text{and} \quad (-1)A = -A.$$

Furthermore, for any scalar $c$ and matrix $A$, we have

$$c0 = 0 \quad \text{and} \quad 0A = 0.$$

Note that the symbol 0 is playing two roles in the equation $0A = 0$. It represents the number zero on the left side, and it represents the zero matrix on the right side. This ambiguity is harmless as it is always clear from the context which meaning is correct.

There are four basic properties of scalar multiplication which we record for reference.

---

**THEOREM** ❷   Let $A$ and $B$ denote matrices and let $c$ and $d$ denote numbers. Then:

    **(1)** $c(A + B) = cA + cB$.
    **(2)** $(c + d)A = cA + dA$.
    **(3)** $c(dA) = (cd)A$.
    **(4)** $1A = A$.

---

[4] In linear algebra, numbers are often referred to as *scalars*; hence the name.

PROOF    We have already mentioned (4). To verify (1), write $A = [a_{ij}]$ and $B = [b_{ij}]$. Then $A + B = [a_{ij} + b_{ij}]$ as before, so the $(i, j)$-entry of the matrix $c(A + B)$ is

$$c(a_{ij} + b_{ij}) = ca_{ij} + cb_{ij}.$$

But $ca_{ij}$ and $cb_{ij}$ are the $(i, j)$-entries of $cA$ and $cB$ respectively, so $ca_{ij} + cb_{ij}$ is the $(i, j)$-entry of $cA + cB$. It follows that $c(A + B) = cA + cB$, proving (1). The verifications of (2) and (3) are similar and are left as exercises.

Properties (1) and (2) in Theorem 2 are called **distributive laws**. They extend to sums of more than two matrices. For example

$$c(A + B + C) = cA + cB + cC,$$
$$(c + d + e)A = cA + dA + eA.$$

Similar expressions hold for more than two summands. These facts, together with the properties in Theorem 1, enable us to simplify matrix expressions by collecting like terms, expanding, and taking out common factors, in exactly the same way that algebraic expressions are simplified. Here are some examples.

**Example 12**    Simplify the matrix expression $3(8A - 5B) + 4(4B - 6A)$.

SOLUTION    The procedure is the same as if $A$ and $B$ were numbers:

$$3(8A - 5B) + 4(4B - 6A) = 24A - 15B + 16B - 24A = B.$$

Hence the expression simplifies a great deal.

**Example 13**    Find $A$ if $\frac{1}{5}\left\{4A - \begin{bmatrix} 1 \\ -2 \end{bmatrix}\right\} = \begin{bmatrix} 0 \\ 5 \end{bmatrix} - 2A$.

SOLUTION    Multiplying both sides by 5 yields

$$4A - \begin{bmatrix} 1 \\ -2 \end{bmatrix} = 5\left\{\begin{bmatrix} 0 \\ 5 \end{bmatrix} - 2A\right\}$$

$$4A - \begin{bmatrix} 1 \\ -2 \end{bmatrix} = \begin{bmatrix} 0 \\ 25 \end{bmatrix} - 10A.$$

Adding $10A$ to both sides gives $14A - \begin{bmatrix} 1 \\ -2 \end{bmatrix} = \begin{bmatrix} 0 \\ 25 \end{bmatrix}$; then adding $\begin{bmatrix} 1 \\ -2 \end{bmatrix}$ to both sides gives

$$14A = \begin{bmatrix} 0 \\ 25 \end{bmatrix} + \begin{bmatrix} 1 \\ -2 \end{bmatrix} = \begin{bmatrix} 1 \\ 23 \end{bmatrix}.$$

Finally, dividing by 14 yields $A = \frac{1}{14}\begin{bmatrix} 1 \\ 23 \end{bmatrix} = \begin{bmatrix} \frac{1}{14} \\ \frac{23}{14} \end{bmatrix}$.

We have already observed that a scalar product $cA$ is the zero matrix if either $c = 0$ is the zero number or $A = 0$ is the zero matrix. The next useful result shows that the converse holds.

**THEOREM 3**  If $cA = 0$, then either $c = 0$ or $A = 0$.

**PROOF**  If $c = 0$, there is nothing to prove; if $c \neq 0$, we show that $A = 0$. Write $A = [a_{ij}]$ where $a_{ij}$ denotes the $(i, j)$-entry of $A$. Since $cA = 0$, we have $c\,a_{ij} = 0$ for each $i$ and $j$ ($c\,a_{ij}$ is the $(i, j)$-entry of $cA$). Since $c \neq 0$, this gives $a_{ij} = 0$ for each $i$ and $j$; that is, $A = 0$.

## 1.1.4 Transposition

Writing column matrices is cumbersome and takes up text space. The idea of the transpose of a matrix helps with this problem by turning columns into rows (it has many other uses as well as we shall see). Before defining it, we must discuss another important notion.

If $A = [a_{ij}]$ is an $m \times n$ matrix, the **main diagonal** of $A$ consists of the entries

$$a_{11}, \ a_{22}, \ a_{33}, \cdots.$$

Thus the main diagonal entries lie on the diagonal line starting from the upper left corner of $A$. For example, in each of the following matrices the main diagonal consists of the entries $a$ and $b$:

$$\begin{bmatrix} a & x \\ y & b \end{bmatrix} \quad \begin{bmatrix} a & x & p \\ y & b & q \end{bmatrix} \quad \begin{bmatrix} a & x \\ y & b \\ r & s \end{bmatrix}$$

Note that the main diagonal of a square matrix extends from the upper left corner to the lower right corner.

We now informally define the **transpose** of an $m \times n$ matrix $A$ to be the $n \times m$ matrix $A^T$ obtained when $A$ is "flipped" over its main diagonal; that is $A$ is rotated $180°$ about the line containing the main diagonal entries.

**Example 14**  Find the transpose of each of the following matrices:

$$A = \begin{bmatrix} 3 \\ 5 \end{bmatrix}, \ B = \begin{bmatrix} 0 & -1 \\ 2 & 11 \end{bmatrix}, \ C = \begin{bmatrix} 5 & -2 & 3 \end{bmatrix}, \ D = \begin{bmatrix} 1 & 6 & 9 \\ 3 & 2 & 7 \end{bmatrix}.$$

**SOLUTION**

$$A^T = \begin{bmatrix} 3 & 5 \end{bmatrix}, \ B^T = \begin{bmatrix} 0 & 2 \\ -1 & 11 \end{bmatrix}, \ C^T = \begin{bmatrix} 5 \\ -2 \\ 3 \end{bmatrix}, \ D^T = \begin{bmatrix} 1 & 3 \\ 6 & 2 \\ 9 & 7 \end{bmatrix}.$$

Clearly the transpose of any row matrix is a column matrix, and the transpose of any column matrix is a row matrix. This gives another way to think of the transpose of a matrix $A$:

The columns of $A^T$ are just the transposes of the rows of $A$, in the same order.

The rows of $A^T$ are just the transposes of the columns of $A$, in the same order.

This is illustrated by the matrices in Example 14, and leads to a more formal description of $A^T$:

If $A = [a_{ij}]$, then $A^T = [b_{ij}]$ where $b_{ij} = a_{ji}$ for all $i$ and $j$.

The following theorem collects four basic properties of transposition which will be used repeatedly and without comment.

**THEOREM 4**    Let $A$ and $B$ denote matrices of the same size, and let $c$ denote a scalar.

    **(1)** If $A$ is an $m \times n$ matrix, then $A^T$ is an $n \times m$ matrix.
    **(2)** $(A^T)^T = A$.
    **(3)** $(cA)^T = c\,A^T$.
    **(4)** $(A + B)^T = A^T + B^T$.

**PROOF**    (1) is part of the definition of $A^T$, and (2) is true because $(A^T)^T$ is the result of flipping twice, and this gets you back to $A$. Similarly, $(cA)^T$ is the result of multiplying each entry by $c$ and then flipping, which is the same as the result of first flipping $A$ and then multiplying by $c$.

While we can prove (4) in the same way, we use the formal definition instead. Write $A = [a_{ij}]$ and $B = [b_{ij}]$, so that $A^T = [a_{ji}]$ and $B^T = [b_{ji}]$ are obtained by interchanging $i$ and $j$ in every entry. Since $A + B = [a_{ij} + b_{ij}]$, this gives

$$(A + B)^T = [a_{ji} + b_{ji}] = [a_{ji}] + [b_{ji}] = A^T + B^T.$$

This is what we wanted.

**Example 15**    Find the matrix $A$ if $\left\{ A^T + 2\begin{bmatrix} 1 & -2 \\ 5 & 0 \end{bmatrix} \right\}^T = 5\begin{bmatrix} 3 & -1 \\ 1 & 7 \end{bmatrix}$.

**SOLUTION**    Using Theorem 4 several times we get

$$5\begin{bmatrix} 3 & -1 \\ 1 & 7 \end{bmatrix} = \left(A^T\right)^T + \left\{ 2\begin{bmatrix} 1 & -2 \\ 5 & 0 \end{bmatrix} \right\}^T$$

$$= A + 2\begin{bmatrix} 1 & -2 \\ 5 & 0 \end{bmatrix}^T$$

$$= A + 2\begin{bmatrix} 1 & 5 \\ -2 & 0 \end{bmatrix}.$$

Hence $A = 5\begin{bmatrix} 3 & -1 \\ 1 & 7 \end{bmatrix} - 2\begin{bmatrix} 1 & 5 \\ -2 & 0 \end{bmatrix} = \begin{bmatrix} 13 & -15 \\ 9 & 35 \end{bmatrix}$.

A matrix $A$ is called **symmetric** if $A^T = A$. Such matrices are necessarily square (if $A$ is $m \times n$ then $A^T$ is $n \times m$, so $A^T = A$ forces $m = n$). The name comes from the fact that the symmetric matrices exhibit a symmetry about the main diagonal. Indeed, the condition $A^T = A$ means that flipping $A$ about the main diagonal does not change it, and so requires that entries directly across the main diagonal from each other are equal. Hence the general $2 \times 2$ and $3 \times 3$ symmetric matrices have the form

$$\begin{bmatrix} a & x \\ x & b \end{bmatrix} \quad \text{and} \quad \begin{bmatrix} a & x & y \\ x & b & z \\ y & z & c \end{bmatrix}$$

respectively. Clearly every $1 \times 1$ matrix is symmetric.

**Example 16**    An undirected graph consists of vertices some of which are connected by edges with no direction assigned (as distinguished from a directed graph). The adjacency matrix of such a graph has a 1 in the $(i, j)$-position if $v_i$ and $v_j$ are connected, and so is a symmetric matrix, as is illustrated for the given graph. Note the contrast with Example 4 in the directed graph situation.

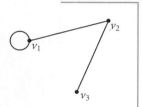

$$\begin{bmatrix} 1 & 1 & 0 \\ 1 & 0 & 1 \\ 0 & 1 & 0 \end{bmatrix}$$

The elementary properties of symmetric matrices come from Theorem 4. Here is an example of how they are used.

**Example 17** If $A$ and $B$ are symmetric, show that $A + B$ is also symmetric.

SOLUTION We are given that $A^T = A$ and $B^T = B$. Hence Theorem 4 gives
$$(A + B)^T = A^T + B^T = A + B.$$
This shows that $A + B$ is symmetric.

---

## Exercises 1.1

**1.** A store sells road bikes and mountain bikes. The manager has the following two data tables available, the first giving sales volumes of each kind for each quarter of 1998, and the second for each quarter of 1999:

| 1998 | Road Bikes | Mountain Bikes |
|---|---|---|
| 1st Quarter | 32 | 27 |
| 2nd Quarter | 65 | 78 |
| 3rd Quarter | 53 | 74 |
| 4th Quarter | 35 | 46 |

| 1999 | Road Bikes | Mountain Bikes |
|---|---|---|
| 1st Quarter | 38 | 43 |
| 2nd Quarter | 72 | 97 |
| 3rd Quarter | 76 | 102 |
| 4th Quarter | 43 | 65 |

**a)** Express each of the tables as a matrix.

**b)** Find a matrix operation that will produce the total sales volumes for these two years combined, and perform the calculations.

**c)** The data for 1997 was given in a slightly different format as follows:

| 1997 | 1st Quarter | 2nd Quarter | 3rd Quarter | 4th Quarter |
|---|---|---|---|---|
| Road Bikes | 24 | 29 | 32 | 21 |
| Mountain Bikes | 23 | 31 | 37 | 29 |

Express this data also as a matrix, find a matrix operation that will produce the total sales volumes for these three years combined, and perform the calculations.

**d)** A second store sold exactly three times as many of each kind of bike as the first store for each quarter in 1997, exactly twice as many for 1998, and exactly four times as many for 1999. Find a matrix operation that will produce the total sales volumes for this second store for these three years combined, and perform the calculations.

**2. a)** If $A$ is an $m \times n$ matrix, what do the $m$ and $n$ represent?

**b)** What are the requirements needed to add two matrices $A$ and $B$?

**3.** How could matrix addition, scalar multiplication, and transposition be helpful in practice?

**4.** Compute the following:

**a)** $3\left(\begin{bmatrix} 1 & -3 & 2 \\ 0 & 4 & -9 \\ 2 & -3 & 1 \end{bmatrix} - 21\begin{bmatrix} 2 & 0 & 12 \\ -15 & -3 & 2 \\ 12 & -8 & -32 \end{bmatrix}\right) + 5\begin{bmatrix} 1 & -5 & -43 \\ 0 & 7 & 0 \\ 21 & 45 & -34 \end{bmatrix}^T$

**b)** $\begin{bmatrix} 21 & 20 & -21 \\ 56 & 11 & 34 \\ 45 & 83 & -34 \end{bmatrix} - 9\left(2\begin{bmatrix} 90 & -45 & 20 \\ -23 & 32 & 72 \\ 65 & 23 & 89 \end{bmatrix} - 13\begin{bmatrix} 19 & -25 & 87 \\ 47 & 40 & 81 \\ 85 & 35 & 24 \end{bmatrix}^T\right)$

**5. a)** Describe when a matrix is the same as its transpose. Such matrices are called *symmetric* matrices.

**b)** Describe when a matrix is equal to the negative of its transpose. Such matrices are called *skew-symmetric* matrices.

**6.** In each case show either that the statement is true or that it is false by giving an example. Assume throughout that $A$, $B$, and $C$ denote matrices.

**a)** If $A + B = A + C$, then $B$ and $C$ have the same size.

**b)** If $A + B = 0$, then $B = 0$.

**c)** If the $(2, 3)$-entry of $A$ is 7, then the $(3, 2)$-entry of $A^T$ is $-7$.

**d)** If $A = -A$, then $A = 0$.

**e)** $A$ and $A^T$ have the same main diagonal for every matrix $A$.

**7.** Find $a$, $b$, $c$, and $d$ if:

**a)** $\begin{bmatrix} a & b \\ c & d \end{bmatrix} = \begin{bmatrix} b - c & c \\ d & 1 \end{bmatrix}$.  **b)** $3\begin{bmatrix} a \\ b \end{bmatrix} - 2\begin{bmatrix} b \\ 0 \end{bmatrix} = \begin{bmatrix} 1 \\ 4 \end{bmatrix}$.

**8.** In each case find the matrix $A$:

**a)** $2A - \begin{bmatrix} 1 & 0 & -2 \\ 4 & 7 & 3 \end{bmatrix}^T = \begin{bmatrix} 2 & 0 \\ -3 & 4 \\ 0 & 8 \end{bmatrix}$.

**b)** $\left(3A^T - 2\begin{bmatrix} 0 & 2 \\ -5 & 2 \\ 1 & 3 \end{bmatrix}\right)^T = \begin{bmatrix} 7 & 0 & -1 \\ 2 & 1 & 5 \end{bmatrix}$.

**9. a)** Simplify $3(5A - 3B) + 6(B - 4A) + 3(2A + B)$.
  **b)** Simplify $2(3A - 5B) + 4(B - 2A) + 2(3B - A)$.

**10.** A matrix $A$ is called *diagonal* if it is square and every entry off the main diagonal is zero. If $A$ and $B$ are both diagonal, show that each of the following matrices is diagonal:
  **a)** $A + B$.   **b)** $cA$ for any scalar $c$.   **c)** $A^T$.

**11. a)** If $A$ is any matrix, show that $(c + d)A = cA + dA$ for all scalars $c$ and $d$. [Part (2) of Theorem 2]
  **b)** If $A$ is any matrix, show that $(cd)A = c(dA)$ for all scalars $c$ and $d$. [Part (3) of Theorem 2]

## 1.2   LINEAR EQUATIONS

One of the historical motivations for the development of mathematics has been to find a way to analyze and solve practical problems. Apart from calculus, the most common method has been to reduce a problem to the solution of a set of linear equations. This has proved to be effective in science, engineering, and social science. Here is a simple example.

### 1.2.1   Linear Equations

**Example 1**   A charity wishes to endow a fund that will provide \$50 000 per year for cancer research. The charity has \$480 000 and, to reduce risk, wants to invest in two banks paying 10% and 11% respectively. How much should be invested in each bank?

**SOLUTION**   If $x$ dollars are invested at 10% and $y$ dollars are invested at 11%, then $x + y = 480\ 000$ and the yearly interest is $\frac{10}{100}x + \frac{11}{100}y$. Hence $x$ and $y$ must satisfy the equations

$$x + \quad y = 480\ 000$$
$$\tfrac{10}{100}x + \tfrac{11}{100}y = 50\ 000$$

If the first equation is multiplied by 10, and the second by 100, the resulting equations are

$$10x + 10y = 4\ 800\ 000$$
$$10x + 11y = 5\ 000\ 000$$

Subtracting the first equation from the second gives $y = 200\ 000$, whence $x = 280\ 000$. In other words, the charity should invest \$280 000 at 10% and \$200 000 at 11%.

An equation of the form $ax + by = c$ is called a linear equation[5] in the variables $x$ and $y$. If more than two variables are present, they will usually be denoted by $x_1, x_2, \cdots, x_n$. Then an equation of the form

$$a_1 x_1 + a_2 x_2 + \cdots + a_n x_n = b \qquad (*)$$

is called a **linear equation** in the **variables** $x_1, x_2, \cdots, x_n$. Here $a_1, a_2, \cdots, a_n$ denote real numbers, called the **coefficients** of the variables $x_1, x_2, \cdots, x_n$, and the number $b$ is called the **constant term** of the equation. The column matrix

$$X = \begin{bmatrix} x_1 \\ x_2 \\ \vdots \\ x_n \end{bmatrix} = [x_1 \quad x_2 \quad \cdots \quad x_n]^T$$

---

[5] The graph of the equation $ax + by = c$ is a straight line if $a$ and $b$ are not both zero; hence the name.

is called the **matrix of variables**. A set of numbers $s_1$, $s_2$, $\cdots$, $s_n$ is called a **solution** to the linear equation (∗) if

$$a_1 s_1 + a_2 s_2 + \cdots + a_n s_n = b,$$

that is, if the equation is satisfied when the substitutions $x_1 = s_1$, $x_2 = s_2$, $\cdots$, $x_n = s_n$ are made. We express this by saying that $X = [s_1 \ \ s_2 \ \ \cdots \ \ s_n]^T$ is a solution[6] to equation (∗).

**Example 2**  Show that $X = [1 \ \ -2]^T$ is a solution to the equation $2x_1 - 3x_2 = 8$, but that $Y = [1 \ \ 1]^T$ is not a solution.

SOLUTION  $[1 \ \ -2]^T$ is a solution because $x_1 = 1$ and $x_2 = -2$ satisfy the equation: $2(1) - 3(-2) = 8$. But $x_1 = 1$ and $x_2 = 1$ do not satisfy the equation since $2(1) - 3(1) \neq 8$, so $[1 \ \ 1]^T$ is not a solution.

## 1.2.2 Systems of Linear Equations

A finite collection of linear equations is called a **system of linear equations**, or simply a **system**. A solution to *every* equation in the system is called a **solution to the system**. The heart of linear algebra is a routine procedure for finding all solutions to any system of linear equations. But first some examples.

**Example 3**  In Example 1, the system of two linear equations

$$\begin{aligned} x + \quad y &= 480\ 000 \\ \tfrac{10}{100}x + \tfrac{11}{100}y &= 50\ 000 \end{aligned} \qquad \text{has solution } X = \begin{bmatrix} x \\ y \end{bmatrix} = \begin{bmatrix} 280\ 000 \\ 200\ 000 \end{bmatrix}.$$

Note that this system has a *unique* solution.

**Example 4**  A system of equations need not have a solution. For example, the system

$$\begin{aligned} x + y \quad &= 1 \\ x \quad - z &= 2 \\ y + z &= 1 \end{aligned}$$

has *no* solution. Indeed adding the last two equations gives $x + y = 3$, contrary to the first equation.

A system of linear equations is called **inconsistent** if it has no solutions, and the system is called **consistent** if it has one or more solutions.

**Example 5**  Verify that $X = [1 + t - s \ \ \ 2 + t + s \ \ \ s \ \ \ t]^T$ is a solution to the system

$$\begin{aligned} x_1 - 2x_2 + 3x_3 + x_4 &= -3 \\ 2x_1 - x_2 + 3x_3 - x_4 &= 0 \end{aligned}$$

for all values of the numbers $s$ and $t$, called **parameters** in this context.

SOLUTION  Simply substitute $x_1 = 1 + t - s$, $x_2 = 2 + t + s$, $x_3 = s$ and $x_4 = t$ into each equation:

$$\begin{aligned} x_1 - 2x_2 + 3x_3 + x_4 &= (1 + t - s) - 2(2 + t + s) + 3s + t = -3 \\ 2x_1 - x_2 + 3x_3 - x_4 &= 2(1 + t - s) - (2 + t + s) + 3s - t = 0 \end{aligned}$$

---

[6] The reason for writing solutions (or variables) as *columns* will be clear later. We will frequently use the transpose notation $[s_1 \ \ s_2 \ \ \cdots \ \ s_n]^T$ to simplify printing.

Because both equations are satisfied, $X$ is a solution for all $s$ and $t$. Note that this system has *infinitely many* solutions since there are infinitely many choices for the parameters $s$ and $t$.

In fact, *every* solution of the system in Example 5 arises as shown for some parameters $s$ and $t$. To see why this is so, and to see how one arrives at sets of solutions like those in Example 5, we develop a general procedure for finding solutions in this form. To simplify the computations we introduce a matrix notation for describing systems of linear equations. Given the system

$$x_1 - 2x_2 + 3x_3 + x_4 = -3$$
$$2x_1 - x_2 + 3x_3 - x_4 = 0$$

of 2 equations in 4 variables, the coefficients of the variables form a $2 \times 4$ matrix

$$\begin{bmatrix} 1 & -2 & 3 & 1 \\ 2 & -1 & 3 & -1 \end{bmatrix}$$

called the **coefficient matrix** for the system. The $2 \times 5$ matrix

$$\begin{bmatrix} 1 & -2 & 3 & 1 & -3 \\ 2 & -1 & 3 & -1 & 0 \end{bmatrix}$$

is called the **augmented matrix** for the system (it is just the coefficient matrix *augmented* by the column of constants). It is evident that the system is completely described by the augmented matrix,[7] so it is not surprising that by manipulating this matrix we can find all solutions to the system.

To see how this is done, it is convenient to call two systems of linear equations **equivalent** if they have the same set of solutions. Starting with a given system of linear equations, we solve it by writing a series of systems, one after the other, each of which is equivalent to the previous one. Since all these systems have the same solutions, the aim is to find one that is easy to solve. Surprisingly enough, there is a simple, routine method for doing this. The following example provides an illustration.

**Example 6**   Find all solutions to the system in Example 5:

$$\begin{cases} x_1 - 2x_2 + 3x_3 + x_4 = -3 \\ 2x_1 - x_2 + 3x_3 - x_4 = 0 \end{cases}$$

SOLUTION   The system is written below together with its augmented matrix:

$$\begin{array}{ll} x_1 - 2x_2 + 3x_3 + x_4 = -3 & \begin{bmatrix} 1 & -2 & 3 & 1 & -3 \\ 2 & -1 & 3 & -1 & 0 \end{bmatrix} \\ 2x_1 - x_2 + 3x_3 - x_4 = 0 & \end{array}$$

We first eliminate $x_1$ from equation 2 by subtracting twice the first equation from the second. The result is the following system (with its augmented matrix).

$$\begin{array}{ll} x_1 - 2x_2 + 3x_3 + x_4 = -3 & \begin{bmatrix} 1 & -2 & 3 & 1 & -3 \\ 0 & 3 & -3 & -3 & 6 \end{bmatrix} \\ 3x_2 - 3x_3 - 3x_4 = 6 & \end{array}$$

This new system is equivalent to the original system (see Theorem 1 below). Note that the new augmented matrix can be obtained directly from the original one by subtracting twice the first row from the second row.

We now multiply the second equation by $\frac{1}{3}$ to obtain another equivalent system

$$\begin{array}{ll} x_1 - 2x_2 + 3x_3 + x_4 = -3 & \begin{bmatrix} 1 & -2 & 3 & 1 & -3 \\ 0 & 1 & -1 & -1 & 2 \end{bmatrix} \\ x_2 - x_3 - x_4 = 2 & \end{array}$$

---

[7] When the system is solved in a computer, it is the augmented matrix that is used.

Again the new augmented matrix comes from the preceding one by multiplying the second row by $\frac{1}{3}$.

Finally, we eliminate $x_2$ from equation 1 by adding twice the second equation to the first. The result is the (equivalent) system

$$\begin{aligned} x_1 \qquad + x_3 - x_4 &= 1 \\ x_2 - x_3 - x_4 &= 2 \end{aligned} \qquad \begin{bmatrix} 1 & 0 & 1 & -1 & 1 \\ 0 & 1 & -1 & -1 & 2 \end{bmatrix}$$

This system is *easy* to solve. Indeed, if the numbers $x_3$ and $x_4$ are chosen arbitrarily, then $x_1$ and $x_2$ can be found so that the equations are satisfied. More precisely, if we set $x_3 = s$ and $x_4 = t$ where $s$ and $t$ are arbitrary parameters, the equations become $x_1 + s - t = 1$ and $x_2 - s - t = 2$, whence

$$x_1 = 1 + t - s \quad \text{and} \quad x_2 = 2 + t + s$$

This gives the solutions exhibited in Example 5 and, because all the systems in the series are equivalent (as will be proved in Theorem 1), we have obtained *all* the solutions to the original system.

Observe that at each stage in the above procedure a certain operation is performed on the system (and thus on the augmented matrix) to produce an equivalent system. The following operations, called **elementary operations**, can routinely be performed on systems to produce equivalent systems:

I     *Interchange two equations.*

II    *Multiply an equation by a nonzero number.*

III   *Add a multiple of one equation to a different equation.*

We only needed operations of type II and III in the computation in Example 6, but type I operations are sometimes useful as well. The next theorem is crucial for the method we are developing.

**THEOREM 1**

If an elementary operation is performed on a system of linear equations, the resulting system is equivalent to the original system.

**PROOF**

We prove it for type III operations; similar arguments work for operations of type I or II.

Suppose we modify the original system by replacing equation $p$ by a new equation, formed by adding a multiple of a different equation $q$ to equation $p$. Then any solution to the original system will satisfy this new equation (it satisfies both equations $p$ and $q$), and so will be a solution to the new system. On the other hand, the new system contains equation $q$ (because $p$ and $q$ are *different* equations). This means that the new system can be transformed *back* to the original system by *subtracting* the same multiple of equation $q$ from the new equation. Hence the same argument shows that every solution to the new system is a solution to the original system. Thus the two systems have the same set of solutions, so operations of type III produce equivalent systems.

Theorem 1 has profound consequences for linear algebra. In particular it allows us to use the procedure in Example 6 on *any* system of linear equations. The idea is to apply a series of elementary operations to the system with the goal of finding a system that is easy to solve. Since all the systems created in this way are equivalent (by Theorem 1), the solutions to the easy system are the solutions to the original system.

As in Example 6, elementary operations performed on a system of equations produce corresponding manipulations of the *rows* of the augmented matrix (regarded as row matrices). In hand calculations (and in computer programs) it is easier to manipulate rows than equations. For this reason we restate these elementary operations for rows:

    I     *Interchange two rows.*
    II    *Multiply a row by a nonzero number.*
    III   *Add a multiple of one row to a different row.*

These are called **elementary row operations**. Here is another example of our method where the entire calculation is carried out by manipulating the augmented matrix.

**Example 7**    Find all solutions to the following system of linear equations:

$$\begin{aligned} x_1 + x_2 - 3x_3 &= 3 \\ -2x_1 - x_2 \phantom{+ 3x_3} &= -4 \\ 4x_1 + 2x_2 + 3x_3 &= 7 \end{aligned}$$

**PROOF**    The augmented matrix of the original system is

$$\begin{bmatrix} 1 & 1 & -3 & 3 \\ -2 & -1 & 0 & -4 \\ 4 & 2 & 3 & 7 \end{bmatrix}.$$

We begin by using the 1 in the upper left corner to "clean up" column 1; that is, to obtain zeros in the other locations (this corresponds to eliminating $x_1$ from equations 2 and 3). More precisely, we add twice row 1 to row 2, and we subtract 4 times row 1 from row 3. The result is

$$\begin{bmatrix} 1 & 1 & -3 & 3 \\ 0 & 1 & -6 & 2 \\ 0 & -2 & 15 & -5 \end{bmatrix}.$$

This completes the work on column 1.

We now use the 1 in the second position of row 2 to clean up column 2; that is, to obtain zeros in the top and bottom positions (this corresponds to eliminating $x_2$ from equations 1 and 3). We do this by subtracting row 2 from row 1 and adding twice row 2 to row 3. This gives

$$\begin{bmatrix} 1 & 0 & 3 & 1 \\ 0 & 1 & -6 & 2 \\ 0 & 0 & 3 & -1 \end{bmatrix}.$$

Note that these row operations have not disturbed column 1 because the first entry of row 2 is a *zero*.

We next divide row 3 by 3 to obtain a 1 in the third position:

$$\begin{bmatrix} 1 & 0 & 3 & 1 \\ 0 & 1 & -6 & 2 \\ 0 & 0 & 1 & -\frac{1}{3} \end{bmatrix}.$$

Finally, we clean up column 3 by subtracting 3 times row 3 from row 1, and adding 6 times row 3 to row 2:

$$\begin{bmatrix} 1 & 0 & 0 & 2 \\ 0 & 1 & 0 & 0 \\ 0 & 0 & 1 & -\frac{1}{3} \end{bmatrix}.$$

The corresponding system of equations is

$$\begin{aligned} x_1 \qquad\qquad &= \ 2 \\ x_2 \qquad &= \ 0 \\ x_3 &= -\tfrac{1}{3} \end{aligned}$$

and the (unique) solution $X = \begin{bmatrix} 2 & 0 & -\tfrac{1}{3} \end{bmatrix}^T$ is apparent. Since this system of equations is equivalent to the original system, this is the solution to the original system.

---

### 1.2.3  Gaussian[8] Elimination

In the computations in Examples 6 and 7, elementary row operations (on the augmented matrix) led to matrices of the form

$$\begin{bmatrix} 1 & 0 & * & * & * \\ 0 & 1 & * & * & * \end{bmatrix} \quad \text{and} \quad \begin{bmatrix} 1 & 0 & 0 & * \\ 0 & 1 & 0 & * \\ 0 & 0 & 1 & * \end{bmatrix}$$

respectively, where each $*$ indicates a number. In both cases the solution was easily obtained from the corresponding system of equations. The matrices that arise in general are described as follows.

A matrix is said to be in **row-echelon form** (and will be called a **row-echelon matrix**) if the following conditions are satisfied:

1. All zero rows are at the bottom.
2. The first nonzero entry from the left in each nonzero row is a 1, called the **leading 1** for that row.
3. Each leading 1 is to the right of all leading 1's in the rows above it.

A row-echelon matrix is said to be in **reduced row-echelon form** if, in addition, it satisfies

4. Each leading 1 is the only nonzero entry in its column.

Thus the row-echelon matrices have a "staircase" form as indicated below (as before, the asterisks indicate arbitrary numbers):

$$\begin{bmatrix} 0 & 1 & * & * & * & * & * \\ 0 & 0 & 0 & 1 & * & * & * \\ 0 & 0 & 0 & 0 & 1 & * & * \\ 0 & 0 & 0 & 0 & 0 & 0 & 1 \\ 0 & 0 & 0 & 0 & 0 & 0 & 0 \end{bmatrix}.$$

The leading 1's proceed down and to the right through the matrix, and every entry below and to the left of a leading 1 is a zero. In a *reduced* row-echelon matrix the additional requirement is that all entries *above* each leading 1 must also be zero. Any row-echelon matrix can be carried to reduced form using a few more row operations (clean up above the leading 1's one after the other).

**Example 8**   The first matrix below is in row-echelon form, and the second matrix is the reduced row-echelon matrix to which it can be carried by row operations:

---

[8] Carl Friedrich Gauss (1777–1855) was one of the greatest mathematicians of all time. He made ground-breaking discoveries in every part of mathematics, and made important contributions to astronomy and physics.

$$\begin{bmatrix} 1 & * & * & * & * \\ 0 & 0 & 1 & * & * \\ 0 & 0 & 0 & 0 & 1 \end{bmatrix} \rightarrow \begin{bmatrix} 1 & * & 0 & * & 0 \\ 0 & 0 & 1 & * & 0 \\ 0 & 0 & 0 & 0 & 1 \end{bmatrix}$$

In general we use an arrow $\rightarrow$ to indicate that row operations have been performed.

Here is a procedure by which any matrix can be carried to row-echelon form (and hence to reduced form if desired) using nothing but elementary row operations.

**GAUSSIAN ALGORITHM**[9]    Every matrix can be carried to row-echelon form as follows:

**Step 1.** If the matrix consists entirely of zeros, stop: it is already in row-echelon form.

**Step 2.** Otherwise, find the first column from the left containing a nonzero entry $k$, and move the row containing that entry to the top of the matrix.

**Step 3.** Multiply the top row by $\frac{1}{k}$ to create the first leading 1.

**Step 4.** Make each entry below the leading 1 zero by subtracting multiples of its row from lower rows.

This completes the first row; all further row operations are carried out on the remaining rows.

**Step 5.** Repeat steps 1–4 on the matrix consisting of the remaining rows.

Note that the Gaussian algorithm is recursive in the sense that, after the first leading 1 has been created, the whole procedure is repeated on the remaining rows. This makes it easy to use on a computer. Note further that at step 4 we can also make every entry *above* the leading 1 zero. Then the algorithm carries the matrix to *reduced* row-echelon form (as in Examples 6 and 7). The reason for distinguishing two row-echelon forms will be discussed later.

The Gaussian algorithm certainly proves the following important theorem.

**THEOREM 2**    Every matrix can be carried to row-echelon form (reduced, if desired) by a sequence of elementary row operations.

**Example 9**    Find all solutions to the following system of linear equations:

$$\begin{aligned} x_1 - 2x_2 - \ \ x_3 + 3x_4 &= 1 \\ 2x_1 - 4x_2 + \ \ x_3 \qquad\quad &= 5 \\ x_1 - 2x_2 + 2x_3 - 3x_4 &= 4 \end{aligned}$$

**SOLUTION**    The augmented matrix is given below. The first leading 1 is in place so we clean up column 1:

$$\begin{bmatrix} 1 & -2 & -1 & 3 & 1 \\ 2 & -4 & 1 & 0 & 5 \\ 1 & -2 & 2 & -3 & 4 \end{bmatrix} \rightarrow \begin{bmatrix} 1 & -2 & -1 & 3 & 1 \\ 0 & 0 & 3 & -6 & 3 \\ 0 & 0 & 3 & -6 & 3 \end{bmatrix}$$

Now subtract the second row from the third, and then[10] multiply the second row by $\frac{1}{3}$, to get the matrix below (now in row-echelon form):

---

[9] While Gauss did use this procedure, the method has been attributed to the Chinese several centuries earlier.

[10] These steps are not strictly in the order specified by the algorithm. However, the point is to carry the matrix to reduced form using *some* sequence of row operations. The sequence in the algorithm will always work, but it may not be the most efficient.

$$\rightarrow \begin{bmatrix} 1 & -2 & -1 & 3 & 1 \\ 0 & 0 & 1 & -2 & 1 \\ 0 & 0 & 0 & 0 & 0 \end{bmatrix}$$

Next use the second leading 1 (in column 3) to clean up column 3 and so achieve reduced row-echelon form:

$$\rightarrow \begin{bmatrix} 1 & -2 & 0 & 1 & 2 \\ 0 & 0 & 1 & -2 & 1 \\ 0 & 0 & 0 & 0 & 0 \end{bmatrix}$$

This is as far as the Gaussian algorithm will take us. The corresponding system of equations is

$$\begin{aligned} x_1 - 2x_2 \quad + \quad x_4 &= 2 \\ x_3 - 2x_4 &= 1 \\ 0 &= 0 \end{aligned}$$

The leading 1's are in columns 1 and 3 here, and the corresponding variables $x_1$ and $x_3$ are called the leading variables. To solve the system the nonleading variables are assigned arbitrary values (called parameters), and then the two equations are used to determine the leading variables in terms of the parameters. More precisely, we set $x_2 = s$, and $x_4 = t$ where $s$ and $t$ are arbitrary parameters, so the equations become $x_1 - 2s + t = 2$ and $x_3 - 2t = 1$. Solving these gives $x_1 = 2 + 2s - t$ and $x_3 = 1 + 2t$. Hence the solutions are given by

$$X = [2 + 2s - t \quad s \quad 1 + 2t \quad t]^T = \begin{bmatrix} 2+2s-t \\ s \\ 1+2t \\ t \end{bmatrix}.$$

The solution $X = [2 + 2s - t \quad s \quad 1 + 2t \quad t]^T$ in Example 9 is called the **general solution** of the system because *every* solution has this form for some values of the parameters $s$ and $t$.

When the augmented matrix of a linear system has been carried to reduced row-echelon form, the variables corresponding to the leading 1's are called the **leading variables**. Then the method of solution in Example 9 provides a way of writing down the solutions of any linear system (assuming that there are solutions).

**Gaussian Elimination**  Assume that a system of linear equations has at least one solution. Then the general solution can be found in parametric form as follows:

**Step 1.**  Carry the augmented matrix of the system to reduced row-echelon form.
**Step 2.**  Assign the nonleading variables as parameters.
**Step 3.**  Use the equations corresponding to the reduced row-echelon form to solve for the leading variables in terms of the parameters.

This procedure solves any system of linear equations which *has* a solution. The following example shows how the method reveals that a system has *no* solution.

**Example 10**  In Example 4 it was shown directly that the system

$$\begin{aligned} x + y \quad\quad &= 1 \\ x \quad - z &= 2 \\ y + z &= 1 \end{aligned}$$

has no solution. The reduction of the augmented matrix to row-echelon form is as follows:

$$\begin{bmatrix} 1 & 1 & 0 & 1 \\ 1 & 0 & -1 & 2 \\ 0 & 1 & 1 & 1 \end{bmatrix} \rightarrow \begin{bmatrix} 1 & 1 & 0 & 1 \\ 0 & -1 & -1 & 1 \\ 0 & 1 & 1 & 1 \end{bmatrix}$$

$$\rightarrow \begin{bmatrix} 1 & 1 & 0 & 1 \\ 0 & 1 & 1 & -1 \\ 0 & 0 & 0 & 2 \end{bmatrix}$$

$$\rightarrow \begin{bmatrix} 1 & 1 & 0 & 1 \\ 0 & 1 & 1 & -1 \\ 0 & 0 & 0 & 1 \end{bmatrix}$$

This last matrix corresponds to a system in which the last equation is

$$0x + 0y + 0z = 1.$$

It is clear that *no* choice of $x$, $y$ and $z$ will satisfy *this* equation, so this last system (and hence the original system) has no solution. This is typical of what happens when there is no solution.

When using Gaussian elimination to solve a large system it is more efficient to carry the augmented matrix only to row-echelon form,[11] assign parameters to the nonleading variables, and then solve for the leading variables using **back-substitution**: Use the last equation to find the last leading variable in terms of the parameters, then substitute that value in the second last equation to solve for the second last leading variable, and so on. This method is more efficient than carrying the matrix directly to reduced row-echelon form, and follows from an operations count (see most numerical analysis books).

## 1.2.4  Rank of a Matrix

It will be proved in Section 1.6 (Theorem 4) that:

*The reduced row-echelon form of a matrix A is uniquely determined by A.*

That is, no matter which series of row operations is used to carry $A$ to a reduced row-echelon matrix, the result will always be the same. By contrast, the same matrix can be taken to *different* row-echelon matrices. For example, if $A = \begin{bmatrix} 1 & 3 & 4 \\ 2 & 7 & 9 \end{bmatrix}$, then

$$A \rightarrow \begin{bmatrix} 1 & 3 & 4 \\ 0 & 1 & 1 \end{bmatrix} \text{ and } A \rightarrow \begin{bmatrix} 1 & 3 & 4 \\ 0 & 1 & 1 \end{bmatrix} \rightarrow \begin{bmatrix} 1 & 2 & 3 \\ 0 & 1 & 1 \end{bmatrix}.$$

Hence $A$ can be carried to two different row-echelon matrices $\begin{bmatrix} 1 & 3 & 4 \\ 0 & 1 & 1 \end{bmatrix}$ and $\begin{bmatrix} 1 & 2 & 3 \\ 0 & 1 & 1 \end{bmatrix}$. However, we will show in Section 4.4 that:

*The number of leading 1's must be the same no matter how A is carried to row-echelon form.*

This number of leading 1's is called the **rank** of the matrix $A$ and is denoted *rankA*.

**Example 11**    Compute the rank of $A = \begin{bmatrix} 1 & 2 & -1 & 3 \\ 2 & 1 & 1 & 5 \\ -1 & 4 & -5 & -1 \end{bmatrix}$.

SOLUTION    The matrix $A$ is carried to row-echelon form as follows:

---

[11] This is often referred to as the Gaussian algorithm, particularly in numerical analysis.

$$A = \begin{bmatrix} 1 & 2 & -1 & 3 \\ 2 & 1 & 1 & 5 \\ -1 & 4 & -5 & -1 \end{bmatrix} \rightarrow \begin{bmatrix} 1 & 2 & -1 & 3 \\ 0 & -3 & 3 & -1 \\ 0 & 6 & -6 & 2 \end{bmatrix} \rightarrow \begin{bmatrix} 1 & 2 & -1 & 3 \\ 0 & 1 & -1 & \frac{1}{3} \\ 0 & 0 & 0 & 0 \end{bmatrix}$$

Because there are two leading 1's, we have $rank A = 2$.

The relationship between rank and systems of equations is given in the following theorem.

**THEOREM 3**  Suppose a system of $m$ equations in $n$ variables has at least one solution. If the rank of the augmented matrix is $r$, the set of solutions has exactly $(n - r)$ parameters.

**PROOF**  Carry the augmented matrix to a reduced row-echelon matrix $R$. Then $R$ has $r$ leading 1's (since the rank is $r$), so there are exactly $r$ leading variables. Hence there are $(n - r)$ nonleading variables, and each of these is assigned a parameter.

Theorem 3 has a surprising number of consequences and will be referred to several times below. We have seen examples of systems with no solution, one solution, or infinitely many solutions; the first application of Theorem 3 is to show that these are the only possibilities.

**THEOREM 4**  For any system of linear equations there are exactly three possibilities:

(1) No solution.
(2) A unique solution.
(3) Infinitely many solutions.

**PROOF**  If there is a solution, then either every variable is a leading variable (unique solution) or there is at least one nonleading variable (infinitely many solutions because a parameter is involved).

A system of linear equations is called **consistent** if it has at least one solution, and the system is called **inconsistent** if it has no solution. Thus a consistent system has either a unique solution or infinitely many solutions; it cannot have (say) exactly two solutions.

The truth of Theorem 4 can be seen graphically for a system of two equations in two variables $x$ and $y$. Recall that the graph of an equation $ax + by = c$ is a straight line if $a$ and $b$ are not both zero, and that $\begin{bmatrix} s \\ t \end{bmatrix}$ is a solution of the equation exactly when the point $P(s, t)$ with coordinates $(s, t)$ lies on the line. Now consider a system

$$\begin{aligned} a_1 x + b_1 y &= c_1 \\ a_2 x + b_2 y &= c_2 \end{aligned} \qquad (**)$$

The graphs of these equations are two straight lines, $L_1$ and $L_2$ assuming that $a_1$ and $b_1$ are not both zero, and that $a_2$ and $b_2$ are not both zero. Geometrically there are three possibilities for these two lines (illustrated in Figure 1.1):

(1) Lines are parallel and distinct | System $(**)$ has no solution because there is no point on both lines

(2) Lines are not parallel | System $(**)$ has a unique solution corresponding to the point of intersection of the lines

(3) Lines are identical | System $(**)$ has infinitely many solutions, one for each point on the (common) line

Clearly these three possibilities correspond to those in Theorem 4.

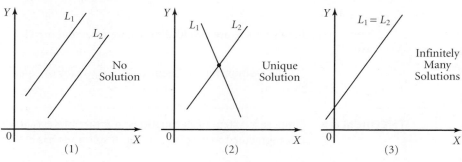

Figure 1.1

The graph of an equation $ax + by + cz = d$ is a plane in space if $a$, $b$, and $c$ are not all zero (this is discussed in detail in Section 3.3). Thus a system of 2 equations in 3 variables must either have no solutions (the planes are parallel) or have infinitely many solutions (the planes coincide or intersect in a line). A unique solution is not possible here. This illustrates Theorem 3 since there are $n = 3$ variables and the augmented matrix has rank $r \leq 2$ (because there are 2 equations) so $n - r \geq 3 - 2 = 1$ and these are parameters.

A similar graphical argument can be given that a system of 3 equations in 3 variables must have zero, one, or infinitely many solutions as in Theorem 4. However, this geometrical argument fails for systems with more than three variables, and we must rely on Theorem 4.

## Exercises 1.2

**1.** In each case show that the column matrix $X$ is a solution to the given system of linear equations; if it involves parameters $s$ and $t$, show that $X$ is a solution for all possible values of these parameters.

**a)**   $x_1 - 2x_2 + 3x_3 + x_4 = -3$
$2x_1 - x_2 + 3x_3 - x_4 = 0$
$X = [1 \ 2 \ 0 \ 0]^T$
$X = [1 \ 4 \ 1 \ 1]^T$
$X = [-s + t + 1 \ \ s \ + t + 2 \ \ s \ \ t]^T$

**b)**   $x_1 - 2x_2 + 3x_3 - 3x_4 = 2$
$2x_1 - x_2 + 3x_3 \quad\quad = 1$
$-x_1 - 4x_2 + 3x_3 - 9x_4 = 4$
$2x_1 - 7x_2 + 9x_3 - 12x_4 = 7$
$X = [-2 \ -2 \ 1 \ 1]^T$
$X = [-3 \ -1 \ 2 \ 1]^T$
$X = [-s - t \ \ -1 + s - 2t \ \ s \ \ t]^T$

**2.** Describe each of the following in your own words: A linear equation, a solution to a linear equation, a system of linear equations, and a solution to a system of linear equations.

**3.** In each case carry the given matrix to reduced row-echelon form. For a methodical approach, follow the Gaussian algorithm.

**a)** $\begin{bmatrix} 3 & -1 & 2 & 1 & 2 & 1 \\ -4 & 1 & -2 & 2 & 7 & 2 \\ 2 & -2 & 4 & 3 & 7 & 1 \\ 0 & 3 & -6 & 1 & 6 & 4 \end{bmatrix}$

**b)** $\begin{bmatrix} 1 & -1 & 3 & 1 & 3 & 2 & 1 \\ -1 & -2 & 6 & 1 & -5 & 0 & -1 \\ 2 & 3 & -9 & 2 & 4 & 1 & -1 \\ 1 & 1 & -3 & -1 & 3 & 0 & 1 \end{bmatrix}$

**4.** For each of the following systems of linear equations, form the augmented matrix, perform elementary row operations to transform it to reduced row-echelon form, and solve the given system this way. For a methodical approach to reducing to row-echelon form, follow the Gaussian algorithm.

**a)** $3x_1 - x_2 = 4$
$2x_1 - \frac{1}{2}x_2 = 1$

**b)** $2x_1 - 3x_2 = 4$
$x_1 - 3x_2 = 1$

**c)** $x_1 + x_2 - x_3 = 1$
$3x_1 - x_2 + x_3 = 0$
$x_1 - 3x_2 + 3x_3 = -2$

**d)** $2x_1 + 2x_2 - 3x_3 = 1$
$x_1 \quad\quad + x_3 = 5$
$3x_1 + 4x_2 - 7x_3 = -3$

**e)** $x_1 - 2x_2 + 2x_3 = 4$
$-2x_1 + x_2 + x_3 = 1$
$x_1 - 5x_2 + 7x_3 = -1$

**f)** $4x_1 + x_2 - 8x_3 = 1$
$3x_1 - 2x_2 + 3x_3 = 5$
$-x_1 + 8x_2 - 25x_3 = -3$

**g)** 
$$x_1 - 3x_2 + x_3 + x_4 - x_5 = 8$$
$$-2x_1 + 6x_2 + x_3 - 2x_4 - 4x_5 = -1$$
$$3x_1 - 9x_2 + 8x_3 + 4x_4 - 13x_5 = 49$$

**h)** 
$$x_1 - 2x_2 + x_3 + 3x_4 - x_5 = 1$$
$$-3x_1 + 6x_2 - 4x_3 - 9x_4 + 3x_5 = -1$$
$$-x_1 + 2x_2 - 2x_3 - 4x_4 - 3x_5 = 3$$
$$x_1 - 2x_2 + 2x_3 + 2x_4 - 5x_5 = 1$$

**5.** In your own words, can you describe an algorithm that will solve a system of linear equations for you?

**6.** In each of the systems of linear equations given in Exercise 4, calculate the rank of the corresponding augmented matrix, and find a connection with the number of variables, the number of solutions, and the number of parameters in the solution.

**7.** If a system of 5 equations in 7 variables has a solution, explain why there is more than one solution.

**8.** Consider the system $ax = b$ of one equation in one variable. Give conditions on $a$ and $b$ such that the system has no solution, a unique solution, and infinitely many solutions.

**9.** In each case find (if possible) conditions on the numbers $a$, $b$, and $c$ that the given system has no solution, a unique solution, or infinitely many solutions.

**a)** 
$$2x_1 - 3x_2 - 3x_3 = a$$
$$-x_1 + x_2 + 2x_3 = b$$
$$x_1 - 3x_2 = c$$

**b)** 
$$x_1 - 2x_2 + 2x_3 = a$$
$$-2x_1 + x_2 + x_3 = b$$
$$x_1 - 5x_2 + 7x_3 = c$$

**c)** 
$$x_1 + ax_2 = 1$$
$$bx_1 + 2x_2 = 5$$

**d)** 
$$ax_1 + x_2 = -1$$
$$2x_1 + x_2 = b$$

**10.** Let $A = \begin{bmatrix} 1 & -2 & 0 & 3 \\ -4 & 1 & 5 & -2 \\ -1 & -5 & 5 & 7 \end{bmatrix}$. Show that $A$ and $A^T$ have the same rank. [Remark: We will see in Section 4.4 that this is true for *every* matrix $A$.]

**11.** In your own words, can you describe an algorithm that will compute the rank of a matrix?

**12.** Four one-way streets feed into a traffic circle as in the diagram, where the traffic flows are measured in cars per minute. In order to design the roads, the engineers must determine the flows $f_1$, $f_2$, $f_3$, and $f_4$ in the circle. Use the fact that the traffic flow into any intersection must equal the traffic flow out of that intersection to obtain four equations relating these flows. Solve the equations and express $f_1$, $f_2$, and $f_3$ in terms of $f_4$. Which part of the circle will carry the most traffic?

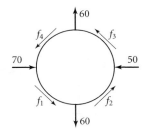

**13.** A man must take 5 units of vitamin A, 13 units of vitamin B, and 23 units of vitamin C each day. Three brands of vitamin pills are available and the number of units of each vitamin per pill are given in the table.

|  | Vitamin A | Vitamin B | Vitamin C |
|---|---|---|---|
| Brand 1 | 1 | 2 | 4 |
| Brand 2 | 1 | 1 | 3 |
| Brand 3 | 0 | 1 | 1 |

**a)** Find all combinations of pills that provide the exact daily requirement (no partial pills).

**b)** If brands 1, 2, and 3 cost $0.90, $0.60, and $1.50 per pill respectively, find the least expensive treatment.

**14.** In each case either show that the statement is true or give an example showing that it is false. Assume that a system of equations is given with augmented matrix $A$, coefficient matrix $C$, and reduced row-echelon form $R$.

**a)** $A$ and $R$ are the same size.

**b)** If $R$ has a row of zeros, there are infinitely many solutions.

**c)** If there is more than one solution, $R$ must have a row of zeros.

**d)** If there exists a solution, there are infinitely many solutions.

**e)** If there are more variables than equations, there are infinitely many solutions.

**f)** If every row of $R$ has a leading 1, the system has at least one solution.

**g)** If the system has a solution, $rank(A) = rank(C)$.

**h)** $rank(C) \leq rank(A)$.

**i)** $rank(A) \leq 1 + rank(C)$.

**j)** If $A$ is $m \times n$ and $rankA = m$, the system is consistent.

## 1.3   HOMOGENEOUS SYSTEMS

In this section, we concentrate on a particular class of systems of linear equations, that is, those systems where the constant term in each equation is 0.

### 1.3.1   Homogeneous Systems

A system of linear equations is called **homogeneous** if all the constant terms are zero. Thus a typical homogeneous linear equation in the $n$ variables $x_1, x_2, \cdots, x_n$ has the form

$$a_1x_1 + a_2x_2 + \cdots + a_nx_n = 0.$$

Because the constants are all zero, any homogeneous system always has the **trivial** solution

$$x_1 = 0, \; x_2 = 0, \; \cdots, \; x_n = 0$$

in which every variable is zero. Many practical problems come down to discovering whether or not some homogeneous system has a **nontrivial** solution, that is, a solution where at least one of the variables is nonzero. The next theorem gives an important situation where this *must* happen.

**THEOREM 1**   If a homogeneous system of linear equations has more variables than equations, then it has nontrivial solutions.

**PROOF**   Suppose there are $m$ equations in $n$ variables, so our assumption is that $n > m$. If $r$ is the rank of the augmented matrix, then $r \leq m$ because the number $r$ of leading 1's cannot exceed the number $m$ of equations. Hence $r \leq m < n$, whence $r < n$. By Theorem 3 §1.2, this means that the number $n - r$ of parameters is not zero, so there are (infinitely many) nontrivial solutions.

The existence of a nontrivial solution is often the desired outcome for a homogeneous system. The following example provides an illustration of how Theorem 1 can be used in geometry.

**Example 1**   The graph of an equation $ax^2 + bxy + cy^2 + dx + ey + f = 0$ is called a **conic** if $a$, $b$, and $c$ are not all zero. (Circles, ellipses, hyperbolas, and parabolas are all examples of conics.) Show that there is at least one conic through any five points in the plane that are not all on a line.

**SOLUTION**   Suppose that the coordinates of the five points are $(p_1, q_1)$, $(p_2, q_2)$, $(p_3, q_3)$, $(p_4, q_4)$, and $(p_5, q_5)$. The graph of the equation $ax^2 + bxy + cy^2 + dx + ey + f = 0$ passes through the point $(p_i, q_i)$ if

$$ap_i^2 + bp_iq_i + cq_i^2 + dp_i + eq_i + f = 0$$

Since there are five points, this gives five homogeneous linear equations in the six variables $a$, $b$, $c$, $d$, $e$, and $f$. Hence there is a nontrivial solution by Theorem 1. If $a = b = c = 0$ in this solution, then the five points all lie on the line with equation $dx + ey + f = 0$, contrary to our assumption. Hence one of $a$, $b$, and $c$ is nonzero and we have a conic.

## 1.3.2  Basic Solutions

Of course, Gaussian elimination also works for homogeneous systems. In fact, it provides a way to write the solutions in a convenient form which will be needed later. This is illustrated in the following example.

**Example 2**  Solve the following homogeneous system

$$x_1 - 2x_2 + x_3 + x_4 = 0$$
$$-x_1 + 2x_2 \quad\quad + x_4 = 0$$
$$2x_1 - 4x_2 + x_3 \quad\quad = 0$$

and express the solutions as sums of scalar multiples of specific solutions.

**SOLUTION**  The augmented matrix is reduced as follows:

$$\begin{bmatrix} 1 & -2 & 1 & 1 & 0 \\ -1 & 2 & 0 & 1 & 0 \\ 2 & -4 & 1 & 0 & 0 \end{bmatrix} \rightarrow \begin{bmatrix} 1 & -2 & 1 & 1 & 0 \\ 0 & 0 & 1 & 2 & 0 \\ 0 & 0 & -1 & -2 & 0 \end{bmatrix}$$
$$\rightarrow \begin{bmatrix} 1 & -2 & 0 & -1 & 0 \\ 0 & 0 & 1 & 2 & 0 \\ 0 & 0 & 0 & 0 & 0 \end{bmatrix}$$

Hence the leading variables are $x_1$ and $x_3$, and the nonleading variables $x_2$ and $x_4$ become parameters: $x_2 = s$ and $x_4 = t$. Then the equations in the final system determine the leading variables in terms of the parameters:

$$x_1 = 2s + t \quad \text{and} \quad x_3 = -2t$$

This means that the general solution is $X = [2s + t \quad s \quad -2t \quad t]^T$. The new idea here is to separate this into parts involving only $s$ or $t$:

$$X = \begin{bmatrix} 2s+t \\ s \\ -2t \\ t \end{bmatrix} = \begin{bmatrix} 2s \\ s \\ 0 \\ 0 \end{bmatrix} + \begin{bmatrix} t \\ 0 \\ -2t \\ t \end{bmatrix} = s\begin{bmatrix} 2 \\ 1 \\ 0 \\ 0 \end{bmatrix} + t\begin{bmatrix} 1 \\ 0 \\ -2 \\ 1 \end{bmatrix}$$

Hence $X_1 = \begin{bmatrix} 2 \\ 1 \\ 0 \\ 0 \end{bmatrix}$ and $X_2 = \begin{bmatrix} 1 \\ 0 \\ -2 \\ 1 \end{bmatrix}$ are specific solutions, and the general solution $X$ has the form $X = sX_1 + tX_2$.

The specific solutions $X_1$ and $X_2$ in Example 2 are called the basic solutions of the homogeneous system. They have the property that every solution $X$ has the form

$$X = sX_1 + tX_2 \quad \text{where } s \text{ and } t \text{ are arbitrary numbers.}$$

This happens for every homogeneous system.

To describe the general situation, the following terminology is useful: If $X_1, X_2, \cdots, X_k$ are columns, an expression of the form

$$s_1X_1 + s_2X_2 + \cdots + s_nX_k \quad \text{where } s_1, s_2, \cdots, s_k \text{ are arbitrary numbers}$$

is called a **linear combination** of the columns $X_1, X_2, \cdots, X_k$. Given *any* homogeneous system of linear equations, a computation like that in Example 2 expresses every solution to the system as a linear combination of certain particular solutions. These particular solutions are called the **basic solutions** produced by the Gaussian algorithm. (Of course, the system may have only the trivial solution; in this case we say that the system has **no basic solutions**.)

**THEOREM 2** Suppose a system of homogeneous linear equations in $n$ variables is given. Assume that the rank of the augmented matrix is $r$. Then:

    **(1)** The Gaussian algorithm produces exactly $n - r$ basic solutions.
    **(2)** Every solution is a linear combination of these basic solutions.

**PROOF** We already have observed that every solution is a linear combination of the basic ones produced by the Gaussian algorithm. By Theorem 3 §1.2 there are exactly $n - r$ parameters in the general solution, and so there are exactly $n - r$ basic solutions (one for each parameter).

**Example 3** Find basic solutions for the following homogeneous system, and illustrate Theorem 2.

$$\begin{aligned}
x_1 - 2x_2 + 4x_3 - x_4 + \phantom{2x_5 +} 5x_6 &= 0 \\
-2x_1 + 4x_2 - 7x_3 + x_4 + 2x_5 - 8x_6 &= 0 \\
3x_1 - 6x_2 + 12x_3 - 3x_4 + x_5 + 15x_6 &= 0 \\
2x_1 - 4x_2 + 9x_3 - 3x_4 + 3x_5 + 12x_6 &= 0
\end{aligned}$$

**SOLUTION** The reduction of the augmented matrix to reduced form is as follows:

$$\begin{bmatrix}
1 & -2 & 4 & -1 & 0 & 5 & 0 \\
-2 & 4 & -7 & 1 & 2 & -8 & 0 \\
3 & -6 & 12 & -3 & 1 & 15 & 0 \\
2 & -4 & 9 & -3 & 3 & 12 & 0
\end{bmatrix}
\rightarrow
\begin{bmatrix}
1 & -2 & 4 & -1 & 0 & 5 & 0 \\
0 & 0 & 1 & -1 & 2 & 2 & 0 \\
0 & 0 & 0 & 0 & 1 & 0 & 0 \\
0 & 0 & 1 & -1 & 3 & 2 & 0
\end{bmatrix}$$

$$\rightarrow
\begin{bmatrix}
1 & -2 & 0 & 3 & 0 & -3 & 0 \\
0 & 0 & 1 & -1 & 0 & 2 & 0 \\
0 & 0 & 0 & 0 & 1 & 0 & 0 \\
0 & 0 & 0 & 0 & 0 & 0 & 0
\end{bmatrix}$$

Hence the nonleading variables are assigned as parameters: $x_2 = s$, $x_4 = t$, and $x_6 = u$. Then the equations give the leading variables: $x_1 = 2s - 3t + 3u$, $x_3 = t - 2u$, and $x_5 = 0$. Thus the general solution $X$ is

$$X = \begin{bmatrix} x_1 \\ x_2 \\ x_3 \\ x_4 \\ x_5 \\ x_6 \end{bmatrix} = \begin{bmatrix} 2s - 3t + 3u \\ s \\ t - 2u \\ t \\ 0 \\ u \end{bmatrix} = s\begin{bmatrix} 2 \\ 1 \\ 0 \\ 0 \\ 0 \\ 0 \end{bmatrix} + t\begin{bmatrix} -3 \\ 0 \\ 1 \\ 1 \\ 0 \\ 0 \end{bmatrix} + u\begin{bmatrix} 3 \\ 0 \\ -2 \\ 0 \\ 0 \\ 1 \end{bmatrix}$$

so the basic solutions are $X_1 = \begin{bmatrix} 2 \\ 1 \\ 0 \\ 0 \\ 0 \\ 0 \end{bmatrix}$, $X_2 = \begin{bmatrix} -3 \\ 0 \\ 1 \\ 1 \\ 0 \\ 0 \end{bmatrix}$ and $X_3 = \begin{bmatrix} 3 \\ 0 \\ -2 \\ 0 \\ 0 \\ 1 \end{bmatrix}$.

Thus there are 3 basic solutions, which agrees with Theorem 2 because the rank of the augmented matrix is $r = 3$ in this case and there are $n = 6$ variables.

The basic solutions of a homogeneous system of linear equations are useful because they give an easy description of *all* solutions: Just take all linear combinations. However, the basic solutions have other uses as well, as we shall see in Section 2.3 when we use them to study diagonalization, one of the most useful techniques in linear algebra. We conclude with an example which applies homogeneous equations to balancing chemical reactions.

**Example 4**  A chemical reaction is a rearrangement of molecules to form new chemical units, where reactants are transformed into products. For example, $2Na + Cl_2 \to 2NaCl$ represents the reaction of two sodium molecules and one chlorine molecule to give two molecules of sodium chloride (table salt). This reaction is balanced in the sense that the number of atoms of each element on one side of the reaction is the same as the number on the other side. As another illustration, consider burning methane:

$$\text{Methane } (CH_4) + \text{Oxygen } (O_2) \to \text{Carbon Dioxide } (CO_2) + \text{Water } (H_2O).$$

To balance this reaction, let $x$, $y$, $z$, and $w$ denote the number of molecules of methane, oxygen, carbon dioxide, and water, respectively, that must be present. Then the reaction can be represented by

$$xCH_4 + yO_2 \to zCO_2 + wH_2O.$$

The number of atoms of each element on each side of the reaction must be the same, giving the equations $x = z$ (to balance C), $4x = 2w$ (to balance H) and $2y = 2z + w$ (to balance O). These equations can be written as a homogeneous linear system

$$
\begin{aligned}
x \quad\quad - z \quad\quad &= 0 \\
4x \quad\quad\quad - 2w &= 0 \\
2y - 2z - \ w &= 0
\end{aligned}
$$

and the general solution is $X = [t \ \ 2t \ \ t \ \ 2t]^T$ where $t$ is arbitrary. Since $x$, $y$, $z$, and $w$ must be positive integers, taking $t = 1$ gives one solution $x = z = 1$ and $y = w = 2$. Hence the balanced reaction is

$$CH_4 + 2O_2 \to CO_2 + 2H_2O.$$

## Exercises 1.3

**1.** In each case, consider the system of homogeneous equations with the given matrix as coefficient matrix, find basic solutions to the system, and express the general solution as a linear combination of these basic solutions.

**a)** $\begin{bmatrix} 1 & 2 & 3 \\ -1 & 3 & 0 \\ -1 & 8 & 3 \end{bmatrix}$

**b)** $\begin{bmatrix} 1 & 2 & 3 \\ -1 & 3 & 0 \\ 1 & 8 & 3 \end{bmatrix}$

**c)** $\begin{bmatrix} 1 & 3 & -2 & -1 \\ -2 & 1 & 0 & 4 \\ 1 & 10 & -6 & 1 \end{bmatrix}$

**d)** $\begin{bmatrix} 1 & -2 & 0 & 3 \\ -3 & 6 & 1 & -4 \\ 2 & -4 & -1 & 1 \end{bmatrix}$

**e)** $\begin{bmatrix} 1 & 2 & 0 \\ 2 & 3 & -1 \\ 3 & 4 & -2 \\ 1 & 3 & 1 \\ 6 & 9 & -3 \end{bmatrix}$

**f)** $\begin{bmatrix} 2 & -1 & 6 & 0 & -3 & 7 \\ -3 & 1 & 3 & -5 & 1 & 4 \end{bmatrix}$

**g)** $\begin{bmatrix} 1 & 2 & 1 & -1 & 3 \\ 1 & 2 & 2 & 1 & 2 \\ 2 & 4 & 2 & -1 & 7 \end{bmatrix}$

**h)** $\begin{bmatrix} 1 & 1 & -2 & 3 & 2 \\ 2 & -1 & 3 & 4 & 1 \\ -1 & -2 & 3 & 0 & 1 \\ 3 & 0 & 1 & 7 & 3 \end{bmatrix}$

**2.** Find a balanced equation for the following chemical reactions:

**a)** Ammonia $(NH_3)$ + Copper Oxide $(CuO)$
    $\to$ Nitrogen $(N_2)$ + Copper $(Cu)$ + Water $(H_2O)$.

**b)** Octane $(C_8H_{18})$ + Oxygen $(O_2)$
    $\to$ Carbon Dioxide $(CO_2)$ + Water $(H_2O)$.

**3.** Explain the difference between general and homogeneous systems of linear equations, and what the implications are for the solutions.

**4.** Suppose a homogeneous system has 4 equations and 6 variables, and let $A$ denote the augmented matrix.

**a)** Can the system have a unique solution? No solution? Support your answers.

**b)** How many parameters are possible? Support your answer.

**c)** How many parameters are possible if a row of $A$ is a multiple of a different row? Support your answer.

**d)** How many parameters are possible if $rankA = 4$? Support your answer.

**e)** How many parameters are possible if $rankA = 2$? Support your answer.

**5.** Show that the converse of Theorem 1 is not true. That is, show that if a homogeneous system has a nontrivial solution, it may *not* have more variables than equations.

**6.** If a system has more equations than variables, can it have a unique solution? Justify your answer.

7. In each case either show that the statement is true or give an example showing that it is false. Assume that a system of equations is given with augmented matrix $A$ and reduced row-echelon form $R$.
   a) If the system is homogeneous, every solution is trivial.
   b) If the system has a nontrivial solution, it cannot be homogeneous.
   c) If the system has a trivial solution, it must be homogeneous.
   d) If the system is homogeneous and has a nontrivial solution, $R$ has a row of zeros.
   e) If the system is homogeneous and has a nontrivial solution, then it has infinitely many nontrivial solutions.
   f) If the system is homogeneous and $R$ has a row of zeros, there are nontrivial solutions.
   g) If $A$ is $m \times n$ and $rankA = m$, the system has only the trivial solution.

8. For a homogeneous system with coefficient matrix $C$ and augmented matrix $A$, show that $rankA = rankC$.

## 1.4   MATRIX MULTIPLICATION

In addition to the operations in Section 1.1, there is another operation by which certain matrices can be multiplied together to yield another matrix. This is useful for systems of linear equations, and turns out to have many other applications. We begin with an example of how matrix multiplication arises in a commercial context.

### 1.4.1   Matrix Multiplication

**Example 1**   A store sells two brands of toasters, brand $X$ and brand $Y$. The manager has the following two data matrices available, the first giving sales volumes for three months, and the second giving the retail price and the dealer cost (in dollars):

$$\begin{array}{c} \\ \text{Jan} \\ \text{Feb} \\ \text{Mar} \end{array} \begin{array}{c} X \quad Y \\ \begin{bmatrix} 11 & 9 \\ 10 & 7 \\ 14 & 8 \end{bmatrix} \end{array} \text{ and } \begin{array}{c} \\ X \\ Y \end{array} \begin{array}{c} \text{Retail} \quad \text{Cost} \\ \begin{bmatrix} 29 & 22 \\ 31 & 26 \end{bmatrix} \end{array}$$

The manager wants a systematic way to calculate the new matrix

$$\begin{array}{c} \\ \text{Jan} \\ \text{Feb} \\ \text{Mar} \end{array} \begin{array}{c} \text{Retail} \quad \text{Cost} \\ \begin{bmatrix} 598 & 476 \\ 507 & 402 \\ 654 & 516 \end{bmatrix} \end{array}$$

of total sales and costs for the three months. For example, for January, the total retail sales were

$$598 = 11 \cdot 29 + 9 \cdot 31$$

because there were 11 brand-$X$ toasters sold at \$29 each, and 9 brand-$Y$ toasters sold at \$31 each. This gives the $(1, \ 1)$-entry 598 of the new matrix. Observe that this entry can be calculated directly from the first two matrices by going along the first row $[11 \ \ 9]$ of the left matrix and the first column $\begin{bmatrix} 29 \\ 31 \end{bmatrix}$ of the right matrix, multiplying corresponding entries, and adding the results. This is called taking the dot product of this row and column. Similarly, the $(3, \ 2)$-entry, 516, of the new matrix is obtained by taking the dot product of row 3 of the left matrix and column 2 of the right matrix:

$$516 = 14 \cdot 22 + 8 \cdot 26$$

Every entry in the new matrix is calculated in the same way: The $(i, \ j)$-entry of the new matrix is the dot product of row $i$ of the left matrix and column $j$ of the right matrix.

The calculation in Example 1 is a special case of general matrix multiplication. To describe it we use the following terminology: If $R = [r_1 \ r_2 \ \cdots \ r_n]$ is a row matrix and $C = [c_1 \ c_2 \ \cdots \ c_n]^T$ is a column matrix, each with $n$ entries, we define their **dot product** to be the number

$$r_1 c_1 + r_2 c_2 + \cdots + r_n c_n.$$

In other words, the dot product of $R$ and $C$ is the number formed by multiplying corresponding entries and adding the results. As in Example 1, this leads to a way to multiply two matrices:

> The **product** *of an* $m \times n$ *matrix* $A$ *and an* $n \times p$ *matrix* $B$ *is the* $m \times p$ *matrix* $AB$ *whose* $(i, j)$*-entry is the dot product of row* $i$ *of* $A$ *and column* $j$ *of* $B$.

Thus computing the $(i, j)$-entry of the product $AB$ means going across row $i$ of $A$ and down column $j$ of $B$, multiplying corresponding entries, and adding the results. This is illustrated in Figure 1.2.

Figure 1.2

**Example 2**  Here are some typical matrix products.

$$\begin{bmatrix} 5 & 1 \\ 7 & 6 \\ -4 & 9 \end{bmatrix} \begin{bmatrix} 2 & -1 \\ 0 & 3 \end{bmatrix} = \begin{bmatrix} 5 \cdot 2 + 1 \cdot 0 & 5(-1) + 1 \cdot 3 \\ 7 \cdot 2 + 6 \cdot 0 & 7(-1) + 6 \cdot 3 \\ (-4)2 + 9 \cdot 0 & (-4)(-1) + 9 \cdot 3 \end{bmatrix} = \begin{bmatrix} 10 & -2 \\ 14 & 11 \\ -8 & 31 \end{bmatrix}$$

$$\begin{bmatrix} p & q & r \end{bmatrix} \begin{bmatrix} 5 & 7 \\ -3 & 9 \\ 2 & -5 \end{bmatrix} = \begin{bmatrix} 5p - 3q + 2r & 7p + 9q - 5r \end{bmatrix}$$

$$\begin{bmatrix} 2 & 7 & -5 \\ 3 & 0 & 1 \end{bmatrix} \begin{bmatrix} x \\ y \\ z \end{bmatrix} = \begin{bmatrix} 2x + 7y - 5z \\ 3x + z \end{bmatrix}$$

Note that forming the product $AB$ is not always possible; it requires that the rows of $A$ have the same length as the columns of $B$ (so the dot products can be computed). The following rule is a useful way to remember when $AB$ can be formed and what the size of $AB$ is when $A$ is $m \times n$ and $B$ is $n' \times p$:

$$\begin{array}{cc} A & B \\ m \times \boxed{n \qquad n'} \times p \end{array}$$

> *The product* $AB$ *is defined if and only if* $n = n'$, *and in this case the matrix* $AB$ *has size* $m \times p$.

When these conditions are met we say that $A$ and $B$ are **compatible for multiplication**. Whenever we write a matrix product $AB$, it is tacitly assumed that this is the case.

The next example shows that matrix multiplication has properties not encountered in numerical arithmetic. If $A$ is a square matrix, we write the **powers** of $A$ as for numbers:

$$A^2 = AA$$
$$A^3 = AAA \quad \text{etc.}$$

**Example 3**　Show that the following can happen for $2 \times 2$ matrices $A$ and $B$.

　　(1) $A^2 = 0$ even though $A \neq 0$.
　　(2) $AB \neq BA$.

**SOLUTION**　Let $A = \begin{bmatrix} 0 & 1 \\ 0 & 0 \end{bmatrix}$ and $B = \begin{bmatrix} 1 & 0 \\ 0 & 0 \end{bmatrix}$. Then $A^2 = \begin{bmatrix} 0 & 1 \\ 0 & 0 \end{bmatrix}\begin{bmatrix} 0 & 1 \\ 0 & 0 \end{bmatrix} = \begin{bmatrix} 0 & 0 \\ 0 & 0 \end{bmatrix} = 0$, even though

$A \neq 0$. Moreover $AB = \begin{bmatrix} 0 & 1 \\ 0 & 0 \end{bmatrix}\begin{bmatrix} 1 & 0 \\ 0 & 0 \end{bmatrix} = 0$ while $BA = \begin{bmatrix} 1 & 0 \\ 0 & 0 \end{bmatrix}\begin{bmatrix} 0 & 1 \\ 0 & 0 \end{bmatrix} = A$.

Hence $AB \neq BA$.

Two matrices $A$ and $B$ are said to **commute** if $AB = BA$. Thus Example 3 exhibits two matrices that do *not* commute. In fact commuting is relatively rare: Given two matrices, the chances are they will not commute.

**Example 4**　If $B = \begin{bmatrix} 1 & 1 \\ 0 & 0 \end{bmatrix}$, find all $2 \times 2$ matrices $A$ that commute with $B$.

**SOLUTION**　The condition on $A$ is that $AB = BA$. If we write $A = \begin{bmatrix} x & y \\ z & w \end{bmatrix}$ in terms of its entries, this condition becomes

$$\begin{bmatrix} x & y \\ z & w \end{bmatrix}\begin{bmatrix} 1 & 1 \\ 0 & 0 \end{bmatrix} = \begin{bmatrix} 1 & 1 \\ 0 & 0 \end{bmatrix}\begin{bmatrix} x & y \\ z & w \end{bmatrix},$$

that is

$$\begin{bmatrix} x & x \\ z & z \end{bmatrix} = \begin{bmatrix} x+z & y+w \\ 0 & 0 \end{bmatrix}.$$

Equating entries gives $z = 0$ and $x = y + w$, so $A$ has the form $A = \begin{bmatrix} y+w & y \\ 0 & w \end{bmatrix}$ where $y$ and $w$ are arbitrary numbers.

## 1.4.2　Properties of Matrix Multiplication

Example 3 shows that matrix arithmetic differs in basic ways from ordinary numerical arithmetic. On the other hand, many similarities do occur (as seen for matrix addition in Theorem 1, §1.1).

The number 1 has the property that $1 \cdot a = a = a \cdot 1$ for every number $a$. An analogous role is played for matrix multiplication by square matrices of the following type:

$$I_2 = \begin{bmatrix} 1 & 0 \\ 0 & 1 \end{bmatrix} \quad I_3 = \begin{bmatrix} 1 & 0 & 0 \\ 0 & 1 & 0 \\ 0 & 0 & 1 \end{bmatrix} \quad I_4 = \begin{bmatrix} 1 & 0 & 0 & 0 \\ 0 & 1 & 0 & 0 \\ 0 & 0 & 1 & 0 \\ 0 & 0 & 0 & 1 \end{bmatrix} \quad \text{etc.}$$

In general the $n \times n$ **identity matrix** $I_n$ is the square $n \times n$ matrix with 1's on the main diagonal and zeros elsewhere. It is often written simply as $I$ when no confusion can result. It is routine to verify that identity matrices enjoy the following properties

$$IA = A \quad \text{and} \quad BI = B$$

whenever the products are defined. These are the first of the following important facts about matrix multiplication.

**THEOREM** ① Let $c$ be a scalar and let $A$, $B$, and $C$ denote matrices of sizes such that the products below can be performed.

    **(1)** $IA = A$ and $BI = B$.
    **(2)** $A(BC) = (AB)C$.
    **(3)** $A(B + C) = AB + AC$    and    $A(B - C) = AB - AC$.
    **(4)** $(B + C)A = BA + CA$    and    $(B - C)A = BA - CA$.
    **(5)** $c(AB) = (cA)B = A(cB)$.
    **(6)** $(AB)^T = B^T A^T$.

The proofs of these properties are routine and are given at the end of this section.

The property that $A(BC) = (AB)C$ is called the **associative law** for matrix multiplication. It asserts that the product of three matrices $A$, $B$, and $C$ (in that order) is the same no matter how it is formed and so is denoted simply as $ABC$. Similarly, the product $ABCD$ of four matrices can be formed in several ways (examples: $A[B(CD)]$, $(AB)(CD)$, $[A(BC)]D$, etc.). However the associative law shows that they are all equal, and so are written simply as $ABCD$. Similar remarks apply to products of five or more matrices.

However, a note of caution is needed here. The fact that $AB$ and $BA$ may not be equal (see Example 3) means that the *order* of the factors in a matrix product *is* important. For example, the products $ADCB$ and $ABCD$ may *not* be equal.

**Warning** If the order of the factors in a product of matrices is changed, the product matrix may change (or may not exist.)

Ignoring this is the source of many errors by linear algebra students! For example, when multiplying a matrix equation
$$B = C$$
by a matrix $A$ care must be taken to multiply $B$ and $C$ on the *same side* by $A$. Hence we speak of *left-multiplying* by $A$ to get $AB = AC$, and *right-multiplying* by $A$ to get $BA = CA$.[12] The next example illustrates how this is used.

**Example 5** If $A$, $B$, and $C$ are matrices such that $AB = I$ and $CA = I$, show that $B = C$.

**SOLUTION** Left multiply the given equation $I = AB$ by $C$, and use (1) and (2) of Theorem 1, to get
$$C = CI = C(AB) = (CA)B = IB = B.$$

Properties (3) and (4) of Theorem 1 are called the **distributive laws**. They extend to sums of more than two terms, and they combine with property (5) to allow algebraic manipulations very much like those in numerical arithmetic. For example:
$$A(2B - C + 3D) = 2AB - AC + 3AD$$
$$(A + 2C - 3E + 5D)B = AB + 2CB - 3EB + 5DB$$
Note that the warning is still in effect: the order of each of the products here must not be changed.

---

[12] It is vital to realize that $AB = CA$ *can happen* even if $B = C$.

**Example 6**    Simplify the expression
$$A(BC - 2CD) + A(2C - B)D + (BA - AB)C.$$

SOLUTION    We use the properties in Theorem 1 and the extended distributive laws
$$A(BC - 2CD) + A(2C - B)D + (BA - AB)C$$
$$= ABC - 2ACD + 2ACD - ABD + BAC - ABC$$
$$= BAC - ABD.$$

Recall that two matrices $A$ and $B$ are said to commute if $AB = BA$. Many formulas that are valid for numbers will fail for noncommuting matrices. For example, the well-known "difference of squares" formula $a^2 - b^2 = (a - b)(a + b)$ of arithmetic may not hold for matrices; the whole thing depends upon whether the matrices commute.

**Example 7**    Let $A$ and $B$ denote $n \times n$ matrices.

    **(a)** If $AB = BA$, show that $A^2 - B^2 = (A - B)(A + B)$.
    **(b)** If $A^2 - B^2 = (A - B)(A + B)$, show that necessarily $AB = BA$

SOLUTION    The distributive laws give
$$(A - B)(A + B) = A(A + B) - B(A + B) = A^2 + AB - BA - B^2. \qquad (*)$$
This holds whether or not $AB = BA$, and so is helpful in both (a) and (b).

    **(a)** If $AB = BA$, the right side of equation $(*)$ equals $A^2 - B^2$, giving (a).
    **(b)** If $A^2 - B^2 = (A - B)(A + B)$, equation $(*)$ becomes
    $A^2 - B^2 = A^2 + AB - BA - B^2$. Subtracting $A^2 - B^2$ from both sides gives
    $0 = AB - BA$, whence $AB = BA$. This proves (b).

Recall that a matrix $A$ is called symmetric if $A^T = A$ where $A^T$ is the transpose of $A$.

**Example 8**    If $A$ and $B$ are symmetric matrices such that $AB = BA$, show that $AB$ is also symmetric.

SOLUTION    We are given $A^T = A$ and $B^T = B$. Compute the transpose of $AB$ using (6) of Theorem 1:
$$(AB)^T = B^T A^T = BA.$$
Since we are assuming that $AB = BA$, this gives $(AB)^T = AB$; that is, $AB$ is symmetric.

## 1.4.3    Matrix Multiplication and Linear Equations

One of the most important motivations for matrix multiplication is in the study of systems of linear equations begun in Section 1.2. Consider the following system of 3 linear equations in 4 variables:
$$x_1 - 2x_2 + 2x_3 - x_4 = 1$$
$$2x_1 - 4x_2 + 3x_3 + x_4 = 3$$
$$3x_1 - 6x_2 + 5x_3 \qquad = 4$$
This system can be written as a single matrix equation
$$\begin{bmatrix} x_1 & - & 2x_2 & + & 2x_3 & - & x_4 \\ 2x_1 & - & 4x_2 & + & 3x_3 & + & x_4 \\ 3x_1 & - & 6x_2 & + & 5x_3 & & \end{bmatrix} = \begin{bmatrix} 1 \\ 3 \\ 4 \end{bmatrix}.$$

The column matrix on the left can be factored as a matrix product

$$\begin{bmatrix} 1 & -2 & 2 & -1 \\ 2 & -4 & 3 & 1 \\ 3 & -6 & 5 & 0 \end{bmatrix} \begin{bmatrix} x_1 \\ x_2 \\ x_3 \\ x_4 \end{bmatrix} = \begin{bmatrix} 1 \\ 3 \\ 4 \end{bmatrix}.$$

Thus the *system* of linear equations becomes a *single* matrix equation

$$AX = B$$

where $A = \begin{bmatrix} 1 & -2 & 2 & -1 \\ 2 & -4 & 3 & 1 \\ 3 & -6 & 5 & 0 \end{bmatrix}$, $X = \begin{bmatrix} x_1 \\ x_2 \\ x_3 \\ x_4 \end{bmatrix}$, and $B = \begin{bmatrix} 1 \\ 3 \\ 4 \end{bmatrix}$. These are called the **coefficient**

**matrix**, the **matrix of variables**, and the **matrix of constants**, respectively. Clearly every system of linear equations can be written in this way.

Given a system $AX = B$ written in matrix form, the new system

$$AX = 0$$

is called the **associated homogeneous system**. It differs from $AX = B$ only in that all the constants have been set equal to zero. The solutions of the two systems are closely related as the next theorem demonstrates.

**THEOREM 2**

Suppose $X_0$ is any particular solution to the system $AX = B$ of linear equations.
   (1) If $X'$ is any solution to the associated homogeneous system $AX = 0$, then $X = X_0 + X'$ is a solution to the system $AX = B$.
   (2) Every solution $X$ to $AX = B$ has the form $X = X_0 + X'$ for some solution $X'$ to $AX = 0$.

**PROOF**

We are assuming that $AX_0 = B$.
   (1) If $AX' = 0$, we have $AX = A(X_0 + X') = AX_0 + AX' = B + 0 = B$.
   (2) If $X$ is *any* solution to the system $AX = B$, write $X' = X - X_0$. Clearly $X = X_0 + X'$, and we have $AX' = A(X - X_0) = AX - AX_0 = B - B = 0$. Thus $X'$ is indeed a solution to the associated homogeneous system $AX = 0$.

The next example shows that the decomposition in (2) is actually accomplished automatically by Gaussian elimination.

**Example 9**

If $A = \begin{bmatrix} 1 & -2 & 2 & -1 \\ 2 & -4 & 3 & 1 \\ 3 & -6 & 5 & 0 \end{bmatrix}$ and $B = \begin{bmatrix} 1 \\ 3 \\ 4 \end{bmatrix}$ as above, use Gaussian elimination to express every

solution to the system $AX = B$ as the sum of a particular solution and the general solution to the associated homogeneous system $AX = 0$.

**SOLUTION**

The reduction of the augmented matrix to reduced form is as follows:

$$\begin{bmatrix} 1 & -2 & 2 & -1 & 1 \\ 2 & -4 & 3 & 1 & 3 \\ 3 & -6 & 5 & 0 & 4 \end{bmatrix} \rightarrow \begin{bmatrix} 1 & -2 & 0 & 5 & 3 \\ 0 & 0 & 1 & -3 & -1 \\ 0 & 0 & 0 & 0 & 0 \end{bmatrix}$$

Hence the leading variables are $x_1$ and $x_3$, and the nonleading variables become parameters: $x_2 = s$ and $x_4 = t$. Thus the reduced form gives the general solution

$$X = \begin{bmatrix} x_1 \\ x_2 \\ x_3 \\ x_4 \end{bmatrix} = \begin{bmatrix} 3+2s-5t \\ s \\ -1+3t \\ t \end{bmatrix} = \begin{bmatrix} 3 \\ 0 \\ -1 \\ 0 \end{bmatrix} + s \begin{bmatrix} 2 \\ 1 \\ 0 \\ 0 \end{bmatrix} + t \begin{bmatrix} -5 \\ 0 \\ 3 \\ 1 \end{bmatrix}$$

$$= X_0 + sX_1 + tX_2.$$

Here $X_0 = \begin{bmatrix} 3 & 0 & -1 & 0 \end{bmatrix}^T$ is a particular solution to $AX = B$ (corresponding to $s = t = 0$) and $X' = sX_1 + tX_2$ is the general solution to the associated homogeneous system $AX = 0$ (where $X_1 = \begin{bmatrix} 2 & 1 & 0 & 0 \end{bmatrix}^T$ and $X_2 = \begin{bmatrix} -5 & 0 & 3 & 1 \end{bmatrix}^T$ are the basic solutions). This last observation also can be seen directly from the algorithm by replacing the last column in the augmented matrix by zeros.

A version of Theorem 2 holds in other contexts as well. For example, it splits the problem of solving a system of linear differential equations into two problems: (1) Finding a particular solution; and (2) Solving the associated homogeneous system. In this situation the two problems are attacked using quite different methods.

## 1.4.4 Block Multiplication

When working with very large matrices which exceed computer memory capacity, it is important to regard a matrix as being made up of smaller matrices (called blocks), and to be able to carry out matrix multiplication by multiplying one pair of blocks at a time. This is only one reason for the importance of block multiplication.

Recall that the augmented matrix of a system $AX = B$ is the matrix $\begin{bmatrix} A & B \end{bmatrix}$ consisting of the coefficient matrix $A$ and the (column) matrix $B$ of constants. This as an example of a **block partitioning** of the augmented matrix, and we say that the augmented matrix $\begin{bmatrix} A & B \end{bmatrix}$ is written in **block form**.

Another common example is to write an $m \times n$ matrix $A$ as a row of columns. If the columns are denoted $C_1, C_2, \cdots, C_n$, we write $A$ as follows:

$$A = \begin{bmatrix} C_1 & C_2 & \cdots & C_n \end{bmatrix}$$

Then the definition of a matrix product $BA$ can be expressed as follows:

$$BA = B\begin{bmatrix} C_1 & C_2 & \cdots & C_n \end{bmatrix} = \begin{bmatrix} BC_1 & BC_2 & \cdots & BC_n \end{bmatrix}$$

In other words, each column of $BA$ is formed by multiplying $B$ times the corresponding column of $A$.

Partitioning a matrix $A$ in this way with its columns as blocks will be used several times below. Here is another way in which it arises.

**Example 10** Write the $2 \times 3$ matrix $A$ as $A = \begin{bmatrix} C_1 & C_2 & C_3 \end{bmatrix}$ where $C_1, C_2,$ and $C_3$ denote the columns of $A$. If $X = \begin{bmatrix} x_1 & x_2 & x_3 \end{bmatrix}^T$, show that

$$AX = \begin{bmatrix} C_1 & C_2 & C_3 \end{bmatrix} \begin{bmatrix} x_1 \\ x_2 \\ x_3 \end{bmatrix} = x_1 C_1 + x_2 C_2 + x_3 C_3.$$

**SOLUTION** Write $A = \begin{bmatrix} a_1 & a_2 & a_3 \\ b_1 & b_2 & b_3 \end{bmatrix}$ so that $C_1 = \begin{bmatrix} a_1 \\ b_1 \end{bmatrix}$, $C_2 = \begin{bmatrix} a_2 \\ b_2 \end{bmatrix}$, and $C_3 = \begin{bmatrix} a_3 \\ b_3 \end{bmatrix}$. Then

$$AX = \begin{bmatrix} a_1 & a_2 & a_3 \\ b_1 & b_2 & b_3 \end{bmatrix} \begin{bmatrix} x_1 \\ x_2 \\ x_3 \end{bmatrix} = \begin{bmatrix} a_1x_1 + a_2x_2 + a_3x_3 \\ b_1x_1 + b_2x_2 + b_3x_3 \end{bmatrix}$$

$$= x_1 \begin{bmatrix} a_1 \\ b_1 \end{bmatrix} + x_2 \begin{bmatrix} a_2 \\ b_2 \end{bmatrix} + x_3 \begin{bmatrix} a_3 \\ b_3 \end{bmatrix}$$

$$= x_1 C_1 + x_2 C_2 + x_3 C_3$$

as required.

The calculation in Example 10 generalizes, and is given as part of the following theorem.

**THEOREM 3**

Let $A = [C_1 \ C_2 \ \cdots \ C_n]$ be an $m \times n$ matrix with columns $C_1, \ C_2, \ \cdots, \ C_n$.

**(1)** If $B$ is a $k \times m$ matrix, then the matrix product $BA$ can be written as
$$BA = B[C_1 \ C_2 \ \cdots \ C_n] = [BC_1 \ BC_2 \ \cdots \ BC_n]$$

**(2)** If $X = [x_1 \ x_2 \ \cdots \ x_n]^T$ is any column, then

$$AX = [C_1 \ C_2 \ \cdots \ C_n] \begin{bmatrix} x_1 \\ x_2 \\ \vdots \\ x_n \end{bmatrix} = x_1 C_1 + x_2 C_2 + \cdots + x_n C_n$$

Theorem 3 will be used frequently throughout the book.

Observe that the products in Theorem 3 are computed using ordinary matrix multiplication, regarding the columns $C_j$ as entries. This observation holds much more generally as the next example illustrates.

**Example 11**  Consider $A = \begin{bmatrix} 1 & -1 & 7 & 8 & 0 \\ 3 & 2 & 5 & 6 & 7 \\ 0 & 0 & -1 & -2 & -3 \end{bmatrix} = \begin{bmatrix} P & X \\ 0 & Q \end{bmatrix}$ and $B = \begin{bmatrix} -2 & 0 \\ 5 & 8 \\ 1 & 1 \\ 2 & 0 \\ -4 & 6 \end{bmatrix} = \begin{bmatrix} U \\ V \end{bmatrix}$ partitioned

as shown. If we use the blocks as entries, the product is

$$AB = \begin{bmatrix} P & X \\ 0 & Q \end{bmatrix} \begin{bmatrix} U \\ V \end{bmatrix} = \begin{bmatrix} PU + XV \\ 0U + QV \end{bmatrix} = \begin{bmatrix} PU + XV \\ QV \end{bmatrix} = \begin{bmatrix} 16 & -1 \\ -7 & 63 \\ 7 & -19 \end{bmatrix}$$

since $PU + XV = \begin{bmatrix} 16 & -1 \\ -7 & 63 \end{bmatrix}$ and $QV = [7 \ -19]$. This agrees with ordinary matrix multiplication as the reader can verify.

Block multiplication can be used to compute *any* matrix product $AB$, subject only to the requirement that the partitioning of $A$ and $B$ are **compatible for multiplication**: The block sizes must be such that every product of blocks that occurs make sense. Then the product can be computed by matrix multiplication using blocks as entries.

**Block Multiplication**  If matrices $A$ and $B$ are partitioned compatibly into blocks, the product $AB$ can be computed by matrix multiplication using the blocks as entries.

An important occurrence of block multiplication is given in the following Theorem, which will be used throughout the book.

**THEOREM 4**   Let $A = \begin{bmatrix} B & X \\ 0 & C \end{bmatrix}$ and $A_1 = \begin{bmatrix} B_1 & X_1 \\ 0 & C_1 \end{bmatrix}$, in block form, where $B$ and $B_1$ are square matrices of the same size, and $C$ and $C_1$ are square matrices of the same size. These are compatible partitionings, and block multiplication gives

$$AA_1 = \begin{bmatrix} B & X \\ 0 & C \end{bmatrix}\begin{bmatrix} B_1 & X_1 \\ 0 & C_1 \end{bmatrix} = \begin{bmatrix} BB_1 & BX_1 + XC_1 \\ 0 & CC_1 \end{bmatrix}.$$

Note that there is a similar theorem for block matrices of the form $\begin{bmatrix} B & O \\ X & C \end{bmatrix}$ where $B$ and $C$ are square.

## 1.4.5  Directed Graphs

Matrix multiplication arises in a variety of situations other than the study of linear equations. In Example 1 it is used to describe how to systematically compute the total sales of retail items. We now describe how it is used to compute the number of paths in a directed graph from one vertex to another.

A **directed graph** is a set of points (called **vertices**) together with arrows (called **edges**) joining various pairs of vertices. Sometimes a double arrow indicates an edge in both directions. For example the vertices could represent cities with arrows indicating direct flights between them. If the graph has $n$ vertices, the **adjacency matrix** $A = [a_{ij}]$ is the $n \times n$ matrix whose $(i, j)$-entry $a_{ij}$ is 1 if there is an edge from vertex $v_i$ to vertex $v_j$, and 0 otherwise. For example, the adjacency matrix of the directed graph below is

$A = \begin{bmatrix} 1 & 1 & 1 \\ 1 & 0 & 0 \\ 0 & 1 & 0 \end{bmatrix}$. Observe that the $(i, j)$-entry $a_{ij}$ of the matrix $A$ is the number of edges

(0 or 1) from vertex $v_i$ to vertex $v_j$.

The next result gives an important extension of this observation. A **path of length** $r$ from vertex $v_i$ to vertex $v_j$ is a sequence of $r$ edges beginning at $v_i$ and leading one after the other to $v_j$. Thus, in the given graph,

$$v_1 \to v_2 \to v_1 \to v_1 \to v_3$$

is a path of length 4 from $v_1$ to $v_3$. The powers of $A$ determine the number of such paths.

**THEOREM 5**   If $A$ is the adjacency matrix of a directed graph, the $(i, j)$-entry of $A^r$ is the number of paths of length $r$ in the graph from vertex $v_i$ to vertex $v_j$.

For example, consider the adjacency matrix $A$ of the graph above. Then

$$A = \begin{bmatrix} 1 & 1 & 1 \\ 1 & 0 & 0 \\ 0 & 1 & 0 \end{bmatrix}, \quad A^2 = \begin{bmatrix} 2 & 2 & 1 \\ 1 & 1 & 1 \\ 1 & 0 & 0 \end{bmatrix} \quad \text{and} \quad A^3 = \begin{bmatrix} 4 & 3 & 2 \\ 2 & 2 & 1 \\ 1 & 1 & 1 \end{bmatrix}.$$

Since the $(1, 2)$-entry of $A^2$ is 2, there must be exactly 2 paths of length 2 from vertex $v_1$ to $v_2$. This can be verified directly in this case: In fact the paths are $v_1 \to v_1 \to v_2$ and $v_1 \to v_3 \to v_2$. The other entries of $A^2$ can also be checked. Similarly, the fact that $A^3$ has no zero entries shows that it is possible to go from *any* vertex to *any* other vertex in exactly three steps.

**PROOF OF THEOREM 5**

Let $b_{ij}$ denote the number of paths of length $r$ from $v_i$ to $v_j$, and write $B_r = [b_{ij}]$. We must show that $B_r = A^r$ for each $r$. This is clear if $r = 1$ because $B_1 = A$. In general we show that, if $B_r = A^r$ holds for some $r \geq 1$, then it follows necessarily that $B_{r+1} = A^{r+1}$. Then $B_1 = A = A^1$ gives $B_2 = A^2$; this in turn gives $B_3 = A^3$; from which we get $B_4 = A^4$, and so on.[13]

So assume that $B_r = A^r$ holds for some $r \geq 1$. Every path of length $r + 1$ from $v_i$ to $v_j$ must pass first through *some* vertex $v_k$:

$$v_i \overset{\text{length } 1}{\to} v_k \overset{\text{length } r}{\to} v_j$$

But the number of such paths is $a_{ik}b_{kj}$ because there are $a_{ik}$ paths of length 1 from $v_i$ to $v_k$, and $b_{kj}$ paths of length $r$ from $v_k$ to $v_j$. Hence the *total* number of paths of length $r + 1$ from $v_i$ to $v_j$ is the sum

$$a_{i1}b_{1j} + a_{i2}b_{2j} + \cdots + a_{in}b_{nj}. \tag{$*$}$$

In other words, $(*)$ is the $(i, j)$-entry of $B_{r+1}$. But this is what we wanted because, by the definition of matrix multiplication, $(*)$ is just the $(i, j)$-entry of the matrix $AB_r = AA^r = A^{r+1}$. (Indeed, row $i$ of $A$ is $[a_{i1} \ a_{i2} \ \cdots \ a_{in}]$ and column $j$ of $B$ is $[b_{1j} \ b_{2j} \ \cdots \ b_{nj}]^T$, and $(*)$ is their dot product.)

## 1.4.6 Proof of Theorem 1

Theorem 1 is a list of the most basic facts about matrix multiplication. Hence, before looking at the proof, we must formalize the definition of matrix multiplication. If $A = [a_{ij}]$ is $m \times n$ and $B = [b_{ij}]$ is $n \times p$, then row $i$ of $A$ and column $j$ of $B$ are, respectively

$$[a_{i1} \ a_{i2} \ \cdots \ a_{in}] \quad \text{and} \quad \begin{bmatrix} b_{1j} \\ b_{2j} \\ \vdots \\ b_{nj} \end{bmatrix}.$$

Hence the $(i, j)$-entry of $AB$ is the dot product

$$a_{i1}b_{1j} + a_{i2}b_{2j} + \cdots + a_{in}b_{nj} = \Sigma_{k=1}^n a_{ik}b_{kj}$$

where we are using **summation notation**[14] for convenience. With this we prove parts 2, 3, and 6 of Theorem 1; the rest is similar and is left as an exercise.

**Proof of (2):** $A(BC) = (AB)C$. Assume $C$ is of size $p \times q$ and write $C = [c_{ij}]$ and $BC = [d_{ij}]$. Then the $(i, j)$-entry of $A(BC)$ is $\Sigma_{k=1}^n a_{ik}d_{kj}$. Now $d_{kj} = \Sigma_{l=1}^p b_{kl}c_{lj}$, so the $(i, j)$-entry of of $A(BC)$ is

$$\Sigma_{k=1}^n a_{ik}(\Sigma_{l=1}^p b_{kl}c_{lj}) = \Sigma_{k=1}^n \Sigma_{l=1}^p a_{ik}b_{kl}c_{lj}.$$

A similar argument shows that this is also the $(i, j)$-entry of $(AB)C$.

**Proof of (3):** $A(B + C) = AB + AC$. Now assume $C$ is of size $n \times p$, and write $C = [c_{ij}]$ so that $B + C = [b_{ij} + c_{ij}]$. Hence the $(i, j)$-entry of $A(B + C)$ is

$$\Sigma_{k=1}^n a_{ik}(b_{kj} + c_{kj}) = \Sigma_{k=1}^n (a_{ik}b_{kj} + a_{ik}c_{kj}) = \Sigma_{k=1}^n a_{ik}b_{kj} + \Sigma_{k=1}^n a_{ik}c_{kj}.$$

This is the $(i, j)$-entry of $AB + AC$ because the sums on the right are the $(i, j)$-entries of $AB$ and $AC$ respectively.

---

[13] This argument uses induction, a proof technique discussed in Appendix A.2.

[14] Summation notation is a convenient shorthand way to denote sums of similar terms. For example: $a_1 + a_2 + a_3 + a_4 = \Sigma_{k=1}^4 a_k$, and $a_3b_3 + a_4b_4 + a_5b_5 = \Sigma_{k=3}^5 a_kb_k$.

**Proof of (6):** $(AB)^T = B^T A^T$. Write $A^T = [a'_{ij}]$ and $B^T = [b'_{ij}]$ where $a'_{ij} = a_{ji}$ and $b'_{ij} = b_{ji}$ for all $i$ and $j$. Then the $(i, j)$-entry of $B^T A^T$ is

$$\Sigma_{k=1}^{n} b'_{ik} a'_{kj} = \Sigma_{k=1}^{n} b_{ki} a_{jk} = \Sigma_{k=1}^{n} a_{jk} b_{ki}.$$

But this is the $(j, i)$-entry of $AB$, that is the $(i, j)$-entry of $(AB)^T$. Hence $B^T A^T = (AB)^T$.

## Exercises 1.4

**1.** In each case compute the indicated matrix products.

**a)** $\begin{bmatrix} 2 & 0 & -3 \\ -1 & 6 & 5 \end{bmatrix} \begin{bmatrix} 0 & 3 \\ -4 & 7 \\ 2 & -5 \end{bmatrix}$    **b)** $\begin{bmatrix} 5 & 0 & -3 \\ 1 & 2 & 1 \end{bmatrix} \begin{bmatrix} 8 \\ -5 \\ 2 \end{bmatrix}$

**c)** $\begin{bmatrix} 1 & 3 & -6 \end{bmatrix} \begin{bmatrix} 2 & -3 \\ 0 & 1 \\ -4 & 5 \end{bmatrix}$    **d)** $\begin{bmatrix} 2 & 7 & -4 \end{bmatrix} \begin{bmatrix} -1 \\ 2 \\ 8 \end{bmatrix}$

**e)** $\begin{bmatrix} 3 \\ -5 \\ 2 \end{bmatrix} \begin{bmatrix} 5 & 9 & -3 \end{bmatrix}$    **f)** $\begin{bmatrix} 4 & 2 \\ -2 & -5 \\ 1 & 5 \end{bmatrix} \begin{bmatrix} 1 & 2 & 3 \\ 0 & -3 & 5 \end{bmatrix}$

**2.** Let $A = \begin{bmatrix} -2 & 5 & 1 \\ -3 & -6 & 5 \end{bmatrix}$, $B = \begin{bmatrix} -2 & -5 \\ 0 & 3 \\ 9 & -4 \end{bmatrix}$, $C = \begin{bmatrix} 1 & 3 \\ -2 & 5 \end{bmatrix}$ and

$D = \begin{bmatrix} -2 \\ 3 \end{bmatrix}$. In each case either compute the indicated product or explain why it is not defined.

**a)** $AB$    **b)** $A^2$    **c)** $CD$    **d)** $DC$
**e)** $BC$    **f)** $AC$    **g)** $C^2$    **h)** $AD$

**3.** In your own words, describe an algorithm to check whether two given matrices can be multiplied, and an algorithm to perform the multiplication when they are compatible.

**4.** If $A$ and $B$ are matrices and $AB$ is square, explain why $BA$ is also square.

**5.** If $AB$ and $BA$ can both be formed, describe the sizes of $A$ and $B$.

**6.** If $AC = CB$, explain why $A$ and $B$ are both square.

**7.** Simplify the following expressions where $A$, $B$, and $C$ represent matrices.
**a)** $A(3B - C) + A^3 C$
**b)** $A(3B - C) + A^2 B + 3A(C - 2B)$
**c)** $(A - B)(A - B) - A^2 + B^2$

**8. a)** If $AB = BA$, show that $(A + B)^2 = A^2 + 2AB + B^2$.
**b)** If $AB \neq BA$, explain why $(A + B)^2 \neq A^2 + 2AB + B^2$.

**9.** If $AB = A$ and $BA = B$, show that $A^2 = A$ and $B^2 = B$.

**10.** In each case either show that the statement is true or give an example showing that it is false. Let $A$ and $B$ denote matrices.

**a)** If $AB$ is defined, then $BA$ is defined.
**b)** If $AB$ is defined and is a square matrix, then $BA$ is defined.
**c)** If $AB = BA$, then $A$ and $B$ are both square and the same size.
**d)** If $A^2$ can be formed, then $A$ must be square.
**e)** If $A$ has a row of zeros, then $AB$ has a row of zeros.
**f)** If $A$ has a column of zeros, then $AB$ has a column of zeros.
**g)** If $AB = 0$, then $A = 0$ or $B = 0$.
**h)** $(AB)^2 = A^2 B^2$ always holds.
**i)** $(A + B)^2 = A^2 + 2AB + B^2$ always holds.
**j)** If $AJ = A$, then $J$ is an identity matrix.
**k)** If $A^2 = A$, then $A = 0$ or $A = I$.
**l)** If $A$ is $m \times n$ and $m > n$, then the system $AX = 0$ has infinitely many solutions.
**m)** If $A$ is $m \times n$ and $m < n$, then the system $AX = B$ has a solution for any column $B$.

**11.** If $A$ is any matrix, show that $AA^T$ and $A^T A$ are both symmetric matrices. [Hint: Theorem 1(6).]

**12.** Suppose that $A$ and $B$ are both symmetric matrices.
**a)** If $AB$ is symmetric, show that $AB = BA$.
**b)** Explain how this relates to Example 8.

**13.** In each case write the system in matrix form $AX = B$ and express the general solution as a sum of a particular solution plus the general solution of the associated homogeneous system $AX = 0$ (See Example 9).

**a)** $\begin{aligned} x_1 + x_2 + x_3 &= 2 \\ 2x_1 + x_2 \quad\;\; &= 3 \\ x_1 - x_2 - 3x_3 &= 0 \end{aligned}$    **b)** $\begin{aligned} x_1 - x_2 - 4x_3 &= -4 \\ x_1 + 2x_2 + 5x_3 &= 2 \\ x_1 + x_2 + 2x_3 &= 0 \end{aligned}$

**c)** $\begin{aligned} x_1 + x_2 - x_3 \quad\quad - 5x_5 &= 2 \\ x_2 + x_3 \quad\quad - 4x_5 &= -1 \\ x_2 + x_3 + x_4 + x_5 &= -1 \\ 2x_1 \quad\quad - 4x_3 + x_4 + x_5 &= 6 \end{aligned}$

**d)** $\begin{aligned} 2x_1 + x_2 - x_3 - x_4 &= -1 \\ 3x_1 + x_2 + x_3 - 2x_4 &= -2 \\ -x_1 - x_2 + 2x_3 + x_4 &= 2 \\ -2x_1 - x_2 \quad\quad + 2x_4 &= 3 \end{aligned}$

**14.** Find an example of an inconsistent system of 4 equations in 4 variables such that the associated homogeneous system has a 2-parameter family of solutions.

**15.** Suppose $A$ is a $3 \times 3$ matrix and $AX = 0$ has a unique solution. Is $AX = B$ consistent for each column $B$? If so, is the solution unique? Support your answer.

**16.** Let $A$ be a $4 \times 6$ matrix with $\text{rank}A = 4$. Explain why $AX = B$ has a solution for any $4 \times 1$ matrix $B$.

**17.** If $A$ is a matrix and the system $AX = 0$ has a nontrivial solution, show that there is no matrix $B$ such that $BA = I$.

**18.** Let $D = \begin{bmatrix} a_1 & 0 & 0 \\ 0 & a_2 & 0 \\ 0 & 0 & a_3 \end{bmatrix}$. A matrix of this type is called a **diagonal matrix**.
  **a)** Calculate the products $D^2$, $D^3$, and $D^4$, and make a guess at what $D^n$ looks like.
  **b)** If $A$ is a $3 \times 3$ matrix, describe the product $AD$ in terms of the columns, and the product $DA$ in terms of the rows, of $A$.

**19.** Let $A = \begin{bmatrix} 3 & -1 & 7 \\ 0 & 2 & 0 \\ 0 & 0 & -1 \end{bmatrix}$. Matrices of this form are called *upper triangular*.
  **a)** Compute $A^2$, $A^3$, and $A^4$, and make a conjecture about $A^n$ for $n \geq 1$.
  **b)** Make a conjecture about the product $AB$ of any two $3 \times 3$ upper triangular matrices.

**20.** Let $A = \left[ \begin{array}{cc|c} 1 & 0 & 2 \\ 1 & 0 & 0 \\ \hline 0 & 0 & -1 \end{array} \right]$. Compute $A^2$, $A^3$, $A^4$, and $A^5$. Make a conjecture about $A^n$ for $n \geq 1$.

**21.** Verify that $A^2 - 5A + 4I = 0$ if:
  **a)** $A = \begin{bmatrix} 1 & 7 \\ 0 & 4 \end{bmatrix}$
  **b)** $A = \begin{bmatrix} -5 & 6 \\ -9 & 10 \end{bmatrix}$

**22.** **a)** If $A = \begin{bmatrix} 0 & 1 \\ 0 & 0 \end{bmatrix}$ show that $A^2 = 0$.
  **b)** If $A$ is a symmetric $2 \times 2$ matrix with $A^2 = 0$, show that $A = 0$.
  **c)** If $A^2 = 0$, show that $(A^T)^2 = 0$, and that $(cA)^2 = 0$ for all constants $c$.

**23.** Compute the matrix product $AB$ using block multiplication where

$A = \left[ \begin{array}{cc|ccc} -2 & 1 & 6 & -2 & 0 \\ 0 & 3 & 1 & 3 & 7 \\ \hline 0 & 0 & -1 & 2 & 3 \end{array} \right]$ and $B = \left[ \begin{array}{cc|cc} 1 & -3 & 0 & 0 \\ 5 & 9 & 0 & 0 \\ \hline -1 & 5 & -1 & -2 \\ 7 & 0 & 3 & 3 \\ 2 & -3 & -1 & 0 \end{array} \right]$.

**24.** In each case compute $A$, $A^2$, $A^3$, and $A^4$. Then find an expression for $A^n$ for any $n \geq 1$.
  **a)** $A = \left[ \begin{array}{c|cc} 1 & 0 & 0 \\ \hline 1 & 1 & -1 \\ 1 & -1 & 1 \end{array} \right]$
  **b)** $A = \left[ \begin{array}{cc|cc} 1 & -1 & 2 & -1 \\ 0 & 1 & 0 & 0 \\ \hline 0 & 0 & -1 & 1 \\ 0 & 0 & 0 & 1 \end{array} \right]$

**25.** Compute the following matrix products using block multiplication. Assume that the partitionings are compatible for multiplication.
  **a)** $\begin{bmatrix} I & X \\ -Y & I \end{bmatrix} \begin{bmatrix} I & 0 \\ Y & I \end{bmatrix}$
  **b)** $\begin{bmatrix} I & X \\ 0 & I \end{bmatrix} \begin{bmatrix} I & -X \\ 0 & I \end{bmatrix}$
  **c)** $\begin{bmatrix} I & X \end{bmatrix} \begin{bmatrix} I \\ Y \end{bmatrix}$
  **d)** $\begin{bmatrix} X & I \end{bmatrix} \begin{bmatrix} I \\ -X \end{bmatrix}$

**26.** Let $A = \begin{bmatrix} 0 & I \\ B & 0 \end{bmatrix}$ where each block is square of the same size. Compute $A$, $A^2$, $A^3$, and $A^4$. What is $A^n$ if $n$ is even? What if $n$ is odd? Support your answer.

**27.** In each of the directed graphs, determine the number of paths of length 4 and 10 from any vertex to any other vertex, and see if you can check the graph itself to verify your findings.
  **a)**
  **b)**†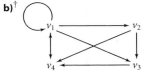

**28.** If $B$ is $n \times n$ and $BC = C$ for all columns $C$, show that $B = I_n$ is the $n \times n$ identity matrix.

**29.** **a)** If the $2 \times 2$ matrix $A$ commutes with every $2 \times 2$ matrix, show that $A = aI_2$ for some number $a$.
  **b)** If $a$ is any number, show that $aI_2$ commutes with every $2 \times 2$ matrix.
  [Remark: The only $n \times n$ matrices that commute with every $n \times n$ matrix are the scalar multiples $aI_n$ of the $n \times n$ identity matrix. You might like to show this as in **a)**.]

**30.** If $A = \begin{bmatrix} a & b & c \\ a' & b' & c' \end{bmatrix}$ and $AA^T = 0$, show that $A = 0$.
  [Remark: More generally, if $A$ is *any* matrix such that $AA^T = 0$, then necessarily $A = 0$.]

**31.** A square matrix $E$ is called an **idempotent** if $E^2 = E$.
  **a)** Show that $0$, $I$, $\begin{bmatrix} 1 & a \\ 0 & 0 \end{bmatrix}$, and $\frac{1}{\sqrt{2}} \begin{bmatrix} 1 & 1 \\ 1 & 1 \end{bmatrix}$ are all idempotents.
  **b)** If $E$ is an idempotent, show that $I - E$ and $E^T$ are both idempotents.
  **c)** If $AB = I$ for some matrices $A$ and $B$, show that $E = BA$ is an idempotent.

† Requires computer assistance (e.g., Matlab).

**32.** If $A = \begin{bmatrix} a & b \\ c & d \end{bmatrix}$ where $a \neq 0$, show that $A$ can be factored in the form $A = \begin{bmatrix} 1 & 0 \\ x & 1 \end{bmatrix}\begin{bmatrix} y & z \\ 0 & w \end{bmatrix}$.

**33.** If $A = \begin{bmatrix} 3 & 2 \\ 2 & 3 \end{bmatrix}$ compute $A^2$, $A^3$, and $A^4$. Describe the pattern. Explain the pattern by examining $\begin{bmatrix} a+1 & a \\ a & a+1 \end{bmatrix}^2$.

**34.** Let $K$ denote the $n \times n$ matrix with ones on the "anti-diagonal"—top right to lower left—and zeros elsewhere. If $A$ is an $n \times n$ matrix, show that $KA^TK$ is the matrix obtained from $A$ by "flipping" $A$ about its anti-diagonal.

## 1.5   MATRIX INVERSES

It is frequently important to "reverse" the effect of multiplying by a matrix. Here is an example.

**Example 1**   A spy plane flies over enemy territory and radios its position $X = \begin{bmatrix} x \\ y \end{bmatrix}$ back to headquarters (here $x$ and $y$ denote the longitude and latitude respectively). These transmissions will likely be intercepted, so they must be encoded to keep the exact location secret. The method chosen is to multiply the position coordinates by the matrix $A = \begin{bmatrix} 3 & -4 \\ 2 & 7 \end{bmatrix}$ to obtain encoded coordinates

$$\begin{bmatrix} x' \\ y' \end{bmatrix} = A\begin{bmatrix} x \\ y \end{bmatrix} = \begin{bmatrix} 3x - 4y \\ 2x + 7y \end{bmatrix}$$

which are sent to headquarters. Devise a method for headquarters to decode these coordinates.

**SOLUTION**   Write the encoded coordinates as $X' = \begin{bmatrix} x' \\ y' \end{bmatrix}$. Headquarters receives $x'$ and $y'$, and must recover the true coordinates $x$ and $y$. They know that $X' = AX$, that is

$$\begin{aligned} 3x - 4y &= x' \\ 2x + 7y &= y' \end{aligned} \qquad\qquad (*)$$

and so they solve this system for $x$ and $y$. The solution is

$$\begin{aligned} x &= \tfrac{7}{29}x' + \tfrac{4}{29}y' \\ y &= -\tfrac{2}{29}x' + \tfrac{3}{29}y' \end{aligned}$$

In matrix form this is $X = CX'$ where $C = \begin{bmatrix} \frac{7}{29} & \frac{4}{29} \\ -\frac{2}{29} & \frac{3}{29} \end{bmatrix} = \frac{1}{29}\begin{bmatrix} 7 & 4 \\ -2 & 3 \end{bmatrix}$. Hence headquarters can decode the received coordinates $X'$ by multiplying by the matrix $C$.

### 1.5.1   Matrix Inverses

The coding matrix $A$ in Example 1 transforms the true coordinates $X$ into coded coordinates $X' = AX$. Then the decoding matrix $C$ transforms $X'$ back to $X = CX'$. If we combine these equations we get $X = CX' = C(AX) = (CA)X$. In fact the reader can verify that

$$CA = I \quad \text{and also} \quad AC = I.$$

These equations reveal the relationship between the matrices $C$ and $A$. Moreover, $C$ is obtained from $A$ by solving a system of equations, so we are led once again to examine how systems of equations are solved.

As we have seen, every system of linear equations can be written in matrix form

$$AX = B$$

where the column $X$ is to be determined. This suggests that we first investigate how we solve *numerical* equations of the form $ax = b$. If $a = 0$, there is no solution (unless $b = 0$). But if $a \neq 0$, we can multiply both sides by the inverse $a^{-1}$ to get $a^{-1}ax = a^{-1}b$, that is, $x = a^{-1}b$. The property of $a^{-1}$ that makes this work is $a^{-1}a = 1$. This condition can be stated for matrices when we remember that the identity matrix $I$ in matrix algebra plays the role of 1 in arithmetic.

With this in mind, we make the following definition: If $A$ is a square matrix, a matrix $C$ is called an **inverse** of $A$ if

$$AC = I \quad \text{and} \quad CA = I.$$

A square matrix that has an inverse is called an **invertible matrix**[15]. Since matrix multiplication is not commutative, we insist that *both $AC$ and $CA$* equal $I$. Note that the inverse matrix $C$ must be the same size as $A$.

**Example 2**  Show that $A = \begin{bmatrix} -1 & 1 \\ 1 & 0 \end{bmatrix}$ has inverse $C = \begin{bmatrix} 0 & 1 \\ 1 & 1 \end{bmatrix}$.

SOLUTION  We verify that $AC = I$ and $CA = I$:

$$AC = \begin{bmatrix} -1 & 1 \\ 1 & 0 \end{bmatrix}\begin{bmatrix} 0 & 1 \\ 1 & 1 \end{bmatrix} = \begin{bmatrix} 1 & 0 \\ 0 & 1 \end{bmatrix} = I$$

and

$$CA = \begin{bmatrix} 0 & 1 \\ 1 & 1 \end{bmatrix}\begin{bmatrix} -1 & 1 \\ 1 & 0 \end{bmatrix} = \begin{bmatrix} 1 & 0 \\ 0 & 1 \end{bmatrix} = I$$

If a square matrix $A$ has a row (or a column) of zeros, then $A$ is not invertible. The next example illustrates why.

**Example 3**  Show that $A = \begin{bmatrix} 2 & -3 \\ 0 & 0 \end{bmatrix}$ has no inverse.

SOLUTION  If $C = \begin{bmatrix} a & b \\ c & d \end{bmatrix}$ is *any* matrix, then $AC = \begin{bmatrix} 2 & -3 \\ 0 & 0 \end{bmatrix}\begin{bmatrix} a & b \\ c & d \end{bmatrix} = \begin{bmatrix} 2a-3c & 2b-3d \\ 0 & 0 \end{bmatrix}$. Since the $(2, 2)$-entry of $AC$ is not 1, $AC$ can never equal $I$ for any choice of the matrix $C$.

Example 3 shows that, contrary to the situation for numbers, *it is possible for a nonzero square matrix to have no inverse.* On the other hand, if a matrix $A$ *has* an inverse, that inverse is uniquely determined by $A$. Indeed, if $C$ and $D$ are both inverses of $A$, then certainly $AC = I$ and $DA = I$, so

$$D = DI = D(AC) = (DA)C = IC = C.$$

Because of this, we speak of *the* inverse of $A$ and denote it by $A^{-1}$ (when it exists). Hence the inverse matrix $A^{-1}$ is determined by the following conditions:

$$AA^{-1} = I \quad \text{and} \quad A^{-1}A = I.$$

The above discussion shows that these equations characterize $A^{-1}$ in the following sense.

---

[15] We will see later that only square matrices can be invertible.

**THEOREM ①**

If a matrix $C$ can be found such that
$$AC = I = CA$$
then $C$ is the inverse of $A$; that is, $C = A^{-1}$. If no such matrix $C$ can be found, then $A$ has no inverse.

**Example 4**　If $A = \begin{bmatrix} -1 & -1 \\ 1 & 0 \end{bmatrix}$, compute $A^3$ and so find $A^{-1}$.

**SOLUTION**　$A^2 = \begin{bmatrix} -1 & -1 \\ 1 & 0 \end{bmatrix}\begin{bmatrix} -1 & -1 \\ 1 & 0 \end{bmatrix} = \begin{bmatrix} 0 & 1 \\ -1 & -1 \end{bmatrix}$, so

$$A^3 = A^2 A = \begin{bmatrix} 0 & 1 \\ -1 & -1 \end{bmatrix}\begin{bmatrix} -1 & -1 \\ 1 & 0 \end{bmatrix} = \begin{bmatrix} 1 & 0 \\ 0 & 1 \end{bmatrix} = I.$$

Hence $A \cdot A^2 = A^3 = I$ and $A^2 \cdot A = A^3 = I$, so $A^2$ is the inverse of $A$ by Theorem 1. In other words, $A^{-1} = A^2 = \begin{bmatrix} 0 & 1 \\ -1 & -1 \end{bmatrix}$.

In contrast to Example 4, the inverse of $A$ is usually *not* a power of $A$. If $A$ is $2 \times 2$, there is a simple formula for the inverse (when it exists) which will be used frequently.

**Example 5**　If $A = \begin{bmatrix} a & b \\ c & d \end{bmatrix}$ and $ad - bc \neq 0$, show that

$$A^{-1} = \frac{1}{ad-bc}\begin{bmatrix} d & -b \\ -c & a \end{bmatrix}.$$

**SOLUTION**　If we write $C = \frac{1}{ad-bc}\begin{bmatrix} d & -b \\ -c & a \end{bmatrix}$, then

$$AC = \frac{1}{ad-bc}\begin{bmatrix} a & b \\ c & d \end{bmatrix}\begin{bmatrix} d & -b \\ -c & a \end{bmatrix} = \frac{1}{ad-bc}\begin{bmatrix} ad-bc & 0 \\ 0 & ad-bc \end{bmatrix} = I.$$

The reader can verify that $CA = I$ also, so $C$ is indeed the inverse of $A$.

The number $ad - bc$ in Example 5 is called the **determinant** of $A = \begin{bmatrix} a & b \\ c & d \end{bmatrix}$, and $\begin{bmatrix} d & -b \\ -c & a \end{bmatrix}$ is called the **adjoint** of $A$. In Chapter 2 we will discuss the determinant and adjoint of an arbitrary $n \times n$ matrix, and use them to give a formula for the inverse (when it exists).

**Example 6**　If $YA = 0$ for some nonzero matrix $Y$, show that $A$ has no inverse.

**SOLUTION**　Suppose $A^{-1}$ does exist. Then right-multiply the equation $YA = 0$ by $A^{-1}$ to obtain $YAA^{-1} = 0A^{-1}$. This becomes $YI = 0$, that is, $Y = 0$, contrary to the assumption that $Y$ is nonzero. So $A^{-1}$ does not exist.

## 1.5.2　Inverses and Systems of Linear Equations

With the notion of the inverse of a matrix in hand, we return to the study of a system of linear equations written as a single matrix equation
$$AX = B.$$

Here $A$ is the coefficient matrix, $B$ is the column of constants, and the column $X$ of variables is to be determined. If $A$ is (square and) invertible, we can solve the system by left-multiplying each side by $A^{-1}$ to get

$$A^{-1}AX = A^{-1}B$$
$$IX = A^{-1}B$$
$$X = A^{-1}B.$$

Thus $X = A^{-1}B$ is a solution to the system. [Check: $AX = A(A^{-1}B) = IB = B.$] Moreover the argument shows that if $X$ is *any* solution to the system, then necessarily $X = A^{-1}B$, so the solution is unique. This discussion is summarized in the following theorem.

**THEOREM 2**    Suppose a system of $n$ equations in $n$ variables is written in matrix form as

$$AX = B.$$

If the $n \times n$ coefficient matrix $A$ is invertible, the system has the unique solution

$$X = A^{-1}B.$$

We show later (Theorem 5) that the converse of Theorem 2 is true: That is, if $A$ is a square matrix and $AX = B$ has a solution for every choice of $B$, then $A$ must be invertible.

**Example 7**    Solve the system of equations $\begin{cases} 5x - 3y = -2 \\ 2x + 7y = \phantom{-}3 \end{cases}$ by finding the inverse of the coefficient matrix and then using Theorem 2.

**SOLUTION**    In matrix form the system is $AX = B$ where $A = \begin{bmatrix} 5 & -3 \\ 2 & 7 \end{bmatrix}$, $X = \begin{bmatrix} x \\ y \end{bmatrix}$, and $B = \begin{bmatrix} -2 \\ 3 \end{bmatrix}$.

The determinant of $A$ is $5 \cdot 7 - (-3) \cdot 2 = 41$. Since this is not zero, Example 5 gives $A^{-1} = \frac{1}{41} \begin{bmatrix} 7 & 3 \\ -2 & 5 \end{bmatrix}$. Hence the solution of the system of equations is

$$X = A^{-1}B = \frac{1}{41} \begin{bmatrix} 7 & 3 \\ -2 & 5 \end{bmatrix} \begin{bmatrix} -2 \\ 3 \end{bmatrix} = \frac{1}{41} \begin{bmatrix} -5 \\ 19 \end{bmatrix}$$

by Theorem 2. In other words, $x = \frac{-5}{41}$ and $y = \frac{19}{41}$.

## 1.5.3    The Matrix Inversion Algorithm

Except for the formula in Example 5 for the inverse of a $2 \times 2$ matrix, we have not given a systematic way to find the inverse of a square matrix. Fortunately, an easy method exists which is based on Gaussian elimination. To illustrate the technique, consider the invertible matrix $A = \begin{bmatrix} 1 & 2 \\ 3 & 7 \end{bmatrix}$. If we write $A^{-1} = \begin{bmatrix} x_1 & x_2 \\ y_1 & y_2 \end{bmatrix}$, the condition $AA^{-1} = I$ becomes

$$A \begin{bmatrix} x_1 & x_2 \\ y_1 & y_2 \end{bmatrix} = \begin{bmatrix} 1 & 0 \\ 0 & 1 \end{bmatrix}.$$

Equating columns, this becomes two systems of linear equations with $A$ as coefficient matrix:

$$A \begin{bmatrix} x_1 \\ y_1 \end{bmatrix} = \begin{bmatrix} 1 \\ 0 \end{bmatrix} \quad \text{and} \quad A \begin{bmatrix} x_2 \\ y_2 \end{bmatrix} = \begin{bmatrix} 0 \\ 1 \end{bmatrix}.$$

Since $A$ is invertible, each of these systems has a unique solution (by Theorem 2). Because $A$ is square, this means that the reduced row-echelon form $R$ of $A$ cannot have a row of zeros (otherwise, there would be at least one nonleading variable and hence at least one parameter). But then $R$ is the identity matrix (it is square and reduced row-echelon), so there exists a sequence of row operations carrying $A$ to the identity matrix. This sequence carries the augmented matrices of both systems to reduced row-echelon form as follows:

$$\begin{bmatrix} A & \begin{matrix} 1 \\ 0 \end{matrix} \end{bmatrix} = \begin{bmatrix} 1 & 2 & 1 \\ 3 & 7 & 0 \end{bmatrix} \rightarrow \begin{bmatrix} 1 & 2 & 1 \\ 0 & 1 & -3 \end{bmatrix} \rightarrow \begin{bmatrix} 1 & 0 & 7 \\ 0 & 1 & -3 \end{bmatrix},$$

$$\begin{bmatrix} A & \begin{matrix} 0 \\ 1 \end{matrix} \end{bmatrix} = \begin{bmatrix} 1 & 2 & 0 \\ 3 & 7 & 1 \end{bmatrix} \rightarrow \begin{bmatrix} 1 & 2 & 0 \\ 0 & 1 & 1 \end{bmatrix} \rightarrow \begin{bmatrix} 1 & 0 & -2 \\ 0 & 1 & 1 \end{bmatrix}.$$

Hence the solutions are $\begin{bmatrix} x_1 \\ y_1 \end{bmatrix} = \begin{bmatrix} 7 \\ -3 \end{bmatrix}$ and $\begin{bmatrix} x_2 \\ y_2 \end{bmatrix} = \begin{bmatrix} -2 \\ 1 \end{bmatrix}$ respectively, so $A^{-1} = \begin{bmatrix} 7 & -2 \\ -3 & 1 \end{bmatrix}$.

Now observe that the two row reductions can be carried out *simultaneously* because the row operations used are the same (the ones carrying $A \rightarrow I$). The result is

$$\begin{bmatrix} A & \begin{matrix} 1 & 0 \\ 0 & 1 \end{matrix} \end{bmatrix} = \begin{bmatrix} 1 & 2 & 1 & 0 \\ 3 & 7 & 0 & 1 \end{bmatrix} \rightarrow \begin{bmatrix} 1 & 2 & 1 & 0 \\ 0 & 1 & -3 & 1 \end{bmatrix} \rightarrow \begin{bmatrix} 1 & 0 & 7 & -2 \\ 0 & 1 & -3 & 1 \end{bmatrix}.$$

This last reduction can be written more compactly as

$$[A \ \ I] \rightarrow [I \ \ A^{-1}].$$

In other words, the set of row operations that carry $A \rightarrow I$ also carry $I \rightarrow A^{-1}$.

This procedure works for any invertible square matrix $A$. In fact all that is needed is that $A$ can be carried to $I$ by row operations (this is equivalent to invertibility as will be seen in Theorem 5). Hence we get the desired algorithm.

**The Matrix Inversion Algorithm**

Suppose $A$ is a square matrix and there exists a sequence of elementary row operations that carry $A \rightarrow I$. Then $A$ is invertible and this same sequence carries $I \rightarrow A^{-1}$. Thus

$$[A \ \ I] \rightarrow [I \ \ A^{-1}]$$

where the row operations on $A$ and $I$ are carried out simultaneously.

**Example 8**  Find the inverse of the matrix $A = \begin{bmatrix} 1 & 4 & -1 \\ 2 & 7 & 1 \\ 1 & 3 & 0 \end{bmatrix}$.

**SOLUTION**  The reduction $[A \ \ I] \rightarrow [I \ \ A^{-1}]$ is as follows:

$$\begin{bmatrix} 1 & 4 & -1 & 1 & 0 & 0 \\ 2 & 7 & 1 & 0 & 1 & 0 \\ 1 & 3 & 0 & 0 & 0 & 1 \end{bmatrix}$$

$$\rightarrow \begin{bmatrix} 1 & 4 & -1 & 1 & 0 & 0 \\ 0 & -1 & 3 & -2 & 1 & 0 \\ 0 & -1 & 1 & -1 & 0 & 1 \end{bmatrix}$$

$$\rightarrow \begin{bmatrix} 1 & 0 & 11 & -7 & 4 & 0 \\ 0 & 1 & -3 & 2 & -1 & 0 \\ 0 & 0 & -2 & 1 & -1 & 1 \end{bmatrix}$$

$$\rightarrow \begin{bmatrix} 1 & 0 & 11 & -7 & 4 & 0 \\ 0 & 1 & -3 & 2 & -1 & 0 \\ 0 & 0 & 1 & -\frac{1}{2} & \frac{1}{2} & -\frac{1}{2} \end{bmatrix}$$

$$\rightarrow \begin{bmatrix} 1 & 0 & 0 & -\frac{3}{2} & -\frac{3}{2} & \frac{11}{2} \\ 0 & 1 & 0 & \frac{1}{2} & \frac{1}{2} & -\frac{3}{2} \\ 0 & 0 & 1 & -\frac{1}{2} & \frac{1}{2} & -\frac{1}{2} \end{bmatrix}$$

Hence $A^{-1} = \begin{bmatrix} -\frac{3}{2} & -\frac{3}{2} & \frac{11}{2} \\ \frac{1}{2} & \frac{1}{2} & -\frac{3}{2} \\ -\frac{1}{2} & \frac{1}{2} & -\frac{1}{2} \end{bmatrix} = \frac{1}{2} \begin{bmatrix} -3 & -3 & 11 \\ 1 & 1 & -3 \\ -1 & 1 & -1 \end{bmatrix}$ by the matrix inversion algorithm. Of course this can be checked by verifying that $AA^{-1} = I$ and $A^{-1}A = I$.

**Remark**  If $A$ is not invertible, then no sequence of row operations can carry $A \to I$ as we shall see in Theorem 5 below. Hence the algorithm breaks down because a row of zeros is encountered. Here is an example.

**Example 9**  Show that the matrix $A = \begin{bmatrix} 1 & 3 & -2 \\ 1 & 2 & 0 \\ 2 & 8 & -8 \end{bmatrix}$ is not invertible.

**SOLUTION**  We try the algorithm on $A$ to get:

$$[A \quad I] = \begin{bmatrix} 1 & 3 & -2 & 1 & 0 & 0 \\ 1 & 2 & 0 & 0 & 1 & 0 \\ 2 & 8 & -8 & 0 & 0 & 1 \end{bmatrix}$$

$$\to \begin{bmatrix} 1 & 3 & -2 & 1 & 0 & 0 \\ 0 & -1 & 2 & -1 & 1 & 0 \\ 0 & 2 & -4 & -2 & 0 & 1 \end{bmatrix}$$

$$\to \begin{bmatrix} 1 & 3 & -2 & 1 & 0 & 0 \\ 0 & -1 & 2 & -1 & 1 & 0 \\ 0 & 0 & 0 & -4 & 2 & 1 \end{bmatrix}$$

Since $A$ will never be transformed to the identity matrix by elementary row operations, $A$ is not invertible.

## 1.5.4  Properties of Inverses

For reference later, the following theorem collects several properties of inverses which are used throughout linear algebra.

**THEOREM ③**  The following properties hold where the matrices are all square.

(1) The identity matrix $I$ is invertible, and $I^{-1} = I$.

(2) If $A$ is invertible, then $A^{-1}$ is also invertible, and $(A^{-1})^{-1} = A$.

(3) If $A$ and $B$ are invertible, then $AB$ is invertible, and $(AB)^{-1} = B^{-1}A^{-1}$.

(4) If $A_1, A_2, \cdots, A_{k-1}, A_k$ are all invertible, their product $A_1 A_2 \cdots A_{k-1} A_k$ is also invertible, and $(A_1 A_2 \cdots A_{k-1} A_k)^{-1} = A_k^{-1} A_{k-1}^{-1} \cdots A_2^{-1} A_1^{-1}$.

(5) If $A$ is invertible, then $A^k$ is invertible for $k \geq 1$, and $(A^k)^{-1} = (A^{-1})^k$.

(6) If $A$ is invertible, then $A^T$ is also invertible, and $(A^T)^{-1} = (A^{-1})^T$.

(7) If $A$ is invertible and $c \neq 0$ is a number, then $cA$ is also invertible, and $(cA)^{-1} = \frac{1}{c} A^{-1}$.

**PROOF**  (1) Clearly $I^2 = I$ so this follows from Theorem 1.

(2) We have $AA^{-1} = I = A^{-1}A$. By Theorem 1, these equations show that $A$ is the inverse of $A^{-1}$, proving (2).

(3) By Theorem 1 we must find a matrix $C$ such that $(AB)C = I$ and $C(AB) = I$. The proposed candidate is $C = B^{-1}A^{-1}$ (which exists because $A^{-1}$ and $B^{-1}$ are assumed to exist). We test it as follows:

$$(AB)C = (AB)(B^{-1}A^{-1}) = A(BB^{-1})A^{-1} = AIA^{-1} = AA^{-1} = I.$$

Similarly, $C(AB) = I$ as the reader can verify. So $C = (AB)^{-1}$ by Theorem 1, proving (3).

(4) We begin with the following result.

*Claim.* If (4) holds for some integer $k \geq 1$, then it holds for $k + 1$.

*Proof.* If $(A_1A_2 \cdots A_k)^{-1} = A_k^{-1} \cdots A_2^{-1}A_1^{-1}$ holds for an integer $k$, then (3) gives

$$(A_1A_2 \cdots A_kA_{k+1})^{-1} = [(A_1A_2 \cdots A_k)(A_{k+1})]^{-1}$$
$$= A_{k+1}^{-1}(A_1A_2 \cdots A_k)^{-1}$$
$$= A_{k+1}^{-1}(A_k^{-1} \cdots A_2^{-1}A_1^{-1}).$$

This proves the Claim.

Now (4) clearly holds if $k = 1$, and it holds for $k = 2$ by (3). Then the Claim shows that (4) holds for $k = 3$; then the Claim again shows it holds for $k = 4$, etc. It follows that (4) holds for every $k \geq 0$.[16]

(5) This restates (4) in the special case that $A_1 = A_2 = \cdots = A_k = A$.

(6) We must find a matrix $C$ such that $A^TC = CA^T$, and a candidate $C = (A^{-1})^T$ is suggested. Using Theorem 1 §1.4, we have

$$A^TC = A^T(A^{-1})^T = (A^{-1}A)^T = I^T = I.$$

Similarly $CA^T = I$, so Theorem 1 shows that $C$ is the inverse of $A^T$, that is $(A^T)^{-1} = C$. This proves (6).

(7) This is left to the reader (Exercise 30).

Part (6) of Theorem 3 yields a useful criterion that a square matrix is invertible.

**COROLLARY**   If $A$ is a square matrix, then $A$ is invertible if and only if $A^T$ is invertible.

**PROOF**   If $A$ is invertible, then $A^T$ is invertible by (6) of Theorem 3. Conversely, if $A^T$ is invertible, then $A = (A^T)^T$ is invertible, again by (6) of Theorem 3 (applied to $A^T$).

We now give four examples showing how the properties in Theorem 3 are used.

**Example 10**   Find $A$ if $(A^{-1} - 3I)^T = 2\begin{bmatrix} -1 & 2 \\ 5 & 4 \end{bmatrix}$.

**SOLUTION**   We use Theorem 3 several times. First transpose both sides to get

$$\left\{(A^{-1} - 3I)^T\right\}^T = \left\{2\begin{bmatrix} -1 & 2 \\ 5 & 4 \end{bmatrix}\right\}^T,$$

that is

$$A^{-1} - 3I = 2\begin{bmatrix} -1 & 5 \\ 2 & 4 \end{bmatrix}.$$

Hence $A^{-1} = 3I + 2\begin{bmatrix} -1 & 5 \\ 2 & 4 \end{bmatrix} = \begin{bmatrix} 1 & 10 \\ 4 & 11 \end{bmatrix}$, and so the fact that $[A^{-1}]^{-1} = A$ gives

$$A = [A^{-1}]^{-1} = \begin{bmatrix} 1 & 10 \\ 4 & 11 \end{bmatrix}^{-1} = \frac{1}{-29}\begin{bmatrix} 11 & -10 \\ -4 & 1 \end{bmatrix}$$

by Example 5, as required.

---

[16] This method of proof is called **mathematical induction**: Suppose that $p_1, p_2, \cdots$ are statements such that $p_1$ is true; and $p_{k+1}$ is true whenever the preceding statement $p_k$ is true. Then $p_k$ is true for every $k \geq 1$. The method is discussed in Appendix A.2.

**Example 11**    If $A$, $B$, and $C$ are invertible, simplify $C^T B (AB)^{-1} [C^{-1} A^T]^T$.

SOLUTION

$$
\begin{aligned}
C^T B (AB)^{-1} [C^{-1} A^T]^T &= C^T B (B^{-1} A^{-1}) [(A^T)^T (C^{-1})^T] \\
&= C^T I A^{-1} [A (C^T)^{-1}] \\
&= C^T I (C^T)^{-1} \\
&= I.
\end{aligned}
$$

**Example 12**    If $A$ and $AB$ are both invertible, show that $B$ is also invertible.

SOLUTION    We have $B = A^{-1}(AB)$. Now $A^{-1}$ is invertible by Theorem 3, because $A$ is invertible. Since $AB$ is invertible by assumption, it follows that $A^{-1}(AB) = B$ is invertible by (3) of Theorem 3.

**Example 13**    A matrix $E$ is called an idempotent if $E^2 = E$. Show that $I$ is the only invertible idempotent.

SOLUTION    $I$ is invertible (by Theorem 3) and $I^2 = I$, so $I$ is indeed an invertible idempotent. Conversely, assume that $E^2 = E$ and that $E^{-1}$ exists. Left multiply the equation $E^2 = E$ by $E^{-1}$ to get $E^{-1}E^2 = E^{-1}E$, that is $E = I$.

A square matrix is called **upper triangular** if every entry below the main diagonal is zero. Similarly, a matrix $A$ is **lower triangular** if every entry above the main diagonal is zero; equivalently, if $A^T$ is upper triangular. A matrix is called **triangular** if it is either upper or lower triangular. One of the reasons for the importance of these matrices is that it is easy to tell if they are invertible.

**THEOREM 4**    Let $A$ denote a square triangular matrix.
    **(1)** $A$ is invertible if and only if no entry on the main diagonal is zero.
    **(2)** If $A$ is upper (lower) triangular, then $A^{-1}$ is also upper (lower) triangular.

PROOF    **(1)** If $A$ is upper triangular and every main diagonal entry is nonzero, it is clear that $A \to I$ by row operations, so $A^{-1}$ exists by the matrix inversion algorithm. Conversely, if some main diagonal entry of $A$ is zero, then $A \to I$ is impossible because the row containing the lowest diagonal zero can be carried to a row of zeros by row operations (verify). Hence the inverse does not exist by Theorem 5 below. Thus (1) holds for upper triangular matrices. But then it holds for a lower triangular matrix $B$ because $B$ is invertible if and only $B^T$ is invertible (Corollary of Theorem 3), and $B^T$ is upper triangular with the same main diagonal as $B$.
    **(2)** If $A$ is upper triangular and invertible, then the matrix inversion algorithm shows that $A^{-1}$ will also be upper triangular because no multiple of any row is added to a lower row. If $B$ is lower triangular, then $B^T$ is upper triangular, whence $(B^{-1})^T = (B^T)^{-1}$ is upper triangular (by the above). Thus $B^{-1} = [(B^{-1})^T]^T$ is lower triangular.

The next example illustrates Theorem 4. The verification (using Example 5) is left to the reader.

**Example 14**   If $A = \begin{bmatrix} a & b \\ 0 & c \end{bmatrix}$ where $a \neq 0$ and $c \neq 0$, show that the inverse is given by $A^{-1} = \begin{bmatrix} \frac{1}{a} & -\frac{b}{ac} \\ 0 & \frac{1}{c} \end{bmatrix}$.

## 1.5.5   Conditions for Invertibility

The following theorem collects a number of conditions on a matrix all equivalent[17] to invertibility. It will be referred to frequently below.

**THEOREM 5**   The following conditions are equivalent for a square matrix $A$:

(1) $A$ is invertible.
(2) The homogeneous system $AX = 0$ has only the trivial solution $X = 0$.
(3) $A$ can be carried to $I$ by elementary row operations.
(4) The system $AX = B$ has a solution $X$ for every choice of column $B$.
(5) There exists a matrix $C$ such that $AC = I$.

**PROOF**   We show that each of these conditions implies the next, and that (5) implies (1).

(1) $\Rightarrow$ (2). If $A^{-1}$ exists, then $AX = 0$ gives $X = IX = A^{-1}AX = A^{-1}0 = 0$. This is (2).

(2) $\Rightarrow$ (3). Assume that (2) is true. Certainly $A$ can be carried to a reduced, row-echelon matrix $R$; we show that $R = I$. Suppose on the contrary that $R \neq I$. Then $R$ has a row of zeros (being square), so the augmented matrix $[A \;\; 0]$ of the system $AX = 0$ also has a row of zeros when carried to row-echelon form. Since $A$ is square, this means that there is at least one nonleading variable, and hence at least one parameter. Thus $AX = 0$ has infinitely many solutions, contrary to (2). So $R = I$ after all.

(3) $\Rightarrow$ (4). Consider the augmented matrix $[A \;\; B]$ of the system $AX = B$. Using (3), let $A \rightarrow I$ by a sequence of row operations. Then these same operations carry $[A \;\; B] \rightarrow [I \;\; C]$ for some column $C$. Hence, the system $AX = B$ has a solution (in fact unique) by Gaussian elimination.

(4) $\Rightarrow$ (5). By (4), assume that the system $AX = B$ has a solution for *any* choice of the column $B$. If we take $B = E_j$ where $E_j$ is column $j$ of the identity matrix $I_n$, this gives a column $C_j$ such that $AC_j = E_j$. Now let $C = [C_1 \;\; C_2 \;\; \cdots \;\; C_n]$ be the $n \times n$ matrix with these matrices $C_j$ as its columns. Then the definition of matrix multiplication gives

$$AC = A[C_1 \;\; C_2 \;\; \cdots \;\; C_n] = [AC_1 \;\; AC_2 \;\; \cdots \;\; AC_n] = [E_1 \;\; E_2 \;\; \cdots \;\; E_n] = I_n.$$

This proves that (5) is true.

(5) $\Rightarrow$ (1). Assume that (5) is true so that $AC = I$ for some matrix $C$.
*Claim.* If $CX = 0$, then $X = 0$.
*Proof.* If $0 = CX$, left multiply by $A$ to get $X = IX = ACX = A0 = 0$, which proves the Claim.

Thus condition (2) holds for the matrix $C$ rather than $A$. Hence the argument above that (2) $\Rightarrow$ (3) $\Rightarrow$ (4) $\Rightarrow$ (5) (with $A$ replaced by $C$) shows that a matrix $C'$ exists such that $CC' = I$. But then

$$A = AI = A(CC') = (AC)C' = IC' = C'.$$

Thus $CA = CC' = I$ which, together with $AC = I$, shows that $C$ is the inverse of $A$. This proves (1).

---

[17] If $p$ and $q$ are statements, we say that $p$ **implies** $q$ (written $p \Rightarrow q$) if $q$ is true whenever $p$ is true. The statements are called **equivalent** if both $p \Rightarrow q$ and $q \Rightarrow p$ (written $p \Leftrightarrow q$, spoken "$p$ if and only if $q$").

Theorem 5 has some wonderful consequences. First, to verify that a square matrix $A$ is invertible, it is enough to find some matrix $C$ such that either $AC = I$ or $CA = I$. Recall that the definition of invertibility requires that *both* conditions are satisfied.

**COROLLARY 1**

If $AC = I$ where $A$ and $C$ are square matrices, then $CA = I$ also. In this case, $A$ and $C$ are both invertible, $C = A^{-1}$, and $A = C^{-1}$.

**PROOF**

If $AC = I$, Theorem 5 shows that $A$ is invertible. But then left multiplication by $A^{-1}$ gives

$$A^{-1} = A^{-1}I = A^{-1}(AC) = (A^{-1}A)C = IC = C.$$

Thus $C$ is invertible (by Theorem 3) and $C^{-1} = (A^{-1})^{-1} = A$. This proves Corollary 1.

Observe that Corollary 1 is false if $A$ and $C$ are not square matrices. For example, we have

$$\begin{bmatrix} 1 & 2 & 1 \\ 1 & 1 & 1 \end{bmatrix}\begin{bmatrix} -1 & 1 \\ 1 & -1 \\ 0 & 1 \end{bmatrix} = I_2 \quad \text{but} \quad \begin{bmatrix} -1 & 1 \\ 1 & -1 \\ 0 & 1 \end{bmatrix}\begin{bmatrix} 1 & 2 & 1 \\ 1 & 1 & 1 \end{bmatrix} \neq I_3.$$

In fact, we can say more. It is quite plausible that $AC = I_m$ and $CA = I_n$ could happen for some $m \times n$ matrix $A$ and some $n \times m$ matrix $C$. We claim that these requirements force $m = n$. Suppose, for example, that $m < n$. Then the system of equations

$$AX = 0$$

has a nontrivial solution $X \neq 0$ by Theorem 1 §1.3. But then $X = I_nX = CAX = C0 = 0$, a contradiction. Thus $m < n$ cannot be true. Similarly we show that $n < m$ cannot happen by considering the system $CX = 0$. So the only possibility is $m = n$. This proves another consequence of Theorem 5:

**COROLLARY 2**

The only invertible matrices are square. That is, if $A$ is an $m \times n$ matrix, and if $AC = I_m$ and $CA = I_n$ hold for some $n \times m$ matrix $C$, then $m = n$.

## Exercises 1.5

**1.** If $A = \begin{bmatrix} -5 & 2 \\ 2 & -1 \end{bmatrix}$, then show that $A^{-1} = \begin{bmatrix} -1 & -2 \\ -2 & -5 \end{bmatrix}$ by verifying that $AA^{-1} = I = A^{-1}A$. What is $(A^{-1})^{-1}$?

**2.** In each case, use the matrix inversion algorithm to find the inverse of the given matrix.

**a)** $\begin{bmatrix} 3 & -7 \\ -2 & 5 \end{bmatrix}$ **b)** $\begin{bmatrix} 3 & 5 \\ 1 & 2 \end{bmatrix}$

**c)** $\begin{bmatrix} 1 & 0 & 1 \\ 2 & 2 & 3 \\ 0 & 3 & 1 \end{bmatrix}$ **d)** $\begin{bmatrix} 1 & 3 & -1 \\ -1 & 0 & 0 \\ 0 & 2 & -1 \end{bmatrix}$

**e)** $\begin{bmatrix} 3 & 5 & 1 \\ 1 & 2 & -1 \\ -1 & 0 & +1 \end{bmatrix}$ **f)** $\begin{bmatrix} 2 & 1 & 2 \\ -3 & -1 & -1 \\ 5 & 2 & 1 \end{bmatrix}$

**g)** $\begin{bmatrix} 1 & 1 & 0 & 1 \\ -1 & 0 & 1 & -1 \\ 5 & 7 & 3 & 5 \\ 2 & 5 & 6 & 1 \end{bmatrix}$ **h)** $\begin{bmatrix} 4 & 0 & 2 & -1 \\ -1 & 0 & 1 & 0 \\ 5 & 0 & -2 & -1 \\ 2 & -1 & 0 & 0 \end{bmatrix}$

**3.** Write your own version of the matrix inverse algorithm.

**4.** If $A = \begin{bmatrix} 3 & -7 & 2 \\ -2 & 5 & -3 \\ 1 & -2 & 0 \end{bmatrix}$, find $A^{-1}$ and $(A^{-1})^{-1}$. Explain your result.

**5.** In each case find the matrix $A$.

**a)** $(3A)^{-1} = \begin{bmatrix} 1 & 0 \\ -5 & -1 \end{bmatrix}$ **b)** $(5A)^T = \begin{bmatrix} 2 & -3 \\ -1 & 4 \end{bmatrix}^{-1}$

**c)** $[2A^T - 3I]^{-1} = \begin{bmatrix} 3 & 2 \\ 1 & 1 \end{bmatrix}$ **d)** $(A^{-1} - 3I)^T = 5\begin{bmatrix} 1 & 2 \\ 3 & 4 \end{bmatrix}$

**e)** $\left[A^T - 3\begin{bmatrix} 1 & 0 \\ 2 & -1 \end{bmatrix}\right]^{-1} = \begin{bmatrix} 3 & 1 \\ 1 & 1 \end{bmatrix}$

**f)** $\left[2\begin{bmatrix} 1 & 1 \\ -2 & 3 \end{bmatrix} - 5A^{-1}\right]^T = (4A^T)^{-1}$

**6.** Write a strategy to solve a system of linear equations $AX = B$ if $A$ is invertible.

**7.** In each case, solve the system by finding the inverse of the coefficient matrix if possible.

**a)** $2x + 3y = -2$
$\quad\ x - y = 5$

**b)** $3x + 4y = 3$
$\quad 2x + 2y = -1$

**c)** 
$$\begin{aligned} x + y + z &= 0 \\ x + y + 2z &= -2 \\ 2x + y + 4z &= 3 \end{aligned}$$
**d)** 
$$\begin{aligned} 3x + 5y &= -5 \\ x + 2y + z &= 1 \\ 3x + 7y + z &= 0 \end{aligned}$$

**8.** What conclusion can you draw about the number of solutions for a system of linear equations $AX = B$ if $A$ is invertible? Does it depend on $B$?

**9.** Assume that you are given that $A^{-1} = \begin{bmatrix} 1 & 0 & 2 \\ 1 & 2 & 1 \\ 3 & 5 & 3 \end{bmatrix}$.

    **a)** Find $X$ such that $AX = \begin{bmatrix} 2 & -1 \\ 1 & 0 \\ 0 & -3 \end{bmatrix}$.

    **b)** Find $X$ such that $XA = \begin{bmatrix} 2 & 3 & -1 \\ -1 & 0 & 5 \end{bmatrix}$.

**10.** If $A$ is an invertible $n \times n$ matrix, show that $AX = B$ has a unique solution for any $n \times k$ matrix $B$. [Hint: Theorem 2.]

**11.** Suppose that $CA = I_n$ where $C$ is $n \times m$ and $A$ is $m \times n$.
    **a)** Show that the system $AX = B$ has a solution for every $n \times 1$ column $B$.
    **b)** Show that the solution in **a)** is uniquely determined by $B$.

**12.** Show that $A = \begin{bmatrix} a & 0 \\ b & 0 \end{bmatrix}$ is not invertible for any choice of $a$ and $b$.

**13.** In each case either show that the statement is true or give an example showing that it is false. Assume throughout that $A$ and $B$ are square matrices.
    **a)** If $A \neq 0$, then $A$ is invertible.
    **b)** If $A$ is invertible, then $A \neq 0$.
    **c)** If $A^3 = 3I$, then $A$ is invertible.
    **d)** If $A^2 = A$ and $A \neq 0$, then $A$ is invertible.
    **e)** If $A$ and $B$ are both invertible, then $A + B$ is invertible.
    **f)** If $AB = 0$ and $A \neq 0$, then $B = 0$.
    **g)** If $A$ is invertible and $AC = I$, then $C = A^{-1}$.
    **h)** If $AX = B$ has no solution for some column $B$, then $AX = 0$ has no solution.
    **i)** If $AX = 0$ has only the trivial solution, then $AX = B$ has a unique solution for every column $B$.
    **j)** If $A^2$ is invertible, then $A$ is invertible.

**14.** If $U = \begin{bmatrix} 3 & -4 \\ 7 & 5 \end{bmatrix}$ and $AU = 0$ for some matrix $A$, show that necessarily $A = 0$.

**15.** Let $A = \begin{bmatrix} 5 & -3 \\ 6 & -4 \end{bmatrix}$.
    **a)** Show that $A^2 - A - 2I = 0$.
    **b)** Use **a)** to obtain a formula for $A^{-1}$ in terms of $A$.

**16.** **a)** Simplify the matrix product $(I - A)(I + A)$. If $A^2 = 0$, show that $I - A$ is invertible and that $(I - A)^{-1} = I + A$.

    **b)** If $A^3 = 0$, show that $I - A$ is invertible and that $(I - A)^{-1} = I + A + A^2$.
    **c)** Generalize.

**17.** Consider the matrix $A = \begin{bmatrix} 2 & 1 & -1 \\ -3 & -3 & 2 \\ 1 & 3 & -1 \end{bmatrix}$.
    **a)** Find the inverse of $A$ using the matrix inversion algorithm.
    **b)** Now that you know that the matrix $A$ is invertible, how could you convince someone that a system of linear equations $AX = B$ has a unique solution for any $B$? If you fix some $B$ and transform the augmented matrix $[A \ B]$ to reduced row-echelon form, what connection do you see with the computations in part **a)**?
    **c)** If $AX = B$ has a unique solution for any $B$, what can you say about the homogeneous system $AX = 0$?
    **d)** Solve $AX = 0$ by transforming the augmented matrix $[A \ 0]$ to reduced row-echelon form; what happens to $A$?
    **e)** Now since $A$ can be carried to $I$ by elementary row operations, what does that say about solving a system $AX = B$? Use this to find a matrix $C$ such that $AC = I$ by successively finding each column of $C$.
    **f)** Now that $AC = I$, verify that $CA = I$ as well. Will this happen all the time?
    **g)** What happens to all these conditions for
$$A = \begin{bmatrix} 1 & 1 & 2 \\ -2 & -3 & -5 \\ -1 & 1 & 0 \end{bmatrix}?$$

**18.** If $A$ is a square matrix, a matrix of the form $P^{-1}AP$ where $P$ is invertible is called a **conjugate** of $A$.
    **a)** Is $P^{-1}AP = A$? Explain.
    **b)** If some conjugate of $A$ is invertible, show that $A$ is also invertible.
    **c)** Show that $(P^{-1}AP)^2 = P^{-1}A^2P$ and $(P^{-1}AP)^3 = P^{-1}A^3P$. Generalize.

**19.** **a)** If $P$ is invertible, solve the equation $P^{-1}XP = B$ for $X$.
    **b)** If $P$ and $Q$ are invertible, solve the equation $P(A + X)Q = B$ for $X$.

**20.** **a)** If $A$ is invertible and $A$ commutes with $C$, show that $A^{-1}$ also commutes with $C$.
    **b)** If $A$ and $B$ are invertible and commute, show that $A^{-1}$ and $B^{-1}$ also commute.

**21.** **a)** If the first row of a matrix $A$ consists of zeros, show that $A^{-1}$ does not exist.
    **b)** If the first column of a matrix $A$ consists of zeros, show that $A^{-1}$ does not exist.

**22.** Find the inverse of $A = \begin{bmatrix} \sin\theta & \cos\theta \\ -\cos\theta & \sin\theta \end{bmatrix}$ for any angle $\theta$.

**23.** A matrix $A = \begin{bmatrix} B & X \\ 0 & C \end{bmatrix}$ in block form is called a **block upper triangular matrix** if $B$ and $C$ are both square matrices.

   **a)** If $B$ and $C$ are invertible show that $A$ is invertible and
   $$A^{-1} = \begin{bmatrix} B^{-1} & -B^{-1}XC^{-1} \\ 0 & C^{-1} \end{bmatrix}$$ in block form.
   [Hint: Theorem 4 §1.4.]

   **b)** If $A$ is invertible, show that $B$ and $C$ must also be invertible.

**24.** If $AB = AC$ where $A$ is an invertible matrix, show that $B = C$.

**25.** **a)** If $AB = BA$, show that $(AB)^2 = A^2B^2$.

   **b)** If $A$ and $B$ are both invertible and $(AB)^2 = A^2B^2$, show that $AB = BA$.

**26.** Suppose that $A$ and $B$ are nonzero square matrices such that $AB = 0$. Show that neither $A$ nor $B$ has an inverse.

**27.** What is wrong with the following solution to Example 12? Support your answer.

"*Solution*": Since $AB$ is invertible, we have $(AB)^{-1} = B^{-1}A^{-1}$ by Theorem 3, so $B^{-1} = (AB)^{-1}A$.

**28.** **a)** If $A$ and $B$ are square matrices and $BA$ is invertible, show that $A$ and $B$ are invertible.

   **b)** If $A$, $B$, and $C$ are square matrices and $ABC$ is invertible, show that $A$, $B$, and $C$ are invertible.

**29.** A matrix $A$ is called **selfinverse** if $A^{-1} = A$. Recall that $E$ is an idempotent if $E^2 = E$.

   **a)** Show that $A$ is selfinverse if and only if $A^2 = I$.

   **b)** If $A$ is selfinverse, show that $E = \frac{1}{2}(I - A)$ is an idempotent.

   **c)** If $E$ is an idempotent, show that $A = I - 2E$ is selfinverse.

**30.** Prove (7) of Theorem 3 by verifying directly that $\frac{1}{c}A^{-1}$ is the inverse of $cA$ and using Theorem 1.

**31.** If $A$ is a square matrix, show that the following conditions are equivalent: (1) $A$ is invertible; (2) If $YA = 0$, $Y$ a row, then $Y = 0$. [Hint: Apply Theorem 5(2) to $A^T$.]

## 1.6  ELEMENTARY MATRICES[18]

By now the reader will be convinced of the importance of elementary row operations. It turns out that these operations can be performed by left multiplication by certain invertible matrices. This gives a new way of looking at matrix inverses, and has other uses as well.

### 1.6.1  Elementary Matrices

A square matrix $E$ that is obtained by doing a single elementary row operation to an identity matrix is called an **elementary matrix**. We say that $E$ is of Type I, II, or III when the corresponding row operation is of type I, II, or III (see page 14). Here are some examples, one of each type.

**Example 1**  $E_1 = \begin{bmatrix} 1 & 0 & 0 \\ 0 & 0 & 1 \\ 0 & 1 & 0 \end{bmatrix}$, $E_2 = \begin{bmatrix} 1 & 0 & 0 \\ 0 & 7 & 0 \\ 0 & 0 & 1 \end{bmatrix}$ and $E_3 = \begin{bmatrix} 1 & 0 & -3 \\ 0 & 1 & 0 \\ 0 & 0 & 1 \end{bmatrix}$ are elementary matrices of type I, II,

and III respectively, obtained by performing the following elementary row operations on the $3 \times 3$ identity matrix: (1) Interchange rows 2 and 3; (2) Multiply row 2 by 7; and (3) Subtract 3 times row 3 from row 1.

Consider the $2 \times 2$ elementary matrices $E_1 = \begin{bmatrix} 0 & 1 \\ 1 & 0 \end{bmatrix}$, $E_2 = \begin{bmatrix} 1 & 0 \\ 0 & 5 \end{bmatrix}$, and $E_3 = \begin{bmatrix} 1 & 3 \\ 0 & 1 \end{bmatrix}$. If a $2 \times 3$ matrix $A = \begin{bmatrix} a & b & c \\ x & y & z \end{bmatrix}$ is left multiplied by these elementary matrices, the results are:

---

[18] This section can be omitted at a first reading as it is used sparingly in the sequel.

$$E_1 A = \begin{bmatrix} 0 & 1 \\ 1 & 0 \end{bmatrix} \begin{bmatrix} a & b & c \\ x & y & z \end{bmatrix} = \begin{bmatrix} x & y & z \\ a & b & c \end{bmatrix}$$

$$E_2 A = \begin{bmatrix} 1 & 0 \\ 0 & 5 \end{bmatrix} \begin{bmatrix} a & b & c \\ x & y & z \end{bmatrix} = \begin{bmatrix} a & b & c \\ 5x & 5y & 5z \end{bmatrix}$$

$$E_3 A = \begin{bmatrix} 1 & 3 \\ 0 & 1 \end{bmatrix} \begin{bmatrix} a & b & c \\ x & y & z \end{bmatrix} = \begin{bmatrix} a+3x & b+3y & c+3z \\ x & y & z \end{bmatrix}$$

Observe that, in each case, $E_i A$ is the matrix obtained from $A$ by performing the elementary operation that created $E_i$ from $I_2$. This works in general.

**LEMMA 1** [19]   If an elementary row operation is performed on an $m \times n$ matrix $A$, the result is the matrix $EA$ where $E$ is the $m \times m$ elementary matrix created by performing the same row operation on $I_m$.

We omit the routine proof.

Any elementary row operation can be reversed by another elementary row operation (called its **inverse**). For example:

The inverse of multiplying row 3 by 7 is to multiply row 3 by $\frac{1}{7}$.

The inverse of adding 3 times row 2 to row 1 is to *subtract* 3 times row 2 from row 1.

The inverse of interchanging rows 2 and 3 is the *same* operation; interchange them again.

This observation translates into the fact that every elementary matrix $E$ is invertible; moreover, it gives a simple way to find the inverse matrix. In fact, since $E$ corresponds to some row operation, we let $F$ be the elementary matrix corresponding to the *inverse* operation. Since $E$ is created by doing a row operation to $I$, then doing the inverse operation will carry $E$ back to $I$. But Lemma 1 shows that $E$ is carried to $FE$ by the inverse operation, whence $FE = I$. This shows that $F$ is the inverse of the matrix $E$, and so proves

**LEMMA 2**   Every elementary matrix $E$ is invertible, and $E^{-1}$ is the elementary matrix (of the same type) obtained from $I$ by the inverse of the operation that produced $E$ from $I$.

**Example 2**   Given elementary matrices

$$E_1 = \begin{bmatrix} 1 & 0 & 5 \\ 0 & 1 & 0 \\ 0 & 0 & 1 \end{bmatrix}, \; E_2 = \begin{bmatrix} 1 & 0 & 0 \\ 0 & 3 & 0 \\ 0 & 0 & 1 \end{bmatrix} \text{ and } E_3 = \begin{bmatrix} 1 & 0 & 0 \\ 0 & 0 & 1 \\ 0 & 1 & 0 \end{bmatrix}, \text{ write down the inverses.}$$

**SOLUTION**   The elementary matrix $E_1$ corresponds to adding 5 times row 3 of $I_3$ to row 1, so the inverse operation is *subtracting* 5 times row 3 from row 1. Hence $E_1^{-1} = \begin{bmatrix} 1 & 0 & -5 \\ 0 & 1 & 0 \\ 0 & 0 & 1 \end{bmatrix}$.

Similarly $E_2^{-1} = \begin{bmatrix} 1 & 0 & 0 \\ 0 & \frac{1}{3} & 0 \\ 0 & 0 & 1 \end{bmatrix}$ because the inverse of multiplying row 2 by 3 is to multiply row 2 by $\frac{1}{3}$. Finally, $E_3^{-1} = E_3$ because the corresponding operation (interchanging rows 2 and 3) is selfinverse.

---

[19] The term "lemma" means an auxiliary proposition used in the demonstration of another proposition.

We note in passing that an argument similar to Lemma 1 shows that if an elementary *column* operation is performed on a matrix $A$, the result is the matrix $AF$ where $F$ is an elementary matrix, and that $F$ can be obtained by doing the same column operation to the identity matrix.

1.6.2 **Applications to Inverses and Rank**

As in Gaussian elimination, suppose that a series of $k$ elementary row operations is applied to an arbitrary matrix $A$, and let $E_i$ denote the elementary matrix corresponding to the $i^{th}$ row operation. Then Lemma 1 asserts that step $i$ of the row-reduction is given by left-multiplication by $E_i$, so the reduction becomes

$$A \rightarrow E_1A \rightarrow E_2E_1A \rightarrow E_3E_2E_1A \rightarrow \cdots \rightarrow E_k\cdots E_2E_1A.$$

Thus

$$A \rightarrow UA \quad \text{where} \quad U = E_k\cdots E_2E_1.$$

This matrix $U$ is invertible because each $E_i$ is invertible (by Lemma 2). Moreover $U$ can be easily constructed *without* finding the $E_i$. Indeed, if we apply the same sequence of row operations to the identity matrix $I$ (in place of $A$), the result is $I \rightarrow UI = U$. Thus these operations carry the "double matrix" $[A \ \ I]$ to $[UA \ \ U]$. If we denote this as $[A \ \ I] \rightarrow [UA \ \ U]$, we get part (2) of the following theorem; the rest of the theorem follows from the above discussion.

**THEOREM 1**

Suppose a series of row operations carries $A \rightarrow B$. Then:

(1) $B = UA$ for some invertible matrix $U$.

(2) $U$ can be constructed by performing the same row operations on the double matrix $[A \ \ I]$:
$$[A \ \ I] \rightarrow [B \ \ U]$$

(3) $U = E_k\cdots E_2E_1$ where $E_1, E_2, \cdots, E_k$ are the elementary matrices corresponding in order to the row operations carrying $A$ to $B$.

**Example 3**

Find the reduced row-echelon form $R$ of $A = \begin{bmatrix} 3 & -2 & 5 \\ 1 & -1 & 0 \end{bmatrix}$, and express it as $R = UA$ where $U$ is invertible.

**SOLUTION**

The reduction of the double matrix $[A \ \ I]$ is as follows:

$$\begin{bmatrix} 3 & -2 & 5 & 1 & 0 \\ 1 & -1 & 0 & 0 & 1 \end{bmatrix} \rightarrow \begin{bmatrix} 1 & -1 & 0 & 0 & 1 \\ 3 & -2 & 5 & 1 & 0 \end{bmatrix}$$

$$\rightarrow \begin{bmatrix} 1 & -1 & 0 & 0 & 1 \\ 0 & 1 & 5 & 1 & -3 \end{bmatrix}$$

$$\rightarrow \begin{bmatrix} 1 & 0 & 5 & 1 & -2 \\ 0 & 1 & 5 & 1 & -3 \end{bmatrix}$$

Hence $R = \begin{bmatrix} 1 & 0 & 5 \\ 0 & 1 & 5 \end{bmatrix}$ and $U = \begin{bmatrix} 1 & -2 \\ 1 & -3 \end{bmatrix}$. The reader can verify that $UA = R$.

Theorem 1 is a generalization of the matrix inversion algorithm given in Section 1.5. Indeed, suppose $A$ is invertible so that $A \rightarrow I$ by row operations by Theorem 5 §1.5. If we take $B = I$, the reduction in Theorem 1 is $[A \ \ I] \rightarrow [I \ \ U]$. But part (1) of Theorem 1 asserts that $UA = B = I$, so $U = A^{-1}$. Thus the reduction becomes

$$[A \ \ I] \rightarrow [I \ \ A^{-1}].$$

This is the matrix inversion algorithm.

However, we can say more. We have $U = E_k \cdots E_2 E_1$ by part (3) of Theorem 1 where $E_1$, $E_2$, $\cdots$, $E_k$ are the elementary matrices corresponding (in order) to the row operations carrying $A$ to $I$. Hence

$$A = (A^{-1})^{-1} = U^{-1} = (E_k \cdots E_2 E_1)^{-1} = E_1^{-1} E_2^{-1} \cdots E_k^{-1}$$

Since each $E_i^{-1}$ is an elementary matrix by Lemma 2, this shows that $A$ is a product of elementary matrices. On the other hand, every product of elementary matrices is invertible (by Lemma 2), so this proves

**THEOREM 2**   A square matrix $A$ is invertible if and only if $A$ is a product of elementary matrices.

**Example 4**   Express the invertible matrix $A = \begin{bmatrix} 2 & 1 \\ 1 & -1 \end{bmatrix}$ as a product of elementary matrices.

**SOLUTION**   Because $A$ is invertible, its reduced row-echelon form is $I$. The reduction is

$$\begin{bmatrix} 2 & 1 \\ 1 & -1 \end{bmatrix} \xrightarrow{E_1} \begin{bmatrix} 1 & -1 \\ 2 & 1 \end{bmatrix} \xrightarrow{E_2} \begin{bmatrix} 1 & -1 \\ 0 & 3 \end{bmatrix} \xrightarrow{E_3} \begin{bmatrix} 1 & -1 \\ 0 & 1 \end{bmatrix} \xrightarrow{E_4} \begin{bmatrix} 1 & 0 \\ 0 & 1 \end{bmatrix}$$

where the corresponding elementary matrices are

$$E_1 = \begin{bmatrix} 0 & 1 \\ 1 & 0 \end{bmatrix} \quad E_2 = \begin{bmatrix} 1 & 0 \\ -2 & 1 \end{bmatrix} \quad E_3 = \begin{bmatrix} 1 & 0 \\ 0 & \frac{1}{3} \end{bmatrix} \quad E_4 = \begin{bmatrix} 1 & 1 \\ 0 & 1 \end{bmatrix}.$$

Thus the reduction is $A \to E_1 A \to E_2 E_1 A \to E_3 E_2 E_1 A \to E_4 E_3 E_2 E_1 A = I$ so, using Lemma 2,

$$A = (E_4 E_3 E_2 E_1)^{-1} = E_1^{-1} E_2^{-1} E_3^{-1} E_4^{-1}$$

$$= \begin{bmatrix} 0 & 1 \\ 1 & 0 \end{bmatrix} \begin{bmatrix} 1 & 0 \\ 2 & 1 \end{bmatrix} \begin{bmatrix} 1 & 0 \\ 0 & 3 \end{bmatrix} \begin{bmatrix} 1 & -1 \\ 0 & 1 \end{bmatrix}.$$

This is the desired factorization.

Theorem 1 has another application. Suppose that an $m \times n$ matrix $A$ is given, and let $A \to R$ by row operations where $R$ is a reduced row-echelon matrix. Then Theorem 1 gives $R = UA$ where the invertible matrix $U$ can be found by

$$[A \quad I_m] \to [R \quad U].$$

If $\text{rank} A = r$, then $R$ has $r$ nonzero rows and, because $R$ is reduced, the top $r$ rows contain each column of the $r \times r$ identity matrix $I_r$. Hence elementary *column* operations will carry $R$ to the block $m \times n$ matrix $\begin{bmatrix} I_r & 0 \\ 0 & 0 \end{bmatrix}$. Furthermore, these column operations can be accomplished by doing the corresponding *row* operations to $R^T$ to obtain its $n \times m$ row-echelon form: $R^T \to U_1 R^T = \begin{bmatrix} I_r & 0 \\ 0 & 0 \end{bmatrix}$ where $U_1$ is $n \times n$ and invertible. If we write $V = U_1^T$, we get

$$UAV = RV = RU_1^T = (U_1 R^T)^T = \begin{bmatrix} I_r & 0 \\ 0 & 0 \end{bmatrix}^T = \begin{bmatrix} I_r & 0 \\ 0 & 0 \end{bmatrix}.$$

By Theorem 1, the matrix $U_1 = V^T$ can be obtained by

$$[R^T \quad I_n] \to \left[ \begin{bmatrix} I_r & 0 \\ 0 & 0 \end{bmatrix} \quad V^T \right].$$

Hence, we get

**THEOREM** ❸ Let $A$ be an $m \times n$ matrix of rank $r$. Then there exist invertible matrices $U$ and $V$ (of sizes $m \times m$ and $n \times n$ respectively) such that

$$UAV = \begin{bmatrix} I_r & 0 \\ 0 & 0 \end{bmatrix}.$$

Moreover, $U$ and $V$ can be computed in two steps:

**Step 1.** $[A \quad I_m] \rightarrow [R \quad U]$ where $R$ is a reduced row-echelon matrix.

**Step 2.** $[R^T \quad I_n] \rightarrow \begin{bmatrix} \begin{bmatrix} I_r & 0 \\ 0 & 0 \end{bmatrix} & V^T \end{bmatrix}.$

The procedure in Theorem 3 to compute $U$ and $V$ is easily adapted for a computer.

**Example 5** If $A = \begin{bmatrix} 1 & -2 & 3 & 1 \\ -1 & 2 & -1 & 1 \\ 2 & -4 & 5 & 1 \end{bmatrix}$, show that $rankA = 2$, and find invertible matrices $U$ and $V$ such that $UAV = \begin{bmatrix} I_2 & 0 \\ 0 & 0 \end{bmatrix}.$

**SOLUTION** The reduction in Step 1 of Theorem 3 is $[A \quad I_3] \rightarrow [R \quad U]$ where, as the reader can verify,

$$R = \begin{bmatrix} 1 & -2 & 0 & -2 \\ 0 & 0 & 1 & 1 \\ 0 & 0 & 0 & 0 \end{bmatrix} \quad \text{and} \quad U = \begin{bmatrix} -5 & 0 & 3 \\ 2 & 0 & -1 \\ -3 & 1 & 2 \end{bmatrix}$$

Thus $rankA = 2$, so Step 2 of Theorem 3 is the reduction

$$[R^T \quad I_4] \rightarrow \begin{bmatrix} \begin{bmatrix} I_2 & 0 \\ 0 & 0 \end{bmatrix} & V^T \end{bmatrix}$$

where

$$V = \begin{bmatrix} 1 & 0 & 2 & 2 \\ 0 & 0 & 1 & 0 \\ 0 & 1 & 0 & -1 \\ 0 & 0 & 0 & 1 \end{bmatrix}.$$ The reader can verify that $UAV = \begin{bmatrix} I_2 & 0 \\ 0 & 0 \end{bmatrix}.$

## 1.6.3 Uniqueness of Reduced Row-echelon Matrices[20]

We stated earlier that the reduced row-echelon form of a matrix is uniquely determined. We conclude this section with a proof of this.

**THEOREM** ❹ If $A$ is a matrix and $A \rightarrow R$ and $A \rightarrow S$ by row operations where $R$ and $S$ are both reduced row-echelon matrices, then $R = S$.

By Theorem 1 there are invertible matrices $P$ and $Q$ such that $R = PA$ and $S = QA$. Then $U = QP^{-1}$ is an invertible matrix with the property that $UR = S$. Hence we prove the following statement:

---

[20] This section is not used in the sequel, and so may be omitted.

**PROPOSITION**    Suppose $UR = S$ where $R$ and $S$ are $m \times n$ reduced row-echelon matrices and $U$ is invertible. Then $R = S$.

**PROOF**    We proceed by induction on $m$. The case $m = 1$ is left to the reader. Let $R_j$ and $S_j$ denote column $j$ of $R$ and $S$ respectively. Then matrix multiplication gives

$$UR_j = S_j \quad \text{for each } j. \tag{$*$}$$

Since $U$ is invertible, it follows that $R_j = 0$ if and only if $S_j = 0$; that is $R$ and $S$ have the same zero columns. Hence, by passing to the matrices obtained by deleting all zero columns from $R$ and $S$, we may assume that there are no zero columns.

But then the first column from the left in $R$ and $S$ is column 1 of $I_m$ (both are row-echelon). Applying $(*)$ to this column shows that column 1 of $U$ equals column 1 of $I_m$. Hence write $U$, $R$, and $S$ in block form as follows:

$$U = \begin{bmatrix} 1 & X \\ 0 & V \end{bmatrix}, \quad R = \begin{bmatrix} 1 & Y \\ 0 & R' \end{bmatrix}, \quad \text{and} \quad S = \begin{bmatrix} 1 & Z \\ 0 & S' \end{bmatrix}$$

Since $UR = S$, block multiplication shows that $VR' = S'$. Hence $R' = S'$ by induction because $V$ is invertible ($U$ is invertible).

It follows that $R$ and $S$ have leading 1's in the same columns, say $k$ of them. Applying $(*)$ to these columns then yields that the first $k$ columns of $U$ are the first $k$ columns of $I_m$. Hence we can write $U$, $R$, and $S$ in block form as follows:

$$U = \begin{bmatrix} I_k & M \\ 0 & W \end{bmatrix}, \quad R = \begin{bmatrix} R_1 & R_2 \\ 0 & 0 \end{bmatrix}, \quad S = \begin{bmatrix} S_1 & S_2 \\ 0 & 0 \end{bmatrix}$$

where $R_1$ and $S_1$ are $k \times k$. But then $UR = R$ by block multiplication, that is $S = R$. This completes the proof.

## Exercises 1.6

**1.** For each of the following elementary matrices, describe the corresponding row operation and write the inverse.

**a)** $E = \begin{bmatrix} 0 & 1 & 0 \\ 1 & 0 & 0 \\ 0 & 0 & 1 \end{bmatrix}$    **b)** $E = \begin{bmatrix} 1 & 0 & 0 \\ 5 & 1 & 0 \\ 0 & 0 & 1 \end{bmatrix}$

**c)** $E = \begin{bmatrix} 2 & 0 & 0 \\ 0 & 1 & 0 \\ 0 & 0 & 1 \end{bmatrix}$    **d)** $E = \begin{bmatrix} 1 & 0 & -3 \\ 0 & 1 & 0 \\ 0 & 0 & 1 \end{bmatrix}$

**2.** In each case find elementary matrices $E$ and $F$ such that $B = EA$ and $A = FB$. What is the relationship between $E$ and $F$?

**a)** $A = \begin{bmatrix} 2 & -1 \\ 5 & 3 \end{bmatrix}$    $B = \begin{bmatrix} 2 & -1 \\ 1 & 5 \end{bmatrix}$

**b)** $A = \begin{bmatrix} 3 & 2 \\ -5 & 7 \end{bmatrix}$    $B = \begin{bmatrix} -2 & 9 \\ -5 & 7 \end{bmatrix}$

**c)** $A = \begin{bmatrix} 3 & -2 & 0 \\ 1 & 5 & 6 \\ -2 & 1 & 3 \end{bmatrix}$    $B = \begin{bmatrix} 3 & -2 & 0 \\ -3 & 7 & 12 \\ -2 & 1 & 3 \end{bmatrix}$

**d)** $A = \begin{bmatrix} 1 & 3 & -7 \\ -1 & 2 & -5 \\ -3 & 0 & 6 \end{bmatrix}$    $B = \begin{bmatrix} 1 & 3 & -7 \\ -1 & 2 & -5 \\ -1 & 6 & -8 \end{bmatrix}$

**3.** In each case find elementary matrices $E_1$ and $E_2$ such that $B = E_2 E_1 A$.

**a)** $A = \begin{bmatrix} 3 & -5 \\ 2 & 4 \end{bmatrix}$    $B = \begin{bmatrix} 0 & -11 \\ 1 & 2 \end{bmatrix}$

**b)** $A = \begin{bmatrix} 5 & -3 \\ 2 & 1 \end{bmatrix}$    $B = \begin{bmatrix} 2 & 1 \\ 1 & -5 \end{bmatrix}$

**4.** Describe the connection between elementary row operations and elementary matrices.

**5.** In each case express the invertible matrix $A$ as a product of elementary matrices.

**a)** $A = \begin{bmatrix} 3 & -1 \\ 1 & 0 \end{bmatrix}$    **b)** $A = \begin{bmatrix} 2 & -1 \\ 1 & -1 \end{bmatrix}$

**6.** In each case find an invertible matrix $U$ such that $UA = R$ is the reduced row-echelon form of $A$.

**a)** $A = \begin{bmatrix} 3 & -1 \\ -2 & 0 \end{bmatrix}$    **b)** $A = \begin{bmatrix} 3 & -1 \\ 1 & -3 \end{bmatrix}$

**c)** $A = \begin{bmatrix} 3 & -1 & 4 \\ -2 & 0 & -3 \\ -1 & 3 & 0 \end{bmatrix}$    **d)** $A = \begin{bmatrix} 1 & -3 & -1 \\ 1 & -3 & 5 \\ 3 & -9 & -1 \end{bmatrix}$

**7.** Explain why a matrix can have many row-echelon forms.

**8.** In each case find an invertible matrix $U$ such that $UA = B$.

**a)** $A = \begin{bmatrix} 3 & 1 \\ 2 & 5 \end{bmatrix}$    $B = \begin{bmatrix} 1 & -4 \\ 4 & -3 \end{bmatrix}$

**b)** $A = \begin{bmatrix} 2 & -3 \\ 5 & 1 \end{bmatrix}$    $B = \begin{bmatrix} 1 & 7 \\ 2 & -3 \end{bmatrix}$

**9.** What is the connection between invertible matrices and elementary matrices?

**10.** Find an elementary matrix $F$ such that $B = AF$, where $A = \begin{bmatrix} 2 & -3 \\ 5 & 7 \end{bmatrix}$ and $B = \begin{bmatrix} 8 & -3 \\ -9 & 7 \end{bmatrix}$.

**11.** Let $E$ be an elementary matrix. Show that $E^T$ is also elementary of the same type as $E$.

**12.** In each case either show that the statement is true or give an example showing that it is false.

**a)** If $B = EA$ where $E$ is elementary, then $A = FB$ for some elementary $F$.

**b)** The product of two elementary matrices is again elementary.

**c)** The transpose of an elementary matrix is again elementary.

**d)** If $A \rightarrow R$ and $B \rightarrow R$ where $R$ is a reduced row-echelon matrix, then $A = B$.

**13.** In each case determine $r = rankA$ and find invertible matrices $U$ and $V$ such that $UAV = \begin{bmatrix} I_r & 0 \\ 0 & 0 \end{bmatrix}$.

**a)** $A = \begin{bmatrix} 1 & -1 & 2 & 1 \\ 2 & -1 & 0 & 3 \\ 0 & 1 & -4 & 1 \end{bmatrix}$    **b)** $A = \begin{bmatrix} 1 & 1 & 0 & -1 \\ 3 & 2 & 1 & 1 \\ 1 & 0 & 1 & 3 \end{bmatrix}$

**14.** While trying to invert $A$, the double matrix $[A \ \ I]$ is carried to $[P \ \ Q]$. Show that $P = QA$.

**15.** Use Theorem 1 to show that the only matrix $A$ that can be carried to 0 by row operations is $A = 0$.

**16.** If $B = UA$ where $U$ is invertible, show that $A \rightarrow B$ by row operations.

**17. a)** Show that every $m \times n$ matrix $A$ can be factored as $A = UR$ where $U$ is invertible and $R$ is in reduced row-echelon form.

**b)** Show that $R$ in the factorization in **a)** is unique, that is, if $A = U_1R_1$ as in **a)**, then $R_1 = R$.

**c)** Show that the factorization in **a)** may not be unique in general.

**18.** Suppose that $A \rightarrow B$ by column operations, and we apply these same column operations to the double matrix $[A \ \ I]$ and get $[A \ \ I] \rightarrow [B \ \ V]$, explain why $V$ is invertible and $AV = B$.

**19.** Two matrices $A$ and $B$ are called **row-equivalent** (written $A \overset{r}{\sim} B$) if $A$ can be carried to $B$ by row operations.

**a)** Show that $A \overset{r}{\sim} B$ if and only if $B = UA$ for some invertible matrix $U$.

**b)** If $A \overset{r}{\sim} B$, show that also $B \overset{r}{\sim} A$.

**c)** If $A \overset{r}{\sim} B$ and $B \overset{r}{\sim} C$, show that $A \overset{r}{\sim} C$.

**d)** If $A$ and $B$ are both row-equivalent to some third matrix, show that $A \overset{r}{\sim} B$.

**e)** If $A \overset{r}{\sim} B$, show that $rankA = rankB$.

**f)** If $A \overset{r}{\sim} B$ and $A$ is invertible, show that $B$ is also invertible.

## 1.7 LU-FACTORIZATION

The solution of a system of linear equations $AX = B$ can be computed much more quickly if the matrix $A$ can be factored in the form $A = LU$ where $L$ and $U$ are matrices of a particularly nice form. Such factorizations play an important role in countless numerical applications. In this section we show that Gaussian elimination can be used to find such factorizations.

### 1.7.1 Triangular Matrices

As for square matrices, an $m \times n$ matrix $A$ is called **upper triangular** if each entry below and to the left of the main diagonal is zero. For example, every row-echelon matrix is upper triangular, as is each of the following matrices:

$$\begin{bmatrix} 2 & -3 & 0 & 1 \\ 0 & 1 & 3 & 0 \\ 0 & 0 & 0 & 2 \end{bmatrix} \qquad \begin{bmatrix} 0 & -1 & 0 & 5 & 2 \\ 0 & 0 & 0 & 5 & -2 \\ 0 & 0 & 1 & 0 & 3 \end{bmatrix} \qquad \begin{bmatrix} 1 & 2 & 3 \\ 0 & -2 & 5 \\ 0 & 0 & 0 \\ 0 & 0 & 0 \end{bmatrix}$$

Similarly, a matrix $A$ is called **lower triangular** if every entry above and to the right of the main diagonal is zero, equivalently if $A^T$ is upper triangular. A matrix is called **triangular** if it is either upper or lower triangular. The following result will be needed, and follows from the definition of matrix multiplication.

**LEMMA 1**   The product of two lower (upper) triangular matrices is again lower (upper) triangular.

One reason for the importance of triangular matrices is the ease with which a linear system $AX = B$ can be solved when the coefficient matrix $A$ is triangular.

**Example 1**   Solve the system

$$-x_1 - 2x_2 + x_3 + 3x_4 + 2x_5 = 2$$
$$x_3 + 5x_4 - 3x_5 = -1$$
$$2x_5 = 3$$

where the coefficient matrix is upper triangular.

**SOLUTION**   As for a row-echelon matrix, let $x_2 = s$ and $x_4 = t$ be parameters. We solve for $x_5$, $x_3$, and $x_1$ in that order. The third equation gives

$$x_5 = \tfrac{3}{2}$$

Then substitution into the second equation gives

$$x_3 = \tfrac{7}{2} - 5t$$

as the reader can verify. Finally substitute these into the first equation to get

$$x_1 = \tfrac{9}{2} - 2s - 2t.$$

This gives the general solution.

The method used in Example 1 is called **back substitution** because later variables are substituted into earlier equations. It works because the coefficient matrix is upper triangular. Similarly, if the matrix of coefficients is lower triangular, the system can be solved by **forward substitution** where earlier variables are substituted in later equations.

Now consider a system $AX = B$ where $A$ is an arbitrary matrix. If $A$ can be factored as $A = LU$ where $L$ is an invertible lower triangular and $U$ is upper triangular, the system $AX = B$ can be solved in two stages as follows:

1. First solve $LY = B$ for $Y$ by forward substitution.
2. Then solve $UX = Y$ for $X$ by back substitution.

Then $X$ is the solution to the system $AX = B$. Indeed, since $A = LU$, we have $AX = LUX = LY = B$ as required. The method adapts easily for use in computer programs.

## 1.7.2   LU-Factorization

Hence if $A$ is any $m \times n$ matrix, we try to factor $A$ as $A = LU$ where $L$ is lower triangular and $U$ is upper triangular. The idea is to carry $A$ by row operations to a row-echelon matrix $U$ (which is upper triangular), and look for ways to find $L$. Suppose that $k$ row operations are needed, and that the corresponding elementary matrices are $E_1$, $E_2$, $\cdots$, $E_k$. Then the reduction takes the form

$$A \to E_1 A \to E_2 E_1 A \to \cdots \to E_k \cdots E_2 E_1 A = U.$$

Thus $PA = U$ where $P = E_k \cdots E_2 E_1$ is invertible because each $E_i$ is invertible.

If we do not insist that $U$ is reduced, the row-reduction can be carried out by adding multiples of a row to rows below it, and possibly by row interchanges. The point is that no row of $A$ need ever be added to a row *above* it. Thus, apart from row-interchanges, the only row operations needed are those for which the corresponding elementary matrix $E_i$ is *lower* triangular. In particular, if no row interchanges are needed, the matrix $P$ is lower triangular by Lemma 1. Since $PA = U$ we obtain $A = P^{-1}U$, and $P^{-1}$ is also lower triangular by Theorem 4 §1.5. If we write $L = P^{-1}$, this proves the first part of the following theorem.

For convenience we say that a matrix $A$ can be **lower reduced** if it can be carried to a row-echelon matrix using no row interchanges (and adding no multiple of a row to a row above it). A square matrix $D$ is called **diagonal** if every entry off the main diagonal is zero.

**THEOREM**  Suppose that a matrix $A$ can be lower reduced to a row-echelon matrix $U$. Then $A$ can be factored as

$$A = LU$$

where $L$ is lower triangular and invertible, and $U$ is upper triangular and row-echelon. Moreover, if $A$ is invertible, this factorization is unique; that is, if $A = L_1U_1$ is another such factorization, then $L_1 = L$ and $U_1 = U$.

PROOF   It remains to prove uniqueness. If $LU = A = L_1U_1$, then $L_1^{-1}L = U_1U^{-1} = D$, say. This matrix $D$ is lower triangular ($L_1^{-1}L$) and upper triangular ($U_1U^{-1}$), and so is actually diagonal. But $D = U_1U^{-1}$ has 1's on the main diagonal (this is true of $U$ and $U_1$ since they are invertible and row-echelon). It follows that $D = I$, and so $L_1^{-1}L = I$ and $U_1U^{-1} = I$. Hence $L_1 = L$ and $U_1 = U$, as asserted.

A factorization $A = LU$ as in Theorem 1 is called an **LU-factorization** of $A$. Such a factorization may not exist (Exercise 6) because $A$ cannot be carried to row-echelon form without using at least one row interchange. A procedure for dealing with this situation will be outlined later.

If an LU-factorization $A = LU$ does exist, the above discussion shows that $U$ can be any row-echelon form for $A$ (obtained by lower reduction), and $L$ is the inverse of the product of elementary matrices needed to carry $A \to U$. However, there is a simpler way to obtain $L$ whereby the columns of $L$ are obtained one by one. The following examples illustrate the technique. For convenience, the first nonzero column from the left in a matrix $A$ is called the **leading column** of $A$.

**Example 2**   Find an LU-factorization for $A = \begin{bmatrix} 0 & 2 & -6 & -2 & 4 \\ 0 & -1 & 3 & 3 & 2 \\ 0 & -1 & 3 & 7 & 10 \end{bmatrix}$.

SOLUTION   We lower reduce $A$ to row echelon form as follows:

$$A = \begin{bmatrix} 0 & 2 & -6 & -2 & 4 \\ 0 & -1 & 3 & 3 & 2 \\ 0 & -1 & 3 & 7 & 10 \end{bmatrix}$$

$$\to \begin{bmatrix} 0 & 1 & -3 & -1 & 2 \\ 0 & 0 & 0 & 2 & 4 \\ 0 & 0 & 0 & 6 & 12 \end{bmatrix}$$

$$\to \begin{bmatrix} 0 & 1 & -3 & -1 & 2 \\ 0 & 0 & 0 & 1 & 2 \\ 0 & 0 & 0 & 0 & 0 \end{bmatrix} = U$$

The shaded columns are determined as follows: The first is the leading column of $A$, and it is used to create the first leading 1 (using lower reduction). At this stage, we are finished with row 1, and we repeat the procedure on the matrix consisting of the remaining rows. Thus the second shaded column is the leading column of this smaller matrix, and we create the second leading 1. As the remaining row is zero in this case, we are finished. Then $A = LU$ where

$$L = \begin{bmatrix} 2 & 0 & 0 \\ -1 & 2 & 0 \\ -1 & 6 & 1 \end{bmatrix}.$$

This matrix $L$ is obtained from the identity matrix by replacing the bottom of columns 1 and 2 by the leading columns (in order) that were shaded in the lower reduction. Note that $rank A = 2$ here, and this is the number of shaded columns.

The calculation in Example 2 works for any $m \times n$ matrix $A$ that can be lower reduced to row-echelon form. The procedure can be stated formally as follows:

**LU-ALGORITHM**    Let $A$ be an $m \times n$ matrix, and suppose $A \to U$ by lower reduction where $U$ is a row-echelon matrix. Then $A = LU$ where the lower triangular, invertible matrix $L$ is created as follows:

(1) If $A = 0$ take $L = I_m$ and $U = 0$.
(2) If $A \neq 0$ let $C_1$ denote the leading column of $A$. Then use $C_1$ to create the first leading 1 and make the rest of the entries in that column zero (using lower reduction). When this is done, let $A_2$ denote the resulting matrix with the first row deleted.
(3) If $A_2 \neq 0$ let $C_2$ denote its leading column and repeat Step 2 on $A_2$ to create $A_3$.
(4) Continue in this way until $U$ is reached where all rows below the last leading 1 consist of zeros.
(5) Create $L$ by placing $C_1, C_2, C_3, \cdots, C_r$ at the bottom of the first $r$ columns of $I_m$, where $r = rank A$.

The proof that $LU = A$ in the LU-algorithm involves induction and block multiplication; we omit the details.

**Example 3**    Find an LU-factorization for $A = \begin{bmatrix} 5 & -5 & 10 & 0 & 5 \\ -3 & 3 & 2 & 2 & 1 \\ -2 & 2 & 0 & -1 & 0 \\ 1 & -1 & 10 & 2 & 5 \end{bmatrix}$.

**SOLUTION**    The lower reduction to row-echelon form is as follows:

$$\begin{bmatrix} 5 & -5 & 10 & 0 & 5 \\ -3 & 3 & 2 & 2 & 1 \\ -2 & 2 & 0 & -1 & 0 \\ 1 & -1 & 10 & 2 & 5 \end{bmatrix} \to \begin{bmatrix} 1 & -1 & 2 & 0 & 1 \\ 0 & 0 & 8 & 2 & 4 \\ 0 & 0 & 4 & -1 & 2 \\ 0 & 0 & 8 & 2 & 4 \end{bmatrix}$$

$$\to \begin{bmatrix} 1 & -1 & 2 & 0 & 1 \\ 0 & 0 & 1 & \frac{1}{4} & \frac{1}{2} \\ 0 & 0 & 0 & -2 & 0 \\ 0 & 0 & 0 & 0 & 0 \end{bmatrix}$$

$$\to \begin{bmatrix} 1 & -1 & 2 & 0 & 1 \\ 0 & 0 & 1 & \frac{1}{4} & \frac{1}{2} \\ 0 & 0 & 0 & 1 & 0 \\ 0 & 0 & 0 & 0 & 0 \end{bmatrix}.$$

Hence $A = LU$ where $L = \begin{bmatrix} 5 & 0 & 0 & 0 \\ -3 & 8 & 0 & 0 \\ -2 & 4 & -2 & 0 \\ 1 & 8 & 0 & 1 \end{bmatrix}$ and $U = \begin{bmatrix} 1 & -1 & 2 & 0 & 1 \\ 0 & 0 & 1 & \frac{1}{4} & \frac{1}{2} \\ 0 & 0 & 0 & 1 & 0 \\ 0 & 0 & 0 & 0 & 0 \end{bmatrix}$ as the reader can verify.

**Example 4**  Find an LU-factorization for the invertible matrix

$$A = \begin{bmatrix} 1 & 2 & 1 \\ 2 & -1 & 1 \\ 1 & 1 & 2 \end{bmatrix}.$$

**SOLUTION**  The lower reduction to row-echelon form is

$$A = \begin{bmatrix} 1 & 2 & 1 \\ 2 & -1 & 1 \\ 1 & 1 & 2 \end{bmatrix} \rightarrow \begin{bmatrix} 1 & 2 & 1 \\ 0 & -5 & -1 \\ 0 & -1 & 1 \end{bmatrix}$$

$$\rightarrow \begin{bmatrix} 1 & 2 & 1 \\ 0 & 1 & \frac{1}{5} \\ 0 & 0 & \frac{6}{5} \end{bmatrix} \rightarrow \begin{bmatrix} 1 & 2 & 1 \\ 0 & 1 & \frac{1}{5} \\ 0 & 0 & 1 \end{bmatrix}$$

Hence $A = LU$ where $L = \begin{bmatrix} 1 & 0 & 0 \\ 2 & -5 & 0 \\ 1 & -1 & \frac{6}{5} \end{bmatrix}$ and $U = \begin{bmatrix} 1 & 2 & 1 \\ 0 & 1 & \frac{1}{5} \\ 0 & 0 & 1 \end{bmatrix}$.

If an invertible symmetric matrix $A$ has an LU-factorization, we can get a better result.[21] More precisely, suppose $A = MU$ where $U$ is a row-echelon form for $A$ and $M$ is lower triangular (the reason for denoting it as $M$ will be apparent soon). Then

$$MU = A = A^T = U^T M^T.$$

Now observe that $M$ and $U$ are both invertible (by Corollary 1 of Theorem 5 §1.5, since $A$ is invertible). Hence we get

$$(U^T)^{-1}M = M^T U^{-1} = D, \text{ say.}$$

This matrix $D$ is lower triangular $((U^T)^{-1}M)$ and upper triangular $(M^T U^{-1})$, and so is actually diagonal. Since $M = U^T D$, the factorization $A = MU$ becomes

$$A = U^T D U.$$

Finally, write $L = U^T$, so that $L$ is lower triangular with 1's on the main triangular (because $U$ is row-echelon and invertible). This proves the following theorem which is important in numerical analysis. (The proof of uniqueness is like that in Theorem 1, and is left as an exercise for the reader.)

**THEOREM 2**  If $A$ is a symmetric, invertible matrix which has an LU-factorization, then

$$A = LDL^T$$

where $D$ is diagonal and $L$ is lower triangular with 1's on the main diagonal. Moreover, this factorization is unique; that is, if $A = L_1 D_1 L_1^T$ is another such factorization, then $L_1 = L$ and $D_1 = D$.

The factorization in Theorem 2 is called an **LDLT-factorization** of the symmetric matrix $A$.

---

[21] But note that $\begin{bmatrix} 0 & 1 \\ 1 & 0 \end{bmatrix}$ does not have an LU-factorization.

**Example 5**   If $A$ is the symmetric invertible matrix in Example 4, we have (in the notation of that example but writing $M$ in place of $L$)

$$D = M^T U^{-1} = \begin{bmatrix} 1 & 2 & 1 \\ 0 & -5 & -1 \\ 0 & 0 & \frac{6}{5} \end{bmatrix} \begin{bmatrix} 1 & -2 & -\frac{3}{5} \\ 0 & 1 & -\frac{1}{5} \\ 0 & 0 & 1 \end{bmatrix} = \begin{bmatrix} 1 & 0 & 0 \\ 0 & -5 & 0 \\ 0 & 0 & \frac{6}{5} \end{bmatrix}.$$

The reader can verify that $A = LDL^T$ where $L = U^T$.

There is one special case of Theorem 2 that deserves mention. A symmetric matrix is called **positive definite** if $X^T A X > 0$ for every column $X \neq 0$ (here we regard the $1 \times 1$ matrix $X^T A X$ as a number). These matrices arise in geometry and in numerical analysis. They will be studied in Section 4.8.2 where several other characterizations are given and, in addition, it is shown that every positive definite matrix is invertible and admits an LU-factorization, and hence an LDLT-factorization.

## 1.7.3   Dealing with Row Interchanges

Every matrix $A$ can be carried to a row-echelon matrix $U$ by row operations. If we do not insist that $U$ is reduced, this can be done without adding any multiples of a row of $A$ to a row above it. If no row interchanges are required, we have seen that $A$ admits an LU-factorization. But some matrices have no LU-factorization (for example $A = \begin{bmatrix} 0 & 1 \\ 1 & 0 \end{bmatrix}$) and so require at least one row interchange. However, it turns out that if all the necessary row interchanges are carried out first, the resulting matrix requires no interchanges and so has an LU-factorization.

**THEOREM 3**   Suppose an $m \times n$ matrix $A$ is carried to a row-echelon matrix $U$ using no row operations that add a multiple of a row to a row above it. Let $P_1, P_2, \cdots, P_{s-1}, P_s$ denote the elementary matrices corresponding (in order) to the row interchanges used. Write $P = P_s P_{s-1} \cdots P_2 P_1$ (if no interchanges are used, take $P = I_m$). Then:

(1) $PA$ is the matrix obtained by doing these interchanges (in order) to $A$.

(2) $PA$ has an LU-factorization.

The proof of Theorem 3 is omitted.

If $P_1, P_2, \cdots, P_{s-1}, P_s$ are elementary matrices corresponding to row interchanges, the matrix $P = P_s P_{s-1} \cdots P_2 P_1$ is called a **permutation matrix**. Such a matrix is obtained from the identity matrix by doing the same interchanges in order to $I$ (starting with the one corresponding to $P_1$). Thus $P$ is obtained from $I$ by placing the rows in a different order (hence the name), and so has exactly one 1 in each row and column. We regard the identity matrix as a permutation matrix.

**Example 6**   If $A = \begin{bmatrix} 0 & 0 & -2 \\ 2 & 4 & 2 \\ 1 & -1 & 4 \end{bmatrix}$, find a permutation matrix $P$ such that $PA$ has an LU-factorization, and then find the factorization.

SOLUTION | First carry $A$ to row-echelon form:

$$A = \begin{bmatrix} 0 & 0 & -2 \\ 2 & 4 & 2 \\ 1 & -1 & 4 \end{bmatrix} \rightarrow \begin{bmatrix} 1 & 2 & 1 \\ 0 & 0 & 1 \\ 1 & -1 & 4 \end{bmatrix}$$

$$\rightarrow \begin{bmatrix} 1 & 2 & 1 \\ 0 & 0 & 1 \\ 0 & -3 & 3 \end{bmatrix} \rightarrow \begin{bmatrix} 1 & 2 & 1 \\ 0 & 1 & -1 \\ 0 & 0 & 1 \end{bmatrix}$$

Only two row interchanges were used, first rows 1 and 2, and then rows 2 and 3. Hence, as in Theorem 3, the required permutation matrix is

$$P = P_2 P_1 = \begin{bmatrix} 1 & 0 & 0 \\ 0 & 0 & 1 \\ 0 & 1 & 0 \end{bmatrix} \begin{bmatrix} 0 & 1 & 0 \\ 1 & 0 & 0 \\ 0 & 0 & 1 \end{bmatrix} = \begin{bmatrix} 0 & 1 & 0 \\ 0 & 0 & 1 \\ 1 & 0 & 0 \end{bmatrix}.$$

If we do these interchanges in order to $A$, the result is $PA$. Applying the LU-algorithm to $PA$ gives

$$PA = \begin{bmatrix} 2 & 4 & 2 \\ 1 & -1 & 4 \\ 0 & 0 & -2 \end{bmatrix} \rightarrow \begin{bmatrix} 1 & 2 & 1 \\ 0 & -3 & 3 \\ 0 & 0 & -2 \end{bmatrix}$$

$$\rightarrow \begin{bmatrix} 1 & 2 & 1 \\ 0 & 1 & -1 \\ 0 & 0 & -2 \end{bmatrix} \rightarrow \begin{bmatrix} 1 & 2 & 1 \\ 0 & 1 & -1 \\ 0 & 0 & 1 \end{bmatrix} = U.$$

Hence $PA = LU$ where $L = \begin{bmatrix} 2 & 0 & 0 \\ 1 & -3 & 0 \\ 0 & 0 & -2 \end{bmatrix}$ and $U = \begin{bmatrix} 1 & 2 & 1 \\ 0 & 1 & -1 \\ 0 & 0 & 1 \end{bmatrix}$.

Note that $A$ can be carried also to row-echelon form using only one row interchange (1 and 3). The reader should obtain the corresponding factorization.

Consider the system $AX = B$ of linear equations where $A$ is $m \times n$. If $P$ is any permutation matrix, the matrix $PA$ is obtained from $A$ by writing the rows in a different order. Hence the system of equations

$$(PA)X = PB$$

is identical to the system $AX = B$ except for the order in which the equations are written, and so the two systems have the same solutions. Since the permutation matrix $P$ can be chosen so that $PA$ has an LU-factorization, the system $(PA)X = PB$ can be solved by forward and backward substitution.

## Exercises 1.7

**1.** In each case use the given LU-factorization of $A$ to solve the system $AX = B$ by solving $LY = B$ and $UX = Y$ by forward and backward substition respectively.

**a)** $A = \begin{bmatrix} 2 & 0 & 0 & 0 \\ 3 & -1 & 0 & 0 \\ 7 & 0 & 1 & 0 \\ 2 & -3 & 5 & 2 \end{bmatrix} \begin{bmatrix} 2 & -1 & 0 & 0 & 3 \\ 0 & 5 & 7 & -3 & 2 \\ 0 & 0 & -1 & 0 & 3 \\ 0 & 0 & 0 & 0 & 0 \end{bmatrix}$, $B = \begin{bmatrix} 8 \\ 1 \\ 30 \\ -15 \end{bmatrix}$

**b)** $A = \begin{bmatrix} 1 & 0 & 0 & 0 \\ 5 & -3 & 0 & 0 \\ 2 & 0 & -1 & 0 \\ -1 & 7 & 0 & 1 \end{bmatrix} \begin{bmatrix} 1 & -3 & 7 & 11 & -4 \\ 0 & 2 & -5 & 6 & 1 \\ 0 & 0 & 0 & 0 & 0 \\ 0 & 0 & 0 & 0 & 0 \end{bmatrix}$, $B = \begin{bmatrix} 2 \\ -1 \\ 7 \\ 2 \end{bmatrix}$

**2.** In each case find a LU-factorization of the given matrix.

**a)** $\begin{bmatrix} 2 & 4 & 2 \\ 1 & -1 & 3 \\ -1 & 7 & -7 \end{bmatrix}$

**b)** $\begin{bmatrix} -3 & 6 & 9 \\ 2 & -4 & 1 \\ -1 & 2 & -7 \end{bmatrix}$

**c)** $\begin{bmatrix} 2 & 6 & -2 & 0 & 2 \\ 3 & 9 & -3 & 3 & 1 \\ -1 & -3 & 1 & -3 & 1 \end{bmatrix}$

**d)** $\begin{bmatrix} 3 & -9 & 6 & 0 & -9 \\ -2 & 6 & 1 & 1 & 1 \\ -1 & 3 & 0 & 4 & 3 \end{bmatrix}$

**e)** $\begin{bmatrix} -5 & 10 & 0 & -15 \\ 2 & -3 & 1 & 5 \\ 1 & -2 & 4 & 0 \\ 1 & 1 & 1 & 1 \end{bmatrix}$

**f)** $\begin{bmatrix} 1 & -3 & 2 & 0 \\ 2 & -5 & -3 & 1 \\ 0 & -1 & 3 & 7 \\ -1 & 3 & 2 & 5 \end{bmatrix}$

**g)** $\begin{bmatrix} 2 & 2 & -2 & 4 & 2 \\ 1 & -1 & 0 & 2 & 1 \\ 3 & 1 & -2 & 6 & 3 \\ 1 & 3 & -2 & 2 & 1 \end{bmatrix}$  **h)** $\begin{bmatrix} -1 & -3 & 1 & 0 & -1 \\ 1 & 4 & 1 & 1 & 1 \\ 1 & 2 & -3 & -1 & 1 \\ 0 & -2 & -4 & -2 & 0 \end{bmatrix}$

**3.** In your own words, write the procedure to find an LU-factorization of a matrix $A$, and describe how it is used to solve a system $AX = B$.

**4.** In each case find a permutation matrix $P$ such that $PA$ has an LU-factorization, and find that factorization.

**a)** $A = \begin{bmatrix} -1 & 4 & 5 & 2 \\ 0 & 0 & 0 & -1 \\ 1 & -2 & -2 & 0 \\ 0 & -1 & -1 & 0 \end{bmatrix}$  **b)** $A = \begin{bmatrix} 0 & 0 & 0 & 3 \\ 0 & 0 & 2 & -4 \\ 0 & -1 & 0 & 5 \\ 1 & 3 & -2 & 3 \end{bmatrix}$

**5.** In each case either show that the statement is true or give an example showing that it is false.
**a)** The sum of two upper triangular matrices is again upper triangular.
**b)** Every invertible matrix has an LU-factorization.
**c)** Every permutation matrix is invertible.
**d)** If $B$ is symmetric, so also is $LBL^T$ for any matrix $L$.

**6.** If $A = \begin{bmatrix} 0 & 1 \\ 1 & 0 \end{bmatrix}$, show that $A = LU$ is impossible for *any* lower triangular $2 \times 2$ matrix $L$ and upper triangular $2 \times 2$ matrix $U$. In particular, show that $A$ has no LU-factorization.

**7.** If $A$ is lower triangular and invertible, give an LU-factorization of $A$.

**8.** Show that any row interchange can be accomplished by row operations of other types.

**9.** Show that any multiple of a row can be added to a row above it by row operations of other types.

**10. a)** Show that every permutation matrix is invertible.
**b)** Show that the inverse of a permutation matrix is again a permutation matrix.

**11.** Show that the $LDLT$ factorization in Theorem 2 is unique.

**12.** A triangular matrix is called *unit triangular* if it is square and every main diagonal element is 1.
**a)** If $A$ has an LU-factorization, show that $A = LU$ where $L$ is unit lower triangular and $U$ is upper triangular.
**b)** Show that the factorization in **a)** is unique if $A$ is invertible.

**13.** If $A = \begin{bmatrix} a & b \\ b & c \end{bmatrix}$ where $a \neq 0$ and $ac \neq b^2$, show that $A$ has an LDLT factorization, $A = LDL^T$, and find $D$ and $L$.

## 1.8 APPLICATION TO MARKOV CHAINS[22]

A Markov chain is a mathematical model that provides insight and information about a wide variety of applications, but is simple enough to be described using only linear equations and matrix multiplication. The model requires an acquaintance with probability theory, but this is no hindrance as we can proceed informally.

The probability of getting a head when flipping a fair coin is $\frac{1}{2}$ because, in a long series of flips, the coin will fall heads half the time. In general, the **probability** of an event is the long-term proportion of the time that the event will occur. For example, the probability of obtaining a 6 on the roll of a die is $\frac{1}{6}$ because each of the 6 faces of the die is equally likely to appear. Probabilities of events are thus numbers between 0 and 1, with impossible events having probability 0, and certain events having probability 1. Our use of probabilities arises in applications like the following.

### 1.8.1 A Model of the Weather

Consider the weather in a certain city. We assume for simplicity that it is always in one of two *states*: sunny or cloudy. The weather evolves in daily stages, changing from one state to another. We are interested in finding the answers to questions of the following type:

---

[22] The material in this section will not be used elsewhere in the text.

> Question 1. *If it is sunny on Monday, what is the probability that it is sunny on Thursday?*
>
> Question 2. *What is the long-term proportion of the time that it is sunny?*

To answer such questions, we need to know the probability that the weather will change or remain the same from one day to the next. Suppose that long-term weather records show that:

**(1)** If it is sunny one day, it is equally likely to be sunny or cloudy next day.

**(2)** If it is cloudy one day, the probabilities it is sunny and cloudy the next day are $\frac{3}{4}$ and $\frac{1}{4}$ respectively.

With the help of matrix algebra, this information provides answers to the above questions.

The given weather data are summarized in the accompanying table. Note that, if it is sunny on a given day, the first column lists the probabilities for the weather on the next day. Similarly, the second column gives the probabilities for the weather the day after a cloudy day. In particular, each column sums to 1 because, in this model, it *must* be either sunny or cloudy the next day. The corresponding matrix of probabilities

Present day

|  | S | C |
|---|---|---|
| Next day S | $\frac{1}{2}$ | $\frac{3}{4}$ |
| C | $\frac{1}{2}$ | $\frac{1}{4}$ |

$$P = \begin{bmatrix} \frac{1}{2} & \frac{3}{4} \\ \frac{1}{2} & \frac{1}{4} \end{bmatrix}$$

is called the *probability transition matrix* for this model of the weather.

Of course the weather each day is not determined; all we can hope to know is the *probability* that it is sunny or cloudy on any given day. Hence we introduce the *state vector* for day $m$:

$$S_m = \begin{bmatrix} s_1 \\ s_2 \end{bmatrix}$$

Here $s_1$ and $s_2$ are the probabilities that the weather is sunny and cloudy, respectively, on day $m$. In particular $S_0$ represents the initial situation. Note that $s_1 + s_2 = 1$ because the weather must be in *some* state on day $m$. The remarkable thing is that we can use matrix multiplication to compute these state vectors. More precisely, the following matrix equations hold:

$$S_{m+1} = PS_m \quad \text{for each} \quad m = 0, 1, \cdots. \tag{$*$}$$

In other words, the state vector $S_{m+1}$ for a given day can be computed by multiplying $S_m$ by the transition matrix $P$. To apply this to our situation, choose the initial day to be Monday. Then $S_0 = \begin{bmatrix} 1 \\ 0 \end{bmatrix}$ because we are assuming that the weather is sunny on Monday. With this, we can apply ($*$) repeatedly to compute

$$S_1 = PS_0 = \begin{bmatrix} \frac{1}{2} & \frac{3}{4} \\ \frac{1}{2} & \frac{1}{4} \end{bmatrix} \begin{bmatrix} 1 \\ 0 \end{bmatrix} = \begin{bmatrix} \frac{1}{2} \\ \frac{1}{2} \end{bmatrix},$$

$$S_2 = PS_1 = \begin{bmatrix} \frac{1}{2} & \frac{3}{4} \\ \frac{1}{2} & \frac{1}{4} \end{bmatrix} \begin{bmatrix} \frac{1}{2} \\ \frac{1}{2} \end{bmatrix} = \begin{bmatrix} \frac{5}{8} \\ \frac{3}{8} \end{bmatrix},$$

$$S_3 = PS_2 = \begin{bmatrix} \frac{1}{2} & \frac{3}{4} \\ \frac{1}{2} & \frac{1}{4} \end{bmatrix} \begin{bmatrix} \frac{5}{8} \\ \frac{3}{8} \end{bmatrix} = \begin{bmatrix} \frac{19}{32} \\ \frac{13}{32} \end{bmatrix}.$$

Since we are taking Monday to be day 0, Thursday is day 3. Hence the probability that it is sunny Thursday is $\frac{19}{32}$, answering Question 1 above. Of course we have also determined that the probability it is cloudy on Thursday is $\frac{13}{32}$.

Turning to Question 2, we compute successively

$$S_4 = \begin{bmatrix} 0.6015625 \\ 0.3984375 \end{bmatrix}, \quad S_5 = \begin{bmatrix} 0.599609375 \\ 0.400390625 \end{bmatrix}, \quad S_6 = \begin{bmatrix} 0.600097656 \\ 0.399902344 \end{bmatrix}, \cdots.$$

Clearly $S_m$ is getting closer and closer to $\begin{bmatrix} 0.6 \\ 0.4 \end{bmatrix}$ as $m$ increases. It follows that, in the long run, the weather is sunny with probability 0.6, and cloudy with probability 0.4. This answers Question 2.

The vector $S = \begin{bmatrix} 0.6 \\ 0.4 \end{bmatrix}$ is called the *steady-state vector* for the weather. There is another way to compute it that is much better (and simpler) than computing several state vectors $S_1$, $S_2$, $\cdots$ and observing their behaviour. Assume that $S_m$ is indeed very close to some fixed vector $S$ for all sufficiently large $m$. Then $S_{m+1}$ is also very close to $S$, and so the equation $S_{m+1} = PS_m$ is closely approximated by $S = PS$. This can be written

$$(I - P)S = 0$$

which is a system of linear homogeneous equations for $S$. In the above model of the weather, reduction of the augmented matrix to reduced form is

$$\begin{bmatrix} \frac{1}{2} & -\frac{3}{4} & 0 \\ -\frac{1}{2} & \frac{3}{4} & 0 \end{bmatrix} \rightarrow \begin{bmatrix} 1 & -\frac{3}{2} & 0 \\ 0 & 0 & 0 \end{bmatrix}$$

so the general solution to the system $(I - P)S = 0$ is $\begin{bmatrix} \frac{3}{2}t \\ t \end{bmatrix}$ where $t$ is a parameter. Since the entries of $S$ must sum to 1, we must have $\frac{3}{2}t + t = 1$, and so $t = \frac{2}{5}$. This gives $S = \begin{bmatrix} \frac{3}{5} \\ \frac{2}{5} \end{bmatrix} = \begin{bmatrix} 0.6 \\ 0.4 \end{bmatrix}$, which is what we had before.

## 1.8.2    Markov Chains

This model of the weather is an example of a Markov chain. In general, a **Markov**[23] **chain** is a system that evolves through a series of **stages** (the consecutive days in the above discussion) and, at any stage, must be in one of a finite number of **states** (sunny or cloudy). To analyze the chain, we must know each **transition probability** $p_{ij}$; that is, the probability that, if the chain is in state $j$ at some stage, it will be in state $i$ after one more transition. If the chain has $n$ states, the $n \times n$ matrix $P = [p_{ij}]$ is called the **transition matrix** for the chain. The $3 \times 3$ case is illustrated as follows:

|  |  | Present stage | | |
|---|---|---|---|---|
|  |  | State 1 | State 2 | State 3 |
| Next | State 1 | $p_{11}$ | $p_{12}$ | $p_{13}$ |
| stage | State 2 | $p_{21}$ | $p_{22}$ | $p_{23}$ |
|  | State 3 | $p_{31}$ | $p_{32}$ | $p_{33}$ |

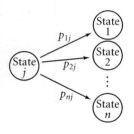

Observe that each column of $P$ sums to 1 because, starting in state $j$ at present, the chain must go to *some* state at the next stage. In general, a square matrix is called **stochastic** if the entries are all nonnegative and each column sum is 1.

---

[23] Named after the Russian mathematician Andreĭ Andreevič Markov (1856–1922).

In general, columns whose entries are probabilities that sum to 1 are called **probability vectors**. At the $m^{\text{th}}$ stage, the chain is described by the **state vector**

$$S_m = \begin{bmatrix} s_1^{(m)} \\ s_2^{(m)} \\ \vdots \\ s_n^{(m)} \end{bmatrix}$$

where $s_k^{(m)}$ is the probability that the chain is in state $k$ after $m$ transitions. Note that each $S_m$ is a probability vector because the chain must be in *some* state after $m$ transitions. A probability vector $S$ is called a **steady-state vector** for the Markov chain if the state vectors $S_m$ get closer and closer to $S$ as $m$ increases. Thus the entries of $S$ are the long-term probabilities that the chain will be in the various states.

It can happen that a Markov chain does not have a steady-state vector. One condition that guarantees that a steady-state vector exists is that the chain is **regular**; that is, some power $P^k$ of the transition matrix has all its entries positive. For small chains this can be checked graphically as follows: Represent the states as the vertices $v_1,\ v_2, \cdots,\ v_n$ of a directed graph (called the **graph** of the chain) and draw an arrow from state $v_j$ to $v_i$ when it is possible to go from state $v_j$ to state $v_i$ (that is, when the probability $p_{ij} > 0$). Then the chain is regular if there exists an integer $k$ such that it is possible to go from any vertex to any other vertex in exactly $k$ transitions.

Thus, for example, the weather chain discussed above is regular because it is possible to go from any state to any other state in *one* transition. The graph is given in the diagram.

The behaviour of regular Markov chains is summarized in the following theorem.

---

**THEOREM 1**

Consider a Markov chain with transition matrix $P$.

**(1)** $S_{m+1} = PS_m$    for all $m = 0,\ 1,\ 2, \cdots$.

**(2)** Suppose the chain is regular. Then the steady state vector $S$ is the unique solution to the homogeneous system

$$(I - P)S = 0$$

whose entries sum to 1.

---

We give a proof of (1) of Theorem 1 at the end of this section which illustrates how it is that matrix multiplication arises here. Part (2) of Theorem 1 will not be proved in this book.[24]

**Example 1**

Consider a mouse in the maze shown. If it is in any cell, assume that it is equally likely to stay in the cell, leave by the first exit, leave by the second exit, etc.

**(a)** If it starts in cell 1, what is the probability it is in each of the cells after three moves?

**(b)** What proportion of the time does it spend in each of the cells?

**SOLUTION**

The states here are the cells 1, 2, and 3, and the stages are the successive moves. If the mouse is in cell 1, it has two options: stay or go to cell 2. By assumption these are equally likely, so both have probability $\frac{1}{2}$. Since it has probability 0 of going to cell 3, this gives the first column of the transition matrix $P = \begin{bmatrix} \frac{1}{2} & \frac{1}{3} & 0 \\ \frac{1}{2} & \frac{1}{3} & \frac{1}{2} \\ 0 & \frac{1}{3} & \frac{1}{2} \end{bmatrix}$.

---

[24] A proof can be found in Kemeny, Mirkil, Snell, and Thompson: *Finite Mathematical Structures*, Prentice-Hall, 1958.

If the mouse is in cell 2 it can stay, go to cell 1, or go to cell 3, and these all have probability $\frac{1}{3}$ (being equally likely). This gives column 2 of $P$, and column 3 is found in a similar way.

If the mouse starts in cell 1, we have $S_0 = [1 \quad 0 \quad 0]^T$. Hence we compute

$$S_1 = PS_0 = \begin{bmatrix} \frac{1}{2} \\ \frac{1}{2} \\ 0 \end{bmatrix}, \quad S_2 = PS_1 = \begin{bmatrix} \frac{5}{12} \\ \frac{5}{12} \\ \frac{2}{12} \end{bmatrix}, \quad S_3 = PS_2 = \begin{bmatrix} \frac{25}{72} \\ \frac{31}{72} \\ \frac{16}{72} \end{bmatrix}.$$

Thus the probabilities the mouse is in cells 1, 2, and 3 after three moves are $\frac{25}{72}$, $\frac{31}{72}$, and $\frac{16}{72}$ respectively.

The graph of this chain is as shown, so any cell is reachable from any other cell in exactly 2 transitions. Hence the chain is regular and we can find the steady-state vector $S$ by solving the system of linear equations

$$(I - P)S = 0.$$

The reduction of the augmented matrix to reduced form is as follows:

$$\begin{bmatrix} \frac{1}{2} & -\frac{1}{3} & 0 & 0 \\ -\frac{1}{2} & \frac{2}{3} & -\frac{1}{2} & 0 \\ 0 & -\frac{1}{3} & \frac{1}{2} & 0 \end{bmatrix} \rightarrow \begin{bmatrix} \frac{1}{2} & -\frac{1}{3} & 0 & 0 \\ 0 & \frac{1}{3} & -\frac{1}{2} & 0 \\ 0 & 0 & 0 & 0 \end{bmatrix}$$

$$\rightarrow \begin{bmatrix} 1 & 0 & -1 & 0 \\ 0 & 1 & -\frac{3}{2} & 0 \\ 0 & 0 & 0 & 0 \end{bmatrix}$$

Thus the general solution to $(I - P)S = 0$ is $\begin{bmatrix} t & \frac{3}{2}t & t \end{bmatrix}^T$. This is a probability vector if the entries sum to 1, that is if $t + \frac{3}{2}t + t = 1$. This gives $t = \frac{2}{7}$, so the steady-state vector is $S = \begin{bmatrix} \frac{2}{7} & \frac{3}{7} & \frac{2}{7} \end{bmatrix}^T$. Thus the mouse spends $\frac{3}{7}$ of his time in cell 2, and $\frac{2}{7}$ of his time in each of cells 1 and 3.

For any regular Markov chain, it can be shown that the solution $S$ of the homogeneous system $(I - P)S = 0$ always involves exactly one parameter $t$ (as in Example 1). Then $t$ is determined by the requirement that the entries of the steady-state vector $S$ must sum to 1.

We conclude with one final observation. Let $P$ be the transition matrix of a Markov chain, and let $S_0$, $S_1$, $S_2$, $\cdots$ $S_m$, $\cdots$ be the state vectors. Then $S_1 = PS_0$ by Theorem 1, so $S_2 = PS_1 = P(PS_0) = P^2 S_0$, again by Theorem 1. Next $S_3 = PS_2 = P(P^2 S_0) = P^3 S_0$. The pattern is clear:

$$S_m = P^m S_0 \quad \text{for each} \quad m = 1, 2, 3, \cdots.$$

Hence the behaviour of the chain is entirely determined by the initial state vector $S_0$ and the powers $P^m$ of the transition matrix. In Section 2.3 we develop methods to calculate the powers of any square matrix.

**1.8.3**  ### Proof of (1) of Theorem 1

Think of the chain as having been run $N$ times where $N$ is large. Recall that the $m^{th}$ state vector is

$$S_m = \begin{bmatrix} s_1^{(m)} & s_2^{(m)} & \cdots & s_n^{(m)} \end{bmatrix}^T$$

where $s_i^{(m)}$ is the probability that it is in state $i$ at stage $m$. Looking at stage $m + 1$, we have that, approximately,

$$s_i^{(m+1)}N \text{ is the number of times it is in state } i \text{ at stage } m + 1.$$

We count this number a different way. To get to state $i$ at stage $m + 1$ the chain had to be in *some* state $j$ at stage $m$. Since the approximate number of times it is in state $j$ at stage $m$ is $s_j^{(m)}N$, and since $p_{ij}$ is the probability that it goes from state $j$ to state $i$ in one transition, the number of times it got to state $i$ via state $j$ is $p_{ij}(s_j^{(m)}N)$. Summing over $j$ gives the total number of times it got to state $i$. This is the number we got before, so

$$s_i^{(m+1)}N = p_{i1}\, s_1^{(m)}N + p_{i2}\, s_2^{(m)}N + \cdots + p_{in}\, s_n^{(m)}N.$$

Cancelling $N$ gives $s_i^{(m+1)} = p_{i1}\, s_1^{(m)} + p_{i2}\, s_2^{(m)} + \cdots + p_{in}\, s_n^{(m)}$. This asserts that entry $i$ of the column matrix $S_{m+1}$ is equal to entry $i$ of the matrix product $PS_m$. Hence $S_{m+1} = PS_m$, which is (1) of Theorem 1.

## Exercises 1.8

**1.** A city has only rainy or sunny days. Assume that there is an 80% probability of having sun the day following a sunny day, and a 60% probability of having rain the day following a rainy day. If there is an 80% probability of having a sunny day today, calculate the probabilities of rain and shine for each of the following 10 days, calculate the long-term probabilities of rain and shine, and find out how many days it will take to reach this steady-state vector with 2 decimal accuracy.

**2.** In each case find the steady-state vector and, assuming the chain starts in state 1, find the probability that it is in state 2 after 3 transitions.

**a)** $P = \begin{bmatrix} 0.5 & 0.3 \\ 0.5 & 0.7 \end{bmatrix}$

**b)** $P = \begin{bmatrix} \frac{1}{2} & 1 \\ \frac{1}{2} & 0 \end{bmatrix}$

**c)** $P = \begin{bmatrix} 0 & \frac{1}{2} & \frac{1}{4} \\ 1 & 0 & \frac{1}{4} \\ 0 & \frac{1}{2} & \frac{1}{2} \end{bmatrix}$

**d)** $P = \begin{bmatrix} 0.4 & 0.1 & 0.5 \\ 0.2 & 0.6 & 0.2 \\ 0.4 & 0.3 & 0.3 \end{bmatrix}$

**e)** $P = \begin{bmatrix} 0.8 & 0.0 & 0.2 \\ 0.1 & 0.6 & 0.1 \\ 0.1 & 0.4 & 0.7 \end{bmatrix}$

**f)** $P = \begin{bmatrix} 0.1 & 0.3 & 0.3 \\ 0.3 & 0.1 & 0.6 \\ 0.6 & 0.6 & 0.1 \end{bmatrix}$

**3.** In each case determine whether the given probability transition matrix is regular by drawing the graph.

**a)** $P = \begin{bmatrix} 0 & 0 & \frac{1}{2} \\ 1 & 0 & \frac{1}{2} \\ 0 & 1 & 0 \end{bmatrix}$

**b)** $P = \begin{bmatrix} \frac{1}{2} & 0 & \frac{1}{3} \\ \frac{1}{4} & 1 & \frac{1}{3} \\ \frac{1}{4} & 0 & \frac{1}{3} \end{bmatrix}$

**c)** $P = \begin{bmatrix} \frac{1}{2} & 0 & 0 \\ \frac{1}{2} & \frac{1}{4} & \frac{2}{3} \\ 0 & \frac{3}{4} & \frac{1}{3} \end{bmatrix}$

**d)** $P = \begin{bmatrix} 0 & 0 & \frac{1}{3} \\ \frac{1}{2} & 0 & \frac{1}{2} \\ \frac{1}{2} & 1 & \frac{1}{6} \end{bmatrix}$

**4.** A man eats one of three soups for lunch each day—beef, chicken, or vegetable. He never eats the same soup two days in a row. If he eats beef soup one day, he is equally likely to eat each of the others next day; if he does not eat beef soup one day, he is twice as likely to eat it next day as the alternative.

**a)** If he has beef soup one day, what is the probability he has it again two days later?

**b)** Find the long-run probabilities he eats each of the soups.

**5.** A wolf pack always hunts in one of four regions, $R_1$, $R_2$, $R_3$, or $R_4$. Its hunting habits are as follows:

(1) If it hunts in some region, it is as likely as not to hunt there the next day.

(2) If it hunts in $R_1$, it never hunts in $R_2$ or in $R_3$ the next day.

(3) If it hunts in $R_2$ or $R_3$, it is equally likely to hunt in each of the other regions the next day.

(4) If it hunts in $R_4$, it is equally likely to hunt in regions $R_2$ and $R_3$ on the next day, but twice as likely to hunt in $R_1$ than in either $R_2$ or $R_3$.

**a)** Find the proportion of the time the pack hunts in $R_1$, $R_2$, $R_3$, and $R_4$.

**b)** If the pack hunts in $R_1$ on Monday, find the probability it hunts there on Thursday.

**6.** A gambler always plays one of three slot machines, $A$, $B$, or $C$. He never plays the same machine on two successive days. If he plays $A$, he plays $C$ the next day. If he plays $B$ or $C$, he is twice as likely to play $A$ the next day as the alternative.

**a)** What proportion of his time does he spend playing $A$, $B$, and $C$?

**b)** If he plays $C$ one day, find the probability he plays it again three days later.

**7.** A lab rat has a choice of three foods $P$, $Q$, and $R$ each day. On any given day it has a 60% chance of choosing the same food as it chose the previous day, and is equally likely to choose either of the other foods.

**a)** If it chooses $P$ one day, find the probability it chooses $Q$ three days later.

**b)** What percentage of its meals are food $P$, $Q$, and $R$?

8. Assume that there are three classes—upper $U$, middle $M$, and lower $L$—and that social mobility is modelled as follows:
   (1) Of children of $U$ parents, 70% remain $U$ while 20% become $L$ and 10% become $M$.
   (2) Of children of $M$ parents, 80% remain $M$ while 10% become $L$ and 10% become $U$.
   (3) Of children of $L$ parents, 60% remain $L$ while 10% become $U$ and 30% become $M$.
   a) Find the probability that the grandchild of $L$ parents becomes $U$.
   b) Find the long-term breakdown of society into classes.

9. John makes it to work on time one Monday out of four. On other work days his behaviour is as follows: If he is late one day, he is twice as likely to be on time next day as to be late. If he is on time one day, he is equally likely to be on time or late next day. Find the probability of his being on time Thursdays.

10. The governor says she will call an election. This gossip is passed from person to person with probability $p \neq 0$ that the information is passed incorrectly at any stage. Assume that when a person hears the news he or she passes it to one person who does not know. Find the long-term probability that a person will hear that there is going to be an election.

11. Suppose you play a game with a friend in which you either win or lose $1 at each round with equal probabilities. You start with $1 and you decide in advance to quit if you ever go broke or get $4. Assume that your friend never quits.
   a) If the states are 0, 1, 2, 3, and 4 (representing your wealth), show that the transition matrix is not regular.

   b) Find the probability that you go broke after exactly 3 matches.
   c) Find a steady-state vector by calculating $S_n$ for large values of $n$, and make an argument whether, as a casino owner, you would profit from playing such a game in the long run against your customers.

12. In each case either show that the statement is true or give an example showing that it is false. Throughout, $P$ denotes a probability transition matrix.
   a) If all rows but the last in $P$ are known, then the last row is also known.
   b) If $P$ has a 1 on the main diagonal, it is regular.
   c) If $P$ is regular, then it is invertible.
   d) If $P$ is invertible, then it is regular.

13. Let $E = \begin{bmatrix} 1 & 1 & \cdots & 1 \end{bmatrix}$ denote the $1 \times n$ matrix with each entry 1.
   a) If an $n \times n$ matrix $P$ has nonnegative entries, show that $P$ is stochastic if and only if $EP = E$.
   b) Show that the product of two stochastic matrices is stochastic.
   c) If $P$ is stochastic, show that there exists a column $S \neq 0$ such that $PS = S$. [In fact $S$ can be chosen to be a probability vector.]

14. Let $P = \begin{bmatrix} 0 & 0 & 1 \\ 1 & 0 & 0 \\ 0 & 1 & 0 \end{bmatrix}$. Is the corresponding chain regular?
   Draw the graph of the chain. Can you get from any vector to any other vector? Why does the graphical test fail? Show that $PS = S$ when $S = \begin{bmatrix} \frac{1}{3} & \frac{1}{3} & \frac{1}{3} \end{bmatrix}^T$. Do the state vectors get closer and closer to $S$?

## 1.9   FINITE FIELDS AND CRYPTOGRAPHY

Up to now, virtually all matrices have had entries that are real (possibly complex) numbers. The real numbers have the properties that they can be added and multiplied, that the usual laws of arithmetic hold, and that we can divide by any nonzero real number. As such the real numbers are an example of an algebraic system called a **field**. The set of complex numbers is also a field, as is the set of rational numbers (fractions). Our motivation for isolating the concept of a field is that nearly everything we have done remains valid if the scalars are restricted to some field: The Gaussian algorithm can be used to solve systems of linear equations with coefficients in the field, the matrix inversion algorithm works in the same way, and so on. The reason is that the field has all the properties used in the proofs of these results, so all the theorems remain valid.

It turns out that there are finite fields—that is, finite sets that satisfy the usual laws of arithmetic and in which every nonzero member has an inverse. This section is an informal introduction to some of these finite fields, with an application to cryptography.

## 1.9.1  Finite Fields

Our construction of a finite field is based on the process of "long division" familiar from arithmetic. For example, we can divide an integer[25] $n$ by 5 and leave a remainder in the set $\{0, 1, 2, 3, 4\}$ (called the remainder of $n$ modulo 5). For example

$$19 = 3 \cdot 5 + 4,$$

so the remainder of 19 modulo 5 is 4. Similarly the remainder of 137 modulo 5 is 2 because $137 = 27 \cdot 5 + 2$. This even works for negative integers: For example,

$$-17 = (-4) \cdot 5 + 3$$

so the remainder of $-17$ modulo 5 is 3.

This process is called the **division algorithm**. More formally, let $p \geq 2$ denote a positive integer. Then every integer $n$ can be written uniquely in the form

$$n = qp + r \quad \text{where } q \text{ and } r \text{ are integers and } 0 \leq r \leq p - 1.$$

Here $q$ is called the **quotient**, and $r$ is called the **remainder** of $n$ **modulo** $p$, and we refer to $p$ as the **modulus**. Thus if $p = 6$, then the fact that $134 = 22 \cdot 6 + 2$ means that 134 has quotient 22 and remainder 2 modulo 6.

Returning to the case $p = 5$, consider the set $\mathbb{F}_5 = \{0, 1, 2, 3, 4\}$ of possible remainders. We are going to turn $\mathbb{F}_5$ into a field by adding and multiplying modulo 5: That is we add or multiply two numbers in $\mathbb{F}_5$ by calculating the usual sum or product and taking the remainder modulo 5. Hence $2 + 4 = 1$ and $2 \cdot 4 = 3$ in $\mathbb{F}_5$, while $3 + 4 = 2$, and $3 \cdot 4 = 2$. In this way we can fill in the addition and multiplication tables for $\mathbb{F}_5$; the result is:

| + | 0 | 1 | 2 | 3 | 4 |
|---|---|---|---|---|---|
| **0** | 0 | 1 | 2 | 3 | 4 |
| **1** | 1 | 2 | 3 | 4 | 0 |
| **2** | 2 | 3 | 4 | 0 | 1 |
| **3** | 3 | 4 | 0 | 1 | 2 |
| **4** | 4 | 0 | 1 | 2 | 3 |

| × | 0 | 1 | 2 | 3 | 4 |
|---|---|---|---|---|---|
| **0** | 0 | 0 | 0 | 0 | 0 |
| **1** | 0 | 1 | 2 | 3 | 4 |
| **2** | 0 | 2 | 4 | 1 | 3 |
| **3** | 0 | 3 | 1 | 4 | 2 |
| **4** | 0 | 4 | 3 | 2 | 1 |

The remarkable thing about this is that addition and multiplication modulo 5 make $\mathbb{F}_5$ into a field. That the usual laws of arithmetic hold is not too surprising since they are true for ordinary addition and multiplication and all we are doing is reducing modulo 5. Perhaps more surprising is the fact that $\mathbb{F}_5$ is a field as the multiplication table shows: We have $1 \cdot 1 = 1$, $2 \cdot 3 = 1$ $(= 3 \cdot 2)$, and $4 \cdot 4 = 1$, so 1 and 4 are selfinverse in $\mathbb{F}_5$, and 2 and 3 are inverses of each other. In other words, *every nonzero element* of $\mathbb{F}_5$ has an inverse!

Note that if we do all of this using 6 as the modulus, we do not get a field. For in that case we have $\mathbb{F}_6 = \{0, 1, 2, 3, 4, 5\}$, and $2 \cdot 3 = 0$ so neither 2 nor 3 has an inverse in $\mathbb{F}_6$ (verify this). The reason, of course, is that 6 factors as 2 times 3 in ordinary arithmetic, and so leaves a remainder of 0 modulo 6. Of course 5 has no such factorization, and it is this fact that makes $\mathbb{F}_5$ into a field. More generally, an integer $p \geq 2$ is called a **prime** if $p$ cannot be factored as $p = ab$ where $a$ and $b$ are positive integers, and neither $a$ nor $b$ equals 1. Thus the first few primes are $2, 3, 5, 7, 11, 13, 17, \cdots$.

If $p$ is a prime, let

$$\mathbb{F}_p = \{0, 1, 2, 3, \cdots, p - 1\}$$

---

[25] The **integers** are the numbers $0, \pm 1, \pm 2, \pm 3, \cdots$.

denote the set of possible remainders modulo $p$. We add and multiply in $\mathbb{F}_p$ by finding the usual sum or product and taking the remainder modulo $p$. The important fact that we need is the following result.

**THEOREM ①**

If $p$ is a prime and we write $\mathbb{F}_p = \{0, 1, 2, 3, \cdots, p-1\}$, then $\mathbb{F}_p$ is a field using addition and multiplication modulo $p$.

The proof can be found in books on abstract algebra.[26]

Note that we will make statements like $-16 = 4$ in $\mathbb{F}_5$; it means that $-16$ and $4$ leave the same remainder when divided by 5, and so are equal in $\mathbb{F}_5$ (and we say that $-16 = 4$ modulo 5). In general, if $p$ is any prime, the operative fact is that $n = m$ modulo $p$ if $n - m$ is a multiple of $p$.

**Example 1**

If $p = 2$, then $\mathbb{F}_2 = \{0, 1\}$ with addition and multiplication modulo 2 given by the tables

| + | 0 | 1 |
|---|---|---|
| 0 | 0 | 1 |
| 1 | 1 | 0 |

and

| × | 0 | 1 |
|---|---|---|
| 0 | 0 | 0 |
| 1 | 0 | 1 |

This is binary arithmetic, the basis of computers.

Arithmetic in $\mathbb{F}_p$ is in a sense simpler than that for the real numbers. For example, consider negatives. Given the element 8 in $\mathbb{F}_{17}$, what is $-8$? The answer lies in the observation that $8 + 9 = 0$ in $\mathbb{F}_{17}$, so $-8 = 9$ (and $-9 = 8$). In the same way, finding negatives is not difficult in $\mathbb{F}_p$ for any prime $p$.

However, as for matrices, it is more difficult to find inverses in $\mathbb{F}_p$. For example, how does one find $\frac{1}{14}$ in $\mathbb{F}_{17}$? Since $\frac{1}{14} \cdot 14 = 1$ in $\mathbb{F}_{17}$, we are looking for an integer $a$ with the property that $a \cdot 14 = 1$ modulo 17. Of course we can try all possibilities (there are only 17 of them!), and the result is $a = 11$ (verify). However this method is of little use for large primes $p$, and it is a comfort to know that there is a systematic procedure (called the **Euclidean algorithm**) for finding inverses in $\mathbb{F}_p$ for any prime $p$. Furthermore, this algorithm is easy to program for a computer. To illustrate the method, let us once again find the inverse of 14 in $\mathbb{F}_{17}$.

**Example 2**

Find the inverse of 14 in $\mathbb{F}_{17}$.

SOLUTION

The idea is to first divide $p = 17$ by 14:

$$17 = 1 \cdot 14 + 3.$$

Now divide 14 by the new remainder 3 to get

$$14 = 4 \cdot 3 + 2,$$

and then divide 3 by the new remainder 2 to get

$$3 = 1 \cdot 2 + 1.$$

It is a theorem of number theory that, because 17 is a prime, this procedure will always lead to a remainder of 1 as is the case here. At this point we eliminate remainders in these equations from the bottom up. In the present example, we get:

[26] For example, W.K. Nicholson, *Introduction to Abstract Algebra*, 2nd ed., Wiley, 1999.

$$1 = 3 - 1 \cdot 2$$
$$= 3 - 1 \cdot (14 - 4 \cdot 3) = 5 \cdot 3 - 1 \cdot 14$$
$$= 5 \cdot (17 - 1 \cdot 14) - 1 \cdot 14 = 5 \cdot 17 - 6 \cdot 14.$$

Hence $(-6) \cdot 14 = 1$ in $\mathbb{F}_{17}$, that is $11 \cdot 14 = 1$. So $\frac{1}{14} = 11$ in $\mathbb{F}_{17}$.

As mentioned above, nearly everything we have done with matrices over the field of real numbers can be done in the same way for matrices with entries from $\mathbb{F}_p$. We illustrate with one example.

**Example 3**

Determine if the matrix $A = \begin{bmatrix} 1 & 4 \\ 6 & 5 \end{bmatrix}$ from $\mathbb{F}_7$ is invertible, and if so find its inverse.

**PROOF**

Working in $\mathbb{F}_7$, $det A = 1 \cdot 5 - 6 \cdot 4 = 5 - 3 = 2 \neq 0$ in $\mathbb{F}_7$, so $A$ is invertible. Hence Example 5 §1.5 gives $A^{-1} = \frac{1}{2}\begin{bmatrix} 5 & -4 \\ -6 & 1 \end{bmatrix}$. Now $2 \cdot 4 = 1$ in $\mathbb{F}_7$, so $\frac{1}{2} = 4$. Also, $-4 = 3$ and $-6 = 1$ in $\mathbb{F}_7$ so, finally, $A^{-1} = 4\begin{bmatrix} 5 & 3 \\ 1 & 1 \end{bmatrix} = \begin{bmatrix} 6 & 5 \\ 4 & 4 \end{bmatrix}$. Thus $\begin{bmatrix} 1 & 4 \\ 6 & 5 \end{bmatrix}\begin{bmatrix} 6 & 5 \\ 4 & 4 \end{bmatrix} = \begin{bmatrix} 1 & 0 \\ 0 & 1 \end{bmatrix}$ in $\mathbb{F}_7$ as the reader can verify.

## 1.9.2 An Application to Cryptography[27]

Cryptography is the science of encoding messages so that they will not be understood if they are intercepted, but which can be easily decoded by the intended recipient. We conclude with one simple example in this large and rapidly growing field.

Suppose a general wants to send the message

ATTACK AT DAWN

to his major in the field, but does not want the message to be understood by the opposing forces. We use the following method of coding the message: Convert every letter and punctuation mark into a number as follows

| a | b | c | d | e | f | g | h | i | j | k | l | m | n | o | p |
|---|---|---|---|---|---|---|---|---|---|---|---|---|---|---|---|
| 0 | 1 | 2 | 3 | 4 | 5 | 6 | 7 | 8 | 9 | 10 | 11 | 12 | 13 | 14 | 15 |

| q | r | s | t | u | v | w | x | y | z | , | . | — |
|---|---|---|---|---|---|---|---|---|---|---|---|---|
| 16 | 17 | 18 | 19 | 20 | 21 | 22 | 23 | 24 | 25 | 26 | 27 | 28 |

where 28 represents a space. Then the message ATTACK AT DAWN becomes

0  19  19  0  2  10  28  0  19  28  3  0  22  13

in numerical form.

Now we encode the message for transmission as follows: Since there are 29 symbols, and since the number 29 is a prime, we choose an invertible $2 \times 2$ matrix over $\mathbb{F}_{29}$, say

$$A = \begin{bmatrix} 2 & 15 \\ 7 & 22 \end{bmatrix}.$$

Then break the message into $1 \times 2$ segments as follows:

$$[0 \ 19] [19 \ 0] [2 \ 10] [28 \ 0] [19 \ 28] [3 \ 0] [22 \ 13].$$

Next encode the message by right multiplying each $1 \times 2$ segment by $A$ to get

$$[17 \ 12] [9 \ 24] [16 \ 18] [27 \ 14] [2 \ 2] [6 \ 16] [19 \ 7]$$

[27] We acknowledge the contribution of Murray Bremner, University of Saskatchewan.

(corresponding to the message RMJYQS.OCCGQTH). It is this sequence that is transmitted to the major who, being in possession of $A$, finds $A^{-1}$ and decodes the message by right multiplying each of these new segments by $A^{-1}$.

More precisely, $det A = -61 = 26$ in $\mathbb{F}_{29}$, so $A$ is invertible. Using the Euclidean algorithm, we find $\frac{1}{26}$ as follows: $29 = 1 \cdot 26 + 3$,  $26 = 8 \cdot 3 + 2$,  $3 = 1 \cdot 2 + 1$; and eliminating remainders gives

$$1 = 3 - 1 \cdot (26 - 8 \cdot 3) = 9 \cdot 3 - 1 \cdot 26 = 9(29 - 1 \cdot 26) - 1 \cdot 26 = 9 \cdot 29 - 10 \cdot 26.$$

Hence $(-10) \cdot 26 = 1$ modulo 29, that is $19 \cdot 26 = 1$. Hence $\frac{1}{26} = 19$ in $\mathbb{F}_{29}$.

Now

$$A^{-1} = \frac{1}{26} \begin{bmatrix} 22 & -15 \\ -7 & 2 \end{bmatrix} = 19 \begin{bmatrix} 22 & 14 \\ 22 & 2 \end{bmatrix} = \begin{bmatrix} 12 & 5 \\ 12 & 9 \end{bmatrix}.$$

Hence the first segment $[17 \ \ 12]$ of the transmitted message is decoded by right multiplication by $A^{-1}$ to give $[17 \ \ 12] \begin{bmatrix} 12 & 5 \\ 12 & 9 \end{bmatrix} = [0 \ \ 19]$, which is the first segment of the original messsage. Similarly each segment is decoded, and so the original messsage is recreated.

Of course this can be varied: The original message can be broken into segments of length 3, say, and then encoded by right multiplication by an invertible $3 \times 3$ matrix $A$. And more symbols can be introduced by choosing a larger prime as the modulus.

## Exercises 1.9

**1.** In each case write the addition and multiplication table of the finite field.

    **a)** $\mathbb{F}_3$         **b)** $\mathbb{F}_7$         **c)** $\mathbb{F}_{11}$

**2.** In each case find $\frac{1}{a}$ in $\mathbb{F}_p$.

    **a)** $a = 12$ in $\mathbb{F}_{41}$.     **b)** $a = 17$ in $\mathbb{F}_{37}$.

    **c)** $a = 11$ in $\mathbb{F}_{23}$.     **d)** $a = 8$ in $\mathbb{F}_{29}$.

**3.** In each case either show that $A$ has no inverse or find $A^{-1}$.

    **a)** $A = \begin{bmatrix} 2 & 1 \\ 5 & 2 \end{bmatrix}$ in $\mathbb{F}_7$.     **b)** $A = \begin{bmatrix} 3 & 4 \\ 7 & 8 \end{bmatrix}$ in $\mathbb{F}_{11}$.

    **c)** $A = \begin{bmatrix} 3 & 11 \\ 2 & 8 \end{bmatrix}$ in $\mathbb{F}_{13}$.     **d)** $A = \begin{bmatrix} 2 & 1 \\ 3 & 4 \end{bmatrix}$ in $\mathbb{F}_5$.

**4.** Encode the message MEET AT NOON using the setup in the text, but using $A = \begin{bmatrix} 6 & 2 \\ 12 & 7 \end{bmatrix}$. What is $A^{-1}$?

**5.** Show that the elements of the last column of the multiplication table for $\mathbb{F}_p$ are $0, \ p-1, \ p-2, \cdots, \ 2, \ 1$ in that order.

# CHAPTER 2

# Determinants and Eigenvalues

## 2.1 COFACTOR EXPANSIONS

In Section 1.5 we defined the determinant of a $2 \times 2$ matrix $A = \begin{bmatrix} a & b \\ c & d \end{bmatrix}$ as follows:

$$det A = det \begin{bmatrix} a & b \\ c & d \end{bmatrix} = ad - bc$$

We then showed (Example 5 §1.5) that $A$ has an inverse if $det A \neq 0$, and gave a formula for the inverse in that case. One goal of this chapter is to do this for *any* square matrix $A$.

There is no problem if $A$ is $1 \times 1$, say $A = [a]$. In that case we define

$$det[a] = a$$

and note that $[a]$ is invertible if (and only if) $a \neq 0$, and that then the formula for the inverse is $[a]^{-1} = \left[\frac{1}{a}\right]$.

So what do we do if $A$ is $n \times n$ where $n \geq 3$? The formula for the determinant of a $3 \times 3$ matrix turns out to be:

$$det \begin{bmatrix} a & b & c \\ d & e & f \\ g & h & i \end{bmatrix} = aei + bfg + cdh - ceg - afh - bdi \qquad (*)$$

and the expression for the inverse is even more involved.[1] The situation is much more complicated for $n \times n$ matrices where $n > 3$.

### 2.1.1  Cofactor Expansion

There is, however, an inductive way to define the determinant $det A$ of any square matrix $A$. The idea is to define the determinant of a $3 \times 3$ matrix in terms of determinants of $2 \times 2$ matrices, then give the determinant of any $4 \times 4$ matrix in terms of determinants of $3 \times 3$ matrices, and so on.

---

[1] One way to remember the formula $(*)$ is to adjoin columns 1 and 2 of $A$ on the right to obtain $\begin{bmatrix} a & b & c & a & b \\ d & e & f & d & e \\ g & h & i & g & h \end{bmatrix}$.

Then the terms $aei$, $bfg$, and $cdh$ in $(*)$ are the products down to the right, starting at $a$, $b$, and $c$, while the terms $-ceg$, $-afh$, and $-bdi$ come from products down to the left, starting at $c$, $a$, and $b$. I hesitate to reveal this rule because it does **not**, repeat **not**, extend to $n \times n$ matrices for $n > 3$.

To motivate the procedure[2], we collect the terms in (∗) involving the first row entries $a$, $b$, and $c$, and observe that the multipliers are $2 \times 2$ determinants:

$$det \begin{bmatrix} a & b & c \\ d & e & f \\ g & h & i \end{bmatrix} = aei + bfg + cdh - ceg - afh - bdi$$

$$= a(ei - fh) - b(di - fg) + c(dh - eg)$$

$$= a \, det \begin{bmatrix} e & f \\ h & i \end{bmatrix} - b \, det \begin{bmatrix} d & f \\ g & i \end{bmatrix} + c \, det \begin{bmatrix} d & e \\ g & h \end{bmatrix}$$

This last expression can be described as follows: To compute the determinant of a $3 \times 3$ matrix $A$, multiply each entry in the first row by $\pm 1$ times the determinant of a certain $2 \times 2$ matrix, namely the one obtained from $A$ by deleting the row and column of the entry in question. Here the signs alternate down the row (starting with $+1$).

**Example 1**   Find the determinant of

$$A = \begin{bmatrix} 2 & -3 & 7 \\ 5 & -1 & 6 \\ 4 & 0 & 8 \end{bmatrix}.$$

**SOLUTION**   Using the above procedure, we get:

$$det A = 2 \, det \begin{bmatrix} -1 & 6 \\ 0 & 8 \end{bmatrix} - (-3) \, det \begin{bmatrix} 5 & 6 \\ 4 & 8 \end{bmatrix} + 7 \, det \begin{bmatrix} 5 & -1 \\ 4 & 0 \end{bmatrix}$$

$$= 2(-8 - 0) + 3(40 - 24) + 7(0 - (-4))$$

$$= -16 + 48 + 28 = 60.$$

In this form the procedure can be generalized. To do so, it is convenient to introduce some terminology. Suppose that we have already defined how to compute the determinant of any $(n - 1) \times (n - 1)$ matrix. Given an $n \times n$ matrix $A$, let $A_{ij}$ denote the $(n - 1) \times (n - 1)$ matrix obtained from $A$ by deleting row $i$ and column $j$. Then we define the $(i, j)$-**cofactor** $C_{ij}(A)$ of $A$ by

$$C_{ij}(A) = (-1)^{i+j} det(A_{ij}) \text{ for each } i \text{ and } j.$$

Here $(-1)^{i+j}$ is called the **sign** of the $(i, j)$-position in the matrix. The following diagram is useful for remembering the sign of a position.

$$\begin{bmatrix} +1 & -1 & +1 & -1 & \cdots \\ -1 & +1 & -1 & +1 & \cdots \\ +1 & -1 & +1 & -1 & \cdots \\ -1 & +1 & -1 & +1 & \cdots \\ \vdots & \vdots & \vdots & \vdots & \end{bmatrix}$$

Note that the signs alternate along each row or column, starting with $+1$ in the upper left corner.

Since we already know how to compute $2 \times 2$ determinants, we can find the cofactors of any $3 \times 3$ matrix.

**Example 2**   Find $C_{12}(A)$, $C_{32}(A)$, and $C_{33}(A)$ for the matrix

$$A = \begin{bmatrix} 2 & -3 & 7 \\ 5 & -1 & 6 \\ 4 & 0 & 8 \end{bmatrix}.$$

---

[2] Another way to generalize (∗) is to notice that the right side is the sum of all products of entries of the matrix which have exactly one factor from each row and column, and which have a sign + or −. When it is clarified which products get which sign, this gives a very satisfactory definition of the determinant.

SOLUTION | The sign of the $(1, 2)$-position is $-1$, and $A_{12}$ is obtained from $A$ by deleting row 1 and column 2. Hence

$$C_{12}(A) = -det(A_{12}) = -det\begin{bmatrix} 5 & 6 \\ 4 & 8 \end{bmatrix} = -(40 - 24) = -16.$$

Similarly,

$$C_{32}(A) = -det(A_{32}) = -det\begin{bmatrix} 2 & 7 \\ 5 & 6 \end{bmatrix} = -(12 - 35) = 23,$$

$$C_{33}(A) = +det(A_{33}) = det\begin{bmatrix} 2 & -3 \\ 5 & -1 \end{bmatrix} = -2 + 15 = 13.$$

Clearly there are $9 = 3^2$ different cofactors for $A$, and they all can be computed in a similar way.

Again, suppose that we have already defined how to compute the determinant of any $(n-1) \times (n-1)$ matrix. Then if $A = [a_{ij}]$ is any $n \times n$ matrix, we define the **determinant** $detA$ of $A$ as follows:

$$detA = a_{11}C_{11}(A) + a_{12}C_{12}(A) + \cdots + a_{1n}C_{1n}(A).$$

In other words, $detA$ is obtained as in $(*)$ by multiplying each entry in row 1 of $A$ by the corresponding cofactor, and adding the results. This is called the cofactor expansion of $A$ along row 1. More generally:

If $A$ is any square matrix, the **cofactor expansion**[3] along any row (or column) is a number defined as follows: *Multiply each entry in the row (or column) by the corresponding cofactor, and add the results.*

Thus a square matrix $A$ has many cofactor expansions, one along each row or column. The astonishing thing about these expansions is that they are all equal. This common value is the determinant $detA$.

---

**THEOREM 1**

**Cofactor Expansion Theorem.** If $A$ is any square matrix, the determinant $detA$ is equal to the Laplace expansion along any row or column of $A$.

---

We omit the proof of Theorem 1.

Note that if a row or column of a square matrix $A$ consists *entirely* of zeros, then $detA = 0$ by the cofactor expansion along that row or column. This observation will be used frequently, and we record it for reference.

---

**COROLLARY**

If a square matrix $A$ has a row or column of zeros, then $detA = 0$.

---

**Example 3** | Compute the determinant of $A = \begin{bmatrix} 5 & -1 & 7 \\ 9 & -3 & 6 \\ 4 & 8 & 0 \end{bmatrix}$ using the cofactor expansion along column 2 and then by using the expansion along row 3.

---

[3] This is sometimes called the *Laplace expansion* because (a more general version of) it was given in 1772 by Pierre-Simon de Laplace (1749–1827).

SOLUTION The expansion along column 2 is as follows:

$$det A = (-1)C_{12}(A) + (-3)C_{22}(A) + 8\,C_{32}(A)$$

$$= (-1)\left\{-det\begin{bmatrix} 9 & 6 \\ 4 & 0 \end{bmatrix}\right\} + (-3)\left\{det\begin{bmatrix} 5 & 7 \\ 4 & 0 \end{bmatrix}\right\} + 8\left\{-det\begin{bmatrix} 5 & 7 \\ 9 & 6 \end{bmatrix}\right\}$$

$$= (-1)(24) + (-3)(-28) + 8(33)$$

$$= 324.$$

Similarly, the expansion along row 3 is

$$det A = 4\,det\begin{bmatrix} -1 & 7 \\ -3 & 6 \end{bmatrix} + 8\left\{-det\begin{bmatrix} 5 & 7 \\ 9 & 6 \end{bmatrix}\right\} + 0\,det\begin{bmatrix} 5 & -1 \\ 9 & -3 \end{bmatrix}$$

$$= 4 \cdot 15 + 8 \cdot 33 + 0 = 324.$$

Of course the two values are equal as Theorem 1 guarantees.

When computing the determinant of a $4 \times 4$ matrix, the cofactor expansion involves computing determinants of $3 \times 3$ matrices. These in turn are evaluated by calculating $2 \times 2$ determinants. Here is an example.

**Example 4**　Compute $det A$ where $A = \begin{bmatrix} 2 & -1 & 0 & 3 \\ 1 & 0 & 5 & 7 \\ 7 & 9 & 0 & 2 \\ 4 & 0 & 0 & 8 \end{bmatrix}$.

SOLUTION Since (by Theorem 1) we may expand $det A$ along any row or column, we choose one that simplifies the calculation. One way to do this is to choose a row or column with as many zeros as possible. Hence we use column 3 in $A$ because it contains three zeros.

$$det A = 0\,C_{13}(A) + 5\,C_{23}(A) + 0\,C_{33}(A) + 0\,C_{43}(A)$$

$$= 5\,C_{23}(A) = -5\,det\begin{bmatrix} 2 & -1 & 3 \\ 7 & 9 & 2 \\ 4 & 0 & 8 \end{bmatrix}$$

where we used the fact that the sign of the $(2, 3)$-position is $-1$. This reduces the calculation to finding a $3 \times 3$ determinant, and again we choose a row or column with as many zeros as possible. Using row 3 gives

$$det A = -5\left\{4\,det\begin{bmatrix} -1 & 3 \\ 9 & 2 \end{bmatrix} + 0 + 8\,det\begin{bmatrix} 2 & -1 \\ 7 & 9 \end{bmatrix}\right\}$$

$$= -5\{4(-29) + 8(25)\} = -420$$

as required.

## 2.1.2　Elementary Operations and Determinants

One way to create zeros in a matrix (and so simplify the computation of the determinant) is by using elementary row operations. The next theorem shows that row operations have a simple effect on the determinant, and that *column* operations can also be used. This can be combined with the cofactor expansion to greatly reduce the computing time involved in finding determinants.

**THEOREM 2**   Let $A$ denote any $n \times n$ matrix.

(1) If $B$ is obtained from $A$ by interchanging two different rows (columns), then $detB = -detA$.

(2) If $B$ is obtained from $A$ by multiplying a row (column) by a number $k$, then $detB = k\,detA$.

(3) If $B$ is obtained from $A$ by adding a multiple of some row (column) of $A$ to a different row (column), then $detB = detA$.

**PROOF**   We prove the theorem for row operations; the proof for column operations is analogous.

(1) We proceed by induction on $n$. The cases $n = 1$ and $n = 2$ are left to the reader. If $n > 2$, expand $detB$ along a row other than the two that were interchanged. The entries of this row are the same for both $A$ and $B$, but the cofactors in $B$ are the negatives of those in $A$ by induction, because the corresponding $(n-1) \times (n-1)$ matrices have two rows interchanged. It follows that $detB = -detA$.

(2) Expand $detB$ along the row that is multiplied by $k$. The entries in this row of $B$ are $k$ times the entries of the same row of $A$, and the corresponding cofactors are the same as those for $A$ (because $B$ and $A$ differ only in this row). It follows that $detB = k\,detA$.

(3) Let $B$ be obtained from $A = [a_{ij}]$ by adding $u$ times row $p$ to row $q$. Thus row $q$ of $B$ is

$$[a_{q1} + u\,a_{p1},\ a_{q2} + u\,a_{p2},\ \cdots\ , a_{qn} + u\,a_{pn}].$$

The cofactors of the positions in this row are the same as those for $A$ (they do not involve row $q$), that is $C_{qj}(B) = C_{qj}(A)$ for each $j$. Hence expanding $detB$ along row $q$ gives

$$detB = (a_{q1} + u\,a_{p1})C_{q1}(B) + \cdots + (a_{qn} + u\,a_{pn})C_{qn}(B)$$

$$= [a_{q1}C_{q1}(B) + \cdots + a_{qn}C_{qn}(B)] + u[a_{p1}C_{q1}(B) + \cdots + a_{pn}C_{qn}(B)]$$

$$= [a_{q1}C_{q1}(A) + \cdots + a_{qn}C_{qn}(A)] + u[a_{p1}C_{q1}(A) + \cdots + a_{pn}C_{qn}(A)]$$

$$= detA + u\,detA_1$$

where $A_1$ is the matrix obtained from $A$ by replacing row $q$ by row $p$, and both expansions are along row $q$. It remains to show that $detA_1 = 0$. But $A_1$ has two identical rows, so $A_1$ is unchanged if these rows are interchanged. Hence $detA_1 = -detA_1$ by (1), whence $detA_1 = 0$. Thus $detB = detA$ as required.

The proof of (3) in Theorem 2 establishes the following useful fact.

**COROLLARY**   If a square matrix $A$ has two identical rows (or columns), then $detA = 0$.

The properties in Theorem 2 can be stated in other words as follows:

(1) If two distinct rows (columns) are interchanged, the determinant changes sign.

(2) A common factor in any row (column) can be "taken out" of the determinant.

(3) Adding a multiple of any row (column) to another row (column) has no effect on the determinant.

These facts are very useful in computing determinants because they reduce the amount of computation required (and for other reasons as well). This is illustrated by the following three examples.

**Example 5**    Compute $detA$ where

$$A = \begin{bmatrix} 5 & -3 & 8 \\ -2 & 0 & 4 \\ 7 & 11 & 1 \end{bmatrix}.$$

SOLUTION    We could expand along row 2 (or column 2) and utilize the zero entry, but we can easily obtain *another* zero in row 2 by adding twice column 1 to column 3. Since this does not change the determinant (by Theorem 2), we obtain

$$detA = det \begin{bmatrix} 5 & -3 & 18 \\ -2 & 0 & 0 \\ 7 & 11 & 15 \end{bmatrix}$$

$$= -2 \left\{ -det \begin{bmatrix} -3 & 18 \\ 11 & 15 \end{bmatrix} \right\} + 0 + 0$$

$$= 2 \, det \begin{bmatrix} -3 & 18 \\ 11 & 15 \end{bmatrix}$$

using the cofactor expansion along row 2. We could simply evaluate at this point to get $detA = 2(-45 - 198) = -486$. However we can simplify the computation by using the fact that there is a common factor 3 in column 2 which we can "take out" by Theorem 2 (2). The result is

$$detA = 2 \, det \begin{bmatrix} -3 & 18 \\ 11 & 15 \end{bmatrix} = 2 \cdot 3 \, det \begin{bmatrix} -3 & 6 \\ 11 & 5 \end{bmatrix} = 6(-15 - 66) = -486$$

as required.

**Example 6**    Compute $detA$ if

$$det \begin{bmatrix} a & b & c \\ p & q & r \\ x & y & z \end{bmatrix} = 5 \quad \text{and} \quad A = \begin{bmatrix} a+2x & b+2y & c+2z \\ 3x+4p & 3y+4q & 3z+4r \\ -2p & -2q & -2r \end{bmatrix}.$$

SOLUTION    We reduce $detA$ to the known determinant using Theorem 2 several times:

$$detA = -2 \, det \begin{bmatrix} a+2x & b+2y & c+2z \\ 3x+4p & 3y+4q & 3z+4r \\ p & q & r \end{bmatrix}$$

$$= -2 \, det \begin{bmatrix} a+2x & b+2y & c+2z \\ 3x & 3y & 3z \\ p & q & r \end{bmatrix}$$

where the last equality results from subtracting 4 times row 3 from row 2. Continuing:

$$detA = -2 \cdot 3 \, det \begin{bmatrix} a+2x & b+2y & c+2z \\ x & y & z \\ p & q & r \end{bmatrix} = -6 \, det \begin{bmatrix} a & b & c \\ x & y & z \\ p & q & r \end{bmatrix}$$

Finally we interchange rows 2 and 3 (and use Theorem 2) to get

$$detA = -6(-1) \, det \begin{bmatrix} a & b & c \\ p & q & r \\ x & y & z \end{bmatrix} = 6 \cdot 5 = 30$$

because the given determinant is 5.

We shall see in Section 2.2 that a square matrix $A$ is *not* invertible if and only if $det A = 0$. Hence it is frequently important to be able to determine when $det A = 0$. If the entries of $A$ are functions of $x$, it is thus helpful to have $det A$ in factored form. Theorem 2 is often useful.

**Example 7**    Find the values of $x$ such that $det A = 0$ where $A = \begin{bmatrix} 1 & x & x \\ x & 1 & x \\ x & x & 1 \end{bmatrix}$.

SOLUTION    As in Gaussian elimination we begin by using the 1 in column 1 to "clean up" column 1:

$$det A = det \begin{bmatrix} 1 & x & x \\ 0 & 1-x^2 & x-x^2 \\ 0 & x-x^2 & 1-x^2 \end{bmatrix} = det \begin{bmatrix} 1-x^2 & x-x^2 \\ x-x^2 & 1-x^2 \end{bmatrix}$$

If we evaluate this we obtain $det A = 1 - 3x^2 + 2x^3$, so finding $x$ such that $det A = 0$ means factoring the cubic $1 - 3x^2 + 2x^3$, a nontrivial task. However, it is often easier to factor each of the entries before evaluating the determinant, and then use Theorem 2. In the present example we use the fact that $1 - x^2 = (1 + x)(1 - x)$ and $x - x^2 = x(1 - x)$ to get

$$det A = det \begin{bmatrix} (1+x)(1-x) & x(1-x) \\ x(1-x) & (1+x)(1-x) \end{bmatrix}$$

$$= (1 - x)^2 det \begin{bmatrix} 1+x & x \\ x & 1+x \end{bmatrix}$$

$$= (1 - x)^2(1 + 2x)$$

where we took the common factor $(1 - x)$ out of each column. Hence the condition $det A = 0$ becomes $(1 - x)^2(1 + 2x) = 0$, that is $x = 1$ or $x = -\frac{1}{2}$.

The fact that a common factor can be taken out of any row or column of a matrix (as in Example 7) leads to the following useful result.

**THEOREM ③**    If $A$ is an $n \times n$ matrix, then $det(kA) = k^n det A$ for any scalar $k$.

PROOF    The matrix $kA$ has $k$ as a common factor in each of its $n$ rows. Hence (by (2) of Theorem 2) we can take out a common factor $k$ from each of the $n$ rows. The result is $det(kA) = k^n det A$.

Recall that a matrix $A$ is called upper (lower) triangular if every entry below (above) the main diagonal is zero, and $A$ is called triangular if it is either upper or lower triangular. The next example reveals that the determinant of an upper triangular matrix is very easy to compute.

**Example 8**    If $A = \begin{bmatrix} a & x & y & z \\ 0 & b & u & v \\ 0 & 0 & c & w \\ 0 & 0 & 0 & d \end{bmatrix}$, show that $det A = abcd$.

SOLUTION    Use column 1 expansions to get

$$det A = a \, det \begin{bmatrix} b & u & v \\ 0 & c & w \\ 0 & 0 & d \end{bmatrix} = ab \, det \begin{bmatrix} c & w \\ 0 & d \end{bmatrix} = abcd.$$

Hence $det A$ is the product of the main diagonal elements.

The procedure in Example 8 works on any triangular matrix (use row expansions in the lower triangular case), and so proves

**THEOREM 4**

If a square matrix $A$ is triangular, then $det A$ is the product of the entries on the main diagonal.

Note that Theorem 4 implies that
$$det I = 1 \text{ for any identity matrix } I$$
because $I$ is upper (and lower) triangular with 1's on the main diagonal.

Theorem 4 shows that finding the determinant of a triangular matrix is easy. This is important in computer calculations because the Gaussian algorithm gives a routine method of carrying a matrix to (upper) triangular form using row operations.

The last result of this section is not surprising in view of the fact that row and column operations play identical roles in the cofactor expansion theorem.

**THEOREM 5**

If $A$ is any $n \times n$ matrix, then $det(A^T) = det A$.

PROOF

We proceed by induction on $n$, the cases $n = 1$ and $n = 2$ being easily verified. If $n \geq 3$, it is enough to show that the cofactor expansion of $det(A^T)$ along row 1 is identical to the cofactor expansion of $det A$ along column 1. But the entries of row 1 of $A^T$ are the same as the entries of column 1 of $A$, and the corresponding cofactors are also the same by induction because they are determinants of $(n-1) \times (n-1)$ matrices which are transposes of each other. It follows that $det(A^T) = det A$.

## Exercises 2.1

**1.** If $A = \begin{bmatrix} 5 & -1 & 4 \\ 2 & 3 & -7 \\ -3 & 9 & 4 \end{bmatrix}$, compute:

**a)** $C_{23}(A)$   **b)** $C_{31}(A)$   **c)** $C_{22}(A)$   **d)** $C_{12}(A)$

**2.** Explain why the sign of the lower right entry of a square matrix $A$ is always $+1$.

**3.** Compute the determinants of the following matrices along various rows and columns, and compare your results.

**a)** $\begin{bmatrix} 3 & -1 \\ 2 & 5 \end{bmatrix}$

**b)** $\begin{bmatrix} 6 & -9 \\ -14 & 21 \end{bmatrix}$

**c)** $\begin{bmatrix} a^2 & ab \\ ab & b^2 \end{bmatrix}$

**d)** $\begin{bmatrix} a+1 & a \\ a & a-1 \end{bmatrix}$

**e)** $\begin{bmatrix} 3 & -1 & 0 \\ 2 & 3 & 5 \\ -2 & 1 & 4 \end{bmatrix}$

**f)** $\begin{bmatrix} 7 & -1 & 2 \\ 5 & 2 & 0 \\ 3 & -4 & 9 \end{bmatrix}$

**g)** $\begin{bmatrix} 1 & b & c \\ b & c & 1 \\ c & 1 & b \end{bmatrix}$

**h)** $\begin{bmatrix} 0 & a & b \\ a & 0 & c \\ b & c & 0 \end{bmatrix}$

**i)** $\begin{bmatrix} -3 & -1 & 5 & 1 \\ 1 & 0 & -3 & 2 \\ 3 & 4 & 0 & 1 \\ -2 & 3 & 4 & 0 \end{bmatrix}$

**j)** $\begin{bmatrix} 4 & 3 & 0 & 1 \\ -1 & 1 & 1 & 2 \\ 3 & 0 & 2 & -1 \\ -1 & 2 & 2 & 1 \end{bmatrix}$

**k)** $\begin{bmatrix} 1 & 3 & -1 & 1 \\ -1 & 1 & -3 & 1 \\ 5 & 2 & 8 & 2 \\ 2 & 4 & 0 & -1 \end{bmatrix}$

**l)** $\begin{bmatrix} 0 & 0 & 0 & a \\ 0 & 0 & b & x \\ 0 & c & y & z \\ d & r & s & t \end{bmatrix}$

**4.** Explain in your own words how to find the determinant of a matrix using the cofactor expansion.

**5.** Explain why a matrix with a row (or column) of zeros has determinant zero.

**6.** In each case, compute the determinant of the given matrix by first adding a multiple of some row (or column) to another row (or column). Explain why this makes the computation by hand a bit simpler.

**a)** $A = \begin{bmatrix} 90 & 71 & 38 \\ -7 & 3 & 5 \\ 29 & -12 & -20 \end{bmatrix}$

**b)** $B = \begin{bmatrix} -35 & 5 & 7 \\ -10 & -11 & 2 \\ 66 & 13 & -13 \end{bmatrix}$

**7.** For various $n \times n$ matrices $A$ and $B$, compute the following pairs of numbers:

**a)** $det(A + B)$ and $det A + det B$

**b)** $det(AB)$ and $det A\, det B$

**c)** $det A$ and $det A^{-1}$

In each case make a conjecture about $det(A + B)$, $det(AB)$, or $det A^{-1}$.

**8.** In each case either show that the statement is true or give an example showing that it is false. Throughout, $A$ and $B$ denote square matrices.

**a)** If $detA = 0$, then $A$ has two identical rows.

**b)** $det(-A) = -detA$.

**c)** $det(A + B) = detA + detB$.

**d)** If $A$ is $2 \times 2$, $det(5A) = 25 \, detA$.

**e)** If $detA = detB$, then $A = B$.

**f)** If the main diagonal of $A$ consists of zeros, then $detA = 0$.

**9.** Prove Theorem 4 by induction on the size of the matrix.

**10.** Consider the matrix $A = \begin{bmatrix} 2 & -4 & 0 \\ 3 & 6 & 5 \\ -2 & 1 & 4 \end{bmatrix}$, and calculate its determinant $detA$. In each of the following cases, first try to guess what $detB$ will be, and verify your claim by a direct calculation.

**a)** $B$ is the matrix obtained from $A$ by interchanging rows 1 and 3.

**b)** $B$ is the matrix obtained from $A$ by multiplying row 2 by $\frac{1}{3}$.

**c)** $B$ is the matrix obtained from $A$ by multiplying row 2 by 3.

**d)** $B$ is the matrix obtained from $A$ by adding row 1 to row 3.

**e)** $B$ is the matrix obtained from $A$ by adding $-2$ times row 1 to row 2.

**f)** $B$ is the matrix obtained by transposing $A$.

**11.** If $B$ is obtained from $A$ by multiplying row 1 by 5, and $detB = 10$, what is $detA$?

**12.** In each case, calculate the determinant by transforming the matrix to upper triangular form by elementary row operations (a row-echelon form for example), and using Theorem 2. Then verify your computation by using the cofactor expansion, and make an estimate about which method is more efficient.

**a)** $A = \begin{bmatrix} -1 & -9 & 6 \\ 2 & -2 & 1 \\ 7 & 6 & -9 \end{bmatrix}$

**b)** $A = \begin{bmatrix} 3 & -6 & 2 \\ 7 & -1 & 0 \\ 7 & 5 & 11 \end{bmatrix}$

**c)** $A = \begin{bmatrix} 2 & -3 & 1 & 4 \\ 12 & -5 & 11 & 3 \\ 7 & 6 & -9 & 4 \\ 2 & -7 & 6 & 9 \end{bmatrix}$

**d)** $A = \begin{bmatrix} 4 & 7 & 2 & -2 \\ 8 & 9 & 2 & -2 \\ 3 & -3 & 1 & -8 \\ 0 & -2 & 3 & 1 \end{bmatrix}$

**13.** Find the real numbers $x$ and $y$ such that $detA = 0$ if:

**a)** $A = \begin{bmatrix} 0 & x & y \\ y & 0 & x \\ x & y & 0 \end{bmatrix}$

**b)** $A = \begin{bmatrix} 1 & -x & -x \\ x & -2 & -x \\ x & x & -3 \end{bmatrix}$

**c)** $A = \begin{bmatrix} 1 & x & x^2 & x^3 \\ x & x^2 & x^3 & 1 \\ x^2 & x^3 & 1 & x \\ x^3 & 1 & x & x^2 \end{bmatrix}$

**d)** $A = \begin{bmatrix} x & 0 & 0 & y \\ y & x & 0 & 0 \\ 0 & y & x & 0 \\ 0 & 0 & y & x \end{bmatrix}$

**14.** Compute $det \begin{bmatrix} \cos\theta & -\sin\theta \\ \sin\theta & \cos\theta \end{bmatrix}$ for any angle $\theta$.

**15.** Find $detA$ if $A$ is $3 \times 3$ and $det(2A) = 6$. What if $A$ is $4 \times 4$?

**16.** Under what conditions is $det(-A) = detA$? Support your answer.

**17.** By direct calculation, show that $detA = detA^T$ for any $2 \times 2$ matrix $A = \begin{bmatrix} a & b \\ c & d \end{bmatrix}$.

**18. a)** Show that $det(A + B^T) = det(A^T + B)$ for any $n \times n$ matrices $A$ and $B$.

**b)** What is wrong with the following argument?
$$det(A + B^T) = detA + detB^T = detA^T + detB$$
$$= det(A^T + B)$$

**19.** Show that $det \begin{bmatrix} 0 & 1 & 1 & 1 \\ 1 & 0 & x & x \\ 1 & x & 0 & x \\ 1 & x & x & 0 \end{bmatrix} = -3x^2$.

**20.** Show that $det \begin{bmatrix} 1 & x & x^2 \\ 1 & y & y^2 \\ 1 & z & z^2 \end{bmatrix} = (y - x)(z - x)(z - y)$. [This is the $3 \times 3$ Vandermonde determinant. There is an $n \times n$ version given in Section 2.7.]

**21. a)** Show that
$$det \begin{bmatrix} x & -1 & 0 & 0 \\ 0 & x & -1 & 0 \\ 0 & 0 & x & -1 \\ a & b & c & x+d \end{bmatrix} = a + bx + cx^2 + dx^3 + x^4.$$
[This matrix is called the **companion matrix** of the polynomial $a + bx + cx^2 + dx^3$].

**b)** Now write the $5 \times 5$ companion matrix for $a + bx + cx^2 + dx^3 + ex^4$ (last row is $[a \ b \ c \ d \ x + e]$) and verify your claim. Can you generalize this to the $n \times n$ case?

## 2.2   DETERMINANTS AND INVERSES

We can now prove the promised theorem that a square matrix $A$ has an inverse if and only if $det A \neq 0$. Furthermore, we derive a formula for $A^{-1}$ using determinants, and give a formula (called Cramer's Rule) for the solution of any set of linear equations with $A$ as coefficient matrix.

### 2.2.1   The Product Theorem

We begin with a remarkable theorem about the determinant of the product of two matrices. The proof is given at the end of this section.

**THEOREM 1**     **Product Theorem.** If $A$ and $B$ are $n \times n$ matrices, then $det(AB) = det A \, det B$.

**Example 1**    For the matrix product $\begin{bmatrix} 2 & -1 \\ 3 & 5 \end{bmatrix} \begin{bmatrix} 4 & 0 \\ -7 & 2 \end{bmatrix} = \begin{bmatrix} 15 & -2 \\ -23 & 10 \end{bmatrix}$, the corresponding determinant product is $13 \cdot 8 = 104$, which is correct as Theorem 1 asserts.

The product theorem extends to products of more than two matrices. For example
$$det(ABC) = det(A(BC)) = det A \, det(BC) = det A \, det B \, det C.$$
In general, if $A_1, A_2, \cdots, A_k$ are square matrices of the same size, then
$$det(A_1 A_2 \cdots A_k) = det A_1 \, det A_2 \cdots det A_k.$$
In particular, taking $A_i = A$ for each $i$ gives
$$det(A^k) = (det A)^k \text{ for each } k \geq 1.$$
We will use these formulas frequently below and record them for reference.

**COROLLARY**     Let $A, A_1, A_2, \cdots, A_k$ denote matrices. Then
   **(1)** $det(A^k) = (det A)^k$.
   **(2)** $det(A_1 A_2 \cdots A_k) = det A_1 \, det A_2 \cdots det A_k$.

Our first application of the product theorem is to give the determinant test for invertibility.

**THEOREM 2**     Let $A$ be an $n \times n$ matrix. Then:
   **(1)** $A$ is invertible if and only if $det A \neq 0$.
   **(2)** If $A$ is invertible, then $det(A^{-1}) = \frac{1}{det A}$.

**PROOF**    If $A$ is invertible, then $AA^{-1} = I$, so the product theorem gives
$$1 = det I = det(AA^{-1}) = det A \, det(A^{-1}).$$

This shows that $det A \neq 0$, proving half of (1), and it also proves (2). For the rest of (1), assume that $det A \neq 0$; we must show that $A$ is invertible. By Theorem 5 §1.5 it suffices to show that $A$ can be carried to the identity matrix by row operations. Certainly $A$ can be carried to a reduced row-echelon matrix $R$ by row operations. Each row operation multiplies the determinant by a nonzero constant by Theorem 2 §2.1. Hence $det A = c \, det R$ for some constant $c$, so $det R \neq 0$. In particular $R$ has no row of zeros, so $R = I$ because it is square and reduced row-echelon. This is what we wanted.

**Example 2**    If $detA = 2$ and $detB = -3$, compute $det(A^3B^{-1}A^TB^2)$.

SOLUTION    We use Theorems 1 and 2 together with Theorem 5 §2.1:

$$det(A^3B^{-1}A^TB^2) = det(A^3)\,det(B^{-1})\,det(A^T)\,det(B^2).$$
$$= (detA)^3\,\tfrac{1}{detB}\,detA\,(detB)^2$$
$$= (detA)^4\,detB$$
$$= -48.$$

**Example 3**    Find the values of $c$ for which $A = \begin{bmatrix} c & -1 & 2 \\ c & c & 1 \\ 1 & c & 1 \end{bmatrix}$ has an inverse.

SOLUTION    We compute $detA$ by first subtracting $c$ times column 3 from column 2:

$$detA = det\begin{bmatrix} c & -1 & 2 \\ c & c & 1 \\ 1 & c & 1 \end{bmatrix} = det\begin{bmatrix} c & -1-2c & 2 \\ c & 0 & 1 \\ 1 & 0 & 1 \end{bmatrix}$$

$$= (1+2c)\,det\begin{bmatrix} c & 1 \\ 1 & 1 \end{bmatrix} = (1+2c)(c-1).$$

Hence $detA = 0$ if and only if $c = -\tfrac{1}{2}$ or $c = 1$, so $A^{-1}$ exists for any value of $c$ *except* $-\tfrac{1}{2}$ and 1.

     Theorem 4 §2.1 asserts that the determinant of any triangular matrix is the product of the main diagonal entries. This extends to **block upper (lower) triangular** matrices, that is matrices which have square blocks on the main diagonal and zero blocks below (respectively above) the main diagonal. Then the extended result asserts that the determinant of a block triangular matrix is the product of the determinants of the main diagonal blocks. We state it here for matrices with two diagonal blocks; the general result follows by induction on the number of diagonal blocks. The proof depends on the product theorem.

**THEOREM ③**    Let $A$ and $B$ be square matrices (possibly of different sizes). Then

$$det\begin{bmatrix} A & X \\ 0 & B \end{bmatrix} = detA\,detB \quad \text{and} \quad det\begin{bmatrix} A & 0 \\ X & B \end{bmatrix} = detA\,detB.$$

PROOF    We factor the first matrix as $\begin{bmatrix} A & X \\ 0 & B \end{bmatrix} = \begin{bmatrix} I & 0 \\ 0 & B \end{bmatrix}\begin{bmatrix} A & X \\ 0 & I \end{bmatrix}$. Then the product theorem gives

$$det\begin{bmatrix} A & X \\ 0 & B \end{bmatrix} = det\begin{bmatrix} I & 0 \\ 0 & B \end{bmatrix}det\begin{bmatrix} A & X \\ 0 & I \end{bmatrix}.$$

The first result follows because $det\begin{bmatrix} I & 0 \\ 0 & B \end{bmatrix} = detB$ by repeated expansions along row 1,

and $det\begin{bmatrix} A & X \\ 0 & I \end{bmatrix} = detA$ by repeated expansions along the last row. The second

formula follows from the first and from the fact that $detC = det(C^T)$ for any matrix $C$.

**Example 4**    Find $detA$ if $A = \begin{bmatrix} a & x & b & y \\ c & z & d & w \\ 0 & p & 0 & q \\ 0 & r & 0 & s \end{bmatrix}$.

SOLUTION    First interchange columns 2 and 3 and then apply Theorem 3:

$$detA = -det\begin{bmatrix} a & b & x & y \\ c & d & z & w \\ 0 & 0 & p & q \\ 0 & 0 & r & s \end{bmatrix}$$

$$= -det\begin{bmatrix} a & b \\ c & d \end{bmatrix} det\begin{bmatrix} p & q \\ r & s \end{bmatrix} = -(ad - bc)(ps - qr)$$

using the block upper triangular partitioning with two $2 \times 2$ diagonal blocks.

## 2.2.2    Adjoint of a Matrix

The cofactor expansion theorem has an important matrix form which we describe next. Let $A$ denote an arbitrary $n \times n$ matrix. The **adjoint**[4] of $A$ is defined to be the transpose of the matrix of cofactors:

$$adjA = [C_{ij}(A)]^T.$$

Note that $adjA$ is also an $n \times n$ matrix.

**Example 5**    Find the adjoint of $A = \begin{bmatrix} 3 & 0 & -1 \\ 4 & 7 & 3 \\ -2 & 8 & 5 \end{bmatrix}$.

SOLUTION    Writing $C_{ij}(A) = C_{ij}$ for simplicity, the adjoint is

$$adjA = \begin{bmatrix} C_{11} & C_{12} & C_{13} \\ C_{21} & C_{22} & C_{23} \\ C_{31} & C_{32} & C_{33} \end{bmatrix}^T = \begin{bmatrix} 11 & -26 & 46 \\ -8 & 13 & -24 \\ 7 & -13 & 21 \end{bmatrix}^T$$

$$= \begin{bmatrix} 11 & -8 & 7 \\ -26 & 13 & -13 \\ 46 & -24 & 21 \end{bmatrix}.$$

**Example 6**    If $A = \begin{bmatrix} a & b \\ c & d \end{bmatrix}$ and we write $C_{ij} = C_{ij}(A)$, we have

$$adjA = \begin{bmatrix} C_{11} & C_{12} \\ C_{21} & C_{22} \end{bmatrix}^T = \begin{bmatrix} det[d] & -det[c] \\ -det[b] & det[a] \end{bmatrix}^T$$

$$= \begin{bmatrix} d & -c \\ -b & a \end{bmatrix}^T = \begin{bmatrix} d & -b \\ -c & a \end{bmatrix}.$$

This agrees with the adjoint used in Example 5 §1.5 to give a formula for the inverse of a $2 \times 2$ matrix.

---

[4] The adjoint is also called the **adjugate** or the **classical adjoint** of the matrix.

The adjoint of a square matrix $A$ is "almost" the inverse of $A$ in that it is off only by a numerical factor (the determinant). This is part (2) of the next theorem.

**THEOREM 4**    Let $A$ be any square matrix.

    (1) (**Adjoint Formula**) $A(adjA) = (detA) I = (adjA)A$.
    (2) If $detA \neq 0$, then $A^{-1} = \frac{1}{detA} adjA$.

**PROOF**    (1) The proof is given at the end of this section.
    (2) If we multiply (1) through by the number $\frac{1}{detA}$, the result is

$$A \cdot \left( \frac{1}{detA} adjA \right) = I = \left( \frac{1}{detA} adjA \right) \cdot A.$$

This shows that $\frac{1}{detA} adjA$ is indeed the inverse of $A$.

**Example 7**    Find the inverse of $A = \begin{bmatrix} 3 & 0 & -1 \\ 4 & 7 & 3 \\ -2 & 8 & 5 \end{bmatrix}$ (see Example 5).

**SOLUTION**    Since $detA = -13$ as the reader can check, Example 5 gives

$$A^{-1} = \frac{1}{-13} \ adjA = -\frac{1}{13} \begin{bmatrix} 11 & -8 & 7 \\ -26 & 13 & -13 \\ 46 & -24 & 21 \end{bmatrix}.$$

**Example 8**    Find the $(2, 3)$-entry of $A^{-1}$ if $A = \begin{bmatrix} 5 & -1 & 0 \\ 6 & 9 & -4 \\ 2 & 7 & 3 \end{bmatrix}$.

**SOLUTION**    We have $detA = 301$ as the reader can verify. The $(2, 3)$-entry of $A^{-1} = \frac{1}{301} \left[ C_{ij}(A) \right]^T$ is the $(3, 2)$-entry of the transpose matrix $\frac{1}{301} \left[ C_{ij}(A) \right]$; that is it equals

$$\frac{1}{301} C_{32}(A) = \frac{1}{301} \left\{ -det \begin{bmatrix} 5 & 0 \\ 6 & -4 \end{bmatrix} \right\} = \frac{20}{301}.$$

**Example 9**    Find a formula for the inverse of the matrix $A = \begin{bmatrix} c & -1 & 2 \\ c & c & 1 \\ 1 & c & 1 \end{bmatrix}$ in Example 3 for those values of $c$ for which it exists.

**SOLUTION**    We have $detA = (1 + 2c)(c - 1)$ by Example 3, and we compute

$$adjA = \begin{bmatrix} C_{11} & C_{12} & C_{13} \\ C_{21} & C_{22} & C_{23} \\ C_{31} & C_{32} & C_{33} \end{bmatrix}^T = \begin{bmatrix} 0 & -c+1 & c^2-c \\ 1+2c & c-2 & -c^2-1 \\ -1-2c & c & c^2+c \end{bmatrix}^T$$

$$= \begin{bmatrix} 0 & 1+2c & -1-2c \\ -c+1 & c-2 & c \\ c^2-c & -c^2-1 & c^2+c \end{bmatrix}.$$

Hence if $c \neq 1$ and $c \neq -\frac{1}{2}$, we have

$$A^{-1} = \frac{1}{(1+2c)(c-1)} \begin{bmatrix} 0 & 1+2c & -1-2c \\ -c+1 & c-2 & c \\ c^2-c & -c^2-1 & c^2+c \end{bmatrix}.$$

Of course this can be checked by multiplying by $A$.

## 2.2.3 Cramer's Rule

We now use the adjoint formula to derive an old theorem about determinants called Cramer's rule[5]. If $A$ is an invertible matrix, we know (from Theorem 2 §1.5) that the system of equations $AX = B$ has the unique solution $X = A^{-1}B$. Cramer's rule uses determinants to give a formula for the solution. In order to state the rule, we need some notation: If $A$ is an $n \times n$ matrix and $B$ is an $n \times 1$ column, let

$A_i(B)$ *denote the $n \times n$ matrix obtained from $A$ by replacing column $i$ by $B$.*

With this we can state Cramer's rule.

**THEOREM 5**

**Cramer's Rule.** Consider the system $AX = B$ of linear equations where $A$ is invertible. If $X = [x_1, x_2, \cdots, x_n]^T$ where $x_1, x_2, \cdots, x_n$ are the variables, then

$$x_i = \frac{det[A_i(B)]}{detA} \text{ for each } i = 1, 2, \cdots, n.$$

**PROOF**

We write $C_{ij} = C_{ij}(A)$ for simplicity, and write $B = [b_1 \ b_2 \ \cdots \ b_n]^T$. The solution to $AX = B$ is given by $X = A^{-1}B$, so the adjoint formula for $A^{-1}$ gives

$$X = A^{-1}B = \left(\frac{1}{detA} adjA\right) B = \frac{1}{detA} \begin{bmatrix} C_{11} & C_{21} & \cdots & C_{n1} \\ C_{12} & C_{22} & \cdots & C_{n2} \\ \vdots & \vdots & & \vdots \\ C_{1n} & C_{2n} & \cdots & C_{nn} \end{bmatrix} \begin{bmatrix} b_1 \\ b_2 \\ \vdots \\ b_n \end{bmatrix}.$$

Since entry $i$ of $X$ is $x_i$, this gives

$$x_i = \frac{1}{detA} [b_1C_{1i} + b_2C_{2i} + \cdots + b_nC_{ni}] = \frac{det[A_i(B)]}{detA}$$

where $det[A_i(B)]$ is expanded along column $i$.

**Example 10**

Use Cramer's rule to find $x_2$ if

$$\begin{cases} 5x_1 & - 7x_2 & + 8x_3 = & 23 \\ 2x_1 & + 6x_2 & - 9x_3 = & 61 \\ -x_1 & - 4x_2 & + 3x_3 = & -19 \end{cases}$$

**SOLUTION**

We have $A = \begin{bmatrix} 5 & -7 & 8 \\ 2 & 6 & -9 \\ -1 & -4 & 3 \end{bmatrix}$ and $A_2(B) = \begin{bmatrix} 5 & 23 & 8 \\ 2 & 61 & -9 \\ -1 & -19 & 3 \end{bmatrix}$ where we are thinking of the

system in matrix form $AX = B$. One verifies that $detA = -127$ and $det[A_2(B)] = 313$, so

$$x_2 = \frac{det[A_2(B)]}{detA} = \frac{313}{-127} = -\frac{313}{127}.$$

Of course, $x_1$ and $x_3$ can be obtained in the same way.

Suppose a system of linear equations is given (as in Example 10) with invertible coefficient matrix. At first appearance it seems to be an efficient procedure to solve for one variable using Cramer's rule without having to find the other variables. However, finding that one variable by Cramer's rule requires computing two determinants, and each of them takes approximately the same amount of computer time as solving the whole system using Gaussian elimination. Except for small systems, Cramer's rule is *not* a practical method for solving systems of linear equations; its virtue is theoretical.

---

[5] Cramer's Rule was known and used long before the Swiss mathematician Gabriel Cramer (1704–1752) popularized it in his book, *Introduction à l'analyse des lignes courbes algébriques*, published in 1750. The rule was widely used for many years but is less important today.

### 2.2.4 Proof of the Product Theorem

If $A$ and $B$ are $n \times n$ matrices, we must show that

$$det(AB) = detA \, detB. \qquad (*)$$

Recall (Section 1.6) that an elementary matrix $E$ is one obtained by doing a single elementary row operation to an identity matrix. Recall further (Lemma 1 §1.6) that doing this operation to a matrix $B$ results in $EB$. By looking at the three types of elementary matrices separately, Theorem 2 §2.1 gives

$$det(EB) = detE \, detB, \text{ whenever } E \text{ is elementary.} \qquad (**)$$

Now suppose that $A \to C$ by row operations with elementary matrices $E_1, E_2, \cdots, E_k$ in order. Using Lemma 1 §1.6 several times, the reduction is $A \to E_1A \to E_2E_1A \to \cdots \to E_k \cdots E_2E_1A = C$. This combines with $(**)$ to give

---

**LEMMA ①** If $A \to C$ by elementary row operations, then $C = E_k \cdots E_2E_1A$ where the $E_i$ are elementary matrices, and $detC = detE_k \cdots detE_2 \, detE_1 \, detA$.

---

**LEMMA ②** If $A$ has no inverse, then $detA = 0$.

**PROOF** Let $A \to R$ by row operations where $R$ is in reduced row-echelon form. Then $R = E_k \cdots E_2E_1A$ where the $E_i$ are elementary by Lemma 1. But $A$ is not invertible, so $R \neq I$ by Theorem 5 §1.5. Thus $R$ has a row of zeros, and so $0 = detR = detE_k \cdots detE_2 \, detE_1 \, detA$, again by Lemma 1. Since $detE_i \neq 0$ for each $i$, this implies that $detA = 0$.

---

Now we can prove $(*)$ by distinguishing two cases.

**Case 1.** $A$ has no inverse. In this case it suffices to show that $AB$ also has no inverse, because then $detA = 0 = det(AB)$ by Lemma 2. But if $(AB)^{-1}$ exists, then $A[B(AB)^{-1}] = I$, so $A$ is invertible by Theorem 5 §1.5, contrary to our assumption.

**Case 2.** $A$ is invertible. Then $A$ is a product of elementary matrices by Theorem 2 §1.6, say $A = E_1E_2 \cdots E_k$. Hence Lemma 1 gives

$$
\begin{aligned}
detA &= det(E_1E_2 \cdots E_kI) \\
&= detE_1 \, detE_2 \cdots detE_k \, detI \\
&= detE_1 \, detE_2 \cdots detE_k.
\end{aligned}
$$

Finally, again using Lemma 1, we obtain

$$det(AB) = det(E_1E_2 \cdots E_kB) = detE_1 \, detE_2 \cdots detE_k \, detB = detA \, detB.$$

Thus $(*)$ holds in this case too.

### 2.2.5 Proof of the Adjoint Formula

If $A$ is any $n \times n$ matrix, we must show that

$$A(adjA) = (detA)I = (adjA)A.$$

We prove the left equality; the other is similar.

Write $A(adjA) = [u_{ij}]$, so $u_{ij}$ denotes the $(i, j)$-entry of the matrix product $A(adjA)$. We must show that

$$u_{ij} = \begin{cases} detA & \text{if } i = j \\ 0 & \text{if } i \neq j \end{cases}$$

Now write $A = [a_{ij}]$, and let $C_{ij} = C_{ij}(A)$ denote the $(i, j)$-cofactor of $A$. By the definition of matrix multiplication, the number $u_{ij}$ is the dot product of row $i$ of $A$ and column $j$ of $adjA$. Now row $i$ of $A$ is $[a_{i1}\ a_{i2}\ \cdots\ a_{in}]$. Since column $j$ of $adjA = [C_{ij}]^T$ is the transpose of row $j$ of the cofactor matrix $[C_{ij}]$, we obtain

$$u_{ij} = [a_{i1}\ a_{i2}\ \cdots\ a_{in}]\begin{bmatrix} C_{j1} \\ C_{j2} \\ \vdots \\ C_{jn} \end{bmatrix} = a_{i1}C_{j1} + a_{i2}C_{j2} + \cdots + a_{in}C_{jn}. \qquad (***)$$

If $i = j$, the sum in $(***)$ is the cofactor expansion of $detA$ along row $i$, so we have $u_{ii} = detA$ for each $i$. If $i \neq j$, let $A'$ denote the matrix obtained from $A$ by replacing row $j$ by row $i$. Then row $j$ of $A'$ is $[a_{i1}\ a_{i2}\ \cdots\ a_{in}]$, and the cofactors are $C_{jk}(A') = C_{jk}(A) = C_{jk}$ for $k = 1, 2, \cdots, n$. Hence, $(***)$ is the expansion of $detA'$ along row $j$; that is, $u_{ij} = detA'$. But $detA' = 0$ because $A'$ has two identical rows, so $u_{ij} = 0$ if $i \neq j$. This proves that $A(adjA) = (detA)I$.

## ■ Exercises 2.2

**1.** Test the product theorem for determinants; that is, create some matrices $A$ and $B$ of your choice; then calculate the matrix product $AB$; calculate the determinants $det(A)$, $det(B)$, $det(AB)$; and verify that $det(AB) = det(A)\,det(B)$.

**2.** Explain in your own words why the product theorem implies that $det(A^{-1}) = 1/det(A)$ whenever $A$ is invertible.

**3.** Test the connection between the determinant of a matrix and that of its inverse; that is, create some square matrix $A$ of your choice, then calculate its inverse $A^{-1}$ and the determinants $det(A)$ and $det(A^{-1})$, and verify that $det(A^{-1}) = 1/det(A)$.

**4.** What can be said about the value of $detA$ where $A$ is an $n \times n$ such that:
**a)** $A^2 = I$    **b)** $A^3 = I$    **c)** $A^2 = 3A$    **d)** $A = -A^T$
**e)** $A^2 + I = 0$   **f)** $A^3 = A$   **g)** $A^{-1} = A^T$?

**5.** If $A^k = 0$ for some $k \geq 1$, show that $A$ has no inverse.

**6.** Assume $detA = -2$, $detB = 3$, and $detC = -1$, where $A$, $B$, and $C$ are $n \times n$ matrices. Compute:
**a)** $det[A^3B^{-1}C^TB^2A^{-1}]$    **b)** $det[B^TA^{-1}B^{-1}CA^2(C^{-1})^T]$

**7.** If $A$ is $3 \times 3$ and $det(2A^{-1}) = 5 = det[A^2(B^T)^{-1}]$, find $detA$ and $detB$.

**8.** Find the adjoint and determinant of the following matrices, and verify your result by verifying that $A\,adj(A) = det(A)I$ in each case. Whenever possible, use this information to calculate the inverse of the matrix.

**a)** $A = \begin{bmatrix} 2 & -1 & 0 \\ 1 & 2 & -2 \\ -1 & -1 & 3 \end{bmatrix}$    **b)** $A = \begin{bmatrix} -2 & 3 & 1 \\ 2 & -3 & 1 \\ 0 & 5 & 1 \end{bmatrix}$

**c)**$^\dagger$ $A = \begin{bmatrix} 8 & -11 & 0 & 4 \\ 11 & 2 & -2 & 9 \\ -1 & 12 & 3 & 3 \\ 12 & 3 & 8 & 0 \end{bmatrix}$    **d)**$^\dagger$ $A = \begin{bmatrix} -2 & 3 & 6 & 2 \\ 7 & -3 & 3 & 9 \\ 3 & 5 & 7 & 12 \\ -5 & 10 & 21 & 20 \end{bmatrix}$

**9.** Explain the difference between the adjoint of a matrix, and the matrix of cofactors.

**10.** In each case either show that the statement is true or give an example showing that it is false. Throughout, $A$ denotes a square matrix.
**a)** If $detA \neq 0$ and $AB = AC$, then $B = C$.
**b)** If $adjA$ exists, then $A$ is invertible.
**c)** If $A^T = -A$, then $detA = -1$.
**d)** $det(3A) = 3\,detA$.
**e)** $det(AB) = det(BA)$.
**f)** If $A$ is invertible and $adj(A) = A^{-1}$, then $det(A) = 1$.
**g)** $det(I^n) = 1^n$.
**h)** If $adj(A) = 0$, then $A = 0$.

**11.** In each case use Theorem 4(2) to find the $(2, 3)$-entry of $A^{-1}$.

**a)** $A = \begin{bmatrix} 3 & -1 & 2 \\ 5 & 5 & -2 \\ 1 & 2 & 3 \end{bmatrix}$    **b)** $A = \begin{bmatrix} -2 & 3 & 6 \\ 7 & -3 & 3 \\ 3 & 5 & 7 \end{bmatrix}$

**12.** If $A$ and $B$ are invertible $n \times n$ matrices, evaluate:
**a)** $det(A^{-1}BA)$    **b)** $det(B^{-1}A^{-1}BA)$.

**13.** By computing the determinant, show that $A = \begin{bmatrix} 1 & a & b \\ -a & 1 & c \\ -b & -c & 1 \end{bmatrix}$ has an inverse for any $a$, $b$, and $c$.

**14.** In each case: (1) Find the values of the number $c$ such that $A$ has an inverse, and (2) Find $A^{-1}$ for those values of $c$.

**a)** $A = \begin{bmatrix} 1 & c & 0 \\ 2 & 0 & c \\ c & -1 & 1 \end{bmatrix}$    **b)** $A = \begin{bmatrix} 1 & -c & c \\ 1 & 1 & -1 \\ c & -c & 1 \end{bmatrix}$

---

$^\dagger$ Requires computer assistance (e.g., Matlab).

**15.** In each case solve for $y$ by Cramer's rule.

**a)** $2x - 5y + 7z = 9$
$-x + 4y + 2z = -2$
$3x + 3y - 6z = 5$

**b)** $3x - 2y + 4z = 5$
$5x + 3y + z = 8$
$-2x + 6y + 7z = -3$

**16.** If $A$, $B$, and $C$ are matrices such that $det A = 3$, $det B = -1$, and $det C = 2$, compute the determinant of:

**a)** $\begin{bmatrix} A & X & Y \\ 0 & B & Z \\ 0 & 0 & C \end{bmatrix}$

**b)** $\begin{bmatrix} A & 0 & 0 \\ X & B & 0 \\ Z & Y & C \end{bmatrix}$

**17.** As in Example 4, use Theorem 3 to find $det A$.

**a)** $A = \begin{bmatrix} 2 & 5 & 1 & -1 \\ -3 & -4 & 2 & 1 \\ 0 & 0 & -1 & -2 \\ 0 & 0 & -2 & 5 \end{bmatrix}$

**b)** $A = \begin{bmatrix} 0 & -1 & 2 & 0 \\ 0 & 3 & 1 & 0 \\ -1 & 2 & -7 & 5 \\ 0 & -3 & 6 & 0 \end{bmatrix}$

**18.** If $A$ is $n \times n$, use Theorem 1 to show that $det(kA) = k^n\, det A$

for all scalars $k$ (This is Theorem 3 §2.1). [Hint: First show that $kA = (kI)A$.]

**19.** **a)** If $A = \begin{bmatrix} 0 & 1 \\ -1 & 0 \end{bmatrix}$, show that $A^2 = -I$.

**b)** Show that there is no $3 \times 3$ matrix $A$ with the property that $A^2 = -I$.

**20.** If $A$ is $n \times n$ and invertible, show that
$$det(adjA) = (detA)^{n-1}.$$

**21.** If $A$ is $3 \times 3$ and $det A = 2$, compute $det[-A^2(adjA)^{-1}]$.

**22.** If $A$ is $n \times n$ and $det A = 2$, evaluate $det[A^{-1} + adjA]$.

**23.** **a)** If $A = UB$ where $det U = 1$, show that $det A = det B$.

**b)** If $A$ and $B$ are invertible and $det A = det B$, show that $A = UB$ for some invertible matrix $U$ such that $det U = 1$.

---

# 2.3 DIAGONALIZATION AND EIGENVALUES

A central problem in the applications of linear algebra is to describe systems that are changing with time. As Example 1 below shows, this often comes down to finding a way to efficiently calculate powers $A$, $A^2$, $A^3$, $\cdots$ of a square matrix $A$. In this section we outline one method of doing this. The technique is called diagonalization, and is one of the most important ideas in linear algebra.

## 2.3.1 A Model of Population Dynamics

We begin with an example that motivates the method.

**Example 1**

We are interested in how the population of an endangered species of birds changes over time. Since the number of males is approximately equal to the number of females, we count only females. Each female remains a juvenile for one year and then becomes an adult. Only adults have offspring. After $k$ years the female population can be expressed as a sum $a_k + j_k$ where

$a_k$   denotes the number of adults after $k$ years,
$j_k$   denotes the number of juveniles after $k$ years.

We assume that we know the initial values $a_0$ and $j_0$, and we are interested in computing $a_k$ and $j_k$ for $k = 1, 2, \cdots$. To do this we model the population growth as follows.

We assume that $j_{k+1} = 2a_k$, that is that the number of juveniles hatched in any year is on average twice the number of adult females alive the year before. We also assume that $\frac{1}{2}$ of the adult birds in any year survive to the next year, but only $\frac{1}{4}$ of the juveniles survive into adulthood. Hence the number $a_{k+1}$ of adults in any year is given by $a_{k+1} = \frac{1}{2}a_k + \frac{1}{4}j_k$. Then the numbers $a_k$ and $j_k$ in successive years are related by the following equations:

$$a_{k+1} = \tfrac{1}{2}a_k + \tfrac{1}{4}j_k$$
$$j_{k+1} = 2a_k$$

If we write $V_k = \begin{bmatrix} a_k \\ j_k \end{bmatrix}$ and $A = \begin{bmatrix} \frac{1}{2} & \frac{1}{4} \\ 2 & 0 \end{bmatrix}$, these equations take the simple matrix form

$$V_{k+1} = AV_k \quad \text{for each } k = 1, 2, \cdots. \tag{$*$}$$

The column $V_k$ is called the *population profile* for the species. Since the initial profile $V_0 = \begin{bmatrix} a_0 \\ j_0 \end{bmatrix}$ is assumed to be known, ($*$) gives successively: $V_1 = AV_0$, $V_2 = AV_1 = A^2V_0$, $V_3 = AV_2 = A^3V_0$, etc. In general we get

$$V_k = A^kV_0 \quad \text{for each } k = 1, 2, \cdots.$$

Thus to find $V_k$ (and hence $a_k$ and $j_k$) it is sufficient to compute the powers $A^k$ of the matrix $A$. In this section we look at an important method of doing this. This example is an illustration of a linear dynamical system, and we return to it in the following section.

Direct computation of the powers $A^k$ of a square matrix $A$ is time-consuming for the large matrices that occur in practice. Furthermore, we are frequently interested only in approximating the long-term behaviour of $V_k = A^kV_0$ as $k$ increases, where $V_0$ is some fixed column. Hence we adopt an indirect method, currently used in applications, which lets us easily compute the powers $A^k$ explicitly, and which also gives information about the long-term behaviour. The idea is to first **diagonalize** the matrix $A$, that is to find an invertible matrix $P$ such that

$$P^{-1}AP = D \quad \text{is a diagonal matrix.} \tag{$**$}$$

The reason this works is two-fold: First, the powers $D^k$ of the diagonal matrix $D$ are easy to compute (more about this later). Second, ($**$) enables us to compute powers $A^k$ of the matrix $A$ in terms of powers $D^k$ of $D$. Indeed, we can solve ($**$) for $A$ to get $A = PDP^{-1}$. Squaring this gives

$$A^2 = (PDP^{-1})(PDP^{-1}) = PD^2P^{-1}.$$

Using this we can compute $A^3$ as follows:

$$A^3 = AA^2 = (PDP^{-1})(PD^2P^{-1}) = PD^3P^{-1}.$$

Continuing in this way (and noting that this works even if $D$ is not diagonal), we obtain

**THEOREM 1**    If $A = PDP^{-1}$, then $A^k = PD^kP^{-1}$ for each $k = 1, 2, \cdots$.

Hence computing $A^k$ comes down to finding the invertible matrix $P$ in equation ($**$). To do this it is necessary to first compute certain numbers (called eigenvalues) associated with the matrix $A$.

## 2.3.2 Eigenvalues and Eigenvectors

If $A$ is an $n \times n$ matrix, a number $\lambda$ is called an **eigenvalue** of $A$ if

$$AX = \lambda X \quad \text{for some column } X \neq 0.$$

Such a nonzero column $X$ is called an **eigenvector** of $A$ corresponding to the eigenvalue $\lambda$. Note that the condition $AX = \lambda X$ is automatically satisfied if $X = 0$, so the requirement that $X \neq 0$ is critical.

**Example 2**　Consider the $2 \times 2$ matrix $A = \begin{bmatrix} 5 & -2 \\ 4 & -1 \end{bmatrix}$. Then $\lambda = 3$ is an eigenvalue of $A$ with

eigenvector $X = \begin{bmatrix} 1 \\ 1 \end{bmatrix}$ because $X \neq 0$ and $AX = \begin{bmatrix} 5 & -2 \\ 4 & -1 \end{bmatrix} \begin{bmatrix} 1 \\ 1 \end{bmatrix} = \begin{bmatrix} 3 \\ 3 \end{bmatrix} = 3 \begin{bmatrix} 1 \\ 1 \end{bmatrix} = 3X$.

The matrix in Example 2 has another eigenvalue in addition to $\lambda = 3$. To find it, we develop the following general procedure which works for any $n \times n$ matrix $A$. By definition, a number $\lambda$ is an eigenvalue of $A$ if and only if

$$AX = \lambda X \text{ for some } X \neq 0.$$

If $I$ is the identity matrix of the same size as $A$, this is equivalent to asking that the homogeneous linear system

$$(\lambda I - A)X = 0$$

has a nontrivial solution $X \neq 0$. By Theorem 5 §1.5 this happens if and only if the matrix $\lambda I - A$ is not invertible, and this in turn holds if and only if the determinant of the coefficient matrix is zero:

$$det(\lambda I - A) = 0.$$

This last condition prompts the following definition: If $A$ is an $n \times n$ matrix, the **characteristic polynomial** $c_A(x)$ of $A$ is defined by

$$c_A(x) = det(xI - A).$$

Note that $c_A(x)$ is indeed a polynomial[6] in the variable $x$, and it has degree $n$ if $A$ is an $n \times n$ matrix (this is illustrated in the examples below).

The above discussion shows that a number $\lambda$ is an eigenvalue of $A$ if and only if $c_A(\lambda) = 0$, that is if and only if $\lambda$ is a **root** of the characteristic polynomial $c_A(x)$. Moreover, the calculation reveals a method for finding the eigenvectors $X$ of $A$ corresponding to $\lambda$. Indeed, $X$ is determined by the condition that $(\lambda I - A)X = 0$, and so the eigenvectors $X$ are the (nonzero) solutions to this system of homogeneous linear equations. We record these observations in

**THEOREM ②**　Let $A$ be an $n \times n$ matrix.

(1) The eigenvalues $\lambda$ of $A$ are the roots of the characteristic polynomial $c_A(x)$ of $A$.

(2) The eigenvectors $X$ corresponding to $\lambda$ are the nonzero solutions to the homogeneous system

$$(\lambda I - A)X = 0$$

of linear equations with $\lambda I - A$ as coefficient matrix.

Observe that once a number $\lambda$ is known to be an eigenvalue of $A$, finding the solutions to the equations $(\lambda I - A)X = 0$ (and hence finding the eigenvectors) is a routine application of Gaussian elimination. Here are two examples.

**Example 3**　Find the characteristic polynomial of the matrix $A = \begin{bmatrix} 5 & -2 \\ 4 & -1 \end{bmatrix}$ discussed in Example 2,

and then find all the eigenvalues and their eigenvectors.

---

[6] A **polynomial** is an expression of the form $a_0 + a_1 x + a_2 x^2 + \cdots + a_n x^n$ where the $a_i$ are numbers and $x$ is a variable. If $a_n \neq 0$, the number $n$ is called the **degree** of the polynomial, and $a_n$ is called the **leading coefficient**. Thus $2 - 5x + 2x^2 - x^4$ is a polynomial of degree 4 with leading coefficient $-1$. Polynomials are discussed in Appendix A.3.

SOLUTION    Since $xI - A = \begin{bmatrix} x & 0 \\ 0 & x \end{bmatrix} - \begin{bmatrix} 5 & -2 \\ 4 & -1 \end{bmatrix} = \begin{bmatrix} x-5 & 2 \\ -4 & x+1 \end{bmatrix}$, we get

$$c_A(x) = det \begin{bmatrix} x-5 & 2 \\ -4 & x+1 \end{bmatrix}$$
$$= (x-5)(x+1) + 8 = x^2 - 4x + 3 = (x-3)(x-1).$$

Hence the roots of $c_A(x)$ are $\lambda_1 = 3$ and $\lambda_2 = 1$, so these are the eigenvalues of $A$. Note that $\lambda_1 = 3$ was the eigenvalue mentioned in Example 2, but we have found a new one: $\lambda_2 = 1$.

To find the eigenvectors corresponding to $\lambda_2 = 1$, observe that in this case

$$\lambda_2 I - A = \begin{bmatrix} \lambda_2 - 5 & 2 \\ -4 & \lambda_2 + 1 \end{bmatrix} = \begin{bmatrix} -4 & 2 \\ -4 & 2 \end{bmatrix}$$

so the solutions to $(\lambda_2 I - A)X = 0$ are $X = t \begin{bmatrix} \frac{1}{2} \\ 1 \end{bmatrix}$ where $t$ is any real number. Hence the eigenvectors $X$ corresponding to $\lambda_2$ are $X = t \begin{bmatrix} \frac{1}{2} \\ 1 \end{bmatrix}$ where $t \neq 0$ is arbitrary. A convenient choice is $X_2 = \begin{bmatrix} 1 \\ 2 \end{bmatrix}$, when $t = 2$. Similarly, $\lambda_1 = 3$ gives rise to the eigenvectors, $X = t \begin{bmatrix} 1 \\ 1 \end{bmatrix}$, $t \neq 0$, which includes the observation in Example 2.

Note that there are *many* eigenvectors of a square matrix $A$ associated with a given eigenvalue $\lambda$, in fact *every* nonzero solution $X$ of $(\lambda I - A)X = 0$ is an eigenvector. Of course the eigenvalue $\lambda$ is chosen so that there *must* be nonzero solutions.

**Example 4**    Find the characteristic polynomial, eigenvalues, and eigenvectors for $A = \begin{bmatrix} 1 & 1 & 1 \\ 0 & 2 & -1 \\ 0 & -3 & 0 \end{bmatrix}$.

SOLUTION    Here the characteristic polynomial is given by

$$c_A(x) = det \begin{bmatrix} x-1 & -1 & -1 \\ 0 & x-2 & 1 \\ 0 & 3 & x \end{bmatrix}$$
$$= (x-1) det \begin{bmatrix} x-2 & 1 \\ 3 & x \end{bmatrix} = (x-1)(x+1)(x-3)$$

so the eigenvalues are $\lambda_1 = 1$, $\lambda_2 = -1$, and $\lambda_3 = 3$. To find the eigenvectors for $\lambda_1 = 1$, compute

$$\lambda_1 I - A = \begin{bmatrix} \lambda_1 - 1 & -1 & -1 \\ 0 & \lambda_1 - 2 & 1 \\ 0 & 3 & \lambda_1 \end{bmatrix} = \begin{bmatrix} 0 & -1 & -1 \\ 0 & -1 & 1 \\ 0 & 3 & 1 \end{bmatrix}.$$

We want the (nonzero) solutions to $(\lambda_1 I - A)X = 0$. The augmented matrix becomes

$$\begin{bmatrix} 0 & -1 & -1 & 0 \\ 0 & -1 & 1 & 0 \\ 0 & 3 & 1 & 0 \end{bmatrix} \rightarrow \begin{bmatrix} 0 & 1 & 1 & 0 \\ 0 & 0 & 2 & 0 \\ 0 & 0 & -2 & 0 \end{bmatrix} \rightarrow \begin{bmatrix} 0 & 1 & 0 & 0 \\ 0 & 0 & 1 & 0 \\ 0 & 0 & 0 & 0 \end{bmatrix}$$

using row operations. Hence the solutions $X$ to $(\lambda_1 I - A)X = 0$ are $X = t \begin{bmatrix} 1 \\ 0 \\ 0 \end{bmatrix}$, where

$t \neq 0$ is arbitrary, so we can use the basic solution $X_1 = \begin{bmatrix} 1 \\ 0 \\ 0 \end{bmatrix}$ as an eigenvector

corresponding to $\lambda_1 = 1$. Similarly, eigenvectors corresponding to $\lambda_2 = -1$ and $\lambda_3 = 3$

are $X_2 = \begin{bmatrix} -2 \\ 1 \\ 3 \end{bmatrix}$ and $X_3 = \begin{bmatrix} 0 \\ 1 \\ -1 \end{bmatrix}$, respectively, as the reader can verify.

There are several comments that must be made at this point:

**Remark 1** The eigenvalues of a real matrix need not be real numbers (as they were in the above examples). For example, the characteristic polynomial of the matrix $A = \begin{bmatrix} 0 & -1 \\ 1 & 0 \end{bmatrix}$ is $x^2 + 1$, so the eigenvalues of $A$ are the nonreal complex roots $\lambda = i$ and $\lambda = -i$. We return to this in Section 2.5.

**Remark 2** An $n \times n$ matrix always has $n$ (possibly complex) eigenvalues, but they may not be distinct (as they are in the above examples). For example, the matrix $A = \begin{bmatrix} 1 & 1 \\ 0 & 1 \end{bmatrix}$ has characteristic polynomial $c_A(x) = (x - 1)^2$, so there is only one eigenvalue $\lambda = 1$. However, $\lambda$ is a double root of $c_A(x)$ and we say that $\lambda$ has multiplicity 2. We shall have more to say about such repeated eigenvalues later (Theorem 5).

**Remark 3** In practice, the eigenvalues of a square matrix $A$ usually are not computed as the roots of the characteristic polynomial. There are iterative numerical methods (for example, the power method or the QR-algorithm) that are more efficient for large matrices. These algorithms (and others) are described in books on computational linear algebra.

## 2.3.3 Diagonalization

An $n \times n$ matrix $D$ is called a **diagonal matrix** if all its entries off the main diagonal are zero; that, is if $D$ has the form

$$D = \begin{bmatrix} \lambda_1 & 0 & \cdots & 0 \\ 0 & \lambda_2 & \cdots & 0 \\ \vdots & \vdots & \ddots & \vdots \\ 0 & 0 & \cdots & \lambda_n \end{bmatrix} = diag(\lambda_1, \lambda_2, \cdots, \lambda_n)$$

where $\lambda_1, \lambda_2, \cdots, \lambda_n$ are numbers. We use the notation $D = diag(\lambda_1, \lambda_2, \cdots, \lambda_n)$ for convenience. Calculations with diagonal matrices are very easy. For example, if $n = 2$ we have

$$\begin{bmatrix} \lambda_1 & 0 \\ 0 & \lambda_2 \end{bmatrix}\begin{bmatrix} \mu_1 & 0 \\ 0 & \mu_2 \end{bmatrix} = \begin{bmatrix} \lambda_1\mu_1 & 0 \\ 0 & \lambda_2\mu_2 \end{bmatrix}$$

and

$$\begin{bmatrix} \lambda_1 & 0 \\ 0 & \lambda_2 \end{bmatrix} + \begin{bmatrix} \mu_1 & 0 \\ 0 & \mu_2 \end{bmatrix} = \begin{bmatrix} \lambda_1 + \mu_1 & 0 \\ 0 & \lambda_2 + \mu_2 \end{bmatrix}.$$

Thus each operation is carried out by doing the same operation on corresponding diagonal entries.

In general, if $D = diag(\lambda_1, \lambda_2, \cdots, \lambda_n)$ and $E = diag(\mu_1, \mu_2, \cdots, \mu_n)$ are two diagonal matrices, their product $DE$ and sum $D + E$ are again diagonal, and are obtained as in the $2 \times 2$ case by doing the same operations to the corresponding diagonal elements:

$$DE = diag(\lambda_1\mu_1,\ \lambda_2\mu_2, \cdots,\ \lambda_n\mu_n)$$

$$D + E = diag(\lambda_1 + \mu_1,\ \lambda_2 + \mu_2, \cdots,\ \lambda_n + \mu_n)$$

The simplicity of these formulas is one reason the following notion is so important.

A square $n \times n$ matrix $A$ is called **diagonalizable** if

$$P^{-1}AP \text{ is diagonal for some invertible } n \times n \text{ matrix } P.$$

In this case the invertible matrix $P$ is called a **diagonalizing matrix** for $A$. To discover when such a matrix $P$ exists, we let $X_1, X_2, \cdots, X_n$ denote the columns of $P$ and look for ways to determine when such $X_i$ exist and how to compute them. To this end, write $P$ in terms of its columns as follows:

$$P = [X_1 \quad X_2 \quad \cdots \quad X_n].$$

The condition $P^{-1}AP = D$ holds for some diagonal matrix $D$ if and only if

$$AP = PD$$

and we proceed by writing each side of this equation in terms of its columns. If we write $D = diag(\lambda_1, \lambda_2, \cdots, \lambda_n)$, where the $\lambda_i$ are numbers to be determined, the equation $AP = PD$ becomes

$$A[X_1 \quad X_2 \quad \cdots \quad X_n] = [X_1 \quad X_2 \quad \cdots \quad X_n]\begin{bmatrix} \lambda_1 & 0 & \cdots & 0 \\ 0 & \lambda_2 & \cdots & 0 \\ \vdots & \vdots & \ddots & \vdots \\ 0 & 0 & \cdots & \lambda_n \end{bmatrix}.$$

By the definition of matrix multiplication each side simplifies as follows

$$[AX_1 \quad AX_2 \quad \cdots \quad AX_n] = [\lambda_1 X_1 \quad \lambda_2 X_2 \quad \cdots \quad \lambda_n X_n].$$

Comparing columns shows that $AX_1 = \lambda_1 X_1$, $AX_2 = \lambda_2 X_2, \cdots$ etc., and so

$$P^{-1}AP = D \quad \text{if and only if} \quad AX_i = \lambda_i X_i \text{ for each } i.$$

In other words, $P^{-1}AP = D$ holds if and only if the diagonal entries of $D$ are eigenvalues of $A$ and the columns of $P$ are corresponding eigenvectors. This proves the following fundamental result.

**THEOREM 3**    Let $A$ be an $n \times n$ matrix.

(1) $A$ is diagonalizable if and only if it has eigenvectors $X_1, X_2, \cdots, X_n$ such that the matrix $P = [X_1 \quad X_2 \quad \cdots \quad X_n]$ is invertible.

(2) When this is the case, $P^{-1}AP = diag(\lambda_1, \lambda_2, \cdots, \lambda_n)$ where, for each $i$, $\lambda_i$ is the eigenvalue of $A$ corresponding to $X_i$.

**Example 5**    Diagonalize the matrix $A = \begin{bmatrix} 1 & 1 & 1 \\ 0 & 2 & -1 \\ 0 & -3 & 0 \end{bmatrix}$ in Example 4.

**SOLUTION**    By Example 4, the eigenvalues of $A$ are $\lambda_1 = 1$, $\lambda_2 = -1$, and $\lambda_3 = 3$ with corresponding eigenvectors $X_1 = \begin{bmatrix} 1 \\ 0 \\ 0 \end{bmatrix}$, $X_2 = \begin{bmatrix} -2 \\ 1 \\ 3 \end{bmatrix}$, and $X_3 = \begin{bmatrix} 0 \\ 1 \\ -1 \end{bmatrix}$. Since the matrix

$$P = [X_1 \quad X_2 \quad X_3] = \begin{bmatrix} 1 & -2 & 0 \\ 0 & 1 & 1 \\ 0 & 3 & -1 \end{bmatrix} \text{ is invertible, Theorem 3 guarantees that}$$

$$P^{-1}AP = \begin{bmatrix} \lambda_1 & 0 & 0 \\ 0 & \lambda_2 & 0 \\ 0 & 0 & \lambda_3 \end{bmatrix} = \begin{bmatrix} 1 & 0 & 0 \\ 0 & -1 & 0 \\ 0 & 0 & 3 \end{bmatrix}.$$

The reader can verify this directly.

**Remark 4**    In Example 5, suppose we let $Q = [X_2 \ X_1 \ X_3]$ be the matrix formed from the eigenvectors $X_1$, $X_2$, and $X_3$ of $A$, but in a *different order* than that used to form $P$. Then $Q^{-1}AQ = D = diag(\lambda_2, \lambda_1, \lambda_3)$ is diagonal by Theorem 3, but the eigenvalues are in the *new* order. This means that we can choose the diagonalizing matrix $P$ so that the eigenvalues $\lambda_i$ appear in any order we want along the main diagonal of $D$.

**Remark 5**    On the other hand, we can multiply any column of $P$ by a nonzero constant without changing $D$ (because the resulting column is still an eigenvector). This is sometimes convenient for simplifying the form of $P$.

Unfortunately, not every square matrix $A$ is diagonalizable. However, if the eigenvalues of $A$ are all distinct (as in Example 5), the matrix $P$ of eigenvectors *must* be invertible (and so $A$ is diagonalizable). We state this result in the next theorem; the proof is deferred to Section 4.7.

**THEOREM ④**    If $A$ is an $n \times n$ matrix with $n$ distinct eigenvalues, then $A$ is diagonalizable.

On the other hand, many matrices with repeated eigenvalues *are* diagonalizable. The following example exhibits such a matrix.

**Example 6**    Diagonalize the matrix $A = \begin{bmatrix} 0 & 1 & 1 \\ 1 & 0 & 1 \\ 1 & 1 & 0 \end{bmatrix}.$

**SOLUTION**    To compute the characteristic polynomial of $A$, first add rows 2 and 3 to row 1:

$$c_A(x) = det\begin{bmatrix} x & -1 & -1 \\ -1 & x & -1 \\ -1 & -1 & x \end{bmatrix} = det\begin{bmatrix} x-2 & x-2 & x-2 \\ -1 & x & -1 \\ -1 & -1 & x \end{bmatrix}$$

$$= det\begin{bmatrix} x-2 & 0 & 0 \\ -1 & x+1 & 0 \\ -1 & 0 & x+1 \end{bmatrix} = (x-2)(x+1)^2.$$

Hence the eigenvalues are $\lambda_1 = 2$ and $\lambda_2 = -1$, with $\lambda_2$ repeated twice (we say that $\lambda_2$ has *multiplicity* two). Thus Theorem 4 does not apply. However, $A$ *is* diagonalizable.

For $\lambda_1 = 2$ the system of equations $(\lambda_1 I - A)X = 0$ has general solution $X = t\begin{bmatrix} 1 \\ 1 \\ 1 \end{bmatrix}$, so

the basic solution $X_1 = \begin{bmatrix} 1 \\ 1 \\ 1 \end{bmatrix}$ is an eigenvector corresponding to $\lambda_1 = 2$. Turning to the

repeated eigenvalue $\lambda_2 = -1$, we must solve $(\lambda_2 I - A)X = 0$. By Gaussian elimination,

the general solution is $X = s\begin{bmatrix} -1 \\ 1 \\ 0 \end{bmatrix} + t\begin{bmatrix} -1 \\ 0 \\ 1 \end{bmatrix}$ where $s$ and $t$ are arbitrary. Hence the basic

solutions $X_2 = \begin{bmatrix} -1 \\ 1 \\ 0 \end{bmatrix}$ and $Y_2 = \begin{bmatrix} -1 \\ 0 \\ 1 \end{bmatrix}$ are *both* eigenvectors of $A$ corresponding to

$\lambda_2 = -1$, and they arise naturally from the Gaussian algorithm. If we take

$$P = [X_1 \quad X_2 \quad Y_2] = \begin{bmatrix} 1 & -1 & -1 \\ 1 & 1 & 0 \\ 1 & 0 & 1 \end{bmatrix}, \text{ we find that } P \text{ is invertible.}$$

Hence $P^{-1}AP = diag(2, -1, -1)$ by Theorem 3.

Example 6 is typical of the situation for any diagonalizable matrix. To describe the general case, we need some terminology. An eigenvalue $\lambda$ of a square matrix $A$ is said to have **multiplicity** $m$ if it occurs $m$ times as a root of the characteristic polynomial $c_A(x)$. When the homogeneous system $(\lambda I - A)X = 0$ is solved, any set of basic solutions is called a set of **basic eigenvectors** corresponding to $\lambda$ (see Remark 5, §2.3). Recall that the number of basic eigenvectors equals the number of parameters involved in the solution of the system $(\lambda I - A)X = 0$.

Thus the eigenvalue $\lambda_2 = -1$ in Example 6 has multiplicity 2. In that case, the Gaussian algorithm yields *two* basic eigenvectors corresponding to $\lambda_2 = -1$, arising from the fact that the general solution to the system $(\lambda_2 I - A)X = 0$ involves *two* parameters. This works in general: A matrix is diagonalizable if and only if every eigenvalue of multiplicity $m$ yields $m$ basic eigenvectors. This is the content of the following theorem; the proof is given in Section 4.7.

**THEOREM 5**

A square matrix $A$ is diagonalizable if and only if it satisfies the following condition:

*The multiplicity of every eigenvalue $\lambda$ of $A$ equals the number of basic eigenvectors corresponding to $\lambda$, which is the number of parameters in the solution of $(\lambda I - A)X = 0$.*

In this case the basic solutions of the system $(\lambda I - A)X = 0$ become columns in the invertible diagonalizing matrix $P$ such that $P^{-1}AP$ is diagonal.

In fact, it turns out that the number of basic eigenvectors corresponding to an eigenvalue $\lambda$ can *never exceed* the multiplicity of $\lambda$. Thus a square matrix is *not* diagonalizable if and only if it has an eigenvalue $\lambda$ of multiplicity $m$ for which the system $(\lambda I - A)X = 0$ has *fewer* than $m$ basic solutions. For this reason, nondiagonalizable matrices are sometimes called **defective**.

The following procedure summarizes Theorem 5.

**DIAGONALIZATION ALGORITHM**

To diagonalize an $n \times n$ matrix $A$:

**Step 1.**  Find the distinct eigenvalues $\lambda$ of $A$.

**Step 2.**  Compute basic eigenvectors corresponding to each of these eigenvalues $\lambda$ from basic solutions of the homogeneous system $(\lambda I - A)X = 0$.

**Step 3.**  The matrix $A$ is diagonalizable if and only if there are $n$ basic eigenvectors in all.

**Step 4.**  If $A$ is diagonalizable, the $n \times n$ matrix $P$ with these basic eigenvectors as columns is a diagonalizing matrix for $A$; that is, $P$ is invertible and $P^{-1}AP$ is diagonal.

**Remark 6**

The diagonalization algorithm is valid even if the eigenvalues are complex numbers that are not real. In this case the eigenvectors also will be complex, but we will not pursue this here.

**Example 7**  Show that the matrix $A = \begin{bmatrix} 1 & 1 \\ 0 & 1 \end{bmatrix}$ is not diagonalizable.

SOLUTION 1  The characteristic polynomial of $A$ is $c_A(x) = (x - 1)^2$, so $A$ has only one eigenvalue $\lambda_1 = 1$, which is of multiplicity 2. But the system of equations $(\lambda_1 I - A)X = 0$ has general solution $X = t\begin{bmatrix} 1 \\ 0 \end{bmatrix}$, so there is only one basic solution $\begin{bmatrix} 1 \\ 0 \end{bmatrix}$. Hence $A$ is not diagonalizable by Theorem 5.

SOLUTION 2  If $A$ were diagonalizable, there would be an invertible matrix $P$ such that $P^{-1}AP = D$ is diagonal. By Theorem 3, the diagonal entries of $D$ must be eigenvalues of $A$, so $D = I$ because 1 is the only eigenvalue of $A$. Hence $P^{-1}AP = I$, so $A = PIP^{-1} = I$, a contradiction. Hence $A$ is not diagonalizable.

## 2.3.4 Similar Matrices

If $A$ and $B$ are $n \times n$ matrices, we say that $A$ and $B$ are *similar* if
$$B = P^{-1}AP \text{ for some invertible matrix } P.$$

When this is the case, we write $A \sim B$. The language of similarity is used throughout linear algebra. For example, in this terminology a matrix $A$ is diagonalizable if and only if it is similar to a diagonal matrix.

If $A \sim B$, then necessarily $B \sim A$. To see why, suppose that $B = P^{-1}AP$. Then $A = PBP^{-1} = Q^{-1}BQ$ where $Q = P^{-1}$. Since $Q$ is invertible, this proves the second of the following properties of similarity (the others are Exercise 18):

  **1.** $A \sim A$ for all square matrices $A$.
  **2.** If $A \sim B$, then $B \sim A$.  $(*)$
  **3.** If $A \sim B$ and $B \sim C$, then $A \sim C$.

These properties are often expressed by saying that the similarity relation $\sim$ is an *equivalence relation* on the set of $n \times n$ matrices. Moreover, similarity is compatible with inverses, transposes, and powers in the following sense:

$$\text{If } A \sim B, \text{ then } \begin{cases} A^{-1} \sim B^{-1} \\ A^T \sim B^T \\ A^k \sim B^k \text{ for all } k \geq 0. \end{cases} \qquad (**)$$

The proofs are routine matrix computations (Exercise 18). In addition, similar matrices share many properties, some of which are collected in the next theorem for reference.

THEOREM **6**  If $A$ and $B$ are similar $n \times n$ matrices, then:
  **(1)** $\det A = \det B$.
  **(2)** $c_A(x) = c_B(x)$.
  **(3)** $A$ and $B$ have the same eigenvalues.

PROOF  Let $B = P^{-1}AP$ for some invertible matrix $P$.
  **(1)** $\det B = \det(P^{-1}) \det A \det P = \det A$ because $\det(P^{-1}) = 1/\det P$.
  **(2)** One verifies that $xI - B = P^{-1}(xI - A)P$. Hence (1) gives
    $c_B(x) = \det(xI - B) = \det\{P^{-1}(xI - A)P\} = \det(xI - A) = c_A(x)$.
  **(3)** The eigenvalues of a matrix are the roots of its characteristic polynomial, so (2) applies.

The class of diagonalizable matrices is well-behaved because these matrices are similar to diagonal matrices. Here are two illustrations.

**Example 8**  If $A$ is similar to $B$ and either $A$ or $B$ is diagonalizable, show that the other is also diagonalizable.

SOLUTION   We have $A \sim B$. Suppose that $A$ is diagonalizable, say $A \sim D$, where $D$ is diagonal. Since $B \sim A$ by ($*$), we have $B \sim A$ and $A \sim D$. Hence $B \sim D$ by ($*$), so $B$ is diagonalizable too. A similar argument works if we assume instead that $B$ is diagonalizable.

**Example 9**  If $A$ is diagonalizable, show that $A^T$, $A^{-1}$ (if it exists), and $A^k$ (for each $k \geq 1$) are all diagonalizable.

SOLUTION   Since $A$ is diagonalizable, we have $A \sim D$ where $D$ is a diagonal matrix. Hence ($**$) gives $A^T \sim D^T$, $A^{-1} \sim D^{-1}$, and $A^k \sim D^k$, and the result follows because $D^T$, $D^{-1}$, and $D^k$ are all diagonal.

Suppose that $A$ is diagonalizable, say $P^{-1}AP = D$ is diagonal. Theorem 6 shows that $A$ and $D$ have the same eigenvalues. But the eigenvalues of the diagonal matrix $D$ are just the main diagonal entries (verify), so the diagonal entries of $D$ must be the eigenvalues of $A$ in some order. This means that $A$ inherits many properties of $D$, and hence of its eigenvalues. The following example illustrates how this happens.

**Example 10**  Let $A$ be a diagonalizable $2 \times 2$ matrix. If $\lambda^4 = 5\lambda$ for each eigenvalue $\lambda$ of $A$, show that $A^4 = 5A$.

SOLUTION   Let $P^{-1}AP = D = \begin{bmatrix} \lambda_1 & 0 \\ 0 & \lambda_2 \end{bmatrix}$ where $\lambda_1$ and $\lambda_2$ are the eigenvalues of $A$, and $P$ is invertible. Then

$$D^4 = \begin{bmatrix} \lambda_1^4 & 0 \\ 0 & \lambda_2^4 \end{bmatrix} = \begin{bmatrix} 5\lambda_1 & 0 \\ 0 & 5\lambda_2 \end{bmatrix} = 5\begin{bmatrix} \lambda_1 & 0 \\ 0 & \lambda_2 \end{bmatrix} = 5D.$$

Since $A = PDP^{-1}$ (obtained by solving $P^{-1}AP = D$ for $A$), this gives

$$A^4 = (PDP^{-1})^4 = PD^4P^{-1} = PD(5D)P^{-1} = 5(PDP^{-1}) = 5A$$

using Theorem 1.

The result in Example 10 can be reformulated as follows: Write $p(x) = x^4 - 5x$. Then the example asserts that, if $p(\lambda) = 0$ for every eigenvalue $\lambda$ of $A$, then $p(A) = 0$. This holds much more generally.

If $p(x) = a_0 + a_1x + a_2x^2 + \cdots + a_kx^k$ is any polynomial, the **evaluation** of $p(x)$ at the (square) matrix $A$ is defined to be

$$p(A) = a_0I + a_1A + a_2A^2 + \cdots + a_kA^k$$

where $I$ is the identity matrix of the same size as $A$. Then the method used in Example 10 shows that if $A$ is diagonalizable, and if $p(x)$ is any polynomial such that $p(\lambda) = 0$ for all eigenvalues $\lambda$ of $A$, then necessarily $p(A) = 0$. In particular, the characteristic polynomial $c_A(x)$ certainly has the property that $c_A(\lambda) = 0$ for each eigenvalue of $A$, so we must have that $c_A(A) = 0$. In other words, this shows that $c_A(A) = 0$ for every diagonalizable matrix $A$. In fact, this holds for *every* square matrix $A$ and, in full generality, is called the

| CAYLEY-HAMILTON THEOREM | If $A$ is a square matrix, then $c_A(A) = 0$. |

The proof is given in Section 5.6.

## Exercises 2.3

**1.** Consider the matrix $A = \begin{bmatrix} 1 & 1 & 1 \\ 1 & 1 & -1 \\ 1 & 1 & 1 \end{bmatrix}$.

**a)**[†] Directly calculate the matrix power $A^{20}$.

**b)** Use the fact that $A$ is diagonalizable to compute $A^{20}$; that is, there are matrices $P$ and $D$ such that $P^{-1}AP = D$

where $P^{-1}AP = \begin{bmatrix} -1 & 0 & 1 \\ 1 & 1 & 0 \\ -1 & -1 & 1 \end{bmatrix}\begin{bmatrix} 1 & 1 & 1 \\ 1 & 1 & -1 \\ 1 & 1 & 1 \end{bmatrix}\begin{bmatrix} -1 & 1 & 1 \\ 1 & 0 & -1 \\ 0 & 1 & 1 \end{bmatrix}$

$= \begin{bmatrix} 0 & 0 & 0 \\ 0 & 2 & 0 \\ 0 & 0 & 1 \end{bmatrix} = D.$

**c)** Explain the benefits of diagonalizing a matrix to compute its powers.

**2.** In each case find the characteristic polynomial, eigenvalues, and eigenvectors and, if possible, find an invertible matrix $P$ such that $P^{-1}AP = D$.

**a)** $A = \begin{bmatrix} 1 & 3 \\ 2 & 2 \end{bmatrix}$

**b)** $A = \begin{bmatrix} 2 & -1 \\ -4 & -1 \end{bmatrix}$

**c)** $A = \begin{bmatrix} 7 & 0 & 5 \\ 0 & 5 & 0 \\ -4 & 0 & -2 \end{bmatrix}$

**d)** $A = \begin{bmatrix} 3 & -4 & 2 \\ 1 & -2 & 2 \\ 1 & -5 & 5 \end{bmatrix}$

**e)** $A = \begin{bmatrix} 1 & 1 & 0 \\ 1 & 1 & -1 \\ 0 & 1 & 1 \end{bmatrix}$

**f)** $A = \begin{bmatrix} 7 & 4 & -16 \\ 2 & 5 & -8 \\ 2 & 2 & -5 \end{bmatrix}$

**3.** Write your own short description of how to find eigenvalues of a matrix $A$, the corresponding eigenvectors, and an invertible matrix $P$ such that $P^{-1}AP$ is a diagonal matrix.

**4.** If $P^{-1}AP = D$ is a diagonal matrix, what are the diagonal entries of $D$ and the columns of $P$?

**5. a)** Show that $A = \begin{bmatrix} 1 & 3 \\ -3 & -5 \end{bmatrix}$ is not diagonalizable.

**b)** Show that $A = \begin{bmatrix} 1 & c \\ 0 & 1 \end{bmatrix}$ is not diagonalizable unless $c = 0$.

**6.** In each case determine whether $A$ is diagonalizable. Give reasons for your answer.

**a)** $A = \begin{bmatrix} 5 & -2 & 2 \\ 4 & 1 & 4 \\ 4 & -4 & 9 \end{bmatrix}$

**b)** $A = \begin{bmatrix} 2 & 2 & -1 \\ 1 & 1 & 0 \\ 1 & -2 & -2 \end{bmatrix}$

**7.** In each case either show that the statement is true or give an example showing that it is false. Throughout, $A$ represents a square matrix.

**a)** If $A$ has real eigenvalues, then it is diagonalizable.

**b)** If $A$ is diagonalizable, then it has distinct eigenvalues.

**c)** If all of the eigenvalues of $A$ are real and distinct, then $A$ is diagonalizable.

**d)** If $A$ is diagonalizable, then its transpose $A^T$ is also diagonalizable.

**e)** Every invertible matrix is diagonalizable.

**f)** Every diagonalizable matrix is symmetric.

**g)** If $A$ has no inverse, then $\lambda = 0$ is an eigenvalue of $A$.

**h)** If $\lambda = 0$ is an eigenvalue of $A$, then $A$ has no inverse.

**8.** If $\lambda$ is an eigenvalue of $A$, and if $\alpha$ is a number, show that $\lambda - \alpha$ is an eigenvalue of $A_1 = A - \alpha I$. How do the eigenvectors compare?

**9.** Find the eigenvalues of $A = \begin{bmatrix} \cos\theta & -\sin\theta \\ \sin\theta & \cos\theta \end{bmatrix}$.

**10.** If $X$ and $Y$ are eigenvectors of $A$ corresponding to $\lambda$, show that the same is true of $X + Y$ and $aX$ for any number $a$ (provided that $X + Y$ and $aX$ are not zero).

**11.** If $\lambda$ is an eigenvalue of $A$, show that:

**a)** $k\lambda$ is an eigenvalue of $kA$ for each real number $k$.

**b)** $\lambda^2$ is an eigenvalue of $A^2$.

**c)** $3 - 2\lambda + 5\lambda^3$ is an eigenvalue of $3I - 2A + 5A^3$.

**12.** Let $A$ denote an invertible matrix.

**a)** If $\lambda$ is an eigenvalue of $A$, show that $\lambda \neq 0$ and that $\frac{1}{\lambda}$ is an eigenvalue of $A^{-1}$.

**b)** Show that every eigenvalue $\mu$ of $A^{-1}$ has the form $\mu = \frac{1}{\lambda}$ where $\lambda$ is some eigenvalue of $A$.

**13.** If $D = diag(\lambda_1, \lambda_2, \cdots, \lambda_n)$ and $E = diag(\mu_1, \mu_2, \cdots, \mu_n)$, show that

**a)** $D + E = diag(\lambda_1 + \mu_1, \lambda_2 + \mu_2, \cdots, \lambda_n + \mu_n)$

**b)** $DE = diag(\lambda_1\mu_1, \lambda_2\mu_2, \cdots, \lambda_n\mu_n)$

**c)** $D^k = diag(\lambda_1^k, \lambda_2^k, \cdots, \lambda_n^k)$ for each integer $k \geq 1$.

**14. a)** Show that $A$ and its transpose $A^T$ have the same characteristic polynomial. [Hint: $xI = (xI)^T$.]

**b)** If $A$ is diagonalizable, show that $A^T$ is diagonalizable.

**15.** If $A$ is diagonalizable with eigenvalues $\lambda_1, \lambda_2, \cdots, \lambda_n$, show that $det A = \lambda_1\lambda_2\cdots\lambda_n$.

---

† Requires computer assistance (e.g., Matlab).

**16. a)** If $A^{-1} = A$ and $\lambda$ is an eigenvalue of $A$, show that $\lambda = \pm 1$.

**b)** If $A$ is diagonalizable and every eigenvalue is $\pm 1$, show that $A^{-1} = A$.

**17. a)** If $A^2 = A$ and $\lambda$ is an eigenvalue of $A$, show that $\lambda = 0, 1$.

**b)** If $A$ is diagonalizable and every eigenvalue is 0 or 1, show that $A^2 = A$.

**18.** If $A$ and $B$ are square matrices, show that:

**a)** If $B = QAQ^{-1}$ for some invertible matrix $Q$, then $A \sim B$.

**b)** $A \sim A$ for all square matrices $A$.

**c)** If $A \sim B$, then $B \sim A$.

**d)** If $A \sim B$ and $B \sim C$, then $A \sim C$.

**e)** If $A \sim B$, then $\det A = \det B$.

**f)** If $A \sim B$, then $A^{-1} \sim B^{-1}$.

**g)** If $A \sim B$, then $A^k \sim B^k$ for all integers $k \geq 1$.

**h)** If $A \sim B$, then $A^T \sim B^T$.

**19.** Assume that $A$ has eigenvalues $\lambda_1$ and $\lambda_2$ with corresponding eigenvectors $X_1$ and $X_2$.

**a)** Find $A$ if $\lambda_1 = 2$, $\lambda_2 = 5$, $X_1 = \begin{bmatrix} 2 \\ 1 \end{bmatrix}$, and $X_2 = \begin{bmatrix} 3 \\ 2 \end{bmatrix}$.

**b)** Find $A$ if $\lambda_1 = -1$, $\lambda_2 = 0$, $X_1 = \begin{bmatrix} 1 \\ 1 \end{bmatrix}$, and $X_2 = \begin{bmatrix} -1 \\ 2 \end{bmatrix}$.

**20.** If $A = \begin{bmatrix} a & b \\ c & d \end{bmatrix}$, show that:

**a)** $c_A(x) = x^2 - (trA)x + \det A$ where $trA = a + d$ is called the **trace** of $A$.

**b)** The eigenvalues of $A$ are
$$\lambda = \tfrac{1}{2}\left[(a + d) \pm \sqrt{(a - d)^2 + 4bc}\right].$$

**21. a)** If $D$ is diagonal with distinct eigenvalues, and if $CD = DC$, show that $C$ is also diagonal. [Hint: Compare $(i, j)$-entries in $CD = DC$.]

**b)** If $A$ is diagonalizable with distinct eigenvalues, and if $BA = AB$, show that $B$ is also diagonalizable. [Hint: Use **a)**.]

## 2.4   LINEAR DYNAMICAL SYSTEMS

We began Section 2.3 with an example from ecology which models the evolution of the population of a species of birds as time goes on. As promised, we now complete the example and extend it to other situations.

### 2.4.1   Linear Dynamical Systems

The bird population was described by computing the female population profile $V_k = \begin{bmatrix} a_k \\ j_k \end{bmatrix}$ of the species, where $a_k$ and $j_k$ represent the number of adult and juvenile females present $k$ years after the initial values $a_0$ and $j_0$ were observed. The model assumes that these numbers are related by the following equations:

$$a_{k+1} = \tfrac{1}{2}a_k + \tfrac{1}{4}j_k$$
$$j_{k+1} = 2a_k$$

If we write $A = \begin{bmatrix} \frac{1}{2} & \frac{1}{4} \\ 2 & 0 \end{bmatrix}$, this system takes the following matrix form

$$V_{k+1} = AV_k \text{ for each } k = 1, 2, \cdots.$$

Hence $V_1 = AV_0$, whence $V_2 = AV_1 = A^2V_0$ and, in general,

$$V_k = A^kV_0 \text{ for } k = 1, 2, \cdots.$$

We can now use our diagonalization techniques to determine the population profile $V_k$ for all values of $k$ in terms of the initial values.

**Example 1**   Assuming that the initial values were $a_0 = 100$ adult females and $j_0 = 40$ juvenile females, compute $a_k$ and $j_k$ for $k = 1, 2, \cdots$.

**SOLUTION**   The characteristic polynomial of the matrix $A = \begin{bmatrix} \frac{1}{2} & \frac{1}{4} \\ 2 & 0 \end{bmatrix}$ is

$c_A(x) = x^2 - \frac{1}{2}x - \frac{1}{2} = (x-1)\left(x+\frac{1}{2}\right)$, so the eigenvalues are $\lambda_1 = 1$ and $\lambda_2 = -\frac{1}{2}$ and Gaussian elimination gives corresponding basic eigenvectors $\begin{bmatrix} \frac{1}{2} \\ 1 \end{bmatrix}$ and $\begin{bmatrix} -\frac{1}{4} \\ 1 \end{bmatrix}$. For convenience (see Remark 5 §2.3), we can use multiples $X_1 = \begin{bmatrix} 1 \\ 2 \end{bmatrix}$ and $X_2 = \begin{bmatrix} -1 \\ 4 \end{bmatrix}$ respectively. Hence a diagonalizing matrix is $P = \begin{bmatrix} 1 & -1 \\ 2 & 4 \end{bmatrix}$ and we obtain

$$P^{-1}AP = D \quad \text{where} \quad D = \begin{bmatrix} 1 & 0 \\ 0 & -\frac{1}{2} \end{bmatrix}.$$

This gives $A = PDP^{-1}$ so, for each $k \geq 0$, we can compute $A^k$ explicitly:

$$A^k = PD^kP^{-1} = \begin{bmatrix} 1 & -1 \\ 2 & 4 \end{bmatrix} \begin{bmatrix} 1 & 0 \\ 0 & (-\frac{1}{2})^k \end{bmatrix} \frac{1}{6}\begin{bmatrix} 4 & 1 \\ -2 & 1 \end{bmatrix}$$

$$= \frac{1}{6}\begin{bmatrix} 4 + 2(-\frac{1}{2})^k & 1 - (-\frac{1}{2})^k \\ 8 - 8(-\frac{1}{2})^k & 2 + 4(-\frac{1}{2})^k \end{bmatrix}.$$

Hence we obtain

$$\begin{bmatrix} a_k \\ j_k \end{bmatrix} = V_k = A^kV_0 = \frac{1}{6}\begin{bmatrix} 4 + 2(-\frac{1}{2})^k & 1 - (-\frac{1}{2})^k \\ 8 - 8(-\frac{1}{2})^k & 2 + 4(-\frac{1}{2})^k \end{bmatrix} \begin{bmatrix} 100 \\ 40 \end{bmatrix}$$

$$= \frac{1}{6}\begin{bmatrix} 440 + 160(-\frac{1}{2})^k \\ 880 - 640(-\frac{1}{2})^k \end{bmatrix}.$$

Equating top and bottom entries, we obtain exact formulas for $a_k$ and $j_k$:

$$a_k = \frac{220}{3} + \frac{80}{3}\left(-\frac{1}{2}\right)^k \quad \text{and} \quad j_k = \frac{440}{3} - \frac{320}{3}\left(-\frac{1}{2}\right)^k \quad \text{for } k = 1, 2, \cdots.$$

In practice, the exact values of $a_k$ and $j_k$ usually are not required. What is needed is a measure of how these numbers behave for large values of $k$. This is easy to obtain here. Since $\left(-\frac{1}{2}\right)^k$ is nearly zero for large $k$, we have the following approximate values

$$a_k \approx \frac{220}{3} \quad \text{and} \quad j_k \approx \frac{440}{3} \quad \text{if } k \text{ is large.}$$

Hence, in the long term, the female population stabilizes with approximately twice as many juveniles as adults.

The population model in Example 1 is an example of a **dynamical system**; that is, a sequence of columns $V_0, V_1, V_2, \cdots$ where the initial column $V_0$ is known and the other columns are determined by the condition

$$V_{k+1} = AV_k \quad \text{for each } k \geq 0$$

where $A$ is a square matrix.[7] The condition $V_{k+1} = AV_k$ is called a **matrix recurrence** for the columns $V_k$. While this example is from ecology, dynamical systems arise in many parts of science and engineering,[8] and in other areas such as economics.

Since $V_0$ is known, the matrix recurrence $V_{k+1} = AV_k$ gives successively

$$V_1 = AV_0,$$
$$V_2 = AV_1 = A(AV_0) = A^2V_0,$$
$$V_3 = AV_2 = A(A^2V_0) = A^3V_0,$$

$$\vdots$$

---

[7] More precisely, this is a *linear discrete* dynamical system. Many models use differential relationships between $V_{k+1}$ and $V_k$, viewed as functions of time.

[8] A Markov Chain (Section 1.8) is a dynamical system where the matrix recursion $S_{k+1} = PS_k$ relates the state vectors $S_k$, and $P$ is the transition matrix.

Continuing in this way, we get

$$V_k = A^k V_0 \text{ for each } k = 1, 2, \cdots. \tag{$*$}$$

Hence we can determine the columns $V_k$ if we can compute the powers $A^k$ of the matrix $A$. As we have seen, this can be done if $A$ is diagonalizable; indeed $(*)$ can be used (as in Example 1) to give a nice "formula" for the columns $V_k$ in this case.

To see what happens in general, assume that $A$ is diagonalizable with eigenvalues $\lambda_1$, $\lambda_2, \cdots, \lambda_n$ and corresponding eigenvectors $X_1, X_2, \cdots, X_n$. If $P = [X_1 \ X_2 \ \cdots \ X_n]$ is the diagonalizing matrix with the $X_i$ as its columns, then $P$ is invertible and

$$P^{-1}AP = D = diag(\lambda_1, \lambda_2, \cdots, \lambda_n)$$

by Theorem 3 §2.3. Hence $A = PDP^{-1}$ so $(*)$ and Theorem 1 §2.3 give

$$V_k = A^k V_0 = (PDP^{-1})^k V_0 = (PD^k P^{-1})V_0 = PD^k(P^{-1}V_0)$$

for each $k = 1, 2, \cdots$. For convenience, we denote the column $P^{-1}V_0$ arising here as follows

$$P^{-1}V_0 = \begin{bmatrix} b_1 \\ b_2 \\ \vdots \\ b_n \end{bmatrix} = [b_1 \ b_2 \ \cdots \ b_n]^T.$$

Then matrix multiplication gives (see Theorem 3 §1.4)

$$V_k = PD^k(P^{-1}V_0)$$

$$= [X_1 \ X_2 \ \cdots \ X_n] \begin{bmatrix} \lambda_1^k & 0 & \cdots & 0 \\ 0 & \lambda_2^k & \cdots & 0 \\ \vdots & \vdots & \ddots & \vdots \\ 0 & 0 & \cdots & \lambda_n^k \end{bmatrix} \begin{bmatrix} b_1 \\ b_2 \\ \vdots \\ b_n \end{bmatrix}$$

$$= \begin{bmatrix} \lambda_1^k X_1 & \lambda_2^k X_2 & \cdots & \lambda_n^k X_n \end{bmatrix} \begin{bmatrix} b_1 \\ b_2 \\ \vdots \\ b_n \end{bmatrix}$$

$$= b_1 \lambda_1^k X_1 + b_2 \lambda_2^k X_2 + \cdots + b_n \lambda_n^k X_n$$

for each $k \geq 0$. This is a useful explicit formula for the columns $V_k$. Note that, in particular, $V_0 = b_1 X_1 + b_2 X_2 + \cdots + b_n X_n$.

However, such an exact formula for $V_k$ is often not required in practice; all that is needed is to *estimate* $V_k$ for large values of $k$ (as in Example 1). This can be done easily if the diagonalizable $n \times n$ matrix $A$ has a largest eigenvalue.

## 2.4.2   Dominant Eigenvalues

An eigenvalue $\lambda$ of an $n \times n$ matrix $A$ is called a **dominant eigenvalue** of $A$ if

$$|\lambda| > |\mu| \text{ for all eigenvalues } \mu \neq \lambda$$

where $|\lambda|$ denotes the absolute value[9] of the number $\lambda$. For example, $\lambda_1 = 1$ is dominant in Example 1 because $\lambda_2 = -\frac{1}{2}$ is smaller in absolute value.

Returning to the above discussion, suppose $A$ has a dominant eigenvalue. By choosing the order in which the columns $X_i$ are placed in $P$, we may assume that $\lambda_1$ is dominant among the eigenvalues $\lambda_1, \lambda_2, \cdots, \lambda_n$ of $A$. In this case, take $\lambda_1^k$ out as a common factor in the expression $V_k = b_1 \lambda_1^k X_1 + b_2 \lambda_2^k X_2 + \cdots + b_n \lambda_n^k X_n$ for $V_k$. The result is

---

[9] The **absolute value** $|a|$ of a number $a$ is defined as follows: $|a| = \begin{cases} a & \text{if } a \geq 0 \\ -a & \text{if } a < 0 \end{cases}$. Thus, for example, $|3| = 3$ and $|-2| = 2$. It is useful to note that the absolute value is also given by the formula $|a| = \sqrt{a^2}$, where $\sqrt{a^2}$ denotes the positive square root of $a^2$.

$$V_k = \lambda_1^k \left[ b_1 X_1 + b_2 \left( \tfrac{\lambda_2}{\lambda_1} \right)^k X_2 + \cdots + b_n \left( \tfrac{\lambda_n}{\lambda_1} \right)^k X_n \right]$$

for each $k \geq 0$. Since $\lambda_1$ is dominant, we have $|\lambda_i| < |\lambda_1|$ for each $i \geq 2$, so each of the numbers $\left( \tfrac{\lambda_i}{\lambda_1} \right)^k$ becomes small in absolute value as $k$ increases. Hence $V_k$ is approximately equal to the first term $\lambda_1^k b_1 X_1$, and we write this as $V_k \approx \lambda_1^k b_1 X_1$. These observations are summarized in the following theorem (together with the above exact formula for $V_k$).

---

**THEOREM** **1**  Consider the dynamical system with matrix recurrence
$$V_{k+1} = AV_k \text{ for } k \geq 0$$
where $A$ and $V_0$ are given. Assume that $A$ is a diagonalizable $n \times n$ matrix with eigenvalues $\lambda_1, \lambda_2, \cdots, \lambda_n$ and corresponding eigenvectors $X_1, X_2, \cdots, X_n$, and let $P = [X_1 \ X_2 \ \cdots \ X_n]$ be the corresponding diagonalizing matrix. Then an exact formula for the $V_k$ is
$$V_k = b_1 \lambda_1^k X_1 + b_2 \lambda_2^k X_2 + \cdots + b_n \lambda_n^k X_n \quad \text{for each} \quad k \geq 0$$
where the coefficients $b_i$ come from $P^{-1} V_0 = [b_1 \ b_2 \ \cdots \ b_n]^T$. Moreover, if $A$ has dominant eigenvalue $\lambda_1$, of multiplicity 1, then[10] $V_k$ is approximated by
$$V_k \approx \lambda_1^k b_1 X_1 \text{ for sufficiently large } k.$$

**Remark**  Suppose the dynamical system in Theorem 1 has a dominant eigenvalue $\lambda_1$. If $\lambda_1 = 1$, Theorem 1 shows that the $V_k$ become constant in the long run. If $|\lambda_1| < 1$, then the powers $\lambda_1^k$ will become close to zero for large $k$, so the columns $V_k$ will converge to 0. On the other hand, if $|\lambda_1| > 1$, then the $V_k$ will become large as $k$ increases. We give some examples of this behaviour below.

In Example 1 the numbers $a_k$ and $j_k$ of adult and juvenile female birds alive were related by the equations
$$a_{k+1} = \tfrac{1}{2} a_k + \tfrac{1}{4} j_k$$
$$j_{k+1} = 2a_k.$$

This means that $\tfrac{1}{2}$ of the adults and $\tfrac{1}{4}$ of the juveniles alive in year $k$ survive to year $k+1$, and that each female produces 2 juveniles per year on average. In Example 1 we showed that the species population becomes stable under these conditions. We now examine what happens if we lower the juvenile survival rate to $\tfrac{1}{8}$. It turns out that this drives the species to extinction.

**Example 2**  Assume that the numbers $a_k$ and $j_k$ of adult and juvenile birds in Example 1 satisfy the equations
$$a_{k+1} = \tfrac{1}{2} a_k + \tfrac{1}{8} j_k$$
$$j_{k+1} = 2a_k.$$
If $a_0 = 100$ and $j_0 = 40$ as in Example 1, estimate the population profile $V_k = \begin{bmatrix} a_k \\ j_k \end{bmatrix}$ for large $k$.

**SOLUTION**  Now $V_{k+1} = AV_k$ where $A = \begin{bmatrix} \tfrac{1}{2} & \tfrac{1}{8} \\ 2 & 0 \end{bmatrix}$. The characteristic polynomial is $c_A(x) = x^2 - \tfrac{1}{2}x - \tfrac{1}{4}$, so the eigenvalues are $\lambda_1 = \tfrac{1}{4} \left[ 1 + \sqrt{5} \right] = 0.809$ and $\lambda_2 = \tfrac{1}{4} \left[ 1 - \sqrt{5} \right] = -0.309$. Corresponding eigenvectors are $X_1 = \begin{bmatrix} 0.809 \\ 2 \end{bmatrix}$ and $X_2 = \begin{bmatrix} -0.309 \\ 2 \end{bmatrix}$, so a diagonalizing matrix

---

[10] More can be said here. Suppose, for example, that $|\lambda_1| = |\lambda_2| > \lambda_i$ for each $i \geq 3$. Then $V_k \approx \lambda_1^k b_1 X_1 + \lambda_2^k b_2 X_2$ for sufficiently large $k$.

is $P = \begin{bmatrix} 0.809 & -0.309 \\ 2 & 2 \end{bmatrix}$. Since we are still assuming that $V_0 = \begin{bmatrix} 100 \\ 40 \end{bmatrix}$, the coefficients $b_1$ and $b_2$ are given by

$$\begin{bmatrix} b_1 \\ b_2 \end{bmatrix} = P^{-1}V_0 = \frac{1}{2.236}\begin{bmatrix} 2 & 0.309 \\ -2 & 0.809 \end{bmatrix}\begin{bmatrix} 100 \\ 40 \end{bmatrix} = \begin{bmatrix} 94.97 \\ -74.97 \end{bmatrix}$$

Since $\lambda_1$ is dominant and of multiplicity 1, an estimate for $V_k$ is given by

$$\begin{bmatrix} a_k \\ j_k \end{bmatrix} = V_k \cong b_1\lambda_1^k X_1 = 94.97(0.809)^k\begin{bmatrix} 0.809 \\ 2 \end{bmatrix}$$

Comparing entries gives

$$a_k \approx 76.83(0.809)^k \quad \text{and} \quad j_k \approx 189.94(0.809)^k$$

for sufficiently large $k$. We thus obtain the total female population $a_k + j_k \cong 266.77\,(0.809)^k$ so, since $0.809 < 1$, the population becomes smaller and smaller as $k$ increases. In other words, the species becomes extinct. Note that this also shows that the breakdown of the (decreasing) female population into adults and juveniles stabilizes with approximately $\frac{76.83}{266.77} = 0.288 = 28.8\%$ adults and $\frac{189.94}{266.77} = 0.712 = 71.2\%$ juveniles.

Much of the computation in Examples 1 and 2 can in fact be carried out for a much more general model. As before, let $a_k$ and $j_k$ denote the number of adult and juvenile female birds alive in year $k$, and assume they satisfy the equations

$$a_{k+1} = \alpha a_k + \beta j_k$$
$$j_{k+1} = m a_k$$

for $k \geq 0$. Here $\alpha$ and $\beta$ are the *adult and juvenile survival rates*; that is, the proportions of adult and juvenile females respectively to survive from one year to the next. Similarly, $m$ is the *reproduction rate*, that is the number of juveniles hatched (per adult female) in any year. We assume that $\alpha$, $\beta$, and $m$ are all positive.

Writing $V_k = \begin{bmatrix} a_k \\ j_k \end{bmatrix}$ and $A = \begin{bmatrix} \alpha & \beta \\ m & 0 \end{bmatrix}$ as before, the above equations become the matrix recurrence $V_{k+1} = AV_k$. The characteristic polynomial of $A$ is

$$c_A(x) = det\begin{bmatrix} x-\alpha & -\beta \\ -m & x \end{bmatrix} = x^2 - \alpha x - \beta m.$$ Hence the quadratic formula gives the eigenvalues

$$\lambda_1 = \tfrac{1}{2}(\alpha + r) \quad \text{and} \quad \lambda_2 = \tfrac{1}{2}(\alpha - r)$$

where, for convenience, we write $r = \sqrt{\alpha^2 + 4\beta m}$. Since the eigenvalues $\lambda_i$ are roots of the characteristic polynomial, we have $\lambda_i^2 = \alpha\lambda_i + \beta m$ for each $i$. This shows that corresponding eigenvectors are

$$X_1 = \begin{bmatrix} \lambda_1 \\ m \end{bmatrix} \quad \text{and} \quad X_2 = \begin{bmatrix} \lambda_2 \\ m \end{bmatrix}$$

(because $AX_i = \lambda_i X_i$ for each $i$ as the reader can verify). Furthermore, $r = \sqrt{\alpha^2 + 4\beta m} > \alpha$, so $\lambda_2 < 0$. Thus

$$|\lambda_2| = -\lambda_2 = \tfrac{1}{2}(r - \alpha) < \tfrac{1}{2}(\alpha + r) = \lambda_1 = |\lambda_1|,$$

so $\lambda_1$ is the dominant eigenvalue no matter what choice of $\alpha$, $\beta$, and $m$ is made. Hence Theorem 1 gives

$$V_k \approx b_1\lambda_1^k X_1 = b_1\lambda_1^k\begin{bmatrix} \lambda_1 \\ m \end{bmatrix}$$

where $b_1$ is computed as in Theorem 1 ($b_1 > 0$ as the reader can verify). Thus, in this model, the magnitude of $\lambda_1 > 0$ determines the ultimate population size as follows:

If $\lambda_1 > 1$, the population becomes very large as $k$ increases.

If $\lambda_1 = 1$, the population stabilizes as $k$ increases.

If $\lambda_1 < 1$, the population becomes very small as $k$ increases.

In the last case, the population will become extinct unless something happens to change the values of $\alpha$, $\beta$, and $m$.

---

## 2.4.3 A Predator-Prey Model

Examples 1 and 2 describe a single population and its evolution over time. The following example uses a matrix recurrence to study the dynamics between two interrelated populations.

**Example 3**

Let $h_k$ and $m_k$ denote the number of hawks and mice respectively in a certain region in year $k$. Assume that they are related as follows:

$$h_{k+1} = \tfrac{1}{2}h_k + \tfrac{1}{100}m_k$$
$$m_{k+1} = -\tfrac{50}{4}h_k + \tfrac{5}{4}m_k \qquad \text{for } k \geq 0.$$

Find the limiting values of the populations of hawks and mice if the populations begin with 50 hawks and 1600 mice.

**SOLUTION**

If we write $V_k = \begin{bmatrix} h_k \\ m_k \end{bmatrix}$ and $A = \begin{bmatrix} \tfrac{1}{2} & \tfrac{1}{100} \\ -\tfrac{50}{4} & \tfrac{5}{4} \end{bmatrix}$, then the equations become $V_{k+1} = AV_k$ and

we have a matrix recurrence. The characteristic polynomial is

$x^2 - \tfrac{7}{4}x + \tfrac{3}{4} = (x - 1)\left(x - \tfrac{3}{4}\right)$, so the eigenvalues are $\lambda_1 = 1$ and $\lambda_2 = \tfrac{3}{4}$ with correspon-

ding eigenvectors $X_1 = \begin{bmatrix} 1 \\ 50 \end{bmatrix}$ and $X_2 = \begin{bmatrix} 1 \\ 25 \end{bmatrix}$. Since $\lambda_1$ is dominant here, we can estimate

the limiting value of $V_k$. Because $P = \begin{bmatrix} 1 & 1 \\ 50 & 25 \end{bmatrix}$ is a diagonalizing matrix, we obtain

$$\begin{bmatrix} b_1 \\ b_2 \end{bmatrix} = P^{-1}V_0 = \begin{bmatrix} -1 & \tfrac{1}{25} \\ 2 & -\tfrac{1}{25} \end{bmatrix}\begin{bmatrix} 50 \\ 1600 \end{bmatrix} = \begin{bmatrix} 14 \\ 36 \end{bmatrix} \text{ as in Theorem 1. Hence, for large } k,$$

$$V_k \approx \lambda_1^k b_1 X_1 = 14\begin{bmatrix} 1 \\ 50 \end{bmatrix}$$

Thus the populations stabilize at approximately 14 hawks and 700 mice.

Of course, if the matrix $A$ is changed in Example 3 the populations may not stabilize. One way to guarantee stability is to insist that $\lambda_1 = 1$ is a dominant eigenvalue. It turns out that this can be done in a whole class of models.

To see what happens in general, assume that the numbers $h_k$ and $m_k$ of hawks and mice are related as follows:

$$h_{k+1} = ph_k + \gamma m_k$$
$$m_{k+1} = -\delta h_k + qm_k \qquad \text{for } k \geq 0 \qquad (**)$$

where $p$ and $q$ are the survival rate parameters, and $\gamma$ and $\delta$ are the correlation parameters. We are looking for a "stable range" for these parameters, that is for values of the parameters that will ensure that the two populations will reach equilibrium levels. The following conditions on the parameters are inherent in the model:

$p > 0$   because more hawks one year means more the next year.

$p < 1$   because few mice one year means the number of hawks decreases.

$q > 1$   because few hawks one year means the number of mice increases.

$\gamma > 0$   because more mice one year means more hawks the next year.

$\delta > 0$   because more hawks one year means fewer mice the next year.

If we write $A = \begin{bmatrix} p & \gamma \\ -\delta & q \end{bmatrix}$ and $V_k = \begin{bmatrix} h_k \\ m_k \end{bmatrix}$, the equations (**) become the matrix recursion

$$V_{k+1} = A\,V_k \text{ for } k = 0,\ 1,\ 2,\cdots$$

where the initial situation $V_0$ is assumed to be known. We compute the characteristic polynomial

$$c_A(x) = det\begin{bmatrix} x-p & -\gamma \\ \delta & x-q \end{bmatrix} = x^2 - (p+q)x + (pq + \gamma\delta).$$

This has roots

$$\tfrac{1}{2}\left\{(p+q) \pm \sqrt{(p+q)^2 - 4(pq+\gamma\delta)}\right\} = \tfrac{1}{2}\left\{(p+q) \pm \sqrt{(q-p)^2 - 4\gamma\delta}\right\}$$

by the quadratic formula so, if we write

$$r = \sqrt{(q-p)^2 - 4\gamma\delta} = \sqrt{(p+q)^2 - 4(pq + \gamma\delta)},$$

the eigenvalues are

$$\lambda_1 = \tfrac{1}{2}\left[(p+q) + r\right] \quad \text{and} \quad \lambda_2 = \tfrac{1}{2}\left[(p+q) - r\right].$$

For $r$ to be a real number, we require that $4\gamma\delta \le (q-p)^2$. In addition, we want $r > 0$ because the eigenvalues should be distinct (so $A$ is diagonalizable). So we insist that

$$4\gamma\delta < (q-p)^2. \tag{***}$$

Hence $\lambda_1 > \lambda_2 > 0$ (because $p + q > r > 0$), so $\lambda_1$ is the dominant eigenvalue. By Theorem 1 the columns $V_k$ become constant for large $k$ if $\lambda_1 = 1$, that is if $p + q + r = 2$. This gives $(p+q-2)^2 = r^2 = (q-p)^2 - 4\gamma\delta$, which simplifies to the equation

$$\gamma\delta = (1-p)(q-1)$$

as the reader can verify. This condition leads to equilibrium.

---

**EQUILIBRIUM CONDITION**     Choose numbers $p$ and $q$ such that $0 < p < 1, q > 1$, and $p + q < 2$, and then choose $\gamma > 0$ and $\delta > 0$ satisfying

$$\gamma\delta = (1-p)(q-1).$$

If we use the model given by (**), the populations of hawks and mice become constant for large $k$.

**PROOF**     In fact, using the condition that $\gamma\delta = (1-p)(q-1)$, we have

$$(q-p)^2 - 4\gamma\delta = [2 - (p+q)]^2.$$

Since $p + q < 2$, it follows that (***) holds and $r = 2 - (p + q)$. This means that $\lambda_1 = \tfrac{1}{2}[(p+q) + r] = 1$, so $V_k = \begin{bmatrix} h_k \\ m_k \end{bmatrix}$ becomes constant for large $k$ by Theorem 1 as required.

The constraints on $p$ and $q$ in the equilibrium condition are:

$$0 < p < 1$$
$$1 < q$$
$$p + q < 2$$

The region of acceptable pairs $(p, q)$ is shown in the diagram. Having chosen such $p$ and $q$, any positive numbers $\gamma$ and $\delta$ can be chosen so that $\gamma\delta = (1 - p)(q - 1)$ and the populations in the resulting model will be in equilibrium.

In fact we can find formulas for the limiting values for $h_k$ and $m_k$. First

$$X_1 = \begin{bmatrix} \gamma \\ \lambda_1 - p \end{bmatrix} \quad \text{and} \quad X_2 = \begin{bmatrix} \gamma \\ \lambda_2 - p \end{bmatrix}$$

are eigenvectors corresponding to $\lambda_1$ and $\lambda_2$ respectively because $AX_i = \lambda_i X_i$ for each $i$. (This is because $\lambda_i$ is a root of the characteristic equation, so $\lambda_i^2 - (p + q)\lambda_i + (pq + \gamma\delta) = 0$.) Then Theorem 1 gives

$$\begin{bmatrix} h_k \\ m_k \end{bmatrix} = V_k \approx \lambda_1^k b_1 X_1 = b_1 X_1$$

where $b_1$ is obtained from $[b_1 \ \ b_2]^T = P^{-1}V_0$, and $P = [X_1 \ \ X_2]$ is the diagonalizing matrix for $A$. Since $\lambda_1 = 1$ for equilibrium, this gives

$$h_k = b_1\gamma \quad \text{and} \quad m_k = b_1(1 - p).$$

Thus, in this model the limiting ratio of hawks to mice is $\frac{\gamma}{1 - p}$.

## 2.4.4 Graphical Description of Solutions

If a dynamical system $V_{k+1} = AV_k$ is given, the sequence $V_0, V_1, V_2, \cdots$ is called the **trajectory** of the system starting at $V_0$. It is instructive to obtain a graphical plot of the system by writing $V_k = \begin{bmatrix} x_k \\ y_k \end{bmatrix}$ and plotting the successive values as points in the plane, identifying $V_k$ with the point $P(x_k, y_k)$. We give several examples which illustrate properties of dynamical systems. For ease of calculation we assume that the matrix $A$ is simple, usually diagonal.

**Example 4**    Let $A = \begin{bmatrix} \frac{1}{2} & 0 \\ 0 & \frac{1}{3} \end{bmatrix}$. Then the eigenvalues are $\frac{1}{2}$ and $\frac{1}{3}$, with corresponding eigenvectors

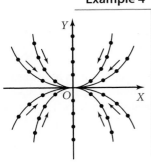

$X_1 = \begin{bmatrix} 1 \\ 0 \end{bmatrix}$ and $X_2 = \begin{bmatrix} 0 \\ 1 \end{bmatrix}$. The general solution is

$$V_k = b_1 \left(\tfrac{1}{2}\right)^k \begin{bmatrix} 1 \\ 0 \end{bmatrix} + b_2 \left(\tfrac{1}{3}\right)^k \begin{bmatrix} 0 \\ 1 \end{bmatrix}$$

for $k = 0, 1, 2, \cdots$ by Theorem 1, where the coefficients $b_1$ and $b_2$ depend[11] on the initial point $V_0$. Several trajectories are plotted in the diagram and, for each choice of $V_0$, the trajectories converge toward the origin because both eigenvalues are less than 1 in absolute value. For this reason, the origin is called an **attractor** for the system.

---

[11] In fact, $P = I$ here, so $V_0 = \begin{bmatrix} b_1 \\ b_2 \end{bmatrix}$.

**Example 5**   Let $A = \begin{bmatrix} \frac{3}{2} & 0 \\ 0 & \frac{4}{3} \end{bmatrix}$. Here the eigenvalues are $\frac{3}{2}$ and $\frac{4}{3}$, with corresponding eigenvectors

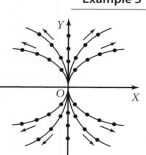

$X_1 = \begin{bmatrix} 1 \\ 0 \end{bmatrix}$ and $X_2 = \begin{bmatrix} 0 \\ 1 \end{bmatrix}$ as before. The general solution is

$$V_k = b_1 \left(\tfrac{3}{2}\right)^k \begin{bmatrix} 1 \\ 0 \end{bmatrix} + b_2 \left(\tfrac{4}{3}\right)^k \begin{bmatrix} 0 \\ 1 \end{bmatrix}$$

for $k = 0, 1, 2, \cdots$. Since both eigenvalues are greater than 1 in absolute value, the trajectories diverge away from the origin for every choice of initial point $V_0$. For this reason, the origin is called a **repellor** for the system.

**Example 6**   Let $A = \begin{bmatrix} 1 & -\frac{1}{2} \\ -\frac{1}{2} & 1 \end{bmatrix}$. Now the eigenvalues are $\frac{3}{2}$ and $\frac{1}{2}$, with corresponding eigenvectors

$X_1 = \begin{bmatrix} -1 \\ 1 \end{bmatrix}$ and $X_2 = \begin{bmatrix} 1 \\ 1 \end{bmatrix}$. The general solution is

$$V_k = b_1 \left(\tfrac{3}{2}\right)^k \begin{bmatrix} -1 \\ 1 \end{bmatrix} + b_2 \left(\tfrac{1}{2}\right)^k \begin{bmatrix} 1 \\ 1 \end{bmatrix}$$

for $k = 0, 1, 2, \cdots$. In this case $\frac{3}{2}$ is the dominant eigenvalue so, if $b_1 \neq 0$, we have $V_k \approx b_1 \left(\tfrac{3}{2}\right)^k \begin{bmatrix} -1 \\ 1 \end{bmatrix}$ for large $k$ and $V_k$ is approaching the line $y = -x$. However, if $b_1 = 0$, then $V_k = b_2 \left(\tfrac{1}{2}\right)^k \begin{bmatrix} 1 \\ 1 \end{bmatrix}$ and so approaches the origin along the line $y = x$. In general the trajectories appear as in the diagram, and the origin is called a **saddle point** for the dynamical system in this case.

**Example 7**   Let $A = \begin{bmatrix} 0 & \frac{1}{2} \\ -\frac{1}{2} & 0 \end{bmatrix}$. Now the characteristic polynomial is $c_A(x) = x^2 + \frac{1}{4}$, so the eigenvalues

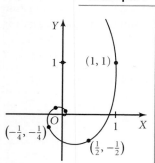

are the complex numbers $\frac{i}{2}$ and $-\frac{i}{2}$ where $i^2 = -1$. Hence $A$ is not diagonalizable as a real matrix[12]. However, the trajectories are not difficult to describe. If we start with $V_0 = \begin{bmatrix} 1 \\ 1 \end{bmatrix}$, then the trajectory begins as

$$V_1 = \begin{bmatrix} \frac{1}{2} \\ -\frac{1}{2} \end{bmatrix}, \; V_2 = \begin{bmatrix} -\frac{1}{4} \\ -\frac{1}{4} \end{bmatrix}, \; V_3 = \begin{bmatrix} -\frac{1}{8} \\ \frac{1}{8} \end{bmatrix},$$

$$V_4 = \begin{bmatrix} \frac{1}{16} \\ \frac{1}{16} \end{bmatrix}, \; V_5 = \begin{bmatrix} \frac{1}{32} \\ -\frac{1}{32} \end{bmatrix}, \; V_6 = \begin{bmatrix} -\frac{1}{64} \\ -\frac{1}{64} \end{bmatrix}, \cdots$$

Five of these points are plotted in the diagram. Here each trajectory spirals in toward the origin, so the origin is an attractor. Note that the two (complex) eigenvalues have absolute value less than 1 here. If they had absolute value greater than 1, the trajectories would spiral out from the origin.

---

[12] Corresponding eigenvectors are $X_1 = \begin{bmatrix} 1 \\ i \end{bmatrix}$ and $X_2 = \begin{bmatrix} 1 \\ -i \end{bmatrix}$ so the theory works for the complex numbers. We return to this in Section 2.5.

## Exercises 2.4

1. Consider the female bird population model where the adult survival rate is $\frac{1}{2}$, the juvenile survival rate is $\frac{1}{3}$, and the reproduction rate is 2; that is, the matrix $A$ is
$$\begin{bmatrix} \frac{1}{2} & \frac{1}{3} \\ 2 & 0 \end{bmatrix}.$$
   a) Using an initial population of 100 adults and 60 juveniles, calculate the population for the next few years.
   b) Does the population become extinct? Stabilize? Diverge?
   c) Does it depend on the initial population?
   d) In the long term, what percentage of the total female population is made up of juveniles?

2. Consider once again the female bird population model where the adult survival rate is $\frac{3}{10}$, the juvenile survival rate is $\frac{1}{3}$, the reproduction rate is 2, and the initial population is 100 adults and 50 juveniles.
   a) Write a matrix recurrence for the bird population in the form $V_{k+1} = AV_k = A^k V_0$, where $A$ is a matrix.
   b) Find the eigenvalues of $A$, corresponding eigenvectors, and an invertible matrix $P$ which diagonalizes the matrix $A$.
   c) Find an exact formula for $V_k$ in term of the eigenvalues and eigenvectors.
   d) Use the formula found in c) to calculate the population 5 years from now.
   e) Find the dominant eigenvalue.
   f) Find an approximating formula for $V_k$ using the dominant eigenvalue.
   g) Use the formula found in f) to approximate the population 10 years from now.
   h) Compare the results found in d) and g).
   i) Does the population become extinct? Stabilize? Diverge?
   j) Does it depend on the initial population?
   k) In the long term, what percentage of the total female population is made up of juveniles?

3. Consider the female bird population model where the adult survival rate is $\frac{2}{5}$, and the reproduction rate is 4. Find

$\beta$, the required value of the juvenile survival rate, to ensure that the population will stabilize in the long term.

4. In the general bird population model, assume that the adult survival rate $\alpha$ and the reproduction rate $m$ are fixed. If $\beta$ denotes the juvenile survival rate, show that the population:
   a) stabilizes if $\beta = \frac{1-\alpha}{m}$;
   b) becomes extinct if $\beta < \frac{1-\alpha}{m}$;
   c) becomes very large if $\beta > \frac{1-\alpha}{m}$.

5. The population of an endangered hawk species and its natural prey the field mouse, originally 30 hawks and 2000 mice, is represented by the following model:
$$h_{k+1} = 0.55h_k + 0.005m_k$$
$$m_{k+1} = -18h_k + 1.2m_k$$
where $h_k$ and $m_k$ represent the hawk and mice population respectively in year $k$.
   a) Make sure you understand what each parameter means.
   b) Express the model in the matrix form $V_{k+1} = AV_k$, and write $V_0 = \begin{bmatrix} 30 \\ 2000 \end{bmatrix}$.
   c) Find the eigenvalues and corresponding eigenvectors of $A$, use them to diagonalize the matrix $A$, and write a general formula to calculate $V_k$.
   d) Calculate the predicted population every few years for the next 30 years or so and, using eigenvalues, reason whether an equilibrium will be reached.
   e) In 5 years, biologists try to improve the hawk population by introducing 1000 mice in the hawk hunting area. Predict and compare the new hawk population 10 years from now.
   f) From that point on, the hawk survival rate drops to 0.50. The biologists are alarmed that this may wipe out the hawk population. Test to see what the model is predicting and explain what is happening.

6. Describe the trajectories of the dynamical system with matrix
$$A = \begin{bmatrix} 0 & 1 \\ -1 & 0 \end{bmatrix}.$$

## 2.5 COMPLEX EIGENVALUES

For the most part the matrices we have been dealing with have had real eigenvalues. However there are important matrices for which this is not the case. For example, the matrix $A = \begin{bmatrix} 0 & -1 \\ 1 & 0 \end{bmatrix}$ has characteristic polynomial $c_A(x) = x^2 + 1$, and this has no real root because $x^2 \geq 0$ for every real number $x$. Nonetheless, we can find eigenvalues for $A$ and diagonalize it, but we have to use complex numbers. This section develops the basic properties of these numbers.

The concept of complex numbers depends on the existence of a number $i$ such that $i^2 = -1$. Because such a number $i$ cannot be real, it is usually called an "imaginary" number. But it is no more imaginary than the number $-1$ or the number $2$ for that matter. Numbers exist only in our minds and so are *all* imaginary. The integers $1, 2, 3, \cdots$ were undoubtedly invented to count things (enemy warriors or animals to hunt), the positive real numbers were used to measure distances and areas, and the negative numbers enabled people to systematically denote phenomena like debt. In each case the numbers were invented (some would say "discovered") to describe some aspect of the world. This proved to be very useful because the algebraic properties of numbers could then be used to describe "laws" of nature. For example, $A = lw$ is the area of a rectangle of length $l$ and width $w$; and $f = ma$ relates the force $f$, mass $m$, and acceleration $a$ of a particle moving in a straight line (Newton's second law of motion). Thus it is important that the new numbers being created satisfy the rules of arithmetic.

## 2.5.1   Complex Numbers

In any event, mathematicians in the 1700s introduced a new (nonreal) number $i$ with the property that

$$i^2 = -1$$

and they called expressions of the form

$$a + ib \quad \text{where } a \text{ and } b \text{ are real}$$

**complex numbers.** The set of all complex numbers is denoted $\mathbb{C}$. The rules of real arithmetic hold for these complex numbers with the single exception that $i^2 = -1$.

Given a complex number $z = a + ib$, the real number $a$ is called the **real part** of $z$, and $b$ is the **imaginary part** of $z$. Every real number $a$ is a complex number, namely $a = a + 0i$, and complex numbers of the form $bi = 0 + bi$ are sometimes called **pure imaginary** numbers. The number $i$ itself is called the **imaginary unit**.

If two complex numbers are equal, they must have the same real and imaginary parts. That is

$$\text{If } a + bi = a' + b'i \quad \text{then} \quad a = a' \text{ and } b = b'.$$

Indeed, if $b$ were not equal to $b'$ then we could solve for $i = \frac{a - a'}{b' - b}$, so $i$ would be a real number, a contradiction. Hence $b = b'$, and so $a = a'$ (because $a + bi = a' + b'i$).

Because of this, it is possible to display the complex numbers geometrically as the points in the plane. This is accomplished by identifying the complex number $z = a + bi$ with the point $P(a, b)$.

Some examples are given in the diagram. When this is done the plane is called the **complex plane.** Note that the real numbers are identified with the points on the $X$-axis in the usual way. For this reason, the $X$-axis is called the **real axis.** Similarly, the $Y$-axis is called the **imaginary axis.**

Complex numbers are added and multiplied in much the same way as linear polynomials $a + bx$, except that $i^2 = -1$. Thus

$$\text{If } z = a + ib \text{ and } w = a' + ib' \text{ then } \begin{cases} z + w = (a + a') + (b + b')i, \\ zw = (aa' - bb') + (ab' + ba')i. \end{cases}$$

**Example 1**   If $z = 3 - 2i$ and $w = -5 + 7i$, compute $z + w$, $z - w$, $5z$, $zw$, and $z^2$.

SOLUTION

$z + w = (3 - 5) + (-2 + 7)i = -2 + 5i$.

$z - w = (3 + 5) + (-2 - 7)i = 8 - 9i$.

$5z = 15 - 10i$.

$zw = (-15 - 14i^2) + (21 + 10)i = -1 + 31i$.

$z^2 = (9 + 4i^2) + (-6 - 6)i = 5 - 12i$.

Before describing the process of dividing one complex number by another we must introduce two important notions. If $z = a + ib$ is any complex number, the **conjugate** of $z$ is another complex number, denoted $\bar{z}$, given by

$$\bar{z} = a - ib.$$

That is, $\bar{z}$ is formed by negating the imaginary part of $z$. Thus, in the complex plane, $z$ and $\bar{z}$ are reflections of each other in the real axis (see the diagram).

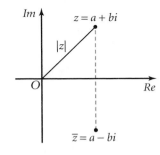

Here are four useful properties of conjugation:

C1   $\overline{(z \pm w)} = \bar{z} \pm \bar{w}$.
C2   $\overline{(zw)} = \bar{z}\,\bar{w}$.
C3   $\overline{(\bar{z})} = z$.
C4   $z$ is real if and only if $\bar{z} = z$.

The routine verifications of C1, C2, and C3 are left as Exercise 6. As to C4, if $z = a + 0i$ is real, then $\bar{z} = a - 0i = z$. Conversely, if $z = a + bi$ and $\bar{z} = z$, then $a - bi = a + bi$. Hence equating imaginary parts gives $-b = b$. This means $b = 0$, whence $z = a + 0i = a$ is real.

Given any complex number $z = a + bi$, the distance from $z$ to the origin 0 is

$$\sqrt{a^2 + b^2}$$

by Pythagoras' theorem[13]. This number is called the **absolute value** or **modulus** of the complex number $z = a + bi$, and is denoted (see the preceding diagram)

$$|z| = \sqrt{a^2 + b^2}.$$

It is routine to verify that

$$z\bar{z} = a^2 + b^2$$

which gives the following useful formula:

$$|z|^2 = z\bar{z} \quad \text{for every complex number } z.$$

Note that, if $z = a + 0i$ is real, then $|z| = \sqrt{a^2}$ is the absolute value of $z$ viewed as a real number.

We list some basic properties of the absolute value.

A1   $|z| \geq 0$ for all complex numbers $z$.
A2   $|z| = 0$ if and only if $z = 0$.
A3   $|zw| = |z|\,|w|$.

A1 and A2 follow immediately from the fact that $|z|$ is the distance from $z$ to the origin $O$. To see why A3 holds, we use the fact (above) that $|z|^2 = z\bar{z}$ holds for every complex number $z$. Combining this with C2, we compute

$$|zw|^2 = (zw)\overline{(zw)} = zw\,\bar{z}\,\bar{w} = (z\bar{z})(w\bar{w}) = |z|^2\,|w|^2 = (|z|\,|w|)^2.$$

Now A3 follows by taking positive square roots.

---

[13] $\sqrt{p}$ denotes the *positive* square root of the real number $p \geq 0$.

We can now describe division by a complex number.

**Example 2**   Compute $\frac{3-5i}{2+7i}$.

SOLUTION   The idea is to multiply top and bottom by the conjugate of the denominator (and so turn the denominator into a real number):

$$\frac{3-5i}{2+7i} = \frac{(3-5i)(2-7i)}{(2+7i)(2-7i)} = \frac{-29-31i}{2^2+7^2} = -\frac{29}{53} - \frac{31}{53}i.$$

## 2.5.2   Quadratics

While the introduction of the imaginary unit $i$ has been motivated by the problem of finding a root of the equation $x^2 + 1 = 0$, the complex numbers easily give the roots of any real quadratic. In fact, if $a$, $b$, and $c$ are real numbers with $a \neq 0$, the quadratic equation $ax^2 + bx + c = 0$ has roots given by the famous **quadratic formula**:

$$x = \frac{-b \pm \sqrt{b^2 - 4ac}}{2a}.$$

The quantity $d = b^2 - 4ac$ is called the **discriminant** of the quadratic $ax^2 + bx + c$. It determines the nature of the roots:

If $d > 0$, there are two real roots.

If $d = 0$, there is one (repeated) real root.

If $d < 0$, there is no real root.

When $d < 0$ the quadratic $ax^2 + bx + c$ is said to be **irreducible**, and the quadratic formula gives two complex roots which are conjugates of each other:

$$\lambda = \tfrac{1}{2a}(-b + i\sqrt{|d|}) \quad \text{and} \quad \overline{\lambda} = \tfrac{1}{2a}(-b - i\sqrt{|d|})$$

The converse of this is also true: Given any nonreal complex number $\lambda$, then $\lambda$ and $\overline{\lambda}$ are the roots of some real irreducible quadratic. Indeed, the quadratic

$$x^2 - (\lambda + \overline{\lambda})x + (\lambda\overline{\lambda}) = (x - \lambda)(x - \overline{\lambda})$$

is irreducible (the roots $\lambda$ and $\overline{\lambda}$ are not real) and has real coefficients ($\lambda\overline{\lambda} = |\lambda|^2$ and $\lambda + \overline{\lambda}$ is twice the real part of $\lambda$).

**Example 3**   Find a real irreducible quadratic with $\lambda = 3 - 4i$ as a root.

SOLUTION   We have $\lambda + \overline{\lambda} = 6$ and $|\lambda|^2 = 25$, so $x^2 - 6x + 25$ is irreducible with roots $\lambda$ and $\overline{\lambda} = 3 + 4i$.

## 2.5.3   Complex Linear Algebra

Everything we have done for real matrices can be done for complex matrices. The methods are the same; the only difference is that the arithmetic is carried out with complex numbers rather than real ones. We give three examples.

**Example 4**   Solve the system

$$\begin{aligned} x + (1+i)y \quad\quad &= 1 - 2i \\ ix - \quad y + iz &= 2 \\ (i-1)x - \quad y - z &= 2i + 3 \end{aligned}$$

SOLUTION | The augmented matrix is reduced as follows:

$$\begin{bmatrix} 1 & 1+i & 0 & 1-2i \\ i & -1 & i & 2 \\ i-1 & -1 & -1 & 2+3i \end{bmatrix} \rightarrow \begin{bmatrix} 1 & 1+i & 0 & 1-2i \\ 0 & -i & i & -i \\ 0 & 1 & -1 & 1 \end{bmatrix}$$

$$\rightarrow \begin{bmatrix} 1 & 1+i & 0 & 1-2i \\ 0 & 1 & -1 & 1 \\ 0 & 0 & 0 & 0 \end{bmatrix}$$

$$\rightarrow \begin{bmatrix} 1 & 0 & 1+i & -3i \\ 0 & 1 & -1 & 1 \\ 0 & 0 & 0 & 0 \end{bmatrix}.$$

Hence we take $z = t$ where $t$ is an arbitrary (complex) parameter, and then solve for $x = -3i - (1 + i)t$ and $y = 1 + t$.

**Example 5**  Find the inverse of $A = \begin{bmatrix} i & -1 \\ -i & 2 \end{bmatrix}$.

SOLUTION | We have $det A = 2i - (-1)(-i) = i \neq 0$. So $A$ has an inverse and (the complex version of) Example 5 §1.5 gives

$$A^{-1} = \frac{1}{i}\begin{bmatrix} 2 & 1 \\ i & i \end{bmatrix} = (-i)\begin{bmatrix} 2 & 1 \\ i & i \end{bmatrix} = \begin{bmatrix} -2i & -i \\ 1 & 1 \end{bmatrix}.$$

The reader can check this directly. Of course the inverse can also be found using the matrix inversion algorithm.

**Example 6**  Diagonalize the matrix $A = \begin{bmatrix} 0 & -1 \\ 1 & 0 \end{bmatrix}$.

SOLUTION | The characteristic polynomial of $A$ is $c_A(x) = det(xI - A) = det\begin{bmatrix} x & 1 \\ -1 & x \end{bmatrix} = x^2 + 1$, so the eigenvalues are $\lambda_1 = i$ and $\lambda_2 = -i$. The augmented matrix of the system $(\lambda_1 I - A)X = 0$ is reduced as follows:

$$\begin{bmatrix} i & 1 & 0 \\ -1 & i & 0 \end{bmatrix} \rightarrow \begin{bmatrix} -1 & i & 0 \\ 0 & 0 & 0 \end{bmatrix} \rightarrow \begin{bmatrix} 1 & -i & 0 \\ 0 & 0 & 0 \end{bmatrix}.$$

Hence an eigenvector corresponding to $\lambda_1 = i$ is $X_1 = \begin{bmatrix} i \\ 1 \end{bmatrix}$. Similarly, $X_2 = \begin{bmatrix} -i \\ 1 \end{bmatrix}$ is an eigenvector for $\lambda_2 = -i$. Hence the complex version of Theorem 3 §2.3 shows that $P = [X_1 \quad X_2] = \begin{bmatrix} i & -i \\ 1 & 1 \end{bmatrix}$ is invertible and

$$P^{-1}AP = \begin{bmatrix} \lambda_1 & 0 \\ 0 & \lambda_2 \end{bmatrix} = \begin{bmatrix} i & 0 \\ 0 & -i \end{bmatrix}.$$

As these examples illustrate, calculations in complex linear algebra are really not much different from their real counterparts, except for the fact that the arithmetic uses complex numbers.

## 2.5.4  Roots of Polynomials

There is, however, one major advantage of the complex case. As we have seen, real matrices can have no real eigenvalues. Perhaps the most important reason for the study of complex linear algebra is that this situation does not arise: Every complex matrix has an eigenvalue (possibly complex). This is a consequence of the following remarkable theorem. (We omit the proof.)

**Fundamental Theorem of Algebra**[14]. Every nonconstant polynomial with complex coefficients has a complex root.

If $f(x)$ is a polynomial with complex coefficients of degree $n \geq 1$, and if $\lambda_1$ is a root, then the factor theorem[15] asserts that

$$f(x) = (x - \lambda_1)\, g(x) \text{ where } g(x) \text{ is a polynomial of degree } n - 1.$$

Suppose that $\lambda_2$ is a root of $g(x)$, again by the fundamental theorem. Then $g(x) = (x - \lambda_2)\, h(x)$, so

$$f(x) = (x - \lambda_1)(x - \lambda_2)\, h(x).$$

This process continues until the last polynomial to appear is linear. This last factor can be written in the form $u \cdot (x - \lambda_n)$, so the fundamental theorem takes the following form:

**THEOREM 1**

Every complex polynomial $f(x)$ of degree $n \geq 1$ has the form

$$f(x) = u \cdot (x - \lambda_1)(x - \lambda_2) \cdots (x - \lambda_n)$$

where the complex numbers $\lambda_1,\ \lambda_2,\ \cdots\ \lambda_n$ are the roots of $f(x)$ and need not be all distinct, and $u \neq 0$ is the coefficient of $x^n$ in $f(x)$.

In the case of a quadratic polynomial $f(x)$, the degree is 2 and the quadratic formula provides the two roots $\lambda_1$ and $\lambda_2$. For polynomials of higher degree it is not so simple to find the roots, but computer programs exist to do the calculations.

Of course, the characteristic polynomial $c_A(x)$ of any complex matrix $A$ factors completely as in Theorem 1. Since the roots of $c_A(x)$ are just the eigenvalues of $A$, we obtain

**THEOREM 2**

Every $n \times n$ complex matrix $A$ has $n$ complex eigenvalues (possibly with some repeated).

We have seen several illustrations of Theorem 2 (with real eigenvalues) in Section 2.3.

### 2.5.5  Symmetric Matrices

Most of the applications of linear algebra involve a real matrix $A$ and, while $A$ will have complex eigenvalues by Theorem 2, it is always of interest to know when the eigenvalues are in fact real. This turns out to be the case whenever $A$ is symmetric:

**THEOREM 3**

Let $A$ be a symmetric real matrix. If $\lambda$ is any eigenvalue of $A$, then $\lambda$ is real.

This important theorem will be used extensively in Chapter 4. Surprisingly enough, the theory of *complex* eigenvalues can be used to prove this useful result about *real* eigenvalues. The proof is given at the end of this section; we verify the $2 \times 2$ case.

**Example 7**  Verify Theorem 3 when $A$ is $2 \times 2$ real and symmetric.

**SOLUTION**  If $A = \begin{bmatrix} a & b \\ b & c \end{bmatrix}$ is real, the characteristic polynomial is

$$c_A(x) = (x - a)(x - c) - b^2 = x^2 - (a + c)x + (ac - b^2).$$

---

[14] This theorem was first proved by Gauss at age 20, and was published in his doctoral dissertation in 1799.

[15] See Appendix A.3.

The discriminant of $c_A(x)$ is

$$d = (a + c)^2 - 4(ac - b^2) = (a - c)^2 + 4b^2.$$

Hence $d \geq 0$ because $a$, $b$, and $c$ are real, so $c_A(x)$ has real roots (the eigenvalues).

## 2.5.6 Polar Form[16]

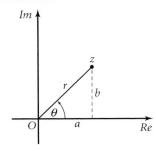

Up to this point we have been describing complex numbers $z$ by writing them in the form $z = a + ib$ where $(a, b)$ are the coordinates of $z$ as a point in the complex plane. However, to describe multiplication of two complex numbers, it is convenient to represent them using polar coordinates in the plane. Write the absolute value of $z = a + ib$ as

$$r = |z| = \sqrt{a^2 + b^2}.$$

If $z \neq 0$, the angle $\theta$ shown in the diagram is called an **argument** of $z$ and is denoted

$$\theta = arg(z).$$

The angle $\theta$ is not uniquely determined by $z$ and the angle $\theta \pm 2\pi k$ would do equally well for any integer $k$.[17] The numbers $r$ and $\theta$ are related to the real and imaginary parts $a$ and $b$ of $z$ by

$$a = r \cos \theta \quad \text{and} \quad b = r \sin \theta.$$

Hence the complex number $z = a + ib$ has the form

$$z = r(\cos \theta + i \sin \theta).$$

There is a famous expression (called **Euler's formula**[18]) for the complex numbers $\cos \theta + i \sin \theta$:

$$e^{i\theta} = \cos \theta + i \sin \theta \qquad \text{for any angle } \theta.$$

Here $e = 2.71828 \cdots$ is the constant familiar from calculus. Then the **polar form** of $z$ is

$$z = re^{i\theta} \text{ where } r = |z| \geq 0 \quad \text{and} \quad \theta = arg(z).$$

Of course it is not unique because the argument of $\theta$ can be changed by adding or subtracting multiples of $2\pi$.

**Example 8** Find the polar form of $z = -1 + i$ and of $w = \sqrt{3} - i$.

**SOLUTION**

These complex numbers are plotted in the diagram. We have $|z| = \sqrt{(-1)^2 + 1^2} = \sqrt{2}$, and the triangle determined by $z$ and the real axis has sides of length 1 and hypotenuse $\sqrt{2}$. Hence the acute angles in the triangle are both $\frac{\pi}{4} = 45°$, so one argument for $z$ is $arg(z) = \frac{3\pi}{4}$ as can be seen in the diagram. Thus the polar form of $z$ is $z = \sqrt{2}e^{3\pi i/4}$. Turning to $w$, we see that $|w| = \sqrt{(\sqrt{3})^2 + (-1)^2} = 2$. The triangle determined by $w$ and the real axis has acute angles $\frac{\pi}{6} = 30°$ and $\frac{\pi}{3} = 60°$, so $arg(w) = -\frac{\pi}{6}$. Hence the polar form of $w$ is $w = 2e^{-\pi i/6}$.

---

[16] This material is not required elsewhere in this book.

[17] Recall that $2\pi$ radians represents one full circle, so the angle $\theta \pm 2\pi k$ is obtained from $\theta$ by adding (or subtracting) $k$ full revolutions. For a review of the properties of a general angle $\theta$ and of $\cos \theta$ and $\sin \theta$, see Appendix A.1.

[18] Named after the great Swiss mathematician Leonhard Euler (1707–1783). We have the Maclaurin expansion $e^x = 1 + x + \frac{1}{2!}x^2 + \frac{1}{3!}x^3 + \cdots$, and substituting $x = i\theta$ gives Euler's formula when we remember that $\sin x = x - \frac{1}{3!}x^3 + \frac{1}{5!}x^5 - \cdots$ and $\cos x = 1 - \frac{1}{2!}x^2 + \frac{1}{4!}x^4 - \cdots$.

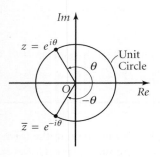

Because $\cos^2\theta + \sin^2\theta = 1$ for any angle $\theta$, the numbers $e^{i\theta} = \cos\theta + i\sin\theta$ all have absolute value 1:

$$\left| e^{i\theta} \right| = 1 \text{ for any angle } \theta.$$

In other words, each $e^{i\theta}$ lies on the **unit circle** (centre at the origin and radius 1). On the other hand, every complex number $z$ on the unit circle has the form $z = e^{i\theta}$ for some angle $\theta$ (see the diagram):

$$|z| = 1 \text{ if and only if } z = e^{i\theta} \text{ for some angle } \theta.$$

Furthermore, if $|z| = 1$, then $z\bar{z} = |z|^2 = 1$, so $\bar{z} = 1/z$. The diagram also shows that:

$$\text{If } z = e^{i\theta}, \text{ then } \quad \bar{z} = e^{-i\theta} = 1/z.$$

Our main purpose for introducing the polar form is to clearly describe the product of two complex numbers. This is achieved in the following result.

**THEOREM 4**

**Multiplication Rule.** If $z = re^{i\theta}$ and $w = se^{i\varphi}$ in polar form, then
$$re^{i\theta} se^{i\varphi} = rs\, e^{i(\theta+\varphi)}.$$

Thus, to multiply two complex numbers, simply multiply the absolute values and add the angles.

PROOF

Since $re^{i\theta}se^{i\varphi} = rs\, e^{i\theta}e^{i\varphi}$, it suffices to show that $e^{i\theta}e^{i\varphi} = e^{i(\theta+\varphi)}$. This is a consequence of the trigonometric identities for $\cos(\theta + \varphi)$ and $\sin(\theta + \varphi)$:

$$\begin{aligned}
e^{i\theta}e^{i\varphi} &= (\cos\theta + i\sin\theta)(\cos\varphi + i\sin\varphi) \\
&= (\cos\theta\cos\varphi - \sin\theta\sin\varphi) + i(\cos\theta\sin\varphi + \sin\theta\cos\varphi) \\
&= \cos(\theta + \varphi) + i\sin(\theta + \varphi) \\
&= e^{i(\theta+\varphi)}.
\end{aligned}$$

**Example 9**　Multiply $(-1 + i)\left(\sqrt{3} - i\right)$ using the multiplication rule.

SOLUTION

In Example 8 we found that polar forms for these numbers are $-1 + i = \sqrt{2}\, e^{3\pi i/4}$ and $\sqrt{3} - i = 2\, e^{-\pi i/6}$. Hence the multiplication rule gives

$$(-1+i)\left(\sqrt{3}-i\right) = \sqrt{2}e^{(3\pi i/4)i}\, 2\, e^{-\pi i/6} = 2\sqrt{2}\, e^{(3\pi/4 - \pi/6)i}$$
$$= 2\sqrt{2}\, e^{7\pi i/12} = 2\sqrt{2}\left(\cos\tfrac{7\pi}{12} + i\sin\tfrac{7\pi}{12}\right)$$

in polar form. Of course, we also can multiply the left side in Cartesian form to obtain $(1 - \sqrt{3}) + (1 + \sqrt{3})i = 2\sqrt{2}\left(\cos\tfrac{7\pi}{12} + i\sin\tfrac{7\pi}{12}\right)$. Equating real and imaginary parts gives the (somewhat unexpected) formulas $\cos\tfrac{7\pi}{12} = \tfrac{1-\sqrt{3}}{2\sqrt{2}}$ and $\sin\tfrac{7\pi}{12} = \tfrac{1+\sqrt{3}}{2\sqrt{2}}$.

The multiplication rule takes a particularly nice form when we use it to compute powers of a complex number. If $z = re^{i\theta}$ in polar form, then

$$z^2 = (re^{i\theta})(re^{i\theta}) = r^2 e^{i2\theta}.$$

Next compute $z^3$ as $z^3 = zz^2$ to get

$$z^3 = zz^2 = (re^{i\theta})(r^2 e^{i2\theta}) = r^3 e^{i3\theta},$$

again by the multiplication rule. The pattern is clear:

**THEOREM 5**  **DeMoivre's Theorem**[19]. If $z = re^{i\theta}$ in polar form, then
$$z^n = r^n e^{in\theta} \text{ for } n = 0, \pm 1, \pm 2, \cdots.$$

**PROOF**  The above argument proves it for $n = 1, 2, \cdots$, and it is clear if $n = 0$. If $n$ is negative, write $n = -m$ where $m > 0$. Then
$$z^n = \tfrac{1}{z^m} = \tfrac{1}{r^m e^{im\theta}} = r^{-m}(e^{im\theta})^{-1} = r^{-m}e^{-(im\theta)} = r^{-m}e^{i(-m)\theta} = r^n e^{in\theta}.$$

**Example 10**  Compute $(1 - i)^{27}$.

**SOLUTION**  We have $1 - i = \sqrt{2}e^{-\pi i/4}$, so DeMoivre's theorem gives
$$(1 - i)^{27} = (\sqrt{2})^{27}e^{-27\pi i/4}.$$
Since $(\sqrt{2})^2 = 2$, we have $(\sqrt{2})^{27} = 2^{13}\sqrt{2}$. We can simplify the expression $e^{-\frac{27\pi}{4}i}$ by using the fact that $e^{2\pi ki} = 1$ for all integers $k$ (even negative values). Observe that $-\frac{27\pi}{4} = -\frac{3\pi}{4} - 6\pi$. Hence
$$e^{-27\pi i/4} = e^{-3\pi i/4}e^{-6\pi i} = e^{-3\pi i/4} \cdot 1 = e^{-3\pi i/4} = -\tfrac{1}{\sqrt{2}}(1 + i).$$
Finally $(1 - i)^{27} = (\sqrt{2})^{27}e^{-27\pi i/4i} = 2^{13}\sqrt{2}\{-\tfrac{1}{\sqrt{2}}(1 + i)\} = -2^{13}(1 + i)$.

DeMoivre's theorem also allows us to take roots of complex numbers. Here is an example.

**Example 11**  Find all fourth roots of $8(1 + \sqrt{3}i)$; that is, find all complex numbers $z$ such that $z^4 = 8(1 + \sqrt{3}i)$.

**SOLUTION**  The idea is to write $z = re^{i\theta}$, and find all possibilities for $r$ and $\theta$. We have $8(1 + \sqrt{3}i) = 8(2e^{\pi i/3})$, so the condition that $z^4 = 8(1 + \sqrt{3}i)$ becomes
$$r^4 e^{4\theta i} = 16e^{\pi i/3} \qquad (*)$$
Equating absolute values gives $r^4 = 16$, so $r = \sqrt[4]{16} = 2$ because $r > 0$. Even though the complex numbers in $(*)$ are equal, their angles may differ by a multiple of $2\pi$:
$$4\theta = \tfrac{\pi}{3} + 2\pi k \text{ for some } k = 0, \pm 1, \pm 2, \pm 3, \cdots.$$
Hence $\theta = \tfrac{\pi}{12} + \tfrac{\pi}{2}k$ for any integer $k$, and it appears that there are infinitely many possibilities for $\theta$, one for each choice of $k$. However, the various possibilities for $\theta$ are obtained from $\tfrac{\pi}{12}$ by adding multiples of $\tfrac{\pi}{2}$. Hence the distinct ones correspond to $k = 0, 1, 2,$ and $3$; all the others differ from one of these by a multiple of $2\pi$ and so yield nothing new (see the diagram).

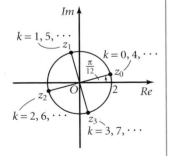

So the four possibilities for $z$ are
$$z_0 = 2e^{\pi i/12}, \quad z_1 = 2e^{7\pi i/12}, \quad z_2 = 2e^{13\pi i/12}, \quad z_3 = 2e^{19\pi i/12}.$$
These are plotted in the diagram. They are equally spaced around the circle $|z| = 2$, starting at $\theta = \tfrac{\pi}{12}$.

The technique in Example 11 works to find all $n^{\text{th}}$ roots of any complex number $w$; that is, all complex numbers $z$ such that $z^n = w$. It turns out that there are $n$ of them, equally spaced on a circle of radius $\sqrt[n]{|w|}$ so the angle between consecutive values is $\tfrac{2\pi}{n}$.

Possibly the most important case is that of the complex numbers $z$ such that $z^n = 1$, the so-called $n^{\text{th}}$ **roots of unity**. Here, if we write $z = re^{i\theta}$, we obtain $r^n e^{n\theta i} = 1e^{0i}$, so equating absolute values gives $r = 1$, and equating angles gives $n\theta = 0 + 2\pi k$ for $k = 0, \pm 1, \pm 2, \pm 3, \cdots$. Thus $\theta = \frac{2\pi}{n}k$, so the roots are all given by taking $k = 0, 1, 2, \cdots, n - 1$.

---

**THEOREM 6**

The $n^{\text{th}}$ roots of unity (that is the solutions to $z^n = 1$) are the complex numbers
$$e^{2\pi ki/n} \text{ where } k = 0, 1, 2, 3, \cdots, n - 1.$$

---

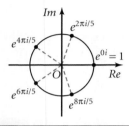

These $n^{\text{th}}$ roots of unity are equally spaced on the unit circle. Every value of $k$ yields a value of $\theta = \frac{2\pi}{n}k$ that differs from one of these by a multiple of $2\pi$. Of course $k = 0$ gives the root $z = 1$, so all the roots are obtained by dividing the unit circle into $n$ equal sectors, starting at $z = 1$. The $5^{\text{th}}$ roots of unity are plotted in the diagram.

## 2.5.7   Proof of Theorem 3

If $Z$ is a complex matrix, the **conjugate matrix** $\overline{Z}$ is defined to be the matrix obtained from $Z$ by conjugating every entry. Thus, if $Z = [z_{ij}]$, then $\overline{Z} = [\overline{z}_{ij}]$. For example, if
$$Z = \begin{bmatrix} -i+2 & 5 \\ i & 3+4i \end{bmatrix}, \text{ then } \overline{Z} = \begin{bmatrix} i+2 & 5 \\ -i & 3-4i \end{bmatrix}.$$

It follows from C1 and C2 in §2.5.1 above that, if $Z$ and $W$ are two complex matrices and $\lambda$ is a complex number, then
$$\overline{Z + W} = \overline{Z} + \overline{W}$$
$$\overline{ZW} = \overline{Z}\,\overline{W}$$
$$\overline{(\lambda Z)} = \overline{\lambda}\,\overline{Z}$$

Now suppose that $A$ is a real symmetric matrix. Observe that $\overline{A} = A$ because $A$ is real (using C4 in §2.5.1 above). If $\lambda$ is an eigenvalue of $A$, we show that $\lambda$ is real by showing that $\overline{\lambda} = \lambda$ (again by C4). Let $X$ be a (possibly complex) eigenvector corresponding to $\lambda$, so that $X \neq 0$ and $AX = \lambda X$. Define $c = X^T \overline{X}$. If we write $X = [z_1 \ z_2 \ \cdots \ z_n]^T$ where the $z_i$ are complex numbers, we have
$$c = X^T \overline{X} = [z_1 \ z_2 \ \cdots \ z_n] \begin{bmatrix} \overline{z}_1 \\ \overline{z}_2 \\ \vdots \\ \overline{z}_n \end{bmatrix}$$
$$= z_1\overline{z}_1 + z_2\overline{z}_2 + \cdots + z_n\overline{z}_n$$
$$= |z_1|^2 + |z_2|^2 + \cdots + |z_n|^2.$$

Thus $c$ is a real number, and $c > 0$ because at least one of the $z_i \neq 0$ (as $X \neq 0$). We show that $\overline{\lambda} = \lambda$ by verifying that $\lambda c = \overline{\lambda} c$. We have
$$\lambda c = \lambda(X^T \overline{X}) = (\lambda X)^T \overline{X} = (AX)^T \overline{X} = X^T A^T \overline{X}.$$

At this point we use the hypothesis that $A$ is symmetric and real. This means $A^T = A = \overline{A}$, so we continue
$$\lambda c = X^T A^T \overline{X} = X^T (\overline{A}\ \overline{X}) = X^T \overline{(AX)} = X^T \overline{(\lambda X)} = X^T (\overline{\lambda}\ \overline{X}) = \overline{\lambda} X^T \overline{X} = \overline{\lambda} c$$

as required. The technique in this proof will be used again in Chapter 4.

## Exercises 2.5

Throughout these exercises, $z$ and $w$ denote complex numbers.

**1.** Write each of the following in the form $a + bi$.
**a)** $(2 - 3i) - 3(5 + 7i) + 4$   **b)** $(4 - i)(2i - 5) + |3 - 2i|$
**c)** $i^{11}$
**d)** $(2 - 3i)^3$
**e)** $\frac{3-i}{2i+5}$
**f)** $\frac{3-5i}{7-i}$
**g)** $\frac{1-i}{2-3i} - \frac{1+2i}{5+i}$
**h)** $\frac{3-i}{2-i} + 3\frac{5i}{2i+1}$
**i)** $\frac{|2-3i|}{|2+3i|}$
**j)** $\left|\frac{3+4i}{|3+4i|}\right|$

**2.** In each case solve for the real number $x$.
**a)** $(2 + xi)(3 - 2i) = 12 + 5i$
**b)** $(2 + xi)^2 = 4$

**3.** In each case solve for the complex number $z$.
**a)** $iz + (1 - i)^2 = 5 - 2i$   **b)** $(i + z) - 3i(2 - z) = iz + 1$
**c)** $z(1 + i) = \bar{z} - (3 + 2i)$   **d)** $z(2 - i) = (\bar{z} + 1)(1 + i)$

**4.** Find all complex numbers $z$ such that $|z| = 1$ and $z/\bar{z} = -1$.

**5. a)** Show that $z + \bar{z}$ is a real number for each complex number $z$.
**b)** What can you say about $z - \bar{z}$?

**6.** Let $z = a + bi$ and $w = a' + b'i$ be arbitrary complex numbers. In each case verify the identity by computing each side separately and comparing.
**a)** C1. $\overline{(z \pm w)} = \bar{z} \pm \bar{w}$
**b)** C2. $\overline{(zw)} = \bar{z}\,\bar{w}$
**c)** C3. $\overline{(\bar{z})} = z$

**7.** In each case either show that the statement is true or give an example showing that it is false.
**a)** If $a^2 + b^2 = 0$ where $a$ and $b$ are real numbers, then $a = 0$ and $b = 0$.
**b)** If $z^2 + w^2 = 0$ where $z$ and $w$ are complex numbers, then $z = 0$ and $w = 0$.
**c)** If $z = 1/z$, then $z = \pm 1$.
**d)** If $|z| = 1$, then $z = 1, -1, i$ or $-i$.
**e)** If $|z| = |w|$, then $z = \pm w$.
**f)** If $\bar{z} = -z$, then $z$ is pure imaginary.

**8.** Find the roots of the following quadratic polynomials:
**a)** $x^2 - 1$
**b)** $x^2 + 1$
**c)** $3x^2 - 5x + 10$
**d)** $3x^2 + 5x + 10$
**e)** $-3x^2 - 5x - 10$
**f)** $-3x^2 + 5x + 10$

**9.** In each case find a real irreducible quadratic with $\lambda$ as a root:
**a)** $\lambda = 1 - 2i$
**b)** $\lambda = 3i$
**c)** $\lambda = -2i$
**d)** $\lambda = 2 - 5i$

**10.** Solve each of the following systems of equations.
**a)**
$$x + \quad iy - \quad iz = 3 + i$$
$$-ix + \quad 2y + \quad iz = 2$$
$$(i - 1)x - (1 + 2i)y + 2z = i - 1$$
**b)**
$$x + (1 + i)y - \quad iz = 2$$
$$(1 + i)x + \quad + (1 - i)z = 3 + i$$
$$-x + (1 - i)y + \quad iz = i - 1$$

**11.** Find the inverse of the following matrices, if possible.
**a)** $\begin{bmatrix} 1+i & -2 \\ 2i & -1-i \end{bmatrix}$
**b)** $\begin{bmatrix} 1 & 1-i \\ 2+i & 3+i \end{bmatrix}$
**c)** $\begin{bmatrix} 1+2i & -i & 3+5i \\ 2 & 1+4i & 3-i \\ -i & 5+i & 4-2i \end{bmatrix}$
**d)** $\begin{bmatrix} 1-i & 1-i & 1+i \\ 2+i & 3+i & 4+i \\ -3i & 2-i & 5+5i \end{bmatrix}$

**12.** In each case find, if possible, an invertible matrix $P$ such that $P^{-1}AP = D$ is diagonal.
**a)** $\begin{bmatrix} -2 & -1 \\ 5 & 2 \end{bmatrix}$
**b)** $\begin{bmatrix} 1 & i \\ -i & 1 \end{bmatrix}$
**c)** $\begin{bmatrix} -3 & 5 & 4 \\ -2 & 3 & 2 \\ 0 & 0 & 1 \end{bmatrix}$
**d)** $\begin{bmatrix} -5 & 6 & 2 \\ -3 & 4 & 1 \\ -5 & 5 & 2 \end{bmatrix}$
**e)** $\begin{bmatrix} 2 & 51 & -53 \\ 10 & -32 & -24 \\ -11 & 19 & -78 \end{bmatrix}$
**f)** $\begin{bmatrix} -91 & 70 & -23 \\ 32 & 14 & 16 \\ 25 & 19 & -19 \end{bmatrix}$
**g)** $\begin{bmatrix} -1-i & -3+i & -2+2i \\ 1+i & -3+2i & -i \\ -3+5i & 2+3i & 3+5i \end{bmatrix}$
**h)** $\begin{bmatrix} 2+i & 2-i & -4+i \\ 3-i & 1+2i & 6+5i \\ 5-i & 6+i & 2+7i \end{bmatrix}$

**13.** Let $A\begin{bmatrix} \cos\theta & -\sin\theta \\ \sin\theta & \cos\theta \end{bmatrix}$ where $\theta$ is any angle.
**a)** Show that the characteristic polynomial of $A$ is $c_A(x) = x^2 - 2\cos\theta\, x + 1$.
**b)** Show that the eigenvalues of $A$ are $\lambda_1 = \cos\theta + i\sin\theta$ and $\lambda_2 = \cos\theta - i\sin\theta$.
**c)** Find an invertible matrix $P$ such that $P^{-1}AP = \begin{bmatrix} \lambda_1 & 0 \\ 0 & \lambda_2 \end{bmatrix}$.

**14.** If $Z$ and $W$ are complex matrices, show that $\overline{ZW} = \bar{Z}\,\bar{W}$.

**15.** If $\lambda$ is an eigenvalue (possibly not real) of a real matrix $A$, show that $\bar{\lambda}$ is also an eigenvalue.

**16.** If $\lambda$ is a root of a real polynomial $f(x)$, show that $\bar{\lambda}$ is also a root.

**17.** Show that $|\bar{z}| = |z|$ for any complex number $z$.

**18.** Describe the set of all complex numbers $z$ such that $|z| = 1$.

**19.** Show that $\left|\frac{z}{w}\right| = \frac{|z|}{|w|}$ whenever $w \neq 0$.

**20.** Express each of the following in polar form.

   **a)** $2 - 2i$        **b)** $-2i$

   **c)** $-\sqrt{3} + i$     **d)** $-3 - \sqrt{3}i$

   **e)** $5i$           **f)** $5(1 + i)$

**21.** Express each of the following in the form $a + bi$.

   **a)** $3\, e^{\pi i}$       **b)** $e^{5\pi i/3}$

   **c)** $2\, e^{\pi i/4}$     **d)** $\sqrt{2}\, e^{-\pi i/4}$

   **e)** $e^{15\pi/4}$     **f)** $2\sqrt{3}\, e^{-\pi i/3}$

**22.** Express each of the following in the form $a + bi$.

   **a)** $(1 - \sqrt{3}i)^5$     **b)** $(\sqrt{3} + i)^{-4}$

   **c)** $(1 + i)^{12}$       **d)** $(1 - i)^{10}$

   **e)** $(1 - i)^6 (\sqrt{3} + i)^3$    **f)** $(\sqrt{3} - i)^9 (2 - 2i)^5$

**23.** Find all:

   **a)** fourth roots of unity.   **b)** cube roots of unity.

**24.** Find all complex numbers $z$ such that:

   **a)** $z^4 = -1$       **b)** $z^4 = 2(\sqrt{3}i - 1)$

   **c)** $z^3 = -27i$     **d)** $z^6 = -64$

**25.** Use DeMoivre's theorem to show that:

   **a)** $\cos(2\theta) = \cos^2\theta - \sin^2\theta$   and
   $\sin(2\theta) = 2\cos\theta\sin\theta$.

   **b)** $\cos(3\theta) = \cos^3\theta - 3\cos\theta\sin^2\theta$   and
   $\sin(2\theta) = 3\cos^2\theta\sin\theta - \sin^3\theta$.

**26.** Show that $\cos\theta = \frac{1}{2}(e^{i\theta} + e^{-i\theta})$ and $\sin\theta = \frac{1}{2i}(e^{i\theta} - e^{-i\theta})$ for any angle $\theta$.

## 2.6    LINEAR RECURRENCES

A sequence $x_0, x_1, \cdots$ of numbers is said to be given **recursively** if each number in the sequence is completely determined by those that come before it. Such sequences arise frequently in mathematics and computer science and in various parts of the natural and social sciences. Here is an example.

**Example 1**

An urban planner wants to determine the number $x_k$ of ways that a row of $k$ parking spaces can be filled with cars and trucks if trucks take up two spaces each. Find $x_{12}$.

**SOLUTION**

Clearly $x_0 = 1$ and $x_1 = 1$, while $x_2 = 2$ because there can be only two cars or one truck. We have $x_3 = 3$ (the 3 configurations are *ccc*, *cT* and *Tc*), and $x_4 = 5$ (*cccc*, *ccT*, *cTc*, *Tcc* and *TT*). However, enumerating the possibilities in this way becomes difficult if the number $k$ of parking spaces is large. Some other method is needed. In this case (and many others) it turns out that the numbers $x_k$ are given recursively, and this solves the problem. In fact, we claim that

$$x_{k+2} = x_k + x_{k+1} \quad \text{for every } k \geq 0.$$

This determines $x_k$ for every $k \geq 2$ since $x_0$ and $x_1$ are given. Indeed, the first few possibilities are

$$x_0 = 1$$
$$x_1 = 1$$
$$x_2 = x_0 + x_1 = 2$$
$$x_3 = x_1 + x_2 = 3$$
$$x_4 = x_2 + x_3 = 5$$
$$x_5 = x_3 + x_4 = 8$$
$$x_6 = x_4 + x_5 = 13$$
$$\vdots \qquad \vdots \qquad \vdots$$

Continuing in this way we obtain $x_{12} = 233$ as the reader can verify. While this method can be used with a computer to quickly calculate $x_k$ for large values of $k$, it would be nice to have a "formula" for $x_k$ in terms of $k$ to get an approximate idea of the size of $x_k$ as $k$ increases (for example $x_{30} = 832\,040$). This can be done using diagonalization, as we shall see.

### 2.6.1 Linear Recurrences

The equation $x_{k+2} = x_k + x_{k+1}$ that occurs in Example 1 is called a **linear recurrence relation** for the numbers $x_0, x_1, x_2, \cdots$. We begin our discussion with a recurrence that is somewhat easier to solve than this one (we return to it in Example 3).

Suppose the numbers $x_0, x_1, x_2, \cdots$ are given by

$$x_{k+2} = x_{k+1} + 2x_k \quad \text{for } k \geq 0$$

where $x_0$ and $x_1$ are specified. If $x_0 = 1$ and $x_1 = 2$, the sequence is uniquely determined:

$$x_0 = 1$$
$$x_1 = 2$$
$$x_2 = x_1 + 2x_0 = 4$$
$$x_3 = x_2 + 2x_1 = 8$$
$$x_4 = x_3 + 2x_2 = 16$$
$$\vdots \qquad \vdots \qquad \vdots$$

It is clear that $x_k = 2^k$ holds for $k = 0, 1, 2, 3,$ and $4$, and this formula satisfies the recurrence $x_{k+2} = x_{k+1} + 2x_k$ as is readily checked.

However if we take $x_0 = 1$ and $x_1 = 1$ instead, the sequence continues $x_2 = 3$, $x_3 = 5$, $x_4 = 11$, $x_5 = 21, \cdots$. In this case, the sequence is uniquely determined but no formula is apparent. Nontheless, a simple device transforms the recurrence into a matrix recurrence, which we can solve easily.

**Example 2**   Suppose the sequence $x_0, x_1, x_2, \cdots$ is given recursively by $x_0 = 1$, $x_1 = 1$, and $x_{k+2} = x_{k+1} + 2x_k$ for each $k \geq 0$. Find a formula for $x_k$ in terms of $k$.

**SOLUTION**   The idea is to consider the sequence $V_0, V_1, V_2, \cdots$ instead of $x_0, x_1, x_2, \cdots$ where

$$V_k = \begin{bmatrix} x_k \\ x_{k+1} \end{bmatrix}.$$

Then $V_0 = \begin{bmatrix} x_0 \\ x_1 \end{bmatrix} = \begin{bmatrix} 1 \\ 1 \end{bmatrix}$ is specified, and the numerical recurrence $x_{k+2} = x_{k+1} + 2x_k$ transforms into a matrix recurrence as follows:

$$V_{k+1} = \begin{bmatrix} x_{k+1} \\ x_{k+2} \end{bmatrix} = \begin{bmatrix} x_{k+1} \\ x_{k+1} + 2x_k \end{bmatrix} = \begin{bmatrix} 0 & 1 \\ 2 & 1 \end{bmatrix}\begin{bmatrix} x_k \\ x_{k+1} \end{bmatrix} = AV_k$$

where $A = \begin{bmatrix} 0 & 1 \\ 2 & 1 \end{bmatrix}$. Hence $V_0, V_1, \cdots$ is a dynamical system, so Theorem 1 §2.4 applies if the matrix $A$ is diagonalizable. We have

$$c_A(x) = det\begin{bmatrix} x & -1 \\ -2 & x-1 \end{bmatrix} = x(x-1) - 2 = (x-2)(x+1).$$

Hence the eigenvalues $\lambda_1 = 2$ and $\lambda_2 = -1$ are distinct, so $A$ is indeed diagonalizable. Corresponding eigenvectors are $X_1 = \begin{bmatrix} 1 \\ 2 \end{bmatrix}$ and $X_2 = \begin{bmatrix} 1 \\ -1 \end{bmatrix}$ as the reader can check, so $P = \begin{bmatrix} 1 & 1 \\ 2 & -1 \end{bmatrix}$ is a diagonalizing matrix for $A$. We compute the coefficients $b_i$ in Theorem 1 §2.4:

$$\begin{bmatrix} b_1 \\ b_2 \end{bmatrix} = P^{-1}V_0 = -\frac{1}{3}\begin{bmatrix} -1 & -1 \\ -2 & 1 \end{bmatrix}\begin{bmatrix} 1 \\ 1 \end{bmatrix} = \begin{bmatrix} \frac{2}{3} \\ \frac{1}{3} \end{bmatrix}.$$

Thus Theorem 1 §2.4 gives

$$\begin{bmatrix} x_k \\ x_{k+1} \end{bmatrix} = V_k = b_1 \lambda_1^k X_1 + b_2 \lambda_2^k X_2 = \tfrac{2}{3} 2^k \begin{bmatrix} 1 \\ 2 \end{bmatrix} + \tfrac{1}{3}(-1)^k \begin{bmatrix} 1 \\ -1 \end{bmatrix}.$$

Equating top entries yields

$$x_k = \tfrac{1}{3}\left(2^{k+1} + (-1)^k\right) \quad \text{for } k \ge 0$$

which is the desired formula for the $x_k$.

Note that we can write $x_k = \tfrac{1}{3}2^{k+1}\left(1 + \tfrac{1}{2}(\tfrac{-1}{2})^k\right) \approx \tfrac{1}{3}2^{k+1}$ for large $k$. (This amounts to the observation that $\lambda_1 = 2$ is the dominant eigenvalue here.) This formula is quite useful in this case because the approximation is good even for small $k$. For example, we get $x_6 \approx \tfrac{1}{3}2^{6+1} = \tfrac{128}{3} = 42.67$, which compares favourably with the true value $x_6 = 43$. In fact, for large $k$, $x_k$ is the integer nearest to $\tfrac{1}{3}2^{k+1}$.

We now apply this diagonalization technique to the combinatorial problem in Example 1.

**Example 3**    An urban planner wants to determine the number $x_k$ of ways that a row of $k$ parking spaces can be filled with cars and trucks if trucks take up two spaces each. Find a formula for $x_k$ and estimate it for large $k$.

SOLUTION    We saw in Example 1 that $x_0 = 1$, $x_1 = 1$, and

$$x_{k+2} = x_k + x_{k+1} \quad \text{for every } k \ge 0. \tag{$*$}$$

If we write $V_k = \begin{bmatrix} x_k \\ x_{k+1} \end{bmatrix}$, the recurrence $(*)$ becomes a matrix recursion for the $V_k$:

$$\begin{aligned} V_{k+1} = \begin{bmatrix} x_{k+1} \\ x_{k+2} \end{bmatrix} &= \begin{bmatrix} x_{k+1} \\ x_k + x_{k+1} \end{bmatrix} \\ &= \begin{bmatrix} 0 & 1 \\ 1 & 1 \end{bmatrix}\begin{bmatrix} x_k \\ x_{k+1} \end{bmatrix} \\ &= A V_k \quad \text{for all } k \ge 0 \end{aligned}$$

where $A = \begin{bmatrix} 0 & 1 \\ 1 & 1 \end{bmatrix}$. Moreover $A$ is diagonalizable here. The characteristic polynomial is

$$c_A(x) = det\begin{bmatrix} x & -1 \\ -1 & x-1 \end{bmatrix} = x^2 - x - 1.$$ This has roots $\tfrac{1}{2}[1 \pm \sqrt{5}]$ by the quadratic formula, so $A$ has distinct eigenvalues

$$\lambda_1 = \tfrac{1}{2}[1 + \sqrt{5}] \quad \text{and} \quad \lambda_2 = \tfrac{1}{2}[1 - \sqrt{5}].$$

Since $\lambda_i^2 = \lambda_i + 1$ for each $i$, corresponding eigenvectors are $X_1 = \begin{bmatrix} 1 \\ \lambda_1 \end{bmatrix}$ and $X_2 = \begin{bmatrix} 1 \\ \lambda_2 \end{bmatrix}$ respectively as the reader can verify. Hence $P = \begin{bmatrix} 1 & 1 \\ \lambda_1 & \lambda_2 \end{bmatrix}$ is a diagonalizing matrix, so we compute the coefficients $b_1$ and $b_2$ (in Theorem 1 §2.4) as follows

$$\begin{bmatrix} b_1 \\ b_2 \end{bmatrix} = P^{-1}V_0 = \tfrac{1}{-\sqrt{5}}\begin{bmatrix} \lambda_2 & -1 \\ -\lambda_1 & 1 \end{bmatrix}\begin{bmatrix} 1 \\ 1 \end{bmatrix} = \tfrac{1}{\sqrt{5}}\begin{bmatrix} \lambda_1 \\ -\lambda_2 \end{bmatrix}$$

where we used the fact that $\lambda_1 + \lambda_2 = 1$. Thus Theorem 1 §2.4 gives

$$\begin{bmatrix} x_k \\ x_{k+1} \end{bmatrix} = V_k = b_1\lambda_1^k X_1 + b_2\lambda_2^k X_2 = \tfrac{\lambda_1}{\sqrt{5}}\lambda_1^k\begin{bmatrix} 1 \\ \lambda_1 \end{bmatrix} - \tfrac{\lambda_2}{\sqrt{5}}\lambda_2^k\begin{bmatrix} 1 \\ \lambda_2 \end{bmatrix}.$$

Comparing first entries gives an exact formula for the numbers $x_k$:

$$x_k = \tfrac{1}{\sqrt{5}}[\lambda_1^{k+1} - \lambda_2^{k+1}] \quad \text{for each } k \ge 0.$$

Finally, observe that $\lambda_1$ is dominant here (in fact, $\lambda_1 = 1.618$ and $\lambda_2 = -0.618$ to three figures) so $\lambda_2^{k+1}$ is negligible compared with $\lambda_1^{k+1}$ if $k$ is large. Thus

$$x_k \approx \tfrac{1}{\sqrt{5}}\lambda_1^{k+1} \quad \text{for each } k \geq 0.$$

(This also follows from Theorem 1 §2.4.) This is a good approximation, even for as small a value as $k = 12$. Indeed, repeated use of the recurrence $x_{k+2} = x_k + x_{k+1}$ gives the exact value $x_{12} = 233$, while the approximation is $x_{12} \approx \frac{(1.618)^{13}}{\sqrt{5}} = 232.94$.

The sequence $x_0, x_1, x_2, \cdots$ in Example 3 was first discussed in 1202 by Leonardo Pisano of Pisa, also known as Fibonacci, and is now called the **Fibonacci sequence**. It is completely determined by the conditions $x_0 = 1$, $x_1 = 1$ and the recurrence $x_{k+2} = x_k + x_{k+1}$ for each $k \geq 0$. These numbers have been studied for centuries, and have many interesting properties (there is even a journal devoted exclusively to them). For example, biologists have discovered that the arrangement of leaves around the stems of some plants follow a Fibonacci pattern.

We conclude with an example showing that nonlinear recurrences can be complicated.

**Example 4**    Suppose a sequence $x_0, x_1, x_2, \cdots$ satisfies the following recurrence:

$$x_{k+1} = \begin{cases} \tfrac{1}{2}x_k & \text{if } x_k \text{ is even} \\ 3x_k + 1 & \text{if } x_k \text{ is odd} \end{cases}$$

If $x_0 = 1$, the sequence is $1, 4, 2, 1, 4, 2, 1, \cdots$ and so continues to cycle indefinitely. The same thing happens if $x_0 = 7$. Then the sequence is

$$7, 22, 11, 34, 17, 52, 26, 13, 40, 20, 10, 5, 16, 8, 4, 2, 1, \cdots$$

and it again cycles. However, it is not known whether every choice of $x_0$ will lead eventually to 1. It is quite possible that, for some $x_0$, the sequence will continue to produce different values indefinitely, or will repeat a value and cycle without reaching 1. No one knows for sure.

---

## Exercises 2.6

**1.** Consider a sequence $x_0, x_1, x_2, \cdots$ satisfying the recurrence $x_{k+2} = 6x_k + x_{k+1}$.
   **a)** If $x_0 = 1$ and $x_1 = 3$, compute the first five terms of the sequence and guess a formula for $x_k$.
   **b)** If $x_0 = 1$ and $x_1 = 1$, find a formula for $x_k$.

**2.** Solve the following linear recurrences.
   **a)** $x_{k+2} = 3x_k + 2x_{k+1}$, where $x_0 = 1$ and $x_1 = 1$.
   **b)** $x_{k+2} = 2x_k - x_{k+1}$, where $x_0 = 1$ and $x_1 = 2$.
   **c)** $x_{k+2} = 2x_k + x_{k+1}$, where $x_0 = 0$ and $x_1 = 1$.
   **d)** $x_{k+2} = 6x_k - x_{k+1}$, where $x_0 = 1$ and $x_1 = 1$.

**3.** Solve the following linear recurrences.
   **a)** $x_{k+3} = 6x_{k+2} - 11x_{k+1} + 6x_k$, where $x_0 = 1$, $x_1 = 0$, and $x_2 = 1$.
   **b)** $x_{k+3} = -2x_{k+2} + x_{k+1} + 2x_k$, where $x_0 = 1$, $x_1 = 0$, and $x_2 = 1$.

**4.**[†] Solve the following linear recurrences.
   **a)** $x_{k+3} = x_{k+2} + x_{k+1} + x_k$, where $x_0 = 1$, $x_1 = 1$, and $x_2 = 1$.
   **b)** $x_{k+3} = 2x_{k+2} - x_{k+1} - 2x_k$, where $x_0 = 1$, $x_1 = 1$, and $x_2 = 1$.
   **c)** $x_{k+5} = 2x_{k+4} - x_{k+3} - 2x_{k+2} - x_{k+1} - x_k$, where $x_0 = 1$, $x_1 = 0$, $x_2 = 1$, $x_3 = 1$, and $x_4 = 1$.

**5.** A man must climb a flight of $k$ steps. He always takes one or two steps at a time. Thus he can climb 3 steps in the following ways: 111, 12, or 21. Find the number $s_k$ of ways he can climb the flight of $k$ steps. [Hint: Examples 1 and 3.]

**6.** How many "words" of $k$ letters can be made from the letters $\{a, b\}$ if there are no adjacent $a$'s?

**7.** How many sequences of $k$ flips of a coin are there with no $HH$ (where $H$ = heads)?

---

† Requires computer assistance (e.g., Matlab).

8. Find the number $x_k$ of ways to make a stack of $k$ poker chips if only red, blue, and gold chips are used and no two gold chips are adjacent.

9. A nuclear reactor contains $\alpha$- and $\beta$-particles. In every second each $\alpha$-particle splits into three $\beta$-particles, and each $\beta$-particle splits into an $\alpha$-particle and two $\beta$-particles. If there is a single $\alpha$-particle in the reactor at time $t = 0$, how many $\alpha$-particles are there at $t = 20$ seconds?

10. The annual yield of wheat in a certain country has been found to equal the average of the yield in the previous two years. If the yields in 1990 and 1991 were 10 and 12 million tons respectively, find a formula for the yield $k$ years after 1990. What is the long-term average yield?

11. Suppose a sequence $x_0, x_1, x_2, \cdots$ is given recursively by $x_{k+1} = ax_k$ for all $k \geq 0$, where $a$ is a fixed number. If $x_0 = b$, find $x_k$ in terms of $k$.

12. **a)** Consider the recurrence $x_{k+2} = ax_k + bx_{k+1}$. Show that the corresponding matrix $A$ has characteristic polynomial $x^2 - bx - a$, eigenvalues $\lambda_1 = \frac{1}{2}\left[b + \sqrt{b^2 + 4a}\right]$ and $\lambda_2 = \frac{1}{2}\left[b - \sqrt{b^2 + 4a}\right]$, and corresponding eigenvectors $X_i = \begin{bmatrix} 1 \\ \lambda_i \end{bmatrix}$ for each $i$.

  **b)** If $x_0 = 0, x_1 = 1$, use the diagonalizing matrix $P = [X_1\ X_2]$ for $A$ to get $V_k = \frac{1}{\sqrt{b^2 + 4a}}\left[\lambda_1^k X_1 - \lambda_2^k X_2\right]$, and hence $x_k = \frac{1}{\sqrt{b^2 + 4a}}\left[\lambda_1^k - \lambda_2^k\right]$ for all $k \geq 0$.

13. Consider $x_{k+2} = -5x_k + 2x_{k+1}$; $x_0 = 0$, $x_1 = 1$. Show that the eigenvalues are $\lambda_1 = 1 + 2i$ and $\lambda_2 = 1 - 2i$, with eigenvectors $X_1 = \begin{bmatrix} 1 \\ 1+2i \end{bmatrix}$ and $X_2 = \begin{bmatrix} 1 \\ 1-2i \end{bmatrix}$. Use Exercise 12 to get the formula
$$x_k = \frac{i}{4}\left[(1 - 2i)^k - (1 + 2i)^k\right] \text{ for } k \geq 0.$$

14. Find the general solution to the recurrence $x_{k+1} = rx_k + c$ where $r$ and $c$ are constants.

15. Consider a linear recurrence $x_{k+2} = ax_{k+1} + bx_k + c$ where $c$ may not be zero.
  **a)** If $a + b \neq 1$, show that $p$ can be found such that, if we set $y_k = x_k + p$, then $y_{k+2} = ay_{k+1} + by_k$. [Hence the sequence $x_k$ can be found provided $y_k$ can be found by the methods of this section (or otherwise).]
  **b)** Use **a)** to solve the recurrence $x_{k+2} = x_{k+1} + 6x_k + 5$ where $x_0 = 1$ and $x_1 = 1$.

16. Consider the linear recurrence
$$x_{k+2} = ax_{k+1} + bx_k + c(k) \tag{*}$$
where $c(k)$ is a function of $k$. Also consider the related recurrence
$$x_{k+2} = ax_{k+1} + bx_k. \tag{**}$$
Suppose that $x_k = p_k$ is a particular solution of (*).
  **a)** If $q_k$ is any solution of (**), show that $q_k + p_k$ is a solution of (*).
  **b)** Show that every solution of (*) arises as in **a)** as the sum of a solution of (**) plus the particular solution $p_k$ of (*).

## 2.7    POLYNOMIAL INTERPOLATION

In experimental science it is frequently necessary to use known data to estimate some quantity. There are many methods of doing this. In this section, we present the method of polynomial interpolation.

### 2.7.1    Polynomial Interpolation

**Example 1**    An engineer wants to estimate the deflection of a wooden beam under a load of 2200 kg. He knows the data in the following table:

| Load (kg) | 1000 | 2000 | 3000 |
|---|---|---|---|
| Deflection (mm) | 1.5 | 2.1 | 3.4 |

Deflection

and wants to "interpolate" to find the estimate he needs. A common way to do this is to find a quadratic $p(x)$ which fits the data exactly in the sense that
$$p(1000) = 1.5$$
$$p(2000) = 2.1$$
$$p(3000) = 3.4$$

Then the estimate would be $p(2200)$. If $p(x) = r_0 + r_1x + r_2x^2$, these conditions become three equations in the unknown coefficients $r_0$, $r_1$, and $r_2$:

$$r_0 + 1000\,r_1 + 1000^2 r_2 = 1.5$$
$$r_0 + 2000\,r_1 + 2000^2 r_2 = 2.1$$
$$r_0 + 3000\,r_1 + 3000^2 r_2 = 3.4$$

The solution is $r_0 = 1.6$, $r_1 = -0.000\,45$, and $r_2 = 0.000\,000\,35$, so the polynomial is

$$p(x) = 1.6 - 0.000\,45x + 0.000\,000\,35x^2.$$

The engineer uses $p(2200) = 2.304$ as the estimate of the deflection.

It often happens that two variables $x$ and $y$ are known to be related but the actual function $y = f(x)$ relating them is not known. Suppose that for certain values $x_1, x_2, \cdots, x_n$ of $x$, the corresponding values $y_1, y_2, \cdots, y_n$ are known (say from tables or experimental measurements), and it is desired to estimate the value of $y$ corresponding to some other value of $x$. As in Example 1, one way to do this is to find a polynomial $p(x)$ that "fits" the data in the sense that $p(x_i) = y_i$ for each $i = 1, 2, \cdots, n$. Then the estimate for $y$ is $p(x)$. Such a polynomial always exists if the $x_i$ are distinct.

**THEOREM 1**

Suppose $n$ data pairs $(x_1, y_1)$, $(x_2, y_2), \cdots, (x_n, y_n)$ are given where the $x_i$ are distinct. Then there exists a unique polynomial

$$p(x) = r_0 + r_1x + r_2x^2 + \cdots + r_{n-1}x^{n-1}$$

such that $p(x_i) = y_i$ for each $i = 1, 2, \cdots, n$.

**PROOF**

We consider the case $n = 4$; the general case is entirely similar. The given conditions that $p(x_i) = y_i$ for each $i = 1, 2, 3$, and $4$ give equations

$$r_0 + r_1x_1 + r_2x_1{}^2 + r_3x_1{}^3 = y_1$$
$$r_0 + r_1x_2 + r_2x_2{}^2 + r_3x_2{}^3 = y_2$$
$$r_0 + r_1x_3 + r_2x_3{}^2 + r_3x_3{}^3 = y_3$$
$$r_0 + r_1x_4 + r_2x_4{}^2 + r_3x_4{}^3 = y_4$$

In matrix form this system can be written as $AR = Y$ where

$$A = \begin{bmatrix} 1 & x_1 & x_1{}^2 & x_1{}^3 \\ 1 & x_2 & x_2{}^2 & x_2{}^3 \\ 1 & x_3 & x_3{}^2 & x_3{}^3 \\ 1 & x_4 & x_4{}^2 & x_4{}^3 \end{bmatrix}, \; R = \begin{bmatrix} r_0 \\ r_1 \\ r_2 \\ r_3 \end{bmatrix}, \text{and } Y = \begin{bmatrix} y_1 \\ y_2 \\ y_3 \\ y_4 \end{bmatrix}.$$

Hence it suffices to show that the matrix $A$ is invertible when the $x_i$ are distinct (then $R = A^{-1}Y$ is the column of coefficients of the polynomial we want).

We show that $A$ is invertible by verifying that the only solution to the homogeneous system $AX = 0$ is the trivial solution $X = 0$ (Theorem 5 §1.5). If we write $X = [t_0 \; t_1 \; t_2 \; t_3]^T$, consider the polynomial $f(x) = t_0 + t_1x + t_2x^2 + t_3x^3$ with the entries of $X$ as coefficients. Then we have

$$AX = \begin{bmatrix} 1 & x_1 & x_1{}^2 & x_1{}^3 \\ 1 & x_2 & x_2{}^2 & x_2{}^3 \\ 1 & x_3 & x_3{}^2 & x_3{}^3 \\ 1 & x_4 & x_4{}^2 & x_4{}^3 \end{bmatrix} \begin{bmatrix} t_0 \\ t_1 \\ t_2 \\ t_3 \end{bmatrix} = \begin{bmatrix} f(x_1) \\ f(x_2) \\ f(x_3) \\ f(x_4) \end{bmatrix}$$

Hence the requirement that $AX = 0$ means $f(x_i) = 0$ for each $i = 1$, 2, 3, and 4. In other words, $f(x)$ is a polynomial of degree at most 3 which has 4 distinct roots $x_1$, $x_2$, $x_3$, and $x_4$. This can happen only if $f(x)$ is identically zero[20] that is if $t_0 = t_1 = t_2 = t_3 = 0$. This means that $X = 0$ as required.

The polynomial $p(x)$ in Theorem 1 is called the **interpolating polynomial** for the data $(x_1, y_1)$, $(x_2, y_2)$, $\cdots$, $(x_n, y_n)$. It is an exact fit in the sense that $p(x_i) = y_i$ holds for each $i$. Finding $p(x)$ may not be feasible if $n$ is large, and polynomials of lower degree which "best fit" the data can be found, for example, by the method of least squares (see Section 4.6.5).

Matrices like $A$ in the proof of Theorem 1 and their determinants, arise in other contexts and so have independent interest. Given numbers $x_1$, $x_2$, $\cdots$, $x_n$, the matrix

$$\begin{bmatrix} 1 & x_1 & x_1{}^2 & \cdots & x_1{}^{n-1} \\ 1 & x_2 & x_2{}^2 & \cdots & x_2{}^{n-1} \\ 1 & x_3 & x_3{}^2 & \cdots & x_3{}^{n-1} \\ \vdots & \vdots & \vdots & & \vdots \\ 1 & x_n & x_n{}^2 & \cdots & x_n{}^{n-1} \end{bmatrix}$$

(or its transpose) is called the **Vandermonde matrix**[21] corresponding to the numbers $x_1$, $x_2$, $\cdots$, $x_n$, and its determinant is called the **Vandermonde determinant**. There is a simple formula for this determinant. The following example treats the case $n = 3$.

**Example 2**    Show that $det \begin{bmatrix} 1 & x_1 & x_1{}^2 \\ 1 & x_2 & x_2{}^2 \\ 1 & x_3 & x_3{}^2 \end{bmatrix} = (x_2 - x_1)(x_3 - x_1)(x_3 - x_2)$.

**SOLUTION**    We begin by using row operations to introduce zeros into column 1:

$$det \begin{bmatrix} 1 & x_1 & x_1^2 \\ 1 & x_2 & x_2^2 \\ 1 & x_3 & x_3^2 \end{bmatrix} = det \begin{bmatrix} 1 & x_1 & x_1^2 \\ 0 & x_2-x_1 & x_2^2-x_1^2 \\ 0 & x_3-x_1 & x_3^2-x_1^2 \end{bmatrix}$$

$$= det \begin{bmatrix} 1 & x_1 & x_1^2 \\ 0 & x_2-x_1 & (x_2-x_1)(x_2+x_1) \\ 0 & x_3-x_1 & (x_3-x_1)(x_3+x_1) \end{bmatrix}$$

Now take the common factor out of each of rows 2 and 3:

$$det \begin{bmatrix} 1 & x_1 & x_1^2 \\ 1 & x_2 & x_2^2 \\ 1 & x_3 & x_3^2 \end{bmatrix} = (x_2 - x_1)(x_3 - x_1)\, det \begin{bmatrix} 1 & x_1 & x_1^2 \\ 0 & 1 & x_2+x_1 \\ 0 & 1 & x_3+x_1 \end{bmatrix}$$

$$= (x_2 - x_1)(x_3 - x_1)\, det \begin{bmatrix} 1 & x_2+x_1 \\ 1 & x_3+x_1 \end{bmatrix}$$

$$= (x_2 - x_1)(x_3 - x_1)(x_3 - x_2).$$

The result in Example 2 is typical of what happens for any Vandermonde determinant. The result is stated in full generality in the following theorem.

---

[20] Since $f(x_1) = 0$, the factor theorem asserts that $f(x) = (x - x_1)g(x)$ where $g(x)$ is another polynomial. Since $f(x)$ has *four* distinct roots, continuing this process means that all four terms $(x - x_1)$, $(x - x_2)$, $(x - x_3)$, and $(x - x_4)$ are factors of $f(x)$, contrary to the fact that $f(x)$ has degree at most 3. See Appendix A.3.

[21] After A.T. Vandermonde (1735–1796), a French mathematician.

THEOREM **2** **Vandermonde Determinant.** Let $x_1, x_2, \cdots, x_n$ be real numbers where $n \geq 2$. Then the corresponding Vandermonde determinant is given by

$$
\begin{bmatrix}
1 & x_1 & x_1^2 & \cdots & x_1^{n-1} \\
1 & x_2 & x_2^2 & \cdots & x_2^{n-1} \\
1 & x_3 & x_3^2 & \cdots & x_3^{n-1} \\
\vdots & \vdots & \vdots & & \vdots \\
1 & x_n & x_n^2 & \cdots & x_n^{n-1}
\end{bmatrix}
= \prod_{1 \leq j < i \leq n} (x_i - x_j)
$$

where $\prod_{1 \leq j < i \leq n}(x_i - x_j)$ means the product of all factors $(x_i - x_j)$ where $j < i$ and both $i$ and $j$ are between 1 and $n$.

We omit the proof. To illustrate the theorem we state the expansion of the $4 \times 4$ Vandermonde determinant:

$$
det \begin{bmatrix}
1 & x_1 & x_1^2 & x_1^3 \\
1 & x_2 & x_2^2 & x_2^3 \\
1 & x_3 & x_3^2 & x_3^3 \\
1 & x_4 & x_4^2 & x_4^3
\end{bmatrix}
= (x_2 - x_1)(x_3 - x_1)(x_3 - x_2)(x_4 - x_1)(x_4 - x_2)(x_4 - x_3).
$$

## Exercises 2.7

1. In each case find the interpolating polynomial for the given data, and use the polynomial to estimate the value of $y$ corresponding to the given value of $x$.
   a) $(0, 2), (1, 3), (3, 8)$; $x = 2$.
   b) $(0, 5), (1, 3), (2, 5)$; $x = 1.5$.
   c) $(0, 1), (1, 1), (-1, 4), (2, 5)$; $x = \frac{1}{2}$.
   d) $(0, 1), (1, 1), (-1, 2), (-2, -3)$; $x = -\frac{1}{2}$.

2. Some observations of the growth of a stock are given in the following table:

   | Week | 1 | 2 | 3 |
   |---|---|---|---|
   | Value ($) | 1.22 | 1.49 | 2.45 |

   a) Use polynomial interpolation to estimate the value of the stock in weeks 4 and 10.
   b)[†] The value turned out to be $3.75 in week 4; now use polynomial interpolation again to estimate the value of the stock in weeks 5 and 10.
   c) Argue whether this method provides valuable information or not.

3. Consider a ski jump starting at a height of 40 m, finishing at a height of 5 m, covering from start to finish a horizontal distance of 35 m.
   a) Using polynomial interpolation, design the jump so that, 30 horizontal metres from the top, the skier is 7.5 m high.
   b) If you know a bit of calculus, design the jump so that the angle at takeoff is 0° with the horizontal.

   c) Specify now that the jump forms a straight line descent for the first 25 horizontal and 20 vertical metres. Design the jump so that the line segment and the remaining curve have the same first derivative (tangent) at the point of intersection.
   d) Do as in c), but require also that the second derivatives also match at the point of intersection.

4. Let $A$ denote the Vandermonde matrix corresponding to the numbers $x_1, x_2, \cdots, x_n$. In the proof of Theorem 1 it was shown that $A$ is invertible if the $x_i$ are distinct. Show, conversely, that if $A$ is invertible, then the $x_i$ must be distinct.

5.[†] *Cubic Spline* In some applications, a better choice of a curve through several points is obtained by joining cubic (i.e., degree 3) polynomials through consecutive points with matching first and second derivatives at the given points.

So a *cubic spline* through data points $(0, 1), (1, 3)$, and $(3, 2)$ for example, is formed by 2 cubic polynomials of the form

$$s_1 = a_1 x^3 + b_1 x^2 + c_1 x + d_1 \text{ and } s_2 = a_2 x^3 + b_2 x^2 + c_2 x + d_2$$

satisfying the 6 equations (in the 8 unknowns $a_1, a_2, b_1, b_2, c_1, c_2, d_1, d_2$):

$$s_1(0) = 1, \ s_1(1) = 3, \ s_2(1) = 3, \ s_2(3) = 2,$$
$$s_1'(1) = s_2'(1), \ s_1''(1) = s_2''(1)$$

[†] Requires computer assistance (e.g., Matlab).

To uniquely determine the two cubics, we also require some conditions at the endpoints; for example, a vanishing second derivative:

$$s_1''(0) = 0 \text{ and } s_2''(2) = 0$$

**a)** Solve the above system of 8 equations in 8 unknowns to find the two required cubic polynomials.

**b)** Find the cubic spline through the data points of Example 1, and use it to estimate the deflection of the beam under a load of 2200 kg. Compare with the results of polynomial interpolation and argue why there is a difference.

## 2.8　SYSTEMS OF DIFFERENTIAL EQUATIONS

Many problems in Science and Engineering come down to solving differential equations. In this brief section we sketch how diagonalization can be used to solve certain systems of first order, ordinary differential equations. Of course some familiarity with calculus is required.

Recall that a real valued function $f$ is a rule assigning a uniquely determined real number $f(x)$ to every real number $x$. Sometimes $f(x)$ is specified by a formula, for example $f(x) = 2x^2 + 3$ for all $x$. But more elaborate functions also arise such as the exponential function $f(x) = e^x$, or the cosine function $f(x) = \cos x$. Note that two functions $f$ and $g$ are *equal* if they have the same effect on every real number $x$, that is if $f(x) = g(x)$ holds for all real $x$. The *sum* $f + g$ and the *scalar product* $af$ (where $a$ is a number) are new functions defined by

$$(f + g)(x) = f(x) + g(x) \quad \text{for all } x,$$
$$af(x) = af(x) \quad \text{for all } x.$$

These are called **pointwise** addition and scalar multiplication, respectively; they are the usual operations that occur in calculus and algebra.

A function $f$ is called **differentiable** if its derivative $f'$ exists; we refer the reader to calculus texts for the precise definition of the derivative. All we need is the fact that, if $f$ and $g$ are both differentiable, so also are $f + g$ and $af$ (for any constant $a$), and the derivatives are given by

$$(f + g)' = f' + g' \quad \text{and} \quad (af)' = af'.$$

These facts are proved in every calculus text.

### 2.8.1 The Diagonalization Method

Let $f$ denote a differentiable function of a real variable $x$. The simplest first order[22] differential equation is

$$f' = af \text{ where } a \text{ is a constant.}$$

The chain rule of differentiation shows that $f(x) = e^{ax}$ satisfies this equation, and some elementary calculus shows that every solution is a multiple of $e^{ax}$.

**LEMMA ❶**　Every solution of the differential equation $f' = af$ is given by

$$f(x) = ce^{ax} \text{ for some constant } c.$$

**PROOF**　If $f$ is any solution, define a function $g$ by $g(x) = f(x)e^{-ax}$. The product rule of differentiation shows that $g'(x) = f'(x)e^{-ax} - af(x)e^{-ax} = 0$, so $g(x) = c$ is a constant function. Hence $f(x) = ce^{ax}$.

---

[22] A *first order* differential equation is one involving only the function and its derivative, and no second or higher derivatives occur.

The problem we want to address here is how to use Lemma 1 to find all solutions of a *system* of first order differential equations. More precisely, if $f_1, f_2, \cdots, f_n$, are differentiable functions, we want to solve the system

$$
\begin{aligned}
f'_1 &= a_{11}f_1 + a_{12}f_2 + \cdots + a_{1n}f_n \\
f'_2 &= a_{21}f_1 + a_{22}f_2 + \cdots + a_{2n}f_n \\
&\;\;\vdots \qquad\qquad\qquad \vdots \\
f'_n &= a_{n1}f_1 + a_{n2}f_2 + \cdots + a_{nn}f_n
\end{aligned}
\tag{$*$}
$$

where the $a_{ij}$ are constants. That is we want to describe all differentiable functions $f_1, f_2, \cdots, f_n$ that satisfy these equations. This system can easily be written in matrix form. Let

$$
\mathbf{f} = \begin{bmatrix} f_1 \\ f_2 \\ \vdots \\ f_n \end{bmatrix} \quad
\mathbf{f}' = \begin{bmatrix} f'_1 \\ f'_2 \\ \vdots \\ f'_n \end{bmatrix} \quad \text{and} \quad
A = \begin{bmatrix} a_{11} & a_{12} & \cdots & a_{1n} \\ a_{21} & a_{22} & \cdots & a_{2n} \\ \vdots & \vdots & & \vdots \\ a_{n1} & a_{n2} & \cdots & a_{nn} \end{bmatrix}.
$$

Then the system $(*)$ of differential equations becomes the single matrix equation

$$
\mathbf{f}' = A\mathbf{f}.
\tag{$**$}
$$

Given the matrix $A$ we ask for all columns $\mathbf{f}$ of functions that satisfy $(**)$. A variation on the technique used on dynamical systems will solve the problem if $A$ is a diagonalizable matrix. We illustrate the method with an example where $n = 2$.

**Example 1**

Find all solutions to the system

$$
\begin{aligned}
f'_1 &= 2f_1 - f_2 \\
f'_2 &= 6f_1 - 5f_2
\end{aligned}.
$$

Then find a solution that satisfies $f_1(0) = 2$ and $f_2(0) = 3$.

**SOLUTION**

This system has the form $\mathbf{f}' = A\mathbf{f}$ where $A = \begin{bmatrix} 2 & -1 \\ 6 & -5 \end{bmatrix}$ and $\mathbf{f} = \begin{bmatrix} f_1 \\ f_2 \end{bmatrix}$. The characteristic polynomial of $A$ is $c_A(x) = (x - 1)(x + 4)$, so the eigenvalues are $\lambda_1 = 1$ and $\lambda_2 = -4$, with corresponding eigenvectors $X_1 = \begin{bmatrix} 1 \\ 1 \end{bmatrix}$ and $X_2 = \begin{bmatrix} 1 \\ 6 \end{bmatrix}$ as is easily verified. Hence the matrix $P = [X_1 \; X_2] = \begin{bmatrix} 1 & 1 \\ 1 & 6 \end{bmatrix}$ diagonalizes $A$, that is

$$
P^{-1}AP = D = diag(\lambda_1, \; \lambda_2) = \begin{bmatrix} \lambda_1 & 0 \\ 0 & \lambda_2 \end{bmatrix}.
$$

The key to our method is to consider new functions $\mathbf{g} = \begin{bmatrix} g_1 \\ g_2 \end{bmatrix}$ given by

$$
\mathbf{f} = P\mathbf{g}, \text{ equivalently } \mathbf{g} = P^{-1}\mathbf{f}.
$$

In other words,

$$
\begin{bmatrix} f_1 \\ f_2 \end{bmatrix} = \begin{bmatrix} 1 & 1 \\ 1 & 6 \end{bmatrix}\begin{bmatrix} g_1 \\ g_2 \end{bmatrix}, \text{ that is } \begin{cases} f_1 = g_1 + g_2 \\ f_2 = g_1 + 6g_2 \end{cases}.
$$

It follows that $f'_1 = g'_1 + g'_2$ and $f'_2 = g'_1 + 6g'_2$, and in matrix form this is

$$
\mathbf{f}' = \begin{bmatrix} f'_1 \\ f'_2 \end{bmatrix} = \begin{bmatrix} g'_1 + g'_2 \\ g'_1 + 6g'_2 \end{bmatrix} = \begin{bmatrix} 1 & 1 \\ 1 & 6 \end{bmatrix}\begin{bmatrix} g'_1 \\ g'_2 \end{bmatrix} = P\mathbf{g}'.
$$

Thus the system $\mathbf{f}' = A\mathbf{f}$ becomes $P\mathbf{g}' = A(P\mathbf{g})$, that is $\mathbf{g}' = P^{-1}AP\mathbf{g} = D\mathbf{g}$. In matrix form this is

$$\begin{bmatrix} g_1' \\ g_2' \end{bmatrix} = \begin{bmatrix} 1 & 0 \\ 0 & -4 \end{bmatrix} \begin{bmatrix} g_1 \\ g_2 \end{bmatrix} \quad \text{or} \quad \begin{matrix} g_1' = g_1 \\ g_2' = -4g_2 \end{matrix}.$$

By Lemma 1, the only solutions to these equations are $g_1 = c_1 e^x$ and $g_2 = c_2 e^{-4x}$ where $c_1$ and $c_2$ are constants. Since $f_1 = g_1 + g_2$ and $f_2 = g_1 + 6g_2$, this gives all solutions to the system $\mathbf{f}' = A\mathbf{f}$:

$$\begin{matrix} f_1(x) = c_1 e^x + c_2 e^{-4x} \\ f_2(x) = c_1 e^x + 6c_2 e^{-4x} \end{matrix} \quad \text{where } c_1 \text{ and } c_2 \text{ are arbitrary constants.}$$

The matrix form of these equations reveals the role of the eigenvectors:

$$\begin{bmatrix} f_1(x) \\ f_2(x) \end{bmatrix} = \begin{bmatrix} c_1 e^x + c_2 e^{-4x} \\ c_1 e^x + 6c_2 e^{-4x} \end{bmatrix}$$

$$= c_1 \begin{bmatrix} 1 \\ 1 \end{bmatrix} e^x + c_2 \begin{bmatrix} 1 \\ 6 \end{bmatrix} e^{-4x}$$

$$= c_1 X_1 e^x + c_2 X_2 e^{-4x}.$$

If we write $\mathbf{f}(x) = \begin{bmatrix} f_1(x) \\ f_2(x) \end{bmatrix}$, this takes the compact form

$$\mathbf{f}(x) = c_1 X_1 e^x + c_2 X_2 e^{-4x}.$$

Finally, if we insist that $f_1(0) = 2$ and $f_2(0) = -3$, we see that $c_1 = 3$ and $c_2 = -1$. Hence $f_1(x) = 3e^x - e^{-4x}$ and $f_2(x) = 3e^x - 6e^{-4x}$ is a solution satisfying these additional requirements.

The technique in Example 1 works in general provided the matrix of coefficients is diagonalizable. For convenience, given a column $\mathbf{f} = \begin{bmatrix} f_1 \\ f_2 \\ \vdots \\ f_n \end{bmatrix}$ of functions, we write

$$\mathbf{f}(x) = \begin{bmatrix} f_1(x) \\ f_2(x) \\ \vdots \\ f_n(x) \end{bmatrix} \quad \text{for all real } x.$$

**THEOREM 1**

Consider a system $\mathbf{f}' = A\mathbf{f}$ of first order differential equations where $\mathbf{f}$ is a column of differentiable functions and $A$ is a diagonalizable matrix. If $\lambda_1,\ \lambda_2,\ \cdots\ \lambda_n$ are the eigenvalues of $A$, with corresponding eigenvectors $X_1,\ X_2,\ \cdots\ X_n$, then every solution to $\mathbf{f}' = A\mathbf{f}$ is given in the form

$$\mathbf{f}(x) = c_1 X_1 e^{\lambda_1 x} + c_2 X_2 e^{\lambda_2 x} + \cdots + c_n X_n e^{\lambda_n x}$$

where the $c_i$ are arbitrary constants.

**PROOF**

If $P = [X_1\ X_2\ \cdots\ X_n]$, then $P^{-1}AP = D = diag(\lambda_1,\ \lambda_2,\ \cdots,\ \lambda_n)$ is diagonal. Define a new column of functions $\mathbf{g}$ by $\mathbf{g} = P^{-1}\mathbf{f}$, equivalently $\mathbf{f} = P\mathbf{g}$. Write $\mathbf{f} = [f_1\ f_2\ \cdots\ f_n]^T$ and $P = [p_{ij}]$. For each $i = 1, 2, \cdots, n$ the equation $\mathbf{f} = P\mathbf{g}$ gives

$$f_i = p_{i1}g_1 + p_{i2}g_2 + \cdots + p_{in}g_n, \quad \text{whence} \quad f_i' = p_{i1}g_1' + p_{i2}g_2' + \cdots + p_{in}g_n'.$$

It follows that $\mathbf{f}' = P\mathbf{g}'$. When this is substituted in the matrix equation $\mathbf{f}' = A\mathbf{f}$, the result is $P\mathbf{g}' = AP\mathbf{g}$, whence $\mathbf{g}' = P^{-1}AP\mathbf{g} = D\mathbf{g}$. If we write $\mathbf{g} = [g_1\ g_2\ \cdots\ g_n]^T$, comparing entries gives

$$g_i' = \lambda_i g_i \text{ for each } i, \text{ so that } g_i(x) = c_i e^{\lambda_i x} \text{ for all } x$$

by Lemma 1. Using Theorem 3 §1.4, it follows that

$$\mathbf{f}(x) = P\mathbf{g}(x) = \begin{bmatrix} X_1 & X_2 & \cdots & X_n \end{bmatrix} \begin{bmatrix} g_1(x) \\ g_2(x) \\ \vdots \\ g_n(x) \end{bmatrix}$$

$$= g_1(x)X_1 + g_2(x)X_2 + \cdots + g_n(x)X_n$$

$$= c_1 X_1 e^{\lambda_1 x} + c_2 X_2 e^{\lambda_2 x} + \cdots + c_n X_n e^{\lambda_n x}.$$

This completes the proof.

For each $i$, the individual function $f_i(x)$ can be read off from the $i^{\text{th}}$ component of $\mathbf{f}(x)$. Here is an example.

**Example 2**   Find all solutions to the system:

$$f_1' = f_2 + f_3$$
$$f_2' = f_1 + f_3$$
$$f_3' = f_1 + f_2$$

Then find the solution such that $f_1(0) = -1$ and $f_2(0) = f_3(0) = 1$.

**SOLUTION**   This system has the form $\mathbf{f}' = A\mathbf{f}$ where $A = \begin{bmatrix} 0 & 1 & 1 \\ 1 & 0 & 1 \\ 1 & 1 & 0 \end{bmatrix}$. This matrix is diagonalizable;

it was analyzed in Example 6 §2.3, where the eigenvalues were found to be $\lambda_1 = 2$ and $\lambda_2 = \lambda_3 = -1$, with corresponding eigenvectors $X_1 = [1\ 1\ 1]^T$, $X_2 = [-1\ 1\ 0]^T$, and $X_3 = [-1\ 0\ 1]^T$. Hence Theorem 1 applies and every solution has the form

$$\mathbf{f}(x) = c_1 \begin{bmatrix} 1 \\ 1 \\ 1 \end{bmatrix} e^{2x} + c_2 \begin{bmatrix} -1 \\ 1 \\ 0 \end{bmatrix} e^{-x} + c_3 \begin{bmatrix} -1 \\ 0 \\ 1 \end{bmatrix} e^{-x}$$

for some choice of $c_1$, $c_2$, and $c_3$. Hence

$$f_1(x) = c_1 e^{2x} - c_2 e^{-x} - c_3 e^{-x}$$
$$f_2(x) = c_1 e^{2x} + c_2 e^{-x}$$
$$f_3(x) = c_1 e^{2x} + c_3 e^{-x}$$

If we require that $f_1(0) = -1$ and $f_2(0) = f_3(0) = 1$, then $c_1 = \frac{1}{3}$ and $c_2 = c_3 = \frac{2}{3}$.

We conclude with a comment about differential equations of higher order. The second derivative of a function $f$ is defined by $f'' = (f')'$. Consider the second order differential equation

$$f'' - af' - bf = 0.$$

If we put $f_1 = f$, and then put $f_2 = f' - af$, then the given equation reads $f_2' - bf_1 = 0$. Hence these functions satisfy the following system of first order equations:

$$f_1' = af_1 + f_2$$
$$f_2' = bf_1$$

Hence solving this system is equivalent to solving $f'' - af' - bf = 0$. Similarly, an $n^{\text{th}}$ order differential equation (one involving differentiating $f$ as many as $n$ times) leads to a system of $n$ first order equations.

## Exercises 2.8

1. In each case find all solutions to the system of differential equations, and find a particular solution that satisfies the given conditions.

   a) $f'_1 = 2f_1 + 4f_2$     $f_1(0) = 0$
   $f'_2 = 3f_1 + 3f_2$     $f_2(0) = 1$

   b) $f'_1 = -f_1 + 5f_2$     $f_1(0) = 1$
   $f'_2 = f_1 + 3f_2$     $f_2(0) = -1$

   c) $f'_1 = 4f_2 + 4f_3$     $f_1(0) = f_2(0) = f_3(0) = 1$
   $f'_2 = f_1 + f_2 - 2f_3$
   $f'_3 = -f_1 + f_2 + 4f_3$

   d) $f'_1 = 2f_1 + f_2 + 2f_3$     $f_1(0) = f_2(0) = f_3(0) = 1$
   $f'_2 = 2f_1 + 2f_2 - 2f_3$
   $f'_3 = 3f_1 + f_2 + f_3$

2. Set up a linear system of three first order differential equations that solves the single third order equation $f''' - af'' - bf' - cf = 0$.

# CHAPTER 3

# Vector Geometry

## 3.1 GEOMETRIC VECTORS

The word geometry in Greek means *measurement of the earth*, and the practical uses of geometry go back to antiquity. In ancient Greece all of mathematics was regarded as geometry, but in this chapter we will be primarily concerned with lines and planes in space. Our approach will be to regard points as column matrices (called vectors in this context) and then use matrix algebra to simplify calculations.

### 3.1.1 Coordinate Systems

Figure 3.1

Figure 3.2

The Greeks practised **synthetic geometry**, which means dealing with geometric figures without using a coordinate system. The use of coordinates was initiated by René Descartes[1] in 1637, and enabled him to use algebraic equations to describe geometric figures. This method, called **analytic geometry**, allows the computational power of arithmetic to be used to simplify many geometric calculations, and was fundamental to the development of analysis (calculus) in the 18th century. On the other hand, the method clarifies the study of equations by applying the intuitive, conceptual power of geometry to their graphs.

In the plane, coordinates are introduced as follows: Choose a point $O$ called the **origin**, choose two perpendicular lines through $O$ called the **X-axis**, and the **Y-axis**, and introduce a number scale on each axis with $O$ at the origin. Then each point $P$ determines a unique pair $(x, y)$ of numbers called the **coordinates** as shown in Figure 3.1. We write $P = P(x, y)$ when the coordinates of $P$ are to be emphasized.

Coordinates in space are introduced in the same way except that *three* mutually perpendicular axes through the origin $O$ are chosen as in Figure 3.2 (called the **X-axis**, the **Y-axis** and the **Z-axis**). Then a point $P$ determines a unique triple $(x, y, z)$ of numbers (the **coordinates** of $P$), and we denote $P$ as $P(x, y, z)$.

### 3.1.2 Vectors and Scalars

We are all familiar with quantities that are completely specified by a single number. Examples are time, area, mass, temperature, and pressure. These are called **scalar** quantities. However some quantities are not scalars. For example, the velocity of an airplane is not completely described by saying that the airplane is going 400 kilometres per hour; its *direction* must be given too, say *east* at 400 km/h. A quantity which requires both a **magnitude** and a **direction** to specify it is called a **vector** quantity. Thus the vector velocity of our airplane has magnitude 400 km/h (usually called the *speed*) and it has

---

[1] René Descartes (1596–1650) was a French philosopher and mathematician. His geometrical methods were first published in his *La géometrié*, Paris, 1637.

Figure 3.3

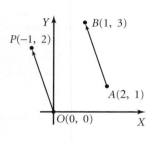

Figure 3.4

easterly direction. In order to say that two airplanes have the same velocity, both the speeds and the directions must be the same. In general:

> *Two vector quantities are the same if and only if they have the same magnitude and the same direction.*

In this chapter we will be concerned only with vectors arising in geometry. If $A$ and $B$ are two geometrical points, the directed line segment from $A$ to $B$ is called the **geometric vector** (or simply the **vector**) from $A$ to $B$, and is denoted $\overrightarrow{AB}$. As the name implies, this is a vector quantity with direction from $A$ to $B$ and magnitude the distance between $A$ and $B$. The vector $\overrightarrow{AB}$ is represented by an arrow from $A$ to $B$ as shown in Figure 3.3. The point $A$ is called the **tail** of the vector $\overrightarrow{AB}$, and $B$ is called the **tip** of $\overrightarrow{AB}$. The magnitude of a geometric vector $\overrightarrow{AB}$ is called its **length** and is denoted $\|\overrightarrow{AB}\|$.

Throughout this chapter, vectors will arise in both two and three dimensions. We will usually discuss the three-dimensional case, but the two situations are analogous.

Two vectors can be the same even though the tips and tails are different. For example $\overrightarrow{AB} = \overrightarrow{OP}$ in Figure 3.4 because they have the same length and the same direction (they both proceed one unit to the left and two units up).[2] Thus the same vector can be translated from one position to another; what is important is that the length and direction remain the same, and not where the tips and tails are located. For this reason, we shall often denote vectors as $\vec{u}, \vec{v}, \vec{w}$, etc. which make no reference to the tips or tails.

<h3>3.1.3 Matrix Form</h3>

Figure 3.5

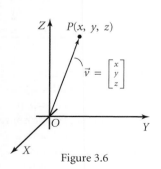

Figure 3.6

Thus the same vector $\vec{v}$ can have any point as its tail, as illustrated in Figure 3.5. In particular, $\vec{v}$ is said to be in **standard position** if its tail is at the origin $O$. Then $\vec{v} = \overrightarrow{OP}$ as in Figure 3.6 where $P(x, y, z)$ is the uniquely determined point at its tip, and we write $\vec{v}$ as the $3 \times 1$ matrix[3]

$$\vec{v} = \begin{bmatrix} x \\ y \\ z \end{bmatrix} = [x \ y \ z]^T.$$

In this case we say that $\vec{v}$ is written[4] in **matrix form**, that $x, y,$ and $z$ are the **components** of $\vec{v}$, and that $\vec{v}$ is the **position vector** of the point $P = P(x, y, z)$. Thus every vector $\vec{v}$ is the position vector of a (uniquely determined) point $P$, and the coordinates of $P$ are the same as the components of $\vec{v}$. Furthermore, every point $P$ *has* a position vector $\vec{v} = \overrightarrow{OP}$.

In the same way, a two-dimensional vector $\vec{v}$ has matrix form $\vec{v} = \begin{bmatrix} x \\ y \end{bmatrix}$ where $\vec{v}$ is the position vector of $P = P(x, y)$. Thus, for example, the vector $\vec{v} = \overrightarrow{AB} = \overrightarrow{OP}$ in Figure 3.4 has matrix form $\vec{v} = \begin{bmatrix} -1 \\ 2 \end{bmatrix}$.

Because we can write vectors as matrices, we can speak of sums and scalar multiples of vectors. This yields a very powerful tool in geometry because these matrix operations have natural geometric meanings. In particular, geometric calculations can be performed easily using matrix algebra. This method is called **vector geometry**, and its study is the main theme of this chapter.

---

[2] Fractions provide another example where quantities can be the same but *look* different: $\frac{6}{9}$ and $\frac{14}{21}$ certainly look different, but they are equal fractions (in fact both equal $\frac{2}{3}$).

[3] Since a vector $\vec{v}$ is a matrix, it should be denoted (as in Section 1.1) by an upper case letter, say $\vec{v} = V$. However the notation $\vec{v}$ for vectors is standard, and it is consistent with the notation $\overrightarrow{AB}$.

[4] The more compact notation $[x \ y \ z]^T$ is used for convenience of printing.

Let $\vec{v} = \begin{bmatrix} x \\ y \\ z \end{bmatrix} = [x \ \ y \ \ z]^T$ and $\vec{w} = \begin{bmatrix} x_1 \\ y_1 \\ z_1 \end{bmatrix} = [x_1 \ \ y_1 \ \ z_1]^T$ be vectors

and let $a$ be a scalar. We define the **zero vector** $\vec{0}$, the **negative** $-\vec{v}$, the **sum** $\vec{v} + \vec{w}$ and the **scalar product** $a\vec{v}$ as the corresponding matrices:

$$\vec{0} = \begin{bmatrix} 0 \\ 0 \\ 0 \end{bmatrix}, \qquad -\vec{v} = \begin{bmatrix} -x \\ -y \\ -z \end{bmatrix}, \qquad \vec{v} + \vec{w} = \begin{bmatrix} x+x_1 \\ y+y_1 \\ z+z_1 \end{bmatrix}, \qquad a\vec{v} = \begin{bmatrix} ax \\ ay \\ az \end{bmatrix}$$

Because they are matrix operations, vector addition and scalar multiplication obey the basic rules of matrix arithmetic (Theorems 1 and 2, Section 1.1). We repeat them here for convenience: If $\vec{u}$, $\vec{v}$, and $\vec{w}$ are (two- or) three-dimensional vectors, and if $a$ and $b$ are scalars, we have:

$$\vec{u} + \vec{v} = \vec{v} + \vec{u} \qquad\qquad (a+b)\vec{v} = a\vec{v} + b\vec{v}$$
$$\vec{u} + (\vec{v} + \vec{w}) = (\vec{u} + \vec{v}) + \vec{w} \qquad a(\vec{u} + \vec{v}) = a\vec{u} + a\vec{v}$$
$$\vec{0} + \vec{v} = \vec{v} \qquad\qquad\qquad a(b\vec{v}) = (ab)\vec{v}$$
$$\vec{v} + (-\vec{v}) = \vec{0} \qquad\qquad\qquad 1\vec{v} = \vec{v}$$

We will use these rules without comment below. They are useful for doing geometric calculations because, as mentioned above, these matrix operations have simple geometric interpretations. In fact the point of view in vector geometry is this: If it is desired to find a point $P(x, y, z)$, find its position vector $\vec{p} = [x \ \ y \ \ z]^T$ instead. This allows us to bring the full power of matrix algebra to bear on the geometric problem. Hence, the rest of this section is devoted to clarifying what these matrix operations mean geometrically.

## 3.1.4 Geometric Descriptions

We are going to describe sums and scalar products of vectors solely in terms of the lengths and directions of the vectors involved. Such descriptions are called **intrinsic** because they make no reference to the coordinate system, a fact which is very important in applications (they were, in fact, the original definitions for vector operations in physics). In addition, we provide other useful translations between geometric and matrix language.

We begin with intrinsic descriptions of vector equality, the zero vector, and the negative of a vector. These come from the following theorem. The theorem also contains two important results about the magnitude of a vector $\vec{v}$, not the least of which is a formula in part (2) for calculating the magnitude $\|\vec{v}\|$ from the components of $\vec{v}$. Recall that the absolute value of a number $a$ is denoted $|a|$.

**THEOREM 1**    Let $\vec{v} = \begin{bmatrix} x \\ y \\ z \end{bmatrix}$ and $\vec{w} = \begin{bmatrix} x_1 \\ y_1 \\ z_1 \end{bmatrix}$ be vectors.

(1) $\vec{v} = \vec{w}$ as vectors if and only if $x = x_1$, $y = y_1$, and $z = z_1$.
(2) $\|\vec{v}\| = \sqrt{x^2 + y^2 + z^2}$.[5]
(3) $\vec{v} = \vec{0}$ if and only if $\|\vec{v}\| = 0$.
(4) $\|a\vec{v}\| = |a| \, \|\vec{v}\|$ for any scalar $a$.

---

[5] Throughout this book $\sqrt{a}$ means the *positive* square root of $a$.

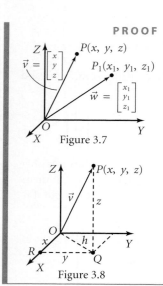

Figure 3.7

Figure 3.8

**PROOF**

**(1)** If $\vec{v}$ and $\vec{w}$ are positioned with their tails at the origin as in Figure 3.7, the tips of $\vec{v}$ and $\vec{w}$ are the points $P(x,\ y,\ z)$ and $P_1(x_1,\ y_1,\ z_1)$ respectively. Then it is clear from Figure 3.7 that $\vec{v} = \vec{w}$ (same length and direction) if and only if $P$ and $P_1$ are equal points, that is if and only if $x = x_1$, $y = y_1$, and $z = z_1$.

**(2)** We have $\vec{v} = \overrightarrow{OP}$ as in Figure 3.8, where $P = P(x,\ y,\ z)$. Since $OQP$ is a right triangle, Pythagoras' theorem[6] gives $\|\vec{v}\|^2 = h^2 + z^2$, and it gives $x^2 + y^2 = h^2$ when applied to the right triangle $ORQ$. Then (2) follows by eliminating $h^2$.

**(3)** Since $\vec{0} = \overrightarrow{OO}$ is the position vector of the origin $O$, it is clear that $\|\vec{0}\| = 0$. Conversely, if $\|\vec{v}\| = 0$ for some vector $\vec{v} = [x\ \ y\ \ z]^T$, then $\sqrt{x^2 + y^2 + z^2} = 0$ by (2). Hence $x = y = z = 0$, and so $\vec{v} = \vec{0}$.

**(4)** If $\vec{v} = [x\ \ y\ \ z]^T$, then $a\vec{v} = [ax\ \ ay\ \ az]^T$ by definition, so (2) gives
$$\|a\vec{v}\|^2 = (ax)^2 + (ay)^2 + (az)^2 = a^2(x^2 + y^2 + z^2) = a^2\|\vec{v}\|^2.$$
Now (4) follows by taking positive square roots because $\sqrt{a^2} = |a|$.

**Example 1**    If $\vec{v} = \begin{bmatrix} 3 \\ -1 \\ 2 \end{bmatrix}$ and $\vec{w} = \begin{bmatrix} 4 \\ -3 \end{bmatrix}$, then $\|\vec{v}\| = \sqrt{3^2 + (-1)^2 + 2^2} = \sqrt{14}$ and

$\|\vec{w}\| = \sqrt{4^2 + (-3)^2} = \sqrt{25} = 5.$

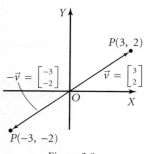

Figure 3.9

Our definition of equality for vectors is intrinsic (same length and direction), and part (1) of Theorem 1 shows that this is consistent with matrix equality:

**Vector Equality.** Two vectors are equal as vectors (intrinsic) if and only if they are equal as matrices.

Part (3) of Theorem 1 shows that the zero vector has an intrinsic description:

**Zero Vector.** $\vec{0}$ is the only vector of length zero. Note that, since $\vec{0}$ has length zero it makes no sense to talk about the direction of the zero vector.

We can also give an intrinsic description of the negative of a vector $\vec{v}$. We have $-\vec{v} = (-1)\vec{v}$ so, by Theorem 1, $\|-\vec{v}\| = |-1|\ \|\vec{v}\| = \|\vec{v}\|$. Hence $\vec{v}$ and $-\vec{v}$ have the same length. Moreover, if $\vec{v} = \overrightarrow{OP}$ then $-\vec{v} = \overrightarrow{PO}$ (see Figure 3.9 for the 2-dimensional case) so $-\vec{v}$ has the opposite direction to $\vec{v}$. Hence:

**Negative of a Vector.** If $\vec{v}$ is a vector, then $-\vec{v}$ is the vector with the same length as $\vec{v}$ but opposite direction.

**3.1.5    Vector Addition**

Figure 3.10

If two vectors $\vec{v}$ and $\vec{w}$ are positioned with a common tail, they create a parallelogram,[7] called the parallelogram **determined** by $\vec{v}$ and $\vec{w}$. Two illustrations of this appear in Figure 3.10. Since this parallelogram is completely determined by the lengths and directions of $\vec{v}$ and $\vec{w}$, the following "law" provides an intrinsic description of the sum $\vec{v} + \vec{w}$.

---

[6] Pythagoras' theorem states that if $a$ and $b$ are the sides of a right-angled triangle with hypotenuse $h$, then $a^2 + b^2 = h^2$. A proof is given at the end of this section.

[7] A **parallelogram** is a four-sided figure with straight sides such that opposite sides are parallel and of equal length. In particular, a rectangle is a parallelogram in which all the interior angles are right angles.

Figure 3.11

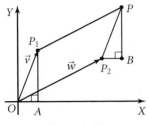

Figure 3.12

**Parallelogram Law of Vector Addition**. In the parallelogram determined by two vectors $\vec{v}$ and $\vec{w}$, the vector $\vec{v} + \vec{w}$ is the diagonal with the same tail as $\vec{v}$ and $\vec{w}$ (see Figure 3.11).

We may assume that $\vec{v}$ and $\vec{w}$ are in standard position because the parallelogram they determine is the same for any choice of common tail. We verify the parallelogram law in the two-dimensional case; the three-dimensional case is analogous. Hence write

$$\vec{v} = \begin{bmatrix} x_1 \\ y_1 \end{bmatrix} \text{ and } \vec{w} = \begin{bmatrix} x_2 \\ y_2 \end{bmatrix}, \text{ so that } \vec{v} + \vec{w} = \begin{bmatrix} x_1 + x_2 \\ y_1 + y_2 \end{bmatrix}. \text{ Then the tips of } \vec{v}, \vec{w}, \text{ and } \vec{v} + \vec{w} \text{ are}$$

$P_1(x_1, y_1)$, $P_2(x_2, y_2)$ and $P(x_1 + x_2, y_1 + y_2)$ respectively in Figure 3.12, and we must verify that the figure $OP_1PP_2$ is a parallelogram. But this follows from the fact that the right triangles $OAP_1$ and $P_2BP$ in Figure 3.12 are identical. In fact, both have vertical and horizontal sides of length $|y_1|$ and $|x_1|$ respectively. (Figure 3.12 depicts the situation where $x_1$ and $y_1$ are positive.) This proves the parallelogram law.

We note in passing that the parallelogram law for vector addition can be shown by experiment to be valid for such nongeometrical vectors as velocity and force. Hence it is a powerful tool in physics.

The parallelogram law provides an intrinsic description of the sum $\vec{v} + \vec{w}$ by describing the length and direction of $\vec{v} + \vec{w}$ in terms of the parallelogram determined by $\vec{v}$ and $\vec{w}$. However, there is another way to look at vector addition, which is more convenient than the parallelogram law in most situations, and is very useful in applications. It is described as follows:

Figure 3.13

**Tip-to-Tail Method of Adding Vectors**. Given two vectors $\vec{v}$ and $\vec{w}$, place the tail of $\vec{w}$ at the tip of $\vec{v}$ as in Figure 3.13. Then $\vec{v} + \vec{w}$ is the vector from the tail of $\vec{v}$ to the tip of $\vec{w}$.

Figure 3.14

This can be described as the "first $\vec{v}$ then $\vec{w}$" method of computing $\vec{v} + \vec{w}$, and it really does give $\vec{v} + \vec{w}$ because the diagram in Figure 3.13 is the "top half" of the parallelogram determined by $\vec{v}$ and $\vec{w}$. Clearly these vectors could also be added as "first $\vec{w}$ then $\vec{v}$" to get $\vec{w} + \vec{v} = \vec{v} + \vec{w}$ as in Figure 3.14.

Figure 3.15

However the most important fact is that two (or more) vectors can be added graphically by placing them tip-to-tail in a sequence. This is illustrated for three vectors in Figure 3.15, and gives a very useful "picture" of the sum of several vectors as a "tip-to-tail" sequence.

**Example 2** Find the net effect of a 1-km walk northeast and a 1-km walk east.

**SOLUTION** Let $\vec{w}_1$ and $\vec{w}_2$ denote the northeast and east walks, respectively. If walk $\vec{w}_1$ is taken first, followed by $\vec{w}_2$, the result is $\vec{w}_1 + \vec{w}_2$ as in the first diagram in Figure 3.16. On the other hand, if $\vec{w}_2$ is taken first, the result is $\vec{w}_2 + \vec{w}_1$ as in the second diagram. The net result is the walk $\vec{w}_1 + \vec{w}_2 = \vec{w}_2 + \vec{w}_1$.

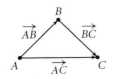

Figure 3.16

The tip-to-tail method of adding vectors takes an easily remembered form for geometric vectors when the $\overrightarrow{AB}$ notation is used. Indeed, if $A$, $B$, and $C$ are points, it asserts that

$$\overrightarrow{AB} + \overrightarrow{BC} = \overrightarrow{AC}$$

as is illustrated in Figure 3.17. Thus adding vectors in this notation becomes a matter of "following the letters". Here is an illustration of how this can be used in geometry.

Figure 3.17

**Example 3**    Show that the diagonals of any parallelogram bisect each other.

**SOLUTION**    Let $A$, $B$, $C$, and $D$ be the four vertices of the parallelogram in order, as in Figure 3.18, and let $M$ be the midpoint of the diagonal $AC$. We must show that $M$ is also the midpoint of the diagonal $BD$. To do this it suffices to show that $\overrightarrow{BM} = \overrightarrow{MD}$. Indeed, the fact that these vectors have the same direction shows that $M$ is *on* the diagonal $BD$; the fact that they have the same length shows that $M$ is the *midpoint* of $BD$.

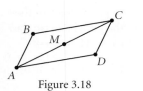

Figure 3.18

Now observe that $\overrightarrow{BA} = \overrightarrow{CD}$ because $ABCD$ is a parallelogram, and $\overrightarrow{AM} = \overrightarrow{MC}$ because $M$ is the midpoint of $AC$. Hence tip-to-tail addition (twice) gives

$$\overrightarrow{BM} = \overrightarrow{BA} + \overrightarrow{AM} = \overrightarrow{CD} + \overrightarrow{MC} = \overrightarrow{MC} + \overrightarrow{CD} = \overrightarrow{MD}.$$

This is what we wanted.

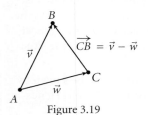

Figure 3.19

There is a simple geometric way to visualize the **difference** $\vec{v} - \vec{w}$ of two vectors $\vec{v}$ and $\vec{w}$. Position $\vec{v}$ and $\vec{w}$ with a common tail $A$ as in Figure 3.19, and let $B$ and $C$ be the respective tips. Then adding tip-to-tail gives $\vec{w} + \overrightarrow{CB} = \vec{v}$ so $\overrightarrow{CB} = \vec{v} - \vec{w}$. Hence we obtain an intrinsic description of vector subtraction.

**Vector Subtraction.** $\vec{v} - \vec{w}$ is the vector from the tip of $\vec{w}$ to the tip of $\vec{v}$.

Thus in the parallelogram determined by $\vec{v}$ and $\vec{w}$, both $\vec{v} - \vec{w}$ and $\vec{v} + \vec{w}$ appear as diagonals (Figure 3.20).

Figure 3.20

One of the reasons for the importance of vector subtraction is that it gives a simple way to find the vector from one point to another in matrix form. This appears in the following theorem together with a formula for the distance between two points.

**THEOREM 2**    Let $P_1(x_1, y_1, z_1)$ and $P_2(x_2, y_2, z_2)$ be two points. Then:

(1) $\overrightarrow{P_1P_2} = \begin{bmatrix} x_2 - x_1 \\ y_2 - y_1 \\ z_2 - z_1 \end{bmatrix}$.

(2) The distance between $P_1$ and $P_2$ is $\sqrt{(x_2 - x_1)^2 + (y_2 - y_1)^2 + (z_2 - z_1)^2}$.

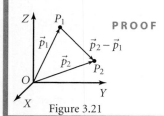

Figure 3.21

**PROOF**    (1) Write $\vec{p}_1 = \overrightarrow{OP_1}$ and $\vec{p}_2 = \overrightarrow{OP_2}$ as in Figure 3.21, so that $\vec{p}_1 = [x_1 \ y_1 \ z_1]^T$ and $\vec{p}_2 = [x_2 \ y_2 \ z_2]^T$. Then the above discussion of vector subtraction gives $\overrightarrow{P_1P_2} = \vec{p}_2 - \vec{p}_1 = [x_2 - x_1 \ y_2 - y_1 \ z_2 - z_1]^T$, as required.

(2) The distance is $\left\|\overrightarrow{P_1P_2}\right\| = \sqrt{(x_2 - x_1)^2 + (y_2 - y_1)^2 + (z_2 - z_1)^2}$ using (1) and Theorem 1.

Of course the two-dimensional version of Theorem 2 is also valid: If $P_1(x_1, y_1)$ and $P_2(x_2, y_2)$ are points, then $\overrightarrow{P_1P_2} = \begin{bmatrix} x_2 - x_1 \\ y_2 - y_1 \end{bmatrix}$, and the distance between $P_1$ and $P_2$ is $\left\|\overrightarrow{P_1P_2}\right\| = \sqrt{(x_2 - x_1)^2 + (y_2 - y_1)^2}$.

**Example 4**    Given points $P_1(3, -1, 2)$ and $P_2(1, 2, 0)$, the vector from $P_1$ to $P_2$ is

$$\overrightarrow{P_1P_2} = [1 - 3 \quad 2 - (-1) \quad 0 - 2]^T = [-2 \ 3 \ -2]^T$$

and the distance between $P_1$ and $P_2$ is $\left\|\overrightarrow{P_1P_2}\right\| = \sqrt{(-2)^2 + 3^2 + (-2)^2}$
$$= \sqrt{4 + 9 + 4} = \sqrt{17}.$$

## 3.1.6 Scalar Multiplication

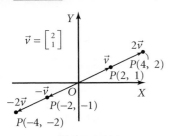

$\vec{v} = \begin{bmatrix} 2 \\ 1 \end{bmatrix}$

Figure 3.22

As for vector addition, we want to give an intrinsic description of the scalar product $a\vec{v}$ by describing its length and direction in terms of the scalar $a$ and the length and direction of the vector $\vec{v}$. The length $\|a\vec{v}\| = |a|\,\|\vec{v}\|$ of the vector $a\vec{v}$ has already been found in Theorem 1. To describe the direction of $a\vec{v}$, we need only deal with the case $a\vec{v} \neq \vec{0}$ because the zero vector $\vec{0}$ has no direction. Hence we restrict our attention to the case $a\vec{v} \neq \vec{0}$ in part (2) of the following result.

**Geometric Description of Scalar Multiplication.** If $a$ is a scalar and $\vec{v}$ is a vector, then:

(1) The length of $a\vec{v}$ is $\|a\vec{v}\| = |a|\,\|\vec{v}\|$.

(2) If $a\vec{v} \neq \vec{0}$, the direction of $a\vec{v}$ is $\begin{cases} \text{the same as } \vec{v} \text{ if } a > 0 \\ \text{opposite to } \vec{v} \text{ if } a < 0 \end{cases}$.

In fact, (1) is part (4) of Theorem 1. To verify (2) first note that each component of $a\vec{v}$ equals $a$ times the corresponding component of $\vec{v}$. If $a > 0$, this means that $a\vec{v}$ has the same direction as $\vec{v}$, as is illustrated in Figure 3.22 for the vector $\vec{v} = [2 \;\; 1]^T$. If $a < 0$, write $b = -a$. Then $b > 0$, so $b\vec{v}$ and $\vec{v}$ have the same direction (by what has just been proved). But $a\vec{v} = -(b\vec{v})$, so $a\vec{v}$ has direction opposite to $b\vec{v}$, and so opposite to $\vec{v}$.

This description of scalar multiplication is illustrated in Figure 3.23 where several scalar multiples of a vector $\vec{v}$ are displayed.

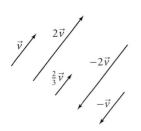

Figure 3.23

We give two examples illustrating how useful the above geometric description of scalar multiplication is in geometry. Recall that the position vector of a point $P(x, y, z)$ is the vector $\vec{p} = \overrightarrow{OP} = [x \;\; y \;\; z]^T$ from the origin to $P$.

**Example 5** Find the coordinates of the midpoint $M(x, y, z)$ between the points $P(3, -1, 5)$ and $Q(-1, 2, 3)$.

**SOLUTION** Let $\vec{p} = [3 \;\; -1 \;\; 5]^T$, $\vec{q} = [-1 \;\; 2 \;\; 3]^T$, and $\vec{m} = [x \;\; y \;\; z]^T$ be the position vectors of $P$, $Q$, and $M$ respectively, as in Figure 3.24. We find the point $M$ by computing its position vector $\vec{m}$. We have $\overrightarrow{PQ} = \vec{q} - \vec{p}$. Hence, as $M$ is the midpoint of $PQ$, we obtain $\overrightarrow{PM} = \frac{1}{2}\overrightarrow{PQ} = \frac{1}{2}(\vec{q} - \vec{p})$. Thus

$$\vec{m} = \overrightarrow{OM} = \vec{p} + \overrightarrow{PM} = \vec{p} + \tfrac{1}{2}(\vec{q} - \vec{p}) = \tfrac{1}{2}(\vec{p} + \vec{q}) = \tfrac{1}{2}\begin{bmatrix} 3-1 \\ -1+2 \\ 5+3 \end{bmatrix} = \begin{bmatrix} 1 \\ \frac{1}{2} \\ 4 \end{bmatrix}.$$

It follows that $M = M(1, \frac{1}{2}, 4)$ is the required midpoint.

Figure 3.24

The method in Example 5 works in general: The position vector $\vec{m}$ of the midpoint between *any* two points $P$ and $Q$ is the average $\vec{m} = \frac{1}{2}(\vec{p} + \vec{q})$ of the position vectors $\vec{p}$ and $\vec{q}$ of the two points.

**Example 6** A *quadrilateral* is a figure with straight sides and four vertices. Show that the midpoints of the sides of any quadrilateral are the vertices of a parallelogram.

**SOLUTION** Let the vertices of the quadrilateral be $A$, $B$, $C$, and $D$ in order, and let $E$, $F$, $G$, and $H$ be the midpoints as shown in Figure 3.25. It suffices to show that $\overrightarrow{EF} = \overrightarrow{HG}$ since then the corresponding sides are parallel (because $\overrightarrow{EF}$ and $\overrightarrow{HG}$ have the same direction) and of equal length (because $\|\overrightarrow{EF}\| = \|\overrightarrow{HG}\|$).

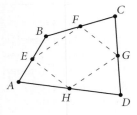

**Figure 3.25**

Now observe that $\overrightarrow{EB} = \frac{1}{2}\overrightarrow{AB}$ because $E$ is the midpoint of $AB$, and similarly $\overrightarrow{BF} = \frac{1}{2}\overrightarrow{BC}$. Hence tip-to-tail vector addition gives

$$\overrightarrow{EF} = \overrightarrow{EB} + \overrightarrow{BF} = \frac{1}{2}\overrightarrow{AB} + \frac{1}{2}\overrightarrow{BC} = \frac{1}{2}\left(\overrightarrow{AB} + \overrightarrow{BC}\right) = \frac{1}{2}\overrightarrow{AC}.$$

A similar argument shows that $\overrightarrow{HG} = \frac{1}{2}\overrightarrow{AC}$ too, so $\overrightarrow{EF} = \frac{1}{2}\overrightarrow{AC} = \overrightarrow{HG}$ as required.

The notion of parallel lines has been a central concept in geometry for centuries. We define it in vector terms as follows: Two nonzero vectors are called **parallel** if they have the same or opposite direction. We conclude this section by giving an important connection between parallel vectors and scalar multiplication.

**THEOREM 3**

The following conditions are equivalent for nonzero vectors $\vec{v}$ and $\vec{w}$:

(1) $\vec{v}$ and $\vec{w}$ are parallel.
(2) Each of $\vec{v}$ and $\vec{w}$ is a scalar multiple of the other.
(3) One of $\vec{v}$ and $\vec{w}$ is a scalar multiple of the other.

**PROOF**

(2)$\Rightarrow$(3) is obvious, and (3)$\Rightarrow$(1) by Part (2) of the above geometric definition of scalar multiplication. So we show that (1)$\Rightarrow$(2). Hence assume that $\vec{v}$ and $\vec{w}$ are parallel; there are two cases:

**Case 1.** *$\vec{v}$ and $\vec{w}$ have the same direction.* In this case we show that $\vec{v} = t\vec{w}$ where $t = \frac{\|\vec{v}\|}{\|\vec{w}\|}$. (Then $\vec{w} = \frac{1}{t}\vec{v}$.) Since $t$ is positive $\vec{v}$ and $t\vec{w}$ have the same direction (the common direction of $\vec{v}$ and $\vec{w}$). Moreover, Theorem 1 gives $\|t\vec{w}\| = |t|\,\|\vec{w}\| = t\,\|\vec{w}\| = \|\vec{v}\|$ so $\vec{v}$ and $t\vec{w}$ also have the same length. Hence $\vec{v} = t\vec{w}$ as required.

**Case 2.** *$\vec{v}$ and $\vec{w}$ have opposite directions.* In this case $\vec{v} = -t\vec{w}$ where $t = \frac{\|\vec{v}\|}{\|\vec{w}\|}$. The details (similar to Case 1) are left to the reader.

**Example 7**

Show that $\vec{v} = [7 \ \ -28 \ \ 35]^T$ and $\vec{w} = [-5 \ \ 20 \ \ -25]^T$ are parallel, while $\vec{u} = [2 \ \ -3 \ \ 6]^T$ and $\vec{z} = [1 \ \ -2 \ \ 3]^T$ are not parallel.

**SOLUTION**

We have $\vec{v} = -\frac{7}{5}\vec{w}$, so $\vec{v}$ and $\vec{w}$ are parallel (and have opposite direction). If $\vec{u}$ and $\vec{z}$ are parallel then each is a scalar multiple of the other, say $\vec{u} = t\vec{z}$ for some scalar $t$. Comparing components gives $2 = t$, $-3 = -2t$, and $6 = 3t$. Clearly, no such number $t$ exists. Hence $\vec{u}$ and $\vec{z}$ are not parallel.

## 3.1.7  A Proof of Pythagoras' Theorem

Pythagoras (c. 550 B.C.) founded a semireligious school whose motto was "Everything is number". Their studies were called *mathema* ("what is learned") and consisted of music, astronomy, geometry, and arithmetic. The Pythagorean theorem was known and used earlier, but Pythagoras is credited with the first logical, deductive proof of the result. Many proofs of the theorem are known; the one we give depends on the basic properties of similar triangles.

**THEOREM 4** **Pythagoras' Theorem.** Given a right-angled triangle with hypotenuse $h$ and sides $a$ and $b$, then

$$a^2 + b^2 = h^2.$$

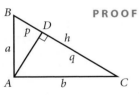

Figure 3.26

**PROOF** Suppose that the triangle has vertices $A$, $B$, and $C$ with the right angle at $A$, as in Figure 3.26. Draw a perpendicular from $A$ to the point $D$ on the hypotenuse and let $p$ and $q$ be the lengths of $BD$ and $DC$ respectively. Then $ABC$ and $DBA$ are similar triangles, so $\frac{p}{a} = \frac{a}{h}$. This means $a^2 = ph$. In the same way, the similarity of $ABC$ and $DAC$ gives $\frac{q}{b} = \frac{b}{h}$, whence $b^2 = qh$. But then

$$a^2 + b^2 = ph + qh = (p + q)h = h^2$$

because $p + q = h$. This proves Pythagoras' theorem.

## Exercises 3.1

**1.** Consider the points $P(6,\ 7)$ and $Q(-12,\ 3)$ in the 2-dimensional $XY$-plane.

  **a)** Locate these two points. Explain how they differ from the points $P'(7,\ 6)$ and $Q'(3,\ -12)$.

  **b)** Form the vector $\overrightarrow{PQ}$ from the point $P$ to the point $Q$. Explain how it differs from the vector $\overrightarrow{QP}$ from the point $Q$ to the point $P$.

  **c)** Translate the vector $\overrightarrow{PQ}$ to standard position, then give the coordinates of its terminal point, its components, and its matrix form.

  **d)** Find the length of the vector $\overrightarrow{PQ}$. What is the connection with the distance between the two points $P$ and $Q$?

  **e)** Repeat **c)** and **d)** with the new points $P(2,\ -1,\ 4)$ and $Q(-8,\ -7,\ 3)$ in 3-dimensional $XYZ$-space.

**2.** Consider the points $P(3,\ -2)$ and $Q(-1,\ 6)$ in the 2-dimensional $XY$-plane.

  **a)** Represent the position vectors $\vec{p}$ and $\vec{q}$ of $P$ and $Q$ in matrix form and calculate their lengths.

  **b)** Draw and compute the matrix forms and lengths of the vectors $\vec{p} + \vec{q}$, $\vec{p} - \vec{q}$, $3\vec{p}$, $-2\vec{q}$, and $3\vec{p} - 2\vec{q}$.

  **c)** Repeat the preceding questions with the new points $P(4,\ 1,\ -3)$ and $Q(-2,\ 4,\ -8)$ in 3-dimensional $XYZ$-space.

**3.** In each case compute $\overrightarrow{PQ}$ and $\|\overrightarrow{PQ}\|$.

  **a)** $P(2,\ -1,\ 3)$ and $Q(0,\ 1,\ 1)$

  **b)** $P(0,\ 2,\ -3)$ and $Q(4,\ -1,\ 2)$

  **c)** $P(2,\ 0,\ -5)$ and $Q(0,\ 5,\ 7)$

  **d)** $P(1,\ 2,\ 3)$ and $Q(3,\ 2,\ 1)$

**4. a)** Given the point $P(3,\ -2,\ 1)$, find the point $Q$ such that $\overrightarrow{PQ} = [-1\ \ 2\ \ 5]^T$.

  **b)** Given the point $Q(0,\ -1,\ 3)$, find the point $P$ such that $\overrightarrow{PQ} = [3\ \ 0\ \ -2]^T$.

**5.** Let $\vec{u} = [2\ \ -1\ \ 3]^T$, $\vec{v} = [0\ \ 2\ \ -3]^T$, and $\vec{w} = [-5\ \ 1\ \ 2]^T$. Compute each of the following.

  **a)** $3\vec{u} - 5\vec{v} + 7\vec{w}$

  **b)** $7\vec{u} + 2\vec{v} - 5\vec{w}$

  **c)** $\left\| 3\vec{u} - \frac{1}{2}\vec{v} + \vec{w} \right\|$

  **d)** $\left\| \frac{1}{2}(\vec{u} + \vec{v} + 2\vec{w}) \right\|$

**6. a)** Find a vector $\vec{u}$ of length 1 in the same direction as $\vec{v} = [3\ \ -1\ \ 2]^T$.

  **b)** If $\vec{v} \neq \vec{0}$, show that $\vec{u} = \frac{1}{\|\vec{v}\|}\vec{v}$ is a vector of length 1 in the same direction as $\vec{v}$.

**7.** In each case simplify the expression.

  **a)** $3(2\vec{w} - 2\vec{u} - \vec{v}) - 2(\vec{u} + \vec{w} - 5\vec{v}) + 2(4\vec{u} + 4\vec{v} - 2\vec{w})$

  **b)** $5(\vec{u} - \vec{v} - 2\vec{w}) + 3(2\vec{v} + 3\vec{w} - 3\vec{u}) - 2(3\vec{u} + \vec{v} + 8\vec{w})$

**8.** Let $\vec{u} = [-2\ \ 3\ \ -5]^T$ and $\vec{v} = [1\ \ -2\ \ 2]^T$. In each case find the vector $\vec{x}$.

  **a)** $3\vec{u} - 5\vec{x} = \frac{3}{2}(4\vec{x} - \vec{v})$

  **b)** $3\vec{x} - \|\vec{v}\|\vec{x} = 5(\vec{u} - 3\vec{x})$

  **c)** $2\vec{x} - 3\vec{v} = \|\vec{u}\|^2(\vec{u} + 4\vec{x})$

  **d)** $4(5\vec{x} - 2\vec{u}) = 3(5\vec{v} + 6\vec{x})$

**9.** Let $\vec{u} = [2\ \ -1\ \ 3]^T$, $\vec{v} = [0\ \ 2\ \ -3]^T$, and $\vec{w} = [6\ \ 1\ \ 3]^T$. In each case determine whether or not numbers $a$, $b$, and $c$ exist such that $\vec{x} = a\vec{u} + b\vec{v} + c\vec{w}$. Support your answer.

  **a)** $\vec{x} = [1\ \ 0\ \ 0]^T$

  **b)** $\vec{x} = [2\ \ -1\ \ 3]^T$

  **c)** $\vec{x} = [9\ \ 3\ \ 0]^T$

  **d)** $\vec{x} = [1\ \ 0\ \ 1]^T$

**10.** In each case determine if $\vec{v}$ and $\vec{w}$ are parallel. Support your answer.

  **a)** $\vec{u} = [10\ \ -15\ \ 25]^T$ and $\vec{v} = [14\ \ -21\ \ 35]^T$

  **b)** $\vec{u} = [3\ \ -2\ \ 1]^T$ and $\vec{v} = [-6\ \ 4\ \ -2]^T$

  **c)** $\vec{u} = [2\ \ 3\ \ -5]^T$ and $\vec{v} = [-8\ \ 12\ \ 20]^T$

  **d)** $\vec{u} = [-3\ \ 5\ \ 7]^T$ and $\vec{v} = [6\ \ -10\ \ 0]^T$

**11.** Find all vectors $\vec{u}$ that are parallel to $\vec{v} = [3\ \ 5\ \ -1]^T$ and satisfy $\|\vec{u}\| = 3\|\vec{v}\|$.

**12.** In each case find all points $Q$ at a distance 5 from $P$ and such that $\overrightarrow{PQ}$ is parallel to $\vec{d}$.

  **a)** $P = P(5,\ 0,\ -3)$, $\vec{d} = [1\ \ 2\ \ -2]^T$

  **b)** $P = P(1,\ 2,\ -5)$, $\vec{d} = [0\ \ -3\ \ 2]^T$

**13.** In each case $P$, $Q$, and $R$ are three vertices of a parallelogram $PQRS$. Find the other vertex $S$.

  **a)** $P(3, -1, -1), Q(1, -2, 0), R(1, -1, 2)$

  **b)** $P(2, 0, -1), Q(-2, 4, 1), R(3, -1, 0)$

**14.** In each case find the midpoint between the given points.

  **a)** $P(2, -3, 5), Q(-1, 7, -3)$

  **b)** $P(1, 2, 3), Q(-3, 5, -3)$

**15.** Given $P(1, 1, 3)$ and $Q(0, -3, 5)$ find:

  **a)** the point $A$ that is $\frac{1}{3}$ the way from $P$ to $Q$.

  **b)** the point $B$ that is $\frac{1}{4}$ the way from $Q$ to $P$.

**16.** Let $P$ and $Q$ be two points with position vectors $\vec{p}$ and $\vec{q}$ respectively. If $0 \leq r \leq 1$, show that the point $A$ that is fraction $r$ of the way from $P$ to $Q$ has position vector $\vec{a} = (1 - r)\vec{p} + r\vec{q}$.

**17.** In each case either show that the statement is true or give an example showing that it is false. Throughout, $\vec{u}$, $\vec{v}$, and $\vec{w}$ denote vectors.

  **a)** The zero vector $\vec{0}$ is the only vector of length 0.

  **b)** If $\|\vec{v} - \vec{w}\| = \vec{0}$, then $\vec{v} = \vec{w}$.

  **c)** If $\vec{v} = -\vec{v}$, then $\vec{v} = \vec{0}$.

  **d)** If $\|\vec{v}\| = \|\vec{w}\|$, then $\vec{v} = \vec{w}$.

  **e)** If $\|\vec{v}\| = \|\vec{w}\|$, then $\vec{v} = \pm\vec{w}$.

  **f)** If $\vec{w} = t\,\vec{v}$ for some scalar $t$, then $\vec{v}$ and $\vec{w}$ have the same direction.

  **g)** If $\vec{v}$ and $\vec{v} + \vec{w}$ are parallel, then $\vec{v}$ and $\vec{w}$ are parallel.

  **h)** $\|(-5)\vec{v}\| = -5\|\vec{v}\|$ for all $\vec{v}$.

  **i)** If $\|\vec{v}\| = \|2\vec{v}\|$, then $\vec{v} = \vec{0}$.

  **j)** $\|\vec{v} + \vec{w}\| = \|\vec{v}\| + \|\vec{w}\|$ for all $\vec{v}$ and $\vec{w}$.

**18.** Consider a triangle with vertices $A$, $B$, and $C$, and let $E$ and $F$ be the midpoints of sides $AB$ and $BC$.

  **a)** Explain why $\overrightarrow{EB} = \frac{1}{2}\overrightarrow{AB}$ and $\overrightarrow{BF} = \frac{1}{2}\overrightarrow{BC}$.

  **b)** Show that the line joining $E$ and $F$ is parallel to $AC$ and half as long.

**19.** Consider a triangle with vertices $A$, $B$, and $C$, where $A = A(1, 0, 0)$ and $\overrightarrow{AC} = [2 \ \ 0 \ \ 1]^T$, $\overrightarrow{BC}$ is parallel to the $Y$-axis and $\|\overrightarrow{AB}\| = 3$; find all possibilities for $B$.

**20.** Let $A$, $B$, $C$, $D$, $E$, and $F$ be the six vertices in order of a regular hexagon. Show that
$$\overrightarrow{AB} + \overrightarrow{AC} + \overrightarrow{AD} + \overrightarrow{AE} + \overrightarrow{AF} = 3\overrightarrow{AD}.$$

**21.** Let $P = P(x, y)$ be an arbitrary point in the plane with position vector $\vec{p} = [x \ \ y]^T$. Let $\mathcal{C}$ denote the circle with centre at the origin and radius $r > 0$.

  **a)** If $P$ is on the circle $\mathcal{C}$, explain geometrically why $\|\vec{p}\| = r$.

  **b)** If $\|\vec{p}\| = r$, explain geometrically why $P$ is on the circle $\mathcal{C}$.

  **c)** Use **a)** and **b)** to show that the equation of the circle $\mathcal{C}$ is $x^2 + y^2 = r^2$.

**21.** Let $P = P(x, y, z)$ be an arbitrary point in space with position vector $\vec{p} = [x \ \ y \ \ z]^T$. Let $\mathcal{S}$ denote the sphere with centre at the origin and radius $r > 0$.

  **a)** If $P$ is on the sphere $\mathcal{S}$, explain geometrically why $\|\vec{p}\| = r$.

  **b)** If $\|\vec{p}\| = r$, explain geometrically why $P$ is on the sphere $\mathcal{S}$.

  **c)** Use **a)** and **b)** to show that the equation of the sphere $\mathcal{S}$ is $x^2 + y^2 + z^2 = r^2$.

## 3.2   DOT PRODUCT AND PROJECTIONS

The concept of perpendicular lines is fundamental in geometry. It is required for the statement of Pythagoras' theorem and it is basic to the study of trigonometry. In order to use vectors to study perpendicularity, we develop a simple test for when two nonzero vectors have perpendicular directions. The test depends on whether a certain number (called the dot product of the two vectors) is zero or not. As is often the case with a good idea, the dot product has other applications: It can be used to find the angle between *any* two nonzero vectors, and it leads to a method of solving certain optimization problems in geometry. For example, we will be able (in Section 3.3) to find the point on a plane that is closest to an arbitrary point.

### 3.2.1   The Dot Product

Given two vectors $\vec{v}$ and $\vec{w}$, the fact that they are column matrices means that the matrix product $\vec{v}^T \vec{w}$ is a $1 \times 1$ matrix, which we regard as a number. This number is called the **dot product** $\vec{v} \cdot \vec{w}$ of $\vec{v}$ and $\vec{w}$. In terms of components, the definition is as follows:

$$\text{If } \vec{v} = \begin{bmatrix} x_1 \\ y_1 \\ z_1 \end{bmatrix} \text{ and } \vec{w} = \begin{bmatrix} x_2 \\ y_2 \\ z_2 \end{bmatrix}, \text{ then } \vec{v} \cdot \vec{w} = \vec{v}^T \vec{w} = x_1 x_2 + y_1 y_2 + z_1 z_2.$$

$$\text{If } \vec{v} = \begin{bmatrix} x_1 \\ y_1 \end{bmatrix} \text{ and } \vec{w} = \begin{bmatrix} x_2 \\ y_2 \end{bmatrix}, \text{ then } \vec{v} \cdot \vec{w} = \vec{v}^T \vec{w} = x_1 x_2 + y_1 y_2.$$

In words, the dot product of two vectors is formed by multiplying corresponding components and adding the results. Because $\vec{v} \cdot \vec{w}$ is a *number* (even though $\vec{v}$ and $\vec{w}$ are vectors), it is sometimes called the **scalar product** of $\vec{v}$ and $\vec{w}$.

**Example 1**   If $\vec{v} = \begin{bmatrix} 6 \\ -2 \\ 5 \end{bmatrix}$ and $\vec{w} = \begin{bmatrix} 2 \\ -1 \\ -3 \end{bmatrix}$, then $\vec{v} \cdot \vec{w} = 6 \cdot 2 + (-2)(-1) + 5(-3) = -1$.

The dot product $\vec{v} \cdot \vec{w}$ is the central concept in this section, and the following theorem collects several of its basic properties.

**THEOREM 1**   Let $\vec{u}$, $\vec{v}$, and $\vec{w}$ denote vectors, and let $a$ be a number. Then:

(1) $\vec{v} \cdot \vec{w}$ is a number.
(2) $\vec{v} \cdot \vec{w} = \vec{w} \cdot \vec{v}$.
(3) $\vec{v} \cdot \vec{0} = 0 = \vec{0} \cdot \vec{w}$.
(4) $\vec{v} \cdot \vec{v} = \|\vec{v}\|^2$.
(5) $(a\vec{v}) \cdot \vec{w} = a(\vec{v} \cdot \vec{w}) = \vec{v} \cdot (a\vec{w})$.
(6) $\vec{v} \cdot (\vec{u} + \vec{w}) = \vec{v} \cdot \vec{u} + \vec{v} \cdot \vec{w}$.
(7) $\vec{v} \cdot (\vec{u} - \vec{w}) = \vec{v} \cdot \vec{u} - \vec{v} \cdot \vec{w}$.

**PROOF**   (1) holds by the definition of $\vec{v} \cdot \vec{w}$, and (4) comes from (2) of Theorem 1, §3.1. To prove (2) write $\vec{v} = [x_1 \ \ y_1 \ \ z_1]^T$ and $\vec{w} = [x_2 \ \ y_2 \ \ z_2]^T$. Then:

$$\vec{v} \cdot \vec{w} = x_1 x_2 + y_1 y_2 + z_1 z_2 = x_2 x_1 + y_2 y_1 + z_2 z_1 = \vec{w} \cdot \vec{v}.$$

The rest of the assertions in the theorem are properties of matrix arithmetic. For example (6) is verified as follows:

$$\vec{v} \cdot (\vec{u} + \vec{w}) = \vec{v}^T(\vec{u} + \vec{w}) = \vec{v}^T \vec{u} + \vec{v}^T \vec{w} = \vec{v} \cdot \vec{u} + \vec{v} \cdot \vec{w}.$$

The proofs of (3), (5), and (7) are similar and are left as Exercise 17.

The properties in Theorem 1 can be combined to enable calculations like the following:

$$3\vec{u} \cdot (\vec{v} - 2\vec{w} + 5\vec{z}) = 3(\vec{u} \cdot \vec{v}) - 6(\vec{u} \cdot \vec{w}) + 15(\vec{u} \cdot \vec{z}).$$

Such computations will be used without comment below. Here is an illustration.

**Example 2**   Verify that $\|\vec{v} + \vec{w}\|^2 = \|\vec{v}\|^2 + 2(\vec{v} \cdot \vec{w}) + \|\vec{w}\|^2$ for any vectors $\vec{v}$ and $\vec{w}$.

**SOLUTION**   We apply Theorem 1 several times:

$$\begin{aligned} \|\vec{v} + \vec{w}\|^2 &= (\vec{v} + \vec{w}) \cdot (\vec{v} + \vec{w}) \\ &= \vec{v} \cdot \vec{v} + \vec{v} \cdot \vec{w} + \vec{w} \cdot \vec{v} + \vec{w} \cdot \vec{w} \\ &= \|\vec{v}\|^2 + 2(\vec{v} \cdot \vec{w}) + \|\vec{w}\|^2. \end{aligned}$$

This is what we wanted.

## 3.2.2  Angles

Figure 3.27

Suppose two nonzero vectors $\vec{v}$ and $\vec{w}$ are positioned with a common tail. Then they determine a unique angle $\theta$ in the range[8]

$$0 \le \theta \le \pi$$

called the **angle between** the vectors $\vec{v}$ and $\vec{w}$. Figure 3.27 illustrates the cases where $\theta$ is acute (less than $\frac{\pi}{2}$) and obtuse (greater than $\frac{\pi}{2}$). If either $\vec{v}$ or $\vec{w}$ is the zero vector, the angle between them is not defined.

There is a close connection between the dot product $\vec{v} \cdot \vec{w}$ and the angle between $\vec{v}$ and $\vec{w}$. The relationship depends upon an extension of Pythagoras' theorem known as the

**LAW OF COSINES**

**PROOF**

Figure 3.28

If $a$, $b$, and $h$ are the sides of a triangle and $\theta$ is the angle opposite $h$ (as in Figure 3.28), then

$$h^2 = a^2 + b^2 - 2ab \cos \theta.$$

To see this, consider the case when $\theta$ is acute; the obtuse case is similar. Referring to Figure 3.29, right triangle trigonometry gives $p = a \sin \theta$ and $q = b - a \cos \theta$, so Pythagoras' theorem gives

$$h^2 = p^2 + q^2 = a^2 \sin^2 \theta + [b^2 - 2ab \cos \theta + a^2 \cos^2 \theta] = a^2 + b^2 - 2ab \cos \theta,$$

where we used the fact that $\sin^2 \theta + \cos^2 \theta = 1$ for any angle $\theta$.

Figure 3.29

Note that the law of cosines reduces to Pythagoras' theorem if $\theta$ is a right angle because then $\theta = \frac{\pi}{2}$ and so $\cos \theta = \cos \frac{\pi}{2} = 0$.

With the law of cosines in hand we can derive an important connection between angles and dot products.

**THEOREM ②**

If $\theta$ is the angle between the nonzero vectors $\vec{v}$ and $\vec{w}$, then

$$\vec{v} \cdot \vec{w} = \|\vec{v}\| \|\vec{w}\| \cos \theta.$$

**PROOF**

Figure 3.30

We compute $\|\vec{v} - \vec{w}\|^2$ two ways and compare the results. If we apply the law of cosines to the triangle in Figure 3.30, the result is

$$\|\vec{v} - \vec{w}\|^2 = \|\vec{v}\|^2 + \|\vec{w}\|^2 - 2\|\vec{v}\| \|\vec{w}\| \cos \theta.$$

On the other hand, Theorem 1 gives

$$\|\vec{v} - \vec{w}\|^2 = (\vec{v} - \vec{w}) \cdot (\vec{v} - \vec{w})$$
$$= \vec{v} \cdot \vec{v} - \vec{v} \cdot \vec{w} - \vec{w} \cdot \vec{v} + \vec{w} \cdot \vec{w}$$
$$= \|\vec{v}\|^2 - 2\vec{v} \cdot \vec{w} + \|\vec{w}\|^2.$$

Comparing these expressions proves Theorem 2.

Theorem 2 gives an intrinsic description of the dot product because the expression $\|\vec{v}\| \|\vec{w}\| \cos \theta$ depends only on the lengths of the vectors $\vec{v}$ and $\vec{w}$ and the angle $\theta$ between them, and does not depend on the coordinate system used to describe the vectors. Moreover, if $\vec{v}$ and $\vec{w}$ are nonzero, it provides a formula for the cosine of the angle $\theta$ between them:

$$\cos \theta = \frac{\vec{v} \cdot \vec{w}}{\|\vec{v}\| \|\vec{w}\|}.$$

---

[8] We use radian measure for angles. Thus $\pi$ equals 180°, $\frac{\pi}{2}$ equals 90°, etc. See Appendix A.1 for a discussion of angles.

Here is how the formula is used.

**Example 3** Find the angle $\theta$ between $\vec{v} = \begin{bmatrix} -1 \\ -2 \\ 1 \end{bmatrix}$ and $\vec{w} = \begin{bmatrix} 2 \\ 1 \\ 1 \end{bmatrix}$.

SOLUTION Using Theorem 2, we compute
$$\cos\theta = \frac{\vec{v}\cdot\vec{w}}{\|\vec{v}\|\|\vec{w}\|} = \frac{-2-2+1}{\sqrt{1+4+1}\sqrt{4+1+1}} = \frac{-3}{\sqrt{6}\sqrt{6}} = -\frac{1}{2}.$$

Now recall that $\cos\theta$ and $\sin\theta$ are defined[9] so that $(\cos\theta,\ \sin\theta)$ is the point on the unit circle determined by the angle $\theta$ (drawn counterclockwise from the positive $X$-axis as in Figure 3.31). In the present example we have $\cos\theta = -\frac{1}{2}$ and $0 \le \theta \le \pi$, so it follows that $\theta = \frac{2\pi}{3}$ or $120°$.

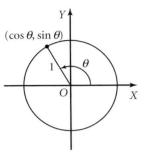

Figure 3.31

Let $\theta$ denote the angle between the nonzero vectors $\vec{v}$ and $\vec{w}$, so that $0 \le \theta \le \pi$. It follows from the formula $\cos\theta = \frac{\vec{v}\cdot\vec{w}}{\|\vec{v}\|\|\vec{w}\|}$ that $\vec{v}\cdot\vec{w}$ and $\cos\theta$ have the same sign. But $\cos\theta$ is the $X$-coordinate of the point on the unit circle determined by $\theta$ (see Figure 3.31), so $\cos\theta$ is positive, zero, or negative if and only if $\theta < \frac{\pi}{2}, \theta = \frac{\pi}{2}$, or $\theta > \frac{\pi}{2}$ respectively. Hence we obtain the following relationship between the sign of $\vec{v}\cdot\vec{w}$ and the size of $\theta$ :

$$\vec{v}\cdot\vec{w} > 0 \text{ if and only if } \theta \text{ is } \textbf{acute } \left(0 \le \theta < \frac{\pi}{2}\right)$$
$$\vec{v}\cdot\vec{w} = 0 \text{ if and only if } \theta \text{ is a } \textbf{right angle } \left(\theta = \frac{\pi}{2}\right)$$
$$\vec{v}\cdot\vec{w} < 0 \text{ if and only if } \theta \text{ is } \textbf{obtuse } \left(\frac{\pi}{2} < \theta \le \pi\right)$$

The second of these three cases is the most important. The nonzero vectors $\vec{v}$ and $\vec{w}$ are called **orthogonal** if the angle $\theta$ between them is a right angle, that is, if $\theta = \frac{\pi}{2}$. The above discussion proves the following fundamental result.

**THEOREM 3** Let $\vec{v}$ and $\vec{w}$ denote nonzero vectors. Then:
$$\vec{v} \text{ and } \vec{w} \text{ are orthogonal if and only if } \vec{v}\cdot\vec{w} = 0.$$

Theorem 3 will be referred to frequently because perpendicular lines play such a basic role in geometry. The reason, of course, is that one can test whether two lines are perpendicular by checking if two nonzero vectors pointing along the lines are orthogonal. The following three examples provide illustrations.

**Example 4** Show that the points $A(3,\ -1,\ 2), B(5,\ -2,\ 5)$, and $C(2,\ 0,\ 3)$ are the vertices of a right-angled triangle.

SOLUTION We have $\overrightarrow{AB} = [2\ \ -1\ \ 3]^T, \overrightarrow{AC} = [-1\ \ 1\ \ 1]^T$, and $\overrightarrow{BC} = [-3\ \ 2\ \ -2]^T$. Now observe that $\overrightarrow{AB}\cdot\overrightarrow{AC} = -2 - 1 + 3 = 0$, so the vectors $\overrightarrow{AB}$ and $\overrightarrow{AC}$ are orthogonal by Theorem 3. Since these vectors point along the sides $AB$ and $AC$ of the triangle (Figure 3.32), this means that the angle at vertex $A$ is a right angle.

Figure 3.32

**Example 5** A **rhombus** is a parallelogram in which all sides are of equal length. Show that the diagonals of any rhombus are perpendicular.

---
[9] A review of basic trigonometry is given in Appendix A.1.

SOLUTION

Figure 3.33

Let $\vec{v}$ and $\vec{w}$ denote the vectors along adjacent sides of the rhombus as in Figure 3.33. Then $\vec{v} + \vec{w}$ and $\vec{v} - \vec{w}$ point along the diagonals, and we compute

$$(\vec{v} - \vec{w}) \cdot (\vec{v} + \vec{w}) = \vec{v} \cdot \vec{v} + \vec{v} \cdot \vec{w} - \vec{w} \cdot \vec{v} - \vec{w} \cdot \vec{w}$$
$$= \|\vec{v}\|^2 - \|\vec{w}\|^2$$
$$= 0$$

because $\|\vec{v}\| = \|\vec{w}\|$ (it is a rhombus). Hence $\vec{v} + \vec{w}$ and $\vec{v} - \vec{w}$ are orthogonal by Theorem 3, and so the diagonals are perpendicular.

**Example 6**

The line through a vertex of a triangle which is perpendicular to the opposite side is called an **altitude** of the triangle. Show that the three altitudes of any triangle are concurrent (that is, they meet at a common point).

SOLUTION

Figure 3.34

Let the vertices be $A$, $B$, and $C$, and let $P$ be the point of intersection of the altitudes through $A$ and $B$, as in Figure 3.34. It suffices to show that $\overrightarrow{PC}$ is orthogonal to $\overrightarrow{AB}$; that is, $\overrightarrow{PC} \cdot \overrightarrow{AB} = 0$. We have $\overrightarrow{AB} = \overrightarrow{AC} - \overrightarrow{BC}$, so

$$\overrightarrow{PC} \cdot \overrightarrow{AB} = \overrightarrow{PC} \cdot (\overrightarrow{AC} - \overrightarrow{BC}) = \overrightarrow{PC} \cdot \overrightarrow{AC} - \overrightarrow{PC} \cdot \overrightarrow{BC}$$

Writing $\overrightarrow{PC}$ two different ways on the right side, we get

$$\overrightarrow{PC} \cdot \overrightarrow{AB} = (\overrightarrow{PB} + \overrightarrow{BC}) \cdot \overrightarrow{AC} - (\overrightarrow{PA} + \overrightarrow{AC}) \cdot \overrightarrow{BC}$$
$$= \overrightarrow{PB} \cdot \overrightarrow{AC} + \overrightarrow{BC} \cdot \overrightarrow{AC} - \overrightarrow{PA} \cdot \overrightarrow{BC} - \overrightarrow{AC} \cdot \overrightarrow{BC}$$
$$= 0 + \overrightarrow{BC} \cdot \overrightarrow{AC} - 0 - \overrightarrow{AC} \cdot \overrightarrow{BC}$$
$$= 0$$

where $\overrightarrow{PB} \cdot \overrightarrow{AC} = 0$ and $\overrightarrow{PA} \cdot \overrightarrow{BC} = 0$ because $P$ lies on both altitudes. This is what we wanted.

## 3.2.3  Projections

In applications of vectors, it is frequently useful to write a vector as the sum of two orthogonal vectors. Here is an example.

**Example 7**

Suppose a ten-kilogram block is placed on a flat surface inclined 30° to the horizontal as in Figure 3.35. Neglecting friction, how much force is required to keep the block from sliding down the surface?

SOLUTION

Figure 3.35

Let $\vec{w}$ denote the weight (force due to gravity) exerted on the block. Then $\|\vec{w}\| = 10$ and the direction of $\vec{w}$ is vertically down as in Figure 3.35.

The idea is to write $\vec{w}$ as a sum $\vec{w} = \vec{w}_1 + \vec{w}_2$ where $\vec{w}_1$ is parallel to the inclined surface, and $\vec{w}_2$ is perpendicular to the surface. Since there is no friction, the force required is $-\vec{w}_1$ because the force $\vec{w}_2$ has no effect parallel to the surface. As the angle between $\vec{w}$ and $\vec{w}_2$ is 30° in Figure 3.35, we have $\frac{\|\vec{w}_1\|}{\|\vec{w}\|} = \sin 30° = \frac{1}{2}$. Hence $\|\vec{w}_1\| = \frac{1}{2}\|\vec{w}\| = \frac{1}{2}10 = 5$. Thus the required force has a magnitude of 5 kilograms, directed up the surface.

The key idea in the solution of Example 7 was to write $\vec{w}$ as a sum of two vectors, one parallel to the surface and one perpendicular to the surface. This type of decomposition arises frequently in applications, and there is a simple way to find it. The general formulation of the problem is as follows.

Suppose that vectors $\vec{v}$ and $\vec{d}$ are given with $\vec{d} \neq \vec{0}$. We want to be able to write $\vec{v}$ as a sum

$$\vec{v} = \vec{v}_1 + \vec{v}_2$$

where $\vec{v}_1$ is parallel to $\vec{d}$ and $\vec{v}_2$ is orthogonal to $\vec{d}$ (see Figure 3.36). Since $\vec{v}_1$ is parallel to $\vec{d}$ it must have the form

$$\vec{v}_1 = t\vec{d}$$

where $t$ is some scalar. This scalar $t$ is determined by the fact that $\vec{v}_2$ is orthogonal to $\vec{d}$, that is $\vec{v}_2 \cdot \vec{d} = 0$. In fact, since $\vec{v}_2 = \vec{v} - \vec{v}_1$, this condition takes the form

$$0 = \vec{v}_2 \cdot \vec{d} = (\vec{v} - \vec{v}_1) \cdot \vec{d} = (\vec{v} - t\vec{d}) \cdot \vec{d} = \vec{v} \cdot \vec{d} - t(\vec{d} \cdot \vec{d}) = \vec{v} \cdot \vec{d} - t\|\vec{d}\|^2.$$

In other words, the requirement is that $t\|\vec{d}\|^2 = \vec{v} \cdot \vec{d}$. Since $\|\vec{d}\|^2 \neq 0$ (because $\vec{d} \neq \vec{0}$) this gives

$$t = \frac{\vec{v} \cdot \vec{d}}{\|\vec{d}\|^2}$$

Hence $\vec{v}_1 = \left( \frac{\vec{v} \cdot \vec{d}}{\|\vec{d}\|^2} \right) \vec{d}$, so $\vec{v}_1$ and $\vec{v}_2 = \vec{v} - \vec{v}_1$ are uniquely determined by $\vec{v}$ and $\vec{d}$.

The vector $\vec{v}_1$ is called the **projection of $\vec{v}$ on $\vec{d}$**, written

$$\vec{v}_1 = proj_{\vec{d}}(\vec{v}).$$

This proves the following useful theorem.

Figure 3.36

---

**THEOREM 4**

**Projection Theorem.** Let $\vec{v}$ and $\vec{d} \neq \vec{0}$ be vectors.

(1) The projection of $\vec{v}$ on $\vec{d}$ is given by

$$proj_{\vec{d}}(\vec{v}) = \left( \frac{\vec{v} \cdot \vec{d}}{\|\vec{d}\|^2} \right) \vec{d}.$$

(2) The vector $\vec{v} - proj_{\vec{d}}(\vec{v})$ is orthogonal to $\vec{d}$.
(3) The vector $\vec{v}$ can be written uniquely in the form $\vec{v} = \vec{v}_1 + \vec{v}_2$ where $\vec{v}_1$ is parallel to $\vec{d}$ and $\vec{v}_2$ is orthogonal to $\vec{d}$. In fact, $\vec{v}_1 = proj_{\vec{d}}(\vec{v})$ and $\vec{v}_2 = \vec{v} - \vec{v}_1$.

---

Observe that the projection $\vec{v}_1 = proj_{\vec{d}}(\vec{v})$ has the same direction as $\vec{d}$ if the angle $\theta$ between $\vec{v}$ and $\vec{d}$ is less than $\frac{\pi}{2}$, and $\vec{v}_1$ has the opposite direction to $\vec{d}$ if $\theta$ is greater than $\frac{\pi}{2}$. These two situations are illustrated in Figure 3.37. Of course, if $\vec{v}$ and $\vec{d}$ are orthogonal (that is, if $\theta = \frac{\pi}{2}$) then $proj_{\vec{d}}(\vec{v}) = \vec{v}_1 = \vec{0}$.

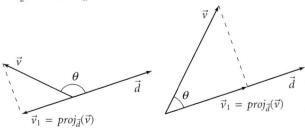

Figure 3.37

**Example 8**   If $\vec{v} = \begin{bmatrix} 1 \\ 2 \\ -3 \end{bmatrix}$ and $\vec{d} = \begin{bmatrix} 4 \\ -1 \\ 1 \end{bmatrix}$, express $\vec{v}$ in the form $\vec{v} = \vec{v}_1 + \vec{v}_2$

where $\vec{v}_1$ is parallel to $\vec{d}$ and $\vec{v}_2$ is orthogonal to $\vec{d}$.

**SOLUTION**   By the Projection Theorem, $\vec{v}_1$ is the projection of $\vec{v}$ on $\vec{d}$; that is,

$$\vec{v}_1 = proj_{\vec{d}}(\vec{v}) = \left( \frac{\vec{v} \cdot \vec{d}}{\|\vec{d}\|^2} \right)\vec{d} = \left( \frac{4 - 2 - 3}{4^2 + (-1)^2 + 1^2} \right)\vec{d} = \left( \frac{-1}{18} \right)\vec{d} = \frac{1}{18}\begin{bmatrix} -4 \\ 1 \\ -1 \end{bmatrix}$$

Since $\vec{v} = \vec{v}_1 + \vec{v}_2$, we can compute $\vec{v}_2$ as follows:

$$\vec{v}_2 = \vec{v} - \vec{v}_1 = \begin{bmatrix} 1 \\ 2 \\ -3 \end{bmatrix} - \frac{1}{18}\begin{bmatrix} -4 \\ 1 \\ -1 \end{bmatrix}$$

$$= \frac{1}{18}\left( \begin{bmatrix} 18 \\ 36 \\ -54 \end{bmatrix} - \begin{bmatrix} -4 \\ 1 \\ -1 \end{bmatrix} \right)$$

$$= \frac{1}{18}\begin{bmatrix} 22 \\ 35 \\ -53 \end{bmatrix}$$

A good check on the arithmetic is to verify that $\vec{v}_2$ is actually orthogonal to $\vec{d}$. In this case, we have $\vec{v}_2 \cdot \vec{d} = \frac{1}{18}(22 \cdot 4 + 35(-1) + (-53) \cdot 1) = 0$, so this vector $\vec{v}_2$ is indeed orthogonal to $\vec{d}$.

## Exercises 3.2

**1.** In each case compute $\vec{v} \cdot \vec{w}$.
   **a)** $\vec{v} = [5 \ -3]^T, \vec{w} = [2 \ 6]^T$
   **b)** $\vec{v} = [4 \ 0]^T, \vec{w} = [1 \ -1]^T$
   **c)** $\vec{v} = [2 \ -1 \ 3]^T, \vec{w} = [1 \ 2 \ -5]^T$
   **d)** $\vec{v} = [3 \ -1 \ 4]^T, \vec{w} = \vec{v}$
   **e)** $\vec{v} = [2 \ 2 \ -1]^T, \vec{w} = -3[1 \ -1 \ 5]^T$
   **f)** $\vec{v} = [3 \ -1 \ 5]^T, \vec{w} = [4 \ 2 \ -2]^T$
   **g)** $\vec{v} = [-5 \ 7 \ 2]^T, \vec{w} = [9 \ 7 \ -2]^T$
   **h)** $\vec{v} = [5 \ -6 \ 2]^T, \vec{w} = [-3 \ 4 \ -2]^T$

**2.** In each case compute $\vec{v} \cdot \vec{v}$ and $\|\vec{v}\|$.
   **a)** $\vec{v} = [2 \ -1 \ 3]^T$     **b)** $\vec{v} = [1 \ -2 \ 2]^T$
   **c)** $\vec{v} = \frac{\sqrt{2}}{10}[3 \ -5 \ 4]^T$    **d)** $\vec{v} = [1 \ -1 \ 1]^T$

**3.** For each of the following pairs of vectors, determine whether the angle between $\vec{v}$ and $\vec{w}$ is acute, obtuse, or a right angle.
   **a)** $\vec{v} = [5 \ -3]^T, \vec{w} = [2 \ 6]^T$
   **b)** $\vec{v} = [4 \ 0]^T, \vec{w} = [1 \ -1]^T$
   **c)** $\vec{v} = [2 \ -1 \ 3]^T, \vec{w} = [1 \ 2 \ -5]^T$
   **d)** $\vec{v} = [3 \ -1 \ 4]^T, \vec{w} = \vec{v}$
   **e)** $\vec{v} = [2 \ 2 \ -1]^T, \vec{w} = -3[1 \ -1 \ 5]^T$
   **f)** $\vec{v} = [3 \ -1 \ 5]^T, \vec{w} = [4 \ 2 \ -2]^T$
   **g)** $\vec{v} = [-5 \ 7 \ 2]^T, \vec{w} = [9 \ 7 \ -2]^T$
   **h)** $\vec{v} = [5 \ -6 \ 2]^T, \vec{w} = [-3 \ 4 \ -2]^T$

**4.** In each case find the angle between $\vec{v}$ and $\vec{w}$.
   **a)** $\vec{v} = [3 \ 1]^T, \vec{w} = [1 \ 2]^T$
   **b)** $\vec{v} = [1 \ 1]^T, \vec{w} = [-1 \ 1]^T$
   **c)** $\vec{v} = [2 \ 1 \ 1]^T, \vec{w} = [1 \ 1 \ 0]^T$
   **d)** $\vec{v} = [2 \ 3 \ 4]^T, \vec{w} = [3 \ -4 \ 7]^T$
   **e)** $\vec{v} = [7 \ 3 \ -1]^T, \vec{w} = [1 \ -1 \ 4]^T$
   **f)** $\vec{v} = [0 \ -1 \ 1]^T, \vec{w} = [1 \ 1 \ 0]^T$

**5.** Find all $a$, $b$, and $c$ such that $\vec{u} = [a \ b \ c]^T$ is orthogonal to both vectors $\vec{v} = [2 \ -1 \ 3]^T$ and $\vec{w} = [1 \ 1 \ -2]^T$.

**6.** Consider the triangle with vertices $A(2, \ -3, \ 0)$, $B(5, \ 1, -2)$, and $C(7, \ 3, \ 5)$.
   **a)** Show that $ABC$ is a right-angled triangle.
   **b)** Find the lengths of the three sides and verify Pythagoras' theorem.

**7.** Show that the line through $A(1, \ -2, \ 1)$ and $B(0, \ 2, \ 5)$ is perpendicular to the line through $C(2, \ 1, \ -1)$ and $D(2, \ 3, \ -3)$.

**8.** In each case consider the line through $A$ and $B$, and the line through $C$ and $D$, and determine if these lines are parallel, perpendicular, or neither. Support your answer.
   **a)** $A(2, \ -1, \ 3)$, $B(3, \ 1, \ -1)$, $C(1, \ 1, \ 0)$, $D(0, \ -1, \ 4)$
   **b)** $A(5, \ 0, \ -2)$, $B(1, \ 1, \ 2)$, $C(3, \ -2, \ 5)$, $D(4, \ -2, \ 3)$

**c)** $A(3, 2, 1)$, $B(1, 1, 1)$, $C(-5, -1, 0)$, $D(4, -3, 2)$
**d)** $A(-1, 2, 5)$, $B(9, -3, 2)$, $C(0, -3, 2)$, $D(1, -4, 7)$
**e)** $A(5, 4, -1)$, $B(8, 3, 3)$, $C(2, 4, 6)$, $D(1, 5, 7)$
**f)** $A(3, 0, -2)$, $B(6, -5, 2)$, $C(4, 4, -3)$, $D(-2, 14, -11)$

**9.** In each case compute the projection of $\vec{v}$ on $\vec{w}$.
**a)** $\vec{v} = [-2 \ \ 3]^T$, $\vec{w} = [1 \ \ 2]^T$
**b)** $\vec{v} = [-1 \ \ 3]^T$, $\vec{w} = [5 \ \ 0]^T$
**c)** $\vec{v} = [3 \ \ -1 \ \ 1]^T$, $\vec{w} = [2 \ \ 0 \ \ -3]^T$
**d)** $\vec{v} = [-2 \ \ 0 \ \ 12]^T$, $\vec{w} = [5 \ \ -4 \ \ 4]^T$
**e)** $\vec{v} = [6 \ \ -1 \ \ 2]^T$, $\vec{w} = [-1 \ \ 2 \ \ 4]^T$
**f)** $\vec{v} = [5 \ \ -1 \ \ 1]^T$, $\vec{w} = [1 \ \ -3 \ \ -2]^T$

**10.** In each case write $\vec{v}$ as a sum $\vec{v} = \vec{v}_1 + \vec{v}_2$ where $\vec{v}_1$ is parallel to $\vec{w}$ and $\vec{v}_2$ is orthogonal to $\vec{w}$.
**a)** $\vec{v} = [1 \ \ -1 \ \ 3]^T$, $\vec{w} = [2 \ \ -1 \ \ 1]^T$
**b)** $\vec{v} = [3 \ \ 1 \ \ 2]^T$, $\vec{w} = [-2 \ \ 1 \ \ 4]^T$
**c)** $\vec{v} = [2 \ \ -1 \ \ 0]^T$, $\vec{w} = [3 \ \ 2 \ \ -1]^T$
**d)** $\vec{v} = [3 \ \ -2 \ \ 1]^T$, $\vec{w} = [1 \ \ 4 \ \ -1]^T$

**11.** Let $A$, $B$, $C$, and $D$ be the vertices in order of a parallelogram $ABCD$. Given $A(1, -1, 2)$, $C(2, 1, 0)$, and the midpoint $M(1, 0, -3)$ of side $AB$, find $\vec{BD}$.

**12.** Let $A$, $B$, and $C$ be the vertices of a triangle, and let $M$ be the midpoint of $BC$. Given $A(2, -1, 3)$, $\vec{BC} = [4 \ \ 1 \ \ -2]^T$, and $\vec{AM} = [2 \ \ 0 \ \ -1]^T$, find $B$ and $C$.

**13.** Show that if the diagonals of a parallelogram are perpendicular, it is necessarily a rhombus. (See Example 5).

**14.** Let $A$ and $B$ be the end-points of a diameter of a circle as in the diagram. If $C$ is any point on the circle, show that the line segments $AC$ and $BC$ are perpendicular.

**15.** Consider the rectangular solid with sides of length 1, 1, and $\sqrt{2}$. Find the angle between a diagonal and one of the longest sides.

**16.** In each case either show that the statement is true or give an example showing that it is false. Throughout, $\vec{u}$, $\vec{v}$, and $\vec{w}$ represent vectors.
**a)** If $\vec{v} \cdot \vec{w} = 0$, then either $\vec{v} = \vec{0}$ or $\vec{w} = \vec{0}$.

**b)** If $\vec{v}$ and $\vec{w}$ are orthogonal, then $2\vec{v}$ and $-3\vec{w}$ are also orthogonal.
**c)** If $-\vec{v}$ is orthogonal to $\vec{w}$, then $\vec{v}$ is parallel to $\vec{w}$.
**d)** If the angle between $\vec{v}$ and $\vec{w}$ is acute, then the angle between $-\vec{v}$ and $\vec{w}$ is obtuse.
**e)** If the projection $proj_{\vec{d}}(\vec{v}) = \vec{0}$, then $\vec{v} = \vec{0}$.
**f)** If $proj_{\vec{d}}(\vec{v}) = \vec{0}$, then $\vec{v}$ is orthogonal to $\vec{d}$.
**g)** If $\vec{v}$ is parallel to $\vec{d}$, then $proj_{\vec{d}}(\vec{v}) = \vec{v}$.

**17.** Prove the following parts of Theorem 1:
**a)** Part (3)    **b)** Part (5)    **c)** Part (7).

**18.** If $\vec{v}$ and $\vec{w}$ are orthogonal, show that
$$\|\vec{v} + \vec{w}\|^2 = \|\vec{v}\|^2 + \|\vec{w}\|^2.$$

**19.** Show that $(\vec{v} + \vec{w}) \cdot (\vec{v} - \vec{w}) = \|\vec{v}\|^2 - \|\vec{w}\|^2$.

**20.** Show that $\|\vec{v} + \vec{w}\|^2 + \|\vec{v} - \vec{w}\|^2 = 2\left(\|\vec{v}\|^2 + \|\vec{w}\|^2\right)$.

**21.** If each of $\vec{u}$, $\vec{v}$, and $\vec{w}$ is orthogonal to the other two, show that $\|\vec{u} + \vec{v} + \vec{w}\|^2 = \|\vec{u}\|^2 + \|\vec{v}\|^2 + \|\vec{w}\|^2$.

**22.** Verify the following two properties of projection.
**a)** $proj_{\vec{d}}(\vec{v} + \vec{w}) = proj_{\vec{d}}(\vec{v}) + proj_{\vec{d}}(\vec{w})$ for all vectors $\vec{v}$ and $\vec{w}$.
**b)** $proj_{\vec{d}}(a\vec{v}) = a\,proj_{\vec{d}}(\vec{v})$ for all scalars $a$ and vectors $\vec{v}$. [Because of these properties, the function $proj_{\vec{d}}(\ )$ is said to be a **linear transformation**; this will be treated further in Section 3.4.]

**23.** If $\vec{v}$ is any nonzero vector, let $\alpha$, $\beta$, and $\gamma$ be the angles $\vec{v}$ makes with the positive $X$-, $Y$-, and $Z$-axes respectively. Then $\cos \alpha$, $\cos \beta$, and $\cos \gamma$ are called the **direction cosines** of the vector $\vec{v}$.
**a)** If $\vec{v} = [a \ \ b \ \ c]^T$, show that $\cos \alpha = \frac{a}{\|\vec{v}\|}$, $\cos \beta = \frac{b}{\|\vec{v}\|}$, and $\cos \gamma = \frac{c}{\|\vec{v}\|}$.
**b)** Show that $\cos^2 \alpha + \cos^2 \beta + \cos^2 \gamma = 1$.

**24.** Consider the line $y = mx$ through the origin with slope $m$.
**a)** Show that $\vec{d} = [1 \ \ m]^T$ is a vector pointing along the line.
**b)** Show that the lines $y = m_1 x$ and $y = m_2 x$ are perpendicular if and only if $m_1 m_2 = -1$.

**25.** Assume that $\vec{d}$ and $\vec{d}'$ are parallel nonzero vectors. Use part (1) of Theorem 4 to show that $proj_{\vec{d}}(\vec{v}) = proj_{\vec{d}'}(\vec{v})$ for every vector $\vec{v}$.

**26.** **a)** Show that $|\vec{v} \cdot \vec{w}| \leq \|\vec{v}\| \|\vec{w}\|$ for all vectors $\vec{v}$ and $\vec{w}$. This is called the **Cauchy-Schwarz inequality**.
**b)** Show that $|\vec{v} \cdot \vec{w}| = \|\vec{v}\| \|\vec{w}\|$ if and only if $\vec{v}$ and $\vec{w}$ are parallel.

## 3.3    LINES AND PLANES

We now turn our attention to describing two of the most commonly encountered objects in geometry: lines and planes. We take a vector approach, and the main tool is the concept of the position vector of a point.

### 3.3.1    Position Vectors

Let $P = P(x, y, z)$ be the point in space with coordinates $(x, y, z)$. The **position vector** $\vec{p}$ of $P$ was defined in Section 3.1 to be the vector from the origin to $P$:

$$\vec{p} = \overrightarrow{OP} = \begin{bmatrix} x \\ y \\ z \end{bmatrix} = [x \ \ y \ \ z]^T.$$

Similarly, the position vector of $P(x, y)$ in the plane is $\vec{p} = \begin{bmatrix} x \\ y \end{bmatrix} = [x \ \ y]^T$. Thus every point has a unique position vector. Conversely every vector is the position vector of a unique point (place the vector with its tail at the origin and use the point at its tip). Hence position vectors are just a different way to describe points.

The introduction of position vectors provides a new perspective in geometry: To compute the *coordinates* $x$, $y$, and $z$ of a point $P(x, y, z)$, we find instead the position vector $\vec{p} = [x \ \ y \ \ z]^T$ of $P$, and read off its *components* $x$, $y$, and $z$. They are the same three numbers, but this change of emphasis allows vector methods to be used in geometry, and so gives a new approach to the subject. This vector perspective is particularly effective for studying lines and planes, and so is central to this section.

### 3.3.2    Lines

Given a point $P$, it is evident geometrically that there is exactly one line[10] through $P$ which is parallel to a given nonzero vector. We are going to use this observation to give a precise description of lines in space. With this in mind, a nonzero vector $\vec{d}$ is called a **direction vector** for the line if it is parallel to the line; that is, if $\vec{d}$ is parallel to $\overrightarrow{AB}$ for some distinct points $A$ and $B$ on the line. Thus $\overrightarrow{AB}$ itself is a direction vector for the line, but there are many others. Indeed any nonzero scalar multiple of a direction vector is again a direction vector.

A direction vector for a line in space describes the orientation of the line. This role is played by the slope for lines in the plane, and the following example relates the two concepts.

**Example 1**    Show that $\vec{d} = \begin{bmatrix} 1 \\ m \end{bmatrix}$ is a direction vector for any line in the plane with slope $m$.

**SOLUTION**    Such a line has equation $y = mx + b$ for some number $b$. Taking $x = 0$ and $x = 1$ gives two points $P(0, b)$ and $Q(1, m + b)$ on the line. Thus the vector $\vec{d} = \overrightarrow{PQ} = \begin{bmatrix} 1 - 0 \\ (m + b) - b \end{bmatrix} = \begin{bmatrix} 1 \\ m \end{bmatrix}$ is parallel to the line (see Figure 3.38) and so will serve as a position vector for the line.

Figure 3.38

---

[10] The term "line" means "straight line" in this book.

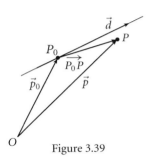

Figure 3.39

Now suppose a point $P_0(x_0, y_0, z_0)$ and a nonzero vector $\vec{d} = [a\ b\ c]^T$ are given. We want to determine the unique line through $P_0$ which has $\vec{d}$ as direction vector; that is, we want to give a condition that an arbitrary point $P = P(x, y, z)$ lies on this line. Let $\vec{p}_0 = [x_0\ y_0\ z_0]^T$ and $\vec{p} = [x\ y\ z]^T$ be the position vectors of $P_0$ and $P$ respectively, so that $\overrightarrow{P_0P} = \vec{p} - \vec{p}_0$ (see Figure 3.39).

The key observation is that $P$ lies on the line if and only if $\vec{p} - \vec{p}_0 = \overrightarrow{P_0P}$ is parallel to $\vec{d}$, that is if and only if

$$\vec{p} - \vec{p}_0 = t\vec{d} \qquad \text{for some scalar } t.$$

In other words, the point $P$ lies on the line if and only if its position vector $\vec{p}$ has the form $\vec{p} = \vec{p}_0 + t\vec{d}$ for some scalar $t$. This equation has a name.

**Vector Equation of a Line.** Consider the line with direction vector $\vec{d} \neq \vec{0}$ which passes through the point with position vector $\vec{p}_0$. The point with position vector $\vec{p}$ lies on this line if and only if

$$\vec{p} = \vec{p}_0 + t\vec{d} \qquad \text{for some scalar } t.$$

In component form, the vector equation of the line becomes

$$\begin{bmatrix} x \\ y \\ z \end{bmatrix} = \begin{bmatrix} x_0 \\ y_0 \\ z_0 \end{bmatrix} + t \begin{bmatrix} a \\ b \\ c \end{bmatrix}$$

Equating components gives another description of the line.

**Scalar Equations of a Line.** The line through the point $P_0(x_0, y_0, z_0)$ with direction vector $\vec{d} = [a\ b\ c]^T \neq \vec{0}$ is given by

$$\begin{aligned} x &= x_0 + ta \\ y &= y_0 + tb \qquad t \text{ any scalar} \\ z &= z_0 + tc \end{aligned}$$

In other words, a point $P(x, y, z)$ lies on the line if and only if a real number $t$ exists such that $x = x_0 + ta$, $y = y_0 + tb$, and $z = z_0 + tc$.

Note that a direction vector $\vec{d}$ can be "read off" from the scalar equations of a line because the components of $\vec{d}$ are the coefficients of the parameter $t$. This is analogous to "reading off" the slope of the line $y = 3 - 4x$ as the coefficient $-4$ of $x$.

The following examples illustrate how these equations are used to describe lines and solve problems about lines.

**Example 2** Determine whether $A(1, -5, 15)$ and $B(-1, 1, 4)$ lie on the line

with scalar equations $\begin{cases} x = -2 + t \\ y = 4 - 3t \\ z = -3 + 6t \end{cases}$

**SOLUTION** Every point $P$ on the line has the form $P(-2 + t, 4 - 3t, -3 + 6t)$ for some $t$. Hence $A$ is on the line if a number $t$ can be found such that $-2 + t = 1$, $4 - 3t = -5$, and $-3 + 6t = 15$. Since $t = 3$ satisfies these equations, the point $A$ is indeed on the line. As to $B$, it lies on the line if $-2 + t = -1$, $4 - 3t = 1$, and $-3 + 6t = 4$ for some $t$. The first two equations each require $t = 1$, but this does not satisfy $-3 + 6t = 4$. Hence no such number $t$ exists, and so $B$ is not on the line.

**Example 3**    If $P(3, -1, 4)$ and $\vec{d} = [5 \ -2 \ 8]^T$, the line through $P$ with direction vector $\vec{d}$ has

vector equation $\vec{p} = \begin{bmatrix} 3 \\ -1 \\ 4 \end{bmatrix} + t \begin{bmatrix} 5 \\ -2 \\ 8 \end{bmatrix}$ and scalar equations $\begin{cases} x = 3 + 5t \\ y = -1 - 2t. \\ z = 4 + 8t \end{cases}$

**Example 4**    Find scalar equations of the line through the points $A(2, 0, -3)$ and $B(1, 1, -5)$.

SOLUTION    Because both $A$ and $B$ lie on the line, the vector

$$\overrightarrow{AB} = \begin{bmatrix} 1-2 \\ 1-0 \\ -5-(-3) \end{bmatrix} = \begin{bmatrix} -1 \\ 1 \\ -2 \end{bmatrix}$$

is a vector along the line. Thus we may use $\vec{d} = \overrightarrow{AB}$ as a direction vector for the line. Using $A(2, 0, -3)$ as the point on the line, the scalar equations are

$$\begin{cases} x = 2 - t \\ y = t \\ z = -3 - 2t \end{cases}.$$

**Example 5**    Determine whether the lines

$$\begin{array}{ll} x = 2 - 5t & x = t \\ y = -1 + t \quad \text{and} & y = -3 + 2t \\ z = 3t & z = 5 + t \end{array}$$

are perpendicular.

SOLUTION    $\vec{d}_1 = [-5 \ 1 \ 3]^T$ and $\vec{d}_2 = [1 \ 2 \ 1]^T$ are direction vectors for these lines, and they are orthogonal because $\vec{d}_1 \cdot \vec{d}_2 = -5 + 2 + 3 = 0$. Hence the lines are perpendicular.

**Example 6**    Find scalar equations of the line through $A(4, -1, 5)$ which is

parallel to the line $\begin{cases} x = 5 \\ y = 2 - t \\ z = -1 + 3t \end{cases}$

SOLUTION    A direction vector of the given line is $\vec{d} = [0 \ -1 \ 3]^T$, and this will do as a direction vector of the line we want because the lines are parallel. Since the point $A(4, -1, 5)$ lies

on the line we want, the scalar equations are $\begin{cases} x = 4 \\ y = -1 - t. \\ z = 5 + 3t \end{cases}$

**Example 7**    Determine whether the lines

$$\begin{array}{ll} x = 2 - t & x = 3 + 2s \\ y = 5 + 2t \quad \text{and} & y = 2 - 3s \\ z = -1 + 3t & z = -6 - 4s \end{array}$$

intersect and, if they do, find the point of intersection.

SOLUTION    Suppose $P(x, y, z)$ is a point of intersection. Then it lies on *both* lines so there must

exist real numbers $t$ and $s$ such that $\begin{bmatrix} 2-t \\ 5+2t \\ -1+3t \end{bmatrix} = \begin{bmatrix} x \\ y \\ z \end{bmatrix} = \begin{bmatrix} 3+2s \\ 2-3s \\ -6-4s \end{bmatrix}.$

Equating components gives three equations in the two variables $s$ and $t$:

$$2 - t = 3 + 2s$$
$$5 + 2t = 2 - 3s$$
$$-1 + 3t = -6 - 4s$$

These equations have a unique solution $s = 1$ and $t = -3$, so the lines meet at the corresponding point $P(5, -1, -10)$. (Of course, if these equations for $s$ and $t$ did not have a solution, the lines would not intersect.)

Two lines in the plane are parallel if they have equal slopes. The next example uses vector methods to derive a well-known test for when two lines in the plane are perpendicular.

**Example 8**

Show that two lines in the plane with slopes $m_1$ and $m_2$ are perpendicular if and only if $m_1 m_2 = -1$.

**SOLUTION**

The two lines have direction vectors $\vec{d}_1 = \begin{bmatrix} 1 \\ m_1 \end{bmatrix}$ and $\vec{d}_2 = \begin{bmatrix} 1 \\ m_2 \end{bmatrix}$ by Example 1. Hence the lines are perpendicular if and only if $\vec{d}_1$ and $\vec{d}_2$ are orthogonal, that is (by Theorem 3 §3.2) if and only if $\vec{d}_1 \cdot \vec{d}_2 = 0$. But this last condition reads $1 + m_1 m_2 = 0$.

In applications of geometry, it is frequently required to solve an optimization problem; that is, to maximize or minimize some quantity. Such problems can sometimes be solved using projections; the following example provides an illustration.

**Example 9**

Find the shortest distance from the point $P(3, 5, 6)$ to the line through $P_0(6, 1, -3)$ with direction vector $\vec{d} = [-1 \ \ 3 \ \ 0]^T$. Then find the point $Q$ on the line that is closest to $P$.

**SOLUTION**

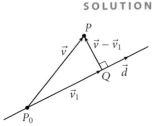

Figure 3.40

Write $\vec{v} = \overrightarrow{P_0P} = [-3 \ \ 4 \ \ 9]^T$. Referring to Figure 3.40, the required distance is $\| \vec{v} - \vec{v}_1 \|$ where $\vec{v}_1$ is the projection of $\vec{v}$ on $\vec{d}$. We can compute $\vec{v}_1$ as follows:

$$\vec{v}_1 = proj_{\vec{d}}(\vec{v}) = \left( \frac{\vec{v} \cdot \vec{d}}{\|\vec{d}\|^2} \right)\vec{d} = \left( \frac{3 + 12 + 0}{1 + 9 + 0} \right)\vec{d} = \frac{3}{2} \begin{bmatrix} -1 \\ 3 \\ 0 \end{bmatrix}.$$

Then the shortest distance from $P$ to the line is

$$\| \overrightarrow{QP} \| = \| \vec{v} - \vec{v}_1 \| = \left\| \begin{bmatrix} -3 \\ 4 \\ 9 \end{bmatrix} - \frac{3}{2} \begin{bmatrix} -1 \\ 3 \\ 0 \end{bmatrix} \right\| = \left\| \frac{1}{2} \begin{bmatrix} -3 \\ -1 \\ 18 \end{bmatrix} \right\| = \frac{1}{2}\sqrt{334}.$$

Observe that we did not have to compute the point $Q$ to find the distance. However, it is easy to find $Q$. Let $\vec{p}_0 = [6 \ \ 1 \ \ -3]^T$ and $\vec{q} = [x \ \ y \ \ z]^T$ be the position vectors of $P_0$ and $Q$ respectively. Then Figure 3.41 gives

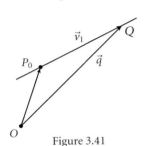

Figure 3.41

$$\vec{q} = \vec{p}_0 + \vec{v}_1 = \begin{bmatrix} 6 \\ 1 \\ -3 \end{bmatrix} + \frac{3}{2} \begin{bmatrix} -1 \\ 3 \\ 0 \end{bmatrix} = \begin{bmatrix} \frac{9}{2} \\ \frac{11}{2} \\ -3 \end{bmatrix}.$$

It follows that $Q = Q(\frac{9}{2}, \frac{11}{2}, -3)$ is the point on the line closest to $P$.

### 3.3.3 Planes

The orientation of a line in space is described by specifying a direction vector for the line. Similarly, the orientation of a plane in space can be described by a vector as follows. A nonzero vector $\vec{n}$ is called a **normal** to a plane if it is orthogonal to every vector in the plane. As for direction vectors, there are many normals to a given plane; in particular, any nonzero scalar multiple of a normal is again a normal.

Given a point $P_0(x_0,\ y_0,\ z_0)$ and a nonzero vector $\vec{n} = [a\ \ b\ \ c]^T$, it is clear geometrically that there is exactly one plane with normal $\vec{n}$ which contains the point $P_0$. Our goal now is to describe this unique plane. In other words, given an arbitrary point $P(x,\ y,\ z)$, we must give a condition (which turns out to be an equation) that $P$ lies in the plane.

Let $\vec{p}_0$ and $\vec{p}$ be the position vectors of $P_0$ and $P$ respectively. Then (see Figure 3.42) $P$ lies in the plane if and only if $\vec{n}$ is orthogonal to $\overrightarrow{P_0P} = \vec{p} - \vec{p}_0$; that is if and only if $\vec{n} \cdot (\vec{p} - \vec{p}_0) = 0$. This is the

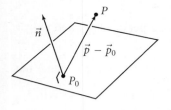

Figure 3.42

**Vector Equation of a Plane.** Consider the plane with normal $\vec{n} \neq \vec{0}$ which contains the point $P_0(x_0,\ y_0,\ z_0)$. The point with position vector $\vec{p}$ lies in this plane if and only if

$$\vec{n} \cdot (\vec{p} - \vec{p}_0) = 0.$$

Now let $\vec{n}$, $\vec{p}_0$, and $\vec{p}$ be given in matrix form, say $\vec{n} = [a\ \ b\ \ c]^T$, $\vec{p}_0 = [x_0\ \ y_0\ \ z_0]^T$, and $\vec{p} = [x\ \ y\ \ z]^T$. Then $\vec{p} - \vec{p}_0 = [x - x_0\ \ y - y_0\ \ z - z_0]^T$, so the vector equation $\vec{n} \cdot (\vec{p} - \vec{p}_0) = 0$ becomes a scalar description of the plane.

**Scalar Equation of a Plane.** The plane through $P_0(x_0,\ y_0,\ z_0)$ with normal $\vec{n} = [a\ \ b\ \ c]^T \neq \vec{0}$ has equation

$$a(x - x_0) + b(y - y_0) + c(z - z_0) = 0.$$

In other words, a point $P(x,\ y,\ z)$ lies in the plane if and only if $x$, $y$, and $z$ satisfy this equation.

**Example 10**

The plane through $P_0(-2,\ 3,\ 5)$ with normal $\vec{n} = [7\ \ -4\ \ 8]^T$ has equation
$$7(x - (-2)) - 4(y - 3) + 8(z - 5) = 0.$$
This simplifies to $7x - 4y + 8z = 14$.

The scalar equation $a(x - x_0) + b(y - y_0) + c(z - z_0) = 0$ of a plane can be written as follows:
$$ax + by + cz = ax_0 + by_0 + cz_0.$$

The quantity $ax_0 + by_0 + cz_0$ on the right side is a constant depending on $P_0$ and $\vec{n}$. If we write it as $k$, we have proved part (1) of the following theorem.

**THEOREM 1**

Let $\vec{n} = [a\ \ b\ \ c]^T \neq \vec{0}$ be a fixed nonzero vector.

(1) Every plane with normal $\vec{n}$ has equation $ax + by + cz = k$ for some constant $k$.

(2) For every constant $k$, the graph of the equation $ax + by + cz = k$ is a plane with normal $\vec{n}$.

**PROOF**

It remains to prove (2). Since $\vec{n} \neq \vec{0}$, one of the components $a$, $b$, and $c$ is nonzero. If $a \neq 0$, the equation $ax + by + cz = k$ can be written
$$a\left(x - \tfrac{k}{a}\right) + b(y - 0) + c(z - 0) = 0$$

and so is the equation of the plane through $P_0\left(\frac{k}{a},\ 0,\ 0\right)$ with normal $\vec{n}$. A similar argument works if $b \neq 0$ or $c \neq 0$.

In particular, if the equation $ax + by + cz = k$ of a plane is given, the components of a normal $\vec{n} = [a\ \ b\ \ c]^T$ of the plane can be "read off" as the coefficients of $x$, $y$, and $z$ in the equation.

**Example 11**

Find the equation of the plane through $P(4,\ 0,\ -1)$ which is parallel to the plane with equation $3x - 2y + z = 9$.

**SOLUTION**

The given plane has normal $\vec{n} = [3\ \ -2\ \ 1]^T$ by Theorem 1, so $\vec{n}$ will do as a normal for the plane we want (because the two planes are parallel). Hence our plane has equation $3x - 2y + z = k$ for some constant $k$, again by Theorem 1. The constant $k$ is determined by the fact that $P(4,\ 0,\ -1)$ lies in the plane. Indeed, this means that $3 \cdot 4 - 2 \cdot 0 + (-1) = k$, whence $k = 11$. So the required equation is $3x - 2y + z = 11$.

**Example 12**

Find the shortest distance from the point $P(1,\ 5,\ 5)$ to the plane with equation $2x - y + 3z = 5$, and find the point $Q$ in the plane closest to $P$.

There are two approaches to this problem, and we present both to illustrate the techniques.

**SOLUTION 1**

Choose any point $P_0$ in the plane, say $P_0(1,\ 0,\ 1)$, and compute $\vec{v} = \overrightarrow{P_0P} = [0\ \ 5\ \ 4]^T$. Position the normal $\vec{n} = [2\ \ -1\ \ 3]^T$ with its tail at $P_0$ as in Figure 3.43, and compute the projection $\vec{v}_1$ of $\vec{v}$ on $\vec{n}$:

$$\vec{v}_1 = proj_{\vec{n}}(\vec{v}) = \frac{\vec{v} \cdot \vec{n}}{\|\vec{n}\|^2}\vec{n} = \frac{0 - 5 + 12}{4 + 1 + 9}\vec{n} = \tfrac{1}{2}\vec{n}.$$

Now observe that $\overrightarrow{QP} = \vec{v}_1$ by Figure 3.43. Hence the required distance is

$$\|\overrightarrow{QP}\| = \|\vec{v}_1\| = \|\tfrac{1}{2}\vec{n}\| = \tfrac{1}{2}\|\vec{n}\| = \tfrac{1}{2}\sqrt{14}.$$

Figure 3.43

Note that we did not have to find $Q$ to be able to compute the shortest distance. However, it is not difficult to find the point $Q$. Let $\vec{p}$ and $\vec{q}$ be the position vectors of $P$ and $Q$ respectively. Then $\vec{p} = \vec{q} + \overrightarrow{QP} = \vec{q} + \vec{v}_1$ by Figure 3.44, so

$$\vec{q} = \vec{p} - \vec{v}_1 = \begin{bmatrix} 1 \\ 5 \\ 5 \end{bmatrix} - \tfrac{1}{2}\begin{bmatrix} 2 \\ -1 \\ 3 \end{bmatrix} = \begin{bmatrix} 0 \\ \frac{11}{2} \\ \frac{7}{2} \end{bmatrix}.$$

Hence $Q = Q\left(0,\ \frac{11}{2},\ \frac{7}{2}\right)$.

Figure 3.44

**SOLUTION 2**

The point $Q$ is the point of intersection of the given plane and the line through $P$ which is perpendicular to the plane (see Figure 3.45). But $\vec{n} = [2\ \ -1\ \ 3]^T$ is a direction vector for this line, so the line is given by the equation

$$[x\ \ y\ \ z]^T = [1\ \ 5\ \ 5]^T + t[2\ \ -1\ \ 3]^T,\ t \text{ a scalar.}$$

Since $Q$ lies on this line, we have $Q = Q(1 + 2t,\ 5 - t,\ 5 + 3t)$ for some $t$. But $Q$ is also on the plane, so its coordinates must satisfy the equation of the plane:

$$2(1 + 2t) - (5 - t) + 3(5 + 3t) = 5$$

Solving this for $t$ gives $t = -\tfrac{1}{2}$, so $Q = Q\left(1 - 1,\ 5 + \tfrac{1}{2},\ 5 - \tfrac{3}{2}\right) = Q\left(0,\ \frac{11}{2},\ \frac{7}{2}\right)$ as before.

Figure 3.45

Moreover, having found $Q$, the required distance is

$$\|\overrightarrow{QP}\| = \|\vec{p} - \vec{q}\| = \left\|[1 \quad \tfrac{-1}{2} \quad \tfrac{3}{2}]^T\right\| = \left\|\tfrac{1}{2}[2 \quad -1 \quad 3]^T\right\| = \tfrac{1}{2}\sqrt{14}.$$

Again this agrees with Solution 1.

## 3.3.4   The Cross Product

It is clear geometrically that there is a unique plane containing any three points $A$, $B$, and $C$ that are not all on a line. In determining the equation of this plane, the problem is finding a normal vector for the plane. Since any nonzero vector which is orthogonal to both $\overrightarrow{AB}$ and $\overrightarrow{AC}$ will do as a normal, what we need is a systematic way to find a nonzero vector orthogonal to any two nonparallel vectors. The cross product solves this problem (and has other uses as well as we shall see in Section 3.5).

Given vectors $\vec{v} = \begin{bmatrix} x_1 \\ y_1 \\ z_1 \end{bmatrix}$ and $\vec{w} = \begin{bmatrix} x_2 \\ y_2 \\ z_2 \end{bmatrix}$, the vector

$$\vec{v} \times \vec{w} = \begin{bmatrix} y_1 z_2 - y_2 z_1 \\ -(x_1 z_2 - x_2 z_1) \\ x_1 y_2 - x_2 y_1 \end{bmatrix} = \begin{bmatrix} \det\begin{bmatrix} y_1 & y_2 \\ z_1 & z_2 \end{bmatrix} \\ -\det\begin{bmatrix} x_1 & x_2 \\ z_1 & z_2 \end{bmatrix} \\ \det\begin{bmatrix} x_1 & x_2 \\ y_1 & y_2 \end{bmatrix} \end{bmatrix} \qquad (*)$$

is called the **cross product** of $\vec{v}$ and $\vec{w}$. This is best remembered as follows: Write $\vec{v}$ and $\vec{w}$ side by side to form a $3 \times 2$ matrix:

$$[\vec{v} \quad \vec{w}] = \begin{bmatrix} x_1 & x_2 \\ y_1 & y_2 \\ z_1 & z_2 \end{bmatrix}.$$

If we delete rows 1, 2, or 3 of this matrix and take the determinant of what remains, we obtain, respectively, components 1, 2, or 3 of $\vec{v} \times \vec{w}$, except for the minus sign on the middle term.[11] This is a good way to remember the definition of $\vec{v} \times \vec{w}$.

The reader can verify directly that $\vec{v} \cdot (\vec{v} \times \vec{w}) = 0$ and $\vec{w} \cdot (\vec{v} \times \vec{w}) = 0$, so $\vec{v} \times \vec{w}$ is indeed orthogonal to both $\vec{v}$ and $\vec{w}$. This proves the first of the assertions in the following Theorem; the proof of the second assertion is deferred to Section 3.5.

**THEOREM 2**   Let $\vec{v}$ and $\vec{w}$ be vectors in space. Then:

(1)  $\vec{v} \times \vec{w}$ is a vector orthogonal to both $\vec{v}$ and $\vec{w}$.
(2)  $\vec{v} \times \vec{w} = \vec{0}$ if and only if $\vec{v}$ and $\vec{w}$ are parallel.

The assertion in (2) of Theorem 2 is a nice dual to Theorem 3 §3.2:

$$\vec{v} \cdot \vec{w} = 0 \quad \text{if and only if} \quad \vec{v} \text{ and } \vec{w} \text{ are orthogonal.}$$

**Example 13**   If $\vec{v} = \begin{bmatrix} 4 \\ -1 \\ 2 \end{bmatrix}$ and $\vec{w} = \begin{bmatrix} 5 \\ 2 \\ -3 \end{bmatrix}$, then $\vec{v} \times \vec{w} = \begin{bmatrix} -1 \\ 22 \\ 13 \end{bmatrix}$, which is orthogonal to both $\vec{v}$ and $\vec{w}$, as Theorem 2 asserts.

---

[11] The connection with determinants that is evident here will be explored in more detail in Section 3.5.

The following three examples illustrate how the cross product is applied to the study of lines and planes.

**Example 14**   Find the equation of the plane containing the points $A(3, 0, -1)$, $B(1, 1, 1)$, and $C(2, -3, 1)$.

**SOLUTION**   The vectors $\overrightarrow{AB} = [-2 \ \ 1 \ \ 2]^T$ and $\overrightarrow{AC} = [-1 \ \ -3 \ \ 2]^T$ lie in the plane (see Figure 3.46) so, if their cross product is nonzero, it will do as a normal. Compute:

$$\overrightarrow{AB} \times \overrightarrow{AC} = \begin{bmatrix} 8 \\ 2 \\ 7 \end{bmatrix}.$$

Using $\overrightarrow{AB} \times \overrightarrow{AC}$ as a normal, the plane has equation $8x + 2y + 7z = k$ for some number $k$. Since the point $A(3, 0, -1)$ lies on the plane, we obtain $k = 8 \cdot 3 + 2 \cdot 0 + 7(-1) = 17$. Thus the required equation is $8x + 2y + 7z = 17$.

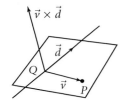

$\overrightarrow{AB} \times \overrightarrow{AC}$

Figure 3.46

**Example 15**   Find the equation of the plane through $P(1, 0, -2)$ which contains the line
$$\begin{cases} x = 2 - t \\ y = -3 \\ z = t \end{cases}.$$

**SOLUTION**   The points $P(1, 0, -2)$ and $Q(2, -3, 0)$ are both in the plane ($Q$ is on the line) so the vector $\vec{v} = \overrightarrow{QP} = [-1 \ \ 3 \ \ -2]^T$ lies in the plane (see Figure 3.47). Since the direction vector $\vec{d} = [-1 \ \ 0 \ \ 1]^T$ of the line also lies in the plane, $\vec{v} \times \vec{d}$ is perpendicular to the plane. We have $\vec{v} \times \vec{d} = \begin{bmatrix} 3 \\ 3 \\ 3 \end{bmatrix}$. Thus we may take $\vec{n} = [1 \ \ 1 \ \ 1]^T$ as the normal to the plane (any nonzero scalar multiple of $\vec{v} \times \vec{d}$ will do), so the equation of the plane is $x + y + z = k$ for some number $k$. Since $P(1, 0, -2)$ is in the plane, we get $k = 1 + 0 - 2 = -1$. Thus the required equation is $x + y + z = -1$.

As a check, note that every point $R$ on the given line has coordinates of the form $R(2 - t, -3, t)$ for some value of $t$, so $R$ lies in the plane $x + y + z = -1$ because $(2 - t) - 3 + t = -1$ for all $t$. Thus the *entire line* lies in the plane, as required.

Figure 3.47

**Example 16**   Find the shortest distance between the lines
$$\begin{cases} x = 3 + t \\ y = -1 + t \\ z = 2 \end{cases} \quad \text{and} \quad \begin{cases} x = t \\ y = 5 - 2t. \\ z = 2 + t \end{cases}$$

and find the points $A_1$ and $A_2$ on the lines which are closest together.

**SOLUTION**   The points $P_1(3, -1, 2)$ and $P_2(0, 5, 2)$ lie on the respective lines, and their direction vectors are $\vec{d}_1 = [1 \ \ 1 \ \ 0]^T$ and $\vec{d}_2 = [1 \ \ -2 \ \ 1]^T$. Hence $\vec{n} = \vec{d}_1 \times \vec{d}_2 = [1 \ \ -1 \ \ -3]^T$ is a vector perpendicular to both lines. We must find the distance between the points $A_1$ and $A_2$ in Figure 3.48. But $A_1$ is in the plane through $P_1$ with normal $\vec{n}$, while $A_2$ is in the plane through $P_2$ with normal $\vec{n}$, so we must compute the distance between these parallel planes. Observe that vector $\vec{v} = \overrightarrow{P_1P_2} = [-3 \ \ 6 \ \ 0]^T$ goes from one plane to the other. Hence if the tail of $\vec{v}$ is moved to $A_1$ as in Figure 3.48, the tip of $\vec{v}$ remains in the other plane. Thus the required distance is the length $\|\vec{v}_1\|$ where $\vec{v}_1$ is the projection of

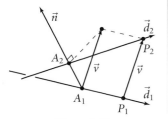

Figure 3.48

$\vec{v}$ along $\vec{n}$. We have $\vec{v}_1 = proj_{\vec{n}}(\vec{v}) = \left(\frac{\vec{v} \cdot \vec{n}}{\|\vec{n}\|^2}\right)\vec{n} = -\frac{9}{11}\vec{n}$, so the distance is
$\|\vec{v}_1\| = \left|-\frac{9}{11}\right| \|\vec{n}\| = \frac{9}{11}\sqrt{11}$.

This computation gives the distance without finding the points $A_1$ and $A_2$. However, these points can be located as follows: Observe that the coordinates of $A_1$ are $A_1 = A_1(3 + t, -1 + t, 2)$ for some $t$ because it lies on the first line, and similarly

$A_2 = A_2(s, 5 - 2s, 2 + s)$ for some $s$. Hence $\overrightarrow{A_1A_2} = \begin{bmatrix} s-3-t \\ 6-2s-t \\ s \end{bmatrix}$, and the parameters

$s$ and $t$ are determined by the fact that $\overrightarrow{A_1A_2}$ is orthogonal to both $\vec{d}_1$ and $\vec{d}_2$. In fact, this means

$$\begin{cases} \vec{d}_1 \cdot (\overrightarrow{A_1A_2}) = 0 \\ \vec{d}_2 \cdot (\overrightarrow{A_1A_2}) = 0 \end{cases} \quad \text{which leads to the equations} \quad \begin{cases} s + 2t = 3 \\ 6s + t = 15 \end{cases}.$$

The solution is $s = \frac{27}{11}$ and $t = \frac{3}{11}$, whence $A_1 = A_1\left(\frac{36}{11}, -\frac{8}{11}, 2\right)$ and $A_2 = A_2\left(\frac{27}{11}, \frac{1}{11}, \frac{49}{11}\right)$. The reader can verify that $\|\overrightarrow{A_1A_2}\| = \frac{9}{11}\sqrt{11}$ as before.

## Exercises 3.3

1. Explain the difference between a point and its position vector; the coordinates of a point and the components of its position vector.

2. In each case find the scalar equations of the line:
   a) Parallel to $[0 \ -1 \ 3]^T$ and passing through $P(4, -3, 5)$.
   b) Parallel to $[3 \ -2 \ 0]^T$ and passing through $P(-3, 0, 7)$.
   c) Passing through $P(5, -7, 2)$ and $Q(0, 0, 4)$.
   d) Passing through $P(9, -3, 1)$ and $Q(-2, 6, -4)$.
   e) Passing through $P(8, 4, -1)$ and parallel to the line $[x \ y \ z]^T = [2 \ 9 \ -3]^T + t[5 \ 1 \ 6]^T$.
   f) Passing through $P(1, 2, 0)$ and parallel to the line $[x \ y \ z]^T = [0 \ 0 \ 1]^T + t[-7 \ 3 \ -5]^T$.
   g) Passing through $P(6, 0, -2)$ and parallel to the line through $Q(1, 1, 1)$ and $R(2, 2, 2)$.
   h) Passing through $P(2, -5, 7)$ and parallel to the line through $Q(0, 1, 2)$ and $R(-5, 7, 3)$.
   i) Passing through $P(2, -5, 7)$ and perpendicular to the plane $3x - 2y + 5z = 7$.
   j) Passing through $P(1, 2, -9)$ and perpendicular to the plane $4y - 3z = 2$.
   k) Passing through $P(6, 0, -3)$, intersecting the line $[x \ y \ z]^T = [1 \ 2 \ -3]^T + t[1 \ -2 \ 0]^T$ and perpendicular to it.
   l) Passing through $P(-7, 2, 7)$, intersecting the line $[x \ y \ z]^T = [0 \ -1 \ 2]^T + t[5 \ 1 \ 1]^T$ and perpendicular to it.

3. Does the line through the points $A(2, -1, 4)$ and $B(-1, 3, 5)$ contain the point $P(1, 2, 4)$? The point $Q(-7, 11, 7)$? Support your answers.

4. In each case determine if the given points all lie on the same line. Support your answer.
   a) $A(2, 1, 3), B(-1, 2, 7), C(5, 0, 4)$
   b) $A(1, 2, -1), B(0, 5, 1), C(2, -1, -3)$.

5. Describe the equations of all lines parallel to the $X$-axis.

6. Find all points $C$ on the line through $A(1, -1, 2)$ and $B(2, 0, 1)$ such that $\|\overrightarrow{AC}\| = 2\|\overrightarrow{BC}\|$.

7. In each case either find the point of intersection of the two lines or show why no such point exists.
   a) $\begin{aligned} x &= 2 - t \\ y &= 3 - t \\ z &= 1 \end{aligned}$  $\begin{aligned} x &= 3 - 2s \\ y &= s \\ z &= 5 - 3s \end{aligned}$
   b) $\begin{aligned} x &= 1 - 3t \\ y &= 7 - 2t \\ z &= 3 + t \end{aligned}$  $\begin{aligned} x &= 1 + s \\ y &= 2 - s \\ z &= 1 - s \end{aligned}$
   c) $[x \ y \ z]^T = [3 \ 1 \ 2]^T + t[-1 \ 1 \ -1]^T$
      $[x \ y \ z]^T = [3 \ 1 \ 5]^T + t[1 \ -1 \ 0]^T$
   d) $[x \ y \ z]^T = [0 \ 1 \ 2]^T + t[3 \ -1 \ -1]^T$
      $[x \ y \ z]^T = [1 \ 3 \ 5]^T + t[-2 \ 1 \ 1]^T$

8. In each case find the shortest distance from the point $P$ to the line, and find the point $Q$ on the line closest to $P$.
   a) $P(1, 2, -3); [x \ y \ z]^T = [1 \ 2 \ 0]^T + t[-3 \ 5 \ 4]^T$
   b) $P(-1, 0, 1); [x \ y \ z]^T = [3 \ -1 \ 4]^T + t[3 \ -2 \ 0]^T$
   c) $P(-3, 0, 5); [x \ y \ z]^T = [3 \ -1 \ 2]^T + t[2 \ -1 \ 2]^T$
   d) $P(1, 1, -1); [x \ y \ z]^T = [4 \ 2 \ -3]^T + t[1 \ -5 \ 0]^T$

9. Does the plane $2x - 3y + z = 6$ contain the point $P(1, -2, -2)$? The point $Q(5, -6, 0)$? Support your answers.

**10.** In each case find the equation of the plane:
  **a)** Passing through $P(3, -7, 5)$ and parallel to the plane with equation $3x - y + 2z = 5$.
  **b)** Passing through $P(1, -2, 4)$ and parallel to the plane with equation $x - 2y + 3z = -6$.
  **c)** Containing $P(2, -3, 0)$ and perpendicular to the line $[x \ y \ z]^T = [2 \ -5 \ 3]^T + t[6 \ -6 \ 5]^T$.
  **d)** Containing $P(1, -1, 4)$ and perpendicular to the line $[x \ y \ z]^T = [7 \ 0 \ 5]^T + t[2 \ 1 \ -7]^T$.
  **e)** Each point of which is the same distance from the points $A(1, -3, 7)$ and $B(3, -5, 9)$.
  **f)** Each point of which is the same distance from the points $A(6, 0, 2)$ and $B(1, 1, -3)$.

**11.** In each case compute $\vec{v} \times \vec{w}$.
  **a)** $\vec{v} = [2 \ 1 \ -3]^T$ and $\vec{w} = [1 \ 2 \ 1]^T$
  **b)** $\vec{v} = [3 \ -5 \ 1]^T$ and $\vec{w} = [1 \ 1 \ 2]^T$
  **c)** $\vec{v} = [3 \ 2 \ -1]^T$ and $\vec{w} = [-6 \ -4 \ 2]^T$
  **d)** $\vec{v} = [1 \ 1 \ 1]^T$ and $\vec{w} = [4 \ 2 \ -3]^T$

**12.** In each case, find the scalar equations of the line:
  **a)** Containing $P(7, 1, 0)$ and perpendicular to the lines $[x \ y \ z]^T = [9 \ -1 \ 4]^T + t[-2 \ 1 \ 3]^T$ and $[x \ y \ z]^T = [1 \ 1 \ 1]^T + t[6 \ -2 \ 3]^T$.
  **b)** Containing $P(2, -1, 3)$ and perpendicular to the lines $[x \ y \ z]^T = [4 \ -1 \ 2]^T + t[7 \ 0 \ 1]^T$ and $[x \ y \ z]^T = [-2 \ 0 \ 1]^T + t[2 \ 3 \ 0]^T$.
  **c)** Of intersection of the planes $2x - 3y + z = 0$ and $3x - 2y + 5z = 4$.
  **d)** Of intersection of the planes $2x - y - 3z = 1$ and $-x - 3y + 5z = 0$.

**13.** In each case find the equation of the plane:
  **a)** Passing through $A(3, 1, 2)$, $B(5, -1, 3)$, and $C(-4, 2, 0)$.
  **b)** Passing through $A(6, -1, 1)$, $B(1, 0, 0)$, and $C(21, 3, 2)$.
  **c)** Passing through $A(6, 1, 0)$ and perpendicular to the line $[x \ y \ z]^T = [4 \ 0 \ -3]^T + t[1 \ -1 \ 4]^T$.
  **d)** Passing through $A(1, -1, 3)$ and perpendicular to the line $[x \ y \ z]^T = [5 \ 5 \ 5]^T + t[8 \ -2 \ 0]^T$.
  **e)** Passing through $P(2, 0, 3)$ and parallel to the plane through the points $A(1, 1, -5)$, $B(0, 1, -2)$, and $C(1, 0, 6)$.
  **f)** Passing through $P(2, -1, 5)$ and parallel to the plane through the points $A(3, -7, 1)$, $B(2, 0, -1)$, and $C(1, 3, 0)$.
  **g)** Containing $A(2, 1, -1)$ and the line $[x \ y \ z]^T = [3 \ -1 \ 0]^T + t[0 \ -1 \ 2]^T$.
  **h)** Containing $A(0, -2, 3)$ and the line $[x \ y \ z]^T = [1 \ -5 \ 2]^T + t[-1 \ 3 \ 5]^T$.
  **i)** Containing the lines $[x \ y \ z]^T = [2 \ 1 \ 0]^T + t[1 \ 1 \ -2]^T$ and $[x \ y \ z]^T = [1 \ 6 \ 2]^T + t[2 \ -1 \ -4]^T$.
  **j)** Containing the lines $[x \ y \ z]^T = [1 \ 1 \ -2]^T + t[3 \ 0 \ 4]^T$ and $[x \ y \ z]^T = [7 \ -1 \ 7]^T + t[3 \ -2 \ 5]^T$.

**14.** In each case determine if the given points all lie on the same plane. Support your answer.
  **a)** $A(1, 0, 1)$, $B(2, 3, 0)$, $C(-1, 1, 2)$, $D(5, 1, 1)$
  **b)** $A(0, 2, 1)$, $B(1, 4, 1)$, $C(2, 0, -1)$, $D(-1, 3, 2)$
  **c)** $A(2, 1, 0)$, $B(3, 0, 3)$, $C(3, 3, 3)$, $D(0, 1, -6)$
  **d)** $A(1, -3, 2)$, $B(2, -4, 4)$, $C(3, -2, 2)$, $D(5, 1, 1)$

**15.** In each case find the shortest distance between the two parallel lines.
  **a)** $[x \ y \ z]^T = [2 \ 1 \ 0]^T + t[1 \ 1 \ -2]^T$ and $[x \ y \ z]^T = [3 \ 3 \ 0]^T + t[1 \ 1 \ -2]^T$.
  **b)** $[x \ y \ z]^T = [1 \ 1 \ 1]^T + t[2 \ 0 \ -3]^T$ and $[x \ y \ z]^T = [3 \ 1 \ 2]^T + t[2 \ 0 \ -3]^T$.

**16.** In each case find the shortest distance between the two nonparallel lines.
  **a)** $[x \ y \ z]^T = [3 \ 0 \ -1]^T + t[1 \ 1 \ 0]^T$ and $[x \ y \ z]^T = [2 \ -3 \ 5]^T + t[2 \ 1 \ -1]^T$.
  **b)** $[x \ y \ z]^T = [3 \ 1 \ -2]^T + t[2 \ 1 \ -3]^T$ and $[x \ y \ z]^T = [2 \ 3 \ -1]^T + t[1 \ 1 \ 1]^T$.
  **c)** $[x \ y \ z]^T = [7 \ 8 \ 1]^T + t[3 \ -2 \ 5]^T$ and $[x \ y \ z]^T = [6 \ 1 \ 5]^T + t[-1 \ -3 \ -2]^T$.
  **d)** $[x \ y \ z]^T = [5 \ 6 \ 7]^T + t[-1 \ -2 \ -3]^T$ and $[x \ y \ z]^T = [3 \ -1 \ 2]^T + t[0 \ 0 \ 1]^T$.

**17.** Give a rule to determine if the plane with equation $ax + by + cz = k$ passes through the origin.

**18.** Describe the equations of all planes perpendicular to the $Z$-axis.

**19.** In each case determine if the given line lies in the given plane. Support your answer.
  **a)** Line through $P(1, 2, -3)$ with direction vector $\vec{d} = [1 \ 1 \ -1]^T$, plane $2x - y + z = -3$.
  **b)** Line through $P(-1, 3, 1)$ with direction vector $\vec{d} = [0 \ 2 \ 5]^T$, plane $x - y + z = 1$.

**20.** In each case determine if the given plane contains the given line. Support your answer.
  **a)** Plane $2x - 3y + z = 1$, line $[x \ y \ z]^T = [3 \ 2 \ 1]^T + t[2 \ 0 \ -4]^T$.
  **b)** Plane $x + 2y - 3z = 5$, line $[x \ y \ z]^T = [3 \ 7 \ 5]^T + t[1 \ 1 \ 1]^T$.

**21.** In each case find the point of intersection of the plane and the line.
  **a)** $x - 3y + 2z = 7$ and $[x \ y \ z]^T = [2 \ 0 \ -1]^T + t[4 \ -1 \ 2]^T$.
  **b)** $2x + 4y - 3z = 1$ and $[x \ y \ z]^T = [5 \ -3 \ 1]^T + t[0 \ 2 \ -3]^T$.

**22.** Show that every plane containing both $A(1, -3, 2)$ and $B(4, 5, 0)$ must also contain $C(7, 13, -2)$.

**23.** In each case find the shortest distance from the point $P$ to the plane, and find the point $Q$ on the plane closest to $P$.
  **a)** $P(1, 0, -2)$; $3x - y + 4z = 5$
  **b)** $P(3, -1, 1)$; $3x - y + 4z = 5$

**c)** $P(5, 1, -3)$; $2x - 3y + z = 1$

**d)** $P(-7, -10, 9)$; $x - 3y + 2z = 6$

**24.** In each case, find the solution of the system of linear equations that is closest to the given point.

**a)** $2x - 3y + 5z = 2$
$x + 2y - z = 1$ and the point $P(2, 3, 0)$.
$3x - y + z = 4$

**b)** $x - y + 3z = 1$
$-x + 3y - z = 3$ and the point $P(-1, 2, 0)$.
$y + z = 2$

**c)** $x - y + 3z = -1$
$-x + 3y - z = 3$ and the point $P(-4, 0, 1)$.
$x + y + 5z = 1$

**d)** $x - 3y + 5z = 1$
$-x + 3y - 5z = -1$ and the point $P(-1, 4, 0)$.

**25.** In each case either show that the statement is true or give an example showing that it is false.

**a)** If a line is parallel to a plane it never intersects the plane.

**b)** Any three planes no two of which are parallel meet in a single point.

**c)** If the plane $ax + by + cz = k$ passes through the origin, then $k = 0$.

**d)** Every plane has exactly one equation of the form $ax + by + cz = k$.

**e)** If two lines do not intersect, they do not both lie in the same plane.

**f)** If a line is parallel to the normal of a plane, then it is parallel to the plane.

**g)** The line of intersection of two (nonparallel) planes is perpendicular to both normals.

**h)** A line perpendicular to the normal of a plane must be parallel to the plane.

**26.** If a plane contains two distinct points $A$ and $B$, show that it contains the line through $A$ and $B$.

**27.** Let $\vec{d} = [a \ b \ c]^T$ be a vector for which $a$, $b$, and $c$ are *all* nonzero. Show that the scalar equations of the line through an arbitrary point $P_0(x_0, y_0, z_0)$ with $\vec{d}$ as direction vector $\vec{d}$ can be written in the form $\frac{x - x_0}{a} = \frac{y - y_0}{b} = \frac{z - z_0}{c}$. This is called the **symmetric form** of the equations.

**28. a)** Show that every plane with normal $\vec{n} \neq \vec{0}$ has vector equation $\vec{p} \cdot \vec{n} = k$ for some scalar $k$, where $\vec{p} = [x \ y \ z]^T$.

**b)** If $\vec{a}$ is the position vector of a point $A$, show that the shortest distance from $A$ to the plane with equation $\vec{p} \cdot \vec{n} = k$ is $\frac{|k - (\vec{a} \cdot \vec{n})|}{\|\vec{n}\|}$.

**29.** Consider the two parallel planes $ax + by + cz = k$ and $ax + by + cz = k_1$. Show that the distance between them is $\frac{|k_1 - k|}{\|\vec{n}\|}$ where $\vec{n} = [a \ b \ c]^T$ is the common normal. [Hint: Example 12.]

## 3.4   MATRIX TRANSFORMATIONS OF $\mathbb{R}^2$

Apart from their use in solving systems of linear equations, matrices have so far been treated rather abstractly with the emphasis on their algebraic properties. In this section, the vector approach to geometry reveals a concrete, geometrical way to view $2 \times 2$ matrices. This discovery not only provides a new interpretation of matrix algebra but also applies matrices to give new geometrical insights.

Throughout this section we denote the set of real numbers by $\mathbb{R}$, and we write $\mathbb{R}^2$ for the Euclidean plane. It is convenient to identify points in $\mathbb{R}^2$ with their position vectors; more precisely, we make no distinction between a point $P(x, y)$ and its position vector $\vec{v} = \begin{bmatrix} x \\ y \end{bmatrix}$, and refer to it simply as a "vector" in $\mathbb{R}^2$.

### 3.4.1   Transformations

To illustrate how $2 \times 2$ matrices arise as geometrical transformations of $\mathbb{R}^2$, consider *reflection* in the $X$-axis. This operation carries the vector $\begin{bmatrix} x \\ y \end{bmatrix}$ to its reflection $\begin{bmatrix} x \\ -y \end{bmatrix}$, and

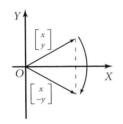

Figure 3.49

so can be described geometrically as in Figure 3.49. Now observe that

$$\begin{bmatrix} x \\ -y \end{bmatrix} = \begin{bmatrix} 1 & 0 \\ 0 & -1 \end{bmatrix}\begin{bmatrix} x \\ y \end{bmatrix}$$

so reflecting $\begin{bmatrix} x \\ y \end{bmatrix}$ in the $X$-axis can be achieved by multiplying by the matrix $\begin{bmatrix} 1 & 0 \\ 0 & -1 \end{bmatrix}$.

Reflection in the $X$-axis is an example of a **transformation** $T$ of $\mathbb{R}^2$, that is a function[12]

$$T : \mathbb{R}^2 \to \mathbb{R}^2$$

which carries every vector $\vec{v} = \begin{bmatrix} x \\ y \end{bmatrix}$ in $\mathbb{R}^2$ to another vector in $\mathbb{R}^2$, denoted $T(\vec{v})$. Thus to describe a transformation $T$ we must specify the vector $T(\vec{v})$ for every vector $\vec{v}$ in $\mathbb{R}^2$. This is referred to as **defining** $T$, or as specifying the **action** of $T$. In particular, if $T$ is reflection in the $X$-axis as above, then $T$ is defined by $T\begin{bmatrix} x \\ y \end{bmatrix} = \begin{bmatrix} x \\ -y \end{bmatrix}$ for all $\begin{bmatrix} x \\ y \end{bmatrix}$ in $\mathbb{R}^2$.

Matrices provide an important way of defining transformations of $\mathbb{R}^2$. If $A$ is any $2 \times 2$ matrix, multiplication by $A$ gives a transformation

$$T : \mathbb{R}^2 \to \mathbb{R}^2 \quad \text{defined by} \quad T(\vec{v}) = A\vec{v} \text{ for every } \vec{v} \text{ in } \mathbb{R}^2 .$$

This is called the **matrix transformation induced** by $A$. Thus reflection in the $X$-axis is the matrix transformation induced by $A = \begin{bmatrix} 1 & 0 \\ 0 & -1 \end{bmatrix}$.

If $A = 0$ is the zero matrix, the corresponding matrix transformation $T$ is given by $T(\vec{v}) = A\vec{v} = \vec{0}$ for every $\vec{v}$ in $\mathbb{R}^2$. This is called the **zero transformation** and is denoted $T = 0$. Another example is the **identity transformation** $1_{\mathbb{R}^2} : \mathbb{R}^2 \to \mathbb{R}^2$ induced by the identity matrix. It is given by

$$1_{\mathbb{R}^2}(\vec{v}) = \vec{v} \quad \text{for all } \vec{v} \text{ in } \mathbb{R}^2.$$

It turns out that many important geometric transformations are in fact matrix transformations. Here is another example.

**Example 1**　Show that the counterclockwise rotation about the origin through a right angle is a matrix transformation of $\mathbb{R}^2$, and find the corresponding matrix.

Figure 3.50

**SOLUTION**　The transformation is shown in Figure 3.50. Since the two right triangles are congruent, the rotation carries $\begin{bmatrix} x \\ y \end{bmatrix}$ to $\begin{bmatrix} -y \\ x \end{bmatrix} = \begin{bmatrix} 0 & -1 \\ 1 & 0 \end{bmatrix}\begin{bmatrix} x \\ y \end{bmatrix}$, so the matrix is $A = \begin{bmatrix} 0 & -1 \\ 1 & 0 \end{bmatrix}$.

We will see later that *any* rotation about the origin is a matrix transformation of $\mathbb{R}^2$.

The line through the origin with slope $m$ has equation $y = mx$. Given $m$, we define two transformations $P_m$ and $Q_m$ of $\mathbb{R}^2$ as follows: Given a vector $\vec{v}$ in $\mathbb{R}^2$, let

$$P_m(\vec{v}) \text{ denote the } \textbf{projection} \text{ of } \vec{v} \text{ on the line } y = mx$$

and let

$$Q_m(\vec{v}) \text{ denote the } \textbf{reflection} \text{ of } \vec{v} \text{ in the line } y = mx$$

as shown in Figure 3.51. Geometrically, $Q_m(\vec{v})$ is the "mirror image" of $\vec{v}$ in the line $y = mx$ while, if $\vec{d}$ is any direction vector for the line, then $P_m(\vec{v}) = proj_{\vec{d}}(\vec{v})$ is the usual projection.

---

[12] Functions $f : \mathbb{R} \to \mathbb{R}$ are basic in high school algebra or calculus. They are often defined by a formula, say $f(x) = x^2 + 1$ for all $x$ in $\mathbb{R}$, but other functions are used (for example, $f(x) = \cos x$ or $f(x) = e^x$). It is conventional to use the term "transformation" for functions $f : \mathbb{R}^2 \to \mathbb{R}^2$.

**THEOREM 1**    Consider the line $y = mx$ through the origin with slope $m$. Then projection $P_m$ and reflection $Q_m$ are both matrix transformations of $\mathbb{R}^2$, indeed

$$P_m \text{ is induced by the matrix } \quad \frac{1}{1+m^2}\begin{bmatrix} 1 & m \\ m & m^2 \end{bmatrix}, \text{ and}$$

$$Q_m \text{ is induced by the matrix } \quad \frac{1}{1+m^2}\begin{bmatrix} 1-m^2 & 2m \\ 2m & m^2-1 \end{bmatrix}.$$

**PROOF**    The line $y = mx$ contains the origin and the point $\begin{bmatrix} 1 \\ m \end{bmatrix}$. Thus $\vec{d} = \begin{bmatrix} 1 \\ m \end{bmatrix}$ is a direction vector for the line. If we write $\vec{v} = \begin{bmatrix} x \\ y \end{bmatrix}$, we have the formula

$$P_m(\vec{v}) = proj_{\vec{d}}(\vec{v}) = \left(\frac{\vec{v} \cdot \vec{d}}{\|\vec{d}\|^2}\right)\vec{d}. \text{ Hence}$$

$$P_m(\vec{v}) = \frac{x+ym}{1+m^2}\begin{bmatrix} 1 \\ m \end{bmatrix} = \frac{1}{1+m^2}\begin{bmatrix} x+my \\ mx+m^2y \end{bmatrix} = \frac{1}{1+m^2}\begin{bmatrix} 1 & m \\ m & m^2 \end{bmatrix}\begin{bmatrix} x \\ y \end{bmatrix}$$

as required. Turning to $Q_m$, Figure 3.51 gives $Q_m(\vec{v}) = \vec{v} + 2[P_m(\vec{v}) - \vec{v}] = 2P_m(\vec{v}) - \vec{v}$. Hence some matrix arithmetic gives

$$Q_m(\vec{v}) = \frac{2}{1+m^2}\begin{bmatrix} 1 & m \\ m & m^2 \end{bmatrix}\begin{bmatrix} x \\ y \end{bmatrix} - \begin{bmatrix} x \\ y \end{bmatrix} = \frac{1}{1+m^2}\begin{bmatrix} 1-m^2 & 2m \\ 2m & m^2-1 \end{bmatrix}\begin{bmatrix} x \\ y \end{bmatrix}.$$

Thus $P_m$ and $Q_m$ are matrix transformations.

Figure 3.51

**Example 2**    Find the projection and reflection of the vector $\vec{v} = \begin{bmatrix} 3 \\ -1 \end{bmatrix}$ on the line of slope $\frac{1}{2}$ through the origin.

**SOLUTION**    Using our derived formulas for $P_m$ and $Q_m$, we have:

$$P_{\frac{1}{2}}\begin{bmatrix} 3 \\ -1 \end{bmatrix} = \frac{1}{1+\left(\frac{1}{2}\right)^2}\begin{bmatrix} 1 & \frac{1}{2} \\ \frac{1}{2} & \left(\frac{1}{2}\right)^2 \end{bmatrix}\begin{bmatrix} 3 \\ -1 \end{bmatrix} = \begin{bmatrix} 2 \\ 1 \end{bmatrix}.$$

and

$$Q_{\frac{1}{2}}\begin{bmatrix} 3 \\ -1 \end{bmatrix} = \frac{1}{1+\left(\frac{1}{2}\right)^2}\begin{bmatrix} 1-\left(\frac{1}{2}\right)^2 & 2\left(\frac{1}{2}\right) \\ 2\left(\frac{1}{2}\right) & \left(\frac{1}{2}\right)^2-1 \end{bmatrix}\begin{bmatrix} 3 \\ -1 \end{bmatrix} = \begin{bmatrix} 1 \\ 3 \end{bmatrix}.$$

We hasten to note that not all transformations of $\mathbb{R}^2$ are matrix transformations. If $\vec{w}$ is a fixed vector, define the transformation $T_{\vec{w}} : \mathbb{R}^2 \to \mathbb{R}^2$ by $T_{\vec{w}}(\vec{v}) = \vec{v} + \vec{w}$ for all $\vec{v}$ in $\mathbb{R}^2$. Then $T_{\vec{w}}$ is called **translation** by $\vec{w}$, and it is not a matrix transformation unless $\vec{w} = \vec{0}$ as the next example shows.

**Example 3**    If $\vec{w} \neq \vec{0}$, show that translation by $\vec{w}$ is not a matrix transformation.

**SOLUTION**    If $T_{\vec{w}}$ were induced by a matrix $A$, then $A\vec{v} = T_{\vec{w}}(\vec{v}) = \vec{v} + \vec{w}$ would hold for each $\vec{v}$ in $\mathbb{R}^2$. In particular, taking $\vec{v} = \vec{0}$ gives $\vec{w} = \vec{0} + \vec{w} = T_{\vec{w}}(\vec{0}) = A\vec{0} = \vec{0}$, contrary to the assumption that $\vec{w} \neq \vec{0}$.

### 3.4.2   Linear Transformations

Let $T : \mathbb{R}^2 \to \mathbb{R}^2$ be the matrix transformation induced by the $2 \times 2$ matrix $A$; that is, $T$ is given by $T(\vec{v}) = A\vec{v}$ for all vectors $\vec{v}$ in $\mathbb{R}^2$. Then $T$ has the following properties:

     T1. $T(\vec{v} + \vec{w}) = T(\vec{v}) + T(\vec{w})$    for all $\vec{v}$ and $\vec{w}$ in $\mathbb{R}^2$.
     T2. $T(a\vec{v}) = aT(\vec{v})$    for all $\vec{v}$ in $\mathbb{R}^2$ and all numbers $a$.

In fact, T1 is just the distributive property of matrix multiplication:

$$T(\vec{v} + \vec{w}) = A(\vec{v} + \vec{w}) = A\vec{v} + A\vec{w} = T(\vec{v}) + T(\vec{w}).$$

Similarly $T(a\vec{v}) = A(a\vec{v}) = a(A\vec{v}) = aT(\vec{v})$, giving T2.

     Transformations $T : \mathbb{R}^2 \to \mathbb{R}^2$ which enjoy properties T1 and T2 are called **linear transformations**. Thus the above discussion shows that every matrix transformation is linear. The remarkable thing is that the converse is true:

*Every linear transformation is actually a matrix transformation.*

Moreover, a simple method exists for finding the corresponding matrix. To derive it, consider the **coordinate vectors**

$$\vec{i} = \begin{bmatrix} 1 \\ 0 \end{bmatrix} \quad \text{and} \quad \vec{j} = \begin{bmatrix} 0 \\ 1 \end{bmatrix}$$

and call the pair of vectors $\{\vec{i}, \vec{j}\}$ the **standard basis** of $\mathbb{R}^2$. The importance of the co-ordinate vectors $\vec{i}$ and $\vec{j}$ (and the reason for the name) lies in the fact that each vector in $\mathbb{R}^2$ can be uniquely expressed in terms of them:

$$\begin{bmatrix} x \\ y \end{bmatrix} = \begin{bmatrix} x \\ 0 \end{bmatrix} + \begin{bmatrix} 0 \\ y \end{bmatrix} = x\vec{i} + y\vec{j} \quad \text{for all} \begin{bmatrix} x \\ y \end{bmatrix} \text{ in } \mathbb{R}^2.$$

In other words, the coordinates of the vector arise as the coefficients of $\vec{i}$ and $\vec{j}$. We will use this observation below.

     Now consider an arbitrary linear transformation $T : \mathbb{R}^2 \to \mathbb{R}^2$. Then $T(\vec{i})$ and $T(\vec{j})$ are columns in $\mathbb{R}^2$, so write them as follows:

$$T(\vec{i}) = \begin{bmatrix} a \\ c \end{bmatrix} \quad \text{and} \quad T(\vec{j}) = \begin{bmatrix} b \\ d \end{bmatrix}.$$

Since $T$ is linear, the properties T1 and T2 apply to $T$ and we obtain

$$\begin{aligned}
T\begin{bmatrix} x \\ y \end{bmatrix} &= T(x\vec{i} + y\vec{j}) = T(x\vec{i}) + T(y\vec{j}) \\
&= xT(\vec{i}) + yT(\vec{j}) \\
&= x\begin{bmatrix} a \\ c \end{bmatrix} + y\begin{bmatrix} b \\ d \end{bmatrix} \\
&= \begin{bmatrix} xa + yb \\ xc + yd \end{bmatrix} = \begin{bmatrix} a & b \\ c & d \end{bmatrix}\begin{bmatrix} x \\ y \end{bmatrix}.
\end{aligned}$$

Hence $T$ is indeed a matrix transformation, the one induced by $A = \begin{bmatrix} a & b \\ c & d \end{bmatrix}$.

Moreover the first and second columns of $A$ are $T(\vec{i}) = \begin{bmatrix} a \\ c \end{bmatrix}$ and $T(\vec{j}) = \begin{bmatrix} b \\ d \end{bmatrix}$

respectively, so it is natural to denote $A$ in terms of its columns as follows:

$$A = [\, T(\vec{i}) \quad T(\vec{j}) \,].$$

This proves Theorem 2.

**THEOREM 2**   A transformation $T : \mathbb{R}^2 \to \mathbb{R}^2$ it is linear if and only if it is a matrix transformation. In this case, if $\{\vec{\imath}, \ \vec{\jmath}\}$ it is the standard basis of $\mathbb{R}^2$, then

$$T \text{ is induced by the matrix } A = [T(\vec{\imath}) \ \ T(\vec{\jmath})].$$

Here $[T(\vec{\imath}) \ \ T(\vec{\jmath})]$ denotes the matrix with columns $T(\vec{\imath})$ and $T(\vec{\jmath})$.

Because of Theorem 2, we will use the phrases "linear transformation" and "matrix transformation" interchangeably.

**Example 4**   Let $T : \mathbb{R}^2 \to \mathbb{R}^2$ be a linear transformation. If $T\begin{bmatrix} 1 \\ 2 \end{bmatrix} = \begin{bmatrix} 3 \\ -1 \end{bmatrix}$ and $T\begin{bmatrix} 1 \\ 0 \end{bmatrix} = \begin{bmatrix} 5 \\ 4 \end{bmatrix}$, find $T\begin{bmatrix} x \\ y \end{bmatrix}$ for any $\begin{bmatrix} x \\ y \end{bmatrix}$ in $\mathbb{R}^2$.

**SOLUTION 1**   We first find the matrix $A = [T(\vec{\imath}) \ \ T(\vec{\jmath})]$ of $T$. We are given the first column $T(\vec{\imath}) = \begin{bmatrix} 5 \\ 4 \end{bmatrix}$. For convenience, write $\vec{w} = \begin{bmatrix} 1 \\ 2 \end{bmatrix}$, so we are also given that $T(\vec{w}) = \begin{bmatrix} 3 \\ -1 \end{bmatrix}$. Now simple matrix arithmetic gives

$$\vec{\jmath} = \tfrac{1}{2}(\vec{w} - \vec{\imath}).$$

If we apply $T$ to both sides, the linearity of $T$ gives

$$T(\vec{\jmath}) = \tfrac{1}{2} T(\vec{w} - \vec{\imath}) = \tfrac{1}{2}[T(\vec{w}) - T(\vec{\imath})] = \tfrac{1}{2}\left[ \begin{bmatrix} 3 \\ -1 \end{bmatrix} - \begin{bmatrix} 5 \\ 4 \end{bmatrix} \right] = \begin{bmatrix} -1 \\ -\tfrac{5}{2} \end{bmatrix}.$$

Hence $A = [T(\vec{\imath}) \ \ T(\vec{\jmath})] = \begin{bmatrix} 5 & -1 \\ 4 & -\tfrac{5}{2} \end{bmatrix}$ by Theorem 2, so

$$T\begin{bmatrix} x \\ y \end{bmatrix} = A\begin{bmatrix} x \\ y \end{bmatrix} = \begin{bmatrix} 5 & -1 \\ 4 & -\tfrac{5}{2} \end{bmatrix}\begin{bmatrix} x \\ y \end{bmatrix} = \begin{bmatrix} 5x - y \\ 4x - \tfrac{5}{2}y \end{bmatrix}.$$

This is what we wanted.

**SOLUTION 2**   We claim that $\begin{bmatrix} x \\ y \end{bmatrix} = s\begin{bmatrix} 1 \\ 2 \end{bmatrix} + t\begin{bmatrix} 1 \\ 0 \end{bmatrix}$ for numbers $s$ and $t$ (depending on $x$ and $y$). In fact, equating entries and solving for $s$ and $t$ gives

$$\begin{bmatrix} x \\ y \end{bmatrix} = \tfrac{y}{2}\begin{bmatrix} 1 \\ 2 \end{bmatrix} + \tfrac{2x - y}{2}\begin{bmatrix} 1 \\ 0 \end{bmatrix} \text{ for all } x, y.$$

Since $T$ has properties T1 and T2, we get

$$T\begin{bmatrix} x \\ y \end{bmatrix} = \tfrac{y}{2}T\begin{bmatrix} 1 \\ 2 \end{bmatrix} + \tfrac{2x-y}{2}T\begin{bmatrix} 1 \\ 0 \end{bmatrix} = \tfrac{y}{2}\begin{bmatrix} 3 \\ -1 \end{bmatrix} + \tfrac{2x-y}{2}\begin{bmatrix} 5 \\ 4 \end{bmatrix} = \begin{bmatrix} 5x - y \\ 4x - \tfrac{5}{2}y \end{bmatrix}$$

as before. Thus, $T\begin{bmatrix} x \\ y \end{bmatrix} = \begin{bmatrix} 5 & -1 \\ 4 & -\tfrac{5}{2} \end{bmatrix}\begin{bmatrix} x \\ y \end{bmatrix}$, and we obtain the matrix $A = \begin{bmatrix} 5 & -1 \\ 4 & -\tfrac{5}{2} \end{bmatrix}$ of $T$.

The rotations about the origin are another important class of transformations of $\mathbb{R}^2$. Given a vector $\vec{v}$ in $\mathbb{R}^2$, let

$R_\theta(\vec{v})$ denote the **counterclockwise rotation** of $\vec{v}$ through the angle $\theta$.

The action of $R_\theta$ is shown in Figure 3.52. It is somewhat surprising that $R_\theta$ is actually a matrix transformation. To see why, the idea is to first show that $R_\theta$ is linear and then apply Theorem 2. Of course this also produces the matrix of the transformation $R_\theta$. (In fact, this is often the most convenient way to obtain the matrix of a geometric transformation.)

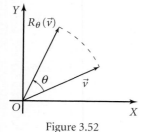

Figure 3.52

**THEOREM 3** The rotation $R_\theta : \mathbb{R}^2 \to \mathbb{R}^2$ is a linear transformation. In fact, $R_\theta$ is the matrix transformation induced by the matrix

$$\begin{bmatrix} \cos\theta & -\sin\theta \\ \sin\theta & \cos\theta \end{bmatrix}.$$

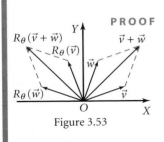

Figure 3.53

**PROOF** To see that $R_\theta$ is linear, consider two vectors $\vec{v}$ and $\vec{w}$ in $\mathbb{R}^2$. Then $\vec{v} + \vec{w}$ is the diagonal of the parallelogram determined by $\vec{v}$ and $\vec{w}$ (see Figure 3.53). The effect of $R_\theta$ is to rotate the entire parallelogram to obtain the parallelogram determined by $R_\theta(\vec{v})$ and $R_\theta(\vec{w})$ with diagonal $R_\theta(\vec{v} + \vec{w})$. But this diagonal is $R_\theta(\vec{v}) + R_\theta(\vec{w})$ by the parallelogram law, and it follows that

$$R_\theta(\vec{v} + \vec{w}) = R_\theta(\vec{v}) + R_\theta(\vec{w}).$$

A similar argument shows that $R_\theta(a\vec{v}) = aR_\theta(\vec{v})$ for any scalar $a$, so $R_\theta$ is a linear transformation.

With this established, we can use Theorem 2. Indeed, from Figure 3.54 we see that

$$R_\theta(\vec{i}) = \begin{bmatrix} \cos\theta \\ \sin\theta \end{bmatrix} \quad \text{and} \quad R_\theta(\vec{j}) = \begin{bmatrix} -\sin\theta \\ \cos\theta \end{bmatrix}$$

so Theorem 2 shows that $R_\theta$ is induced by the matrix

$$[R_\theta(\vec{i}) \quad R_\theta(\vec{j})] = \begin{bmatrix} \cos\theta & -\sin\theta \\ \sin\theta & \cos\theta \end{bmatrix}.$$

Figure 3.54

In particular, the rotations through $\frac{\pi}{2}$ and $\pi$ respectively are given by[13]

$$R_{\frac{\pi}{2}}\begin{bmatrix} x \\ y \end{bmatrix} = \begin{bmatrix} \cos\frac{\pi}{2} & -\sin\frac{\pi}{2} \\ \sin\frac{\pi}{2} & \cos\frac{\pi}{2} \end{bmatrix}\begin{bmatrix} x \\ y \end{bmatrix} = \begin{bmatrix} 0 & -1 \\ 1 & 0 \end{bmatrix}\begin{bmatrix} x \\ y \end{bmatrix} = \begin{bmatrix} -y \\ x \end{bmatrix}$$

and

$$R_{\pi}\begin{bmatrix} x \\ y \end{bmatrix} = \begin{bmatrix} \cos\pi & -\sin\pi \\ \sin\pi & \cos\pi \end{bmatrix}\begin{bmatrix} x \\ y \end{bmatrix} = \begin{bmatrix} -1 & 0 \\ 0 & -1 \end{bmatrix}\begin{bmatrix} x \\ y \end{bmatrix} = \begin{bmatrix} -x \\ -y \end{bmatrix}$$

using Theorem 3. The first of these again confirms the result in Example 1, and the second shows that rotating a vector through an angle $\pi$ is the same as negating the vector, a fact that is clear without Theorem 3.

## 3.4.3 Effect on the Unit Square

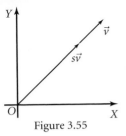

Figure 3.55

If $\vec{v}$ is a vector in $\mathbb{R}^2$, then every vector $\vec{u}$ between $\vec{0}$ and $\vec{v}$ has the form $\vec{u} = s\vec{v}$ for some scalar $s$ with $0 \leq s \leq 1$ (see Figure 3.55). Thus if $\vec{v}$ and $\vec{w}$ are two vectors, Figure 3.56 shows that each vector in the parallelogram determined by $\vec{v}$ and $\vec{w}$ has the form

$$s\vec{v} + t\vec{w} \quad \text{where} \quad 0 \leq s \leq 1 \quad \text{and} \quad 0 \leq t \leq 1.$$

This observation has useful consequences.

Indeed, let $T : \mathbb{R}^2 \to \mathbb{R}^2$ be the matrix transformation induced by $A$. If $s\vec{v} + t\vec{w}$ is any vector in the parallelogram determined by $\vec{v}$ and $\vec{w}$, then (since $T$ is linear)

$$T(s\vec{v} + t\vec{w}) = T(s\vec{v}) + T(t\vec{w}) = s\,T(\vec{v}) + t\,T(\vec{w}).$$

Geometrically this means that

$T(s\vec{v} + t\vec{w})$ is a point in the parallelogram determined by $T(\vec{v})$ and $T(\vec{w})$.

The parallelogram determined by $T(\vec{v})$ and $T(\vec{w})$ is called the **image** under $T$ of the parallelogram determined by $\vec{v}$ and $\vec{w}$. The situation is illustrated in Figure 3.57 on the next page.

Figure 3.56

---

[13] Angles and the trigonometric functions are discussed in Appendix A.1.

$T(\vec{v}) = A\vec{v}$        $T(\vec{w}) = A\vec{w}$

Figure 3.57

The parallelogram determined by the coordinate vectors $\vec{\imath}$ and $\vec{\jmath}$ is actually a square, called the **unit square**. Its image under the matrix transformation $T$ is the parallelogram determined by $T(\vec{\imath})$ and $T(\vec{\jmath})$. This image is important because (as Theorem 2 implies) $T$ is completely determined by $T(\vec{\imath})$ and $T(\vec{\jmath})$. Thus the image of the unit square gives a graphic sense of how the transformation acts. The next two examples illustrate this.

**Example 5**    If $a > 0$, the matrix transformation $T\begin{bmatrix} x \\ y \end{bmatrix} = \begin{bmatrix} ax \\ y \end{bmatrix}$ induced by the matrix $A = \begin{bmatrix} a & 0 \\ 0 & 1 \end{bmatrix}$ is called an $X$-**expansion** if $a > 1$ and an $X$-**compression** if $0 < a < 1$. Here $T(\vec{\imath}) = A\vec{\imath} = a\vec{\imath}$ and $T(\vec{\jmath}) = A\vec{\jmath} = \vec{\jmath}$, so the effect of $T$ on the unit square is shown in Figure 3.58.

Similarly, if $b > 0$, the matrix $\begin{bmatrix} 1 & 0 \\ 0 & b \end{bmatrix}$ gives rise to $Y$-**expansions** and $Y$-**compressions**.

Unit Square                X-expansion                X-compression

Figure 3.58

**Example 6**    If $a$ is a number, the matrix transformation $T\begin{bmatrix} x \\ y \end{bmatrix} = \begin{bmatrix} x + ay \\ y \end{bmatrix}$ induced by the matrix $A = \begin{bmatrix} 1 & a \\ 0 & 1 \end{bmatrix}$ is called an $X$-**shear**, (**positive** if $a > 0$ and **negative** if $a < 0$). In this case, $T(\vec{\imath}) = A\vec{\imath} = \vec{\imath}$ and $T(\vec{\jmath}) = A\vec{\jmath} = \begin{bmatrix} a \\ 1 \end{bmatrix}$. The effect of $T$ on the unit square is as illustrated in Figure 3.59. Similarly, the matrix $\begin{bmatrix} 1 & 0 \\ a & 1 \end{bmatrix}$ induces a $Y$-**shear** for any number $a$.

Unit Square                Positive X-shear                Negative X-shear

Figure 3.59

Theorem 2 shows that a matrix transformation is completely determined by its effect on the unit square (equivalently by its effect on the standard basis). The next theorem reveals the role of the determinant.

**THEOREM 4**    Let $T : \mathbb{R}^2 \to \mathbb{R}^2$ be the matrix transformation induced by the $2 \times 2$ matrix $A$. Then the image under $T$ of the unit square has area $|det(A)|$.

**PROOF**    Write $A = \begin{bmatrix} a & b \\ c & d \end{bmatrix}$. For the moment, identify the vector $\begin{bmatrix} x \\ y \end{bmatrix}$ in $\mathbb{R}^2$ with the vector $\begin{bmatrix} x \\ y \\ 0 \end{bmatrix}$

in $\mathbb{R}^3$. Thus the image of the unit square under $T$ is the parallelogram determined by

$$T(\vec{\imath}) = A\vec{\imath} = \begin{bmatrix} a \\ c \end{bmatrix} = \begin{bmatrix} a \\ c \\ 0 \end{bmatrix} \text{ and } T(\vec{\jmath}) = A\vec{\jmath} = \begin{bmatrix} b \\ d \end{bmatrix} = \begin{bmatrix} b \\ d \\ 0 \end{bmatrix}. \text{ It is proved in Theorem 5 in §3.5}$$

that this parallelogram has area

$$\left\| A\vec{\imath} \times A\vec{\jmath} \right\| = \left\| \begin{bmatrix} 0 \\ 0 \\ ad - bc \end{bmatrix} \right\| = \sqrt{(ad - bc)^2} = |(ad - bc)| = |det(A)|.$$

This is what we wanted.

**Example 7**    The image of the unit square under a rotation $R_\theta$ or a reflection $Q_m$ is again a square of side 1, and so has area 1. Hence the determinant of the matrix of each of these transformations must have absolute value 1 by Theorem 4. In fact the matrices are

$\begin{bmatrix} \cos\theta & -\sin\theta \\ \sin\theta & \cos\theta \end{bmatrix}$ and $\frac{1}{1+m^2}\begin{bmatrix} 1-m^2 & 2m \\ 2m & m^2-1 \end{bmatrix}$ respectively, so the determinants are 1 and $-1$

respectively, as is easily verified. (We return to this in Theorem 7 below.)

On the other hand, the image of the unit square under the projection $P_m$ on the line $y = mx$ is the parallelogram determined by $P_m(\vec{\imath})$ and $P_m(\vec{\jmath})$. But these vectors are parallel (both are parallel to the line $y = mx$), and so the parallelogram they determine has area 0. Thus the matrix $\frac{1}{1+m^2}\begin{bmatrix} 1 & m \\ m & m^2 \end{bmatrix}$ of $P_m$ must have determinant 0, a fact that is easily verified directly.

## 3.4.4    Composition of Matrix Transformations

It often happens that the effect of a transformation can be achieved by doing a sequence of other (often simpler) transformations one after the other. The result is described as follows. If $S : \mathbb{R}^2 \to \mathbb{R}^2$ and $T : \mathbb{R}^2 \to \mathbb{R}^2$ are two transformations of $\mathbb{R}^2$, their **composite** $S \circ T$ is the transformation

$$S \circ T : \mathbb{R}^2 \to \mathbb{R}^2 \quad \text{defined by } [S \circ T](\vec{v}) = S[T(\vec{v})] \quad \text{for all } \vec{v} \text{ in } \mathbb{R}^2.$$

Thus the effect of $S \circ T$ can be described as "first $T$, then $S$" (note the order!). This is illustrated in Figure 3.60.

Figure 3.60

Composition is closely related to matrix multiplication.

**THEOREM 5**    Let $S : \mathbb{R}^2 \to \mathbb{R}^2$ and $T : \mathbb{R}^2 \to \mathbb{R}^2$ be matrix transformations with matrices $A$ and $B$ respectively. Then

$$S \circ T \text{ is the matrix transformation with matrix } AB.$$

**PROOF**    Given any vector $\vec{v}$ in $\mathbb{R}^2$, we have $S(\vec{v}) = A\vec{v}$ and $T(\vec{v}) = B\vec{v}$. Hence

$$[S \circ T](\vec{v}) = S[T(\vec{v})] = S[B(\vec{v})] = A[B(\vec{v})] = (AB)\vec{v}$$

for all $\vec{v}$ in $\mathbb{R}^2$. Hence $S \circ T$ is the matrix transformation induced by the matrix $AB$.

**Example 8**    Show that the effect of first reflecting in the $X$-axis and then reflecting in the line $y = x$ is the same as rotating through the angle $\frac{\pi}{2}$.

**SOLUTION**    Let $T : \mathbb{R}^2 \to \mathbb{R}^2$ and $S : \mathbb{R}^2 \to \mathbb{R}^2$ denote reflection in the $X$-axis and reflection in the line $y = x$, respectively. Then the effect of $T$ is $\begin{bmatrix} x \\ y \end{bmatrix} \to \begin{bmatrix} x \\ -y \end{bmatrix}$, while the effect of $S$ is $\begin{bmatrix} x \\ y \end{bmatrix} \to \begin{bmatrix} y \\ x \end{bmatrix}$. Hence the combined effect is

$$\begin{bmatrix} x \\ y \end{bmatrix} \xrightarrow{S} \begin{bmatrix} x \\ -y \end{bmatrix} \xrightarrow{T} \begin{bmatrix} -y \\ x \end{bmatrix},$$

which is rotation through $\frac{\pi}{2}$ by Example 1. We also can see this using matrix multiplication. Theorem 1 shows that the matrices of $S$ and $T$ are $A = \begin{bmatrix} 0 & 1 \\ 1 & 0 \end{bmatrix}$ and $B = \begin{bmatrix} 1 & 0 \\ 0 & -1 \end{bmatrix}$, respectively. It follows by Theorem 5 that the matrix of $S \circ T$ is

$$AB = \begin{bmatrix} 0 & 1 \\ 1 & 0 \end{bmatrix} \begin{bmatrix} 1 & 0 \\ 0 & -1 \end{bmatrix} = \begin{bmatrix} 0 & -1 \\ 1 & 0 \end{bmatrix},$$

which is the matrix of rotation through $\frac{\pi}{2}$, again by Example 1. Since $S \circ T$ is "first $T$, then $S$," this is what we wanted.

Theorem 5 is useful in two ways. First, it enables the effect of the composite of two geometric transformations to be computed using matrix multiplication. But it also provides a very useful geometric interpretation of matrix multiplication: The matrix product $AB$ corresponds to the transformation resulting from first applying $B$, and then applying $A$ (again, note the order). Thus the study of matrices can cast light on geometrical transformations and vice versa.

**Example 9**    Consider the rotations $R_\theta$ and $R_\psi$ through the angles $\theta$ and $\psi$. The composite $R_\theta \circ R_\psi$ is the transformation resulting from first rotating through $\psi$ and then rotating through $\theta$, and so is the rotation through the angle $\theta + \psi$ as is illustrated in Figure 3.61. Thus

$$R_\theta \circ R_\psi = R_{\theta + \psi}$$

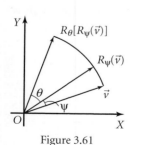

Figure 3.61

Hence Theorem 5 shows that the corresponding relationship holds for the matrices of these transformations:

$$\begin{bmatrix} \cos\theta & -\sin\theta \\ \sin\theta & \cos\theta \end{bmatrix} \begin{bmatrix} \cos\psi & -\sin\psi \\ \sin\psi & \cos\psi \end{bmatrix} = \begin{bmatrix} \cos(\theta+\psi) & -\sin(\theta+\psi) \\ \sin(\theta+\psi) & \cos(\theta+\psi) \end{bmatrix}.$$

If we perform the matrix multiplication on the left, and then compare first column entries, we obtain

$$\cos\theta\cos\psi - \sin\theta\sin\psi = \cos(\theta + \psi)$$
$$\sin\theta\cos\psi + \cos\theta\sin\psi = \sin(\theta + \psi)$$

These are the two basic identities from which most of trigonometry can be derived.

**3.4.5** **Inverse of a Matrix Transformation**

Another application of Theorem 5 provides a geometrical interpretation of the inverse of a matrix. The matrix transformation $T$ induced by a matrix $A$ is called an **invertible** linear transformation if $A$ is an invertible matrix and, in this case, the matrix transformation induced by the inverse matrix $A^{-1}$ is called the **inverse** of the transformation $T$ and is denoted $T^{-1}$. Hence $T^{-1}$ is defined by

$$T^{-1}(\vec{v}) = A^{-1}\vec{v} \quad \text{for all } \vec{v} \text{ in } \mathbb{R}^2.$$

Moreover Theorem 5 shows that $T \circ T^{-1}$ has matrix $A A^{-1} = I_2$, and it follows that $T[T^{-1}(\vec{v})] = (T \circ T^{-1})(\vec{v}) = I_2 \vec{v} = \vec{v}$ for all $\vec{v}$ in $\mathbb{R}^2$. Similarly $T^{-1} \circ T$ has matrix $A^{-1} A = I_2$, and we obtain

$$T[T^{-1}(\vec{v})] = \vec{v} \quad \text{and} \quad T^{-1}[T(\vec{v})] = \vec{v} \quad \text{for all } \vec{v} \text{ in } \mathbb{R}^2.$$

These are called the **fundamental identities** relating $T$ and $T^{-1}$. They show that $T^{-1}$ "reverses" the action of the transformation $T$ in the following sense: If $T$ carries a vector $\vec{v}$ to $\vec{w} = T(\vec{v})$, then $T^{-1}$ carries $\vec{w}$ right back to $\vec{v}$. Similarly $T$ reverses the action of $T^{-1}$. Thus the basic relationship between $T$ and $T^{-1}$ is that each reverses the action of the other.

Recall that the matrix transformation induced by the identity matrix $I$ is called the identity transformation $1_{\mathbb{R}^2} : \mathbb{R}^2 \to \mathbb{R}^2$. It is given by $1_{\mathbb{R}^2}(\vec{v}) = I\vec{v} = \vec{v}$ for each $\vec{v}$ in $\mathbb{R}^2$. Using this, the fundamental identities can by reformulated as follows:

$$T \circ T^{-1} = 1_{\mathbb{R}^2} \quad \text{and} \quad T^{-1} \circ T = 1_{\mathbb{R}^2}$$

This leads to a useful characterization of when a linear transformation is invertible.

**THEOREM 6**

Let $T : \mathbb{R}^2 \to \mathbb{R}^2$ be the linear transformation induced by a $2 \times 2$ matrix $A$. The following conditions are equivalent:

(1) $T$ is invertible (that is $A$ is invertible).
(2) There exists a linear transformation $S : \mathbb{R}^2 \to \mathbb{R}^2$ such that
$T \circ S = 1_{\mathbb{R}^2}$ and $S \circ T = 1_{\mathbb{R}^2}$.

In this case, the matrix of $S$ is $A^{-1}$, so $S = T^{-1}$ is the inverse of $T$.

**PROOF**

If (1) holds, then (2) follows with $S = T^{-1}$. Conversely, suppose that (2) is true. Let $S$ be induced by the matrix $B$. By Theorem 5, the equations $T \circ S = 1_{\mathbb{R}^2}$ and $S \circ T = 1_{\mathbb{R}^2}$ translate to $AB = I$ and $BA = I$. Thus $A$ is invertible and $A^{-1} = B$. This proves (2) and the last statement.

This geometric view of the inverse of a linear transformation provides a way to find the inverse of a matrix. More precisely, if $A$ is an invertible matrix:

(1) Let $T$ be the linear transformation induced by $A$.
(2) Obtain the linear transformation $T^{-1}$ which "reverses" the action of $T$.
(3) Then $A^{-1}$ is the matrix of $T^{-1}$.

Here is an example.

**Example 10**

If $\theta$ is any angle, find the inverse of the matrix

$$A = \begin{bmatrix} \cos \theta & -\sin \theta \\ \sin \theta & \cos \theta \end{bmatrix}$$

by examining the corresponding matrix transformation.

SOLUTION | By Theorem 3, $A$ is the matrix of the rotation $R_\theta$ through the angle $\theta$. Hence $A^{-1}$ will be the matrix of the inverse transformation $[R_\theta]^{-1}$. But $[R_\theta]^{-1}$ must be the rotation $R_{-\theta}$ through the angle $-\theta$ because $R_{-\theta}$ "reverses" the action of $R_\theta$. Hence $A^{-1}$ is the matrix of $R_{-\theta}$, so (by Theorem 3 again with $\theta$ replaced by $-\theta$)

$$A^{-1} = \begin{bmatrix} \cos(-\theta) & -\sin(-\theta) \\ \sin(-\theta) & \cos(-\theta) \end{bmatrix} = \begin{bmatrix} \cos\theta & \sin\theta \\ -\sin\theta & \cos\theta \end{bmatrix}.$$

Of course this can be verified directly by checking that $AA' = I = A'A$.

## 3.4.6  Isometries[14]

A linear transformation $T : \mathbb{R}^2 \to \mathbb{R}^2$ is called an **isometry** if it preserves the distance between any two vectors; that is, if the distance between $T(\vec{v})$ and $T(\vec{w})$ equals the distance between $\vec{v}$ and $\vec{w}$ for all vectors $\vec{v}$ and $\vec{w}$. Since the distance between vectors $\vec{v}$ and $\vec{w}$ is $\|\vec{v} - \vec{w}\|$, this means that

$$\|T(\vec{v}) - T(\vec{w})\| = \|\vec{v} - \vec{w}\| \text{ for all } \vec{v} \text{ and } \vec{w}.$$

If $T$ is induced by the matrix $A$, this condition becomes

$$\|A\vec{v} - A\vec{w}\| = \|\vec{v} - \vec{w}\| \text{ for all } \vec{v} \text{ and } \vec{w}. \tag{$*$}$$

It is clear geometrically that every rotation about the origin is an isometry, as is every reflection about a line through the origin. The surprising fact is that these are the only possibilities.

To see why, let $T : \mathbb{R}^2 \to \mathbb{R}^2$ be an isometry and let $A = \begin{bmatrix} a & b \\ c & d \end{bmatrix}$ be the matrix of $T$. If we set $\vec{w} = \vec{0}$ in equation $(*)$, the result is $\|A\vec{v}\| = \|\vec{v}\|$ for all $\vec{v}$. Taking $\vec{v} = \begin{bmatrix} 1 \\ 0 \end{bmatrix}$, $\vec{v} = \begin{bmatrix} 0 \\ 1 \end{bmatrix}$, and $\vec{v} = \begin{bmatrix} 1 \\ 1 \end{bmatrix}$ results in the equations

$$a^2 + c^2 = 1, \quad b^2 + d^2 = 1 \quad \text{and} \quad ab + cd = 0$$

respectively. These in turn show that $A^T A = I$ as the reader can verify, whence

$$1 = det(I) = det(A^T A) = det(A^T)\,det(A) = (det A)^2.$$

It follows that $det A = 1$ or $det A = -1$, and it turns out that $T$ is a rotation if $det A = 1$ and $T$ is a reflection if $det A = -1$. We examine each case separately.

**Case 1.** $det A = 1$. Then $1 = ad - bc$ and multiplying this by $c$ gives

$$c = acd - bc^2 = a(-ab) - bc^2 = -b(a^2 + c^2) = -b.$$

In the same way, multiplying $1 = ad - bc$ by $d$ yields $d = a$. Thus $A$ has the form

$$A = \begin{bmatrix} a & -c \\ c & a \end{bmatrix} \quad \text{where } a^2 + c^2 = 1.$$

The condition $a^2 + c^2 = 1$ means that the point $P(a,\ c)$ lies on the unit circle. Hence $a = \cos\theta$ and $c = \sin\theta$ where $\theta$ is the angle determined by $P(a,\ c)$, so we have

$$A = \begin{bmatrix} \cos\theta & -\sin\theta \\ \sin\theta & \cos\theta \end{bmatrix} \quad \text{for some angle } \theta.$$

Hence $T$ is the rotation $R_\theta$ through the angle $\theta$ by Theorem 3.

**Case 2.** $det A = -1$. Now $1 = bc - ad$ and an argument analogous to that in Case 1 yields that $A = \begin{bmatrix} a & c \\ c & -a \end{bmatrix}$ where $a^2 + c^2 = 1$. Again, $a = \cos\theta$ and $c = \sin\theta$ where $0 \le \theta < 2\pi$, so

---
[14] While this section is mentioned later, it can be omitted without loss of continuity.

$$A = \begin{bmatrix} \cos\theta & \sin\theta \\ \sin\theta & -\cos\theta \end{bmatrix} \quad \text{for some angle } \theta.$$

If $\theta = \pi$, then $A = \begin{bmatrix} -1 & 0 \\ 0 & 1 \end{bmatrix}$ so $T$ is reflection in the $Y$-axis. If $\theta \neq \pi$, then $\cos\left(\frac{\theta}{2}\right) \neq 0$ because $\frac{\theta}{2} \neq \frac{\pi}{2}$ and $0 \leq \frac{\theta}{2} < \pi$. Hence $\tan\left(\frac{\theta}{2}\right) = \frac{\sin\left(\frac{\theta}{2}\right)}{\cos\left(\frac{\theta}{2}\right)}$ is defined, and we claim that $T$ is the reflection in the line through the origin with slope $m = \tan\left(\frac{\theta}{2}\right)$. In fact, Theorem 1 (and some standard trigonometry) shows that this reflection has matrix

$$\frac{1}{1+m^2}\begin{bmatrix} 1-m^2 & 2m \\ 2m & m^2-1 \end{bmatrix} = \begin{bmatrix} \cos\theta & \sin\theta \\ \sin\theta & -\cos\theta \end{bmatrix} = A.$$

It is worth noting that, in this case, $y = mx$ is the line through the origin which makes an angle $\frac{\theta}{2}$ with the positive $X$-axis.

This discussion is summarized as part of the following theorem.

**THEOREM 7**

Every isometry $T : \mathbb{R}^2 \to \mathbb{R}^2$ is either a rotation or a reflection. Moreover, if $T$ is induced by the matrix $A$, then either $detA = 1$ or $detA = -1$, and

$$T \text{ is a rotation} \quad \text{if and only if} \quad detA = 1,$$
$$T \text{ is a reflection} \quad \text{if and only if} \quad detA = -1.$$

**PROOF**

All that remains is to show that $detA = 1$ if $T$ is a rotation, and $detA = -1$ if $T$ is a reflection. But these follow by Theorem 3 and Theorem 1, respectively.

## 3.4.7 Computer Graphics[15]

Figure 3.62

Figure 3.63

Computer graphics deals with images displayed on a computer screen, and so arise in a variety of applications, ranging from word processors, to *Star Wars* animations, to video games, to wire-frame images of an airplane. These images consist of a number of points on the screen, together with instructions on how to fill in areas bounded by lines and curves. Often curves are approximated by a set of short straight-line segments, so that the curve is specified by a series of points on the screen. Matrix transformations are important here because matrix images of straight line segments are again[16] line segments. Note that a colour image requires that three images are sent, one to each of the red, green, and blue phosphorus dots on the screen, in varying intensities.

Consider displaying the letter K. In reality, it is depicted on the screen as in Figure 3.62 by specifying the coordinates of the 12 corners and filling in the interior. For simplicity, we will disregard the thickness of the letter, so we require only 6 coordinates as in Figure 3.63. This simplified letter can then be stored as a data matrix

$$\begin{array}{cccccc} \textit{Vertex} & {\scriptstyle 1\ \ 2\ \ 3\ \ \ \ 4\ \ \ \ 5\ \ \ 6} \\ K_d = & \begin{bmatrix} 0 & 7 & 3 & 7 & 0 & 0 \\ 0 & 0 & 6 & 10 & 10 & 3 \end{bmatrix} \end{array}$$

where the columns are the coordinates of the vertices in order (and the letter is 10 units high and 7 units wide). We can slant the letter to the right by multiplying by an $X$-shear matrix $A = \begin{bmatrix} 1 & 0.2 \\ 0 & 1 \end{bmatrix}$. The result is the letter with data matrix

$$AK_d = \begin{bmatrix} 1 & 0.2 \\ 0 & 1 \end{bmatrix}\begin{bmatrix} 0 & 7 & 3 & 7 & 0 & 0 \\ 0 & 0 & 6 & 10 & 10 & 3 \end{bmatrix} = \begin{bmatrix} 0 & 7 & 4.2 & 9 & 2 & 0.6 \\ 0 & 0 & 6 & 10 & 10 & 3 \end{bmatrix}$$

---

[15] This section is not referred to elsewhere in this book.

[16] If a vector $\vec{v}$ lies on the line through $\vec{v}_0$ with direction vector $\vec{d}$, then $\vec{v} = \vec{v}_0 + t\vec{d}$ for some number $t$. Hence the matrix image $A\vec{v} = A\vec{v}_0 + tA\vec{d}$ lies on the line through $A\vec{v}_0$ with direction vector $A\vec{d}$. (Note that if $A\vec{d} = \vec{0}$, the image is the single point $A\vec{v}_0$.)

origin

Figure 3.64

origin

Figure 3.65

origin

Figure 3.66

which is shown in Figure 3.64. If we want to make the matrix narrower, we can now apply an $X$-scale matrix $B = \begin{bmatrix} 0.8 & 0 \\ 0 & 1 \end{bmatrix}$ to get

$$BAK_d = \begin{bmatrix} 0.8 & 0 \\ 0 & 1 \end{bmatrix}\begin{bmatrix} 1 & 0.2 \\ 0 & 1 \end{bmatrix}\begin{bmatrix} 0 & 7 & 3 & 7 & 0 & 0 \\ 0 & 0 & 6 & 10 & 10 & 3 \end{bmatrix}$$
$$= \begin{bmatrix} 0 & 5.6 & 3.36 & 7.2 & 1.6 & 0.48 \\ 0 & 0 & 6 & 10 & 10 & 3 \end{bmatrix}$$

and which is drawn in Figure 3.65.

On the other hand, we can rotate the letter about the origin through $\frac{\pi}{6}$ (or 30°) by multiplying by the matrix $R_{\frac{\pi}{6}} = \begin{bmatrix} \cos(\frac{\pi}{6}) & -\sin(\frac{\pi}{6}) \\ \sin(\frac{\pi}{6}) & \cos(\frac{\pi}{6}) \end{bmatrix} = \begin{bmatrix} 0.866 & -0.5 \\ 0.5 & 0.866 \end{bmatrix}$.

This gives

$$R_{\frac{\pi}{6}}K_d = \begin{bmatrix} 0.866 & -0.5 \\ 0.5 & 0.866 \end{bmatrix}\begin{bmatrix} 0 & 7 & 3 & 7 & 0 & 0 \\ 0 & 0 & 6 & 10 & 10 & 3 \end{bmatrix}$$
$$= \begin{bmatrix} 0 & 6.062 & -0.402 & 1.062 & -5.0 & -1.5 \\ 0 & 3.5 & 6.696 & 12.16 & 8.66 & 2.598 \end{bmatrix}$$

and is plotted in Figure 3.66.

This poses a problem: How do we rotate about a point other than the origin? It turns out that we can do this when we have solved another more basic problem. It is clearly important to be able to translate a screen image by a fixed vector $\vec{w}$, that is apply the transformation $T_{\vec{w}} : \mathbb{R}^2 \to \mathbb{R}^2$ given by $T_{\vec{w}}(\vec{v}) = \vec{v} + \vec{w}$ for all $\vec{v}$ in $\mathbb{R}^2$. The problem is that these translations are not matrix transformations (see Example 3). However, there is a clever way around this.

The idea is to represent a point $\vec{v} = \begin{bmatrix} x \\ y \end{bmatrix}$ as a $3 \times 1$ column $\begin{bmatrix} x \\ y \\ 1 \end{bmatrix}$, called the **homogeneous coordinates** of $\vec{v}$. Then translation by $\vec{w} = \begin{bmatrix} p \\ q \end{bmatrix}$ can be achieved by multiplying by a $3 \times 3$ matrix:

$$\begin{bmatrix} 1 & 0 & p \\ 0 & 1 & q \\ 0 & 0 & 1 \end{bmatrix}\begin{bmatrix} x \\ y \\ 1 \end{bmatrix} = \begin{bmatrix} x+p \\ y+q \\ 1 \end{bmatrix} = \begin{bmatrix} T_{\vec{w}}(\vec{v}) \\ 1 \end{bmatrix}.$$

Thus, by using homogeneous coordinates we can implement the translation $T_{\vec{w}}$ in the top two coordinates. On the other hand, the matrix transformation induced by $A = \begin{bmatrix} a & b \\ c & d \end{bmatrix}$ is also given by a $3 \times 3$ matrix:

$$\begin{bmatrix} a & b & 0 \\ c & d & 0 \\ 0 & 0 & 1 \end{bmatrix}\begin{bmatrix} x \\ y \\ 1 \end{bmatrix} = \begin{bmatrix} ax+by \\ cx+dy \\ 1 \end{bmatrix} = \begin{bmatrix} A\vec{v} \\ 1 \end{bmatrix}.$$

So everything can be accomplished at the expense of using $3 \times 3$ matrices and homogeneous coordinates.

**Example 11**   Rotate the letter K in Figure 3.63 through $\frac{\pi}{6}$ about the point $\begin{bmatrix} 4 \\ 5 \end{bmatrix}$.

**SOLUTION**   Using homogeneous coordinates for the vertices of the letter results in a data matrix with three rows:

$$K_d = \begin{bmatrix} 0 & 7 & 3 & 7 & 0 & 0 \\ 0 & 0 & 6 & 10 & 10 & 3 \\ 1 & 1 & 1 & 1 & 1 & 1 \end{bmatrix}.$$

Figure 3.67

If we write $\vec{w} = \begin{bmatrix} 4 \\ 5 \end{bmatrix}$, the idea is to first translate the letter by $-\vec{w}$ so that the point $\vec{w}$ moves to the origin, then rotate this translated letter, and then translate it by $\vec{w}$ back to its original position. The matrix arithmetic is as follows:

$$\begin{bmatrix} 1 & 0 & 4 \\ 0 & 1 & 5 \\ 0 & 0 & 1 \end{bmatrix} \begin{bmatrix} 0.866 & -0.5 & 0 \\ 0.5 & 0.866 & 0 \\ 0 & 0 & 1 \end{bmatrix} \begin{bmatrix} 1 & 0 & -4 \\ 0 & 1 & -5 \\ 0 & 0 & 1 \end{bmatrix} \begin{bmatrix} 0 & 7 & 3 & 7 & 0 & 0 \\ 0 & 0 & 6 & 10 & 10 & 3 \\ 1 & 1 & 1 & 1 & 1 & 1 \end{bmatrix}$$

$$= \begin{bmatrix} 3.036 & 9.098 & 2.634 & 4.098 & -1.964 & 1.536 \\ -1.33 & 2.17 & 5.366 & 10.83 & 7.33 & 1.268 \\ 1 & 1 & 1 & 1 & 1 & 1 \end{bmatrix}.$$

This is plotted in Figure 3.67.

This discussion merely touches the surface of computer graphics, and the reader is referred to specialized books on the subject. Realistic graphic rendering requires an enormous number of matrix calculations. In fact, matrix multiplication algorithms are now embedded in microchip circuits, and can perform over 100 million matrix multiplications per second. This is particularly important in the field of three-dimensional graphics where the homogeneous coordinates have four components and $4 \times 4$ matrices are required.

## Exercises 3.4

**1.** In each case show that $T : \mathbb{R}^2 \to \mathbb{R}^2$ is a matrix transformation, and find the matrix of $T$:

**a)** $T \begin{bmatrix} x \\ y \end{bmatrix} = \begin{bmatrix} 2x+y \\ 3x-2y \end{bmatrix}$ 

**b)** $T \begin{bmatrix} x \\ y \end{bmatrix} = \begin{bmatrix} 3x-y \\ 2y+x \end{bmatrix}$

**c)** $T \begin{bmatrix} x \\ y \end{bmatrix} = \begin{bmatrix} 0 \\ 5y+6x \end{bmatrix}$ 

**d)** $T \begin{bmatrix} x \\ y \end{bmatrix} = \begin{bmatrix} y-x \\ x+y \end{bmatrix}$

**2.** In each case, assume that $T : \mathbb{R}^2 \to \mathbb{R}^2$ is a linear transformation, use the information to determine $T \begin{bmatrix} x \\ y \end{bmatrix}$ for all $\begin{bmatrix} x \\ y \end{bmatrix}$ in $\mathbb{R}^2$, and find the matrix of $T$.

**a)** $T \begin{bmatrix} 1 \\ 0 \end{bmatrix} = \begin{bmatrix} 2 \\ -3 \end{bmatrix}$ and $T \begin{bmatrix} 0 \\ 1 \end{bmatrix} = \begin{bmatrix} 4 \\ -2 \end{bmatrix}$

**b)** $T \begin{bmatrix} 1 \\ 2 \end{bmatrix} = \begin{bmatrix} 1 \\ -3 \end{bmatrix}$ and $T \begin{bmatrix} 3 \\ -1 \end{bmatrix} = \begin{bmatrix} 2 \\ 1 \end{bmatrix}$

**c)** $T \begin{bmatrix} 1 \\ 1 \end{bmatrix} = \begin{bmatrix} 2 \\ -1 \end{bmatrix}$ and $T \begin{bmatrix} 1 \\ 0 \end{bmatrix} = \begin{bmatrix} -1 \\ 1 \end{bmatrix}$

**d)** $T \begin{bmatrix} 1 \\ -1 \end{bmatrix} = \begin{bmatrix} 1 \\ 0 \end{bmatrix}$ and $T \begin{bmatrix} 0 \\ 1 \end{bmatrix} = \begin{bmatrix} 2 \\ -3 \end{bmatrix}$

**3.** Give the matrix of each of the following transformations and sketch the image of the unit square.

**a)** Reflection in the line $y = -x$.

**b)** Reflection in the line $y = 2x$.

**c)** Rotation through $\frac{\pi}{4}$.

**d)** Rotation through $-\frac{\pi}{2}$.

**e)** Projection on the $Y$-axis.

**f)** Projection on the line $y = -x$.

**4.** In each case show that $T$ is not a matrix transformation.

**a)** $T \begin{bmatrix} x \\ y \end{bmatrix} = \begin{bmatrix} xy \\ 0 \end{bmatrix}$ 

**b)** $T \begin{bmatrix} x \\ y \end{bmatrix} = \begin{bmatrix} 0 \\ y^2 \end{bmatrix}$

**5.** In each case show that $T : \mathbb{R}^2 \to \mathbb{R}^2$ is a matrix transformation, and find the matrix of $T$.

**a)** $T(\vec{v}) = a\vec{v}$ for all $\vec{v}$ in $\mathbb{R}^2$. (Here $a$ is a fixed number.)

**b)** $T \begin{bmatrix} x \\ y \end{bmatrix} = x\vec{u} + y\vec{w}$ for all $\begin{bmatrix} x \\ y \end{bmatrix}$ in $\mathbb{R}^2$. (Here $\vec{u} = \begin{bmatrix} p \\ q \end{bmatrix}$ and $\vec{w} = \begin{bmatrix} r \\ s \end{bmatrix}$ are fixed in $\mathbb{R}^2$.)

**6.** In each case, find the matrix of $T$, $T \begin{bmatrix} 1 \\ 1 \end{bmatrix}$, and $T \begin{bmatrix} 2 \\ -1 \end{bmatrix}$, and sketch the image of the unit square.

**a)** $T$ is reflection in the line $y = -2x$.

**b)** $T$ is reflection in the line $y = 3x$.

**c)** $T$ is rotation through $\frac{\pi}{3}$.

**d)** $T$ is rotation through $-\frac{\pi}{4}$.

**e)** $T$ is projection on the line $y = 3x$.

**f)** $T$ is projection on the line $x + 2y = 0$.

**7.** In each case show that $T$ is a projection on a line, or a reflection in a line, or a rotation through an angle, and find the line or angle.

**a)** $T \begin{bmatrix} x \\ y \end{bmatrix} = \frac{1}{5} \begin{bmatrix} x+2y \\ 2x+4y \end{bmatrix}$ 

**b)** $T \begin{bmatrix} x \\ y \end{bmatrix} = \frac{1}{2} \begin{bmatrix} x-y \\ y-x \end{bmatrix}$

**c)** $T \begin{bmatrix} x \\ y \end{bmatrix} = \frac{1}{\sqrt{2}} \begin{bmatrix} -x-y \\ x-y \end{bmatrix}$ 

**d)** $T \begin{bmatrix} x \\ y \end{bmatrix} = \frac{1}{5} \begin{bmatrix} -3x+4y \\ 4x+3y \end{bmatrix}$

**e)** $T \begin{bmatrix} x \\ y \end{bmatrix} = \begin{bmatrix} -y \\ -x \end{bmatrix}$ 

**f)** $T \begin{bmatrix} x \\ y \end{bmatrix} = \frac{1}{2} \begin{bmatrix} x-\sqrt{3}y \\ \sqrt{3}x+y \end{bmatrix}$

**8.** Let $R$ and $S$ be the matrix transformations induced by matrices $A$ and $B$ respectively. In each case show that $T$ is a matrix transformation and find its matrix in terms of $A$ and $B$.
   **a)** $T(\vec{v}) = R(\vec{v}) + S(\vec{v})$ for all $\vec{v}$ in $\mathbb{R}^2$.
   **b)** $T(\vec{v}) = k\,R(\vec{v})$ for all $\vec{v}$ in $\mathbb{R}^2$. (Here $k$ is a fixed number.)

**9.** Given $a > 0$, let $T : \mathbb{R}^2 \to \mathbb{R}^2$ be the transformation which carries each point at distance $r$ from the origin radially out (or in if $a < 1$) to the point at distance $ar$ from the origin. [$T$ is called a **dilation** if $a > 1$; a **contraction** if $0 \le a \le 1$.]
   **a)** Show that $T$ is a matrix transformation, find its matrix, and sketch the image of the unit square.
   **b)** Show that $T \circ S = S \circ T$ holds for every matrix transformation $S : \mathbb{R}^2 \to \mathbb{R}^2$.
   **c)** If $R : \mathbb{R}^2 \to \mathbb{R}^2$ is a matrix transformation such that $R \circ S = S \circ R$ holds for every matrix transformation $S : \mathbb{R}^2 \to \mathbb{R}^2$, show that $R$ is a dilation or a contraction. [Hint: See Exercise 29, §1.4.]

**10.** Let $L$ be the line through the origin with direction vector $\vec{d} \ne \vec{0}$, and let $P_L$ denote the projection on $L$. Consider the formula $P_L(\vec{v}) = \frac{\vec{v} \cdot \vec{d}}{\|\vec{d}\|^2}\vec{d}$ for all $\vec{v}$ in $\mathbb{R}^2$.
   **a)** Use the formula to show that $P_L(\vec{v})$ is the same no matter which (nonzero) direction vector $\vec{d}$ is used for $L$.
   **b)** Use the formula to show directly that $P_L$ is linear.

**11.** Show that the following hold for all matrix transformations $T : \mathbb{R}^2 \to \mathbb{R}^2$.
   **a)** $T(\vec{0}) = \vec{0}$.    **b)** $T(-\vec{v}) = -T(\vec{v})$ for all $\vec{v}$ in $\mathbb{R}^2$.
   **c)** $T(a\vec{v} + b\vec{w}) = aT(\vec{v}) + bT(\vec{w})$ for all $\vec{v}$ and $\vec{w}$ in $\mathbb{R}^2$.

**12.** Let $L$ denote the line through the origin with (nonzero) direction vector $\vec{d} = \begin{bmatrix} a \\ b \end{bmatrix}$.
   **a)** Show that the matrix of the projection on $L$ is
   $$\frac{1}{a^2+b^2}\begin{bmatrix} a^2 & ab \\ ab & b^2 \end{bmatrix}.$$
   **b)** Show that the matrix of the reflection in $L$ is
   $$\frac{1}{a^2+b^2}\begin{bmatrix} a^2-b^2 & 2ab \\ 2ab & b^2-a^2 \end{bmatrix}.$$

**13.** Let $T$ be the matrix transformation induced by an invertible $2 \times 2$ matrix $A$. In each case, interpret $T^{-1}$ geometrically and so (as in Example 10) find $A^{-1}$.
   **a)** $A = \begin{bmatrix} 2 & 0 \\ 0 & 1 \end{bmatrix}$    **b)** $A = \begin{bmatrix} 1 & 5 \\ 0 & 1 \end{bmatrix}$
   **c)** $A = \frac{1}{5}\begin{bmatrix} -3 & 4 \\ 4 & 3 \end{bmatrix}$    **d)** $A = \frac{1}{\sqrt{2}}\begin{bmatrix} 1 & -1 \\ 1 & 1 \end{bmatrix}$
   **e)** $A = \begin{bmatrix} 0 & -1 \\ -1 & 0 \end{bmatrix}$    **f)** $A = \frac{1}{1+m^2}\begin{bmatrix} 1-m^2 & 2m \\ 2m & m^2-1 \end{bmatrix}$

**14.** In each case find a rotation or reflection which equals the given transformation.
   **a)** Reflection in the $Y$-axis followed by rotation through $\frac{\pi}{2}$.

   **b)** Rotation through $\pi$ followed by reflection in the $X$-axis.
   **c)** Rotation through $\frac{\pi}{2}$ followed by reflection in the line $y = x$.
   **d)** Reflection in the $X$-axis followed by rotation through $\frac{\pi}{2}$.
   **e)** Reflection in the line $y = x$ followed by reflection in the $X$-axis.

**15.** Express reflection $R$ in the line $y = -x$ as the composite of a rotation $T$ followed by reflection $S$ in the line $y = x$. [Hint: If $R = S \circ T$, then $T = S^{-1} \circ R$.]

**16.** Determine the effect of the following transformations.
   **a)** Rotation through $\frac{\pi}{2}$, followed by projection on the $Y$-axis, followed by reflection in the line $y = x$.
   **b)** Reflection in the line $y = x$ followed by reflection in the line $y = -x$.
   **c)** Projection on the $X$-axis followed by reflection in the line $y = x$.
   **d)** $Y$-expansion by 2 followed by the rotation through $\frac{\pi}{2}$.

**17.** Given any line $y = mx$ through the origin, show that $P_m \circ Q_m = P_m = Q_m \circ P_m$ in two ways: **a)** geometrically; **b)** using Theorem 1.

**18.** Given two perpendicular lines $y = mx$ and $y = m_1x$ through the origin, show that $P_m \circ P_{m_1} = 0$ in two ways: **a)** geometrically; **b)** using Theorem 1.

**19.** If $T$ is any projection, show that $T \circ T = T$ in two ways: **a)** geometrically; **b)** using Theorem 1.

**20.** If $T$ is any reflection, show that $T^{-1} = T$ in two ways: **a)** geometrically; **b)** using Theorem 1.

**21. a)** Show that the composite of two isometries is an isometry.
   **b)** Show that the composite of two reflections is a rotation.

**22.** Show that every reflection is a reflection in the $X$-axis followed by a rotation. [Hint: Compose it with the reflection in the $X$-axis, and use Exercises 20 and 21.]

**23.** Let $area(\vec{v}, \vec{w})$ denote the area of the parallelogram determined by the vectors $\vec{v}$ and $\vec{w}$.
   **a)** Show that $area(A\vec{v}, A\vec{w}) = |det(A)|\,\big|det[\vec{v}, \vec{w}]\big|$ where $[\vec{v}, \vec{w}]$ denotes the matrix with columns $\vec{v}$ and $\vec{w}$. [Hint: Write $\vec{v} = [p \ \ q]^T$ and $\vec{w} = [r \ \ s]^T$, and view them as $\vec{v} = [p \ \ q \ \ 0]^T$ and $\vec{w} = [r \ \ s \ \ 0]^T$ in $\mathbb{R}^3$. Proceed as in the proof of Theorem 4.]
   **b)** Use the result in **a)** to deduce Theorem 4.

**24.** A $2 \times 2$ matrix $A$ is called **orthogonal** if $A$ is invertible and $A^{-1} = A^T$.
   **a)** Show that the matrix of any isometry is orthogonal. [Hint: Theorem 7.]
   **b)** If $A$ is orthogonal, show that the corresponding matrix transformation is an isometry.

**25.** In each case, find the $3 \times 3$ matrix for the transformation:

   **a)** Reflection in $y = -x$ followed by translation by $\begin{bmatrix} 2 \\ -1 \end{bmatrix}$.

   **b)** Rotation through $\frac{\pi}{2}$ followed by translation by $\begin{bmatrix} 1 \\ 1 \end{bmatrix}$.

   **c)** Translation by $\begin{bmatrix} 1 \\ -1 \end{bmatrix}$ followed by projection on $y = 2x$.

   **d)** Translation by $\begin{bmatrix} 1 \\ 2 \end{bmatrix}$ followed by rotation through $\frac{\pi}{3}$.

**26.** In each case, find the $3 \times 3$ matrix for the transformation:

   **a)** Reflection in the line $y = mx + b$.

   **b)** Projection on the line $y = mx + b$.

---

## 3.5 THE CROSS PRODUCT[17]

In this section we derive some properties of the cross product not given in Section 3.3.

---

### 3.5.1 The Cross Product

Let $\vec{v} = \begin{bmatrix} x_1 \\ y_1 \\ z_1 \end{bmatrix}$ and $\vec{w} = \begin{bmatrix} x_2 \\ y_2 \\ z_2 \end{bmatrix}$ be vectors. In Section 3.3 we introduced the **cross product** $\vec{v} \times \vec{w}$ defined by

$$\vec{v} \times \vec{w} = \begin{bmatrix} y_1 z_2 - y_2 z_1 \\ -(x_1 z_2 - x_2 z_1) \\ x_1 y_2 - x_2 y_1 \end{bmatrix}. \tag{$*$}$$

Observe that $\vec{v} \times \vec{w}$ is a *vector* (as opposed to the *scalar* $\vec{v} \cdot \vec{w}$) and, for this reason, $\vec{v} \times \vec{w}$ is sometimes called the **vector product** of $\vec{v}$ and $\vec{w}$. The motivation for introducing $\vec{v} \times \vec{w}$ was that it is orthogonal to both $\vec{v}$ and $\vec{w}$. This is verified below (Theorem 2) using a description of $\vec{v} \times \vec{w}$ which employs determinants.

The determinant description of the cross product utilizes the **coordinate vectors**

$$\vec{\imath} = \begin{bmatrix} 1 \\ 0 \\ 0 \end{bmatrix}, \ \vec{\jmath} = \begin{bmatrix} 0 \\ 1 \\ 0 \end{bmatrix} \text{ and } \vec{k} = \begin{bmatrix} 0 \\ 0 \\ 1 \end{bmatrix}.$$

These are vectors of length $1$[18] pointing along the positive $X$-, $Y$- and $Z$-axes respectively, and the name comes from that fact that any vector $\begin{bmatrix} x \\ y \\ z \end{bmatrix}$ can be written in the form

$$\begin{bmatrix} x \\ y \\ z \end{bmatrix} = x\vec{\imath} + y\vec{\jmath} + z\vec{k}.$$

This enables us to give a useful description of the cross product using determinants. It comes from the observation that the components of $\vec{v} \times \vec{w}$ in $(*)$ can be written as $2 \times 2$ determinants: $det\begin{bmatrix} y_1 & y_2 \\ z_1 & z_2 \end{bmatrix}$, $-det\begin{bmatrix} x_1 & x_2 \\ z_1 & z_2 \end{bmatrix}$, and $det\begin{bmatrix} x_1 & x_2 \\ y_1 & y_2 \end{bmatrix}$. This prompts the following way of writing the cross product.

---

[17] This section can be omitted as it is used very little in the sequel.

[18] Vectors of length 1 are called **unit vectors**. They will reappear in Chapter 4.

**Determinant Description of the Cross Product.**

If $\vec{v} = \begin{bmatrix} x_1 \\ y_1 \\ z_1 \end{bmatrix}$ and $\vec{w} = \begin{bmatrix} x_2 \\ y_2 \\ z_2 \end{bmatrix}$ are vectors, the cross product is given by

$$\vec{v} \times \vec{w} = det \begin{bmatrix} \vec{\imath} & x_1 & x_2 \\ \vec{\jmath} & y_1 & y_2 \\ \vec{k} & z_1 & z_2 \end{bmatrix} = det \begin{bmatrix} y_1 & y_2 \\ z_1 & z_2 \end{bmatrix} \vec{\imath} - det \begin{bmatrix} x_1 & x_2 \\ z_1 & z_2 \end{bmatrix} \vec{\jmath} + det \begin{bmatrix} x_1 & x_2 \\ y_1 & y_2 \end{bmatrix} \vec{k}.$$

Here the determinant is expanded along the first column by cofactors, so the components of $\vec{v} \times \vec{w}$ are the coefficients of $\vec{\imath}$, $\vec{\jmath}$, and $\vec{k}$ respectively.

**Example 1**    If $\vec{v} = \begin{bmatrix} 4 \\ -2 \\ 1 \end{bmatrix}$ and $\vec{w} = \begin{bmatrix} 5 \\ 0 \\ 3 \end{bmatrix}$, then $\vec{v} \times \vec{w} = det \begin{bmatrix} \vec{\imath} & 4 & 5 \\ \vec{\jmath} & -2 & 0 \\ \vec{k} & 1 & 3 \end{bmatrix} = \begin{bmatrix} -6 \\ -7 \\ 10 \end{bmatrix}$.

Note that $\vec{v} \times \vec{w}$ is orthogonal to both $\vec{v}$ and $\vec{w}$ in Example 1 because, as is easily verified, $\vec{v} \cdot (\vec{v} \times \vec{w}) = 0$ and $\vec{w} \cdot (\vec{v} \times \vec{w}) = 0$. This happens in every case as we shall see. In fact there is an easy way to compute the number $\vec{u} \cdot (\vec{v} \times \vec{w})$ for *any* vector $\vec{u}$: It is the determinant of the matrix $[\vec{u} \ \ \vec{v} \ \ \vec{w}]$ which has $\vec{u}$, $\vec{v}$, and $\vec{w}$ as its columns.

**THEOREM 1**    If $\vec{u}$, $\vec{v}$, and $\vec{w}$ are vectors, then $\vec{u} \cdot (\vec{v} \times \vec{w}) = det[\vec{u} \ \ \vec{v} \ \ \vec{w}]$.

**PROOF**    Write $\vec{u} = \begin{bmatrix} x_0 \\ y_0 \\ z_0 \end{bmatrix}$, $\vec{v} = \begin{bmatrix} x_1 \\ y_1 \\ z_1 \end{bmatrix}$, and $\vec{w} = \begin{bmatrix} x_2 \\ y_2 \\ z_2 \end{bmatrix}$ in matrix form.

Then $\vec{v} \times \vec{w} = \begin{bmatrix} det \begin{bmatrix} y_1 & y_2 \\ z_1 & z_2 \end{bmatrix} & -det \begin{bmatrix} x_1 & x_2 \\ z_1 & z_2 \end{bmatrix} & det \begin{bmatrix} x_1 & x_2 \\ y_1 & y_2 \end{bmatrix} \end{bmatrix}^T$, so

$$\vec{u} \cdot (\vec{v} \times \vec{w}) = x_0 \, det \begin{bmatrix} y_1 & y_2 \\ z_1 & z_2 \end{bmatrix} - y_0 \, det \begin{bmatrix} x_1 & x_2 \\ z_1 & z_2 \end{bmatrix} + z_0 \, det \begin{bmatrix} x_1 & x_2 \\ y_1 & y_2 \end{bmatrix} = det \begin{bmatrix} x_0 & x_1 & x_2 \\ y_0 & y_1 & y_2 \\ z_0 & z_1 & z_2 \end{bmatrix}$$

where this last determinant is expanded along the first column.

The number $\vec{u} \cdot (\vec{v} \times \vec{w})$ is called the **scalar triple product** of the vectors $\vec{u}$, $\vec{v}$ and $\vec{w}$, and has a geometric meaning that we will return to later. For now we use it to confirm that the cross product solves the problem of finding a vector orthogonal to two given vectors.

**THEOREM 2**    For any nonzero vectors $\vec{v}$ and $\vec{w}$, the cross product $\vec{v} \times \vec{w}$ is orthogonal to both $\vec{v}$ and $\vec{w}$.

**PROOF**    Using Theorem 1 with $\vec{u} = \vec{v}$ we obtain $\vec{v} \cdot (\vec{v} \times \vec{w}) = det[\vec{v} \ \ \vec{v} \ \ \vec{w}] = 0$ because the matrix $[\vec{v} \ \ \vec{v} \ \ \vec{w}]$ has two identical columns.[19] Hence $\vec{v}$ and $\vec{v} \times \vec{w}$ are orthogonal. Similarly, $\vec{w} \cdot (\vec{v} \times \vec{w}) = 0$, so $\vec{w}$ and $\vec{v} \times \vec{w}$ are also orthogonal vectors.

We saw in Section 3.3 that Theorem 2 has many applications to the study of planes and lines. In this section we examine other important aspects of the cross product, including some applications to finding areas and volumes.

---

[19] As noted earlier, the fact that $\vec{v} \cdot (\vec{v} \times \vec{w}) = 0$ can be verified directly without any reference to determinants.

### 3.5.2 Properties of the Cross Product

We begin with a theorem which collects several algebraic properties of the cross product.

**THEOREM 3**  Let $\vec{u}$, $\vec{v}$, and $\vec{w}$ be vectors. Then:

(1) $\vec{v} \times \vec{w}$ is a vector.
(2) $\vec{v} \times \vec{0} = \vec{0} = \vec{0} \times \vec{v}$.
(3) $\vec{v} \times \vec{v} = \vec{0}$.
(4) $\vec{w} \times \vec{v} = -(\vec{v} \times \vec{w})$.
(5) $(a\vec{v}) \times \vec{w} = a(\vec{v} \times \vec{w}) = \vec{v} \times (a\vec{w})$ for any scalar $a$.
(6) $\vec{v} \times (\vec{u} + \vec{w}) = (\vec{v} \times \vec{u}) + (\vec{v} \times \vec{w})$.
(7) $(\vec{u} + \vec{w}) \times \vec{v} = (\vec{u} \times \vec{v}) + (\vec{w} \times \vec{v})$.

**PROOF**  We have already noted (1). Now recall that if $\vec{v} = [x_1 \ \ y_1 \ \ z_1]^T$ and $\vec{w} = [x_2 \ \ y_2 \ \ z_2]^T$, then

$$\vec{v} \times \vec{w} = det \begin{bmatrix} \vec{i} & x_1 & x_2 \\ \vec{j} & y_1 & y_2 \\ \vec{k} & z_1 & z_2 \end{bmatrix}.$$

Hence (2) and (3) follow because a determinant is zero if a column is zero or if two columns are equal; (4) follows because a determinant changes sign if two columns are interchanged; and (5) follows because multiplying a column of a determinant by a scalar $a$ multiplies the value of the determinant by $a$. Finally, (6) and (7) follow from the definition of the cross product and the verifications are left as Exercise 11.

We have not yet given an intrinsic description of the cross product that does not depend on the choice of a coordinate system. The following theorem leads to such a description and provides a fundamental connection between the dot and cross products.

**THEOREM 4**  **The Lagrange[20] Identity**. If $\vec{v}$ and $\vec{w}$ are any two vectors, then
$$\|\vec{v} \times \vec{w}\|^2 = \|\vec{v}\|^2 \|\vec{w}\|^2 - (\vec{v} \cdot \vec{w})^2.$$

**PROOF**  Write $\vec{v} = [x_1 \ \ y_1 \ \ z_1]^T$ and $\vec{w} = [x_2 \ \ y_2 \ \ z_2]^T$ in component form. Then the identity becomes
$$\left(y_1 z_2 - y_2 z_1\right)^2 + (x_1 z_2 - x_2 z_1)^2 + \left(x_1 y_2 - x_2 y_1\right)^2$$
$$= \left(x_1^2 + y_1^2 + z_1^2\right)\left(x_2^2 + y_2^2 + z_2^2\right) - \left(x_1 x_2 + y_1 y_2 + z_1 z_2\right)^2$$
We leave the routine verification as an exercise.

If $\vec{v}$ and $\vec{w}$ are nonzero vectors, the Lagrange identity can be used to give an intrinsic description of the length of the vector $\vec{v} \times \vec{w}$. Recall (Theorem 2 §3.2) that $\vec{v} \cdot \vec{w} = \|\vec{v}\| \|\vec{w}\| \cos \theta$ where $\theta$ is the angle between $\vec{v}$ and $\vec{w}$. If this is substituted into the right side of the Lagrange identity, the result is
$$\|\vec{v} \times \vec{w}\|^2 = \|\vec{v}\|^2 \|\vec{w}\|^2 - (\|\vec{v}\| \|\vec{w}\| \cos \theta)^2$$
$$= \|\vec{v}\|^2 \|\vec{w}\|^2 \left(1 - \cos^2 \theta\right)$$
$$= (\|\vec{v}\| \|\vec{w}\| \sin \theta)^2$$
where we have used the fact that $\sin^2 \theta = 1 - \cos^2 \theta$. Because $\sin \theta$ is positive on the range $0 \le \theta \le \pi$, taking positive square roots gives
$$\|\vec{v} \times \vec{w}\| = \|\vec{v}\| \|\vec{w}\| \sin \theta$$

---

[20] Joseph Louis Lagrange (1736–1813) was one of the great mathematicians. He contributed to many parts of mathematics and is best remembered for his work in mechanics.

Figure 3.68

This is an important (intrinsic) formula for the magnitude of $\vec{v} \times \vec{w}$. In addition, it has a nice geometric interpretation in terms of the parallelogram determined by $\vec{v}$ and $\vec{w}$ (see Figure 3.68). In fact, the base of the parallelogram has length $\|\vec{v}\|$ while the height is $\|\vec{w}\| \sin \theta$. Hence the area of the parallelogram is $\|\vec{w}\| \, (\|\vec{v}\| \sin \theta) = \|\vec{v} \times \vec{w}\|$, and we have proved part (1) of

**THEOREM 5**    If $\theta$ is the angle between the nonzero vectors $\vec{v}$ and $\vec{w}$, then:

(1) $\|\vec{v} \times \vec{w}\| = \|\vec{v}\|\|\vec{w}\| \sin \theta =$ The area of the parallelogram determined by $\vec{v}$ and $\vec{w}$.

(2) $\vec{v}$ and $\vec{w}$ are parallel if and only if $\vec{v} \times \vec{w} = \vec{0}$.

**PROOF**    Part (1) is proved above. We have $\vec{v} \times \vec{w} = \vec{0}$ if and only if $\|\vec{v} \times \vec{w}\| = 0$. By (1), this holds if and only if the parallelogram determined by $\vec{v}$ and $\vec{w}$ has zero area. But this happens if and only if $\vec{v}$ and $\vec{w}$ have the same or opposite direction; that is, if and only if $\vec{v}$ and $\vec{w}$ are parallel vectors.

Note the analogue between (2) of Theorem 5 and the fact that vectors $\vec{v}$ and $\vec{w}$ are orthogonal if and only if $\vec{v} \cdot \vec{w} = 0$ (Theorem 3, Section 3.2).

**Example 2**    Determine the area of the triangle with vertices $P(1, -1, 1), Q(2, 0, 3)$, and $R(1, 1, -3)$.

**SOLUTION**    We have $\overrightarrow{PQ} = [1 \;\; 1 \;\; 2]^T$ and $\overrightarrow{PR} = [0 \;\; 2 \;\; -4]^T$, so $\overrightarrow{PQ} \times \overrightarrow{PR} = [-8 \;\; 4 \;\; 2]^T = 2[-4 \;\; 2 \;\; 1]^T$. The area $A$ of the triangle $PQR$ is half the area of the parallelogram determined by $\overrightarrow{PQ}$ and $\overrightarrow{PR}$ (see Figure 3.69), so Theorem 5 gives

$$A = \tfrac{1}{2}\|\overrightarrow{PQ} \times \overrightarrow{PR}\| = \tfrac{1}{2}\left[2\sqrt{16 + 4 + 1}\right] = \sqrt{21}.$$

Figure 3.69

If three vectors $\vec{u}, \vec{v}$, and $\vec{w}$ are given, they determine a geometric solid called a **parallelepiped** (see Figure 3.70), whose six faces are parallelograms. It is often of interest to find the volume of such a solid. The base of the parallelepiped is the parallelogram determined by $\vec{v}$ and $\vec{w}$ and so the base has area $A = \|\vec{v} \times \vec{w}\|$ by Theorem 5. By Figure 3.70 the volume of the parallelepiped is $hA$ where $h$ is the length of the projection of $\vec{u}$ on $\vec{v} \times \vec{w}$. Hence Theorem 4 §3.2 gives

$$h = \left|\frac{\vec{u} \cdot (\vec{v} \times \vec{w})}{\|\vec{v} \times \vec{w}\|^2}\right| \|\vec{v} \times \vec{w}\| = \frac{|\vec{u} \cdot (\vec{v} \times \vec{w})|}{\|\vec{v} \times \vec{w}\|} = \frac{|\vec{u} \cdot (\vec{v} \times \vec{w})|}{A}$$

Figure 3.70

Thus the volume is $hA = |\vec{u} \cdot (\vec{v} \times \vec{w})|$. This, combined with Theorem 1 proves an important formula.

**THEOREM 6**    The volume of the parallelepiped determined by three vectors $\vec{u}, \vec{v}$, and $\vec{w}$ is

$$|\vec{u} \cdot (\vec{v} \times \vec{w})| = |det[\vec{u} \;\; \vec{v} \;\; \vec{w}]|$$

where $[\vec{u} \;\; \vec{v} \;\; \vec{w}]$ is the matrix with $\vec{u}, \vec{v}$, and $\vec{w}$ as its columns.

### 3.5.3 Geometrical Meaning of the Cross Product

Left-Hand System

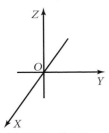

Right-Hand System

Figure 3.71

As for the dot product, our definition of $\vec{v} \times \vec{w}$ is given in terms of the components of $\vec{v}$ and $\vec{w}$ and so appears to depend on the choice of coordinate system. In fact, this is not so. To begin with, Theorem 5 gives an intrinsic description of the magnitude of $\vec{v} \times \vec{w}$:

$\|\vec{v} \times \vec{w}\|$ is the area of the parallelogram determined by $\vec{v}$ and $\vec{w}$.

Note that this holds even if $\vec{v}$ or $\vec{w}$ is zero.

To give a coordinate-free description of the of the *direction* of $\vec{v} \times \vec{w}$ we must first clarify how coordinate axes are chosen in space. The procedure is as follows: An origin is selected and two perpendicular lines (the $X$- and $Y$-axes) are chosen through the origin. Then a positive direction is chosen arbitrarily on each of these axes. Now the $Z$-axis is taken to be the (unique) line through the origin perpendicular to both the $X$- and $Y$-axes. All that remains is to select the positive direction on this $Z$-axis. There are two possibilities (shown in Figure 3.71) and it is a convention to choose a **right-hand system**. This means that the positive $Z$-direction is so chosen that, if the $Z$-axis is grasped in the right hand with the thumb pointing in the positive direction, the fingers will curl around from the positive $X$-axis to the positive $Y$-axis (through a right angle).

With the convention about right-hand systems established, the direction of $\vec{v} \times \vec{w}$ can be given by the following rule:

**Right-Hand Rule.** Let $\vec{v}$ and $\vec{w}$ be nonzero vectors which are not parallel (so $\vec{v} \times \vec{w} \neq \vec{0}$), and let $\theta$ be the angle between $\vec{v}$ and $\vec{w}$. If the vector $\vec{v} \times \vec{w}$ is grasped in the right hand and the fingers curl around from $\vec{v}$ to $\vec{w}$ through the angle $\theta$, the thumb points in the direction of $\vec{v} \times \vec{w}$.

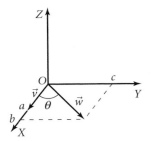

Figure 3.72

To indicate why this is so, choose coordinate axes in such a way that the initial points of $\vec{v}$ and $\vec{w}$ are at the origin, $\vec{v}$ points along the positive $X$-axis, and $\vec{w}$ is in the $XY$-plane on the same side of the $X$-axis as the positive $Y$-axis (see Figure 3.72). In this coordinate system we have $\vec{v} = [a \ \ 0 \ \ 0]^T$ and $\vec{w} = [b \ \ c \ \ 0]^T$ where $a > 0$ and $c > 0$. Hence

$$\vec{v} \times \vec{w} = det \begin{bmatrix} \vec{i} & a & b \\ \vec{j} & 0 & c \\ \vec{k} & 0 & 0 \end{bmatrix} = \begin{bmatrix} 0 \\ 0 \\ ac \end{bmatrix}$$

and so $\vec{v} \times \vec{w}$ has the positive $Z$-direction because $ac > 0$. This is the direction given by the right-hand rule.

### Exercises 3.5

**1.** In each case find the area of the triangle with vertices $A$, $B$, and $C$.
   **a)** $A(1, -1, 2)$, $B(3, 3, 2)$, $C(5, 0, -4)$
   **b)** $A(1, 1, 1)$, $B(2, 3, -5)$, $C(0, 2, 2)$

**2.** If $\vec{i}$, $\vec{j}$, and $\vec{k}$ are the coordinate vectors, show that $\vec{i} \times \vec{j} = \vec{k}$, $\vec{j} \times \vec{k} = \vec{i}$, and $\vec{k} \times \vec{i} = \vec{j}$.

**3.** In each case find the volume of the parallelepiped determined by $\vec{u}$, $\vec{v}$, and $\vec{w}$.

   **a)** $\vec{u} = [1 \ \ 1 \ \ 2]^T$, $\vec{v} = [2 \ \ 0 \ \ 1]^T$, $\vec{w} = [-1 \ \ 1 \ \ 2]^T$.
   **b)** $\vec{u} = [3 \ \ 0 \ \ 1]^T$, $\vec{v} = [-3 \ \ 1 \ \ 2]^T$, $\vec{w} = [1 \ \ 1 \ \ 1]^T$.

**4.** In each case either show that the statement is true or give an example showing that it is false.
   **a)** If $\vec{v} \times \vec{w} = \vec{0}$, then either $\vec{v} = \vec{0}$ or $\vec{w} = \vec{0}$.
   **b)** If $\vec{v} \times \vec{w} = \vec{w} \times \vec{v}$ where $\vec{v} \neq \vec{0}$ and $\vec{w} \neq \vec{0}$, then $\vec{v}$ and $\vec{w}$ are parallel.

**c)** $\vec{u} \cdot (\vec{v} \times \vec{w}) = -\vec{v} \cdot (\vec{u} \times \vec{w})$

**d)** $\vec{u} \cdot (\vec{v} \times \vec{w}) = \vec{w} \cdot (\vec{u} \times \vec{v})$

**5.** Simplify $(a\vec{v} + b\vec{w}) \times (c\vec{v} + d\vec{w})$.

**6.** If $\vec{u} + \vec{v} + \vec{w} = \vec{0}$, show that $\vec{u} \times \vec{v} = \vec{v} \times \vec{w} = \vec{w} \times \vec{u}$.

**7.** Show that the volume of the parallelepiped determined by $\vec{v}$, $\vec{w}$, and $\vec{v} \times \vec{w}$ is $\|\vec{v} \times \vec{w}\|^2$.

**8.** Show that three points $A$, $B$, and $C$ all lie on some line if and only if $\overrightarrow{AB} \times \overrightarrow{AC} = \vec{0}$.

**9.** Show that four points $A$, $B$, $C$, and $D$ all lie in some plane if and only if $\overrightarrow{AB} \cdot (\overrightarrow{AC} \times \overrightarrow{AD}) = 0$.

**10.** If $\vec{u}$, $\vec{v}$, and $\vec{w}$ are vectors, show that
$$|\vec{u} \cdot (\vec{v} \times \vec{w})| = |\vec{v} \cdot (\vec{w} \times \vec{u})| = |\vec{w} \cdot (\vec{u} \times \vec{v})|.$$

**11.** Use the definition of the cross product to: **a)** Prove (6) of Theorem 3. **b)** Prove (7) of Theorem 3.

**12.** Show that the shortest distance from a point $P$ to the line through $P_0$ with direction vector $\vec{d}$ is $\dfrac{\|\overrightarrow{P_0P} \times \vec{d}\|}{\|\vec{d}\|}$.

# The Vector Space $\mathbb{R}^n$

## 4.1   SUBSPACES AND SPANNING

In Chapter 3 we studied 3-dimensional geometry by regarding each point $P = P(x,\ y,\ z)$ in space as a $3 \times 1$ column matrix $\vec{p} = [x\ \ y\ \ z]^T$. This enabled us to give simple descriptions of planes and lines in space, and to study orthogonality and projections. In the same way, we regarded the plane as the set of all $2 \times 1$ column matrices, and described geometric transformations such as rotations and reflections using matrix multiplication. In this chapter we extend these geometrical ideas to $n$-dimensional space where $n > 3$. This leads to some of the most important concepts in linear algebra, and has a number of important applications as the next example illustrates.

### 4.1.1   An Example

**Example 1**

An engineer would like to be able to predict the response of an amplifier for any input voltage $x$ between 2 and 6 millivolts. She inputs trial voltages $x_i$ of 2, 3, 4, 5, and 6 mv and measures the response $y_i$ in the table. When the data pairs are plotted it is apparent (see the diagram) that the points $(x_i,\ y_i)$ are nearly on a straight line. So the question is: What is the straight line that best fits the data?

| Trial $i$ | 1 | 2 | 3 | 4 | 5 |
|---|---|---|---|---|---|
| Input $x_i$ | 2 | 3 | 4 | 5 | 6 |
| Response $y_i$ | 4.8 | 7.5 | 8.6 | 11.1 | 12.3 |

We know that every (nonvertical) line has the form $y = mx + b$ for some constants $m$ and $b$. Hence, given any choice of $m$ and $b$, we get predicted responses $y'_1,\ \cdots,\ y'_5$, where we write $y'_i = mx_i + b$ for convenience. Thus the task is to find the values of $m$ and $b$ such that the observed responses $y_1,\ \cdots,\ y_5$ are as close as possible to the predicted responses $y'_1,\ \cdots,\ y'_5$.

But what does "as close as possible" mean?

At this point recall that the distance between two vectors $\vec{v} = [v_1\ \ v_2\ \ v_3]^T$ and $\vec{w} = [w_1\ \ w_2\ \ w_3]^T$ in $\mathbb{R}^3$ is given by

$$\|\vec{v} - \vec{w}\| = \sqrt{(v_1 - w_1)^2 + (v_2 - w_2)^2 + (v_3 - w_3)^2}.$$

Now think of the observed responses $y_i$ and the predicted values $y'_i$ as columns $Y = [y_1\ \ y_2\ \ y_3\ \ y_4\ \ y_5]^T$ and $Y' = [y'_1\ \ y'_2\ \ y'_3\ \ y'_4\ \ y'_5]^T$. If we think of the "distance" between $Y$ and $Y'$ as the number

$$\sqrt{(y_1 - y'_1)^2 + (y_2 - y'_2)^2 + (y_3 - y'_3)^2 + (y_4 - y'_4)^2 + (y_5 - y'_5)^2},$$

then we are asking for the values of $m$ and $b$ that will make this distance as small as possible. In this form the problem can be solved using matrix theory. For obvious reasons, this technique is called the "method of least squares". The idea can be traced back to Gauss and we will return to it in Section 4.6 below.

---

[1] This chapter is self-contained except for matrix algebra and diagonalizations. Geometric notions from Chapter 3 are primarily used for motivating examples.

Example 1 suggests that geometric notions like distance should be extended to the set of all $n \times 1$ columns. This is the purpose of this chapter.

## 4.1.2  The Vector Space $\mathbb{R}^n$

As in Section 3.4, let $\mathbb{R}$ denote the set of real numbers. Recall that points in the plane and in space are represented by ordered pairs $(x, y)$ and ordered triples $(x, y, z)$ respectively. If $n \geq 1$, an ordered $n$-tuple

$$(a_1, a_2, \cdots, a_n)$$

of $n$ real numbers is called an *$n$-vector*, or simply a **vector**. The set of *all $n$-vectors* is called *$n$-space*, and a special notation is used for it:

$$\mathbb{R}^n \text{ denotes the set of all } n\text{-vectors.}$$

As for ordered pairs and triples, two $n$-vectors are equal if and only if corresponding entries are equal; that is, if they are equal as matrices. Hence we write these $n$-vectors as column matrices (and frequently as row matrices), and denote them as matrices $X$, $Y$, etc. However, in $\mathbb{R}^2$ and $\mathbb{R}^3$ we will continue to use the notation $\vec{v}$ from Chapter 3.

Thus $\mathbb{R}^1$ is just the set $\mathbb{R}$ of real numbers, while $\mathbb{R}^2$ and $\mathbb{R}^3$ represent the plane and space respectively. The fact that we are writing points in $\mathbb{R}^3$ as column matrices is consistent with our convention in Section 3.4 to identify a point $P = P(x, y, z)$ in $\mathbb{R}^3$ with its position vector $\vec{p} = [x \ \ y \ \ z]^T$. Writing $n$-vectors as column matrices amounts to adopting this convention in $\mathbb{R}^n$ for every $n$. Hence, for example, we will speak of the *vector $\vec{p} = [x \ \ y \ \ z]^T$* in $\mathbb{R}^3$, rather than the *point $P(x, y, z)$*.

## 4.1.3  Subspaces of $\mathbb{R}^n$

Consider the line $L$ through the origin in $\mathbb{R}^3$ with direction vector $\vec{d} \neq \vec{0}$. Then every vector $\vec{p}$ on the line $L$ has the form $\vec{p} = \vec{0} + t\vec{d}$ for some scalar $t$, so we have[2]

$$L = \{t\vec{d} \mid t \text{ any real number}\}.$$

Of course the zero vector $\vec{0}$ is in $L$ (because $\vec{0} = 0\vec{d}$). However $L$ has two other properties that are worthy of note: Sums and scalar multiples of vectors in $L$ are again in $L$. Indeed, if $\vec{v} = t\vec{d}$ and $\vec{v}_1 = t_1\vec{d}$ are in $L$, then

$$\vec{v} + \vec{v}_1 = t\vec{d} + t_1\vec{d} = (t + t_1)\vec{d} \quad \text{and} \quad r\vec{v} = r(t\vec{d}) = (rt)\vec{d}$$

are both in $L$ for all scalars $r$. For this reason, the line $L$ is an example of a *subspace* of $\mathbb{R}^3$. This is an important concept in linear algebra, and the general definition is as follows:

A set $U$ of vectors in $\mathbb{R}^n$ is called a **subspace** of $\mathbb{R}^n$ if it has the following three properties:

S1. *The zero vector 0 is in U.*
S2. *If X and Y are in U, then X + Y is also in U.*
S3. *If X is in U, then rX is also in U for any scalar r.*

When property S2 holds for $U$ we say that $U$ is **closed under addition**, and we say that $U$ is **closed under scalar multiplication** if S3 holds.

**Example 2**    Clearly $\mathbb{R}^n$ is a subspace of itself. The set $\{0\}$ consisting of only the zero vector in $\mathbb{R}^n$ is also a subspace of $\mathbb{R}^n$ because $0 + 0 = 0$ and $r0 = 0$ for any scalar $r$. This is called the **zero subspace** of $\mathbb{R}^n$. Any subspace of $\mathbb{R}^n$ other than $\{0\}$ or $\mathbb{R}^n$ is called a **proper subspace** of $\mathbb{R}^n$.

---

[2] We are using set notation here. In general, the notation $\{p \mid q\}$ means the set of all objects $p$ which satisfy condition $q$.

**Example 3**   Show that $U = \left\{ \begin{bmatrix} 2s-t \\ s-3t \\ -s+t \\ t \end{bmatrix} \;\middle|\; s \text{ and } t \text{ real} \right\}$ is a subspace of $\mathbb{R}^4$.

**SOLUTION**   If $X_1 = \begin{bmatrix} 2s_1-t_1 \\ s_1-3t_1 \\ -s_1+t_1 \\ t_1 \end{bmatrix}$ and $X_2 = \begin{bmatrix} 2s_2-t_2 \\ s_2-3t_2 \\ -s_2+t_2 \\ t_2 \end{bmatrix}$ are in $U$, then $X_1 + X_2 = \begin{bmatrix} 2s_3-t_3 \\ s_3-3t_3 \\ -s_3+t_3 \\ t_3 \end{bmatrix}$

where $s_3 = s_1 + s_2$ and $t_3 = t_1 + t_2$. Hence $X_1 + X_2$ is in $U$. The verification that $rX_1$ is in $U$ for any scalar $r$ is similar and is left to the reader. Clearly 0 is in $U$ (take $s = t = 0$), so $U$ is a subspace of $\mathbb{R}^4$.

The reader should not get the impression that every subset of $\mathbb{R}^n$ is a subspace.

**Example 4**   Consider the subsets $U_1 = \{[x \;\; y \;\; z]^T \mid x \ge 0\}$ and $U_2 = \{[x \;\; y \;\; z]^T \mid x^2 = y^2\}$ of $\mathbb{R}^3$. Show that neither $U_1$ nor $U_2$ is a subspace of $\mathbb{R}^3$.

**SOLUTION**   Both $U_1$ and $U_2$ contain $0 = [0 \;\; 0 \;\; 0]^T$. However $U_1$ is not closed under scalar multiplication: For example, $[1 \;\; 0 \;\; 0]^T$ is in $U_1$ but $(-1)[1 \;\; 0 \;\; 0]^T = [-1 \;\; 0 \;\; 0]^T$ is not in $U_1$. Hence $U_1$ is not a subspace, even though it is closed under addition (as the reader can verify). As to $U_2$, it is not closed under addition because, for example, $[1 \;\; 1 \;\; 0]^T$ and $[-1 \;\; 1 \;\; 0]^T$ are both in $U_2$ but their sum $[1 \;\; 1 \;\; 0]^T + [-1 \;\; 1 \;\; 0]^T = [0 \;\; 2 \;\; 0]^T$ is not in $U_2$, so $U_2$ is not a subspace. (Note that $U_2$ is actually closed under scalar multiplication.)

**Example 5**   Show that lines through the origin and planes through the origin are all subspaces of $\mathbb{R}^3$.

**SOLUTION**   The line $L = \{t\vec{d} \mid t \text{ real}\}$ through the origin with direction vector $\vec{d}$ was verified above to be a subspace of $\mathbb{R}^3$.

Let $P$ denote the plane through the origin with normal $\vec{n} = [a \;\; b \;\; c]^T$ (using the notation of Chapter 3). Then $P$ has equation $ax + by + cz = 0$, where the right side is zero because $P$ contains the origin. Now let $\vec{v} = [x \;\; y \;\; z]^T$ and $\vec{v}_1 = [x_1 \;\; y_1 \;\; z_1]^T$ be two vectors in the plane $P$. Then $\vec{v} + \vec{v}_1 = [x+x_1 \;\; y+y_1 \;\; z+z_1]^T$ is also in the plane because

$$a(x + x_1) + b(y + y_1) + c(z + z_1) = (ax + by + cz) + (ax_1 + by_1 + cz_1)$$
$$= 0 + 0 = 0.$$

Hence S2 holds for $P$. Similarly $r\vec{v} = [rx \;\; ry \;\; rz]^T$ is in the plane for any scalar $r$, as the reader can verify, so S3 also holds for $P$. Since S1 clearly holds, $P$ is a subspace of $\mathbb{R}^3$.

It turns out (Example 8 §4.3) that lines and planes through the origin are the *only* proper subspaces of $\mathbb{R}^3$. Thus the geometry of planes and lines through the origin in space is captured by the subspace concept.

However, subspaces have other uses, one of which is describing various features of an $m \times n$ matrix $A$. In fact there are two important subspaces associated with $A$, the **null space** of $A$, denoted *nullA*, and the **image** of $A$, denoted *imA*. They are defined as follows:

$$nullA = \{X \text{ in } \mathbb{R}^n \mid AX = 0\}.$$
$$imA = \{Y \text{ in } \mathbb{R}^m \mid Y = AX \text{ for some } X \text{ in } \mathbb{R}^n\}.$$

Thus *nullA* is the set of all solutions to the homogeneous system $AX = 0$, and *imA* is the set of all columns $Y$ in $\mathbb{R}^m$ such that the system $AX = Y$ has a solution. Observe that *nullA* is a subset of $\mathbb{R}^n$ while *imA* is a subset of $\mathbb{R}^m$. In fact they are both subspaces.

**Example 6**  For any $m \times n$ matrix $A$, show that *nullA* is a subspace of $\mathbb{R}^n$, and that *imA* is a subspace of $\mathbb{R}^m$.

**SOLUTION**  A column $X$ is in *nullA* if it satisfies the *condition* that $AX = 0$. Hence $0 \in nullA$ because it satisfies the condition: $A0 = 0$. Thus S1 holds for *nullA*. Suppose that $X$ and $X_1$ are in *nullA*, that is $AX = 0$ and $AX_1 = 0$. Then

$$A(X + X_1) = AX + AX_1 = 0 + 0 = 0 \quad \text{and} \quad A(rX) = r(AX) = r0 = 0$$

show, respectively, that $X + X_1$ and $rX$ are in *nullA* for all scalars $r$. Hence S2 and S3 are also satisfied for *nullA*.

Turning to *imA*, note that it consists of the vectors $Y$ of the *form* $Y = AX$ for some $X \in \mathbb{R}^n$. Thus $0 \in imA$ because it has the required form: $0 = A0$, so S1 is satisfied for *imA*. Let $Y$ and $Y_1$ be vectors in *imA*, so that $Y = AX$ and $Y_1 = AX_1$ for some $X$ and $X_1$ in $\mathbb{R}^n$. Then

$$Y + Y_1 = AX + AX_1 = A(X + X_1) \quad \text{and} \quad rY = r(AX) = A(rX)$$

show that $Y + Y_1$ and $rY$ are both in *imA* for all $r$. Thus S2 and S3 are both satisfied for *imA*.

The image and null space of a matrix $A$ are essential features of $A$ and will arise frequently in this chapter.

**Example 7**  Let $A$ denote an $n \times n$ matrix. If $\lambda$ is a number, let

$$E_\lambda(A) = \{X \text{ in } \mathbb{R}^n \mid AX = \lambda X\}.$$

This is a subspace of $\mathbb{R}^n$ as is easily verified; in fact, it is the null space of the matrix $(\lambda I - A)$. The subspace $E_\lambda(A)$ is called the **eigenspace** of $A$ corresponding to $\lambda$, and several definitions in Section 2.3 can be stated in terms of $E_\lambda(A)$. The number $\lambda$ is called an **eigenvalue** of $A$ if $E_\lambda(A) \neq \{0\}$, and in that case the nonzero vectors in $E_\lambda(A)$ are called the **eigenvectors** of $A$ corresponding to $\lambda$. Recall that these eigenvalues and eigenvectors are closely associated with the process of diagonalizing $A$; we return to this in Section 4.7.

## 4.1.4  Spanning Sets

The Gaussian algorithm can be regarded as a procedure for finding the null space of a matrix $A$. To illustrate, consider the matrix

$$A = \begin{bmatrix} 1 & -2 & 1 & 1 \\ -1 & 2 & 0 & 1 \\ 2 & -4 & 1 & 0 \end{bmatrix}.$$

The null space of $A$ consists of all columns $X$ in $\mathbb{R}^4$ such that $AX = 0$, that is all solutions $X = [x_1 \ x_2 \ x_3 \ x_4]^T$ to the system $AX = 0$ of (homogeneous) equations with $A$ as coefficient matrix:

$$\begin{aligned} x_1 - 2x_2 + x_3 + x_4 &= 0 \\ -x_1 + 2x_2 \qquad\quad + x_4 &= 0 \ . \\ 2x_1 - 4x_2 + x_3 \qquad &= 0 \end{aligned} \qquad (*)$$

This system is solved in Example 2 §1.3, and the general solution $X$ is presented there as a sum of scalar multiples of the basic solutions $X_1 = [2 \ \ 1 \ \ 0 \ \ 0]^T$ and $X_2 = [1 \ \ 0 \ \ -2 \ \ 1]^T$:

$$X = sX_1 + tX_2 \text{ where } s \text{ and } t \text{ are parameters.}$$

This way of combining the vectors $X_1$ and $X_2$ is important, and some terminology is commonly used to describe it.

Given vectors $X_1,\ X_2,\ \cdots,\ X_k$ in $\mathbb{R}^n$, a vector, of the form

$$X = t_1X_1 + t_2X_2 + \cdots + t_kX_k$$

where the $t_i$ are scalars is called a **linear combination** of the $X_i$. The scalar $t_i$ is called the **coefficient** of $X_i$ in the linear combination. The set of *all* such linear combinations is called the **span** of the $X_i$, and is denoted $span\{X_1,\ X_2,\ \cdots,\ X_k\}$. More formally,

$$span\{X_1,\ X_2,\ \cdots,\ X_k\} = \{t_1X_1 + t_2X_2 + \cdots + t_kX_k \mid t_i \text{ in } \mathbb{R}\}.$$

In particular, if $X$ and $Y$ are two vectors in $\mathbb{R}^n$, then

$$span\{X,\ Y\} = \{sX + tY \mid s \text{ and } t \text{ real numbers}\}$$

consists of all sums of scalar multiples of $X$ and $Y$. Similarly

$$span\{X\} = \{sX \mid s \text{ a real number}\}.$$

**Example 8**

Determine whether $X = [3 \ \ -1 \ \ 2]^T$ and $Y = [-1 \ \ 13 \ \ -3]^T$ are in $span\{X_1,\ X_2\}$ where $X_1 = [1 \ \ 5 \ \ 0]^T$ and $X_2 = [1 \ \ -1 \ \ 1]^T$.

**SOLUTION**

To test if $X$ is in $span\{X_1,\ X_2\}$ we must discover whether $X$ is a linear combination of $X_1$ and $X_2$, that is whether scalars $s$ and $t$ exist such that $X = sX_1 + tX_2$. This condition is

$$X = [3 \ \ -1 \ \ 2]^T = s[1 \ \ 5 \ \ 0]^T + t[1 \ \ -1 \ \ 1]^T$$

and comparing entries gives equations $3 = s + t, -1 = 5s - t$, and $2 = t$. These equations clearly have no solution, so $X$ is not in $span\{X_1,\ X_2\}$.

On the other hand, a similar procedure using $Y$ in place of $X$ yields equations $-1 = s + t, 13 = 5s - t$, and $-3 = t$. These have solution $s = 2$ and $t = -3$, whence $Y = 2X_1 - 3X_2$ as the reader can verify. Hence $Y$ is in $span\{X_1,\ X_2\}$.

If $\vec{d}$ is a nonzero vector in $\mathbb{R}^3$, then (as discussed prior to Example 2 above)

$$span\{\vec{d}\} = \{t\vec{d} \mid t \text{ real}\}$$

is the line through the origin with direction vector $\vec{d}$. In particular $span\{\vec{d}\}$ is a *subspace* of $\mathbb{R}^3$. In fact the span of *any* set of vectors is a subspace as the next theorem shows.

We need the following notation. If $U$ and $V$ are sets, we say that $U$ is **contained** in $V$, and write $U \subseteq V$, if every element of $U$ is also an element of $V$ (that is, if $U$ is a subset of $V$). This gives a useful test for set equality: $U = V$ if and only if both $U \subseteq V$ and $V \subseteq U$ (that is $U$ and $V$ have the same elements).

**THEOREM 1**

Let $X_1,\ X_2,\ \cdots,\ X_k$ be any vectors in $\mathbb{R}^n$.

(1) $span\{X_1,\ X_2,\ \cdots,\ X_k\}$ is a subspace of $\mathbb{R}^n$ which contains each of the vectors $X_1,\ X_2,\ \cdots, X_k$.

(2) If the vectors $X_1,\ X_2,\ \cdots,\ X_k$ all lie in some subspace $V$, then $span\{X_1,\ X_2,\ \cdots,\ X_k\} \subseteq V$.

**PROOF**    **(1)** For convenience, write $U = span\{X_1, X_2, \cdots, X_k\}$. Condition S1 clearly holds for $U$ since $0 = 0X_1 + \cdots + 0X_k$ is in $U$. Given two vectors $X = t_1X_1 + t_2X_2 + \cdots + t_kX_k$ and $Y = s_1X_1 + s_2X_2 + \cdots + s_kX_k$ in $U$, their sum is

$$X + Y = (s_1 + t_1)X_1 + (s_2 + t_2)X_2 + \cdots + (s_k + t_k)X_k$$

which is again in $U$. Hence S2 holds for $U$, and the proof for S3 is similar. Thus $U$ is a subspace. It contains each vector $X_i$ because $X_i$ is the linear combination with the coefficient of $X_i$ equal to 1, and all the other coefficients zero.

**(2)** If $V$ is a subspace containing all the $X_i$, then $V$ contains every linear combination of the $X_i$ by S2 and S3. In other words, $V$ contains $span\{X_1, X_2, \cdots, X_k\}$.

Note that (2) of Theorem 1 asserts that $U = span\{X_1, X_2, \cdots, X_k\}$ is the *smallest* subspace containing all the vectors $X_i$ because $U$ is contained in every subspace that contains each of the $X_i$.

The next example illustrates how Theorem 1 is used.

**Example 9**    If $X$ and $Y$ are vectors in $\mathbb{R}^n$, show that $span\{X, Y\} = span\{X + Y, X - Y\}$.

**SOLUTION**    Clearly both $X + Y$ and $X - Y$ are in $span\{X, Y\}$, so $span\{X + Y, X - Y\} \subseteq span\{X, Y\}$ by Theorem 1. On the other hand, we have

$$X = \tfrac{1}{2}(X + Y) + \tfrac{1}{2}(X - Y) \text{ and } Y = \tfrac{1}{2}(X + Y) - \tfrac{1}{2}(X - Y)$$

which show that both $X$ and $Y$ are in $span\{X + Y, X - Y\}$.

Thus $span\{X, Y\} \subseteq span\{X + Y, X - Y\}$, again by Theorem 1. Since we have verified both containments, $span\{X, Y\} = span\{X + Y, X - Y\}$.

If a subspace $U$ has the form $U = span\{X_1, X_2, \cdots, X_k\}$, we say that the vectors $X_1, X_2, \cdots, X_k$ are a **spanning set** for $U$, or that $U$ is **spanned** by the $X_i$. Many important subspaces have naturally occurring spanning sets. Here are three examples.

**Example 10**    Show that $\mathbb{R}^n = span\{E_1, E_2, \cdots, E_n\}$ where $E_1, E_2, \cdots, E_n$ are the columns of the $n \times n$ identity matrix.

**SOLUTION**    Given $X = [x_1 \ x_2 \ \cdots \ x_n]^T$ in $\mathbb{R}^n$, we can write $X$ as follows:

$$X = \begin{bmatrix} x_1 \\ x_2 \\ \vdots \\ x_n \end{bmatrix} = x_1 \begin{bmatrix} 1 \\ 0 \\ \vdots \\ 0 \end{bmatrix} + x_2 \begin{bmatrix} 0 \\ 1 \\ \vdots \\ 0 \end{bmatrix} + \cdots + x_n \begin{bmatrix} 0 \\ 0 \\ \vdots \\ 1 \end{bmatrix}.$$

In other words, $X = x_1E_1 + x_2E_2 + \cdots + x_nE_n$ is in $span\{E_1, E_2, \cdots, E_n\}$. Since $X$ could be any vector in $\mathbb{R}^n$, this shows that $\mathbb{R}^n \subseteq span\{E_1, E_2, \cdots, E_n\}$. On the other hand, each $E_i \in \mathbb{R}^n$ so Theorem 1 gives $span\{E_1, E_2, \cdots, E_n\} \subseteq \mathbb{R}^n$. Hence we have equality: $span\{E_1, E_2, \cdots, E_n\} = \mathbb{R}^n$.

If $A$ is an $m \times n$ matrix, recall that we defined $nullA = \{X \mid AX = 0\}$ to be the set of all solutions to the homogeneous system $AX = 0$ of linear equations with $A$ as coefficient matrix. The next example gives a systematic way to find a spanning set for $nullA$.

**Example 11** | Let $A$ be an $m \times n$ matrix, and let $\{X_1, X_2, \cdots, X_k\}$ denote the basic solutions of the homogeneous system $AX = 0$ given by Gaussian elimination (Section 1.3). Show that the set $nullA$ of all solutions of the system is

$$nullA = span\{X_1, X_2, \cdots, X_k\}.$$

SOLUTION | Theorem 2 §1.3 shows that every solution $X$ of $AX = 0$ is a linear combination of the basic solutions $X_1, X_2, \cdots, X_k$. In our present terminology, it shows that $nullA \subseteq span\{X_1, X_2, \cdots, X_k\}$. On the other hand, each $X_i$ is in the subspace $nullA$ (being a solution), so $span\{X_1, X_2, \cdots, X_k\} \subseteq nullA$ by Theorem 1. Hence we have equality.

As our third example of a spanning set, we show that $imA$ is spanned by the columns of $A$.

**Example 12** | If $A$ is any $m \times n$ matrix, show that $imA = span\{C_1, C_2, \cdots, C_n\}$ where $C_1, C_2, \cdots, C_n$ are the columns of $A$.

SOLUTION | Write $A = [C_1 \ C_2 \ \cdots \ C_n]$ in block form as a row of columns. If $X = [x_1 \ x_2 \ \cdots \ x_n]^T$ is any vector in $\mathbb{R}^n$, then

$$AX = [C_1 \ C_2 \ \cdots \ C_n]\begin{bmatrix} x_1 \\ x_2 \\ \vdots \\ x_n \end{bmatrix} = x_1C_1 + x_2C_2 + \cdots + x_nC_n$$

by block multiplication (Theorem 3 §1.4). Since $imA = \{AX \mid X \in \mathbb{R}^n\}$, this shows that each vector in $imA$ is in $span\{C_1, C_2, \cdots, C_n\}$, and vice versa. In other words, $imA = span\{C_1, C_2, \cdots, C_n\}$.

## Exercises 4.1

**1.** Give four examples of vectors belonging to each space.
   **a)** $\mathbb{R}^2$.   **b)** $\mathbb{R}^3$.   **c)** $\mathbb{R}^5$.   **d)** $\mathbb{R}^1$.

**2.** Which space do the following vectors belong to? That is, what is the value of $n$ of $\mathbb{R}^n$ for the following vectors?
   **a)** $(-2, -3)$   **b)** $(-2, -3, 0, 0, 0, 0)$
   **c)** $(0, 0, 8, -9, 3)$   **d)** $100$

**3.** Identify which pairs of vectors are equal. If the two vectors are not equal, describe why.
   **a)** $(4, 5, -9)$ and $(4, -5, -9)$
   **b)** $(10, 2, 7)$ and $(10, 2, 7, 1)$
   **c)** $(-1, 4)$ and $(4, -1)$
   **d)** $(3, -9, -1, 0)$ and $(3, -9, -1, 0)$

**4.** Find $nullA$ and $imA$ for the following matrices.

**a)** $A = \begin{bmatrix} 1 & 0 & 0 & 0 \\ 0 & 1 & 0 & 0 \\ 0 & 0 & 1 & 0 \\ 0 & 0 & 0 & 1 \end{bmatrix}$

**b)** $A = \begin{bmatrix} 0 & 0 & 0 & 0 \\ 0 & 0 & 0 & 0 \\ 0 & 0 & 0 & 0 \\ 0 & 0 & 0 & 0 \end{bmatrix}$

**c)** $A = \begin{bmatrix} 0 & 2 & 0 & -5 \\ 0 & 1 & 4 & 0 \\ 0 & 0 & 1 & 0 \\ 0 & 0 & 0 & 1 \end{bmatrix}$

**d)** $A = \begin{bmatrix} 5 & 0 & -7 & 1 \\ 0 & 1 & 4 & 0 \\ 0 & 0 & 1 & 0 \\ 5 & 0 & 0 & 1 \end{bmatrix}$

**e)** $A = \begin{bmatrix} -1 & -4 & 1 \\ 7 & -9 & 0 \\ 10 & 3 & -3 \\ -9 & 1 & 2 \end{bmatrix}$

5. In each case determine whether $U$ is a subspace of $\mathbb{R}^3$. Support your answer.
   a) $U = \{[s \ \ t \ \ 1]^T \mid s \text{ and } t \text{ in } \mathbb{R}\}$.
   b) $U = \{[0 \ \ t \ \ 0]^T \mid t \text{ in } \mathbb{R}\}$.
   c) $U = \{[r \ \ s \ \ t]^T \mid r, s, \text{ and } t \text{ in } \mathbb{R}, 3r - 2s + t = 0\}$.
   d) $U = \{[r \ \ 0 \ \ s]^T \mid r^2 + s^2 = 1, r \text{ and } s \text{ in } \mathbb{R}\}$.
   e) $U = \{[r - 1 \ \ 3s \ \ t]^T \mid r, s, \text{ and } t \text{ in } \mathbb{R}\}$.
   f) $U = \{[r \ \ r \ \ t]^T \mid r \text{ and } t \text{ in } \mathbb{R}\}$.

6. In each case determine if $X$ lies in $U = span\{Y, \ Z\}$. If $X$ is in $U$, write it as a linear combination of $Y$ and $Z$; if $X$ is not in $U$, show why not.
   a) $X = [2 \ \ {-1} \ \ 0 \ \ 1]^T$, $Y = [1 \ \ 0 \ \ 0 \ \ 1]^T$ and $Z = [0 \ \ 1 \ \ 0 \ \ 1]^T$.
   b) $X = [1 \ \ 2 \ \ 15 \ \ 11]^T$, $Y = [2 \ \ {-1} \ \ 0 \ \ 2]^T$ and $Z = [1 \ \ {-1} \ \ {-3} \ \ 1]^T$.
   c) $X = [8 \ \ 3 \ \ {-13} \ \ 20]^T$, $Y = [2 \ \ 1 \ \ {-3} \ \ 5]^T$ and $Z = [{-1} \ \ 0 \ \ 2 \ \ {-3}]^T$.
   d) $X = [2 \ \ 5 \ \ 8 \ \ 3]^T$, $Y = [2 \ \ {-1} \ \ 0 \ \ 5]^T$ and $Z = [{-1} \ \ 2 \ \ 2 \ \ {-3}]^T$.

7. In each case determine if the given vectors span $\mathbb{R}^4$. Support your answer.
   a) $\{[1 \ \ 1 \ \ 1 \ \ 1]^T, [0 \ \ 1 \ \ 1 \ \ 1]^T, [0 \ \ 0 \ \ 1 \ \ 1]^T, [0 \ \ 0 \ \ 0 \ \ 1]^T\}$.
   b) $\{[1 \ \ 3 \ \ {-1} \ \ 0]^T, [{-2} \ \ 1 \ \ 0 \ \ 0]^T, [0 \ \ 2 \ \ 1 \ \ {-1}]^T, [3 \ \ 6 \ \ {-3} \ \ {-2}]^T\}$.

8. In each case either show that the statement is true or give an example showing that it is false.
   a) If $U$ is a subspace of $\mathbb{R}^n$ and $X + Y$ is in $U$, then $X$ and $Y$ are both in $U$.
   b) If $U$ is a subspace of $\mathbb{R}^n$ and $rX$ is in $U$ for all $r$ in $\mathbb{R}$, then $X$ is in $U$.
   c) If $U$ is a nonempty set and $sX + tY$ is in $U$ for all $s$ and $t$ in $\mathbb{R}$ whenever $X$ and $Y$ are in $U$, then $U$ is a subspace.
   d) If $U$ is a subspace of $\mathbb{R}^n$ and $X$ is in $U$, then $-X$ is in $U$.
   e) If $Y$ is in $U = span\{X_1, X_2\}$, then $U = span\{Y, X_1, X_2\}$.
   f) The empty set of vectors is a subspace of $\mathbb{R}^n$.

9. Is it possible that $\{[1 \ \ {-1} \ \ 0]^T, \ [2 \ \ 0 \ \ 3]^T\}$ can span the subspace $U = \{[r \ \ 0 \ \ s]^T \mid r \text{ and } s \text{ in } \mathbb{R}\}$? Defend your answer.

10. If $X$ is any vector in $\mathbb{R}^n$, show that $span\{X\} = span\{aX\}$ for any $a \neq 0$ in $\mathbb{R}$.

11. Describe $span\{0\}$.

12. Is $\mathbb{R}^2$ a subspace of $\mathbb{R}^3$? Defend your answer.

13. Suppose $X_1, X_2, \cdots, X_k$ are vectors in $\mathbb{R}^n$. If $Y = a_1X_1 + a_2X_2 + \cdots + a_kX_k$ where $a_1 \neq 0$, show that $span\{X_1, X_2, \cdots, X_k\} = span\{Y, X_2, \cdots, X_k\}$.

14. Suppose that $U = span\{X_1, X_2, \cdots, X_k\}$ where each $X_i$ is in $\mathbb{R}^n$. If $AX_i = 0$ for each $i$ for some $m \times n$ matrix $A$, show that $AY = 0$ for each $Y$ in $U$.

15. Let $A$ be an $m \times n$ matrix.
    a) If $U$ is an invertible $m \times m$ matrix, show that $null(UA) = nullA$.
    b) If $V$ is an invertible $n \times n$ matrix, show that $im(AV) = imA$.

16. Let $U$ be a subspace of $\mathbb{R}^n$.
    a) If $aX$ is in $U$ where $a \neq 0$ is a number and $X$ is in $\mathbb{R}^n$, show that $X$ is in $U$.
    b) If $X$ and $X + Y$ are in $U$ where $X$ and $Y$ are in $\mathbb{R}^n$, show that $Y$ is in $U$.

17. Let $\mathcal{X}$ and $\mathcal{Y}$ denote finite sets of vectors in $\mathbb{R}^n$. If $\mathcal{X} \subseteq \mathcal{Y}$, show that $span\{\mathcal{X}\} \subseteq span\{\mathcal{Y}\}$.

18. Let $A$ and $B$ denote two $m \times n$ matrices. If $U = \{X \text{ in } \mathbb{R}^n \mid AX = BX\}$, show that $U$ is a subspace of $\mathbb{R}^n$.

19. If $A$ is an $n \times n$ matrix and $\lambda$ is a number, write $E_\lambda(A) = \{X \text{ in } \mathbb{R}^n \mid AX = \lambda X\}$ as in Example 7.
    a) Show that $E_\lambda(A)$ is a subspace of $\mathbb{R}^n$ for any choice of $A$ and $\lambda$.
    b) Explain how these subspaces $E_\lambda(A)$ are related to the eigenvalues and eigenvectors of $A$.

20. Let $A$ be an $m \times n$ matrix. For which columns $B$ is $U = \{X \text{ in } \mathbb{R}^n \mid AX = B\}$ a subspace of $\mathbb{R}^n$? Support your answer.

21. Show that $\mathbb{R}^1$ has no proper subspace.

22. If $V \neq 0$ is a vector in $\mathbb{R}^n$, determine all subspaces of $span\{V\}$.

23. Let $U$ be a *nonempty* subset of $\mathbb{R}^n$. Show that $U$ is a subspace of $\mathbb{R}^n$ if and only if S2 and S3 hold.

24. If $U$ and $W$ are subspaces of $\mathbb{R}^n$, define their **intersection** $U \cap W$ and **sum** $U + W$ as follows:
    $U \cap W = \{X \text{ in } \mathbb{R}^n \mid X \text{ is in both } U \text{ and } W\}$.
    $U + W = \{X \text{ in } \mathbb{R}^n \mid X \text{ is a sum of a vector in } U \text{ and a vector in } W\}$.
    Show that:
    a) $U \cap W$ is a subspace of $\mathbb{R}^n$.
    b) $U + W$ is a subspace of $\mathbb{R}^n$.
    c) If $U \cap W = \{0\}$, show that each vector $X$ in $U + W$ has a unique representation as a sum $X = Y + Z$ where $Y$ is in $U$ and $Z$ is in $W$.

## 4.2 LINEAR INDEPENDENCE

Some spanning sets for a subspace $U$ of $\mathbb{R}^n$ are better than others.

If $U = span\{X_1, X_2, \cdots, X_k\}$, then every vector in $U$ can be written as a linear combination, of the $X_i$ in at least one way. Our interest here is in spanning sets for which each vector in $U$ has *exactly one* representation as a linear combination of these vectors.

### 4.2.1 Independent Sets of Vectors

Suppose that two linear combinations are equal:

$$r_1X_1 + r_2X_2 + \cdots + r_kX_k = s_1X_1 + s_2X_2 + \cdots + s_kX_k.$$

We are looking for a condition on the vectors $X_i$ such that this equation will imply that $r_i = s_i$ for each $i$. Taking all terms to the left side gives

$$(r_1 - s_1)X_1 + (r_2 - s_2)X_2 + \cdots + (r_k - s_k)X_k = 0$$

so the required condition is that this equation forces all the coefficients $r_i - s_i$ to be zero. With this in mind we make the following definition.

A set $\{X_1, X_2, \cdots, X_k\}$ of vectors is called **linearly independent** (or simply **independent**) if it satisfies the following condition:

$$\text{If } t_1X_1 + t_2X_2 + \cdots + t_kX_k = 0, \text{ then } t_1 = t_2 = \cdots = t_k = 0.$$

We record the result of the above discussion for reference.

**THEOREM ①** If $\{X_1, X_2, \cdots, X_k\}$ is an independent set of vectors, then every vector $X$ in $span\{X_1, X_2, \cdots, X_k\}$ has a unique representation as a linear combination of the $X_i$.

It is useful to state the definition of independence in a slightly different form. Call a linear combination of vectors **trivial** if every coefficient is zero. Then:

> *A set of vectors is independent if and only if the only linear combination which vanishes is the trivial one.*

(Here we say that a linear combination **vanishes** if it equals the zero vector.) Hence the procedure for checking that a set of vectors is independent is as follows:

**INDEPENDENCE TEST** To verify that a set $\{X_1, X_2, \cdots, X_k\}$ of vectors in $\mathbb{R}^n$ is linearly independent, proceed as follows:

**Step 1.** Set a linear combination of the vectors equal to zero:

$$t_1X_1 + t_2X_2 + \cdots + t_kX_k = 0.$$

**Step 2.** Show that the only way this can happen is trivially—with all coefficients $t_i = 0$.

Of course, if some nontrivial linear combination exists in Step 2, the vectors $X_i$ are *not* linearly independent.

The following example illustrates how this condition is used to verify independence in a concrete situation.

**Example 1**    Show that the set of vectors $\{[2 \ \ 0 \ \ -2 \ \ 3]^T, [2 \ \ 1 \ \ 0 \ \ 7]^T, [1 \ \ 1 \ \ 3 \ \ 0]^T\}$ in $\mathbb{R}^4$ is linearly independent.

**SOLUTION**    Suppose a linear combination vanishes:

$$r[2 \ \ 0 \ \ -2 \ \ 3]^T + s[2 \ \ 1 \ \ 0 \ \ 7]^T + t[1 \ \ 1 \ \ 3 \ \ 0]^T = 0 = [0 \ \ 0 \ \ 0 \ \ 0]^T.$$

Equating corresponding entries gives four equations:

$$
\begin{aligned}
2r + 2s + \ \ t &= 0 \\
s + \ \ t &= 0 \\
-2r \quad\quad\ + 3t &= 0 \\
3r + 7s \quad\quad &= 0
\end{aligned}
$$

The only solution is the trivial one: $r = s = t = 0$, by Gaussian elimination. Hence the vectors are linearly independent by the Independence test.

The solution of Example 1 is typical of the way the Independence test shows that a set of vectors (columns or rows) in $\mathbb{R}^n$ is linearly independent. However, in many situations matrix algebra can be used to verify that a set of vectors is independent. In every case the idea is to set a linear combination of the vectors equal to zero, and show somehow that this forces all the coefficients to be zero. Here are six examples.

**Example 2**    If $E_1, \ E_2, \ \cdots, \ E_n$ are the columns of the $n \times n$ identity matrix, show that $\{E_1, \ E_2, \ \cdots, \ E_n\}$ is an independent set of vectors in $\mathbb{R}^n$.

**SOLUTION**    Suppose a linear combination vanishes: $t_1E_1 + t_2E_2 + \cdots + t_nE_n = 0$ where $t_1, \ t_2, \ \cdots, \ t_n$ are scalars. We have $t_1E_1 + t_2E_2 + \cdots + t_nE_n = [t_1 \ \ t_2 \ \ \cdots \ \ t_n]^T$ as the reader can verify, and it follows that $[t_1 \ \ t_2 \ \ \cdots \ \ t_n]^T = 0$. This means that $t_1 = t_2 = \cdots = t_n = 0$, so $\{E_1, \ E_2, \ \cdots, \ E_n\}$ is independent by the Independence test.

**Example 3**    Show that $\{X\}$ is an independent set for any vector $X \neq 0$ in $\mathbb{R}^n$.

**SOLUTION**    Suppose $tX = 0$ where $t$ is a scalar. Then $t = 0$ because $X \neq 0$ (Theorem 3 §1.1). In other words, the only linear combination from $\{X\}$ that vanishes is the trivial one.

**Example 4**    If $\{X, \ Y\}$ is linearly independent, show that $\{2X + Y, 3X - 5Y\}$ is also independent.

**SOLUTION**    Suppose a linear combination of the vectors $2X + Y$ and $3X - 5Y$ vanishes, say

$$s(2X + Y) + t(3X - 5Y) = 0.$$

We must show that this is the trivial linear combination; that is, $s = t = 0$. If we collect terms in $X$ and $Y$ on the left side, the result is a linear combination of $X$ and $Y$ that vanishes:

$$(2s + 3t)X + (s - 5t)Y = 0.$$

Hence the independence of $\{X, \ Y\}$ requires that this is the trivial linear combination; that is, both coefficients must be zero. Thus $2s + 3t = 0$ and $s - 5t = 0$, and the reader can verify that these equations have only the trivial solution $s = t = 0$. This is what we wanted.

**Example 5**    Show that no set of vectors containing the zero vector can be linearly independent.

SOLUTION    Given a set $\{0, X_1, X_2, \cdots, X_k\}$ of vectors containing the zero vector 0, we have a nontrivial linear combination

$$1 \cdot 0 + 0X_1 + 0X_2 + \cdots + 0X_k = 0$$

which vanishes. Hence $\{0, X_1, X_2, \cdots, X_k\}$ is not linearly independent.

**Example 6**    Let $\{X_1, X_2, \cdots, X_k\}$ be a linearly independent set in $\mathbb{R}^n$. If $U$ is any invertible $n \times n$ matrix, show that the set $\{UX_1, UX_2, \cdots, UX_k\}$ is also linearly independent.

SOLUTION    Suppose $t_1(UX_1) + t_2(UX_2) + \cdots + t_k(UX_k) = 0$. Then $U(t_1X_1 + t_2X_2 + \cdots + t_kX_k) = 0$, so left multiplication by $U^{-1}$ gives $t_1X_1 + t_2X_2 + \cdots + t_kX_k = 0$.
Hence $t_1 = t_2 = \cdots = t_k = 0$ by the independence of $\{X_1, X_2, \cdots, X_k\}$.

The next example reveals another reason for the importance of row-echelon matrices, and will be referred to later.

**Example 7**    Show that the nonzero rows of any row-echelon matrix $R$ are linearly independent.

SOLUTION    Let $Y_1, Y_2, \cdots, Y_r$ denote the nonzero rows of $R$, and suppose a linear combination vanishes:

$$t_1Y_1 + t_2Y_2 + \cdots + t_rY_r = 0.$$

For convenience write $C = t_1Y_1 + t_2Y_2 + \cdots + t_rY_r$. We must show that each coefficient $t_i$ is zero. Suppose the first leading 1 in $R$ lies in column $j$. Then the $j^{\text{th}}$ entry of $C$ is $t_1$ because all the other $Y$'s have $j^{\text{th}}$ entry zero (the first leading 1 has zeros below it in $R$). Hence the condition $C = 0$ forces $t_1 = 0$.

Thus $C = t_2Y_2 + \cdots + t_rY_r = 0$. Now regard $Y_2, \cdots, Y_r$ as the rows of the (smaller) row echelon matrix obtained from $R$ by deleting row 1. Then the same argument shows that $t_2 = 0$. Continue in this way to conclude that $t_i = 0$ for each $i = 1, 2, \cdots, r$. This is what we wanted.

**4.2.2**   **Invertibility of Matrices**

The invertibility of an $n \times n$ matrix $A$ is closely related to the independence (spanning) of the columns or rows of $A$. The whole thing depends on the following observation. Write $A = [C_1 \ C_2 \ \cdots \ C_n]$ where $C_1, C_2, \cdots, C_n$ are the columns of $A$ and, given scalars $x_1, x_2, \cdots, x_n$, write $X = [x_1 \ x_2 \ \cdots \ x_n]^T$. Then block multiplication (Theorem 3 §1.4) gives

$$AX = [C_1 \ C_2 \ \cdots \ C_n]\begin{bmatrix} x_1 \\ x_2 \\ \vdots \\ x_n \end{bmatrix} = x_1C_1 + x_2C_2 + \cdots + x_nC_n. \tag{$*$}$$

This equation gives the equivalence of (1), (2), and (3) in the following useful theorem.

**THEOREM 2**     The following are equivalent for an $n \times n$ matrix $A$:
  **(1)** $A$ is invertible.
  **(2)** The columns of $A$ are linearly independent in $\mathbb{R}^n$.
  **(3)** The columns of $A$ span $\mathbb{R}^n$.
  **(4)** The rows of $A$ are linearly independent in $\mathbb{R}^n$.
  **(5)** The rows of $A$ span $\mathbb{R}^n$.

PROOF     As above, let $C_1$, $C_2$, $\cdots$, $C_n$ denote the columns of $A$.
$(1) \Leftrightarrow (2)$. By Theorem 5 §1.5, $A$ is invertible if and only if the system $AX = 0$ has only the trivial solution $X = 0$. By $(*)$ this holds if and only if $\{C_1, C_2, \cdots, C_n\}$ is linearly independent.

$(1) \Leftrightarrow (3)$. By Theorem 5 §1.5, $A$ is invertible if and only if $AX = B$ has a solution for every $B$ in $\mathbb{R}^n$. By $(*)$ this holds if and only if $span\{C_1, C_2, \cdots, C_n\} = \mathbb{R}^n$.

$(1) \Leftrightarrow (4)$. The independence of the rows of $A$ in condition (4) is equivalent to the independence of the *columns* of $A^T$, and hence to the invertibility of $A^T$ (by $(1) \Leftrightarrow (2)$ applied to $A^T$). But $A^T$ is invertible if and only if $A$ is invertible (by the Corollary to Theorem 3 §1.5).

$(1) \Leftrightarrow (5)$. This is analogous to the proof of $(1) \Leftrightarrow (3)$.

**Example 8**     Show that the set $\{[1 \ \ 0 \ \ -2 \ \ 4]^T, [5 \ \ -3 \ \ 7 \ \ 0]^T, [2 \ \ 8 \ \ -1 \ \ 6]^T, [1 \ \ 3 \ \ 0 \ \ 2]^T\}$ is linearly independent.

SOLUTION     The matrix $A = \begin{bmatrix} 1 & 5 & 2 & 1 \\ 0 & -3 & 8 & 3 \\ -2 & 7 & -1 & 0 \\ 4 & 0 & 6 & 2 \end{bmatrix}$ with these vectors as its columns is invertible because $detA = 36 \neq 0$. Hence the result follows from Theorem 2.

### 4.2.3   Linear Dependence

A set $\{X_1, X_2, \cdots, X_k\}$ of vectors in $\mathbb{R}^n$ is called **linearly dependent** (or simply **dependent**) if it is *not* linearly independent. Hence our condition for dependence becomes:

> *A set of vectors is linearly dependent if and only if some nontrivial linear combination vanishes.*

There is a useful test for linear dependence which we record for reference.

**THEOREM 3**     A set $\{X_1, X_2, \cdots, X_k\}$ of vectors in $\mathbb{R}^n$ is linearly dependent if and only if one of the vectors $X_i$ is a linear combination of the others.

PROOF     Suppose that $\{X_1, X_2, \cdots, X_k\}$ is dependent. Then some nontrivial linear combination vanishes, say $t_1 X_1 + t_2 X_2 + \cdots + t_k X_k = 0$ where some $t_i \neq 0$. If $t_1 \neq 0$, then $X_1 = -\frac{t_2}{t_1} X_2 - \cdots - \frac{t_k}{t_1} X_k$, so $X_1$ is a linear combination of the rest. Similarly, if $t_i \neq 0$, then $X_i$ is a linear combination of the others.

  Conversely, assume that one of the vectors is a linear combination of the others, say $X_1 = r_2 X_2 + \cdots + r_k X_k$. Then $1X_1 - r_2 X_2 - \cdots - r_k X_k = 0$ is a nontrivial linear combination that vanishes, so $\{X_1, X_2, \cdots, X_k\}$ is dependent. A similar argument works if any $X_i$ is a linear combination of the rest.

Theorem 3 has a geometric interpretation in $\mathbb{R}^3$ which clarifies the notion of linear independence. Let $\vec{v}$ and $\vec{w}$ be two nonzero vectors in $\mathbb{R}^3$. Then Theorem 3 asserts that the set $\{\vec{v}, \vec{w}\}$ is linearly dependent if and only if one of $\vec{v}$ and $\vec{w}$ is a scalar multiple of the other; that is, if and only if they are parallel. Hence we have

**COROLLARY**     Let $\vec{v}$ and $\vec{w}$ be nonzero vectors in $\mathbb{R}^3$.

**(1)** $\{\vec{v}, \vec{w}\}$ is linearly dependent if and only if $\vec{v}$ and $\vec{w}$ are parallel.
**(2)** $\{\vec{v}, \vec{w}\}$ is linearly independent if and only if $\vec{v}$ and $\vec{w}$ are not parallel.

Now we can give a complete geometrical description of the span of two nonzero vectors in $\mathbb{R}^3$.

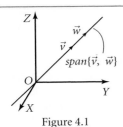

Figure 4.1

Assume that $\vec{v}$ and $\vec{w}$ are two nonzero vectors in $\mathbb{R}^3$. There are two cases:

**Case 1.** If $\vec{v}$ and $\vec{w}$ are parallel (that is, $\{\vec{v}, \vec{w}\}$ is linearly dependent in $\mathbb{R}^3$) then $span\{\vec{v}, \vec{w}\}$ is the line through the origin with direction vector $\vec{v}$ (or $\vec{w}$). This is illustrated in Figure 4.1.

**Case 2.** If $\vec{v}$ and $\vec{w}$ are not parallel (that is, $\{\vec{v}, \vec{w}\}$ is linearly independent in $\mathbb{R}^3$) then $span\{\vec{v}, \vec{w}\}$ is the plane through the origin with normal $\vec{n} = \vec{v} \times \vec{w}$. In this case $span\{\vec{v}, \vec{w}\}$ is the unique plane through the origin containing $\vec{v}$ and $\vec{w}$. This is illustrated in Figure 4.2.

**PROOF**     **Case 1.** In this case, each of $\vec{v}$ and $\vec{w}$ is a scalar multiple of the other by Theorem 3, and so $span\{\vec{v}, \vec{w}\} = span\{\vec{v}\} = span\{\vec{w}\}$ is the line through the origin with direction vector $\vec{v}$ (or $\vec{w}$).

**Case 2.** We have $\vec{n} = \vec{v} \times \vec{w} \neq \vec{0}$ because $\vec{v}$ and $\vec{w}$ are not parallel (Theorem 2 §3.3). For convenience, write $U = span\{\vec{v}, \vec{w}\}$ and let $P$ denote the plane through the origin with normal $\vec{n}$. Then $P = \{\vec{p} \mid \vec{n} \cdot \vec{p} = 0\}$, so $P$ is a subspace of $\mathbb{R}^3$ (verify). Moreover, $\vec{v}$ and $\vec{w}$ are in $P$ by Theorem 2 §3.3 so $U \subseteq P$ by Theorem 1 §4.1. It remains to show that $P \subseteq U$. Let $\vec{p}$ be a vector in $P$; we must show that $\vec{p}$ is in $U$. If $[\vec{p} \ \vec{v} \ \vec{w}]$ denotes the matrix with $\vec{p}$, $\vec{v}$, and $\vec{w}$ as its columns, then Theorem 1 §3.5 gives

$$det[\vec{p} \ \vec{v} \ \vec{w}] = \vec{p} \cdot (\vec{v} \times \vec{w}) = \vec{p} \cdot \vec{n} = 0.$$

Hence $[\vec{p} \ \vec{v} \ \vec{w}]$ is not an invertible matrix, so $\{\vec{p}, \vec{v}, \vec{w}\}$ is linearly dependent by Theorem 2. This means that there is a linear combination $a\vec{p} + b\vec{v} + c\vec{w} = \vec{0}$ where $a$, $b$, and $c$ are not all zero. But $a \neq 0$ because $\{\vec{v}, \vec{w}\}$ is linearly independent, so $\vec{p} = \frac{-b}{a}\vec{v} + \frac{-c}{a}\vec{w}$ is in $U$. This shows that $P \subseteq U$, as required.

Figure 4.2

# Exercises 4.2

**1.** Show that each of the following sets of vectors is independent.
**a)** $\{[-1 \ 2]^T, [3 \ 4]^T\}$ in $\mathbb{R}^2$.
**b)** $\{[1 \ 1 \ 1]^T, [2 \ 1 \ 0]^T\}$ in $\mathbb{R}^3$.
**c)** $\{[3 \ 1 \ -2]^T, [-2 \ -3 \ -1]^T, [1 \ 3 \ 3]^T\}$ in $\mathbb{R}^3$.
**d)** $\{[1 \ 0 \ 1 \ 0]^T, [0 \ 1 \ 0 \ 1]^T, [3 \ -1 \ 2 \ -1]^T\}$ in $\mathbb{R}^4$.

**2.** Which of the following subsets are independent? If the set is dependent, give an example of a nontrivial linear combination that equals zero.

**a)** $\{[1 \ 5]^T, [0 \ -2]^T\}$ in $\mathbb{R}^2$.
**b)** $\{[2 \ -6]^T, [-1 \ 3]^T\}$ in $\mathbb{R}^2$.
**c)** $\{[1 \ -1 \ 0]^T, [3 \ 2 \ -1]^T, [5 \ 0 \ -1]^T\}$ in $\mathbb{R}^3$.
**d)** $\{[3 \ 0 \ -1]^T, [2 \ 1 \ -1]^T\}$ in $\mathbb{R}^3$.
**e)** $\{[1 \ -1 \ 1 \ -1]^T, [2 \ 0 \ 1 \ 0]^T, [0 \ -2 \ 1 \ -2]^T\}$ in $\mathbb{R}^4$.
**f)** $\{[1 \ 2 \ 3 \ 4]^T, [-1 \ 0 \ 2 \ 2]^T, [-4 \ 2 \ 0 \ -1]^T\}$ in $\mathbb{R}^4$.
**g)** $\{[1 \ 1 \ 0 \ 0]^T, [1 \ 0 \ 1 \ 0]^T, [0 \ 0 \ 1 \ 1]^T, [0 \ 1 \ 0 \ 1]^T\}$ in $\mathbb{R}^4$.

**3.** Use Theorem 2 to determine if $A$ is invertible.

**a)** $A = \begin{bmatrix} 5 & 10 \\ 1 & 2 \end{bmatrix}$

**b)** $A = \begin{bmatrix} 1 & 1 & 0 \\ 0 & -2 & 5 \\ 3 & 0 & 4 \end{bmatrix}$

**c)** $A = \begin{bmatrix} 3 & 4 & -1 & 1 \\ -3 & -3 & 0 & 0 \\ 2 & -2 & 1 & -2 \\ 1 & -8 & -6 & 1 \end{bmatrix}$

**d)** $A = \begin{bmatrix} 3 & -1 & 4 & -1 & -4 \\ 5 & 6 & 6 & 13 & 1 \\ 6 & 6 & -6 & 0 & 0 \\ 3 & 0 & 0 & -3 & -3 \\ 2 & 0 & 9 & 7 & -2 \end{bmatrix}$

**4.** Use Theorem 2 to determine whether or not these sets of vectors are independent.
  **a)** $\{[3 \ \ 1]^T, \ [-2 \ \ 2]^T\}$ in $\mathbb{R}^2$.
  **b)** $\{[1 \ \ 1 \ \ -1]^T, \ [1 \ \ -1 \ \ 1]^T, \ [0 \ \ 0 \ \ 1]^T\}$ in $\mathbb{R}^3$.
  **c)** $\{[5 \ \ -2 \ \ 1]^T, \ [-5 \ \ 0 \ \ -5]^T, \ [4 \ \ -2 \ \ 0]^T\}$ in $\mathbb{R}^3$.
  **d)** $\{[1 \ \ 0 \ \ -2 \ \ 5]^T, \ [4 \ \ 4 \ \ -3 \ \ 2]^T, \ [0 \ \ 1 \ \ 0 \ \ -3]^T,$
  $[1 \ \ 3 \ \ 3 \ \ 1]^T\}$ in $\mathbb{R}^4$.

**5.** Using Theorem 2, determine if the following sets of vectors span the indicated space.
  **a)** $\{[4 \ \ -7]^T, \ [2 \ \ -5]^T\}, \mathbb{R}^2$.
  **b)** $\{[0 \ \ -6 \ \ -6]^T, \ [8 \ \ -3 \ \ 5]^T, \ [-9 \ \ 7 \ \ -2]^T\}, \mathbb{R}^3$.
  **c)** $\{[-8 \ \ -1 \ \ 3 \ \ 6]^T, \ [-5 \ \ -5 \ \ 4 \ \ -9]^T, \ [7 \ \ 5 \ \ 9 \ \ 0]^T,$
  $[6 \ \ 1 \ \ -16 \ \ 3]^T\}, \mathbb{R}^4$.
  **d)** $\{[2 \ \ 1 \ \ 7 \ \ -2]^T, \ [3 \ \ 5 \ \ 4 \ \ 5]^T, \ [4 \ \ -4 \ \ -3 \ \ -3]^T,$
  $[-5 \ \ 0 \ \ 6 \ \ -4]^T\}, \mathbb{R}^4$.

**6.** For which sets of vectors in Exercises 1 and 2 could you use Theorem 2 to determine their independence?

**7.** In each case either show that the statement is true or give an example showing that it is false. Throughout, $X, Y, Z, X_1, X_2, \cdots, X_n$ denote vectors in $\mathbb{R}^n$.
  **a)** If $\{X, Y\}$ is independent, then $\{X, Y, X + Y\}$ is independent.
  **b)** If $\{X, Y, Z\}$ is independent, then $\{X, Y\}$ is independent.
  **c)** If $\{X, Y\}$ is dependent, then $\{X, Y, Z\}$ is dependent.
  **d)** If all of $X_1, X_2, \cdots, X_n$ are nonzero, then $\{X_1, X_2, \cdots, X_n\}$ is independent.
  **e)** If one of $X_1, X_2, \cdots, X_n$ is zero, then $\{X_1, X_2, \cdots, X_n\}$ is dependent.
  **f)** If $aX + bY + cZ = 0$ where $a$, $b$, and $c$ are in $\mathbb{R}$, then $\{X, Y, Z\}$ is independent.
  **g)** If $\{X, Y, Z\}$ is independent, then $aX + bY + cZ = 0$ for some $a$, $b$, and $c$ in $\mathbb{R}$.

**h)** If $\{X_1, X_2, \cdots, X_n\}$ is dependent, then $t_1X_1 + t_2X_2 + \cdots + t_nX_n = 0$ for $t_i$ in $\mathbb{R}$ not all zero.
  **i)** If $\{X_1, X_2, \cdots, X_n\}$ is independent, then $t_1X_1 + t_2X_2 + \cdots + t_nX_n = 0$ for some $t_i$ in $\mathbb{R}$.

**8.** Suppose that $\vec{u}$, $\vec{v}$, and $\vec{w}$ are nonzero vectors in $\mathbb{R}^3$ with the property that $\vec{u} \cdot \vec{v} = \vec{u} \cdot \vec{w} = \vec{v} \cdot \vec{w} = 0$. Show that $\{\vec{u}, \vec{v}, \vec{w}\}$ is linearly independent.

**9.** Let $\{X, Y, Z, W\}$ be an independent set in $\mathbb{R}^n$. Which of the following sets is independent? Support your answer.
  **a)** $\{X - Y, \ Y - Z, \ Z - X\}$
  **b)** $\{X + Y, \ Y + Z, \ Z + X\}$
  **c)** $\{X - Y, \ Y - Z, \ Z - W, \ W - X\}$
  **d)** $\{X + Y, \ Y + Z, \ Z + W, \ W + X\}$

**10.** If $A$ is an $n \times n$ matrix, show that $det A = 0$ if and only if some column of $A$ is a linear combination of the other columns.

**11.** Let $A$ be any $m \times n$ matrix, and let $B_1, B_2, B_3, \cdots, B_k$ be columns in $\mathbb{R}^m$ such that the system $AX_i = B_i$ has a solution $X_i$ for each $i$. If $\{B_1, B_2, B_3, \cdots, B_k\}$ is independent in $\mathbb{R}^m$, show that $\{X_1, X_2, X_3, \cdots, X_k\}$ is independent in $\mathbb{R}^n$.

**12.** If $\{X_1, X_2, X_3, \cdots, X_k\}$ is independent, show that $\{X_1, X_1 + X_2, X_1 + X_2 + X_3, \cdots, X_1 + X_2 + \cdots + X_k\}$ is also independent.

**13.** If $\{Y, X_1, X_2, X_3, \cdots, X_k\}$ is independent, show that $\{Y + X_1, Y + X_2, Y + X_3, \cdots, Y + X_k\}$ is also independent.

**14.** Suppose $\{X_1, X_2, X_3, \cdots, X_k\}$ is independent in $\mathbb{R}^n$. If $Y$ is not in $span\{X_1, X_2, X_3, \cdots, X_k\}$, show that $\{Y, X_1, X_2, X_3, \cdots, X_k\}$ is independent.

**15. a)** If $\{X_1, X_2, X_3, X_4, X_5, X_6\}$ is an independent set of vectors, show that the subset $\{X_2, X_3, X_5\}$ is also independent.
  **b)** If $\mathcal{X}$ is any independent set in $\mathbb{R}^n$, show that any non-empty subset $\mathcal{Y} \subseteq \mathcal{X}$ is also independent.

**16. a)** If $\{X_1, X_2, X_3, X_4\}$ is a dependent set of vectors, show that any superset $\{X_1, X_2, X_3, X_4, Y, Z\}$ is also dependent.
  **b)** If $\mathcal{X}$ is any dependent set in $\mathbb{R}^n$, show that any superset $\mathcal{Y} \supseteq \mathcal{X}$ is also dependent.

## 4.3   DIMENSION

Consider the following four classes of subspaces of $\mathbb{R}^3$:

The zero subspace $\{0\}$.

The lines through the origin.

The planes through the origin.

The space $\mathbb{R}^3$ itself.

It is common geometrical language to say that $\mathbb{R}^3$ is 3-dimensional, that planes are all 2-dimensional, and that lines are all 1-dimensional. Thus the idea of "dimension" gives a measure of the "size" of a subspace of $\mathbb{R}^3$. In this section we show how to define the dimension of any subspace of $\mathbb{R}^n$ and, as a byproduct, show that the above list contains *all* the subspaces of $\mathbb{R}^3$. In doing so, we introduce one of the most important concepts in linear algebra—the idea of a basis of a subspace.

## 4.3.1 Fundamental Theorem

The notion of a basis rests on the following remarkable relationship between the sizes of spanning and independent sets: *The number of independent vectors in a subspace U cannot exceed the number of vectors in a spanning set for U.* The importance of this theorem is difficult to exaggerate.

**THEOREM 1**

**Fundamental Theorem.** Let $U$ be a subspace of $\mathbb{R}^n$.

*If $U$ is spanned by $m$ vectors and $U$ contains $k$ linearly independent vectors, then $k \leq m$.*

Because of its importance, we give two proofs of this theorem at the end of this section.

As its name suggests, the Fundamental theorem has a wide variety of applications. Since $\mathbb{R}^n$ can be spanned by the columns of the identity matrix (Example 10 §4.1) and these columns are independent (Example 2 §4.2), taking $U = \mathbb{R}^n$ in Theorem 1 gives:

**COROLLARY**

No linearly independent set in $\mathbb{R}^n$ can contain more than $n$ vectors.
No spanning set for $\mathbb{R}^n$ can contain fewer than $n$ vectors.

Our main use of Theorem 1 depends on the following concept. If $U$ is a subspace of $\mathbb{R}^n$, a set $\{X_1, X_2, \cdots, X_k\}$ of vectors in $U$ is called a **basis** of $U$ if:

**(1)** $\{X_1, X_2, \cdots, X_k\}$ is linearly independent.

**(2)** $U = span\{X_1, X_2, \cdots, X_k\}$.

Possibly the most important consequence of the Fundamental theorem is the following important result: Any two bases[3] of a subspace must have the same number of elements.

**THEOREM 2**

**Invariance Theorem.** If $\{X_1, X_2, \cdots, X_k\}$ and $\{Y_1, Y_2, \cdots, Y_m\}$ are two bases of a subspace $U$ of $\mathbb{R}^n$, then $k = m$.

**PROOF**

We have $k \leq m$ by the Fundamental theorem because $\{X_1, X_2, \cdots, X_k\}$ is independent in $U$ and $\{Y_1, Y_2, \cdots, Y_m\}$ spans $U$. Similarly, $m \leq k$, whence $m = k$.

The whole point of Theorem 2 is that it allows us to make the following definition: If $U$ is a subspace of $\mathbb{R}^n$, the number of vectors in a basis of $U$ is called the **dimension** of $U$, and is denoted $dimU$. The remarkable thing about the dimension is that (by Theorem 2) it can be determined by counting the number of vectors in *any* basis of $U$. Because of this, the dimension is called an **invariant** of the subspace $U$.

---

[3] The plural of "basis" is "bases".

If $\{E_1, E_2, \cdots, E_n\}$ is the set of columns of the $n \times n$ identity matrix, then $\mathbb{R}^n = span\{E_1, E_2, \cdots, E_n\}$ by Example 10 §4.1, and $\{E_1, E_2, \cdots, E_n\}$ is linearly independent by Example 2 §4.2. Hence $\{E_1, E_2, \cdots, E_n\}$ is a basis of $\mathbb{R}^n$, called the **standard basis** of $\mathbb{R}^n$. In particular, this shows:

**Example 1**    $dim(\mathbb{R}^n) = n$ for each $n \geq 1$, and $\{E_1, E_2, \cdots, E_n\}$ is a basis.

For $n = 2$ and $n = 3$ this is compatible with our sense that the plane $\mathbb{R}^2$ is two-dimensional and space $\mathbb{R}^3$ is three-dimensional. Note that if $n = 1$ it says that $dim\mathbb{R} = 1$. In other words, the line $\mathbb{R}$ has dimension 1, and the standard basis is $\{1\}$.

Returning to subspaces of $\mathbb{R}^n$, we define the dimension of the zero subspace $\{0\}$ to be zero:

$$dim\{0\} = 0.$$

This amounts to saying that $\{0\}$ has an empty basis, that is a basis containing no vectors. (This makes sense because the zero vector 0 cannot belong to *any* linearly independent set by Example 5 §4.2.)

**Example 2**    Show that the lines through the origin in $\mathbb{R}^3$ are precisely the subspaces of dimension 1.

SOLUTION    If $L$ is the line through the origin with direction vector $\vec{d}$, then $L = \{t\vec{d} \mid t$ in $\mathbb{R}\} = span\{\vec{d}\}$. Since $\vec{d} \neq \vec{0}$, the set $\{\vec{d}\}$ is linearly independent, and so is a basis of $L$. Thus $dimL = 1$. Conversely, if $U \subseteq \mathbb{R}^3$ is a subspace of dimension 1, let $\{\vec{v}\}$ be a basis. Then $U = span\{\vec{v}\} = \{t\vec{v} \mid t$ in $\mathbb{R}\}$, and so $U$ is the line through the origin with direction vector $\vec{v}$.

We will see below (Example 8) that the planes through the origin are precisely the subspaces of $\mathbb{R}^3$ of dimension 2.

Note that the argument in Example 2 also shows more generally that the 1-dimensional subspaces $U$ of $\mathbb{R}^n$ are precisely the subspaces of the form $U = span\{X\} = \{tX \mid t$ in $\mathbb{R}\}$ for some $X \neq 0$ in $\mathbb{R}^n$.

When working with a subspace $U$, it is often convenient to simplify a basis by multiplying some of the basis vectors by nonzero scalars. The next example shows that this will always produce another basis.

**Example 3**    Assume that $\{X_1, X_2, \cdots, X_k\}$ is a basis of a subspace $U$ of $\mathbb{R}^n$. If $a_1, a_2, \cdots, a_k$ are nonzero scalars, show that $\{a_1X_1, a_2X_2, \cdots, a_kX_k\}$ is also a basis of $U$.

SOLUTION    Suppose that a linear combination of the new set of vectors vanishes:

$$t_1(a_1X_1) + t_2(a_2X_2) + \cdots + t_k(a_kX_k) = 0.$$

Then $t_1a_1 = t_2a_2 = \cdots = t_ka_k = 0$ because $\{X_1, X_2, \cdots, X_k\}$ is independent. Since each $a_i \neq 0$, this forces $t_1 = t_2 = \cdots = t_k = 0$. Hence $\{a_1X_1, a_2X_2, \cdots, a_kX_k\}$ is linearly independent.

It remains to show that $U = span\{a_1X_1, a_2X_2, \cdots, a_kX_k\}$. Write $V = span\{a_1X_1, a_2X_2, \cdots, a_kX_k\}$ for convenience. Then $V \subseteq U$ by Theorem 1 §4.1 because each $a_iX_i$ is in $U$. Moreover, the same theorem shows that $U \subseteq V$ because $X_i = \frac{1}{a_i}(a_iX_i)$ is in $V$ for each $i$. Hence $U = V$ and we are done.

## 4.3.2    **Existence of Bases**

In order to calculate the dimension of a nonzero subspace $U$ of $\mathbb{R}^n$ it is necessary to find a basis of $U$. However, at this point we do not know whether $U$ *has* a basis. In fact it does, and one way to verify this is to find a way to create larger and larger linearly independent sets in $U$ until a basis is reached. The next result contains the key idea.

**LEMMA**    **Independent Lemma.** Suppose $\{X_1, X_2, \cdots, X_k\}$ is a linearly independent set of vectors in $\mathbb{R}^n$. If $Y$ is any vector of $\mathbb{R}^n$ which is not in $span\{X_1, X_2, \cdots, X_k\}$, then the larger set $\{Y, X_1, X_2, \cdots, X_k\}$ is also linearly independent.

**PROOF**    Suppose that

$$tY + t_1X_1 + t_2X_2 + \cdots + t_kX_k = 0 \tag{$*$}$$

is a linear combination from $\{Y, X_1, X_2, \cdots, X_k\}$ that vanishes; we must show that $t = t_1 = t_2 = \cdots = t_k = 0$. First we have $t = 0$ because, otherwise, $Y = \frac{-1}{t}(t_1X_1 + t_2X_2 + \cdots + t_kX_k)$ is in $span\{X_1, X_2, \cdots, X_k\}$, contrary to our assumption. Hence $t = 0$ and $(*)$ becomes $t_1X_1 + t_2X_2 + \cdots + t_kX_k = 0$. But then $t_1 = t_2 = \cdots = t_k = 0$ because $\{X_1, X_2, \cdots, X_k\}$ is independent.

The Independent lemma gives a geometrical interpretation of what it means for three vectors $\vec{u}$, $\vec{v}$, and $\vec{w}$ in $\mathbb{R}^3$ to be linearly independent.

**Example 4**

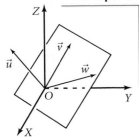

Observe first that if $\{\vec{u}, \vec{v}, \vec{w}\}$ is any linearly independent set of vectors, then $\{\vec{v}, \vec{w}\}$ must also be linearly independent (Exercise 15 §4.2). Hence $\vec{v}$ and $\vec{w}$ are nonparallel vectors in $\mathbb{R}^3$ as in the diagram. Then $span\{\vec{v}, \vec{w}\}$ is the plane through the origin containing $\vec{v}$ and $\vec{w}$ (by Theorem 4 §4.2). Hence the Independent lemma asserts that if $\vec{u}$ is any vector not in the plane containing $\vec{v}$ and $\vec{w}$, then $\{\vec{u}, \vec{v}, \vec{w}\}$ is a linearly independent set of vectors.

With the Independent lemma in hand, we can show that every nonzero subspace of $\mathbb{R}^n$ does indeed have a basis.

**THEOREM 3**    Let $U \neq \{0\}$ be any nonzero subspace of $\mathbb{R}^n$.

(1) $U$ has a basis, and $dim\,U \leq n$.

(2) Any linearly independent subset of $U$ can be enlarged to a basis of $U$.

(3) Any spanning set of $U$ contains a basis of $U$.

**PROOF**    We first prove (2), then (1) and (3).

(2) Suppose that $\{X_1, X_2, \cdots, X_k\}$ is any independent subset of $U$; we must show that it is part of a basis of $U$. If $U = span\{X_1, X_2, \cdots, X_k\}$, we are done because $\{X_1, X_2, \cdots, X_k\}$ is itself a basis. Otherwise, choose a vector $X_{k+1}$ in $U$ which is not in $span\{X_1, X_2, \cdots, X_k\}$. Then $\{X_1, X_2, \cdots, X_k, X_{k+1}\}$ is a larger independent subset of $U$ by the Independent lemma. Now repeat the process.

If $U = span\{X_1, X_2, \cdots, X_k, X_{k+1}\}$, we are done. Otherwise, enlarge it again. Since the Corollary of the Fundamental theorem shows that $U$ contains no independent set with more than $n$ vectors, this enlarging process cannot continue indefinitely. So at some stage we reach an independent set which spans $U$; that is, we reach a basis of $U$. This proves (2).

**(1)** As $U \neq \{0\}$, it contains at least one linearly independent subset (for example, $\{X\}$ where $X$ is any nonzero vector in $U$). Hence $U$ has a basis by (2), and this basis cannot contain more than $n$ vectors by the Corollary of the Fundamental theorem. In other words, $dimU \leq n$.

**(3)** Let $U = span\{X_1, X_2, \cdots, X_k\}$. If $\{X_1, X_2, \cdots, X_k\}$ is independent, it is a basis of $U$ and we are done. Otherwise, one of the $X_i$ is a linear combination of the others by Theorem 3 §4.2. After possible relabelling, assume that $X_1$ is in $span\{X_2, X_3, \cdots, X_k\}$. Hence $U = span\{X_2, X_3, \cdots, X_k\}$ by Theorem 1 §4.1. Again, if $\{X_2, X_3, \cdots, X_k\}$ is independent, we are done. If not, continue the process. Either we are done at some stage or, after possible relabelling, we reach $U = span\{X_k\}$. But then $\{X_k\}$ is a basis because $X_k \neq 0$ (since $U \neq \{0\}$).

**Example 5**    If $\{C_1, C_2, \cdots, C_k\}$ is a linearly independent set of columns in $\mathbb{R}^n$, show that the $C_i$ are the first $k$ columns of some invertible $n \times n$ matrix.

**SOLUTION**    Observe that $k \leq n$ by the Corollary to Theorem 1. Since $\{C_1, C_2, \cdots, C_k\}$ is independent in $\mathbb{R}^n$, (2) of Theorem 3 shows that it can be enlarged to a basis $\{C_1, C_2, \cdots, C_k, C_{k+1}, \cdots, C_n\}$ of $\mathbb{R}^n$. Hence the matrix $A = [C_1 \quad C_2 \quad \cdots \quad C_k \quad C_{k+1} \quad \cdots \quad C_n]$ is invertible by Theorem 2 §4.2, as required.

Note that, apart from the Independent lemma, Theorem 3 does not specify how to construct the new columns $C_{k+1}, \cdots, C_n$ in Example 5. In fact there are many choices. Theorem 3 asserts only the existence of such columns.

In the following sections we will make frequent use of the next result, each part of which is an easy consequence of Theorem 3.

**THEOREM 4**    Let $U$ and $V$ denote subspaces of $\mathbb{R}^n$.

**(1)** If $U \subseteq V$, then $dimU \leq dimV$.

**(2)** If $U \subseteq V$ and $dimU = dimV$, then $U = V$.

**(3)** If $dimU = d$, then any set of $d$ linearly independent vectors in $U$ is automatically a basis of $U$.

**(4)** If $dimU = d$, then any spanning set for $U$ containing $d$ vectors is automatically a basis of $U$.

**PROOF**    **(1)** If $U = \{0\}$, then $dimU = 0 \leq dimV$. Otherwise, any basis of $U$ is an independent set in $V$, and so (by the Fundamental theorem) contains no more vectors than a basis of $V$. In other words, $dimU \leq dimV$.

**(2)** Write $d = dimU = dimV$, and let $\{X_1, X_2, \cdots, X_d\}$ be a basis of $U$. If $U \neq V$, choose $Y$ in $V$ such that $Y$ is not in $U = span\{X_1, X_2, \cdots, X_d\}$. By the Independent lemma the set $\{X_1, X_2, \cdots, X_d, Y\}$ is a linearly independent set in $V$ containing $d + 1$ vectors. But this contradicts the Fundamental theorem because $V$ has a spanning set (in fact a basis) of $d$ vectors. Hence $U \neq V$ is impossible, so $U = V$ after all.

(3) Let $\{X_1, X_2, \cdots, X_d\}$ be an independent set in $U$; we must show that it spans $U$. Write $W = span\{X_1, X_2, \cdots, X_d\}$. Then $W \subseteq U$ by Theorem 1 §4.1 because each $X_i$ is in $U$, and $dim W = d$ because $\{X_1, X_2, \cdots, X_d\}$ is a basis of $W$. Hence $W = U$ by (2). Thus $U = W = span\{X_1, X_2, \cdots, X_d\}$, so $\{X_1, X_2, \cdots, X_d\}$ is a basis of $U$.

(4) If $U = span\{X_1, X_2, \cdots, X_d\}$, then $\{X_1, X_2, \cdots, X_d\}$ contains a basis of $U$ by Theorem 3. This basis contains $d = dim U$ vectors, and so must be all of $\{X_1, X_2, \cdots, X_d\}$. In other words, $\{X_1, X_2, \cdots, X_d\}$ is a basis of $U$.

Parts (3) and (4) of Theorem 4 are "laboursaving" results since they show that it is sufficient to verify either independence or spanning in order to verify that a set of $d$ vectors (in a subspace $U$ of dimension $d$) is a basis; the other is then automatic. Of course it is necessary that the number of vectors in the set equals $dim U$.

**Example 6**  Show that $\{[2 \ \ 0 \ \ -1]^T, [1 \ \ 4 \ \ 9]^T, [-1 \ \ 5 \ \ 6]^T\}$ is a basis of $\mathbb{R}^3$.

**SOLUTION**  By Theorem 4 it suffices to show that $\{[2 \ \ 0 \ \ -1]^T, [1 \ \ 4 \ \ 9]^T, [-1 \ \ 5 \ \ 6]^T\}$ is linearly independent (spanning would do as well). This follows from Theorem 2 §4.2 if we can show that the matrix $A$ with these vectors as columns is invertible. But, using row operations as in Chapter 2,

$$det A = det \begin{bmatrix} 2 & 1 & -1 \\ 0 & 4 & 5 \\ -1 & 9 & 6 \end{bmatrix} = det \begin{bmatrix} 0 & 19 & 11 \\ 0 & 4 & 5 \\ -1 & 9 & 6 \end{bmatrix} = -51 \neq 0.$$

Hence $A$ is indeed invertible.

Part (2) of Theorem 4 is also a "laboursaving" result, and is frequently the easiest way to show that two subspaces of the same dimension are equal: Simply show that one is contained in the other. This is illustrated in the following example (where we give a different proof of part of Theorem 4 §4.2).

**Example 7**  Let $\vec{v}$ and $\vec{w}$ be linearly independent (that is, nonparallel) vectors in $\mathbb{R}^3$, and let $P = span\{\vec{v}, \vec{w}\}$ denote the plane spanned by $\vec{v}$ and $\vec{w}$. Show that $P$ is the plane (through the origin) with normal $\vec{n} = \vec{v} \times \vec{w}$.

**SOLUTION**  Note first that $\vec{n} \neq \vec{0}$ by Theorem 2 §3.3 because $\vec{v}$ and $\vec{w}$ are not parallel. Let $Q$ denote the plane through the origin with $\vec{n}$ as normal, and write $\vec{n} = [a \ \ b \ \ c]^T$. Then $Q$ has equation $ax + by + cz = 0$, that is

$$Q = \{[x \ y \ z]^T \mid ax + by + cz = 0\} = \{\vec{x} \mid \vec{n} \cdot \vec{x} = 0\}$$

where we denote $\vec{x} = [x \ \ y \ \ z]^T$. Since $\vec{n}$ is orthogonal to both $\vec{v}$ and $\vec{w}$ (Theorem 2 §3.3), we have $\vec{n} \cdot \vec{v} = 0$ and $\vec{n} \cdot \vec{w} = 0$. This means that both $\vec{v}$ and $\vec{w}$ are in $Q$, and hence that $P \subseteq Q \subseteq \mathbb{R}^3$. Since $dim P = 2$, Theorem 4 gives

$$dim P \leq dim Q \leq dim \mathbb{R}^3, \text{ that is, } 2 \leq dim Q \leq 3.$$

Now observe that $dim Q$ is an integer, so this means that $dim Q = 2$ or $dim Q = 3$, whence $Q = P$ or $Q = \mathbb{R}^3$ by (2) of Theorem 4. But $Q \neq \mathbb{R}^3$ because $\vec{n}$ is not in $Q$ (as $\vec{n} \neq \vec{0}$), so $Q = P$ as required.

We conclude with a characterization (mentioned earlier) of all subspaces of $\mathbb{R}^3$.

**Example 8**     Show that the only subspaces of $\mathbb{R}^3$ are the zero space $\{0\}$, lines through the origin, planes through the origin, and $\mathbb{R}^3$ itself.

**SOLUTION**     Let $U$ denote a subspace of $\mathbb{R}^3$. Then $dimU \leq dim(\mathbb{R}^3) = 3$ so, since $dimU$ is a nonnegative integer, $dimU$ is one of 0, 1, 2, or 3. We examine each case separately:

$dimU = 0.$     Then $U = \{0\}$.

$dimU = 1.$     If $\{\vec{d}\}$ is a basis of $U$, then $U = span\{\vec{d}\} = \{t\vec{d} \mid t$ a real number$\}$ is the line through the origin with direction vector $\vec{d}$.

$dimU = 2.$     If $\{\vec{v},\ \vec{w}\}$ is a basis of $U$, then $U$ is the plane through the origin with normal $\vec{v} \times \vec{w}$ containing $\vec{v}$ and $\vec{w}$ by Example 7.

$dimU = 3.$     Since $U \subseteq \mathbb{R}^3$ and $dimU = 3 = dim(\mathbb{R}^3)$, we have $U = \mathbb{R}^3$ by (2) of Theorem 4.

## 4.3.3    Proof of the Fundamental Theorem

If $U$ is a subspace of $\mathbb{R}^n$ which is spanned by $m$ vectors, and if $U$ contains $k$ linearly independent vectors, we must show that $k \leq m$. Let $U = span\{X_1,\ X_2,\ \cdots,\ X_m\}$, and let $\{Y_1,\ Y_2,\ \cdots,\ Y_k\}$ be an independent set in $U$.

**PROOF 1**     Since each $Y_j$ in $span\{X_1,\ X_2,\ \cdots,\ X_m\}$, it is a linear combination of $X_1,\ X_2,\ \cdots,\ X_m$, say[4]

$$Y_j = a_{1j}X_1 + a_{2j}X_2 + \cdots + a_{mj}X_m = \sum_{i=1}^{m} a_{ij}X_i \text{ for each } j = 1,\ 2,\ \cdots,\ k.$$

The coefficients here form an $m \times k$ matrix $A = [a_{ij}]$. We must show that $k \leq m$; we do so by showing that $k > m$ leads to a contradiction. Indeed, if $k > m$ the system $AX = 0$ has a nontrivial solution $X = [x_1\ \ x_2\ \ \cdots\ \ x_k]^T \neq 0$ by Theorem 1 §1.3. Using the entries $x_j$ of $X$ as coefficients, we compute the following linear combination of the $Y_j$:

$$\sum_{j=1}^{k} x_j Y_j = \sum_{j=1}^{k} x_j \left( \sum_{i=1}^{m} a_{ij}X_i \right)$$

$$= \sum_{j=1}^{k} \sum_{i=1}^{m} a_{ij}x_j X_i$$

$$= \sum_{i=1}^{m} \left( \sum_{j=1}^{k} a_{ij}x_j \right) X_i$$

$$= \sum_{i=1}^{m} (0)X_i = 0$$

where the sum $\sum_{j=1}^{k} a_{ij}x_j = 0$ for each $i$ because it is entry $i$ of $AX = 0$. Thus $\sum_{j=1}^{k} x_j Y_j = 0$, a contradiction because the $Y_j$ are independent and not all the $x_j$ are zero (since $X \neq 0$).

**PROOF 2**     We assume that $k > m$ and show that this leads to a contradiction.
As $U = span\{X_1,\ X_2,\ \cdots,\ X_m\}$, let $Y_1 = a_1X_1 + a_2X_2 + \cdots + a_mX_m$. As $Y_1 \neq 0$ not all of the $a_i$ are zero, say $a_1 \neq 0$ (after relabelling the $X_i$). Then $U = span\{Y_1,\ X_2,\ X_3,\ \cdots,\ X_m\}$ as the reader can verify. Now write $Y_2 = b_1Y_1 + c_2X_2 + c_3X_3 + \cdots + c_mX_m$. Then some $c_i \neq 0$ because $\{Y_1,\ Y_2\}$ is independent so, as before, $U = span\{Y_1,\ Y_2,\ X_3,\ \cdots,\ X_m\}$ after possible relabelling of the $X_i$. As $k > m$ this procedure continues until all

---

[4] We again use summation notation for convenience. For example $a_1 + a_2 + a_3 + a_4 = \sum_{i=1}^{4} a_i$ and $a_3X_3 + a_4X_4 + a_5X_5 = \sum_{i=3}^{5} a_iX_i$.

the vectors $X_1$, $X_2$, $\cdots$, $X_m$ are replaced by $Y_1$, $Y_2$, $\cdots$, $Y_m$. In particular $U = span\{Y_1, Y_2, \cdots, Y_m\}$. But then $Y_{m+1}$ is a linear combination of $Y_1$, $Y_2$, $\cdots$, $Y_m$ contrary to the independence of the $Y_j$.

The second proof above gives even more. If $U = span\{X_1, X_2, \cdots, X_m\}$, and if $\{Y_1, Y_2, \cdots, Y_k\}$ is independent in $U$, Proof 2 shows not only that $k \leq m$ but also that $k$ of the (spanning) vectors $X_1$, $X_2$, $\cdots$, $X_m$ can be replaced by the (independent) vectors $Y_1$, $Y_2$, $\cdots$, $Y_k$ and the resulting set will still span $U$. In this form the result is called the **Steinitz Exchange Lemma**.

## Exercises 4.3

**1.** The following sets of vectors are all bases of the subspace they span. Give the dimension of each subspace.
  **a)** $\{[8 \ \ 4 \ \ 0]^T, [-7 \ \ 7 \ \ 3]^T\}$.
  **b)** $\{[3 \ \ 0 \ \ -5]^T, [2 \ \ -1 \ \ 0]^T, [-2 \ \ 4 \ \ -9]^T\}$.
  **c)** $\{[-9 \ \ 0 \ \ -1 \ \ 6]^T, [5 \ \ 7 \ \ -3 \ \ -2]^T\}$.
  **d)** $\{[-1 \ \ 6 \ \ 2 \ \ 6]^T, [-2 \ \ 4 \ \ 0 \ \ -4]^T, [-6 \ \ 5 \ \ 0 \ \ 1]^T\}$.
  **e)** $\{[1 \ \ 0 \ \ 0 \ \ 0]^T, [0 \ \ 0 \ \ 1 \ \ 0]^T, [0 \ \ 1 \ \ 0 \ \ 0]^T,$ $[0 \ \ 0 \ \ 0 \ \ 1]^T\}$.

**2.** Show that the following sets of vectors are bases of the indicated space.
  **a)** $\{[1 \ \ 1 \ \ 0]^T, [1 \ \ 0 \ \ 1]^T, [0 \ \ 1 \ \ 1]^T\}$ in $\mathbb{R}^3$.
  **b)** $\{[-1 \ \ 1 \ \ 1]^T, [1 \ \ -1 \ \ 1]^T, [1 \ \ 1 \ \ -1]^T\}$ in $\mathbb{R}^3$.
  **c)** $\{[1 \ \ 2 \ \ 1 \ \ 2]^T, [-1 \ \ 0 \ \ 2 \ \ 1]^T, [0 \ \ 0 \ \ 1 \ \ 1]^T,$ $[0 \ \ 1 \ \ -1 \ \ 1]^T\}$ in $\mathbb{R}^4$.
  **d)** $\{[1 \ \ 1 \ \ 0 \ \ 0]^T, [1 \ \ 0 \ \ 0 \ \ 1]^T, [0 \ \ 1 \ \ 1 \ \ 0]^T,$ $[0 \ \ 1 \ \ 0 \ \ 1]^T\}$ in $\mathbb{R}^4$.

**3.** Verify that the following sets of vectors span the given space. Could they form a basis for the space? Support your answer.
  **a)** $\{[-5 \ \ -2 \ \ -2]^T, [7 \ \ -9 \ \ 3]^T, [4 \ \ -8 \ \ 9]^T, [8 \ \ 4 \ \ 7]^T\}$ in $\mathbb{R}^3$.
  **b)** $\{[2 \ \ 3 \ \ 5 \ \ -4]^T, [6 \ \ 1 \ \ -4 \ \ -7]^T, [3 \ \ 6 \ \ 3 \ \ 2]^T,$ $[1 \ \ 0 \ \ 0 \ \ 0]^T, [6 \ \ -3 \ \ -2 \ \ 0]^T, [7 \ \ -4 \ \ 2 \ \ 9]^T\}$ in $\mathbb{R}^4$.
  **c)** $\{[-4 \ \ 3 \ \ 2]^T, [-1 \ \ 1 \ \ 11]^T, [-3 \ \ 2 \ \ -9]^T\}$ in $\mathbb{R}^3$.
  **d)** $\{[1 \ \ -6 \ \ 3 \ \ -7]^T, [5 \ \ 1 \ \ 5 \ \ 9]^T, [3 \ \ 9 \ \ 0 \ \ 2]^T,$ $[2 \ \ 0 \ \ 3 \ \ 0]^T\}$ in $\mathbb{R}^4$.

**4.** The following pairs of vector sets claim to be two bases of the same subspace. Can you determine if this is false without doing any calculations?
  **a)** $\{[3 \ \ -1]\}$ and $\{[2 \ \ 0], [-5 \ \ 3]\}$ of a subspace in $\mathbb{R}^2$.
  **b)** $\{[6 \ \ 0 \ \ -9], [-4 \ \ 2 \ \ 3], [-3 \ \ 8 \ \ 7]\}$ and $\{[6 \ \ 4 \ \ 3],$ $[2 \ \ 9 \ \ -1]\}$ of a subspace in $\mathbb{R}^3$.
  **c)** $\{[2 \ \ 1 \ \ 0], [-3 \ \ -7 \ \ 3]\}$ and $\{[1 \ \ 6 \ \ -3], [5 \ \ 8 \ \ -3]\}$ of a subspace in $\mathbb{R}^3$.
  **d)** $\{[6 \ \ -2 \ \ -8 \ \ 0], [7 \ \ 0 \ \ 9 \ \ 0]\}$ and $\{[-2 \ \ -5 \ \ -4 \ \ 6],$ $[6 \ \ 3 \ \ 7 \ \ -8], [9 \ \ 9 \ \ 0 \ \ 2], [-7 \ \ -1 \ \ 0 \ \ 9]\}$ of a subspace in $\mathbb{R}^4$.

**5.** Generate the standard basis for the given space.
  **a)** $\mathbb{R}^2$.      **b)** $\mathbb{R}^3$.      **c)** $\mathbb{R}^5$.      **d)** $\mathbb{R}$.

**6.** Find a basis and calculate the dimension of the following subspaces of $\mathbb{R}^4$. Can you find more than one basis for each subspace?
  **a)** $span\{[1 \ \ -1 \ \ 2 \ \ 0]^T, [2 \ \ 3 \ \ 0 \ \ 3]^T, [1 \ \ 9 \ \ -6 \ \ 6]^T\}$.
  **b)** $span\{[2 \ \ 1 \ \ 0 \ \ -1]^T, [-1 \ \ 1 \ \ 1 \ \ 2]^T, [2 \ \ 7 \ \ 4 \ \ 5]^T\}$.
  **c)** $span\{[-1 \ \ 2 \ \ 1 \ \ 0]^T, [2 \ \ 0 \ \ 3 \ \ -1]^T, [4 \ \ 4 \ \ 11 \ \ -3]^T,$ $[3 \ \ -2 \ \ 2 \ \ -1]^T\}$.
  **d)** $span\{[-2 \ \ 0 \ \ 3 \ \ 1]^T, [1 \ \ 2 \ \ -1 \ \ 0]^T, [-2 \ \ 8 \ \ 5 \ \ 3]^T,$ $[1 \ \ 2 \ \ 2 \ \ 1]^T\}$.

**7.** Find a basis and calculate the dimension of the following subspaces of $\mathbb{R}^4$.
  **a)** $U = \{[a \ \ a+b \ \ a-b \ \ b]^T \mid a \text{ and } b \text{ in } \mathbb{R}\}$.
  **b)** $U = \{[a+b \ \ a-b \ \ b \ \ a]^T \mid a \text{ and } b \text{ in } \mathbb{R}\}$.
  **c)** $U = \{[a \ \ b \ \ c+a \ \ c]^T \mid a, b, \text{ and } c \text{ in } \mathbb{R}\}$.
  **d)** $U = \{[a-b \ \ b+c \ \ a \ \ b+c]^T \mid a, b, \text{ and } c \text{ in } \mathbb{R}\}$.
  **e)** $U = \{[a \ \ b \ \ c \ \ d]^T \mid a+b-c+d = 0 \text{ in } \mathbb{R}\}$.

**8.** In each case find a basis for and calculate the dimension of $nullA$:
  **a)** $A = \begin{bmatrix} 1 & 2 & -3 & 4 & 0 \\ -1 & 0 & 2 & -4 & 1 \\ -1 & 4 & 0 & -4 & 3 \end{bmatrix}$

  **b)** $A = \begin{bmatrix} 1 & 1 & -2 & 0 & 4 \\ 2 & 1 & 0 & 1 & -5 \\ 7 & 2 & 6 & -10 & -7 \end{bmatrix}$

**9.** In each case show that $U$ is a subspace of $\mathbb{R}^4$, find a basis of $U$ and calculate the dimension of $U$.
  **a)** $U = \{X \text{ in } \mathbb{R}^4 \mid X^T A = 0\}$ where $A = \begin{bmatrix} 1 & 1 & 0 & 3 \\ -1 & 2 & 2 & -1 \end{bmatrix}^T$.
  **b)** $U = \{X \text{ in } \mathbb{R}^4 \mid X^T A = 0\}$ where $A = \begin{bmatrix} 0 & 1 & 2 & -1 \\ 1 & -1 & 0 & 3 \end{bmatrix}^T$.
  **c)** $U = \{X \text{ in } \mathbb{R}^4 \mid X^T A = (BX)^T\}$ where $A = \begin{bmatrix} 1 & 0 & 1 & 0 \\ 0 & 1 & 0 & 1 \end{bmatrix}^T$ and $B = \begin{bmatrix} 1 & 2 & 3 & 4 \\ 0 & 1 & 1 & -1 \end{bmatrix}$.
  **d)** $U = \{X \text{ in } \mathbb{R}^4 \mid X^T A = (BX)^T\}$ where $A = \begin{bmatrix} 2 & 0 & 1 & -3 \\ -1 & 1 & 1 & 0 \end{bmatrix}^T$ and $B = \begin{bmatrix} 2 & 1 & 0 & -1 \\ 1 & 1 & -1 & -1 \end{bmatrix}$.

**10.** In each case either show that the statement is true or give an example showing that it is false.

**a)** Every set of four nonzero vectors in $\mathbb{R}^4$ is a basis.

**b)** No basis of $\mathbb{R}^3$ can contain a vector with a component 0.

**c)** $\mathbb{R}^3$ has a basis of the form $\{X, X + Y, Y\}$ where $X$ and $Y$ are vectors.

**d)** Every basis of $\mathbb{R}^5$ contains one column of $I_5$.

**e)** Every nonempty subset of a basis of $\mathbb{R}^3$ is again a basis of $\mathbb{R}^3$.

**f)** If $\{X_1, X_2, X_3, X_4\}$ and $\{Y_1, Y_2, Y_3, Y_4\}$ are bases of $\mathbb{R}^4$, then $\{X_1 + Y_1, X_2 + Y_2, X_3 + Y_3, X_4 + Y_4\}$ is also a basis of $\mathbb{R}^4$.

**11.** Given that $\{[3 \;\; -1 \;\; 2 \;\; 1], [0 \;\; 2 \;\; -1 \;\; 0], [4 \;\; -1 \;\; 3 \;\; -3]\}$ is a basis for some subspace $U$ of $\mathbb{R}^4$, is $\{[-6 \;\; 2 \;\; -4 \;\; -2], [0 \;\; -4 \;\; 2 \;\; 0], [-8 \;\; 2 \;\; -6 \;\; 6]\}$ also a basis of $U$?

**12.** Show that every nonzero vector in $\mathbb{R}^n$ is in some basis of $\mathbb{R}^n$.

**13. a)** Find a basis of $\mathbb{R}^4$ containing the vector $[11 \;\; -19 \;\; 203 \;\; -131]^T$.

**b)** If $X$ is *any* vector in $\mathbb{R}^4$, can you find a basis containing $X$? Defend your answer.

**c)** Find a condition on vectors $X$ and $Y$ in $\mathbb{R}^4$ such that there is a basis of $\mathbb{R}^4$ containing both $X$ and $Y$ if and only if the condition is satisfied. Defend your answer.

**14.** Let $\{X, Y, Z, W\}$ denote vectors in $\mathbb{R}^3$. Find a condition on this set of vectors such that it contains a basis of $\mathbb{R}^3$ if and only if the condition is satisfied. Defend your answer.

**15. a)** Find a basis of vectors on $\mathbb{R}^3$ consisting of vectors whose components sum to 1.

**b)** Can you find a basis as in **a)** if we ask that the components sum to 0? Defend your answer.

**16.** Suppose that $\{X, Y, Z, W\}$ is a basis of $\mathbb{R}^4$. Show that:

**a)** $\{X + aW, Y, Z, W\}$ is also a basis of $\mathbb{R}^4$ for any choice of the scalar $a$.

**b)** $\{X + W, Y + W, Z + W, W\}$ is also a basis of $\mathbb{R}^4$.

**c)** $\{X, X + Y, X + Y + Z, X + Y + Z + W\}$ is also a basis of $\mathbb{R}^4$.

**17.** Suppose that the columns $\{X, Y\}$ constitute a basis of $\mathbb{R}^2$, and let $A = \begin{bmatrix} a & c \\ b & d \end{bmatrix}$ denote a matrix.

**a)** If $A$ is invertible, show that $\{aX + bY, cX + dY\}$ is a basis of $\mathbb{R}^2$.

**b)** If $\{aX + bY, cX + dY\}$ is a basis of $\mathbb{R}^2$, show that $A$ is invertible.

**18. a)** If $\{X, Y\}$ is independent in $\mathbb{R}^4$, show that $\{X, Y, E_l, E_t\}$ is a basis of $\mathbb{R}^4$ for some choice of $E_l$ and $E_t$ in the standard basis of $\mathbb{R}^4$.

**b)** If $\{X_1, X_2, \cdots, X_k\}$ is independent in $\mathbb{R}^n$ with $k < n$, show that $\{X_1, X_2, \cdots, X_k, E_{l_{k+1}}, \cdots, E_{l_n}\}$ is a basis of $\mathbb{R}^n$ for some choice of $E_{l_{k+1}}, \cdots, E_{l_n}$ in the standard basis of $\mathbb{R}^n$.

**c)** Let $\{F_1, F_2, \cdots, F_m\}$ be a basis of a subspace $U$ of $\mathbb{R}^n$. If $\{X_1, X_2, \cdots, X_k\}$ is independent in $U$ with $k < m$, show that $\{X_1, X_2, \cdots, X_k, F_{l_{k+1}}, \cdots, F_{l_n}\}$ is a basis of $U$ for some choice of $F_{l_{k+1}}, \cdots, F_{l_n}$.

**19.** Suppose that $\{X_1, X_2, X_3, \cdots, X_n\}$ is a basis of $\mathbb{R}^n$. If $A$ is an invertible matrix, show that $\{AX_1, AX_2, AX_3, \cdots, AX_n\}$ is also a basis of $\mathbb{R}^n$.

**20.** Let $U$ and $W$ denote subspaces of $\mathbb{R}^n$, and assume that $U \subseteq W$. If $dim\,W = 1$, show that either $U = \{0\}$ or $U = W$.

**21.** Let $U$ and $W$ denote subspaces of $\mathbb{R}^n$, and assume that $U \subseteq W$. If $dim\,U = n - 1$, show that either $W = U$ or $W = \mathbb{R}^n$.

**22.** Let $U$ and $W$ denote subspaces of $\mathbb{R}^n$, and suppose that $dim\,U = 2$. Show that either $U \subseteq W$ or $dim(U \cap W) \le 1$. (See Exercise 24 §4.1.) [Hint: $U \cap W$ is a subspace of $U$. Use Theorem 4.]

**23.** Let $A$ denote an $m \times n$ matrix. Show that:

**a)** $dim(null\,A) = dim(null(AV))$ for every invertible $n \times n$ matrix $V$.

**b)** $dim(im\,A) = dim(im(UA))$ for every invertible $m \times m$ matrix $U$.

**24.** Let $U$ be a subspace of $\mathbb{R}^n$, and let $\{X_1, X_2, \cdots, X_k\}$ be a maximal independent subset of $U$ (that is, no independent subset of $U$ contains more than $k$ vectors). Show that $\{X_1, X_2, \cdots, X_k\}$ is a basis of $U$. [Hint: If $span\{X_1, X_2, \cdots, X_k\} \ne U$, choose $Y$ in $U$ which is not in $span\{X_1, X_2, \cdots, X_k\}$, and apply Lemma 1.]

**25.** Let $U \ne \{0\}$ be a subspace of $\mathbb{R}^n$, and let $\{X_1, X_2, \cdots, X_k\}$ be a minimal spanning subset of $U$ (that is, $U = span\{X_1, X_2, \cdots, X_k\}$ and no spanning set of $U$ contains fewer than $k$ vectors). Show that $\{X_1, X_2, \cdots, X_k\}$ is a basis of $U$. [Hint: If $\{X_1, X_2, \cdots, X_k\}$ is not independent, apply Theorem 3 §4.2.]

**26.** Let $A = \begin{bmatrix} B & 0 \\ 0 & C \end{bmatrix}$ where $B$ and $C$ are $n \times n$ and $m \times m$ matrices, respectively.

**a)** If $B$ and $C$ are diagonalizable matrices with diagonalizing matrices $P$ and $Q$ respectively, show that $\begin{bmatrix} B & 0 \\ 0 & C \end{bmatrix}$ is diagonalizable with diagonalizing matrix $\begin{bmatrix} P & 0 \\ 0 & Q \end{bmatrix}$.

**b)** If $A = \begin{bmatrix} B & 0 \\ 0 & C \end{bmatrix}$ is diagonalizable, show that both $B$ and $C$ are diagonalizable. [Hint: Let $\{X_1, \cdots, X_{n+m}\}$ be a basis of $\mathbb{R}^{n+m}$ consisting of eigenvectors of $A$, and write $X_i = \begin{bmatrix} Y_i \\ Z_i \end{bmatrix}$ where each $Y_i$ is in $\mathbb{R}^n$ and each $Z_i$ is in $\mathbb{R}^m$. Show that $\{Y_1, \cdots, Y_n\}$ and $\{Z_1, \cdots, Z_m\}$ span $\mathbb{R}^n$ and $\mathbb{R}^m$ respectively, and so contain bases by Theorem 3. Use Theorem 3 §2.3.]

**27.** Let $U$ and $W$ be any two subspaces of $\mathbb{R}^n$, and consider the subspaces

$$U \cap W = \{X \text{ in } \mathbb{R}^n \mid X \text{ is in both } U \text{ and } W\}$$
$$U + W = \{X \text{ in } \mathbb{R}^n \mid X \text{ is a sum of a vector in } U$$
$$\text{and a vector in } W\} \text{ as in Exercise 23 §4.1.}$$

Show that $dim(U + W) = dimU + dimW - dim(U \cap W)$.

[Hint: Let $\{X_1, \cdots, X_d\}$ be a basis of $U \cap W$ and, by Theorem 3, extend it to bases $\{X_1, \cdots, X_d, Y_1, \cdots, Y_k\}$ and $\{X_1, \cdots, X_d, Z_1, \cdots, Z_m\}$ of $U$ and $W$ respectively. Show that $\{X_1, \cdots, X_d, Y_1, \cdots, Y_k, Z_1, \cdots, Z_m\}$ is a basis of $U + W$.]

## 4.4   RANK

In this section we complete some unfinished business from Section 1.2 by giving a precise definition of the rank of an $m \times n$ matrix, and showing that the rank of $A$ equals the rank of $A^T$. We then relate the rank to the dimensions of the image and null spaces of $A$.

### 4.4.1   Row and Column Spaces

If $A$ is an $m \times n$ matrix, we have already discussed the null space, *nullA*, and the image, *imA*. We are going to find natural bases for both these spaces, and so calculate their dimension. To do this, and for other reasons as well, it is essential to consider two other subspaces associated with $A$. They are defined as follows:

The **column space**, *colA*, of an $m \times n$ matrix $A$ is the subspace of $\mathbb{R}^m$ spanned by the columns of $A$.

The **row space**, *rowA*, of an $m \times n$ matrix $A$ is the subspace of $\mathbb{R}^n$ spanned by the rows of $A$.

Observe that in discussing *rowA* we are regarding the elements of $\mathbb{R}^n$ as rows.

One reason for the importance of the spaces *rowA* and *colA* is that they are unchanged when row (respectively column) operations are performed on the matrix $A$.

LEMMA **1**   Let $A$ denote an $m \times n$ matrix.

(1) If $A \to B$ using a sequence of row operations, then $rowB = rowA$.
(2) If $A \to B$ using a sequence of column operations, then $colB = colA$.

PROOF   (1) It is enough to prove it for a single row operation (since $B$ is obtained from $A$ by a sequence of row operations). We do this for a row operation of type III; a similar argument works for the other types.

Hence, suppose that $B$ is obtained by adding $u$ times row $p$ of $A$ to row $q$. Then each row of $B$ is a linear combination of the rows of $A$, so $rowB \subseteq rowA$ by Theorem 1 §4.1. On the other hand we can carry $B \to A$ by *subtracting* $u$ times row $p$ of $B$ from row $q$. Hence $rowA \subseteq rowB$ follows in the same way. Thus $rowA = rowB$ as required.

(2) This is proved in the same way using column operations.

In particular, suppose $A \to R$ by row operations where $R$ is a row-echelon matrix, and suppose further that $R$ has $k$ leading 1's. Then $R$ has $k$ nonzero rows, say $Y_1, Y_2, \cdots, Y_k$, so $rowR = span\{Y_1, Y_2, \cdots, Y_k\}$. Moreover $\{Y_1, Y_2, \cdots, Y_k\}$ is linearly independent by Example 7 §4.2, so it is a basis of *rowR*.
Since $rowR = rowA$ by Lemma 1, we have

$$k = dim(rowR) = dim(rowA).$$

Hence the number $k$ of leading 1's is the same *no matter how A is carried to row-echelon form*[5]. We express this by saying that the number of leading 1's is an *invariant* of $A$, called the **rank** of $A$, and denoted *rankA*. This is consistent with the usage in Section 1.2, and the above discussion is summarized in the following theorem.

**THEOREM** ❶ 
Let $A$ be an $m \times n$ matrix. Then

$$rankA = dim(rowA).$$

Moreover, if $A \rightarrow R$ by row operations where $R$ is a row-echelon matrix, the nonzero rows of $R$ are a basis of *rowA*.

Theorem 1 has an additional virtue: It provides a routine method for finding a basis for any subspace $U$ of $\mathbb{R}^n$ when a spanning set is given. Indeed, if $U = span\{X_1, X_2, \cdots, X_k\}$ where the $X_i$ are written as rows, simply carry the matrix with the $X_i$ as its rows to row-echelon form $R$. Then the nonzero rows of $R$ are a basis of $U$.

**Example 1**   Find a basis of $U = span\{[1 \ \ 1 \ \ -2 \ \ 4], [2 \ \ 5 \ \ 4 \ \ -2], [1 \ \ 7 \ \ 14 \ \ -16]\}$.

**SOLUTION**   Observe that $U = rowA$ where $A = \begin{bmatrix} 1 & 1 & -2 & 4 \\ 2 & 5 & 4 & -2 \\ 1 & 7 & 14 & -16 \end{bmatrix}$. The reduction of $A$ to row-echelon form is

$$\begin{bmatrix} 1 & 1 & -2 & 4 \\ 2 & 5 & 4 & -2 \\ 1 & 7 & 14 & -16 \end{bmatrix} \rightarrow \begin{bmatrix} 1 & 1 & -2 & 4 \\ 0 & 3 & 8 & -10 \\ 0 & 6 & 16 & -20 \end{bmatrix}$$

$$\rightarrow \begin{bmatrix} 1 & 1 & -2 & 4 \\ 0 & 3 & 8 & -10 \\ 0 & 0 & 0 & 0 \end{bmatrix}$$

$$\rightarrow \begin{bmatrix} 1 & 1 & -2 & 4 \\ 0 & 1 & \frac{8}{3} & -\frac{10}{3} \\ 0 & 0 & 0 & 0 \end{bmatrix}.$$

Hence $\{[1 \ \ 1 \ \ -2 \ \ 4], [0 \ \ 1 \ \ \frac{8}{3} \ \ -\frac{10}{3}]\}$ is a basis of $rowA = U$. We note that $\{[1 \ \ 1 \ \ -2 \ \ 4], [0 \ \ 3 \ \ 8 \ \ -10]\}$ is also a basis (see Example 3 §4.3), possibly more convenient.

### 4.4.2   The Rank Theorem

Theorem 1 can also be used to find a basis of *colA*: Carry $A^T$ to row-echelon form $R$ and use the transposes of the nonzero rows of $R$. (Equivalently, carry $A$ to "column-echelon form" $C$ and use the nonzero columns of $C$.) However, there is an easy way to select a basis of *colA* directly from the columns of $A$. This is part of the following very important theorem: If $A$ is any $m \times n$ matrix, the subspace *rowA* of $\mathbb{R}^n$ has the *same dimension* as the subspace *colA* of $\mathbb{R}^m$.

---

[5] This was asserted without proof in Section 1.2.

**THEOREM 2**

**Rank Theorem.** Let $A$ be an $m \times n$ matrix. Then
$$dim(rowA) = dim(colA) = rankA.$$

Moreover, if $A \rightarrow R$ by row operations where $R$ is a row-echelon matrix, and if the leading 1's are in columns $j_1, j_2, \cdots, j_r$ of $R$, then the corresponding columns $j_1, j_2, \cdots, j_r$ of $A$ are a basis of $colA$.

**PROOF**

Since $r = rankA$, the displayed equations follow from the last sentence and Theorem 1. To prove the last sentence, write $A = [C_1 \ C_2 \ \cdots \ C_n]$ where $C_j$ denotes column $j$ of $A$. We use the fact that $R = UA$ for some invertible matrix $U$ (Theorem 1 §1.6). Hence
$$R = UA = U[C_1 \ C_2 \ \cdots \ C_n]$$
$$= [UC_1 \ UC_2 \ \cdots \ UC_n]$$

so $UC_j$ is column $j$ of $R$. Since $R$ contains $r$ leading 1's, let $UC_{j_1}, UC_{j_2}, \cdots, UC_{j_r}$ be the columns of $R$ that contain a leading 1. Then $\{UC_{j_1}, UC_{j_2}, \cdots, UC_{j_r}\}$ is a basis of $colR$ as the reader can verify. We use this to show that $\{C_{j_1}, C_{j_2}, \cdots, C_{j_r}\}$ is a basis of $colA$, as required.

*Independence:* Suppose $t_1 C_{j_1} + t_2 C_{j_2} + \cdots + t_r C_{j_r} = 0$ where each $t_i$ is a real number. Left multiplying by $U$ gives
$$0 = U0 = U(t_1 C_{j_1} + t_2 C_{j_2} + \cdots + t_r C_{j_r}) = t_1 UC_{j_1} + t_2 UC_{j_2} + \cdots + t_r UC_{j_r}.$$

Hence each $t_i = 0$ by the independence of $\{UC_{j_1}, UC_{j_2}, \cdots, UC_{j_r}\}$, and we have proved the independence of $\{C_{j_1}, C_{j_2} \cdots, C_{j_r}\}$.

*Spanning:* Given $X$ in $colA$, write $X = r_1 C_1 + r_2 C_2 + \cdots + r_n C_n$ where each $r_j$ is a real number. Then $UX = r_1 UC_1 + r_2 UC_2 + \cdots + r_n UC_n$ is in $colR$ because each $UC_j$ is in $colR$. But $\{UC_{j_1}, UC_{j_2}, \cdots, UC_{j_r}\}$ spans $colR$, so there exist $t_1, t_2, \cdots, t_r$ such that
$$UX = t_1 UC_{j_1} + t_2 UC_{j_2} + \cdots + t_r UC_{j_r}.$$

Hence $UX = U(t_1 C_{j_1} + t_2 C_{j_2} + \cdots + t_r C_{j_r})$, so left multiplication by $U^{-1}$ gives $X = t_1 C_{j_1} + t_2 C_{j_2} + \cdots + t_r C_{j_r}$. Thus $X$ is in $span\{C_{j_1}, C_{j_2} \cdots, C_{j_r}\}$, as required.

**Example 2**

If $A = \begin{bmatrix} 1 & -1 & 3 & -2 \\ 2 & -2 & 2 & -1 \\ -1 & 1 & 5 & -4 \end{bmatrix}$, find a basis for $rowA$ and $colA$.

**SOLUTION**

One reduction of $A$ to row-echelon form is as follows:

$$A = \begin{bmatrix} 1 & -1 & 3 & -2 \\ 2 & -2 & 2 & -1 \\ -1 & 1 & 5 & -4 \end{bmatrix}$$

$$\rightarrow \begin{bmatrix} 1 & -1 & 3 & -2 \\ 0 & 0 & -4 & 3 \\ 0 & 0 & 8 & -6 \end{bmatrix}$$

$$\rightarrow \begin{bmatrix} 1 & -1 & 3 & -2 \\ 0 & 0 & 1 & -\frac{3}{4} \\ 0 & 0 & 0 & 0 \end{bmatrix} = R.$$

Hence $\{[1 \ -1 \ 3 \ -2], [0 \ 0 \ 1 \ -\frac{3}{4}]\}$ is a basis of $rowA$ by Theorem 1. Since the leading ones are in columns 1 and 3 of the row-echelon matrix $R$, Theorem 2 shows that columns 1 and 3 of $A$ constitute a basis $\left\{ \begin{bmatrix} 1 \\ 2 \\ -1 \end{bmatrix}, \begin{bmatrix} 3 \\ 2 \\ 5 \end{bmatrix} \right\}$ of $colA$. Of course both bases contain $2 = rankA$ vectors as the Rank theorem asserts.

The Rank theorem has many consequences. First, if $A$ is an $m \times n$ matrix, then $rowA$ is spanned by the $m$ rows of $A$, whence $rankA = dim(rowA) \leq m$ because the dimension of a subspace cannot exceed the number of vectors in any spanning set. Similarly, $rankA \leq n$ because $rankA = dim(colA)$. We record this as

**COROLLARY 1**    If $A$ is an $m \times n$ matrix, then $rankA \leq m$ and $rankA \leq n$.

If $A$ is an $n \times n$ matrix, then $rowA$ is spanned by $n$ rows, so $dim(rowA) = n$ if and only if the rows are independent (using Theorem 4 §4.3). But the rows are independent if and only if $A$ is invertible (by Theorem 2 §4.2), so we have

**COROLLARY 2**    An $n \times n$ matrix $A$ is invertible if and only if $rankA = n$.

The fact that the columns of $A$ are just the transposes of the rows of $A^T$ means that $dim(colA) = dim(row(A^T))$, that is

**COROLLARY 3**    $rankA^T = rankA$ for any matrix $A$.

**COROLLARY 4**    If $A$ is an $m \times n$ matrix, then $rankA = rank(UAV)$ for any invertible matrices $U$ and $V$.

PROOF    Recall (Lemma 1 §1.6) that doing an elementary row operation to $A$ is the same as left multiplying $A$ by an elementary matrix. Since the invertible matrix $U$ is a product of elementary matrices (Theorem 1 §1.6), it follows that $UA$ is obtained from $A$ by a series of row operations. Hence $rank(UA) = rankA$ by Lemma 1. Now, write $UA = B$ for convenience, and apply Corollary 3 to get

$$rank(UAV) = rank(BV) = rank(BV)^T = rank(V^T B^T).$$

But $V^T$ is invertible so the first part of this proof gives $rank(V^T B^T) = rankB^T = rankB = rankA$. Thus $rank(UAV) = rankA$ as required.

The result in Corollary 4 should be compared to Theorem 3 §1.6.

If $A$ is an $n \times n$ matrix, then $A$ is invertible if and only if $rankA = n$ by Corollary 2. In view of Corollary 1, this asserts that $A$ is invertible if and only if $rankA$ is as large as possible. This raises a natural question about an arbitrary $m \times n$ matrix $A$: When do the two extreme cases $rankA = n$ and $rankA = m$ arise? The next two theorems show that the cases are naturally related to the independence and spanning properties of the columns of $A$ (and hence to the systems of linear equations with $A$ as coefficient matrix), and also to the invertibility of the (square and symmetric) matrices $A^T A$ and $A A^T$.

Recall that if $A = [C_1 \ \ C_2 \ \ \cdots \ \ C_n]$ is an $m \times n$ matrix with columns $C_1, \ C_2, \ \cdots, \ C_n$, and if $X = [x_1 \ \ x_2 \ \ \cdots \ \ x_n]^T$ is a column of scalars, then

$$AX = [C_1 \ \ C_2 \ \ \cdots \ \ C_n] \begin{bmatrix} x_1 \\ x_2 \\ \vdots \\ x_n \end{bmatrix} = x_1C_1 + x_2C_2 + \cdots + x_nC_n. \qquad (*)$$

We will refer to this several times.

---

**THEOREM 3**

Let $A$ be an $m \times n$ matrix. The following are equivalent:

(1) $AX = 0$, $X$ a column, has only the trivial solution $X = 0$.
(2) The columns of $A$ are linearly independent.
(3) $rankA = n$.
(4) $A^TA$ is an invertible $n \times n$ matrix.

**PROOF**

Let $C_1, \ C_2, \ \cdots, \ C_n$ denote the columns of $A$.

$(1) \Rightarrow (2)$. If $x_1C_1 + x_2C_2 + \cdots + x_nC_n = 0$, then $AX = 0$ by $(*)$, so each $x_i = 0$ by (1).

$(2) \Rightarrow (3)$. Since $colA = span\{C_1, \ C_2, \ \cdots, \ C_n\}$, (2) implies that $n = dim(colA) = rankA$.

$(3) \Rightarrow (4)$. To prove (4) we verify that $(A^TA)X = 0$ implies that $X = 0$ (by Theorem 5 §1.5). So assume that $A^TAX = 0$, write $AX = [y_1 \ \ y_2 \ \ \cdots \ \ y_m]^T$, and compute:

$$y_1^2 + y_2^2 + \cdots + y_m^2 = (AX)^T(AX) = X^T(A^TAX) = X^T0 = 0.$$

It follows that each $y_j = 0$; that is, $AX = 0$. If $X = [x_1 \ \ x_2 \ \ \cdots \ \ x_n]^T$, this means that $x_1C_1 + x_2C_2 + \cdots + x_nC_n = 0$ by $(*)$. But $\{C_1, \ C_2, \ \cdots, \ C_n\}$ is independent by Theorem 4 §4.3 (because $span\{C_1, \ C_2, \ \cdots, \ C_n\} = colA$ has dimension $n$ by (3)). Hence every $x_i = 0$, whence $X = 0$. This is what we wanted.

$(4) \Rightarrow (1)$. If $AX = 0$, then $A^TAX = 0$, so $X = 0$ by (4).

---

Theorem 3 describes what happens when an $m \times n$ matrix has rank $n$. The following companion theorem discusses the situation when the rank is $m$.

---

**THEOREM 4**

The following are equivalent for an $m \times n$ matrix $A$:

(1) $AX = B$ has a solution for every column $B$ in $\mathbb{R}^m$.
(2) The columns of $A$ span $\mathbb{R}^m$.
(3) $rankA = m$.
(4) $AA^T$ is an invertible $m \times m$ matrix.

**PROOF**

Again let $C_1, \ C_2, \ \cdots, \ C_n$ denote the columns of $A$.

$(1) \Rightarrow (2)$. If $B$ is in $\mathbb{R}^m$, then (1) and equation $(*)$ (preceding Theorem 3) give $B = AX = x_1C_1 + x_2C_2 + \cdots + x_nC_n$ for scalars $x_i$, so $B$ is in $colA$.

$(2) \Rightarrow (3)$. $colA = \mathbb{R}^m$ by (2), so $rankA = dim(colA) = dim(\mathbb{R}^m) = m$.

$(3) \Rightarrow (1)$. Since $colA \subseteq \mathbb{R}^m$ and $dim(colA) = m$ by (3), we have $colA = \mathbb{R}^m$ by Theorem 4 §4.3. Hence if $B$ is in $\mathbb{R}^m$, then $B = x_1C_1 + x_2C_2 + \cdots + x_nC_n$ for scalars $x_i$, that is $B = AX$ where $X = [x_1 \ \ x_2 \ \ \cdots \ \ x_n]^T$. This proves (1).

$(3) \Leftrightarrow (4)$. Write $M = A^T$. If we apply $(3) \Leftrightarrow (4)$ in Theorem 3 to the $n \times m$ matrix $M$, we obtain $M^TM$ is invertible if and only if $rankM = m$; that is, $AA^T$ is invertible if and only if $rankA = m$. This is $(3) \Leftrightarrow (4)$ in the present theorem.

It is instructive to compare conditions (1) and (2) in Theorems 3 and 4 with Theorem 5 §1.5. In that earlier theorem these conditions are all equivalent to the square matrix $A$ being invertible, while the present theorems isolate the roles played by these conditions in an arbitrary matrix. Let $A$ be any $m \times n$ matrix. Theorems 3 and 4 relate several important properties of $A$ to the invertibility of the square matrices $A^T A$ and $A A^T$. Even if the columns of $A$ are not independent or do not span $\mathbb{R}^m$ (as in these theorems), the matrices $A^T A$ and $A A^T$ are both symmetric and as such have real eigenvalues by Theorem 3 §2.5. This focuses our attention on the study of symmetric matrices. In fact the eigenvalues of $A^T A$ and $A A^T$ are all nonnegative, and this leads to the so-called singular value decomposition of $A$, which provides basic information about $A$ and is important in numerical analysis. We investigate this in Section 4.11.

## 4.4.3   Null Space and Image

In Section 4.1 we discussed two subspaces associated with an $m \times n$ matrix $A$, the null space $nullA = \{X \mid X \text{ in } \mathbb{R}^n \text{ and } AX = 0\}$ and the image $imA = \{AX \mid X \text{ in } \mathbb{R}^n\}$. Using the rank, there are simple ways to find bases for these spaces. The next theorem does this for $imA$ because the Rank theorem does it for $colA$.

**THEOREM 5**

If $A$ is any $m \times n$ matrix of rank $r$, then
$$colA = imA = \{AX \mid X \text{ in } \mathbb{R}^n\}.$$
Hence the Rank theorem provides a basis for $imA$. In particular,
$$dim(imA) = rankA.$$

**PROOF**

Equation $(*)$ preceding Theorem 3 shows that the vectors $AX$ in $imA$ are just the linear combinations of the columns of $A$, that is $imA = colA$.[6] In particular, $dim(imA) = dim(colA) = rankA$.

Turning to the null space, a natural basis can be found using the Gaussian algorithm. We first discuss a specific example which exhibits most of the salient features of the situation.

**Example 3**

If $A = \begin{bmatrix} 1 & -2 & 1 & 1 \\ -1 & 2 & 0 & 1 \\ 2 & -4 & 1 & 0 \end{bmatrix}$, find a basis of $nullA$ and so find its dimension.

**SOLUTION**

If $X$ is in $nullA$, then $AX = 0$, so $X$ is given by solving the system $AX = 0$. Using Gaussian elimination, the augmented matrix is carried to reduced row-echelon form as follows:

$$\begin{bmatrix} 1 & -2 & 1 & 1 & 0 \\ -1 & 2 & 0 & 1 & 0 \\ 2 & -4 & 1 & 0 & 0 \end{bmatrix} \rightarrow \begin{bmatrix} 1 & -2 & 1 & 1 & 0 \\ 0 & 0 & 1 & 2 & 0 \\ 0 & 0 & -1 & -2 & 0 \end{bmatrix}$$

$$\rightarrow \begin{bmatrix} 1 & -2 & 0 & -1 & 0 \\ 0 & 0 & 1 & 2 & 0 \\ 0 & 0 & 0 & 0 & 0 \end{bmatrix}.$$

Hence, writing $X = [x_1 \ x_2 \ x_3 \ x_4]^T$, the leading variables are $x_1$ and $x_3$, and the non-leading variables $x_2$ and $x_4$ become parameters: $x_2 = s$ and $x_4 = t$. Then the equations corresponding to the reduced matrix determine the leading variables in terms of the parameters:

---

[6] This observation was made earlier in Example 12 §4.1.

$$x_1 = 2s + t \quad \text{and} \quad x_3 = -2t.$$

This means that the general solution is

$$X = [2s + t \quad s \quad -2t \quad t]^T = s[2 \quad 1 \quad 0 \quad 0]^T + t[1 \quad 0 \quad -2 \quad 1]^T. \qquad (**)$$

Hence $X$ is in $span\{X_1, X_2\}$ where $X_1 = [2 \quad 1 \quad 0 \quad 0]^T$ and $X_2 = [1 \quad 0 \quad -2 \quad 1]^T$ are the basic solutions, and we have shown that $null(A) \subseteq span\{X_1, X_2\}$. But each of $X_1$ and $X_2$ are in $nullA$ (they are solutions of the system $AX = 0$), so we have

$$nullA = span\{X_1, X_2\}$$

by Theorem 1 §4.1. We claim further that $\{X_1, X_2\}$ is linearly independent. To see this, let $sX_1 + tX_2 = 0$ be a linear combination that vanishes. Then $(**)$ shows that $[2s + t \quad s \quad -2t \quad t]^T = 0$, whence $s = t = 0$. Thus $\{X_1, X_2\}$ is a basis of $null(A)$, and so $dim(nullA) = 2$.

The calculation in Example 3 is typical of what happens in general. If $A$ is an $m \times n$ matrix, it was shown in Theorem 2 §1.3 that the system $AX = 0$ has exactly $n - r$ basic solutions $X_1, X_2, \cdots, X_{n-r}$ where $r = rankA$, and that

$$nullA = span\{X_1, X_2, \cdots, X_{n-r}\}.$$

Moreover, the general solution is a linear combination $X = t_1X_1 + t_2X_2 + \cdots + t_{n-r}X_{n-r}$ where each coefficient $t_i$ is a parameter equal to a nonleading variable. Thus if this linear combination vanishes, that is, if $X = 0$, then each $t_i = 0$ (as for $s$ and $t$ in Example 3, each $t_i$ is a coefficient when $X$ is expressed as a linear combination of the standard basis of $\mathbb{R}^n$). This proves that $\{X_1, X_2, \cdots, X_{n-r}\}$ is linearly independent, and so is a basis of $nullA$. Theorem 6 summarizes this discussion.

**THEOREM 6**

Let $A$ denote an $m \times n$ matrix of rank $r$, and let $X_1, X_2, \cdots, X_{n-r}$ be the basic solutions of the homogeneous system $AX = 0$ that are produced by the Gaussian algorithm. Then $\{X_1, X_2, \cdots, X_{n-r}\}$ is a basis of $nullA$. In particular,

$$dim(nullA) = n - r.$$

Combining Theorems 5 and 6, we obtain a useful equation relating the dimensions of the image and null space of a matrix.

**COROLLARY**

If $A$ is any $m \times n$ matrix, then

$$dim(imA) + dim(nullA) = n.$$

## Exercises 4.4

**1.** Give the row space and column space of the following matrices.

**a)** $A = \begin{bmatrix} 4 & -6 \\ 7 & 6 \end{bmatrix}$

**b)** $A = \begin{bmatrix} -5 & -4 \\ 8 & 0 \\ -1 & -3 \end{bmatrix}$

**c)** $A = \begin{bmatrix} 0 & -4 & -8 & -4 \\ 8 & 2 & -8 & 0 \\ -3 & 1 & -1 & -2 \end{bmatrix}$

**d)** $A = \begin{bmatrix} 9 & -5 & 1 & -1 \\ 8 & -5 & -3 & 1 \\ 0 & 4 & 7 & 6 \\ 0 & 2 & -7 & 7 \end{bmatrix}$

**2.** Match each matrix on the left with a matrix on the right sharing the same row or column space.

**a)** $A = \begin{bmatrix} 0 & 1 & -3 & 2 \\ 0 & 2 & 1 & -3 \\ 2 & -5 & 0 & 5 \end{bmatrix}$

**b)** $A = \begin{bmatrix} 1 & 0 & 0 & 0 \\ 0 & 1 & 0 & 0 \\ 2 & 1 & 0 & 0 \end{bmatrix}$

**c)** $A = \begin{bmatrix} 1 & 0 & 1 & 1 \\ 0 & 1 & 1 & -1 \\ 2 & 1 & 3 & 1 \end{bmatrix}$

**d)** $A = \begin{bmatrix} 0 & -2 & 6 & -4 \\ 2 & -3 & 1 & 2 \\ 2 & -5 & 0 & 5 \end{bmatrix}$

**e)** $A = \begin{bmatrix} 0 & 0 & 1 & 1 \\ -1 & 1 & 0 & 2 \\ -1 & 1 & 2 & 4 \end{bmatrix}$

**f)** $A = \begin{bmatrix} 1 & -1 & 0 & -2 \\ 0 & 0 & 1 & 1 \\ 0 & 0 & 0 & 0 \end{bmatrix}$

3. Given $rowA$ for some matrix $A$, what is the rank of $A$?
   a) $rowA = span\{[1 \ \ 0 \ \ 0 \ \ 1], [0 \ \ 1 \ \ 1 \ \ 0], [0 \ \ 0 \ \ 0 \ \ 1]\}$.
   b) $rowA = span\{[1 \ \ 2 \ \ -1 \ \ 3], [2 \ \ -3 \ \ 1 \ \ 4],$
      $[4 \ \ -13 \ \ 5 \ \ 6]\}$.
   c) $rowA = span\{[-1 \ \ 0 \ \ 3 \ \ 2], [3 \ \ 2 \ \ 0 \ \ -4],$
      $[9 \ \ 8 \ \ 9 \ \ -10], [4 \ \ 2 \ \ -3 \ \ -6]\}$.

4. In each case find a basis of the subspace $U$.
   a) $U = span\{[1 \ \ 2 \ \ -1 \ \ 3]^T, [2 \ \ -3 \ \ 1 \ \ 4]^T,$
      $[4 \ \ -13 \ \ 5 \ \ 6]^T\}$.
   b) $U = span\{[1 \ \ 2 \ \ -1 \ \ 0 \ \ 4]^T, [-3 \ \ 1 \ \ 0 \ \ 2 \ \ 5]^T,$
      $[-7 \ \ 7 \ \ -2 \ \ 6 \ \ 13]^T, [-2 \ \ 3 \ \ -1 \ \ 2 \ \ 9]^T\}$.
   c) $U = span\{[-1 \ \ 0 \ \ 3 \ \ 2]^T, [3 \ \ 2 \ \ 0 \ \ -4]^T,$
      $[9 \ \ 8 \ \ 9 \ \ -10]^T, [4 \ \ 2 \ \ -3 \ \ -6]^T\}$.

5. In each case find bases for the row and column spaces of $A$ and determine the rank of $A$.

   a) $A = \begin{bmatrix} 1 & -1 & 3 & 2 & 1 \\ -2 & 3 & 1 & 1 & 1 \\ -4 & 7 & 9 & 7 & 5 \\ 3 & -4 & 2 & 1 & 0 \end{bmatrix}$   b) $A = \begin{bmatrix} 1 & 2 & 0 & 5 & 7 & 1 \\ 2 & 1 & 1 & -1 & 0 & 3 \\ -5 & 2 & -4 & 19 & 21 & 9 \\ 1 & -1 & 1 & -6 & -7 & 2 \end{bmatrix}$

6. In each case find a basis of the nullspace of $A$, compute $rankA$, and verify Theorem 6.

   a) $A = \begin{bmatrix} 3 & 1 & 1 \\ 2 & 0 & 1 \\ 4 & 2 & 1 \\ 1 & -1 & 1 \end{bmatrix}$   b) $A = \begin{bmatrix} 3 & 5 & 5 & 2 & 0 \\ 1 & 0 & 2 & 2 & 1 \\ 1 & 1 & 1 & -2 & -2 \\ 2 & 0 & 4 & 4 & 2 \end{bmatrix}$

7. In each case either show that the statement is true or give an example showing that it is false. Throughout let $A$ denote an $m \times n$ matrix.
   a) If $A$ has a row of zeros, then $rowA = \{0\}$.
   b) If the rows of $A$ are linearly independent, then $rowA = \mathbb{R}^n$.
   c) If the rows of $A$ span $\mathbb{R}^n$, then $rowA = \mathbb{R}^n$.
   d) If $A$ has a row of zeros, then $rankA = 0$.
   e) If $A$ has a row of zeros, then $rankA < m$.
   f) If $B$ is $m \times n$ and $rankB = r = rankA$, then $rank(A + B) = r$.

8. Can a $3 \times 4$ matrix have independent rows? Independent columns? Explain.

9. If $A$ is $4 \times 3$ and $rankA = 2$, can $A$ have independent columns? Independent rows? Explain.

10. Can a nonsquare matrix $A$ have its rows independent and its columns independent? Explain.

11. Can the null space of a $3 \times 6$ matrix have dimension 2? Explain.

12. If a matrix $A$ is not square, show that either the rows or the columns of $A$ are dependent.

13. If $A$ is an $m \times n$ matrix and $rankA = m$, show that $m \le n$.

14. Let $A$ and $B$ denote matrices of size $k \times m$ and $m \times n$ respectively.
   a) Show that each column of $AB$ is a linear combination of the columns of $A$. [Hint: Equation (∗) preceding Theorem 3].
   b) Show that $rank(AB) \le rankA$.
   c) Show that $rank(AB) \le rankB$. [Hint: Corollary 3 of Theorem 2.]

15. Let $A$ and $B$ denote matrices of sizes $m \times n$ and $n \times k$ respectively.
   a) If $AB = 0$, show that $colB \subseteq nullA$. [Hint: Theorem 3 §1.4 (2).]
   b) If $colB \subseteq nullA$, show that $AB = 0$. [Hint: Theorem 3 §1.4 (1).]

16. Consider the system of equations $AX = B$ where $A$ is $m \times n$ and $B$ is $m \times 1$.
   a) Show that the system $AX = B$ has a solution if and only if $B$ is in $colA$. [Hint: Equation (∗) preceding Theorem 3.]
   b) Show that the system $AX = B$ has a solution if and only if $rankA = rank[A \ \ B]$, where $[A \ \ B]$ is the augmented matrix of the system.

17. If $I$ is the $m \times m$ identity matrix, show that $I + ZZ^T$ is invertible for every $m \times k$ matrix $Z$. [Hint: Factor
   $I + ZZ^T = [I \ \ Z]\begin{bmatrix} I \\ Z^T \end{bmatrix}$ in block form and use Theorem 4.]

18. Let $A$ denote an $m \times n$ matrix.
   a) If $rankA = m$, show that $AB = I_m$ for some $n \times m$ matrix $B$. [Hint: Theorem 4 (4).]
   b) If $AB = I_m$ for some $n \times m$ matrix $B$, show that $rankA = m$. [Hint: Exercise 14 b).]

19. Let $A$ denote an $m \times n$ matrix.
   a) Show that $null(A^TA) = null(A)$. [Hint: Use the proof of (3)$\Rightarrow$(4) in Theorem 3.]
   b) Conclude that $rank(A^TA) = rankA$.
   c) Show that $im(A^T) = im(A^TA)$. [Hint: Show that $im(A^TA) \subseteq im(A^T)$, and apply Theorem 4 §4.3.]

20. If $A$ is an $m \times n$ matrix, it can be proved that there exists a unique $n \times m$ matrix $A^\#$ such that $AA^\#A = A$, $A^\#AA^\# = A^\#$, and both $AA^\#$ and $A^\#A$ are symmetric. The matrix $A^\#$ is called the **Moore-Penrose** inverse of $A$.
   a) If $A$ is square and invertible, show that $A^\# = A^{-1}$.
   b) If $rankA = m$, show that $A^\# = A^T(AA^T)^{-1}$.
   c) If $rankA = n$, show that $A^\# = (A^TA)^{-1}A^T$.

## 4.5 ORTHOGONALITY

Length and orthogonality are two of the most important concepts in geometry. In $\mathbb{R}^2$ and $\mathbb{R}^3$ these notions can both be defined using the dot product, and this observation is exploited in this section to extend these concepts to $\mathbb{R}^n$. This in turn leads to a natural extension of the notion of distance to $\mathbb{R}^n$, to an extension of Pythagoras' theorem, and to the idea of an orthogonal basis—one of the most useful concepts in linear algebra.

### 4.5.1 Dot Product, Length, and Distance

Recall that, given $\vec{v} = [x_1 \ x_2 \ x_3]^T$ and $\vec{w} = [y_1 \ y_2 \ y_3]^T$ in $\mathbb{R}^3$, their dot product $\vec{v} \cdot \vec{w}$ and the length $\|\vec{v}\|$ of $\vec{v}$ are defined by

$$\vec{v} \cdot \vec{w} = x_1 y_1 + x_2 y_2 + x_3 y_3 \quad \text{and} \quad \|\vec{v}\| = \sqrt{x_1^2 + x_2^2 + x_3^2}.$$

These formulas extend in the obvious way to columns in $\mathbb{R}^n$. Accordingly, given columns $X = [x_1 \ x_2 \ \cdots \ x_n]^T$ and $Y = [y_1 \ y_2 \ \cdots \ y_n]^T$ in $\mathbb{R}^n$, we define their **dot product** $X \cdot Y$ as follows:

$$X \cdot Y = x_1 y_1 + x_2 y_2 + \cdots + x_n y_n = X^T Y.$$

Note that $X^T Y$ is technically a $1 \times 1$ matrix, which we take to be a number.[7] The **length** of the column vector $X = [x_1 \ x_2 \ \cdots \ x_n]^T$ is defined to be

$$\|X\| = \sqrt{X \cdot X} = \sqrt{x_1^2 + x_2^2 + \cdots + x_n^2}$$

where $\sqrt{(\ )}$ indicates the positive square root.[8]

**Example 1**  Given $X = [3 \ -2 \ 0 \ 4 \ -1]$ and $Y = [1 \ 1 \ -1 \ 0 \ 6]$ in $\mathbb{R}^5$, we have $X \cdot Y = 3 - 2 + 0 + 0 - 6 = -5$, and $\|X\| = \sqrt{9 + 4 + 0 + 16 + 1} = \sqrt{30}$.

The following facts will be used frequently below.

**THEOREM ①**  Let $X$, $Y$, and $Z$ denote columns in $\mathbb{R}^n$. Then:

(1) $X \cdot Y = Y \cdot X$.
(2) $X \cdot (Y + Z) = X \cdot Y + X \cdot Z$.
(3) $(aX) \cdot Y = a(X \cdot Y) = X \cdot (aY)$ for all scalars $a$.
(4) $\|X\|^2 = X \cdot X$.
(5) $\|X\| \geq 0$, and $\|X\| = 0$ if and only if $X = 0$.
(6) $\|aX\| = |a| \, \|X\|$ for all scalars $a$.

**PROOF**  Let $X = [x_1 \ x_2 \ \cdots \ x_n]^T$ and $Y = [y_1 \ y_2 \ \cdots \ y_n]^T$ be columns in $\mathbb{R}^n$. Then (1) follows from $X \cdot Y = x_1 y_1 + x_2 y_2 + \cdots + x_n y_n$, (2) and (3) follow from matrix arithmetic because $X \cdot Y = X^T Y$, (4) is clear from the definition, and (6) follows because $|a| = \sqrt{a^2}$ for all real numbers $a$. As for (5), $\|X\| \geq 0$ is clear, as is $\|0\| = 0$; finally, if $\|X\| = 0$, then $x_1^2 + x_2^2 + \cdots + x_n^2 = 0$, so each $x_i = 0$ because the $x_i$ are real numbers. Thus $X = 0$.

---

[7] If $X$ and $Y$ are written as rows rather than columns, this becomes $X \cdot Y = XY^T$.

[8] Since $X \cdot X = x_1^2 + x_2^2 + \cdots + x_n^2 \geq 0$ for all choices of the real numbers $x_i$, the length $\|X\|$ is a nonnegative real number for every column $X$ in $\mathbb{R}^n$.

Because of Theorem 1, computations with dot products in $\mathbb{R}^n$ are similar to those in $\mathbb{R}^3$. In particular it follows from (1) and (2) that the dot product of two sums $X_1 + \cdots + X_k$ and $Y_1 + \cdots + Y_m$ in $\mathbb{R}^n$ is the sum of the dot products of every $X_i$ with every $Y_j$:

$$(X_1 + \cdots + X_k) \bullet (Y_1 + \cdots + Y_m)$$
$$= X_1 \bullet Y_1 + X_1 \bullet Y_2 + X_2 \bullet Y_1 + \cdots + X_k \bullet Y_m$$
$$= \Sigma_{i,j} X_i \bullet Y_j.$$

If this is combined with (3) in Theorem 1, we can do calculations like

$$(2X_1 - 3X_2) \bullet (5Y_1 + 4Y_2)$$
$$= 10(X_1 \bullet Y_1) + 8(X_1 \bullet Y_2) - 15(X_2 \bullet Y_1) - 12(X_2 \bullet Y_2).$$

Such computations will be carried out without comment below.

**Example 2**    If $X$ and $Y$ are in $\mathbb{R}^n$, show that

$$\|X + Y\|^2 = \|X\|^2 + 2X \bullet Y + \|Y\|^2.$$

SOLUTION    We use Theorem 1 several times:

$$\|X + Y\|^2 = (X + Y) \bullet (X + Y)$$
$$= X \bullet X + X \bullet Y + Y \bullet X + Y \bullet Y$$
$$= \|X\|^2 + 2X \bullet Y + \|Y\|^2.$$

**Example 3**    Let $\mathbb{R}^n = span\{F_1, F_2, \cdots, F_m\}$. If $X$ in $\mathbb{R}^n$ satisfies $X \bullet F_i = 0$ for each $i$, show that $X = 0$.

SOLUTION    We show $X = 0$ by showing that $\|X\| = 0$ and applying (5) of Theorem 1. By hypothesis, write $X = t_1 F_1 + t_2 F_2 + \cdots + t_m F_m$ for some $t_i$ in $\mathbb{R}$. Then

$$\|X\|^2 = X \bullet X = X \bullet (t_1 F_1 + t_2 F_2 + \cdots + t_m F_m)$$
$$= t_1(X \bullet F_1) + t_2(X \bullet F_2) + \cdots + t_m(X \bullet F_m)$$
$$= 0$$

by hypothesis. Hence $\|X\| = 0$, and so $X = 0$ by Theorem 1.

A vector $E$ in $\mathbb{R}^n$ is called a **unit vector** if $\|E\| = 1$. Thus if $\{E_1, E_2, \cdots, E_n\}$ is the standard basis of $\mathbb{R}^n$ then each $E_i$ is a unit vector ($E_i$ is column $i$ of the $n \times n$ identity matrix). We will use the following observation several times below.

**Example 4**    If $X \neq 0$ in $\mathbb{R}^n$, show that $E = \frac{1}{\|X\|} X$ is the unique positive scalar multiple of $X$ that is a unit vector.

SOLUTION    If $E = tX$ where $t > 0$, then $\|E\| = |t| \|X\| = t\|X\|$ by Theorem 1. Hence if $\|E\| = 1$, we have $t\|X\| = 1$, so $t = \frac{1}{\|X\|}$.

If $\vec{v}$ and $\vec{w}$ are nonzero vectors in $\mathbb{R}^3$, we showed (Theorem 2 §3.2) that the angle $\theta$ between them is given by $\cos\theta = \frac{\vec{v} \bullet \vec{w}}{\|\vec{v}\|\|\vec{w}\|}$. Since $|\cos\theta| \leq 1$ for any angle $\theta$, it follows that $\|\vec{v} \bullet \vec{w}\| \leq \|\vec{v}\|\|\vec{w}\|$, and this holds even if $\vec{v} = \vec{0}$ or $\vec{w} = \vec{0}$. This inequality has an important analogue in $\mathbb{R}^n$.

**THEOREM** ❷ **Cauchy Inequality.**[9] If $X$ and $Y$ are vectors in $\mathbb{R}^n$, then

$$|X \bullet Y| \leq \|X\|\|Y\|.$$

Moreover, equality holds if and only if one of $X$ and $Y$ is a scalar multiple of the other.

**PROOF** If either $X = 0$ or $Y = 0$, the inequality holds (in fact, it is equality). Otherwise, write $\|X\| = a$ and $\|Y\| = b$ for convenience. Then $a > 0$ and $b > 0$, and a computation like that in Example 2 gives

$$\|bX - aY\|^2 = 2ab(ab - X \bullet Y) \quad \text{and} \quad \|bX + aY\|^2 = 2ab(ab + X \bullet Y). \qquad (*)$$

It follows that $ab - X \bullet Y \geq 0$ and $ab + X \bullet Y \geq 0$, whence $-ab \leq X \bullet Y \leq ab$. This gives the Cauchy inequality. If equality holds, we have $|X \bullet Y| = ab$, so either $X \bullet Y = ab$ or $X \bullet Y = -ab$. Using $(*)$, these imply that $bX - aY = 0$ or $bX + aY = 0$, respectively. Thus one of $X$ and $Y$ is a scalar multiple of the other (even if $a = 0$ or $b = 0$).

The Cauchy inequality is often written in the equivalent form $(X \bullet Y)^2 \leq \|X\|^2\|Y\|^2$. Thus, for example, in $\mathbb{R}^4$ we have

$$(a_1b_1 + a_2b_2 + a_3b_3 + a_4b_4)^2 \leq (a_1^2 + a_2^2 + a_3^2 + a_4^2)(b_1^2 + b_2^2 + b_3^2 + b_4^2)$$

for all choices of the numbers $a_i$ and $b_i$.

There is an important consequence of the Cauchy inequality. Given $X$ and $Y$ in $\mathbb{R}^n$, we use the inequality and Example 2 to compute

$$\begin{aligned}
\|X + Y\|^2 &= \|X\|^2 + 2X \bullet Y + \|Y\|^2 \\
&\leq \|X\|^2 + 2\|X\|\|Y\| + \|Y\|^2 \\
&= (\|X\| + \|Y\|)^2
\end{aligned}$$

If we take positive square roots, this gives the

**COROLLARY** **Triangle Inequality.** If $X$ and $Y$ are vectors in $\mathbb{R}^n$, then $\|X + Y\| \leq \|X\| + \|Y\|$.

The reason for the name will be clear when we have introduced the notion of distance into $\mathbb{R}^n$.

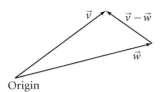

If $\vec{v}$ and $\vec{w}$ denote two vectors in $\mathbb{R}^3$, the distance between $\vec{v}$ and $\vec{w}$ (regarded as points) is $\|\vec{v} - \vec{w}\|$ as in the diagram. The analogue for $\mathbb{R}^n$ is as follows: If $X$ and $Y$ are vectors in $\mathbb{R}^n$, the **distance** between $X$ and $Y$ is defined to be

$$d(X, Y) = \|X - Y\|.$$

This distance function has all the intuitive properties of distance that are familiar from $\mathbb{R}^2$ and $\mathbb{R}^3$, including another version of the triangle inequality.

**THEOREM** ❸ Let $X$, $Y$, and $Z$ denote vectors in $\mathbb{R}^n$. Then:

(1) $d(X, Y) \geq 0$ for all $X$ and $Y$.
(2) $d(X, Y) = 0$ if and only if $X = Y$.
(3) $d(X, Y) = d(Y, X)$.
(4) **Triangle Inequality.** $d(X, Y) \leq d(X, Z) + d(Z, Y)$.

---

[9] Augustin Louis Cauchy (1789–1857) was born in Paris and became a professor at the École Polytechnique at the age of 26. He was one of the great mathematicians and is best remembered for his work in analysis where he established standards of rigour that carry down to today's calculus texts.

PROOF   (1) and (2) are translations of part (5) of Theorem 1, and (3) holds because
$\|X\| = \|-X\|$ by part (6) of Theorem 1. To prove (4), we use the Corollary
to Theorem 2:

$$
\begin{aligned}
d(X, Y) &= \|X - Y\| \\
&= \|(X - Z) + (Z - Y)\| \leq \|(X - Z)\| + \|(Z - Y)\| \\
&= d(X, Z) + d(Z, Y).
\end{aligned}
$$

This proves (4).

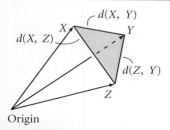

Now the name "Triangle Inequality" in (4) makes sense: Consider the "triangle" in $\mathbb{R}^n$
with vertices $X$, $Y$, and $Z$, as depicted in the diagram. Then the length $d(X, Y)$ of one
side of the triangle is smaller than the sum $d(X, Z) + d(Z, Y)$ of the lengths of the
other two sides. Note that, as in $\mathbb{R}^3$, we regard a vector $X$ in $\mathbb{R}^n$ as an arrow from the
origin to the point $X$. While not totally accurate, this method of visualizing $\mathbb{R}^n$ does not
lead one astray.

## 4.5.2   Orthogonal Sets and the Expansion Theorem

The concept of orthogonality is a central idea in geometry, occurring as it does in
Pythagoras' theorem, in the notion of the normal to a plane, and in the way we find
the point on a line that is closest to a given point. Orthogonality turns out to play an
even more important role in the geometry of $\mathbb{R}^n$ as we shall see. Moreover, as for length
and distance, the definition is a natural extension of the situation in $\mathbb{R}^3$.

As in $\mathbb{R}^3$, two vectors $X$ and $Y$ in $\mathbb{R}^n$ are said to be **orthogonal** if

$$X \cdot Y = 0$$

More generally, a set of vectors $\{X_1, X_2, \cdots, X_k\}$ is called an **orthogonal set** of vectors if

$$
\begin{aligned}
X_i \cdot X_j &= 0 \quad \text{for all } i \neq j \\
X_i &\neq 0 \quad \text{for all } i.
\end{aligned}
$$

(The reason for insisting that each $X_i$ is nonzero is that we will be concerned primarily
with orthogonal bases.) Note that $\{X\}$ is an orthogonal set for any $X \neq 0$ in $\mathbb{R}^n$. An
orthogonal set of vectors $\{X_1, X_2, \cdots, X_k\}$ is called **orthonormal** if $\|X_i\| = 1$ for each $i$,
that is if each $X_i$ is a unit vector.

**Example 5**   The standard basis of $\mathbb{R}^n$ is an orthonormal set of vectors.

If $\{X_1, X_2, \cdots, X_k\}$ is an orthogonal set in $\mathbb{R}^n$, it is easy to verify that
$\{a_1 X_1, a_2 X_2, \cdots, a_k X_k\}$ is also an orthogonal set for any choice of scalars $a_i \neq 0$.
In particular, $\left\{ \frac{1}{\|X_1\|} X_1, \frac{1}{\|X_2\|} X_2, \cdots, \frac{1}{\|X_k\|} X_k \right\}$ is an orthonormal set by Example 4,
and we say it is the result of **normalizing** the orthogonal set $\{X_1, X_2, \cdots, X_k\}$.

**Example 6**   The set $\{[1 \ -1 \ 3 \ 0]^T, [-2 \ 1 \ 1 \ 1]^T, [1 \ 1 \ 0 \ 1]^T, [1 \ 10 \ 3 \ -11]^T\}$ is orthogonal
in $\mathbb{R}^4$ as is easily verified. After normalizing, the corresponding orthonormal set is
$\left\{ \frac{1}{\sqrt{11}}[1 \ -1 \ 3 \ 0]^T, \frac{1}{\sqrt{7}}[-2 \ 1 \ 1 \ 1]^T, \frac{1}{\sqrt{3}}[1 \ 1 \ 0 \ 1]^T, \frac{1}{\sqrt{231}}[1 \ 10 \ 3 \ -11]^T \right\}$.

No discussion of geometry can omit mention of the most famous theorem about orthogonality: Pythagoras' theorem. Suppose orthogonal vectors $\vec{v} \neq \vec{0}$ and $\vec{w} \neq \vec{0}$ are given in $\mathbb{R}^3$. Then Pythagoras' theorem asserts that $\|\vec{v} + \vec{w}\|^2 = \|\vec{v}\|^2 + \|\vec{w}\|^2$ (see the diagram). More generally, if $\{X, Y\}$ is an orthogonal set in $\mathbb{R}^n$, then $X \cdot Y = 0$ so

$$\|X + Y\|^2 = \|X\|^2 + 2(X \cdot Y) + \|Y\|^2$$
$$= \|X\|^2 + \|Y\|^2.$$

In fact, this works for orthogonal sets of more than two vectors in $\mathbb{R}^n$.

**THEOREM 4**

**Pythagoras' Theorem.** Let $\{X_1, X_2, \cdots, X_k\}$ be an orthogonal set[10] of vectors in $\mathbb{R}^n$. Then

$$\|X_1 + X_2 + \cdots + X_k\|^2 = \|X_1\|^2 + \|X_2\|^2 + \cdots + \|X_k\|^2.$$

**PROOF**

The fact that $X_i \cdot X_j = 0$ whenever $i \neq j$ gives

$$\|X_1 + X_2 + \cdots + X_k\|^2$$
$$= (X_1 + X_2 + \cdots + X_k) \cdot (X_1 + X_2 + \cdots + X_k)$$
$$= X_1 \cdot X_1 + X_2 \cdot X_2 + \cdots + X_k \cdot X_k$$
$$+ (X_1 \cdot X_2 + X_1 \cdot X_3 + X_2 \cdot X_3 + \cdots)$$
$$= \|X_1\|^2 + \|X_2\|^2 + \cdots + \|X_k\|^2 + (0 + 0 + 0 + \cdots).$$

This proves the theorem.

If $\{\vec{v}, \vec{w}\}$ is an orthogonal set in $\mathbb{R}^3$, then $\vec{v}$ and $\vec{w}$ are certainly not parallel (being orthogonal), so $\{\vec{v}, \vec{w}\}$ is linearly independent by the Corollary of Theorem 3 §4.2. The next theorem shows that this observation holds much more generally.

**THEOREM 5**

Every orthogonal set of vectors in $\mathbb{R}^n$ is linearly independent.

**PROOF**

If $\{X_1, X_2, \cdots, X_k\}$ is orthogonal, suppose a linear combination vanishes: $t_1 X_1 + t_2 X_2 + \cdots + t_k X_k = 0$. We must show that each $t_i = 0$. If we take the dot product with $X_1$, the result is

$$0 = X_1 \cdot 0 = X_1 \cdot (t_1 X_1 + t_2 X_2 + \cdots + t_k X_k)$$
$$= t_1 (X_1 \cdot X_1) + t_2 (X_1 \cdot X_2) + \cdots + t_k (X_1 \cdot X_k)$$
$$= t_1 \|X_1\|^2 + t_2 (0) + \cdots + t_k (0)$$
$$= t_1 \|X_1\|^2.$$

Since $\|X_1\|^2 \neq 0$ (because $X_1 \neq 0$), it follows that $t_1 = 0$. Similarly, $t_i = 0$ for each $i$.

Theorem 5 points immediately to the consideration of orthogonal bases. These turn out to be much more convenient than ordinary bases for the following reason. If $\{X_1, X_2, \cdots, X_k\}$ is a basis of a subspace $U$, then $U = span\{X_1, X_2, \cdots, X_k\}$ so every vector $X$ in $U$ can be (uniquely) expressed as a linear combination $X = t_1 X_1 + t_2 X_2 + \cdots + t_k X_k$. The problem is that, in general, it is tedious to compute the coefficients $t_i$ (this requires solving $k$ equations in $k$ variables). However, explicit formulas exist for the coefficients if the basis $\{X_1, X_2, \cdots, X_k\}$ is orthogonal.

---

[10] This is valid even if some of the $X_i$ are zero.

**THEOREM 6** **Expansion Theorem.** Let $\{X_1, X_2, \cdots, X_k\}$ be an orthogonal basis of a subspace $U$ of $\mathbb{R}^n$. If $X$ is any vector in $U$, then

$$X = \frac{X \cdot X_1}{\|X_1\|^2}X_1 + \frac{X \cdot X_2}{\|X_2\|^2}X_2 + \cdots + \frac{X \cdot X_k}{\|X_k\|^2}X_k.$$

**PROOF** We have $X = t_1X_1 + t_2X_2 + \cdots + t_kX_k$ for some coefficients $t_i$ because $U = span\{X_1, X_2, \cdots, X_k\}$. Take the dot product of both sides with $X_1$:

$$\begin{aligned} X \cdot X_1 &= (t_1X_1 + t_2X_2 + \cdots + t_kX_k) \cdot X_1 \\ &= t_1(X_1 \cdot X_1) + t_2(X_2 \cdot X_1) + \cdots + t_k(X_k \cdot X_1) \\ &= t_1\|X_1\|^2 + t_2(0) + \cdots + t_k(0) \\ &= t_1\|X_1\|^2. \end{aligned}$$

Since $\|X_1\|^2 \neq 0$, this gives $t_1 = \frac{X \cdot X_1}{\|X_1\|^2}$. Similarly, $t_i = \frac{X \cdot X_i}{\|X_i\|^2}$ for each $i$. ∎

**Example 7** As in Example 6, let $X_1 = [1 \;\; -1 \;\; 3 \;\; 0]^T$, $X_2 = [-2 \;\; 1 \;\; 1 \;\; 1]^T$, $X_3 = [1 \;\; 1 \;\; 0 \;\; 1]^T$, and $X_4 = [1 \;\; 10 \;\; 3 \;\; -11]^T$. Then $\{X_1, X_2, X_3, X_4\}$ is orthogonal in $\mathbb{R}^4$, hence it is independent by Theorem 5, and so it is a basis because $dim(\mathbb{R}^4) = 4$. Given $X = [a \;\; b \;\; c \;\; d]^T$ in $\mathbb{R}^4$, we can express $X$ as a linear combination of $X_1, X_2, X_3$, and $X_4$. In fact the Expansion theorem gives $X = t_1X_1 + t_2X_2 + t_3X_3 + t_4X_4$ where

$$t_1 = \frac{X \cdot X_1}{\|X_1\|^2} = \frac{a - b + 3c}{11} \qquad t_2 = \frac{X \cdot X_2}{\|X_2\|^2} = \frac{-2a + b + c + d}{7}$$

$$t_3 = \frac{X \cdot X_3}{\|X_3\|^2} = \frac{a + b + d}{3} \qquad t_4 = \frac{X \cdot X_4}{\|X_4\|^2} = \frac{a + 10b + 3c - 11d}{231}$$

The reader may wish to verify that in fact $X = t_1X_1 + t_2X_2 + t_3X_3 + t_4X_4$.

## 4.5.3 The Gram-Schmidt Algorithm

The Expansion theorem demonstrates the desirability of orthogonal bases. More evidence will be given in the following section. However, first we must show that every subspace of $\mathbb{R}^n$ *has* an orthogonal basis; we accomplish this by giving a method for converting any basis into an orthogonal basis.

In showing that every subspace of $\mathbb{R}^n$ has a basis, it was essential in Section 4.3 to be able to enlarge any independent set. The key result was the Independent lemma (Lemma 1 §4.3): If $\{X_1, X_2, \cdots, X_m\}$ is independent in $\mathbb{R}^n$ and $Y$ is not in $span\{X_1, X_2, \cdots, X_m\}$, then $\{Y, X_1, X_2, \cdots, X_m\}$ is also an independent set. The analogue for orthogonal sets is

**LEMMA 1** **Orthogonal Lemma.** Let $\{F_1, F_2, \cdots, F_k\}$ be an orthogonal set in $\mathbb{R}^n$. Given $X$ in $\mathbb{R}^n$, write

$$F_{k+1} = X - \left(\frac{X \cdot F_1}{\|F_1\|^2}F_1 + \frac{X \cdot F_2}{\|F_2\|^2}F_2 + \cdots + \frac{X \cdot F_k}{\|F_k\|^2}F_k\right).$$

Then:

(1) $F_{k+1}$ is orthogonal to each of $F_1, F_2, \cdots, F_k$.

(2) If $X$ is not in $span\{F_1, F_2, \cdots, F_k\}$, then $F_{k+1} \neq 0$ and the larger set $\{F_1, F_2, \cdots, F_k, F_{k+1}\}$ is an orthogonal set.

PROOF    For convenience, write $c_i = \frac{X \cdot F_i}{\|F_i\|^2}$ for each $i$. Given $1 \leq i \leq k$, we have

$$F_{k+1} \cdot F_i = (X - c_1 F_1 - c_2 F_2 - \cdots - c_k F_k) \cdot F_i$$
$$= X \cdot F_i - c_1(F_1 \cdot F_i) - c_2(F_2 \cdot F_i) - \cdots - c_k(F_k \cdot F_i)$$
$$= X \cdot F_i - c_i \|F_i\|^2$$
$$= 0.$$

This proves (1), and (2) follows because $F_{k+1} \neq 0$ if $X$ is not in $span\{F_1, F_2, \cdots, F_k\}$.

It is worth observing that if $X$ *is* in $span\{F_1, F_2, \cdots, F_k\}$ in Lemma 1, then $F_{k+1} = 0$ by the Expansion theorem.

The Orthogonal lemma gives a method of extending an orthogonal set to a larger orthogonal set. In this way it provides a version, for orthogonal sets, of the fact that every independent subset of a subspace $U$ is part of an basis of $U$.

---

**THEOREM  7**    If $U$ is a nonzero subspace of $\mathbb{R}^n$, then every orthogonal subset $\{F_1, F_2, \cdots, F_k\}$ of $U$ is part of an orthogonal basis of $U$. In particular, $U$ has an orthogonal basis.

PROOF    If $span\{F_1, F_2, \cdots, F_k\} = U$, then $\{F_1, F_2, \cdots, F_k\}$ is already an orthogonal basis. Otherwise, there exists $X$ in $U$ which is not in $span\{F_1, F_2, \cdots, F_k\}$. If $F_{k+1}$ is as in the Orthogonal lemma, then $F_{k+1}$ is in $U$ and $\{F_1, F_2, \cdots, F_k, F_{k+1}\}$ is orthogonal. If $span\{F_1, F_2, \cdots, F_k, F_{k+1}\} = U$, we are done. Otherwise, the process continues to create larger and larger orthogonal subsets of $U$, each containing $\{F_1, F_2, \cdots, F_k\}$. They are all independent by Theorem 5, so we have an orthogonal basis when we reach such a set containing $dim U$ vectors. Finally, as $U \neq 0$, choose $F_1 \neq 0$ in $U$. Then $\{F_1\}$ is an orthogonal set, so the foregoing procedure produces an orthogonal basis of $U$ (containing the vector $F_1$).

The method of proof of Theorem 7 actually gives more information. It leads to a procedure by which any basis $\{X_1, X_2, \cdots, X_m\}$ of a subspace $U$ of $\mathbb{R}^n$ can be systematically modified to produce an orthogonal basis $\{F_1, F_2, \cdots, F_m\}$ of $U$. These new vectors $F_i$ are constructed one at a time from the $X_i$. To start the process, take

$$F_1 = X_1.$$

Then $X_2$ is not in $span\{F_1\}$ because $\{X_1, X_2\}$ is independent so, with an eye on the Orthogonal lemma, take

$$F_2 = X_2 - \frac{X_2 \cdot F_1}{\|F_1\|^2} F_1.$$

Thus $\{F_1, F_2\}$ is an orthogonal set by the Orthogonal lemma. Moreover, $span\{F_1, F_2\} = span\{X_1, X_2\}$ as is easily verified. Hence $X_3$ is not in $span\{F_1, F_2\}$ because $\{X_1, X_2, X_3\}$ is independent. Now another application of the Orthogonal lemma shows that $\{F_1, F_2, F_3\}$ is orthogonal where

$$F_3 = X_3 - \frac{X_3 \cdot F_1}{\|F_1\|^2} F_1 - \frac{X_3 \cdot F_2}{\|F_2\|^2} F_2.$$

Again one verifies that $span\{F_1, F_2, F_3\} = span\{X_1, X_2, X_3\}$, and the process continues. At the $m$th iteration we construct an orthogonal set $\{F_1, F_2, \cdots, F_m\}$ such that

$$span\{F_1, F_2, \cdots, F_m\} = span\{X_1, X_2, \cdots, X_m\} = U.$$

Hence $\{F_1,\ F_2,\ \cdots,\ F_m\}$ is the desired orthogonal basis of $U$. The procedure can be summarized as follows.

---

**THEOREM 8**

**Gram-Schmidt Algorithm.**[11] If $\{X_1,\ X_2,\ \cdots,\ X_m\}$ is any basis of a subspace $U$ of $\mathbb{R}^n$, construct $F_1,\ F_2,\ \cdots,\ F_m$ in $U$ successively as follows:

$$F_1 = X_1$$
$$F_2 = X_2 - \frac{X_2 \cdot F_1}{\|F_1\|^2} F_1$$
$$F_3 = X_3 - \frac{X_3 \cdot F_1}{\|F_1\|^2} F_1 - \frac{X_3 \cdot F_2}{\|F_2\|^2} F_2$$
$$\vdots$$
$$F_k = X_k - \frac{X_k \cdot F_1}{\|F_1\|^2} F_1 - \frac{X_k \cdot F_2}{\|F_2\|^2} F_2 - \cdots - \frac{X_k \cdot F_{k-1}}{\|F_{k-1}\|^2} F_{k-1}$$
$$\vdots$$

for each $k = 2,\ 3,\ \cdots,\ m$. Then:

**(1)** $\{F_1,\ F_2,\ \cdots,\ F_m\}$ is an orthogonal basis of $U$.
**(2)** $span\{F_1,\ F_2,\ \cdots,\ F_k\} = span\{X_1,\ X_2,\ \cdots,\ X_k\}$ for each $k = 1,\ 2,\ \cdots,\ m$.

---

**Example 8**  Find an orthogonal basis for $span\{[1\ \ 1\ \ 0\ \ 1],\ [0\ \ 1\ \ -1\ \ 1],\ [1\ \ 0\ \ 1\ \ 1]\}$.

**SOLUTION**  Write $X_1 = [1\ \ 1\ \ 0\ \ 1]$, $X_2 = [0\ \ 1\ \ -1\ \ 1]$, and $X_3 = [1\ \ 0\ \ 1\ \ 1]$, and observe that $\{X_1,\ X_2,\ X_3\}$ is linearly independent. The Gram-Schmidt Algorithm gives:

$$F_1 = X_1 = [1\ \ 1\ \ 0\ \ 1].$$
$$F_2 = X_2 - \frac{X_2 \cdot F_1}{\|F_1\|^2} F_1 = [0\ \ 1\ \ -1\ \ 1] - \frac{2}{3}[1\ \ 1\ \ 0\ \ 1]$$
$$= \frac{1}{3}[-2\ \ 1\ \ -3\ \ 1].$$
$$F_3 = X_3 - \frac{X_3 \cdot F_1}{\|F_1\|^2} F_1 - \frac{X_3 \cdot F_2}{\|F_2\|^2} F_2$$
$$= [1\ \ 0\ \ 1\ \ 1] - \frac{2}{3}[1\ \ 1\ \ 0\ \ 1] - \frac{(-4/3)}{(15/9)}\frac{1}{3}[-2\ \ 1\ \ -3\ \ 1]$$
$$= \frac{1}{5}[-1\ \ -2\ \ 1\ \ 3].$$

Hence the algorithm gives the orthogonal basis $\{[1\ \ 1\ \ 0\ \ 1],\ \frac{1}{3}[-2\ \ 1\ \ -3\ \ 1],$ $\frac{1}{5}[-1\ \ -2\ \ 1\ \ 3]\}$. In hand calculations, or for other reasons, it may be convenient to eliminate fractions and common factors, so $\{[1\ \ 1\ \ 0\ \ 1],\ [-2\ \ 1\ \ -3\ \ 1],$ $[-1\ \ -2\ \ 1\ \ 3]\}$ is also an orthogonal basis. (See the following Note.)

**Note on Computation.** Observe that the vector $\frac{X \cdot F_i}{\|F_i\|^2} F_i$ in the Gram-Schmidt algorithm is unchanged if $F_i$ is replaced by $aF_i$ for any nonzero scalar $a$. Hence if a newly constructed $F_i$ is multiplied by a nonzero scalar at some stage of the Gram-Schmidt algorithm, the subsequent $F$'s will be unchanged. This is useful in actual calculations. As an illustration, in Example 8 we could have used $F_2 = [-2\ \ 1\ \ -3\ \ 1]$ at the second stage of the algorithm. This would have made the hand calculation of $F_3$ easier, but there is another reason: Because computing $F_2$ involved a division (by 3 in Example 8), a computer would have carried some round-off error into the calculation of $F_3$. In a large calculation, this round-off error can accumulate and become a serious problem.

---

[11] Erhardt Schmidt (1876–1959) was a German mathematician who first described the algorithm in 1907. Jörgen Pederson Gram (1850–1916) was a Danish actuary.

**QR-Factorization**

We conclude this section with the matrix version of the Gram-Schmidt process. The result is a factorization of any matrix with independent columns as the product of a matrix with *orthonormal* columns and an invertible upper triangular matrix with positive diagonal entries. This is important in applications, and is particularly useful in calculations because there are computer algorithms that accomplish the decomposition with good control over round-off error.

Suppose $A = [C_1 \ C_2 \ \cdots, \ C_n]$ is an $m \times n$ matrix with linearly independent columns $C_1, \ C_2, \ \cdots, \ C_n$. The Gram-Schmidt algorithm can be applied to these columns to provide orthogonal columns $F_1, \ F_2, \ \cdots, \ F_n$ where

$$F_1 = C_1$$
$$F_2 = C_2 - \frac{C_2 \cdot F_1}{\|F_1\|^2} F_1$$
$$F_3 = C_3 - \frac{C_3 \cdot F_1}{\|F_1\|^2} F_1 - \frac{C_3 \cdot F_2}{\|F_2\|^2} F_2$$
$$\vdots$$
$$F_n = C_n - \frac{C_n \cdot F_1}{\|F_1\|^2} F_1 - \frac{C_n \cdot F_2}{\|F_2\|^2} F_2 - \cdots - \frac{C_n \cdot F_{n-1}}{\|F_{n-1}\|^2} F_{n-1}.$$

Now write $Q_j = \frac{1}{\|F_j\|} F_j$ for each $j$. Then $Q_1, \ Q_2, \ \cdots, \ Q_n$ are orthonormal columns, and the above equations become

$$\|F_1\|Q_1 = C_1$$
$$\|F_2\|Q_2 = C_2 - (C_2 \cdot Q_1)Q_1$$
$$\|F_3\|Q_3 = C_3 - (C_3 \cdot Q_1)Q_1 - (C_3 \cdot Q_2)Q_2$$
$$\vdots$$
$$\|F_n\|Q_n = C_n - (C_n \cdot Q_1)Q_1 - (C_n \cdot Q_2)Q_2 - \cdots - (C_n \cdot Q_{n-1})Q_{n-1}.$$

Finally, use these equations to express each $C_k$ as a linear combination of $Q_1, \ Q_2, \ \cdots, \ Q_n$:

$$C_1 = \|F_1\|Q_1$$
$$C_2 = (C_2 \cdot Q_1)Q_1 + \|F_2\|Q_2$$
$$C_3 = (C_3 \cdot Q_1)Q_1 + (C_3 \cdot Q_2)Q_2 + \|F_3\|Q_3$$
$$\vdots$$
$$C_n = (C_n \cdot Q_1)Q_1 + (C_n \cdot Q_2)Q_2 + (C_n \cdot Q_3)Q_3 + \cdots + \|F_n\|Q_n.$$

Using block multiplication (Theorem 3 §1.4), these equations have a matrix form that gives the required factorization:

$$A = [C_1 \ C_2 \ C_3 \ \cdots \ C_n]$$
$$= [Q_1 \ Q_2 \ Q_3 \ \cdots \ Q_n] \begin{bmatrix} \|F_1\| & C_2 \cdot Q_1 & C_3 \cdot Q_1 & \cdots & C_n \cdot Q_1 \\ 0 & \|F_2\| & C_3 \cdot Q_2 & \cdots & C_n \cdot Q_2 \\ 0 & 0 & \|F_3\| & \cdots & C_n \cdot Q_3 \\ \vdots & \vdots & \vdots & \ddots & \vdots \\ 0 & 0 & 0 & \cdots & \|F_n\| \end{bmatrix} \quad (*)$$

Here the first factor $Q = [Q_1 \ Q_2 \ Q_3 \ \cdots \ Q_n]$ has orthonormal columns, and the second factor is an $n \times n$ upper triangular matrix $R$ with positive diagonal entries (and so is invertible). We record this in the following theorem (the proof of uniqueness is deferred to the end of this section).

**THEOREM 9**    **QR-Factorization.** If $A$ is an $m \times n$ matrix with linearly independent columns, then $A$ can be factored as

$$A = QR$$

where $Q$ is an $m \times n$ matrix with orthonormal columns, and $R$ is an invertible, upper triangular matrix with positive entries on the main diagonal. Furthermore, this factorization is unique in the sense that if $A = Q_1 R_1$ is another such factorization, then $Q_1 = Q$ and $R_1 = R$.

If a matrix $A$ has independent rows and we apply QR-factorization to $A^T$, the result is:

**COROLLARY**    If $A$ has independent rows, then $A = LP$ where $P$ has orthonormal rows and $L$ is an invertible lower triangular matrix with positive main diagonal entries.

**Example 9**    Find the QR-factorization of $A = \begin{bmatrix} 1 & 1 & 0 \\ -1 & 0 & 1 \\ 0 & 1 & 1 \\ 0 & 0 & 1 \end{bmatrix}$.

**SOLUTION**    The columns of $A$ are $C_1 = [1 \ -1 \ 0 \ 0]^T$, $C_2 = [1 \ 0 \ 1 \ 0]^T$, and $C_3 = [0 \ 1 \ 1 \ 1]^T$. It is easy to verify that $\{C_1, \ C_2, \ C_3\}$ is independent so, if we apply the Gram-Schmidt algorithm to these columns $C_i$, the result is

$$\begin{aligned} F_1 &= C_1 & &= [1 \ -1 \ 0 \ 0]^T \\ F_2 &= C_2 - \tfrac{1}{2}F_1 & &= \left[\tfrac{1}{2} \ \tfrac{1}{2} \ 1 \ 0\right]^T \\ F_3 &= C_3 + \tfrac{1}{2}F_1 - F_2 &&= [0 \ 0 \ 0 \ 1]^T \end{aligned}$$

Hence let $Q_j = \frac{1}{\|F_j\|} F_j$ for each $j$. Then equation $(*)$ preceding Theorem 9 gives $A = QR$ where

$$Q = [Q_1 \ \ Q_2 \ \ Q_2] = \begin{bmatrix} \frac{1}{\sqrt{2}} & \frac{1}{\sqrt{6}} & 0 \\ \frac{-1}{\sqrt{2}} & \frac{1}{\sqrt{6}} & 0 \\ 0 & \frac{2}{\sqrt{6}} & 0 \\ 0 & 0 & 1 \end{bmatrix} = \frac{1}{\sqrt{6}}\begin{bmatrix} \sqrt{3} & 1 & 0 \\ -\sqrt{3} & 1 & 0 \\ 0 & 2 & 0 \\ 0 & 0 & \sqrt{6} \end{bmatrix},$$

$$R = \begin{bmatrix} \|F_1\| & C_2 \cdot Q_1 & C_3 \cdot Q_1 \\ 0 & \|F_2\| & C_3 \cdot Q_2 \\ 0 & 0 & \|F_3\| \end{bmatrix} = \begin{bmatrix} \sqrt{2} & \frac{1}{\sqrt{2}} & \frac{-1}{\sqrt{2}} \\ 0 & \frac{\sqrt{3}}{\sqrt{2}} & \frac{\sqrt{3}}{\sqrt{2}} \\ 0 & 0 & 1 \end{bmatrix}$$

$$= \frac{1}{\sqrt{2}}\begin{bmatrix} 2 & 1 & -1 \\ 0 & \sqrt{3} & \sqrt{3} \\ 0 & 0 & \sqrt{2} \end{bmatrix}.$$

The reader can verify that indeed $A = QR$.

**Remark.** If an $m \times n$ matrix $A$ has independent columns, then $A^T A$ is invertible (by Theorem 3 §4.4), and it is often desirable to compute the inverse of $A^T A$ (see Section 4.6 for example). This is simplified if we have a QR-factorization of $A$ (and is one of the main reasons for the importance of Theorem 9). For if $A = QR$ is such a factorization, then $Q^T Q = I_n$ because $Q$ has orthonormal columns (verify), so we obtain

$$A^T A = R^T Q^T Q R = R^T R.$$

Hence computing $(A^T A)^{-1}$ amounts to finding $R^{-1}$, and this is a routine matter because $R$ is upper triangular. Thus the difficulty in computing $(A^T A)^{-1}$ lies in obtaining the QR-factorization of $A$.

### 4.5.5 Proof of Uniqueness in the QR-Factorization

Suppose $A = QR = Q_1R_1$ are two QR-factorizations of the $m \times n$ matrix $A$. Write $Q = [C_1 \ C_2 \ \cdots \ C_n]$ and $Q_1 = [D_1 \ D_2 \ \cdots \ D_n]$ in terms of their columns, and observe first that $Q^TQ = I_n = Q_1^TQ_1$ because $Q$ and $Q_1$ have orthonormal columns. Hence it suffices to show that $Q_1 = Q$ (then $R_1 = Q_1^TA = Q^TA = R$). Since $Q_1^TQ_1 = I_n$, the equation $QR = Q_1R_1$ gives $Q_1^TQ = R_1R^{-1}$; for convenience we write this matrix as

$$Q_1^TQ = R_1R^{-1} = [t_{ij}].$$

This matrix is upper triangular with positive diagonal elements (since this is true for $R$ and $R_1$), so $t_{ii} > 0$ for each $i$ and $t_{ij} = 0$ if $i > j$. On the other hand, the $(i, j)$-entry of $Q_1^TQ$ is $D_i^TC_j = D_i \cdot C_j$, so we have $D_i \cdot C_j = t_{ij}$ for all $i$ and $j$. But $C_1, C_2, \cdots, C_n$ is an orthonormal basis of $\mathbb{R}^n$, so the Expansion theorem gives

$$D_i = (D_i \cdot C_1)C_1 + (D_i \cdot C_2)C_2 + \cdots + (D_i \cdot C_i)C_i + \cdots + (D_i \cdot C_n)C_n$$
$$= t_{i1}C_1 + t_{i2}C_2 + \cdots + t_{ii}C_i + \cdots + t_{in}C_n$$
$$= t_{ii}C_i + \cdots + t_{in}C_n$$

because $D_i \cdot C_j = t_{ij} = 0$ if $i > j$. The first few equations here (in reverse order) are

$$D_n = t_{nn}C_n$$
$$D_{n-1} = t_{n-1 \ n-1}C_{n-1} + t_{n-1 \ n}C_n$$
$$D_{n-2} = t_{n-2 \ n-2}C_{n-2} + t_{n-2 \ n-1}C_{n-1} + t_{n-2 \ n}C_n$$
$$\vdots$$

The first of these equations gives $1 = \|D_n\| = \|t_{nn}C_n\| = |t_{nn}| \, \|C_n\| = t_{nn}$, whence $D_n = C_n$. But then

$$t_{n-1 \ n} = D_{n-1} \cdot C_n = D_{n-1} \cdot D_n = 0,$$

so the second equation becomes $D_{n-1} = t_{n-1 \ n-1}C_{n-1}$. As before it follows that $D_{n-1} = C_{n-1}$, and hence that $t_{n-2 \ n} = 0$ and $t_{n-2 \ n-1} = 0$. Hence the third equation reads $D_{n-2} = t_{n-2 \ n-2}C_{n-2}$. Again this gives $D_{n-2} = C_{n-2}$, and the process continues in this way to get $D_i = C_i$ for all $i$. This means that $Q_1 = Q$, which is what we wanted.

## Exercises 4.5

**1.** If $\|X\| = 2$, $\|Y\| = 1$ and $X \cdot Y = -1$, compute:
**a)** $\|2X - 5Y\|$  **b)** $\|3X - 4Y\|$
**c)** $(2X - Y) \cdot (3X + 7Y)$  **d)** $(3X + 4Y) \cdot (X - 5Y)$

**2.** In each case show that the set of vectors is orthogonal in $\mathbb{R}^4$.
**a)** $\{[1 \ 1 \ -1 \ 4]^T, [-1 \ 2 \ 5 \ 1]^T, [28 \ 5 \ 5 \ -7]^T\}$.
**b)** $\{[4 \ 5 \ 2 \ -1]^T, [1 \ -1 \ 0 \ -1]^T, [2 \ -1 \ 0 \ 3]^T\}$.

**3.** By normalizing the following sets, obtain an orthonormal basis of $\mathbb{R}^3$.
**a)** $\{[1 \ 0 \ 2]^T, [-2 \ 5 \ 1]^T, [2 \ 1 \ -1]^T\}$.
**b)** $\{[-3 \ 1 \ 2]^T, [1 \ 1 \ 1]^T, [1 \ -5 \ 4]^T\}$.

**4.** Let $U = span\{[1 \ -2 \ 1 \ -1]^T, [2 \ 1 \ -1 \ 1]^T\}$. Show that $Y = [1 \ 3 \ -2 \ 2]^T$ is in $U$, and find all $Z$ such that $\{Y, Z\}$ is an orthogonal basis of $U$.

**5.** In each case find all vectors $[a \ b \ c \ d]^T$ in $\mathbb{R}^4$ such that the given set is orthogonal.
**a)** $\{[1 \ -1 \ 1 \ 0]^T, [-1 \ 3 \ 4 \ 1]^T, [2 \ 1 \ -1 \ 3]^T, [a \ b \ c \ d]^T\}$.
**b)** $\{[1 \ 2 \ 1 \ -1]^T, [1 \ 1 \ -2 \ 1]^T, [5 \ -2 \ 4 \ 5]^T, [a \ b \ c \ d]^T\}$.

**6.** In each case show that the set of vectors is an orthogonal basis of $\mathbb{R}^3$, and use the Expansion theorem to write $[a \ b \ c]^T$ as a linear combination of the basis vectors.
**a)** $\{[3 \ 1 \ -1]^T, [1 \ -2 \ 1]^T, [1 \ 4 \ 7]^T\}$.
**b)** $\{[-4 \ 1 \ 5]^T, [-1 \ 1 \ -1]^T, [2 \ 3 \ 1]^T\}$.

**7.** In each case use the Gram-Schmidt algorithm to convert the given basis into an orthogonal basis.
**a)** $\{[1 \ -1]^T, [2 \ 1]^T\}$ in $\mathbb{R}^2$
**b)** $\{[2 \ 1]^T, [1 \ 2]^T\}$ in $\mathbb{R}^2$
**c)** $\{[1 \ -1 \ 1]^T, [1 \ 0 \ 1]^T, [1 \ 1 \ 2]^T\}$ in $\mathbb{R}^3$
**d)** $\{[0 \ 1 \ 1]^T, [1 \ 1 \ 1]^T, [1 \ -2 \ 2]^T\}$ in $\mathbb{R}^3$

**8.** Find the QR-factorization of each of the following matrices.

**a)** $\begin{bmatrix} 1 & -1 \\ -1 & 0 \end{bmatrix}$  **b)** $\begin{bmatrix} 2 & 1 \\ 1 & 1 \end{bmatrix}$

**c)** $\begin{bmatrix} 1 & 1 & 1 \\ 1 & 1 & 0 \\ 1 & 0 & 0 \\ 0 & 0 & 0 \end{bmatrix}$  **d)** $\begin{bmatrix} 1 & 1 & 0 \\ -1 & 0 & 1 \\ 0 & 1 & 1 \\ 1 & -1 & 0 \end{bmatrix}$

**9.** In each case either show that the statement is true or give an example showing that it is false.
  **a)** Every independent set in $\mathbb{R}^n$ is orthogonal.
  **b)** If $\{X, Y\}$ is an orthogonal set in $\mathbb{R}^n$, then $\{X, X + Y\}$ is also orthogonal.
  **c)** If $\{X, Y\}$ and $\{Z, W\}$ are both orthogonal in $\mathbb{R}^n$, then $\{X, Y, Z, W\}$ is also orthogonal.
  **d)** If $\{X_1, X_2\}$ and $\{Y_1, Y_2, Y_3\}$ are orthogonal and $X_i \cdot Y_j = 0$ for all $i$ and $j$, then $\{X_1, X_2, Y_1, Y_2, Y_3\}$ is orthogonal.
  **e)** If $\{X_1, X_2, \cdots, X_n\}$ is orthogonal in $\mathbb{R}^n$, then $\mathbb{R}^n = span\{X_1, X_2, \cdots, X_n\}$.
  **f)** If $E$ is a unit vector in $\mathbb{R}^n$, then $(E \cdot X)X = X$ for all $X$ in $\mathbb{R}^n$.
  **g)** If $Q$ is $m \times n$ with orthonormal columns, then $Q^T Q = I$.
  **h)** If $Q$ is $m \times n$ and $Q^T Q = I$, then $Q$ has orthonormal columns.
  **i)** If $X$ is orthogonal to each of $X_1, X_2,$ and $X_3$, then $X$ is orthogonal to $aX_1 + bX_2 + cX_3$ for any scalars $a$, $b$, and $c$.

**10.** If $\{X_1, \cdots, X_m\}$ is an orthogonal set in $\mathbb{R}^n$, show that $\{a_1 X_1, \cdots, a_m X_m\}$ is orthogonal for any numbers $a_i \neq 0$.

**11.** Let $X$ and $Y$ denote vectors in $\mathbb{R}^n$.
  **a)** If $X$ and $Y$ are orthogonal, show that $\|X - Y\| = \|X + Y\|$.
  **b)** If $\|X - Y\| = \|X + Y\|$, show that $X$ and $Y$ are orthogonal.

**12.** Let $X$, $Y$, and $Z$ be vectors in $\mathbb{R}^n$.
  **a)** If $\|X + Y\|^2 = \|X\|^2 + \|Y\|^2$, show that $X$ and $Y$ are orthogonal.
  **b)** Show that $\|X + Y + Z\|^2 = \|X\|^2 + \|Y\|^2 + \|Z\|^2$ does not imply that $\{X, Y, Z\}$ is orthogonal by considering $X = [1 \ 1 \ 0]^T$, $Y = [0 \ -1 \ 1]^T$, and $Z = [1 \ 1 \ 0]^T$.

**13.** Let $X$ and $Y$ denote vectors in $\mathbb{R}^n$.
  **a)** Show that $X \cdot Y = \frac{1}{4}\left\{\left[\|X + Y\|^2 - \|X - Y\|^2\right]\right\}$.
  **b)** Show that $\|X\|^2 + \|Y\|^2 = \frac{1}{2}\left\{\left[\|X + Y\|^2 + \|X - Y\|^2\right]\right\}$.

**14.** Let $\lambda$ be an eigenvalue of $A^T A$ where $A$ is an $n \times n$ matrix. Since $A^T A$ is symmetric, we know that $\lambda$ is real from Theorem 3 §2.5. Show that $\lambda \geq 0$. [Hint: Compute $\|AX\|^2$ where $X$ is an eigenvector.]

**15.** If $Q$ is an $m \times n$ matrix with orthonormal columns, explain why $Q^T Q = I_n$.

**16.** If $\mathbb{R}^n = span\{F_1, F_2, \cdots, F_m\}$ and $X$ in $\mathbb{R}^n$ satisfies $X \cdot F_i = 0$ for all $i$, show that $X = 0$. [Hint: Show that $\|X\| = 0$.]

**17.** If $\mathbb{R}^n = span\{F_1, F_2, \cdots, F_m\}$, and if $X$ and $Y$ in $\mathbb{R}^n$ satisfies $X \cdot F_i = Y \cdot F_i$ for all $i$, show that $X = Y$.

**18.** Let $U$ be a subspace of $\mathbb{R}^n$ with $dim U = m \geq 2$, and assume that $\{F_1, F_2, \cdots, F_{m-1}\}$ is an orthonormal set in $U$.
  **a)** Show that $U$ has an orthonormal basis $\{F_1, F_2, \cdots, F_{m-1}, G\}$ for some $G$ in $U$.
  **b)** If $\{F_1, F_2, \cdots, F_{m-1}, H\}$ is an orthonormal basis of $U$, determine $H$ in terms of $G$.

**19.** Let $\{X_1, X_2, \cdots, X_m\}$ be an orthogonal basis of a subspace $U$ of $\mathbb{R}^n$. If the Gram-Schmidt algorithm is applied to this basis to produce an orthogonal basis $\{F_1, F_2, \cdots, F_m\}$ of $U$, describe the $F_i$ in terms of the $X_i$.

**20.** Let $\{X_1, X_2, \cdots, X_n\}$ be a basis of $\mathbb{R}^n$. If $\{Y_1, Y_2, \cdots, Y_n\}$ are vectors in $\mathbb{R}^n$ which satisfy

$$X_i \cdot Y_j = \begin{cases} 1 \text{ if } i = j \\ 0 \text{ if } i \neq j \end{cases},$$

show that $\{Y_1, Y_2, \cdots, Y_n\}$ is also a basis of $\mathbb{R}^n$. What does this say about matrix inverses?

**21.** Let $\{G_1, G_2, \cdots, G_n\}$ be an orthogonal set in $\mathbb{R}^n$. Given $X$ and $Y$ in $\mathbb{R}^n$, show that
$$X \cdot Y = \frac{(X \cdot G_1)(Y \cdot G_1)}{\|G_1\|^2} + \frac{(X \cdot G_2)(Y \cdot G_2)}{\|G_2\|^2} + \cdots + \frac{(X \cdot G_n)(Y \cdot G_n)}{\|G_n\|^2}.$$

**22.** Let $A$ be an $n \times n$ matrix of rank $r$. Show that there is an invertible $n \times n$ matrix $U$ such that $UA$ is a row-echelon matrix with orthogonal rows.

**23.** Suppose $A = Q_0 R_0$ where $Q_0$ has orthonomal columns and $R_0$ is invertible and upper triangular. Show that a QR-factorization can be obtained as follows: Obtain $R$ from $R_0$ by negating certain rows, then obtain $Q$ from $Q_0$ by negating the corresponding columns. [Hint: $R = DR_0$ for a diagonal matrix $D$.]

# 4.6 PROJECTIONS AND APPROXIMATION

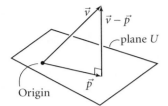

Let $U$ be a plane in $\mathbb{R}^3$ containing the origin, and let $\vec{v}$ be a vector in $\mathbb{R}^3$ (recall that we are identifying points with their position vectors). It is evident geometrically that there is a vector $\vec{p}$ in the plane $U$ which is closest to $\vec{v}$ (see the diagram). Indeed the vector $\vec{p}$ is determined by the fact that the line through $\vec{v}$ and $\vec{p}$ must be perpendicular to the plane.

Now we make two observations. First, the plane is a *subspace* of $\mathbb{R}^3$; and second, the condition that $\vec{v} - \vec{p}$ is perpendicular to the plane means that $\vec{v} - \vec{p}$ is *orthogonal* to every vector in the plane. In this form the whole discussion makes sense in $\mathbb{R}^n$. In fact, we show in this section that, given a vector $X$ and a subspace $U$ of $\mathbb{R}^n$, there exists a vector $P$ in $U$ that is closest to $X$. This vector $P$ is called the projection of $X$ on the subspace $U$, and the result leads to a general method for solving certain minimization problems. This technique will be used throughout this chapter.

## 4.6.1 Orthogonal Complements

If $U$ is a subspace of $\mathbb{R}^n$, the **orthogonal complement** $U^\perp$ is defined[12] to be the set of all vectors that are orthogonal to every vector in $U$:
$$U^\perp = \{X \ \text{ in } \ \mathbb{R}^n \mid X \cdot Y = 0 \text{ for all } Y \text{ in } U\}.$$

**Example 1**  If $L$ is a line through the origin in $\mathbb{R}^3$, then $L^\perp$ is the plane through the origin perpendicular to $L$. Similarly, if $P$ is a plane through the origin in $\mathbb{R}^3$ then $P^\perp$ is the line through the origin perpendicular to $P$.

In Example 1, note that $L^\perp$ and $P^\perp$ are both *subspaces* of $\mathbb{R}^3$. This holds much more generally. In fact, if $U$ is any subspace of $\mathbb{R}^n$, then $U^\perp$ is again a subspace. For if $X$ and $X_1$ are both in $U^\perp$, then $(X + X_1) \cdot Y = X \cdot Y + X_1 \cdot Y = 0 + 0 = 0$ for all $Y$ in $U$, so $X + X_1$ is also in $U^\perp$. Similarly $aX$ is in $U^\perp$ for all scalars $a$, so $U^\perp$ is indeed a subspace. Moreover there is a routine way to determine $U^\perp$ when a spanning set for $U$ is known.

**THEOREM 1**  Let $U$ denote a subspace of $\mathbb{R}^n$. Then:
(1) $U^\perp$ is also a subspace of $\mathbb{R}^n$.
(2) If $U = span\{X_1, \cdots, X_k\}$, then $U^\perp = \{X \mid X \cdot X_i = 0 \text{ for each } i = 1, 2, \cdots, k\}$.

**PROOF**  We verified (1) above. As to (2), suppose that $X \cdot X_i = 0$ for each $i$. If $Y$ is in $U$, write it as $Y = t_1 X_1 + \cdots + t_k X_k$ where the $t_i$ are scalars. Then
$$\begin{aligned} X \cdot Y &= X \cdot (t_1 X_1 + \cdots + t_k X_k) \\ &= t_1(X \cdot X_1) + \cdots + t_k(X \cdot X_k) \\ &= t_1(0) + \cdots + t_k(0) \\ &= 0. \end{aligned}$$
It follows that $\{X \mid X \cdot X_i = 0 \text{ for each } i\} \subseteq U^\perp$. The other inclusion follows because, if $X$ is in $U^\perp$, then $X \cdot X_i = 0$ for each $i$ because $X_i$ is in $U$.

---

[12] When speaking, $U^\perp$ is pronounced "$U$-perp".

When $U$ has a spanning set of $k$ vectors, the second part of Theorem 1 reduces the description of $U^\perp$ to solving a system of $k$ linear equations. Here is an example.

**Example 2**    Find $U^\perp$ if $U = span\{[1 \ \ -2 \ \ -2 \ \ 1 \ \ 0], \ [-2 \ \ 4 \ \ 3 \ \ -1 \ \ 2]\}$ in $\mathbb{R}^5$.

**SOLUTION**    By Theorem 1, a vector $X = [x_1 \ \ x_2 \ \ x_3 \ \ x_4 \ \ x_5]$ is in $U^\perp$ if and only if $X$ is orthogonal to both $[1 \ \ -2 \ \ -2 \ \ 1 \ \ 0]$ and $[-2 \ \ 4 \ \ 3 \ \ -1 \ \ 2]$; that is, if and only if

$$x_1 - 2x_2 - 2x_3 + x_4 \qquad \ = 0$$
$$-2x_1 + 4x_2 + 3x_3 - x_4 + 2x_5 = 0$$

Gaussian elimination gives $X = r[2 \ \ 1 \ \ 0 \ \ 0 \ \ 0] + s[1 \ \ 0 \ \ 1 \ \ 1 \ \ 0] + t[4 \ \ 0 \ \ 2 \ \ 0 \ \ 1]$ where $r$, $s$, and $t$ are parameters. Hence
$U^\perp = span\{[2 \ \ 1 \ \ 0 \ \ 0 \ \ 0], \ [1 \ \ 0 \ \ 1 \ \ 1 \ \ 0], \ [4 \ \ 0 \ \ 2 \ \ 0 \ \ 1]\}$.

## 4.6.2    Projections

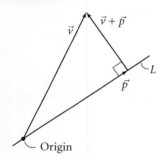

Let $L$ be a line through the origin in $\mathbb{R}^3$, and let $\vec{v}$ be a vector in $\mathbb{R}^3$. If $\vec{d}$ is any direction vector for $L$, let $\vec{p}$ denote the projection of $\vec{v}$ on $\vec{d}$ as discussed in Section 3.2 (see the diagram). Clearly $\vec{p}$ can be characterized as the unique vector in $L$ such that $\vec{v} - \vec{p}$ is perpendicular to $L$, that is such that

$$\vec{v} - \vec{p} \text{ is in } L^\perp.$$

The fact that $L$ is a subspace of $\mathbb{R}^3$ means that we can formulate this in $\mathbb{R}^n$.

Thus if $U$ is a subspace of $\mathbb{R}^n$ and $X$ is a vector in $\mathbb{R}^n$, we look for a vector $P$ in $U$ such that

$$X - P \text{ is in } U^\perp.$$

Remarkably, not only does such a vector $P$ exist, there is an explicit formula for it. Indeed, given an orthogonal basis $\{F_1, \ F_2, \ \cdots, \ F_m\}$ of $U$, consider the vector

$$P = \frac{X \cdot F_1}{\|F_1\|^2} F_1 + \frac{X \cdot F_2}{\|F_2\|^2} F_2 + \cdots + \frac{X \cdot F_m}{\|F_m\|^2} F_m. \qquad (*)$$

Then $P$ is in $U$ because each $F_i$ is in $U$, and $X - P$ is in $U^\perp$ because it is orthogonal to each $F_i$ by the Orthogonal lemma (Lemma 1 §4.5).

Even more is true: This vector $P$ is the *same* for every choice of orthogonal basis of $U$. More precisely, if $\{G_1, \ G_2, \ \cdots, \ G_m\}$ is any other orthogonal basis of $U$, write

$$P' = \frac{X \cdot G_1}{\|G_1\|^2} G_1 + \frac{X \cdot G_2}{\|G_2\|^2} G_2 + \cdots + \frac{X \cdot G_m}{\|G_m\|^2} G_m$$

for the vector formed as in $(*)$ but using the $G_i$ in place of the $F_i$. We claim that $P' = P$. To see this, note first that $P'$ is in U and $X - P'$ is in $U^\perp$ by the Orthogonal lemma, as before. Now write the vector $P - P'$ as follows:

$$P - P' = (X - P') - (X - P).$$

Then $P - P'$ is in $U^\perp$ (because $X - P$ and $X - P'$ are in $U^\perp$), and $P - P'$ is also in $U$ (because $P$ and $P'$ are in $U$). But then $P - P'$ is orthogonal to *itself*! This means that

$$\|P - P'\|^2 = (P - P') \cdot (P - P') = 0.$$

Thus $P - P' = 0$, that is $P = P'$ as claimed. We record this observation as

**LEMMA 1**

Let $U$ be a nonzero subspace of $\mathbb{R}^n$, and let $X$ be any vector in $\mathbb{R}^n$. If $\{F_1, F_2, \cdots, F_m\}$ is an orthogonal basis of $U$, the vector

$$P = \frac{X \cdot F_1}{\|F_1\|^2}F_1 + \frac{X \cdot F_2}{\|F_2\|^2}F_2 + \cdots + \frac{X \cdot F_m}{\|F_m\|^2}F_m$$

depends only on $U$ and $X$; that is, given $X$, $P$ is the same for every choice of orthogonal basis $\{F_1, F_2, \cdots, F_m\}$.

The vector $P$ in Lemma 1 is called[13] the **projection** of $X$ on $U$, and is denoted

$$proj_U(X).$$

If $U = 0$, we define

$$proj_0(X) = 0.$$

The next example shows that this definition agrees with that in Section 3.2 when $U$ is a line through the origin.

**Example 3**

Let $L$ be the line through the origin in $\mathbb{R}^3$ with direction vector $\vec{d}$. Given $\vec{v}$ in $\mathbb{R}^3$, Theorem 4 §3.2 gives a formula for the projection of $\vec{v}$ on $\vec{d}$:

$$proj_{\vec{d}}(\vec{v}) = \frac{\vec{v} \cdot \vec{d}}{\|\vec{d}\|^2}\vec{d}.$$

This agrees with $proj_L(\vec{v})$ if we use $\{\vec{d}\}$ as the orthogonal basis of $L$.

The above discussion proves the most important theorem of this section.

**THEOREM 2**

**Projection Theorem.** Let $U$ be a subspace of $\mathbb{R}^n$, let $X$ be a vector in $\mathbb{R}^n$. Then $proj_U(X)$ is in $U$ and $X - proj_U(X)$ is in $U^\perp$.

Let $X$ be a vector in $\mathbb{R}^n$, let $U$ be a subspace of $\mathbb{R}^n$, and write $P = proj_U(X)$. Then the situation in the Projection theorem can be depicted as in the diagram (where the subspace $U$ is shown as a plane). In particular, we see that

$$X = P + (X - P)$$

is one way to write a vector $X$ in $\mathbb{R}^n$ as a sum of a vector $P$ in $U$ and a vector $X - P$ in $U^\perp$. In fact this is the *only* way to write $X$ as such a sum. This observation is often useful and we record it in the following form.

**COROLLARY**

Let $U$ be a subspace of $\mathbb{R}^n$, and let $X$ be a vector in $\mathbb{R}^n$. If

$$X = Y + Z$$

where $Y$ is in $U$ and $Z$ is in $U^\perp$, then necessarily $Y = proj_U(X)$ and $Z = X - proj_U(X)$.

**PROOF**

Write $P = proj_U(X)$. We have $Y + Z = X = P + (X - P)$. Hence the vector $Y - P = (X - P) - Z$, is in both $U$ and $U^\perp$; in fact, $Y - P$ is in $U$ while $(X - P) - Z$ is in $U^\perp$. It follows that this vector is zero, so $Y = P$ and $(X - P) = Z$.

---

[13] This is sometimes called the **orthogonal projection** of $X$ on $U$, but we prefer the simpler terminology.

**Example 4**

Let $U = span\{X_1, X_2\}$ where $X_1 = [1 \ -1 \ 0 \ 1]$ and $X_2 = [1 \ 0 \ 1 \ 0]$ in $\mathbb{R}^4$. If $X = [2 \ 1 \ -1 \ 1]$, express $X$ as the sum of a vector in $U$ and a vector in $U^\perp$.

**SOLUTION**

Note first that, while $\{X_1, X_2\}$ is a basis of $U$, it is not orthogonal. The Gram-Schmidt algorithm gives an orthogonal basis $\{F_1, F_2\}$ of $U$ where

$$F_1 = X_1 = [1 \ -1 \ 0 \ 1], \text{ and}$$
$$F_2 = X_2 - \frac{X_2 \cdot F_1}{\|F_1\|^2}F_1 = X_2 - \tfrac{1}{3}F_1 = \tfrac{1}{3}[2 \ 1 \ 3 \ -1].$$

For convenience, we use the orthogonal basis $\{G_1, G_2\}$ where we take $G_1 = F_1 = [1 \ -1 \ 0 \ 1]$ and $G_2 = 3F_2 = [2 \ 1 \ 3 \ -1]$. We compute the projection $P$ of $X$ on $U$ using $\{G_1, G_2\}$:

$$P = proj_U(X) = \frac{X \cdot G_1}{\|G_1\|^2}G_1 + \frac{X \cdot G_2}{\|G_2\|^2}G_2 = \tfrac{2}{3}G_1 + \tfrac{1}{15}G_2 = \tfrac{1}{5}[4 \ -3 \ 1 \ 3].$$

Then we compute $X - P = \tfrac{2}{5}[3 \ 4 \ -3 \ 1]$. Theorem 2 asserts that this vector $X - P$ is in $U^\perp$ (this can be verified by checking that it is orthogonal to the two generators $X_1$ and $X_2$ of $U$, and this is a good test of the arithmetic). Finally, since $P$ is in $U$,

$$X = P + (X - P) = \tfrac{1}{5}[4 \ -3 \ 1 \ 3] + \tfrac{2}{5}[3 \ 4 \ -3 \ 1]$$

is the required decomposition of $X$.

The Projection theorem has many consequences, among them the following two important facts about the orthogonal complement $U^\perp$ of a subspace $U$ of $\mathbb{R}^n$. Since $U^\perp$ is also a subspace, it too has an orthogonal complement which, for convenience, we write as $U^{\perp\perp} = (U^\perp)^\perp$. Surprisingly, $U^{\perp\perp} = U$.

**THEOREM 3**

Let $U$ be any subspace of $\mathbb{R}^n$. Then:

**(1)** $dim(U^\perp) = n - dimU$.

**(2)** $U^{\perp\perp} = U$.

**PROOF**

**(1)** If $U = 0$, then $U^\perp = \mathbb{R}^n$ and (1) holds. If $U^\perp = 0$ and $X$ is in $\mathbb{R}^n$, then $X - proj_U(X)$ is in $U^\perp$ by the Projection theorem, so $X = proj_U(X)$ is in $U$. Hence $U = \mathbb{R}^n$, and again (1) holds.

Thus we may assume that $U \neq 0$ and $U^\perp \neq 0$, so let $\{F_1, F_2, \cdots, F_m\}$ and $\{G_1, G_2, \cdots, G_k\}$ be orthogonal bases of $U$ and $U^\perp$ respectively. Then $\{F_1, F_2, \cdots, F_m, G_1, G_2, \cdots, G_k\}$ is an orthogonal set in $\mathbb{R}^n$ (verify), and it spans $\mathbb{R}^n$ by the Projection theorem. It is thus a basis of $\mathbb{R}^n$, so $dim(\mathbb{R}^n) = m + k$. Hence $n = dimU + dim(U^\perp)$, and (1) holds in this case too.

**(2)** We have $U \subseteq U^{\perp\perp}$ because every vector in $U$ is orthogonal to every vector in $U^\perp$. Conversely, let $Y$ be in $U^{\perp\perp}$, and write $P = proj_U(Y)$. Then $P$ is in $U \subseteq U^{\perp\perp}$, so $P$ is in $U^{\perp\perp}$. Hence, $Y - P$ is also in $U^{\perp\perp}$. But $Y - P$ is in $U^\perp$ by the Projection theorem, and it follows that $Y - P$ is orthogonal to itself. Thus $Y - P = 0$, whence $Y = P$ is in $U$. This proves that $U^{\perp\perp} \subseteq U$, so $U^{\perp\perp} = U$.

Let $X$ be a vector in $\mathbb{R}^n$, and let $P = proj_U(X)$ be its projection on a subspace $U$. Then $X = (X - P) + P$ is a decomposition of $X$ as the sum of a vector $X - P$ in $U^\perp$ and a vector $P$ in $U = (U^\perp)^\perp$. Hence the Corollary to the Projection theorem (with $U^\perp$ in place of $U$) shows that $X - P = proj_{U^\perp}(X)$. This proves the

**COROLLARY**  Let $U$ be a subspace of $\mathbb{R}^n$, and let $X$ be a vector in $\mathbb{R}^n$. Then
$$X = proj_U(X) + proj_{U^\perp}(X)$$
is the unique decomposition of $X$ as the sum of a vector in $U$ and a vector in $U^\perp$.

## 4.6.3  Approximation

There is another facet of the Projection theorem that is very important in applications as we shall see. If $X$ is a vector in $\mathbb{R}^n$ and $U$ is a subspace of $\mathbb{R}^n$, then the projection $P$ of $X$ on $U$ is the vector in $U$ that is *closest* to $X$. This is intuitively clear from the diagram where $P$ and $Y$ are vectors in $U$ and, as before, the subspace $U$ is depicted as a plane. The following theorem makes it more precise. Recall that the distance between two vectors $X$ and $Y$ is $\|X - Y\|$.

**THEOREM 4**  **Approximation Theorem**. Let $U$ be a subspace of $\mathbb{R}^n$, let $X$ be a vector in $\mathbb{R}^n$, and write $P = proj_U(X)$. Then $P$ is the vector in $U$ closest to $X$ in the sense that
$$\|X - P\| < \|X - Y\| \text{ for all } Y \neq P \text{ in } U.$$

**PROOF**  Given $Y$ in $U$ where $Y \neq P$, write $X - Y$ in the form
$$X - Y = (X - P) + (P - Y).$$
Then $X - P$ is in $U^\perp$ by the Projection theorem, and $P - Y$ is in $U$ ($P$ and $Y$ are in $U$). Hence $X - P$ and $P - Y$ are orthogonal, so Pythagoras' theorem gives
$$\|X - Y\|^2 = \|X - P\|^2 + \|P - Y\|^2 > \|X - P\|^2$$
because $P - Y \neq 0$. The theorem follows.

**Example 5**  Consider the plane $U$ with equation $2x + 3y - z = 0$. Find the point in $U$ which is closest to the point $X = [3 \ -1 \ 2]^T$.

**SOLUTION**  The plane $U$ is the subspace of $\mathbb{R}^3$ whose points $[x \ y \ z]^T$ satisfy $z = 2x + 3y$. Hence
$$U = \{[s \ t \ 2s + 3t]^T \mid s, t \text{ in } \mathbb{R}\} = span\{[1 \ 0 \ 2]^T, [0 \ 1 \ 3]^T\}.$$
The Gram-Schmidt process gives an orthogonal basis $\{F_1, F_2\}$ for $U$ where $F_1 = [1 \ 0 \ 2]^T$ and $F_2 = [-6 \ 5 \ 3]^T$. Hence the vector in $U$ closest to $X$ is
$$P = proj_U(X) = \frac{X \cdot F_1}{\|F_1\|^2}F_1 + \frac{X \cdot F_2}{\|F_2\|^2}F_2 = \frac{7}{5}F_1 + \frac{-17}{70}F_2 = \frac{1}{14}[40 \ -17 \ 29]^T.$$
As noted before, a good check on the arithmetic is to compute $X - P$ and check that it is in $U^\perp$. In the present case, $X - P = \frac{1}{14}[2 \ 3 \ -1]^T$, which is indeed in $U^\perp$ because it is orthogonal to both $[1 \ 0 \ 2]^T$ and $[0 \ 1 \ 3]^T$.

## 4.6.4  Inconsistent Systems

A system of linear equations may be inconsistent; that is, it may not have a solution. However, even when there is no solution, "best approximations" to a solution always exist, and finding them is one of the most useful applications of the Approximation theorem. The theorem is applied by carefully choosing the subspace $U$ so that the "best approximations" are the vectors in $U$ closest to some particular vector.

More precisely, if $A$ is an $m \times n$ matrix and $B$ is a column in $\mathbb{R}^m$, consider the system

$$AX = B$$

of $m$ linear equations in $n$ variables. This system may have no solution; that is, it may happen that $AZ \ne B$ for every column $Z$ in $\mathbb{R}^n$. On the other hand, the distance $\|B - AZ\|$ between $B$ and $AZ$ is a measure of how close $AZ$ is to $B$. Hence it is natural to ask if there are columns $Z$ in $\mathbb{R}^n$ that are as close as possible to a solution in the sense that

$$\|B - AZ\|$$

is the minimum value of $\|B - AX\|$ as $X$ ranges over all columns in $\mathbb{R}^n$.
The Approximation theorem gives an affirmative answer. To see how, write

$$U = \{AX \mid X \text{ in } \mathbb{R}^n\} = imA.$$

This is a subspace of $\mathbb{R}^n$ as we have seen, and we want a vector in $U$ as close as possible to $B$. But the Approximation theorem shows that $proj_U(B)$ has exactly this property. Moreover, since $proj_U(B)$ is in $U$ we have

$$proj_U(B) = AZ$$

for some $Z$ in $\mathbb{R}^n$. Any such vector $Z$ is called a **best approximation** to a solution of the system $AX = B$.

This appears to solve the problem, but there are two difficulties. First we need an orthogonal basis of $U$ to compute $proj_U(B)$, and second we end up with $AZ$ rather than $Z$. Fortunately, there is a routine way to find all best approximations $Z$ directly. The key observation is that $B - AZ$ lies in $U^\perp$ (by the Projection theorem), so it is orthogonal to every column $AX$ in $U$. Thus, for all $X$ in $\mathbb{R}^n$, we have

$$
\begin{aligned}
0 &= (AX) \cdot (B - AZ) \\
&= (AX)^T (B - AZ) \\
&= X^T A^T (B - AZ) \\
&= X \cdot [A^T (B - AZ)].
\end{aligned}
$$

In other words, $A^T(B - AZ)$ is orthogonal to every vector $x$ in $\mathbb{R}^n$, and so must be zero (being orthogonal to itself). Hence $(A^TA)Z = A^TB$, so every best approximation $Z$ to a solution to the system $AX = B$ is a solution to the matrix equation

$$(A^TA)Z = A^TB.$$

This is a system of linear equations called the **normal equations** for $Z$, and it turns out that *every* solution is actually a best approximation.

**THEOREM 5**

Let $A$ be an $m \times n$ matrix, let $B$ be any column in $\mathbb{R}^m$, and consider the system

$$AX = B$$

of $m$ linear equations in $n$ variables.

(1) The solutions $Z$ to the normal equations

$$(A^TA)Z = A^TB$$

are exactly the best approximations to a solution to $AX = B$ in the sense that $\|B - AZ\|$ is the minimum value of $\|B - AX\|$ as $X$ ranges over all columns of $\mathbb{R}^n$.

(2) The best approximation $Z$ is unique if and only if $A^TA$ is invertible.

**PROOF** (1) We showed above that every best approximation $Z$ is a solution to the normal equations. On the other hand, if $Z$ satisfies these equations then, by reversing the above argument, we see that $B - AZ$ is in $U^\perp$, where $U = \{AX \mid X \text{ in } \mathbb{R}^n\}$. Thus we have $B = AZ + (B - AZ)$ where $AZ$ is in $U$ and $B - AZ$ is in $U^\perp$. This means that $AZ = proj_U(B)$ by the Corollary to the Projection theorem, so $Z$ is a best approximation to a solution of the system $AX = B$.

(2) If $A^TA$ is invertible, then $Z = (A^TA)^{-1}A^TB$ is uniquely determined by $A$ and $B$. Conversely, if $Z$ is unique, we show that $A^TA$ is invertible by showing (Theorem 5 §1.5) that $(A^TA)X = 0$ implies $X = 0$. But if $(A^TA)X = 0$, then $(A^TA)(Z + X) = (A^TA)Z + 0 = A^TB$, so $Z + X = Z$ by the uniqueness. Hence $X = 0$, which proves (2).

In connection with condition (2) of Theorem 5, recall Theorem 3 §4.4: If $A$ is any $m \times n$ matrix, then $A^TA$ is invertible if and only if $A$ has independent columns (equivalently, if $rankA = n$, or if $AX = 0$ has unique solution $X = 0$). Hence in this case there is a unique best approximation $Z$ to a solution of the system $AX = B$.

Of course, having many best approximations is not necessarily a bad thing. It means that you have room to choose a best approximation that optimizes some other feature of the solution to the system.

**Example 6** Find the best approximations to a solution of the following inconsistent system of linear equations:

$$
\begin{aligned}
x_1 - x_2 + 3x_3 \quad\quad &= 2 \\
-2x_1 + 2x_2 \quad\quad + 4x_4 &= -1 \\
-x_1 + x_2 + 9x_3 + 8x_4 &= 3
\end{aligned}
$$

**SOLUTION** Here $A = \begin{bmatrix} 1 & -1 & 3 & 0 \\ -2 & 2 & 0 & 4 \\ -1 & 1 & 9 & 8 \end{bmatrix}$, $A^TA = \begin{bmatrix} 6 & -6 & -6 & -16 \\ -6 & 6 & 6 & 16 \\ -6 & 6 & 90 & 72 \\ -16 & 16 & 72 & 80 \end{bmatrix}$, and $B = [2 \;\; -1 \;\; 3]^T$, so that

$A^TB = [1 \;\; -1 \;\; 33 \;\; 20]^T$. Solving the normal equations $(A^TA)Z = A^TB$ by Gaussian elimination gives the general solution

$$
Z = \begin{bmatrix} \frac{4}{7} & 0 & \frac{17}{42} & 0 \end{bmatrix}^T + s[1 \;\; 1 \;\; 0 \;\; 0]^T + t[6 \;\; 0 \;\; -2 \;\; 3]^T
$$

where $s$ and $t$ are parameters. Hence $Z$ is a best approximation to a solution of $AX = B$ for any value of $s$ and $t$, one possibility being $Z = \begin{bmatrix} \frac{4}{7} & 0 & \frac{17}{42} & 0 \end{bmatrix}^T = \frac{1}{42}[24 \;\; 0 \;\; 17 \;\; 0]^T$. However, for all values of $s$ and $t$ we obtain

$$
AZ = \begin{bmatrix} \frac{25}{14} & \frac{-8}{7} & \frac{43}{14} \end{bmatrix}^T \approx [1.79 \;\; -1.14 \;\; 3.07]^T.
$$

This is as close as possible to the given constants $[2 \;\; -1 \;\; 3]^T$.

**Example 7** It is thought that the steel production $s$ (in thousands of tons) in a country is linearly related to the number $x_1$ of mines and the number $x_2$ of steel mills in that country, that is that there exist constants $a$ and $b$ such that

$$s = ax_1 + bx_2$$

holds for every country. The data in the table were collected for four countries $A$, $B$, $C$, and $D$. Find the best approximation to the coefficients $a$ and $b$.

| Country | A | B | C | D |
|---------|---|---|---|---|
| $x_1$ | 4 | 3 | 5 | 2 |
| $x_2$ | 3 | 2 | 3 | 1 |
| $s$ | 2.9 | 2.2 | 3.6 | 1.6 |

**SOLUTION**   The given data yield four equations (one for each country) in the variables $a$ and $b$:

$$4a + 2b = 2.9$$
$$3a + 2b = 2.2$$
$$5a + 3b = 3.6$$
$$2a + \phantom{2}b = 1.6$$

that is, $AX = B$ where

$$A = \begin{bmatrix} 4 & 2 \\ 3 & 2 \\ 5 & 3 \\ 2 & 1 \end{bmatrix}, X = \begin{bmatrix} a \\ b \end{bmatrix}, \text{ and } B = \begin{bmatrix} 2.9 \\ 2.2 \\ 3.6 \\ 1.6 \end{bmatrix}.$$

Here the system $AX = B$ has no solution. But the matrix $A^TA$ is invertible because $A$ has independent columns, and the unique solution $Z$ to the normal equations $(A^TA)Z = A^TB$ is $Z = \begin{bmatrix} 0.782 \\ -0.091 \end{bmatrix}$, to three decimals. Hence $a = 0.782$, and $b = -0.091$ are the best approximation, whence $s = 0.782x_1 - 0.091x_2$.

## 4.6.5   Least Squares Approximation

We began this chapter with an example where an engineer wants to be able to predict the response of an amplifier for any input voltage $x$ between 2 and 6 millivolts. She inputs trial voltages $x_i$ of 2, 3, 4, 5, and 6 mv and measures the response values $y_i$ in the table.

| Trial $i$ | 1 | 2 | 3 | 4 | 5 |
|---|---|---|---|---|---|
| Input $x_i$ | 2 | 3 | 4 | 5 | 6 |
| Response $y_i$ | 4.8 | 7.5 | 8.6 | 11.1 | 12.3 |

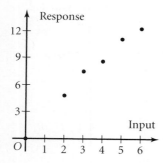

When the data pairs are plotted it is apparent (see the diagram) that the points $(x_i, y_i)$ are nearly on a straight line. Theorem 5 enables us to find the straight line that best fits the data. Recall that every (nonvertical) line has the form $y = a_0 + a_1x$ for some constants $a_0$ and $a_1$. Hence, given $a_0$ and $a_1$, we get predicted responses $y_1', \cdots, y_5'$, where we write $y_i' = a_0 + a_1x_i$ for convenience. Thus the task is to find the values of $a_0$ and $a_1$ such that the observed responses $y_1, \cdots, y_5$ are as close as possible to the predicted responses $y_1', \cdots, y_5'$. This is where Theorem 5 comes in.

If we write $Y = [y_1\ y_2\ y_3\ y_4\ y_5]^T$ and $Y' = [y_1'\ y_2'\ y_3'\ y_4'\ y_5']^T$, the idea is to choose $a_0$ and $a_1$ such that

$$\|Y - Y'\| = \sqrt{(y_1 - y_1')^2 + (y_2 - y_2')^2 + (y_3 - y_3')^2 + (y_4 - y_4')^2 + (y_5 - y_5')^2}$$

is as small as possible (hence the name "least squares" approximation). Now observe that if we define

$$M = \begin{bmatrix} 1 & x_1 \\ 1 & x_2 \\ 1 & x_3 \\ 1 & x_4 \\ 1 & x_5 \end{bmatrix} \text{ and } Z = \begin{bmatrix} a_0 \\ a_1 \end{bmatrix},$$

then

$$MZ = \begin{bmatrix} y_1' \\ y_2' \\ y_3' \\ y_4' \\ y_5' \end{bmatrix} = Y'.$$

Hence we are to choose $Z$ such that $\|Y - Y'\| = \|Y - MZ\|$ is as small as possible. In other words, $Z$ is to be a best approximation to the system $MZ = Y$. Since $Y$ and $M$ are known matrices, this is solved by Theorem 5. Thus $Z$ is given as any solution of the normal equations

$$(M^T M)Z = M^T Y.$$

In the present situation, solving these equations gives a unique best approximation $Z = [1.42 \quad 1.86]^T$, so the best fitting line has equation $y = 1.42 + 1.86x$. Using this, the predicted values are

$$Y' = [5.14 \quad 7.00 \quad 8.86 \quad 10.72 \quad 12.58]^T$$

compared to the observed data

$$Y = [4.8 \quad 7.5 \quad 8.6 \quad 11.1 \quad 12.3]^T.$$

These predicted values $Y'$ are as close as possible (in the least squares sense) to the observed data $Y$ using a straight line approximation.

The general situation is dealt with in the same way. Suppose that $n$ data pairs have been collected giving corresponding values of two variables $x$ and $y$:

$$(x_1, y_1), \quad (x_2, y_2), \quad (x_3, y_3), \quad \cdots, \quad (x_n, y_n).$$

The method used above will best fit a line $y = a_0 + a_1 x$ to these data. In fact, with very little extra effort, we can find a *polynomial*

$$y = a_0 + a_1 x + a_2 x^2 + \cdots + a_m x^m$$

of degree at most $m$ that best fits the data (so the line is the case $m = 1$). For convenience write

$$f(x) = a_0 + a_1 x + a_2 x^2 + \cdots + a_m x^m.$$

Given any choice of the coefficients $a_0, a_1, \cdots, a_m$, we obtain predicted values $y_i' = f(x_i)$. If we write $Y = [y_1 \ y_2 \ y_3 \ \cdots \ y_n]^T$ and $Y' = [y_1' \ y_2' \ y_3' \ \cdots \ y_n']^T$ then, as before, the idea is to choose the coefficients $a_0, a_1, \cdots, a_m$ so that these predicted values $y_i'$ are as close as possible to the observed values $y_i$. In other words, we choose the $a_i$ so that the quantity

$$\|Y - Y'\| = \sqrt{(y_1 - y_1')^2 + (y_2 - y_2')^2 + (y_3 - y_3')^2 + \cdots + (y_n - y_n')^2}$$

is as small as possible. This "least squares" fit can always be achieved.

By analogy with the above computation, write

$$M = \begin{bmatrix} 1 & x_1 & x_1^2 & \cdots & x_1^m \\ 1 & x_2 & x_2^2 & \cdots & x_2^m \\ \vdots & \vdots & \vdots & \vdots & \vdots \\ 1 & x_n & x_n^2 & \cdots & x_n^m \end{bmatrix} \text{ and } Z = \begin{bmatrix} a_0 \\ a_1 \\ \vdots \\ a_m \end{bmatrix},$$

so that (as before) $MZ = Y'$. Hence $\|Y - Y'\| = \|Y - MZ\|$, so we want to find $Z$ such that $\|Y - MZ\|$ is as small as possible. Thus the desired column $Z$ is a best approximation to the linear system $MZ = Y$ where $Y$ and $M$ are known matrices. Hence Theorem 5 shows that any solution $Z = [a_0 \ a_1 \ \cdots \ a_m]^T$ of the normal equations $(M^T M)Z = M^T Y$ will provide a best approximation. The corresponding polynomial $a_0 + a_1 x + a_2 x^2 + \cdots + a_m x^m$ is called a **least squares approximating polynomial of degree** $m$ for the data pairs. This discussion proves most of

**THEOREM 6**　**Least Squares Theorem.** Given $n$ data pairs $(x_1,\ y_1),\ (x_2,\ y_2),\ (x_3,\ y_3),\ \cdots,\ (x_n,\ y_n)$, put

$$
Y = \begin{bmatrix} y_1 \\ y_2 \\ \vdots \\ y_n \end{bmatrix} \quad \text{and} \quad M = \begin{bmatrix} 1 & x_1 & x_1^2 & \cdots & x_1^m \\ 1 & x_2 & x_2^2 & \cdots & x_2^m \\ \vdots & \vdots & \vdots & \vdots & \vdots \\ 1 & x_n & x_n^2 & \cdots & x_n^m \end{bmatrix}.
$$

Then a least squares approximating polynomial of degree $m$ for these data is given by

$$y = a_0 + a_1 x + a_2 x^2 + \cdots + a_m x^m$$

where $Z = [a_0 \ \ a_1 \ \cdots \ a_m]^T$ is any solution of the normal equations

$$(M^T M)Z = M^T Y.$$

Furthermore, if at least $m + 1$ of the $x_i$ are distinct, then $Z$ is unique and so the least squares polynomial is unique. (In fact, $M^T M$ is invertible in this case, so $Z = (M^T M)^{-1} M^T Y$.) [14]

**PROOF**　It remains to verify that $Z$ is unique when at least $m + 1$ of the $x_i$ are distinct. It suffices to show that $M^T M$ is invertible (then $Z = (M^T M)^{-1} M^T Y$), and this follows if $M$ has independent columns (Theorem 3 §4.4). Denote the columns of $M$ by $C_0,\ C_1,\ C_2,\ \cdots,\ C_m$, and suppose a linear combination vanishes: $t_0 C_0 + t_1 C_1 + t_2 C_2 + \cdots + t_m C_m = 0$. If we write $g(x) = t_0 + t_1 x + t_2 x^2 + \cdots + t_m x^m$, this gives $g(x_i) = 0$ for each $i$. If at least $m + 1$ of the $x_i$ are distinct, this means that $g(x)$ is a polynomial of degree at most $m$ which has more than $m$ distinct roots. It is a theorem of algebra (see Appendix A.3), that this cannot happen unless all the coefficients $t_i$ of $g(x)$ are zero. Hence $\{C_0,\ C_1,\ C_2,\ \cdots,\ C_m\}$ is independent.

It is worth noting that the condition in Theorem 6 that $m + 1$ of the $x_i$ are distinct is not a severe restriction. For example, it can be arranged in many experimental situations where the data pairs $(x_i,\ y_i)$ are found by *choosing* the $x_i$ and then observing the corresponding values $y_i$.

**Example 8**　Newton's Laws of Motion imply that an object thrown vertically at a velocity of $v$ metres per second will be at a height of $h = vt - \frac{1}{2}gt^2$ metres after $t$ seconds, where $g$ is the acceleration due to gravity. The values in the table were observed. By fitting a quadratic $h = a_0 + a_1 t + a_2 t^2$ to these data, estimate the values of $v$ and $g$.

| $t$ | 0.5 | 1.0 | 1.5 | 2.0 |
|---|---|---|---|---|
| $h$ | 23.7 | 45.1 | 64.0 | 80.4 |

**SOLUTION**　Here $Y = \begin{bmatrix} 23.7 \\ 45.1 \\ 64.0 \\ 80.4 \end{bmatrix}$ and $M = \begin{bmatrix} 1 & 0.5 & 0.25 \\ 1 & 1.0 & 1.0 \\ 1 & 1.5 & 2.25 \\ 1 & 2.0 & 4.0 \end{bmatrix}$, where $t$ and $h$ are playing the role of $x$ and $y$ in

Theorem 6. Solving the normal equations gives $h = -0.175 + 50.25t - 4.98t^2$, whence the estimates are $v = 50.25$ and $-\frac{1}{2}g = -4.98$. The latter equation gives $g = 9.96$ (the true value is 9.81).

---

[14] Even if $M^T M$ is invertible, it is rarely advisable to calculate $Z$ from the formula $Z = (M^T M)^{-1} M^T Y$. It is nearly always more efficient numerically to solve the normal equations $(M^T M)Z = M^T Y$ by Gaussian elimination.

**Remark**. There is an extension of Theorem 6 that should be mentioned. Given the data pairs $(x_1, y_1), (x_2, y_2), \cdots, (x_n, y_n)$, the theorem shows how to find a polynomial

$$p(x) = a_0 + a_1x + a_2x^2 + \cdots + a_mx^m$$

which approximates the data in the least squares sense. In fact the theorem describes how to find the coefficients $a_i$ of the best fitting polynomial. Now view $p(x)$ as a linear combination of the functions $1, x, x^2, \cdots, x^m$. This suggests that perhaps this will work for linear combinations of other functions. If functions $f_0(x), f_1(x), f_2(x), \cdots, f_m(x)$ are given, we ask whether coefficients $a_i$ can be found such that

$$f(x) = a_0f_0(x) + a_1f_1(x) + a_2f_2(x) + \cdots + a_mf_m(x)$$

is a best approximation in the sense that $\|Y - Y'\|$ is as small as possible where $Y = [y_1 \ y_2 \ y_3 \ \cdots \ y_n]$ and $Y' = [f(x_1) \ f(x_2) \ f(x_3) \ \cdots \ f(x_n)]$. The answer is "yes," and the result is the analogue of Theorem 6 where we use a new matrix

$$M = \begin{bmatrix} f_0(x_1) & f_1(x_1) & \cdots & f_m(x_1) \\ f_0(x_2) & f_1(x_2) & \cdots & f_m(x_2) \\ \vdots & \vdots & & \vdots \\ f_0(x_n) & f_1(x_n) & \cdots & f_m(x_n) \end{bmatrix}.$$

As before, the coefficients $Z = [a_0 \ a_1 \ \cdots \ a_m]^T$ are found as the solution to the normal equations $(M^TM)Z = M^TY$. However, the statement (at the end of Theorem 6) about the uniqueness of $Z$ may fail for other choices of the functions $f_i(x)$.

## Exercises 4.6

**1.** In each case write the vector $X$ as the sum of a vector in $U$ and a vector in $U^\perp$.

**a)** $X = [3 \ -1 \ 2]^T$, $U = span\{[1 \ 2 \ -1]^T, [2 \ 0 \ 1]^T\}$.

**b)** $X = [1 \ 1 \ 3]^T$, $U = span\{[-1 \ 0 \ 2]^T, [3 \ 1 \ 5]^T\}$.

**c)** $X = [3 \ 0 \ 2 \ 1]^T$,
$U = span\{[1 \ 1 \ 1 \ 1]^T, [1 \ 1 \ -1 \ -1]^T, [1 \ -1 \ 1 \ -1]^T\}$.

**d)** $X = [2 \ 1 \ 4 \ -2]^T$,
$U = span\{[1 \ 0 \ 1 \ 1]^T, [0 \ 1 \ -1 \ 1]^T, [-2 \ 0 \ 1 \ 1]^T\}$.

**e)** $X = [a \ b \ c \ d]^T$,
$U = span\{[1 \ 0 \ 0 \ 0]^T, [0 \ 1 \ 0 \ 0]^T, [0 \ 0 \ 1 \ 0]^T\}$.

**f)** $X = [a \ b \ c \ d]^T$,
$U = span\{[1 \ -1 \ 2 \ 0]^T, [-1 \ 1 \ 1 \ 1]^T\}$.

**2.** Let $X = [1 \ 1 \ 0 \ 0]^T$ and
$U = span\{[1 \ -1 \ 0 \ 2]^T, [2 \ 2 \ -1 \ 0]^T\}$.

**a)** Compute $proj_U(X)$.

**b)** Show that $\{[1 \ 3 \ -1 \ -2]^T, [7 \ 1 \ -2 \ 6]^T\}$ is another orthogonal basis of $U$, and use it to compute $proj_U(X)$.

**3.** In each case find the vector in the subspace $U$ that is closest to $X$.

**a)** $U = span\{[1 \ 2 \ -1]^T, [0 \ 1 \ 3]^T\}$, $X = [5 \ -2 \ 1]^T$.

**b)** $U = span\{[1 \ -1 \ 0]^T, [1 \ 0 \ -1]^T\}$, $X = [2 \ 1 \ 3]^T$.

**c)** $U = span\{[1 \ 0 \ 1 \ 0]^T, [1 \ 1 \ 1 \ 0]^T, [1 \ 1 \ 0 \ 0]^T\}$,
$X = [3 \ -1 \ 2 \ 1]^T$.

**d)** $U = span\{[1 \ -1 \ 1 \ 0]^T, [1 \ 1 \ 0 \ 0]^T, [1 \ 1 \ 0 \ 1]^T\}$,
$X = [3 \ 0 \ 5 \ -1]^T$.

**4.** Find the best approximation to a solution of each of the following systems of equations.

**a)**
$$\begin{aligned} x - \ \ y &= \ \ 3 \\ 2x + \ \ y &= -1 \\ x + 5y &= -4 \end{aligned}$$

**b)**
$$\begin{aligned} 2x + \ \ y &= 1 \\ x - \ \ y &= 2 \\ x + 5y &= 3 \end{aligned}$$

**c)**
$$\begin{aligned} x + \ \ y + \ \ z &= 1 \\ 2x - \ \ y - \ \ z &= 0 \\ -x + 2y + 3z &= 2 \\ 2x + 2y + 3z &= 1 \end{aligned}$$

**d)**
$$\begin{aligned} x - \ \ y + 2z &= \ \ 0 \\ 3x + \ \ y + \ \ z &= -1 \\ 2x + \ \ y - 3z &= \ \ 3 \\ 3x + 4y - 4z &= \ \ 1 \end{aligned}$$

**5.** Find the least squares approximating quadratic for the following sets of data pairs.

**a)** $(-1, 1), (0, 0), (2, 3), (3, 4)$

**b)** $(-1, 0), (0, 2), (1, 1), (3, -1)$

**6.** Find the least squares approximating cubic for the following sets of data pairs.

**a)** $(-1, 1), (0, -1), (1, 3), (3, -4)$

**b)** $(-2, 0), (0, 3), (1, 2), (3, -1), (4, 1)$

**7.** Find the least squares approximating function of the form $a_0x + a_1x^2 + a_2(-1)^x$ for the following set of data pairs:
$(-1, 1), (0, 3), (1, 1), (2, 0)$. [Hint: See the Remark following Example 8.]

**8.** Let $U = span\{Y_1, Y_2, \cdots, Y_m\}$ in $\mathbb{R}^n$ where the $Y_i$ are written as rows, and let $A$ denote the $m \times n$ matrix with the $Y_i$ as its rows.
 **a)** Show that $U^\perp = \{X \text{ in } \mathbb{R}^n \mid AX^T = 0\}$.
 **b)** Use **a)** to find $U^\perp$ if
 $U = span\{[1 \;\; -2 \;\; 1 \;\; 1]^T, [0 \;\; -1 \;\; 3 \;\; -2]^T\}$.

**9.** In each case either show that the statement is true or give an example showing that it is false.
 **a)** If $proj_U(X)$ is in $U$, then $X$ is in $U$.
 **b)** If $proj_U(X)$ is in $U^\perp$, then $X$ is in $U^\perp$.
 **c)** There is no vector $X$ in both $U$ and $U^\perp$.
 **d)** If $proj_U(X) = 0$, then $X$ is in $U^\perp$.
 **e)** If $X$ is in $U^\perp$, then $proj_U(X) = 0$.
 **f)** If $U \subseteq W$, then $U^\perp \subseteq W^\perp$.
 **g)** If $U^\perp = W^\perp$, then $U = W$.

**10.** Let $U$ denote a subspace of $\mathbb{R}^n$. Show that a vector $X$ in $\mathbb{R}^n$ is in $U$ if and only if $proj_U(X) = X$.

**11.** If $U = span(X)$ where $X \neq 0$ in $\mathbb{R}^n$, show that $dim(U^\perp) = n - 1$.

**12.** If $U \subseteq W$ are subspaces of $\mathbb{R}^n$, show that $W^\perp \subseteq U^\perp$.

**13.** If $A$ is an $m \times n$ matrix, show that $null(A^T) = (colA)^\perp$ in $\mathbb{R}^m$.

**14.** Let $U$ denote a subspace of $\mathbb{R}^n$.
 **a)** Show that $U^\perp = \mathbb{R}^n$ if and only if $U = \{0\}$.
 **b)** Show that $U^\perp = \{0\}$ if and only if $U = \mathbb{R}^n$.

**15.** If $U$ is a subspace of $\mathbb{R}^n$, show that $U^\perp = \{X \text{ in } \mathbb{R}^n \mid proj_U(X) = 0\}$.

**16.** Let $U$ be a subspace of $\mathbb{R}^n$. Show that:
 **a)** $proj_U(X + Y) = proj_U(X) + proj_U(Y)$ for all $X$ and $Y$ in $\mathbb{R}^n$.
 **b)** $proj_U(aX) = a\, proj_U(X)$ for all numbers $a$ and all $X$ in $\mathbb{R}^n$.
 **c)** Show that $proj_U(proj_U(X)) = proj_U(X)$ for all $X$ in $\mathbb{R}^n$.

**17.** If $\{F_1, F_2, \cdots, F_n\}$ is an orthogonal basis of $\mathbb{R}^n$ and $U = span\{F_1, F_2, \cdots, F_k\}$, show that $U^\perp = span\{F_{k+1}, \cdots, F_n\}$. [Hint: Apply the Expansion theorem to a vector $X$ in $\mathbb{R}^n$.]

**18.** If $U$ is a subspace of $\mathbb{R}^n$, show how to find an $n \times n$ matrix $A$ such that $U = \{X \text{ in } \mathbb{R}^n \mid AX = 0\}$. [Hint: Theorem 3(2) and the preceding exercise.]

**19.** Let $U$ and $W$ be subspaces of $\mathbb{R}^n$ and consider the subspaces
 $U \cap W = \{X \text{ in } \mathbb{R}^n \mid X \text{ is in both } U \text{ and } W\}$.
 $U + W = \{X \text{ in } \mathbb{R}^n \mid X \text{ is the sum of the vector in } U \text{ and a vector in } W\}$.
 **a)** Show that $(U + W)^\perp = U^\perp \cap W^\perp$.
 **b)** Show that $(U \cap W)^\perp \supseteq U^\perp + W^\perp$.
 **c)** Show that in fact $(U \cap W)^\perp = U^\perp + W^\perp$. [Hint: Use **a)**, Theorem 3, and Exercise 25 §4.3 to show that $dim(U^\perp + W^\perp) = dim((U \cap W)^\perp)$. Then use **b)** and Theorem 4 §4.3.]

**20.** A square matrix $E$ is called a **projection matrix** if $E^2 = E = E^T$.
 **a)** If $A$ is $m \times n$ and $AA^T$ is invertible, show that $E = A^T(AA^T)^{-1}A$ is a projection matrix.
 **b)** If $E$ is a projection matrix, show that $I - E$ is also a projection matrix.
 **c)** If $E$ and $F$ are projection matrices such that $EF = 0 = FE$, show that $E + F$ is a projection matrix.

## 4.7    ORTHOGONAL DIAGONALIZATION

In Section 2.3 we studied diagonalization of a square matrix $A$, and found important applications (for example to dynamical systems). We can now utilize the concepts of subspace, basis and dimension to clarify the diagonalization process. Then we integrate orthogonality into the discussion by introducing the idea of orthogonal diagonalization. This leads to the Principal Axis theorem, one of the most useful results in linear algebra. The section concludes by proving some results that could not be demonstrated in Section 2.3 (Theorems 4 and 5).

Recall that a square matrix $A$ is *diagonalizable* if there exists an invertible matrix $P$ such that $P^{-1}AP = D$ is a diagonal matrix. Unfortunately, not all matrices are diagonalizable, for example $\begin{bmatrix} 1 & 1 \\ 0 & 1 \end{bmatrix}$. Determining whether $A$ is diagonalizable is closely related to the eigenvalues and eigenvectors of $A$. Recall that a number $\lambda$ is called an *eigenvalue* of $A$ if $AX = \lambda X$ for some nonzero vector $X$ in $\mathbb{R}^n$, and any such $X$ is called an *eigenvector* of $A$ corresponding to $\lambda$. Theorem 3 §2.3 asserts (in part) that $A$ is diagonalizable if and only if it has $n$ eigenvectors $X_1, \cdots, X_n$ such that the matrix $P = [X_1 \cdots X_n]$ with the $X_i$ as columns is invertible. This is equivalent to asking that $\{X_1, \cdots, X_n\}$ is a basis of $\mathbb{R}^n$ consisting of eigenvectors of $A$. Hence we can restate Theorem 3 §2.3 as follows:

**THEOREM 1**

Let $A$ be an $n \times n$ matrix.

(1) $A$ is diagonalizable if and only if $\mathbb{R}^n$ has a basis $\{X_1, X_2, \cdots, X_n\}$ of eigenvectors of $A$.

(2) When this is the case, the matrix $P = [X_1 \cdots X_n]$ is invertible and $P^{-1}AP = diag(\lambda_1, \lambda_2, \cdots, \lambda_n)$ where, for each $i$, $\lambda_i$ is the eigenvalue of $A$ corresponding to $X_i$.

## 4.7.1 Orthogonal Matrices

Theorem 1 relates diagonalizability to the existence of a basis of eigenvectors. Since the most convenient bases of $\mathbb{R}^n$ are the *orthogonal* ones, this suggests two questions: Which $n \times n$ matrices admit an orthogonal basis $\{X_1, \cdots, X_n\}$ of eigenvectors? And what is special about $P = [X_1 \cdots X_n]$ in this case? Both these questions have interesting and important answers which we develop in this section.

We know (Theorem 2 §4.2) that an $n \times n$ matrix $A$ is invertible if and only if its columns (respectively its rows) form a basis of $\mathbb{R}^n$. It is thus natural to investigate the situation where the columns of $A$ are actually an orthogonal basis of $\mathbb{R}^n$. It turns out, however, that the appropriate condition is that the columns are *orthonormal* (that is, they are orthogonal and each is a unit vector). Surprisingly, this is equivalent to the condition that the *rows* are orthonormal.

**THEOREM 2**

The following conditions are equivalent for an $n \times n$ matrix $P$:

(1) $P$ is invertible and $P^{-1} = P^T$.

(2) The columns of $P$ are orthonormal.

(3) The rows of $P$ are orthonormal.

PROOF

First note that (1) is equivalent to $P^T P = I$ by Corollary 1 to Theorem 5 §1.5. Now write $P = [C_1 \quad C_2 \quad \cdots \quad C_n]$ where $C_1, C_2, \cdots, C_n$ denote the columns of $P$. Since row $i$ of $P^T$ is $C_i^T$, it follows that the $(i, j)$-entry of $P^T P$ is $C_i^T C_j = C_i \cdot C_j$. Thus $P^T P = I$ holds if and only if $C_i \cdot C_j = 0$ when $i \neq j$ and $C_i \cdot C_j = 1$ when $i = j$, that is if and only if the $C_i$ are orthonormal. This proves $(1) \Leftrightarrow (2)$, and $(1) \Leftrightarrow (3)$ follows in a similar way.

A square matrix $P$ is called an **orthogonal matrix**[15] if it satisfies one (and hence all) of the conditions in Theorem 2. Clearly every identity matrix is orthogonal. We begin by listing some simple properties of these matrices that will be referred to below. The routine verifications are left to the reader.

COROLLARY

Let $P$ and $Q$ denote $n \times n$ matrices. Then:

(1) If $P$ is an orthogonal matrix, then either $det P = 1$ or $det P = -1$.

(2) If $P$ is an orthogonal matrix, then $P^{-1} = P^T$ is an orthogonal matrix.

(3) If $P$ and $Q$ are orthogonal matrices, then $PQ$ is an orthogonal matrix.

---

[15] In view of Theorem 2, *orthonormal* matrix might be a better name. However the term *orthogonal matrix* is standard.

The next example exhibits orthogonal $2 \times 2$ matrices with determinant 1 and $-1$; it also shows that orthogonal matrices have geometrical significance.

**Example 1**    The matrices $A = \begin{bmatrix} \cos\theta & -\sin\theta \\ \sin\theta & \cos\theta \end{bmatrix}$ and $B = \begin{bmatrix} \cos\theta & \sin\theta \\ \sin\theta & -\cos\theta \end{bmatrix}$ are orthogonal for any angle $\theta$ because $\sin^2\theta + \cos^2\theta = 1$. In fact these are the *only* $2 \times 2$ orthogonal matrices. This is verified in the discussion leading to Theorem 7 §3.5 where it is also shown that, if these matrices are viewed as transformations of $\mathbb{R}^2$, then $A$ is rotation around the origin through the angle $\theta$, and $B$ is reflection in the line through the origin making an angle $\frac{\theta}{2}$ with the positive $X$-axis.

A $2 \times 2$ matrix $P$, viewed as a transformation of $\mathbb{R}^2$, is called an *isometry* if it preserves distance, that is if $\|P\vec{v} - P\vec{w}\| = \|\vec{v} - \vec{w}\|$ for all vectors $\vec{v}$ and $\vec{w}$ in $\mathbb{R}^2$. Hence Theorem 7 §3.5 shows that $P$ is an isometry if and only if $P$ is a rotation or a reflection, and Example 1 shows that this happens if and only if $P$ is orthogonal. Similarly, the isometries of $\mathbb{R}^n$ are precisely the orthogonal matrices (see Section 4.9). In the case of $\mathbb{R}^3$, the isometries are either rotations (about a line through the origin), reflections (in a plane through the origin), or a composition of two of these.

**Example 2**    It is not enough that the columns are merely orthogonal for a matrix to be an orthogonal matrix. For example, $A = \begin{bmatrix} 1 & 1 & 1 \\ 1 & -1 & 1 \\ -1 & 0 & 2 \end{bmatrix}$ has orthogonal columns, but the rows are not orthogonal. However, if the columns are normalized the result is the orthogonal matrix $P = \begin{bmatrix} \frac{1}{\sqrt{3}} & \frac{1}{\sqrt{2}} & \frac{1}{\sqrt{6}} \\ \frac{1}{\sqrt{3}} & -\frac{1}{\sqrt{2}} & \frac{1}{\sqrt{6}} \\ -\frac{1}{\sqrt{3}} & 0 & \frac{2}{\sqrt{6}} \end{bmatrix}$. Note that now the rows are *also* orthonormal.

The eigenvalues of an orthogonal matrix $P$ need not be real. Indeed, the rotation $P = \begin{bmatrix} \cos\theta & -\sin\theta \\ \sin\theta & \cos\theta \end{bmatrix}$ in Example 1 has eigenvalues $\cos\theta \pm i\sin\theta$. These eigenvalues have absolute value $|\cos\theta \pm i\sin\theta| = 1$ as complex numbers, and the next theorem shows that this is a general property of orthogonal matrices.

**THEOREM 3**    If $P$ is an orthogonal matrix, then the (possibly complex) eigenvalues $\lambda$ of $P$ have absolute value $|\lambda| = 1$. In particular, the only *real* eigenvalues of $P$ are 1 and $-1$.

**PROOF**    Recall (see §2.5) that if $Z = [z_{ij}]$ is a matrix with complex entries $z_{ij}$, the conjugate $\overline{Z} = [\overline{z}_{ij}]$ is defined to be the matrix obtained from $Z$ by conjugating every entry. Recall further that if $Z$ and $W$ are two complex matrices then $\overline{Z + W} = \overline{Z} + \overline{W}$, $\overline{ZW} = \overline{Z}\,\overline{W}$, and $\overline{\lambda Z} = \overline{\lambda}\,\overline{Z}$ for each complex number $\lambda$.

Now suppose that $\lambda$ is a complex eigenvalue of $P$, say $PX = \lambda X$ where $X \neq 0$ is a column in $\mathbb{C}^n$. Since $P$ is real, we have $\overline{P} = P$ and so we compute

$$|\lambda|^2 (X^T \overline{X}) = (\lambda\overline{\lambda})X^T\overline{X} = (\lambda X)^T\overline{(\lambda X)} = (PX)^T\overline{(PX)}$$
$$= X^T P^T \overline{P}\,\overline{X} = X^T P^T P\overline{X} = X^T I\overline{X} = X^T\overline{X}.$$

Hence it suffices to show that $X^T\overline{X} \neq 0$. But if $X = [x_1 \ \ x_2 \ \ \cdots \ \ x_n]^T$ where the $x_i$ are complex numbers, then $X^T\overline{X} = |x_1|^2 + |x_2|^2 + \cdots + |x_n|^2 \neq 0$ because some $x_i \neq 0$ (as $X \neq 0$).

A major virtue of orthogonal matrices is that they are easy to invert—simply take the transpose. This makes the $n \times n$ version of the QR-factorization (Theorem 9 §4.5) even more useful. We restate it here for reference.

**THEOREM** **4**

**QR-Factorization.** Every invertible matrix $A$ can be factored as
$$A = QR$$
where $Q$ is an orthogonal matrix and $R$ is an invertible, upper triangular matrix with positive entries on the main diagonal. Furthermore, this factorization is unique in the sense that if $A = Q_1 R_1$ is another such factorization then $Q_1 = Q$ and $R_1 = R$.

It is important to note that a systematic procedure for finding QR-factorizations is available using the Gram-Schmidt algorithm—see the discussion preceding Theorem 9 §4.5, and other good numerical algorithms which are available.

## 4.7.2 The Principal Axis Theorem

Recall that an $n \times n$ matrix $A$ is said to be diagonalizable if $P^{-1}AP$ is a diagonal matrix for some invertible matrix $P$. By analogy, $A$ is called **orthogonally diagonalizable** if an orthogonal matrix $P$ can be found such that
$$P^{-1}AP = P^{T}AP \text{ is a diagonal matrix.}$$

The Principal Axis theorem asserts that every symmetric matrix is orthogonally diagonalizable (and conversely). The proof requires an important lemma which formalizes a technique that is basic to diagonalization, and which will be used three times below.

**LEMMA** **1**

Let $\{X_1, X_2, \cdots, X_k\}$ be a linearly independent set of eigenvectors of an $n \times n$ matrix $A$, extend it to a basis $\{X_1, X_2, \cdots, X_k, \cdots, X_n\}$ of $\mathbb{R}^n$, and let
$$P = [X_1 \ X_2 \ \cdots \ X_n]$$
be the (invertible) matrix with the $X_i$ as its columns. If $\lambda_1, \lambda_2, \cdots, \lambda_k$ are the (not necessarily distinct) eigenvalues of $A$ corresponding to $X_1, X_2, \cdots, X_k$ respectively, then $P^{-1}AP$ has block form
$$P^{-1}AP = \begin{bmatrix} diag(\lambda_1, \lambda_2, \cdots, \lambda_k) & B \\ 0 & A_1 \end{bmatrix}$$
where $B$ and $A_1$ are matrices of size $k \times (n-k)$ and $(n-k) \times (n-k)$ respectively.

**PROOF**

If $\{E_1, E_2, \cdots, E_n\}$ is the standard basis of $\mathbb{R}^n$, note that $PE_i$ is column $i$ of $P$ for each $i$, that is $PE_i = X_i$. Hence $P^{-1}X_i = E_i$ for each $1 \le i \le n$. On the other hand, observe that
$$P^{-1}AP = P^{-1}A[X_1 \ X_2 \ \cdots \ X_n] = [P^{-1}AX_1 \ P^{-1}AX_2 \ \cdots \ P^{-1}AX_n].$$
Hence, if $1 \le i \le k$, column $i$ of $P^{-1}AP$ is
$$(P^{-1}A)X_i = P^{-1}(\lambda_i X_i) = \lambda_i (P^{-1}X_i) = \lambda_i E_i.$$
This describes the first $k$ columns of $P^{-1}AP$, and Lemma 1 follows.

Note that Lemma 1 (with $k = n$) shows that an $n \times n$ matrix $A$ is diagonalizable if $\mathbb{R}^n$ has a basis of eigenvectors of $A$, as in part (1) of Theorem 1.

We can now prove one of the most useful results in linear algebra.

**THEOREM 5**   **Principal Axis Theorem.** The following are equivalent for an $n \times n$ matrix $A$:

    **(1)** $A$ is orthogonally diagonalizable.
    **(2)** $A$ has an orthonormal set of $n$ eigenvectors.
    **(3)** $A$ is symmetric.

**PROOF**   $(1) \Leftrightarrow (2)$. Let $P = [X_1 \ \cdots \ X_n]$ be an invertible $n \times n$ matrix with columns $X_i$. As in the discussion preceeding Theorem 3 §2.3, $(1) \Leftrightarrow (2)$ follows from the following statements: $P$ is orthogonal if and only if $\{X_1, \cdots, X_n\}$ is an orthonormal set in $\mathbb{R}^n$; and $P^{-1}AP$ is diagonal if and only if $\{X_1, \cdots, X_n\}$ is a basis of eigenvectors of $A$.

$(1) \Rightarrow (3)$. Let $P^T AP = D$ where $D$ is diagonal and $P^{-1} = P^T$. Then $A = PDP^T$ so, since $D$ is symmetric, we have $A^T = P^{TT}D^T P^T = PDP^T = A$. This proves (3).

$(3) \Rightarrow (1)$. Let $A$ be an $n \times n$ symmetric matrix; we prove (1) by induction on $n$. If $n = 1$, then (1) is clear. In general, since $A$ is symmetric let $\lambda_1$ be a real eigenvalue of $A$ (Theorem 3 §2.5), and let $AX_1 = \lambda_1 X_1$ for some eigenvector $X_1$ in $\mathbb{R}^n$ which we may assume satisfies $\|X_1\| = 1$. Use the Gram-Schmidt algorithm to find an orthonormal basis $\{X_1, X_2, \cdots, X_n\}$ of $\mathbb{R}^n$ containing this eigenvector $X_1$. Then $P_1 = [X_1 \ X_2 \ \cdots \ X_n]$ is an orthogonal matrix and

$$P_1^T AP_1 = P_1^{-1}AP_1 = \begin{bmatrix} \lambda_1 & B \\ 0 & A_1 \end{bmatrix}$$

in block form by Lemma 1. But $P_1^T AP_1$ is symmetric (because $A$ is), so it follows that $B = 0$ and $A_1$ is symmetric of size $(n-1) \times (n-1)$. Hence, by induction, there exists an orthogonal $(n-1) \times (n-1)$ matrix $Q$ such that $Q^T A_1 Q = D_1$ is diagonal. But then $P_2 = \begin{bmatrix} 1 & 0 \\ 0 & Q \end{bmatrix}$ is an orthogonal matrix, and

$$(P_1 P_2)^T A (P_1 P_2) = P_2^T (P_1^T AP_1) P_2$$

$$= \begin{bmatrix} 1 & 0 \\ 0 & Q^T \end{bmatrix} \begin{bmatrix} \lambda_1 & 0 \\ 0 & A_1 \end{bmatrix} \begin{bmatrix} 1 & 0 \\ 0 & Q \end{bmatrix}$$

$$= \begin{bmatrix} \lambda_1 & 0 \\ 0 & D_1 \end{bmatrix}$$

is diagonal. Since $P_1 P_2$ is orthogonal (Corollary to Theorem 2), this proves (1).

A set of $n$ orthonormal eigenvectors for a symmetric matrix $A$ is is called a set of **principal axes** for $A$. The name comes from geometry, and will be clarified in Section 4.8 when we discuss quadratic forms.

**Example 3**   If $A = \begin{bmatrix} 1 & 0 & 2 \\ 0 & 1 & -1 \\ 2 & -1 & 5 \end{bmatrix}$, find an orthogonal matrix $P$ such that $P^T AP$ is diagonal.

**SOLUTION**   The procedure is exactly the same as for diagonalizing $A$. The characteristic polynomial of $A$ is $c_A(x) = x(x-1)(x-6)$, so $A$ has eigenvalues $\lambda_1 = 0$, $\lambda_2 = 1$ and $\lambda_3 = 6$, with corresponding eigenvectors $X_1 = [-2 \ 1 \ 1]^T$, $X_2 = [1 \ 2 \ 0]^T$ and $X_3 = [2 \ -1 \ 5]^T$. Surprisingly, these eigenvectors are orthogonal. Hence, after normalizing, we obtain a set $\left\{ \frac{1}{\sqrt{6}}X_1, \frac{1}{\sqrt{5}}X_2, \frac{1}{\sqrt{30}}X_3 \right\}$ of principal axes for $A$. These yield the orthogonal matrix

$$P = \left[ \tfrac{1}{\sqrt{6}}X_1 \ \ \tfrac{1}{\sqrt{5}}X_2 \ \ \tfrac{1}{\sqrt{30}}X_3 \right] = \frac{1}{\sqrt{30}} \begin{bmatrix} -2\sqrt{5} & \sqrt{6} & 2 \\ \sqrt{5} & 2\sqrt{6} & -1 \\ \sqrt{5} & 0 & 5 \end{bmatrix}.$$

This matrix $P$ is a diagonalizing matrix for $A$ (which happens to be orthogonal), and so $P^T AP = P^{-1}AP = diag(\lambda_1, \lambda_2, \lambda_3) = diag(0, 1, 6)$.

The fact that the eigenvectors in Example 3 were orthogonal is no coincidence: The fact that the matrix is *symmetric* forces them to be orthogonal. To see why, we need the following fact about symmetric matrices.

**LEMMA 2**  If $A$ is a symmetric $n \times n$ matrix in $\mathbb{R}^n$, then $(AX) \cdot Y = X \cdot (AY)$ for all columns $X$ and $Y$ in $\mathbb{R}^n$.

**PROOF**  We have $A^T = A$ so $(AX) \cdot Y = (AX)^T Y = X^T A^T Y = X^T (AY) = X \cdot (AY)$.

**THEOREM 6**  If $A$ is a symmetric matrix, then eigenvectors of $A$ corresponding to distinct eigenvalues are orthogonal.

**PROOF**  Suppose that $AX = \lambda X$ and $AY = \mu Y$ where $\lambda \neq \mu$ are distinct eigenvalues. Then Lemma 2 gives

$$\lambda(X \cdot Y) = (\lambda X) \cdot Y = (AX) \cdot Y = X \cdot (AY) = X \cdot (\mu Y) = \mu(X \cdot Y).$$

Hence $(\lambda - \mu)(X \cdot Y) = 0$, so $X \cdot Y = 0$ because $\lambda \neq \mu$.

If $\lambda$ is an eigenvalue of an $n \times n$ matrix $A$, write

$$E_\lambda(A) = \{X \text{ in } \mathbb{R}^n \mid AX = \lambda X\}.$$

This is a subspace of $\mathbb{R}^n$ called the eigenspace of $A$ corresponding to $\lambda$, and the eigenvectors corresponding to $\lambda$ are just the nonzero members of $E_\lambda(A)$.[16] In fact, $E_\lambda(A)$ is the null space of the matrix $\lambda I - A$:

$$E_\lambda(A) = \{X \text{ in } \mathbb{R}^n \mid (\lambda I - A)X = 0\} = null(\lambda I - A),$$

so (by Theorem 6 §4.4) the basic solutions of the linear system $(\lambda I - A)X = 0$ form a basis of $E_\lambda(A)$. Of course the eigenvalues $\lambda$ of $A$ are the roots of the characteristic polynomial $c_A(x)$ of $A$, defined by

$$c_A(x) = det(xI - A).$$

With this and Theorem 6, the process for orthogonally diagonalizing a symmetric $n \times n$ matrix is a straightforward extension of the diagonalization procedure given in Section 2.3.

**Orthogonal Diagonalization Algorithm**

To diagonalize an $n \times n$ symmetric matrix $A$:

**Step 1.**  Find the distinct eigenvalues $\lambda$.

**Step 2.**  Obtain a basis for each eigenspace $E_\lambda(A)$ as the basic solutions of $(\lambda I - A)X = 0$.

**Step 3.**  Obtain an orthonormal basis of each $E_\lambda(A)$ using the Gram-Schmidt procedure if necessary.

**Step 4.**  The set of all the eigenvectors obtained in Step 3 is an orthonormal basis of $\mathbb{R}^n$ by Theorem 5.

**Step 5.**  If $P$ is the orthogonal matrix with the basis in Step 4 as its columns, then $P^T A P$ is diagonal.

**Example 4**  Orthogonally diagonalize the symmetric matrix

$$A = \begin{bmatrix} 5 & -2 & 4 \\ -2 & 8 & 2 \\ 4 & 2 & 5 \end{bmatrix}.$$

**SOLUTION**  The characteristic polynomial is $c_A(x) = x(x - 9)^2$, so the eigenvalues are $\lambda_1 = 0$ and $\lambda_2 = 9$. Bases for the eigenspaces are

---

[16] Note that $E_\lambda(A)$ is a subspace for *any* number $\lambda$; the eigenvalues are the numbers $\lambda$ for which $E_\lambda(A) \neq 0$.

$$E_{\lambda_1}(A) = span\{[2 \ \ 1 \ \ -2]^T\} \quad \text{and} \quad E_{\lambda_2}(A) = span\{[1 \ \ -2 \ \ 0]^T, \ [1 \ \ 0 \ \ 1]^T\}.$$

Note that $[2 \ \ 1 \ \ -2]^T$ from $E_{\lambda_1}(A)$ is orthogonal to both vectors $[1 \ \ -2 \ \ 0]^T$ and $[1 \ \ 0 \ \ 1]^T$ in $E_{\lambda_2}(A)$, as Theorem 6 guarantees, but this basis of $E_{\lambda_2}(A)$ is not orthogonal. Applying the Gram-Schmidt algorithm yields an orthogonal basis: $E_{\lambda_2}(A) = span\{[1 \ \ -2 \ \ 0]^T, \ [4 \ \ 2 \ \ 5]^T\}$. Hence $\{[2 \ \ 1 \ \ -2]^T, \ [1 \ \ -2 \ \ 0]^T, \ [4 \ \ 2 \ \ 5]^T\}$ is an orthogonal basis of $\mathbb{R}^3$ consisting of eigenvectors of $A$, whence normalization gives the orthonormal basis (principal axes for $A$)

$$\left\{ \tfrac{1}{3}[2 \ \ 1 \ \ -2]^T, \ \tfrac{1}{\sqrt{5}}[1 \ \ -2 \ \ 0]^T, \ \tfrac{1}{3\sqrt{5}}[4 \ \ 2 \ \ 5]^T \right\}.$$

Hence $P = \begin{bmatrix} \frac{2}{3} & \frac{1}{\sqrt{5}} & \frac{4}{3\sqrt{5}} \\ \frac{1}{3} & \frac{-2}{\sqrt{5}} & \frac{2}{3\sqrt{5}} \\ \frac{-2}{3} & 0 & \frac{5}{3\sqrt{5}} \end{bmatrix} = \frac{1}{3\sqrt{5}} \begin{bmatrix} 2\sqrt{5} & 3 & 4 \\ \sqrt{5} & -6 & 2 \\ -2\sqrt{5} & 0 & 5 \end{bmatrix}$ is an orthogonal matrix such that $P^T A P = P^{-1} A P = diag(0, \ 9, \ 9)$.

### 4.7.3   Triangulation

If we substitute "*upper triangular*" for "*diagonal*" in the Principal Axis theorem, we can weaken the requirement that $A$ is symmetric to asking only that all the eigenvalues are real. The proof is virtually the same.

**THEOREM 7**

**Triangulation Theorem.** If $A$ is a square matrix with real eigenvalues, there exists an orthogonal matrix $P$ such that $P^T A P$ is upper triangular.

**PROOF**

If $A$ is an $n \times n$ matrix with real eigenvalues, the proof of $(3) \Rightarrow (1)$ of Theorem 5 goes through except that $B$ need not be zero and $D_1$ is upper triangular, not diagonal. We leave the details to the reader.

Unfortunately, Theorem 7 provides no systematic way to find the matrix $P$. An algorithm exists which is a refinement of the method for orthogonal diagonalization.[17] Even so, the theorem is useful as it stands and reveals important information about the determinant and trace of a matrix.

Here the **trace** $trA$ of an $n \times n$ matrix $A$ is defined to be the sum of the main diagonal elements of $A$, in other words:

If $A = [a_{ij}]$, then $trA = a_{11} + a_{22} + \cdots + a_{nn}$.

It is evident that $tr(A + B) = trA + trB$ and that $tr(cA) = c \, trA$ holds for all $n \times n$ matrices $A$ and $B$, and all scalars $c$. The following fact is more surprising.

**LEMMA 3**

Let $A$ and $B$ be $n \times n$ matrices. Then:

(1) $tr(AB) = tr(BA)$.

(2) If $P$ is an invertible matrix, then $tr(P^{-1}AP) = trA$.

**PROOF**

Write $A = [a_{ij}]$ and $B = [b_{ij}]$. Then the $(i, \ i)$-entry of $AB$ is $d_i = a_{i1}b_{1i} + a_{i2}b_{2i} + \cdots + a_{in}b_{ni} = \Sigma_j a_{ij} b_{ji}$. Hence

$$tr(AB) = d_1 + d_2 + \cdots + d_n = \Sigma_i d_i = \Sigma_i (\Sigma_j a_{ij} b_{ji}).$$

Similarly we have $tr(BA) = \Sigma_i (\Sigma_j b_{ij} a_{ji})$. Since these two double sums are the same, this proves (1). Then (2) follows from (1) because $tr(P^{-1}AP) = tr[P^{-1}(AP)] = tr[(AP)P^{-1}] = trA$.

---

[17] See W.K. Nicholson, *Linear Algebra with Applications*, 4th ed., McGraw-Hill Ryerson, Section 8.7.

Our reason for looking at Lemma 3 here is that it gives the first part of

**THEOREM** 8 — Let $A$ be an $n \times n$ matrix with real eigenvalues $\lambda_1$, $\lambda_2$, $\cdots$, $\lambda_n$ (possibly not all distinct). Then
$$tr A = \lambda_1 + \lambda_2 + \cdots + \lambda_n \quad \text{and} \quad det A = \lambda_1 \lambda_2 \cdots \lambda_n.$$

**PROOF** By Theorem 7, let $P$ be a matrix (in fact, orthogonal) such that $P^{-1}AP = U$ is upper triangular. Then $tr A = tr U$ (by Lemma 3) and $det A = det U$. The result now follows because $U$ has the same eigenvalues as $A$ (Theorem 6 §2.3), so they appear on the main diagonal of $U$.

## 4.7.4 Diagonalization Revisited

There are two results in Section 2.3 (Theorems 4 and 5) that could not be proved using the methods available there. In this concluding section we use our present techniques to derive (and, in one case, extend) these theorems.

Theorem 4 §2.3 asserts (without proof) that an $n \times n$ matrix $A$ is diagonalizable if it has $n$ distinct eigenvalues. The next result explains (and extends) this by revealing an important connection between eigenvalues and linear independence: Eigenvectors corresponding to distinct eigenvalues are necessarily linearly independent.

**THEOREM** 9 — Let $X_1$, $X_2$, $\cdots$, $X_k$ be eigenvectors corresponding to distinct eigenvalues $\lambda_1$, $\lambda_2$, $\cdots$, $\lambda_k$ of an $n \times n$ matrix $A$. Then $\{X_1, X_2, \cdots, X_k\}$ is a linearly independent set.

**PROOF** We use induction on $k$. If $k = 1$, then $\{X_1\}$ is independent because $X_1 \neq 0$. In general, suppose the theorem is true for some $k > 0$. Given eigenvectors $\{X_1, X_2, \cdots, X_{k+1}\}$, suppose a linear combination vanishes:
$$t_1 X_1 + t_2 X_2 + \cdots + t_{k+1} X_{k+1} = 0. \tag{$*$}$$
We must show that each $t_i = 0$. Left multiply $(*)$ by $A$ and use the fact that $AX_i = \lambda_i X_i$ to get
$$t_1 \lambda_1 X_1 + t_2 \lambda_2 X_2 + \cdots + t_{k+1} \lambda_{k+1} X_{k+1} = 0. \tag{$**$}$$
If we multiply $(*)$ by $\lambda_1$ and subtract the result from $(**)$, the first terms cancel and we obtain
$$t_2(\lambda_2 - \lambda_1)X_2 + t_3(\lambda_3 - \lambda_1)X_3 + \cdots + t_{k+1}(\lambda_{k+1} - \lambda_1)X_{k+1} = 0.$$
Since $\{X_2, X_3, \cdots, X_{k+1}\}$ is independent by the induction hypothesis, this gives
$$t_2(\lambda_2 - \lambda_1) = 0, \ t_3(\lambda_3 - \lambda_1) = 0, \ \cdots, \ t_{k+1}(\lambda_{k+1} - \lambda_1) = 0,$$
and so $t_2 = t_3 = \cdots = t_{k+1} = 0$ because the $\lambda_i$ are distinct. Hence $(*)$ becomes $t_1 X_1 = 0$, which implies that $t_1 = 0$ because $X_1 \neq 0$. This is what we wanted.

Theorem 9 completes some unfinished business from Section 2.3.

**COROLLARY** (**Theorem 4, Section 2.3.**) If $A$ is an $n \times n$ matrix with $n$ distinct eigenvalues, then $A$ is diagonalizable.

**PROOF** Choose one eigenvector for each of the $n$ distinct eigenvalues. Then these eigenvectors are independent by Theorem 9, and so are a basis of $\mathbb{R}^n$ by Theorem 3 §4.3. Now use Theorem 1.

Recall that the *multiplicity* of an eigenvalue $\lambda$ of $A$ is the number of times $\lambda$ occurs as a root of the characteristic polynomial $c_A(x)$. In other words,[18] the multiplicity of $\lambda$ is

---

[18] See Appendix A.3.

the largest integer $m \geq 1$ such that $c_A(x) = (x - \lambda)^m g(x)$ for some polynomial $g(x)$. The assertion (without proof) in Theorem 5 §2.3 is that a square matrix is diagonalizable if and only if the multiplicity of each eigenvalue $\lambda$ equals $dim[E_\lambda(A)]$. We are going to prove this, and the proof requires the following result which is valid for *any* square matrix, diagonalizable or not.

**LEMMA 4**    Let $\lambda$ be an eigenvalue of multiplicity $m$ of a square matrix $A$. Then $dim[E_\lambda(A)] \leq m$.

**PROOF**    Write $dim[E_\lambda(A)] = d$. By the definition of multiplicity, it suffices to show that $c_A(x) = (x - \lambda)^d g(x)$ for some polynomial $g(x)$. To this end, let $\{X_1, X_2, \cdots, X_d\}$ be a basis of $E_\lambda(A)$. Then Lemma 1 shows that an invertible $n \times n$ matrix $P$ exists such that

$$P^{-1}AP = \begin{bmatrix} \lambda I_d & B \\ 0 & A_1 \end{bmatrix}$$

in block form, where $I_d$ denotes the $d \times d$ identity matrix. Now write $A' = P^{-1}AP$ and observe that $c_{A'}(x) = c_A(x)$ by Theorem 6 §2.3. But Theorem 3 §2.2 gives

$$c_{A'}(x) = det(xI_n - A')$$
$$= det \begin{bmatrix} (x-\lambda)I_d & -B \\ 0 & xI_{n-d} - A_1 \end{bmatrix}$$
$$= det[(x - \lambda)I_d] det[xI_{n-d} - A_1]$$
$$= (x - \lambda)^d c_{A_1}(x).$$

Hence $c_A(x) = c_{A'}(x) = (x - \lambda)^d g(x)$ where $g(x) = c_{A_1}(x)$. This is what we wanted.

It is impossible to ignore the question when equality holds in Lemma 4 for each eigenvalue $\lambda$. It turns out that this characterizes the diagonalizable matrices. This is the second piece of unfinished business from Section 2.3.

**THEOREM 10**    (**Theorem 5, Section 2.3.**) The following are equivalent for a square matrix $A$:

(1) $A$ is diagonalizable.
(2) $dim[E_\lambda(A)]$ equals the multiplicity of $\lambda$ for every eigenvalue $\lambda$ of $A$.

**PROOF**    Let $\lambda_1, \lambda_2, \cdots, \lambda_k$ be the distinct eigenvalues of $A$ and, for each $i$, let $m_i$ denote the multiplicity of $\lambda_i$ and write $d_i = dim[E_{\lambda_i}(A)]$. Then $c_A(x) = (x - \lambda_1)^{m_1}(x - \lambda_2)^{m_2} \cdots (x - \lambda_k)^{m_k}$ so $m_1 + \cdots + m_k = n$. Moreover, $d_i \leq m_i$ for each $i$ by Lemma 4.

$(1) \Rightarrow (2)$. By (1), $\mathbb{R}^n$ has a basis of $n$ eigenvectors of $A$, so let $t_i$ of them lie in $E_{\lambda_i}(A)$ for each $i$. Then $t_i \leq d_i$ for each $i$ and so

$$n = t_1 + \cdots + t_k \leq d_1 + \cdots + d_k \leq m_1 + \cdots + m_k = n.$$

Hence $d_1 + \cdots + d_k = m_1 + \cdots + m_k$ so, since $d_i \leq m_i$ for each $i$, we must have $d_i = m_i$. This is (2).

$(2) \Rightarrow (1)$. Let $B_i$ denote a basis of $E_{\lambda_i}(A)$ for each $i$, and let $B = B_1 \cup \cdots \cup B_k$. Since each $B_i$ contains $m_i$ vectors by (2), and since the $B_i$ are pairwise disjoint (the $\lambda_i$ are distinct), it follows that $B$ contains $n$ vectors. So it suffices to show that $B$ is linearly independent (then $B$ is a basis of $\mathbb{R}^n$). Suppose a linear combination of the vectors in $B$ vanishes, and let $Y_i$ denote the sum of all terms that come from $B_i$. Then $Y_i$ lies in $E_{\lambda_i}(A)$ for each $i$, so the nonzero $Y_i$ are independent by Theorem 2 (as the $\lambda_i$ are distinct). Since the sum of the $Y_i$ is zero, it follows that $Y_i = 0$ for each $i$. Hence all coefficients of terms in $Y_i$ are zero (because $B_i$ is independent). This shows that $B$ is independent.

Examples of how Theorem 3 is used can be found in Section 2.3.

## Exercises 4.7

1. In each case either show that the statement is true or give an example showing that it is false.
   a) Every eigenvalue of a matrix has a unit eigenvector (that is one of length 1).
   b) If $P$ has orthogonal columns and $det\ P = \pm 1$, then $P$ is orthogonal.
   c) Every orthogonal matrix is invertible.
   d) Every orthogonal matrix is symmetric.
   e) Every diagonal matrix is orthogonal.
   f) If $A$ is symmetric, then $P^T A P$ is also symmetric for all matrices $P$.

2. By normalizing the columns, make each of the following matrices orthogonal.

   a) $\begin{bmatrix} 2 & -2 \\ 1 & 4 \end{bmatrix}$
   b) $\begin{bmatrix} 3 & 4 \\ -4 & 3 \end{bmatrix}$

   c) $\begin{bmatrix} \cos\theta & -\sin\theta & 0 \\ \sin\theta & \cos\theta & 0 \\ 0 & 0 & 3 \end{bmatrix}$
   d) $\begin{bmatrix} 1 & 0 & 2 \\ -1 & 1 & 1 \\ 1 & 1 & -1 \end{bmatrix}$

   e) $\begin{bmatrix} -1 & 2 & 2 \\ 2 & -1 & 2 \\ 2 & 2 & -1 \end{bmatrix}$
   f) $\begin{bmatrix} 2 & 3 & -6 \\ 6 & 2 & 3 \\ -3 & 6 & 2 \end{bmatrix}$

3. If the first two rows of an orthogonal matrix are $\begin{bmatrix} \frac{1}{3} & \frac{2}{3} & \frac{2}{3} \end{bmatrix}$ and $\begin{bmatrix} \frac{2}{3} & \frac{-2}{3} & \frac{1}{3} \end{bmatrix}$, find all possible third rows.

4. If $\{X_1, X_2, \cdots, X_n\}$ is an orthonormal basis of $\mathbb{R}^n$, find all $Y$ in $\mathbb{R}^n$ such that $\{Y, X_2, \cdots, X_n\}$ is an orthonormal basis of $\mathbb{R}^n$. [Hint: Expand $Y$ in terms of $\{X_1, X_2, \cdots, X_n\}$ by the Expansion theorem.]

5. For each matrix $A$, find an orthogonal matrix $P$ such that $P^T A P$ is diagonal.

   a) $A = \begin{bmatrix} 3 & 0 & 0 \\ 0 & 2 & 2 \\ 0 & 2 & 5 \end{bmatrix}$
   b) $A = \begin{bmatrix} 3 & 0 & 7 \\ 0 & 5 & 0 \\ 7 & 0 & 3 \end{bmatrix}$

   c) $A = \begin{bmatrix} 3 & 3 & 0 \\ 3 & 3 & 0 \\ 0 & 0 & 6 \end{bmatrix}$
   d) $A = \begin{bmatrix} 1 & 3 & 0 \\ 3 & 13 & 2 \\ 0 & 2 & 1 \end{bmatrix}$

   e) $A = \begin{bmatrix} 3 & 5 & 0 & 0 \\ 5 & 3 & 0 & 0 \\ 0 & 0 & 1 & 7 \\ 0 & 0 & 7 & 1 \end{bmatrix}$
   f) $A = \begin{bmatrix} 3 & 5 & -1 & 1 \\ 5 & 3 & 1 & -1 \\ -1 & 1 & 3 & 5 \\ 1 & -1 & 5 & 3 \end{bmatrix}$

6. If $P$ is an orthogonal matrix and $c$ is a scalar, show that $cP$ is orthogonal if and only if $c = \pm 1$.

7. If $P$ is a triangular, orthogonal matrix, show that $P$ is diagonal and each diagonal entry is 1 or $-1$. [Hint: If $P$ is $n \times n$, induct on $n$. Write $P = \begin{bmatrix} a & X \\ 0 & Q \end{bmatrix}$ in block form.]

8. An $n \times n$ matrix $P$ is called a **permutation matrix** if the columns of $P$ are just the columns of $I_n$ in some order.

   a) Show that every permutation matrix is orthogonal.
   b) Show that the rows of a permutation matrix are the rows of $I_n$ in some order.

9. Show that an $n \times n$ matrix $P$ is orthogonal if and only if $\|PX\| = \|X\|$ for every column $X$ in $\mathbb{R}^n$. [Hint: If $\|PX\| = \|X\|$, replace $X$ by $X + Y$ to show that $(PX) \cdot (PY) = X \cdot Y$ for all $X$ and $Y$.]

10. Let $A$ be an $m \times n$ matrix with columns $C_1, C_2, \cdots, C_n$. Explain why the $(i, j)$-entry of $A^T A$ is $C_i \cdot C_j$.

11. Show that the following are equivalent for an $m \times n$ matrix $A$:
    i) $A$ has orthogonal columns.
    ii) $A = PD$ where $P$ has orthonormal columns and $D$ is diagonal and invertible.
    iii) $A^T A$ is invertible and diagonal.
    [Hint: For (i)$\Rightarrow$(ii) take $D = diag(\|C_1\|, \|C_2\|, \cdots, \|C_n\|)$ where $C_j$ is column $j$ of $A$. Exercise 10 is useful for **iii)**$\Rightarrow$**i)**.]

12. Show that the following are equivalent for a symmetric matrix $A$:
    i) $A$ is orthogonal.
    ii) $A^2 = I$.
    iii) The only eigenvalues of $A$ are $\pm 1$.
    [Hint: For **iii)**$\Rightarrow$**ii)** use the Principal Axis theorem.]

13. If the columns $C_1, C_2, \cdots, C_n$ of the $n \times n$ matrix $A = [a_{ij}]$ are orthogonal, show that the $(i, j)$-entry of $A^{-1}$ is $\frac{a_{ji}}{\|C_i\|^2}$.
    [Hint: Use Exercise 10 to show that $A^T A = D = diag(\|C_1\|^2, \|C_2\|^2, \cdots, \|C_n\|^2)$.]

14. If $A$ is symmetric, show that $\lambda \geq 0$ for every eigenvalue of $A$ if and only if $A = B^2$ for some symmetric matrix $B$. [Hint: If $P^T A P = D = diag(\lambda_1, \cdots, \lambda_n)$ where $P^{-1} = P^T$ and each $\lambda_i > 0$, take $B = P\ diag(\sqrt{\lambda_1}, \cdots, \sqrt{\lambda_n})P^T$.]

15. Extend Theorem 6 as follows: Let $\lambda$ and $\mu$ be distinct eigenvectors of a square matrix $A$, and hence of $A^T$. If $X$ is an eigenvector of $A$ corresponding to $\lambda$ and $Y$ is an eigenvector of $A^T$ corresponding to $\mu$, show that $X$ and $Y$ are orthogonal.

16. Matrices $A$ and $B$ are called **orthogonally similar** (written $A \overset{\circ}{\sim} B$) if $B = P^T A P$ for some orthogonal matrix $P$. Hence the Principal Axis theorem asserts that a matrix $A$ is symmetric if and only if $A \overset{\circ}{\sim} D$ for some diagonal matrix $D$.
    a) Show: (i) $A \overset{\circ}{\sim} A$ for all $A$; (ii) If $A \overset{\circ}{\sim} B$ then $B \overset{\circ}{\sim} A$; (iii) If $A \overset{\circ}{\sim} B$ and $B \overset{\circ}{\sim} C$, then $A \overset{\circ}{\sim} C$.
    b) If $A \overset{\circ}{\sim} B$, show that $A^T \overset{\circ}{\sim} B^T$.
    c) If $A$ and $B$ are invertible and $A \overset{\circ}{\sim} B$, show that $A^{-1} \overset{\circ}{\sim} B^{-1}$.
    d) If $A \overset{\circ}{\sim} B$, show that $A$ and $B$ have the same eigenvalues.
    [Hint: Theorem 6 §2.3.]
    e) If $A$ and $B$ are symmetric, show that $A \overset{\circ}{\sim} B$ if and only if $A$ and $B$ have the same eigenvalues.

**17.** Let $A$ be an $n \times n$ matrix. If $(AX) \cdot Y = X \cdot (AY)$ holds for all columns $X$ and $Y$ in $\mathbb{R}^n$, show that $A$ is symmetric. (Converse to Lemma 2.) [Hint: Show that $(AX) \cdot Y = X \cdot (A^T Y)$ holds for all columns $X$ and $Y$ in $\mathbb{R}^n$. If $X^T BY = 0$ for all columns $X$ and $Y$, show that $B = 0$ by taking $X$ and $Y$ to be various columns of $I_n$.]

**18.** A square matrix $E$ is called a projection matrix if $E^2 = E = E^T$. (See Exercise 20 §4.6.)
   **a)** If $E$ is a projection matrix, show that $P = I - 2E$ is orthogonal and symmetric.
   **b)** If $P$ is an orthogonal, symmetric matrix, show that $E = \frac{1}{2}(I - P)$ is a projection matrix.
   **c)** If $Q$ has orthonormal columns, show that $E = QQ^T$ is a projection matrix.

**19.** Show that every $2 \times 2$ orthogonal matrix has the form $\begin{bmatrix} \cos\theta & -\sin\theta \\ \sin\theta & \cos\theta \end{bmatrix}$ or $\begin{bmatrix} \cos\theta & \sin\theta \\ \sin\theta & -\cos\theta \end{bmatrix}$ for some angle $\theta$. [Hint: See the discussion preceding Theorem 7 §3.5.]

**20.** A square matrix $A$ is said to be **nilpotent** if $A^k = 0$ for some $k \geq 1$.
   **a)** If $N$ is a square upper triangular matrix with 0's on the main diagonal, show that $N$ is nilpotent. [Hint: Induct on $n$ where $A$ is $n \times n$.]
   **b)** If $A$ is nilpotent, show that $P^T AP$ has the form in **a)** for some orthogonal matrix $P$. [Hint: Theorem 7.]

   **c)** If $A$ is a square matrix with real eigenvalues, show that $A = S + N$ for some symmetric matrix $S$ and some nilpotent matrix $N$. [Hint: Theorem 7 and **a)**.]

**21. a)** If $A$ is nilpotent, show that every eigenvalue of $A$ is zero.
   **b)** If every eigenvalue of $A$ is zero, use Theorem 7 to show that $A$ is nilpotent.
   **c)** Show that the only nilpotent, diagonalizable matrix $A$ is $A = 0$.

**22.** If $A$ is a square matrix with real eigenvalues, show that $Q^T AQ$ is lower triangular for some orthogonal matrix $Q$.

**23.** Let $P = \begin{bmatrix} a & -b \\ b & a \end{bmatrix}$ and $Q = \begin{bmatrix} a & -b & -c & -d \\ b & a & -d & c \\ c & d & a & -b \\ d & -c & b & a \end{bmatrix}$. Compute $P^T P$ and $Q^T Q$ and, for each of $P$ and $Q$, give a necessary and sufficient condition that the matrix is orthogonal. (Matrices of the form $P$ are a representation of the complex numbers—identify $a + bi = \begin{bmatrix} a & -b \\ b & a \end{bmatrix}$. Matrices of the form $Q$ are related in the same way to the **quaternions**, an extension of the complex numbers.)

**24.** If $C$ is an $n \times n$ matrix with its nonzero columns orthogonal, show that $C = PD$ where $P$ is orthogonal and $D$ is diagonal with nonnegative entries.

## 4.8   QUADRATIC FORMS [19]

The Principal Axis theorem states that every symmetric matrix can be orthogonally diagonalized. This result has far reaching consequences, some of which are investigated in this section. We begin with quadratic forms which provide a geometrical view (and the name!) of the Principal Axis theorem. This leads to the study of positive definite matrices, and to the Cholesky decomposition. Then we look at constrained optimization where a quantity is maximized (or minimized) subject to a condition on the variables.

### 4.8.1   Quadratic Forms

The equation of a curve in the plane can often be made simpler by a change of coordinates, and the same is true for the equation of a surface in $\mathbb{R}^3$. In this section we show how to use the Principal Axis theorem to accomplish this for graphs of quadratics in $n$ variables.

**Example 1**   Consider the graph of the equation $x_1 x_2 = 1$ as shown in the diagram. The curve is clearly symmetric about the line $x_2 = x_1$. Hence we are going to look at the equation of this curve in terms of new variables $y_1$ and $y_2$, where the $Y_1$-axis is the line $x_2 = x_1$. To do this, we obtain the $Y_1$- and $Y_2$-axes from the $X_1$- and $X_2$-axes by a counterclockwise rotation of $\frac{\pi}{4}$. Recall that the matrix of the rotation through an angle $\theta$ is $\begin{bmatrix} \cos\theta & -\sin\theta \\ \sin\theta & \cos\theta \end{bmatrix}$

---
[19] This section can be omitted without loss of continuity.

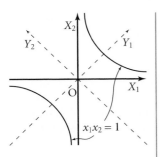

$x_1x_2 = 1$

by Theorem 3 §3.4. Hence, if the vector $[x_1 \ x_2]^T$ is rotated through $\frac{\pi}{4}$ the resulting vector $[y_1 \ y_2]^T$ is given by

$$
\begin{bmatrix} y_1 \\ y_2 \end{bmatrix} = \begin{bmatrix} \cos(\frac{\pi}{4}) & -\sin(\frac{\pi}{4}) \\ \sin(\frac{\pi}{4}) & \cos(\frac{\pi}{4}) \end{bmatrix} \begin{bmatrix} x_1 \\ x_2 \end{bmatrix}
$$

$$
= \begin{bmatrix} \frac{1}{\sqrt{2}} & -\frac{1}{\sqrt{2}} \\ \frac{1}{\sqrt{2}} & \frac{1}{\sqrt{2}} \end{bmatrix} \begin{bmatrix} x_1 \\ x_2 \end{bmatrix}
$$

$$
= \frac{1}{\sqrt{2}} \begin{bmatrix} x_1 - x_2 \\ x_1 + x_2 \end{bmatrix}.
$$

The matrix $\begin{bmatrix} \frac{1}{\sqrt{2}} & -\frac{1}{\sqrt{2}} \\ \frac{1}{\sqrt{2}} & \frac{1}{\sqrt{2}} \end{bmatrix}$ is orthogonal, so

$$
\begin{bmatrix} x_1 \\ x_2 \end{bmatrix} = \begin{bmatrix} \frac{1}{\sqrt{2}} & -\frac{1}{\sqrt{2}} \\ \frac{1}{\sqrt{2}} & \frac{1}{\sqrt{2}} \end{bmatrix}^T \begin{bmatrix} y_1 \\ y_2 \end{bmatrix} = \frac{1}{\sqrt{2}} \begin{bmatrix} y_1 + y_2 \\ -y_1 + y_2 \end{bmatrix}.
$$

Hence $x_1 = \frac{1}{\sqrt{2}}(y_1 + y_2)$ and $x_2 = \frac{1}{\sqrt{2}}(-y_1 + y_2)$, so the equation $x_1x_2 = 1$ becomes $y_2^2 - y_1^2 = 2$ in terms of these new variables. In this form we recognize the equation as that of a hyperbola in the $Y_1$-$Y_2$-coordinate system. The new $Y_1$- and $Y_2$-axes are called the **principal axes** of the hyperbola. This is the source of the name Principal Axis theorem.

Note that we chose the new $Y_1$- and $Y_2$-axes in Example 1 by observing a symmetry of the graph of the original equation $x_1x_2 = 1$. While this is still possible for equations in three variables (the graph is then a surface in $\mathbb{R}^3$), this type of geometric insight is not available when more than three variables are involved. The approach we adopt is as follows: The change of variables in Example 1 carried the equation $x_1x_2 = 1$ to the equation $y_2^2 - y_1^2 = 2$ which has no cross term $y_1y_2$. It is in this form that the Principal Axis theorem applies to a much wider class of equations and actually *reveals* symmetries in their graphs.

An expression like $x_1^2 - x_2^2 + 2x_3^2 + 3x_1x_2 - x_1x_3$ is an example of a quadratic form in the variables $x_1$, $x_2$ and $x_3$. In general, a **quadratic form** $q$ in the $n$ variables $x_1$, $x_2$, $\cdots$, $x_n$ is a linear combination of the terms $x_1^2$, $x_2^2$, $\cdots$, $x_n^2$, $x_1x_2$, $x_1x_3$, $x_2x_3$, $\cdots$:

$$
q = r_1x_1^2 + r_2x_2^2 + \cdots + r_nx_n^2 + r_{12}x_1x_2 + r_{13}x_1x_3 + r_{23}x_2x_3 + \cdots
$$

where the coefficients $r_i$ and $r_{ij}$ are real numbers. Even though $q$ is not a linear function of the $x_i$, it can be easily described using matrix multiplication. Indeed if we write $X = [x_1 \ x_2 \ \cdots \ x_n]^T$, $q$ can be written in **matrix form**

$$
q = X^TAX \text{ where } A \text{ is a } symmetric \ n \times n \text{ matrix.}
$$

In fact, the main diagonal entries of the matrix $A$ are the coefficients (possibly zero) of $x_1^2$, $x_2^2$, and $x_n^2$ in order and, if $i \neq j$, the $(i, \ j)$- and $(j, \ i)$-entries of $A$ are both equal to half the coefficient of $x_ix_j$. Here is an example.

**Example 2**    Write $q = x_1^2 + 3x_3^2 + 2x_1x_2 - x_1x_3$ in matrix form $q = X^TAX$ where $A$ is symmetric.

SOLUTION    Applying the above rule, we get

$$
q = [x_1 \ x_2 \ x_3] \begin{bmatrix} 1 & 1 & -\frac{1}{2} \\ 1 & 0 & 0 \\ -\frac{1}{2} & 0 & 3 \end{bmatrix} \begin{bmatrix} x_1 \\ x_2 \\ x_3 \end{bmatrix}.
$$

The reader should verify that this really works.

The reason for writing quadratic forms as in Example 2 is that the Principal Axis theorem can be applied to the symmetric matrix $A$. Suppose that a quadratic form $q$ is written in matrix form as

$$q = X^T A X$$

where $A$ is a symmetric $n \times n$ matrix and $X = [x_1 \ x_2 \ \cdots \ x_n]^T$. We want to find new variables $Y = [y_1 \ y_2 \ \cdots \ y_n]^T$ such that, when $q$ is written in terms of $y_1, \ y_2, \ \cdots, \ y_n$, there are no cross terms $y_i y_j$ where $i \neq j$. This amounts to asking that $q = Y^T D Y$ where $D$ is a *diagonal* matrix. Since $A$ is symmetric, the Principal Axis theorem guarantees the existence of an orthogonal matrix $P$ such that

$$P^T A P = D = \begin{bmatrix} \lambda_1 & 0 & \cdots & 0 \\ 0 & \lambda_2 & \cdots & 0 \\ \vdots & \vdots & \ddots & \vdots \\ 0 & 0 & \cdots & \lambda_n \end{bmatrix}$$

where the $\lambda_i$ are the (real but not necessarily distinct) eigenvalues of the symmetric matrix $A$. Note that $P^{-1} = P^T$. Now define the new variables $Y$ by

$$Y = P^T X \quad \text{equivalently} \quad X = PY.$$

Then substitution in the equation $q = X^T A X$ gives

$$q = (PY)^T A (PY) = Y^T (P^T A P) Y = Y^T D Y = \lambda_1 y_1^2 + \lambda_2 y_2^2 + \cdots + \lambda_n y_n^2.$$

Hence this change of variables has produced the desired simplification of $q$ (and we say that the quadratic form $q$ has been **diagonalized**). The following theorem summarizes this discussion.

**THEOREM 1**

Let $q = X^T A X$ be a quadratic form where $A$ is a symmetric $n \times n$ matrix and $X = [x_1 \ x_2 \ \cdots \ x_n]^T$. Let $P$ be an orthogonal matrix such that $P^T A P = diag(\lambda_1, \ \lambda_2, \ \cdots, \ \lambda_n)$ where the $\lambda_i$ are the (real) eigenvalues of $A$ repeated according to their multiplicities. Define new variables $Y = [y_1 \ y_2 \ \cdots \ y_n]^T$ by

$$Y = P^T X \quad \text{equivalently} \quad X = PY.$$

Then

$$q = \lambda_1 y_1^2 + \lambda_2 y_2^2 + \cdots + \lambda_n y_n^2$$

in terms of these new variables $y_i$.

**Example 3**

The quadratic form $q = x_1 x_2$ in Example 1 has matrix form $q = X^T A X$ where $X = [x_1 \ x_2]^T$ and $A = \begin{bmatrix} 0 & \frac{1}{2} \\ \frac{1}{2} & 0 \end{bmatrix}$. The eigenvalues are $-\frac{1}{2}$ and $\frac{1}{2}$, with corresponding eigenvectors $\begin{bmatrix} 1 \\ -1 \end{bmatrix}$ and $\begin{bmatrix} 1 \\ 1 \end{bmatrix}$, respectively. They are (of course) orthogonal and, after normalizing, the orthogonal matrix $P$ is $P = \frac{1}{\sqrt{2}} \begin{bmatrix} 1 & 1 \\ -1 & 1 \end{bmatrix}$. Hence the new variables are $\begin{bmatrix} y_1 \\ y_2 \end{bmatrix} = Y = P^T X = \frac{1}{\sqrt{2}} \begin{bmatrix} 1 & -1 \\ 1 & 1 \end{bmatrix} \begin{bmatrix} x_1 \\ x_2 \end{bmatrix}$. Thus $y_1 = \frac{1}{\sqrt{2}}(x_1 - x_2)$ and $y_2 = \frac{1}{\sqrt{2}}(x_1 + x_2)$ as in Example 1.

Note that this "diagonalizing" change of variables is not unique. For example, if we take $[-1 \ 1]^T$ as the eigenvector corresponding to the eigenvalue $-\frac{1}{2}$, the resulting formulas are $y_1' = \frac{1}{\sqrt{2}}(-x_1 + x_2)$ and $y_2' = \frac{1}{\sqrt{2}}(x_1 + x_2)$. Then $y_2'$ is the same as $y_2$, but $y_1'$ is the negative of the preceding $y_1$. Geometrically, this amounts to pointing the $y_1$-axis in the opposite direction in Example 1.

**Example 4**  Diagonalize the quadratic form $q = x_1^2 + 2x_2^2 + x_3^2 - 2x_1x_2 - 2x_2x_3$.

**SOLUTION**  Here $q = X^TAX$ where $A = \begin{bmatrix} 1 & -1 & 0 \\ -1 & 2 & -1 \\ 0 & -1 & 1 \end{bmatrix}$ and $X = \begin{bmatrix} x_1 \\ x_2 \\ x_3 \end{bmatrix}$. The eigenvalues of $A$ are $\lambda_1 = 0$,

$\lambda_2 = 1$, and $\lambda_3 = 3$, with (orthogonal) eigenvectors $G_1 = \begin{bmatrix} 1 & 1 & 1 \end{bmatrix}^T$, $G_2 = \begin{bmatrix} -1 & 0 & 1 \end{bmatrix}^T$,

and $G_3 = \begin{bmatrix} 1 & -2 & 1 \end{bmatrix}^T$ respectively. Normalizing gives the principal axes $F_1 = \frac{1}{\sqrt{3}}G_1$,

$F_2 = \frac{1}{\sqrt{2}}G_2$, and $F_3 = \frac{1}{\sqrt{6}}G_3$, and so we obtain an orthogonal matrix

$$P = [F_1 \ F_2 \ F_3] = \begin{bmatrix} \frac{1}{\sqrt{3}} & -\frac{1}{\sqrt{2}} & \frac{1}{\sqrt{6}} \\ \frac{1}{\sqrt{3}} & 0 & -\frac{2}{\sqrt{6}} \\ \frac{1}{\sqrt{3}} & \frac{1}{\sqrt{2}} & \frac{1}{\sqrt{6}} \end{bmatrix}$$

such that $P^TAP = diag(\lambda_1, \lambda_2, \lambda_3) = diag(0, 1, 3)$. Thus the new variables
$[y_1 \ y_2 \ y_3]^T = Y = P^TX$ are

$$y_1 = \frac{1}{\sqrt{3}}(x_1 + x_2 + x_3), \ y_2 = \frac{1}{\sqrt{2}}(-x_1 + x_3), \text{ and } y_3 = \frac{1}{\sqrt{6}}(x_1 - 2x_2 + x_3).$$

If we compute $X = PY$ and substitute in $q$, we diagonalize the quadratic form in the sense that $q = \lambda_1y_1^2 + \lambda_2y_2^2 + \lambda_3y_3^2 = y_2^2 + 3y_3^2$.

These methods shed light on graphs of the form $q = 1$ where $q$ is a quadratic form in two variables $x_1$ and $x_2$ (as in Example 1). The graph of the equation $rx_1^2 + sx_2^2 = 1$ is called an **ellipse** if $rs > 0$ and a **hyperbola** if $rs < 0$. Our theory guarantees that every equation $ax_1^2 + bx_1x_2 + cx_2^2 = 1$ can be transformed into such a diagonal form by a change of coordinates. The next theorem shows that this can always be achieved by a rotation (as in Example 1), and it also gives a simple way to decide from the coefficients $a$, $b$, and $c$ whether it is an ellipse or a hyperbola.

**THEOREM 2**  Consider the quadratic form $q = ax_1^2 + bx_1x_2 + cx_2^2$.
(1) There is a counterclockwise rotation of the coordinate axes about the origin such that $q$ has the form $q = ry_1^2 + sy_2^2$ in the new variables $y_1$ and $y_2$.
(2) The graph of the equation $ax_1^2 + bx_1x_2 + cx_2^2 = 1$ is an ellipse if $b^2 - 4ac < 0$ and it is a hyperbola if $b^2 - 4ac > 0$.

**PROOF**  The matrix of $q$ is $A = \begin{bmatrix} a & \frac{1}{2}b \\ \frac{1}{2}b & c \end{bmatrix}$. Let $\lambda_1$ and $\lambda_2$ be the eigenvalues of $A$ with

orthonormal eigenvectors $X_1$ and $X_2$. Hence $P = [X_1 \ X_2]$ is an orthogonal matrix and $P^TAP = diag(\lambda_1, \lambda_2)$. We have $detP = \pm 1$ so, by interchanging $X_1$ and $X_2$ if necessary, we can ensure that $detP = 1$ and hence (by Theorem 7 §3.4) that multiplication by $P$ is a rotation of $\mathbb{R}^2$. Since the new coordinates $y_1$ and $y_2$ are given by $[y_1 \ y_2]^T = P^T[x_1 \ x_2]^T$, the new axes are obtained by a rotation (since $P^T$ is also orthogonal and $det(P^T) = 1$). This proves (1). Furthermore, the fact that $P^T = P^{-1}$ gives

$$\lambda_1\lambda_2 = det\begin{bmatrix} \lambda_1 & 0 \\ 0 & \lambda_2 \end{bmatrix} = det(P^TAP) = detA = -\frac{1}{4}(b^2 - 4ac).$$

Since $q = \lambda_1y_1^2 + \lambda_2y_2^2$ in the new coordinate system, the graph of the equation $q = 1$ is an ellipse if $\lambda_1\lambda_2 > 0$ (that is $b^2 - 4ac < 0$), and it is a hyperbola if $\lambda_1\lambda_2 < 0$ (that is $b^2 - 4ac > 0$). This proves (2).

There are many ways to diagonalize a given quadratic form $q$. For example, consider $q = 3x_1^2 + 4x_1x_2 + 2x_2^2$. Then $q = X^TAX$ where $A = \begin{bmatrix} 3 & 2 \\ 2 & 2 \end{bmatrix}$. This matrix has eigenvalues $\lambda_1 = \frac{1}{2}(5 + \sqrt{17})$ and $\lambda_2 = \frac{1}{2}(5 - \sqrt{17})$, and so $q$ can be written in the form $q = \lambda_1 y_1^2 + \lambda_2 y_2^2$ for appropriate new variables $y_1$ and $y_2$. On the other hand, $q$ can be written as follows:

$$q = x_1^2 + 2(x_1 + x_2)^2 \quad \text{and} \quad q = 3(x_1 + \tfrac{2}{3}x_2)^2 + \tfrac{2}{3}x_2^2.$$

These formulas reveal two other changes of variables that diagonalize $q$:

$$\begin{aligned} y'_1 &= x_1 \\ y'_2 &= x_1 + x_2 \end{aligned} \quad \text{and} \quad \begin{aligned} y''_1 &= x_1 + \tfrac{2}{3}x_2 \\ y''_2 &= x_2 \end{aligned}.$$

The general problem of how such changes of variables are related has been studied in great detail; however we shall not pursue this here.

## 4.8.2 Positive Definite Matrices

If $A$ is an $n \times n$ symmetric matrix, the quadratic form $q = X^TAX$ can be viewed as a function $q : \mathbb{R}^n \to \mathbb{R}$ given by $q(X) = X^TAX$. If $n = 2$ and $X = [x_1 \ x_2]^T$, then $q = q(x_1, x_2)$ is a function of two variables and we can look at the graph of $q$ in $\mathbb{R}^3$. Three examples are shown in the diagram.

In this section we examine the important special case where the graph stays strictly above the $X_1$-$X_2$-plane except at the origin (as in the first example in the diagram). Moreover, we treat the general case where $q = q(x_1, x_2, \cdots, x_n)$ is a function of $n$ variables.

If $q = q(X) = X^TAX$ is a quadratic form on $\mathbb{R}^n$, it is clear that $q(0) = 0$, and $q$ is said to be a **positive definite form** if $q(X) > 0$ whenever $X \neq 0$. For example, it is clear that $q(x_1, x_2) = x_1^2 + 2x_2^2$ is positive definite; however it is sometimes less clear.

**Example 5**    If $A = \begin{bmatrix} 2 & -1 \\ -1 & 1 \end{bmatrix}$ and $X = \begin{bmatrix} x_1 \\ x_2 \end{bmatrix}$, show that the corresponding form

$q = X^TAX = 2x_1^2 + x_2^2 - 2x_1x_2$ is positive definite.

**SOLUTION**    We can write $q = x_1^2 + (x_1 - x_2)^2$, so the only way that $q$ can be zero is if $x_1 = 0$ and $x_1 - x_2 = 0$; that is, if $X = 0$.

The Principal Axis theorem provides the following important characterization of these positive definite forms $q = X^TAX$ in terms of the (real) eigenvalues of the symmetric matrix $A$.

**THEOREM 3**  If $A$ is a symmetric matrix, the quadratic form $q(X) = X^TAX$ is positive definite if and only if every eigenvalue of $A$ is positive.

PROOF  By the Principal Axis theorem let $P$ be an orthogonal matrix such that $P^TAP = D = diag(\lambda_1, \lambda_2, \cdots, \lambda_n)$ where the $\lambda_i$ are the eigenvalues of $A$. Suppose $\lambda_i > 0$ for each $i$. If $X \neq 0$, write $Y = P^TX = [y_1 \ y_2 \ \cdots \ y_n]^T$. Then $Y \neq 0$ because $P^T$ is invertible. Since $X = PY$, this means that

$$q(X) = X^TAX = Y^T(P^TAP)Y = Y^TDY = \lambda_1 y_1^2 + \lambda_2 y_2^2 + \cdots + \lambda_n y_n^2 > 0$$

because each $\lambda_i > 0$ and some $y_i \neq 0$. Conversely, if $q(X) > 0$ for all $X \neq 0$, let $X_j = PE_j$ where $E_j$ is column $j$ of $I_n$. Then $0 < q(X_j) = X_j^TAX_j = E_j^T(P^TAP)E_j = E_j^TDE_j = \lambda_j$, as required.

Motivated by Theorem 3, a square matrix is called **positive definite** if it is symmetric and all its eigenvalues are positive. Hence Theorem 3 asserts that a symmetric matrix $A$ is positive definite if and only if the corresponding quadratic form $q(X) = X^TAX$ is a positive definite form. In particular, Example 5 shows that $\begin{bmatrix} 2 & -1 \\ -1 & 1 \end{bmatrix}$ is a positive definite matrix.

COROLLARY  Every positive definite matrix $A$ is invertible; indeed, $det A > 0$.

PROOF  Since $A$ is symmetric, we have $P^TAP = D = diag(\lambda_1, \lambda_2, \cdots, \lambda_n)$ for some orthogonal matrix $P$. Hence $det A = det D = \lambda_1 \lambda_2 \cdots \lambda_n$, so $det D > 0$ because each $\lambda_i > 0$.

These positive definite matrices arise frequently when optimization (maximum or minimum) problems are encountered, and so have applications throughout Science and Engineering. They are also used in factor analysis in statistics, and we will encounter them when discussing general inner products in Chapter 5.

**Example 6**  If $U$ is any invertible $n \times n$ matrix, show that $A = U^TU$ is positive definite.

Let $q(X) = X^TAX$ be the quadratic form associated with $A$. If $X \neq 0$ in $\mathbb{R}^n$, then $q(X) = X^TAX = X^T(U^TU)X = (UX)^T(UX) = \|UX\|^2 > 0$ because $UX \neq 0$ ($U$ is invertible).

If $A$ is an $m \times n$ matrix, it follows from Theorem 3 §4.4 that the $n \times n$ symmetric matrix $A^TA$ is positive definite if and only if $A$ has linearly independent columns (see Exercise 12).

It is remarkable that the converse to Example 6 is also true. In fact every positive definite matrix $A$ can be factored as $A = U^TU$ where $U$ is an upper triangular matrix with positive elements on the main diagonal. However, before verifying this, we introduce another concept that is central to any discussion of positive definite matrices.

If $A$ is any $n \times n$ matrix, let $^rA$ denote the $r \times r$ submatrix of $A$ in the upper left corner of $A$; that is $^rA$ is the matrix obtained from $A$ by deleting the last $n - r$ rows and columns. The matrices $^1A$, $^2A$, $^3A$, $\cdots$, $^nA = A$ are called the **principal submatrices** of $A$.

**Example 7**  If $A = \begin{bmatrix} 10 & 5 & 2 \\ 5 & 3 & 2 \\ 2 & 2 & 3 \end{bmatrix}$ then $^1A = [10]$, $^2A = \begin{bmatrix} 10 & 5 \\ 5 & 3 \end{bmatrix}$ and $^3A = A$.

**LEMMA 1**    Let $A$ be $n \times n$. If $A$ is positive definite, so also is each principal submatrix $^rA$ for $r = 1, 2, \cdots, n$.

**PROOF**    Write $A = \begin{bmatrix} ^rA & P \\ Q & R \end{bmatrix}$ in block form. If $Y \neq 0$ in $\mathbb{R}^r$, write $X = \begin{bmatrix} Y \\ 0 \end{bmatrix}$ in $\mathbb{R}^n$. Then $X \neq 0$ so the fact that $A$ is positive definite gives

$$0 < X^TAX = [Y^T \ \ 0] \begin{bmatrix} ^rA & P \\ Q & R \end{bmatrix} \begin{bmatrix} Y \\ 0 \end{bmatrix} = Y^T (^rA) Y.$$

This shows that $^rA$ is positive definite.[20]

If $A$ is positive definite, Lemma 1 and the Corollary to Theorem 3 show that $det(^rA) > 0$ for every $r$. This proves part of the following theorem which contains the converse to Example 6, and characterizes the positive definite matrices among the symmetric ones.

**THEOREM 4**    The following conditions are equivalent for a symmetric $n \times n$ matrix $A$:

    **(1)** $A$ is positive definite.
    **(2)** $det(^rA) > 0$ for each $r = 1, 2, \cdots, n$.
    **(3)** $A = U^TU$ where $U$ is an upper triangular matrix with positive entries on the main diagonal.

Furthermore, the factorization in (3) is unique (called the **Cholesky Factorization** of $A$).

**PROOF**    $(3) \Rightarrow (1)$ is Example 6, and $(1) \Rightarrow (2)$ follows from Lemma 1 and the Corollary to Theorem 3.

$(2) \Rightarrow (3)$. Assume (2) and proceed by induction on $n$. If $n = 1$, then $A = [a]$ where $a > 0$ by (2), so take $U = [\sqrt{a}]$. If $n > 1$, write $B = {}^{(n-1)}A$. Then $B$ is symmetric and satisfies (2) so, by induction, we have $B = U^TU$ as in (3) where $U$ is of size $(n-1) \times (n-1)$. Then, as $A$ is symmetric, it has block form $A = \begin{bmatrix} B & P \\ P^T & a \end{bmatrix}$ where $P$ is a column in $\mathbb{R}^{n-1}$ and $a$ is in $\mathbb{R}$. If we write $X = (U^T)^{-1}P$ and $b = a - X^TX$, block multiplication gives

$$A = \begin{bmatrix} U^TU & P \\ P^T & a \end{bmatrix} = \begin{bmatrix} U^T & 0 \\ X^T & 1 \end{bmatrix} \begin{bmatrix} U & X \\ 0 & b \end{bmatrix},$$

as the reader can verify. Taking determinants and applying Theorem 3 §2.2 gives $detA = det(U^T) \, detU \cdot b = b(detU)^2$. Hence $b > 0$ because $detA > 0$ by (2). But then the above factorization can be written $A = \begin{bmatrix} U^T & 0 \\ X^T & \sqrt{b} \end{bmatrix} \begin{bmatrix} U & X \\ 0 & \sqrt{b} \end{bmatrix}$. Since $U$ has positive diagonal entries, this is a Cholesky factorization of $A$. This proves (3).

As to the uniqueness, suppose that $A = U^TU = U_1^TU_1$ are two Cholesky factorizations. Write $D = U U_1^{-1} = (U^T)^{-1}U_1^T$. Then $D$ is upper triangular (because $D = U U_1^{-1}$) and lower triangular (because $D = (U^T)^{-1}U_1^T$) and so is a diagonal matrix. Thus $U = DU_1$ and $U_1 = DU$ (since $U_1^T = U^TD$), so it suffices to show that $D = I$. But eliminating $U_1$ gives $U = D^2U$, so $D^2 = I$ because $U$ is invertible. Since the diagonal entries of $D$ are positive (this is true of $U$ and $U_1$), it follows that $D = I$.

The remarkable thing is that the matrix $U$ in the Cholesky factorization is easy to obtain from $A$ using row operations. The key is that Step 1 of the following algorithm is possible for any positive definite matrix $A$. A proof of the algorithm is given following Example 8.

---

[20] A similar argument shows that, if $B$ is any matrix obtained from a positive definite matrix $A$ by deleting certain rows and deleting the *same* columns, then $B$ is also positive definite.

**Algorithm for the Cholesky Factorization**. If $A$ is a positive definite matrix, the Cholesky factorization $A = U^T U$ can be obtained as follows:

**Step 1.** Carry $A$ to an upper triangular matrix $U_1$ with positive diagonal entries by row operations each of which adds a multiple of a row to a lower row.

**Step 2.** Obtain $U$ from $U_1$ by dividing each row of $U_1$ by the square root of the diagonal entry in that row.

**Example 8**    Find the Cholesky factorization of $A = \begin{bmatrix} 10 & 5 & 2 \\ 5 & 3 & 2 \\ 2 & 2 & 3 \end{bmatrix}$.

**SOLUTION**    The matrix $A$ is positive definite by Theorem 4 because $det\,^1(A) = 10 > 0$, $det\,^2(A) = 5 > 0$, and $det\,^3(A) = detA = 3 > 0$. Hence Step 1 of the Algorithm is carried out as follows:

$$A = \begin{bmatrix} 10 & 5 & 2 \\ 5 & 3 & 2 \\ 2 & 2 & 3 \end{bmatrix} \rightarrow \begin{bmatrix} 10 & 5 & 2 \\ 0 & \frac{1}{2} & 1 \\ 0 & 1 & \frac{13}{5} \end{bmatrix} \rightarrow \begin{bmatrix} 10 & 5 & 2 \\ 0 & \frac{1}{2} & 1 \\ 0 & 0 & \frac{3}{5} \end{bmatrix} = U_1.$$

Now carry out Step 2 on $U_1$ to obtain $U = \begin{bmatrix} \sqrt{10} & \frac{5}{\sqrt{10}} & \frac{2}{\sqrt{10}} \\ 0 & \frac{1}{\sqrt{2}} & \sqrt{2} \\ 0 & 0 & \frac{\sqrt{3}}{\sqrt{5}} \end{bmatrix}$. The reader can verify that $U^T U = A$.

**Proof of the Cholesky Algorithm**

If $A$ is positive definite, let $A = U^T U$ be the Cholesky factorization, and let $D = diag(d_1, \cdots, d_n)$ be the common diagonal of $U$ and $U^T$. Then $U^T D^{-1}$ is lower triangular with ones on the diagonal (call such matrices LT-1). Hence $L = (U^T D^{-1})^{-1}$ is also LT-1, and so $I_n \rightarrow L$ by a sequence of row operations each of which adds a multiple of a row to a lower row (verify; modify columns right to left). But then $A \rightarrow LA$ by the same sequence of row operations (see the discussion preceding Theorem 1 §1.6). Since $LA = [D(U^T)^{-1}][U^T U] = DU$ is upper triangular with positive entries on the diagonal, this shows that Step 1 of the Algorithm is possible.

Turning to Step 2, let $A \rightarrow U_1$ as in Step 1 so that $U_1 = L_1 A$ where $L_1$ is LT-1. Since $A$ is symmetric, we get

$$L_1 U_1^T = L_1 (L_1 A)^T = L_1 A^T L_1^T = L_1 A L_1^T = U_1 L_1^T. \tag{$*$}$$

Let $D_1 = diag(e_1, \cdots, e_n)$ where $e_1, \cdots, e_n$ denote the diagonal entries of $U_1$. Then $(*)$ gives $L_1(U_1^T D_1^{-1}) = U_1 L_1^T D_1^{-1}$, and this is both upper triangular (right side) and LT-1 (left side), and so must equal $I_n$. In particular, $U_1^T D_1^{-1} = L_1^{-1}$. Now let $D_2 = diag(\sqrt{e_1}, \cdots, \sqrt{e_n})$, so that $D_2^2 = D_1$. If we write $U = D_2^{-1} U_1$, we have

$$U^T U = (U_1^T D_2^{-1})(D_2^{-1} U_1) = U_1^T (D_2^2)^{-1} U_1 = (U_1^T D_1^{-1}) U_1 = (L_1^{-1}) U_1 = A.$$

This proves Step 2 because $U = D_2^{-1} U_1$ is formed by dividing each row of $U_1$ by the square root of its diagonal entry (verify).

### 4.8.3   Constrained Optimization

It is a frequent occurrence in applications that a function $q = q(x_1, x_2, \cdots, x_n)$ of $n$ variables, called an **objective function**, is to be made as large or as small as possible among all vectors $X = [x_1 \ x_2 \ \cdots \ x_n]^T$ lying in a certain region of $\mathbb{R}^n$ called the **feasible region**. A wide variety of objective functions $q$ arise in practice; our primary concern here is to examine one important situation where $q$ is a quadratic form. The next example gives some indication of how such problems arise.

**Example 9**

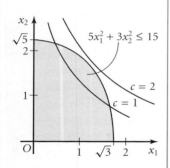

A politician proposes to spend $x_1$ dollars annually on health care and $x_2$ dollars annually on education. She is constrained in her spending by various budget pressures, and one model of this is that the expenditures $x_1$ and $x_2$ should satisfy a constraint like

$$5x_1^2 + 3x_2^2 \leq 15.$$

Since $x_i \geq 0$ for each $i$, the feasible region is the shaded area shown in the diagram. Any choice of feasible point $[x_1 \ x_2]^T$ in this region will satisfy the budget constraints. However, these choices have different effects on voters, and the politician wants to choose $X = [x_1 \ x_2]^T$ to maximize some measure $q = q(x_1, x_2)$ of voter satisfaction. Thus the assumption is that, for any value of $c$, all points on the graph of $q(x_1, x_2) = c$ have the same appeal to voters. Hence the goal is to find the largest value of $c$ for which the graph of $q(x_1, x_2) = c$ contains a feasible point.

The choice of the function $q$ depends upon many factors; we will show how to solve the problem for any quadratic form $q$ (even with more than two variables). In the diagram the function $q$ is given by

$$q(x_1, x_2) = x_1 x_2,$$

and the graphs of $q(x_1, x_2) = c$ are shown for $c = 1$ and $c = 2$. As $c$ increases the graph of $q(x_1, x_2) = c$ moves up and to the right. From this it is clear that there will be a solution for some value of $c$ between 1 and 2 (in fact the largest value is $c = \frac{1}{2}\sqrt{15} = 1.94$ to two decimal places).

The constraint $5x_1^2 + 3x_2^2 \leq 15$ in Example 9 can be put in a standard form. If we divide through by 15, it becomes $\left(\frac{x_1}{\sqrt{3}}\right)^2 + \left(\frac{x_2}{\sqrt{5}}\right)^2 \leq 1$. This suggests that we introduce new variables $Y = [y_1 \ y_2]^T$ where $y_1 = \frac{x_1}{\sqrt{3}}$ and $y_2 = \frac{x_2}{\sqrt{5}}$. Then the constraint becomes $\|Y\|^2 \leq 1$, equivalently $\|Y\| \leq 1$. In terms of these new variables, the objective function is $q = \sqrt{15}y_1 y_2$, and we want to maximize this subject to $\|Y\| \leq 1$. When this is done, the maximizing values of $x_1$ and $x_2$ are obtained from $x_1 = \sqrt{3}\,y_1$ and $x_2 = \sqrt{5}\,y_2$.

Hence, for constraints like that in Example 9, there is no real loss in generality in assuming that the constraint takes the form $\|X\| \leq 1$. In this case the Principal Axis theorem solves the problem. Recall that a vector in $\mathbb{R}^n$ of length 1 is called a *unit vector*.

**THEOREM 5**

Consider the quadratic form $q = q(X) = X^T A X$ where $A$ is an $n \times n$ symmetric matrix, and let $\lambda_1$ and $\lambda_n$ denote the largest and smallest eigenvalues of $A$, respectively. Then:

(1) $max\{q(X) \mid \|X\| \leq 1\} = \lambda_1$, and $q(F_1) = \lambda_1$ where $F_1$ is any unit eigenvector corresponding to $\lambda_1$.

(2) $min\{q(X) \mid \|X\| \leq 1\} = \lambda_n$, and $q(F_n) = \lambda_n$ where $F_n$ is any unit eigenvector corresponding to $\lambda_n$.

**PROOF**

Since $A$ is symmetric, let the (real) eigenvalues $\lambda_i$ of $A$ be ordered as to size as follows: $\lambda_1 \geq \lambda_2 \geq \cdots \geq \lambda_n$. By the Principal Axis Theorem, let $P$ be an orthogonal matrix such that $P^T A P = D = diag(\lambda_1, \lambda_2, \cdots, \lambda_n)$. Define $Y = P^T X$, equivalently $X = PY$, and observe that $\|Y\| = \|X\|$ because $\|Y\|^2 = Y^T Y = X^T(PP^T)X = X^T X = \|X\|^2$. If we write $Y = [y_1 \ y_2 \ \cdots \ y_n]^T$, then

$$\begin{aligned} q(X) = q(PY) &= (PY)^T A(PY) \\ &= Y^T(P^T A P)Y = Y^T D Y \\ &= \lambda_1 y_1^2 + \lambda_2 y_2^2 + \cdots + \lambda_n y_n^2 \,. \end{aligned} \qquad (**)$$

Now assume that $\|X\| \leq 1$. Since $\lambda_i \leq \lambda_1$ for each $i$, $(**)$ gives

$$q(X) = \lambda_1 y_1^2 + \lambda_2 y_2^2 + \cdots + \lambda_n y_n^2 \leq \lambda_1 y_1^2 + \lambda_1 y_2^2 + \cdots + \lambda_1 y_n^2 = \lambda_1 \|Y\|^2 \leq \lambda_1$$

because $\|Y\| = \|X\| \leq 1$. This shows that $q(X)$ cannot exceed $\lambda_1$ when $\|X\| \leq 1$. To see that this maximum is actually achieved, let $F_1$ be a unit eigenvector corresponding to $\lambda_1$. Then

$$q(F_1) = F_1^T A F_1 = F_1^T(\lambda_1 F_1) = \lambda_1(F_1^T F_1) = \lambda_1 \|F_1\|^2 = \lambda_1.$$

Hence $\lambda_1$ is the maximum value of $q(X)$ when $\|X\| \leq 1$, proving (1). The proof of (2) is analogous.

The set of all vectors $X$ in $\mathbb{R}^n$ such that $\|X\| \leq 1$ is called the **unit ball**. If $n = 2$, it is often called the unit disk and consists of the unit circle and its interior; if $n = 3$, it is the unit sphere and its interior. It is worth noting that the maximum value of a quadratic form $q(X)$ as $X$ ranges *throughout* the unit ball is (by Theorem 5) actually attained for a unit vector $X$ on the *boundary* of the unit ball.

Theorem 5 is important for applications involving vibrations in areas as diverse as aerodynamics and particle physics, and the maximum and minimum values in the theorem are often found using advanced calculus to minimize the quadratic form on the unit ball. The algebraic approach using the Principal Axis theorem gives a geometrical interpretation of the optimal values because they are eigenvalues.

**Example 10**    Maximize and minimize the form $q(X) = 3x_1^2 + 14x_1x_2 + 3x_2^2$ subject to $\|X\| \leq 1$.

SOLUTION    The matrix of $q$ is $A = \begin{bmatrix} 3 & 7 \\ 7 & 3 \end{bmatrix}$, with eigenvalues $\lambda_1 = 10$ and $\lambda_2 = -4$, and corresponding unit eigenvectors $F_1 = \frac{1}{\sqrt{2}}[1 \ \ 1]^T$ and $F_2 = \frac{1}{\sqrt{2}}[1 \ \ {-1}]^T$. Hence, among all unit vectors $X$ in $\mathbb{R}^2$, $q(X)$ takes its maximal value 10 at $X = F_1$, and the minimum value of $q(X)$ is $-4$ when $X = F_2$.

As noted above, the objective function in a constrained optimization problem need not be a quadratic form. We conclude with an example where the objective function is linear, and the feasible region is determined by linear constraints.

**Example 11**    A manufacturer makes $x_1$ units of product 1, and $x_2$ units of product 2, at a profit of \$70 and \$50 per unit respectively, and wants to choose $x_1$ and $x_2$ to maximize the total profit $p(x_1, x_2) = 70x_1 + 50x_2$. However $x_1$ and $x_2$ are not arbitrary; for example, $x_1 \geq 0$ and $x_2 \geq 0$. Other conditions also come into play.

Each unit of product 1 costs \$1200 to produce and requires 2000 square metres of warehouse space; each unit of product 2 costs \$1300 to produce and requires 1100 square metres of space. If the total warehouse space is 11 300 square metres, and if the total production budget is \$8700, $x_1$ and $x_2$ must also satisfy the conditions

$$2000x_1 + 1100x_2 \leq 11\,300,$$
$$1200x_1 + 1300x_2 \leq 8700.$$

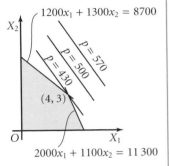

The feasible region in the plane satisfying these constraints (and $x_1 \geq 0$, $x_2 \geq 0$) is shaded in the diagram. If the profit equation $70x_1 + 50x_2 = p$ is plotted for various values of $p$, the resulting lines are parallel, with $p$ increasing with distance from the origin. Hence the best choice occurs for the line $70x_1 + 50x_2 = 430$ that touches the shaded region at the point $(4, 3)$. So the profit $p$ has a maximum of $p = 430$ for $x_1 = 4$ units and $x_2 = 3$ units.

Example 11 is a simple case of the general **linear programming** problem[21] which arises in economic, management, network, and scheduling applications. Here the objective function is a linear combination $q = a_1x_1 + a_2x_2 + \cdots + a_nx_n$ of the variables, and the feasible region consists of the vectors $X = [x_1 \ \ x_2 \ \ \cdots \ \ x_n]^T$ in $\mathbb{R}^n$ which satisfy a set of linear inequalities of the form $b_1x_1 + b_2x_2 + \cdots + b_nx_n \leq b$. There is a good method (an extension of the Gaussian algorithm) called the **simplex algorithm** for finding the maximum and minimum values of $q$ when $X$ ranges over such a feasible set. As Example 11 suggests, the optimal values turn out to be vertices of the feasible set. In particular, they are on the boundary of the feasible region, as is the case in Theorem 5.

## 4.8.4    Statistical Principal Component Analysis

Linear algebra is important in multivariate analysis in statistics, and we conclude with a very short look at one application of diagonalization in this area. A main feature of probability and statistics is the idea of a *random variable X*, that is a real valued function which takes its values according to a probability law (called its *distribution*). Random variables occur in a wide variety of contexts; examples include the number of meteors falling per square kilometre in a given region, the price of a share of Microsoft, or the duration of a long distance telephone call from a certain city.

The values of a random variable $X$ are distributed about a central number $\mu$, called the *mean* of $X$. The mean can be calculated from the distribution as the *expectation* $E(X) = \mu$ of the random variable $X$. Functions of a random variable are again random variables. In particular, $(X - \mu)^2$ is a random variable, and the *variance* of the random variable $X$, denoted $var(X)$, is defined to be the number

$$var(X) = E\{(X - \mu)^2\} \text{ where } \mu = E(X).$$

It is not difficult to see that $var(X) \geq 0$ for every random variable $X$. The number $\sigma = \sqrt{var(X)}$ is called the *standard deviation* of $X$, and is a measure of how much the values of $X$ are spread about the mean $\mu$ of $X$. A main goal of statistical inference is finding reliable methods for estimating the mean and the standard deviation of a random variable $X$ by sampling the values of $X$.

If two random variables $X$ and $Y$ are given, and their joint distribution is known, then functions of $X$ and $Y$ are also random variables. In particular, $X + Y$ and $aX$ are random variables for any real number $a$, and we have

$$E(X + Y) = E(X) + E(Y) \quad \text{and} \quad E(aX) = aE(X). \text{ [22]}$$

An important question is how much the random variables $X$ and $Y$ depend on each other. One measure of this is the *covariance* of $X$ and $Y$, denoted $cov(X, \ Y)$, defined by

$$cov(X, \ Y) = E\{(X - \mu)(Y - v)\} \text{ where } \mu = E(X) \text{ and } v = E(Y).$$

Clearly, $cov(X, X) = var(X)$. If $cov(X, Y) = 0$ then $X$ and $Y$ have little relationship to each other and are said to be *uncorrelated*.[23]

Multivariate statistical analysis deals with a family $X_1, \ X_2, \ \cdots, \ X_n$ of random variables with means $\mu_i = E(X_i)$ and variances $\sigma_i^2 = var(X_i)$ for each $i$. We denote the covariance of $X_i$ and $X_j$ by $\sigma_{ij} = cov(X_i, \ X_j)$. Then the *covariance matrix* of the random variables $X_1, \ X_2, \ \cdots, \ X_n$ is defined to be the $n \times n$ matrix

$$\Sigma = [\sigma_{ij}]$$

---

[21] A good introduction can be found at http://www.mcgrawhill.ca/college/nicholson, and more information is available in "Linear Programming and Extensions" by N. Wu and R. Coppins, McGraw-Hill, 1981.

[22] Because of this we say that $E(\ )$ is a linear transformation from the vector space of all random variables to the space of real numbers. In Section 5.3 we will have more to say about such general linear transformations.

[23] If $X$ and $Y$ are independent in the sense of probability theory, then they are uncorrelated; however, the converse is not true in general.

whose $(i, j)$-entry is $\sigma_{ij}$. The matrix $\Sigma$ is clearly symmetric; in fact it can be shown that $\Sigma$ is **positive semidefinite** in the sense that $\lambda \geq 0$ for every eigenvalue $\lambda$ of $\Sigma$. (In reality, $\Sigma$ is positive definite in most cases of interest.) So suppose that the eigenvalues of $\Sigma$ are $\lambda_1 \geq \lambda_2 \geq \cdots \geq \lambda_n \geq 0$, with corresponding eigenvectors $\overline{E}_1, \overline{E}_2, \cdots, \overline{E}_n$.[24] If $P = [\overline{E}_1 \ \overline{E}_2 \ \cdots \ \overline{E}_n]$, the Principal Axis theorem shows that $P$ is an orthogonal matrix and

$$P^T \Sigma P = diag(\lambda_1, \lambda_2, \cdots, \lambda_n).$$

If we write $\overline{X} = [X_1 \ X_2 \ \cdots \ X_n]^T$, the procedure for diagonalizing a quadratic form (Theorem 1) gives new variables $\overline{Y} = [Y_1 \ Y_2 \ \cdots \ Y_n]^T$ defined by

$$\overline{Y} = P^T \overline{X}. \qquad (***)$$

These new random variables $Y_1, Y_2, \cdots, Y_n$ are called the **principal components** of the original random variables $X_i$, and are linear combinations of the $X_i$. Furthermore, it can be shown that

$$cov(Y_i, Y_j) = 0 \text{ if } i \neq j \quad \text{and} \quad var(Y_i) = \lambda_i \text{ for each } i.$$

Of course the principal components $Y_i$ point along the principal axes of the quadratic form $q = \overline{X}^T \Sigma \overline{X}$.

The sum of the variances of a set of random variables is called the **total variance** of the variables, and determining the source of this total variance is one of the benefits of principal component analysis. The fact that the matrices $\Sigma$ and $diag(\lambda_1, \lambda_2, \cdots, \lambda_n)$ are similar means that they have the same trace, that is

$$\sigma_{11} + \sigma_{22} + \cdots + \sigma_{nn} = \lambda_1 + \lambda_2 + \cdots + \lambda_n$$

This means that the principal components $Y_i$ have the same total variance as the original random variables $X_i$. Moreover, the fact that $\lambda_1 \geq \lambda_2 \geq \cdots \geq \lambda_n \geq 0$ means that most of this variance resides in the first few $Y_i$. In practice, statisticians find that studying these first few $Y_i$ (and ignoring the rest) gives an accurate analysis of the total system variability. This results in substantial data reduction since often only a few $Y_i$ suffice for all practical purposes. Furthermore, these $Y_i$ are easily obtained as linear combinations of the $X_i$. Finally, the analysis of the principal components often reveals relationships among the $X_i$ that were not previously suspected, and so results in interpretations that would not otherwise have been made.[25]

## Exercises 4.8

**1.** In each case either show that the statement is true or give an example showing that it is false. Throughout, $A$ denotes a square matrix.

**a)** Every positive definite matrix is orthogonally diagonalizable.

**b)** If all eigenvalues of $A$ are positive, then $A$ is symmetric.

**c)** Every symmetric matrix with positive determinant is positive definite.

**d)** If $A$ is positive definite and $X^T A X = 0$ where $X$ is in $\mathbb{R}^n$, then $X = 0$.

**e)** If $A$ is positive definite and $det \ A = 1$, then $A$ is orthogonal.

**f)** If $U$ is invertible and all its eigenvalues are positive, then $U$ is positive definite.

**2.** In each case, if $q = X^T B X$ for all $X$ in $\mathbb{R}^n$, find a symmetric matrix $A$ such that $q = X^T A X$ for all $X$.

**a)** $B = \begin{bmatrix} 2 & -1 & 1 \\ 1 & -1 & 3 \\ 0 & 2 & 1 \end{bmatrix}$

**b)** $B = \begin{bmatrix} 2 & 2 & 7 \\ 4 & 1 & -2 \\ -1 & 0 & 5 \end{bmatrix}$

**3.** In each case find a change of variables that will diagonalize the quadratic form $q$, and express $q$ in terms of these new variables.

**a)** $q = x_1^2 + 4x_1x_2 + x_2^2.$    **b)** $q = 2x_1^2 + 4x_1x_2 - x_2^2.$

**c)** $q = x_1^2 + 4x_1x_2 + 3x_2^2.$    **d)** $q = -x_1^2 + 4x_1x_2 + 3x_2^2.$

**e)** $q = x_1^2 + 2x_2^2 + x_3^2 - 2x_1x_2 - 2x_2x_3.$

**f)** $q = 5x_1^2 + 8x_2^2 + 5x_3^2 - 4x_1x_2 - 8x_1x_3 - 4x_2x_3.$

---

[24] In probability and statistics random variables are commonly denoted by upper case letters. Although it is in conflict with our usage of upper case letters for *matrices*, we will bow to convention and stay with the upper case notation, and denote column matrices of random variables as $\overline{X}$.

[25] A more detailed analysis of principal components can be found for example in R.A. Johnson and D.W. Wichern, *Applied Multivariate Statistical Analysis*, 4th ed., Prentice-Hall, 1998.

**4.** In each case decide if the graph of the equation is an ellipse or a hyperbola, and find a change of variables that transforms the equation to one with no cross terms.

**a)** $3x_1^2 - 4x_1x_2 = 2.$  **b)** $2x_1^2 + 4x_1x_2 + 5x_2^2 = 1.$

**5.** In each case find the maximum and minimum of $q(X) = X^TAX$ among all columns $X$ such that $\|X\| \le 1$.

**a)** $q(X) = 3x_1^2 + 4x_1x_2 + x_2^2,$ $X = [x_1 \ x_2]^T.$

**b)** $q(X) = 2x_1^2 - 5x_1x_2 - 3x_2^2,$ $X = [x_1 \ x_2]^T.$

**c)** $q(X) = 3x_1^2 - x_2^2 - 2x_3^2 - 2x_1x_2 + 8x_2x_3,$ $X = [x_1 \ x_2 \ x_3]^T.$

**d)** $q(X) = x_1^2 - 3x_2^2 - x_3^2 - 4x_1x_3 + 4x_2x_3,$ $X = [x_1 \ x_2 \ x_3]^T.$

**6.** In each case find the Cholesky factorization of the matrix $A$.

**a)** $A = \begin{bmatrix} 2 & 3 \\ 3 & 5 \end{bmatrix}$  **b)** $A = \begin{bmatrix} 3 & -2 \\ -2 & 2 \end{bmatrix}$

**c)** $A = \begin{bmatrix} 10 & 4 & 3 \\ 4 & 2 & 2 \\ 3 & 2 & 3 \end{bmatrix}$  **d)** $A = \begin{bmatrix} 5 & 3 & -2 \\ 3 & 2 & -1 \\ -2 & -1 & 4 \end{bmatrix}$

**7.** If $A$ is positive definite, show that every diagonal entry of $A$ is positive. [Hint: Consider $E_j^TAE_j$ where $E_j$ is column $j$ of the identity matrix.]

**8. a)** If $A$ is symmetric and invertible, show that $A^2$ is positive definite. [Hint: Show that $X^TA^2X = \|AX\|^2$ for all $X$ in $\mathbb{R}^n$.]

**b)** If $A = \begin{bmatrix} 3 & 2 \\ 2 & 1 \end{bmatrix}$, show that $A^2$ is positive definite but that $A$ is not.

**c)** If $A$ is positive definite, show that $A^k$ is positive definite for all $k \ge 1$. [Hint: $A$ is symmetric and invertible. Use induction on $k$.]

**d)** If $A^k$ is positive definite where $k$ is odd, show that $A$ is positive definite.

**9.** If $A$ is positive definite, show that $A^{-1}$ is also positive definite. [Hint: Theorem 3.]

**10.** If $A$ and $B$ are positive definite $n \times n$ matrices, show that $A + B$ is positive definite.

**11.** If $A$ and $B$ are positive definite matrices (possibly different sizes), show that $\begin{bmatrix} A & 0 \\ 0 & B \end{bmatrix}$ is positive definite. [Hint: See the proof of Lemma 1.]

**12.** Let $A$ denote an $m \times n$ matrix.

**a)** Show that $A^TA$ is positive definite if and only if $A$ has independent columns. [Hint: Theorem 3 §4.4.]

**b)** Show that $\lambda \ge 0$ for every eigenvalue $\lambda$ of $A^TA$.

**13.** If $A$ is an $n \times n$ positive definite matrix, and if $U$ is an $n \times m$ matrix of rank $m$, show that $U^TAU$ is positive definite. [Hint: Theorem 3 §4.4.]

**14.** If $A$ is positive definite, show that $A = B^2$ where $B$ is positive definite. [Hint: Let $P^TAP = D = diag(\lambda_1, \cdots, \lambda_n)$, $\lambda_i > 0$, and consider $D_0 = diag(\sqrt{\lambda_1}, \cdots, \sqrt{\lambda_n})$.]

**15.** If $A$ is positive definite, show that $A = CC^T$ for some matrix $C$ with orthogonal columns. [Hint: Let $P^TAP = D = diag(\lambda_1, \cdots, \lambda_n)$, $\lambda_i > 0$. Write $D_0 = diag(\sqrt{\lambda_1}, \cdots, \sqrt{\lambda_n})$ and take $C = PD_0$.]

**16.** Call a square matrix **unit triangular** if it is triangular with 1's on the main diagonal.

**a)** Suppose that an invertible matrix $A$ can be factored as $A = LDU$ where $L$ is unit lower triangular, $U$ is unit upper triangular, and $D$ is diagonal with positive diagonal entries. Show that this factorization is unique: that is, if $A = L_1D_1U_1$ is another such factorization, show that $L_1 = L, D_1 = D$ and $U_1 = U$. [Hint: If $L_1D_1U_1 = LDU$, scrutinize $L^{-1}L_1D_1 = DUU_1^{-1}$.]

**b)** Show that a matrix $A$ is positive definite if and only if it has a factorization of the form $A = U^TDU$, which is a special form of **a)**. [Hint: Cholesky.]

**17.** Consider the equation $ax_1^2 + bx_1x_2 + cx_2^2 = 1$ where $b \ne 0$. Write $X = [x_1 \ x_2]^T$, and let $Y = [y_1 \ y_2]^T$ denote the new variables obtained by rotating through an angle $\theta$, so that (Theorem 3 §3.5) $Y = RX$ where $R = \begin{bmatrix} \cos\theta & -\sin\theta \\ \sin\theta & \cos\theta \end{bmatrix}$. If $\theta$ is an angle such that $\cos 2\theta = \frac{c-a}{\sqrt{b^2+(a-c)^2}}$ and $\sin 2\theta = \frac{b}{\sqrt{b^2+(a-c)^2}}$, show that the resulting equation has no cross term. [Hint: $R$ is orthogonal, so $X = R^TY$. Substitute into the equation $ax_1^2 + bx_1x_2 + cx_2^2 = 1$.]

**18.** Let $q = X^TAX$ be a quadratic form where $A$ is an $n \times n$ symmetric matrix.

**a)** Show that $A$ is uniquely determined by the function $q$; that is, if $X^TBX = X^TAX$ holds for all columns $X$ in $\mathbb{R}^n$, show that $B = A$. [Hint: If $C = B - A$ show that $X^TCX = 0$ for all $X$. Replace $X$ by $X + Y$ to conclude that $X^YCY = 0$ for all $X$ and $Y$. Then consider the columns of the identity matrix.]

**b)** If new variables $Y$ are defined by $Y = U^{-1}X$ where $U$ is an invertible matrix, show that $q$ is given in terms of the new variables by $q = Y^TBY$ where $B = U^TAU$.

**c)** Two square matrices $A$ and $B$ are said to be **congruent** (written $A \overset{c}{\sim} B$) if $B = U^TAU$ for some invertible matrix $U$. Show that:

**i)** $A \overset{c}{\sim} A$ for all $A$; If $A \overset{c}{\sim} B$, then $B \overset{c}{\sim} A$; and, If $A \overset{c}{\sim} B$ and $B \overset{c}{\sim} C$, then $A \overset{c}{\sim} C$.

**ii)** If $A \overset{c}{\sim} B$, then $A$ is symmetric (invertible) if and only if $B$ is symmetric (invertible).

**iii)** $A$ is symmetric if and only $A \overset{c}{\sim} D$ for some diagonal matrix $D$.

**iv)** If $A \overset{c}{\sim} B$, then $rankA = rankB$ and $A^T \overset{c}{\sim} B^T$.

## 4.9 LINEAR TRANSFORMATIONS

In Section 3.4 we investigated $2 \times 2$ matrices by viewing them as transformations of the plane $\mathbb{R}^2$. Not only does this give a graphic interpretation of a matrix which is useful in analyzing the matrix, it also provides a way of studying certain geometrical properties of the plane (for example, we found that rotations about the origin correspond to simple matrices depending on the angle of rotation). In this section we do the same thing for $n \times n$ matrices, and view them as linear transformations of $\mathbb{R}^n$.

### 4.9.1 Transformations $\mathbb{R}^n \to \mathbb{R}^m$

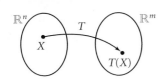

A transformation $T : \mathbb{R}^n \to \mathbb{R}^m$ is a rule[26] that associates with every vector $X$ in $\mathbb{R}^n$ a uniquely determined vector $T(X)$ in $\mathbb{R}^m$. The situation can be visualized as in the diagram. To specify a transformation $T : \mathbb{R}^n \to \mathbb{R}^m$, *every* vector $X$ in $\mathbb{R}^n$ must be assigned to a *uniquely determined* vector $T(X)$ in $\mathbb{R}^m$, called the **image** of $X$ under $T$. This is referred to as specifying the **action** of the transformation $T$, or as **defining** $T$.

Most transformations $\mathbb{R}^n \to \mathbb{R}^m$ are of little interest in linear algebra; the important ones for us are those that preserve addition and scalar multiplication. More precisely, a transformation $T : \mathbb{R}^n \to \mathbb{R}^m$ is called a **linear transformation** if it satisfies the following two properties:

**T1.** $T(X + Y) = T(X) + T(Y)$ for all $X$ and $Y$ in $\mathbb{R}^n$.
($T$ preserves addition)

**T2.** $T(aX) = a\,T(X)$ for all $X$ in $\mathbb{R}^n$ and all scalars $a$.
($T$ preserves scalar multiplication)

A linear transformation $\mathbb{R}^n \to \mathbb{R}^n$ is called a **linear operator** on $\mathbb{R}^n$.

**Example 1**

As we saw in Section 3.4, rotation about the origin, reflection about a line through the origin, and projection on a line through the origin are all examples of linear operators on $\mathbb{R}^2$.

**Example 2**

The transformation $T : \mathbb{R}^n \to \mathbb{R}^m$ given by $T(X) = 0$ for all $X$ in $\mathbb{R}^n$ is a linear transformation because $0 + 0 = 0$. It is called the **zero transformation**, and is usually denoted $T = 0$.

**Example 3**

The transformation $1_{\mathbb{R}^n} : \mathbb{R}^n \to \mathbb{R}^n$ defined by $1_{\mathbb{R}^n}(X) = X$ for all $X$ in $\mathbb{R}^n$ is a linear transformation (verify), called the **identity operator** on $\mathbb{R}^n$. More generally, if $a$ is any scalar, the operator $a : \mathbb{R}^n \to \mathbb{R}^n$ defined by $a(X) = aX$ for all $X$ in $\mathbb{R}^n$ is called a **scalar operator**. In this notation, $1_{\mathbb{R}^n} = 1$ and the zero operator is $0 : \mathbb{R}^n \to \mathbb{R}^n$.

**Example 4**

If $U$ is a subspace of $\mathbb{R}^n$, define the projection operator $T : \mathbb{R}^n \to \mathbb{R}^n$ by $T(X) = proj_U(X)$ for all $X$ in $\mathbb{R}^n$. Show that $T$ is a linear transformation.

**SOLUTION**

If $\{F_1, F_2, \cdots F_k\}$ is any orthogonal basis of $U$, we have

$$T(X) = \frac{X \cdot F_1}{\|F_1\|^2}F_1 + \frac{X \cdot F_2}{\|F_2\|^2}F_2 + \cdots + \frac{X \cdot F_k}{\|F_k\|^2}F_k$$

for each $X$ in $\mathbb{R}^n$. As the reader can verify, the result follows because, for each $i$, we have $(X + Y) \cdot F_i = X \cdot F_i + Y \cdot F_i$ and $(aX) \cdot F_i = a\,(X \cdot F_i)$ for all $X$ and $Y$ in $\mathbb{R}^n$ and all scalars $a$.

---

[26] Transformations $\mathbb{R} \to \mathbb{R}$ are usually called *functions* or *maps*. However the term "transformation" is commonly used in linear algebra.

**Example 5** If $Y$ is a fixed vector in $\mathbb{R}^n$, the transformation $T : \mathbb{R}^n \to \mathbb{R}^n$ defined by $T(X) = X + Y$ for all $X$ in $\mathbb{R}^n$ is called $Y$-**translation**. Thus $T(0) = 0 + Y = Y$, so the map $T$ is *not* a linear transformation if $Y \neq 0$. (If $T$ were linear, then $T(0) = 0$ by T2.)

We now come to the prototype example of a linear transformation $\mathbb{R}^n \to \mathbb{R}^m$.

**Example 6** If $A$ is any $m \times n$ matrix, define a transformation $T : \mathbb{R}^n \to \mathbb{R}^m$ by $T(X) = AX$ for all $X$ in $\mathbb{R}^n$. This is a linear transformation for any choice of $A$, called the **matrix transformation** induced by $A$. Indeed, T1 and T2 are basic properties of matrices.

It turns out that *every* linear transformation $\mathbb{R}^n \to \mathbb{R}^m$ is a matrix transformation. This fact depends in part on the following general properties of linear transformations: As well as preserving addition and scalar multiplication, they also preserve the zero vector, negatives, and linear combinations.

**THEOREM** ❶ Let $T : \mathbb{R}^n \to \mathbb{R}^m$ be a linear transformation, let $X$, $X_1$, $X_2$, $\cdots$, $X_k$ denote vectors in $\mathbb{R}^n$, and let $a_1$, $a_2$, $\cdots$, $a_k$ denote scalars. Then:

(1) $T(0) = 0$.

(2) $T(-X) = -T(X)$.

(3) $T(a_1X_1 + a_2X_2 + \cdots + a_kX_k) = a_1T(X_1) + a_2T(X_2) + \cdots + a_kT(X_k)$.

**PROOF** (1) and (2) follow from $T(aX) = aT(X)$ by taking $a = 0$ and $a = -1$, respectively. We prove (3) by induction on $k$. If $k = 1$, it is T2. If it holds for some $k \geq 1$, then

$$T(a_1X_1 + a_2X_2 + \cdots + a_kX_k + a_{k+1}X_{k+1})$$
$$= T(a_1X_1 + a_2X_2 + \cdots + a_kX_k) + T(a_{k+1}X_{k+1})$$
$$= [a_1T(X_1) + a_2T(X_2) + \cdots + a_kT(X_k)] + a_{k+1}T(X_{k+1})$$

where we used T1 at step 1, and we used T2 and the induction hypothesis at step 2. This completes the proof.

**Example 7** Suppose $T : \mathbb{R}^4 \to \mathbb{R}^3$ is a linear transformation such that
$$T\left([1 \ \ 0 \ -1 \ \ 2]^T\right) = [2 \ \ 1 \ \ 0]^T \text{ and } T\left([1 \ \ 1 \ -2 \ \ 0]^T\right) = [-1 \ \ 2 \ \ 4]^T.$$
Compute $T\left([-2 \ -5 \ \ 7 \ \ 6]^T\right)$.

**SOLUTION** For convenience, write $X_1 = [1 \ \ 0 \ -1 \ \ 2]^T$, $X_2 = [1 \ \ 1 \ -2 \ \ 0]^T$, and $X = [-2 \ -5 \ \ 7 \ \ 6]^T$. At first glance the problem seems impossible: We are given $T(X_1)$ and $T(X_2)$, but no way is apparent to find $T(X)$. However, since $T$ is linear, part (3) of Theorem 1 shows that we can find $T(X)$ if $X$ is *any* linear combination of $X_1$ and $X_2$. So we try to express $X$ in the form $X = rX_1 + sX_2$. Equating components gives two linear equation for $r$ and s, and the solution is $X = 3X_1 - 5X_2$. Hence Theorem 1 gives

$$T(X) = T(3X_1 - 5X_2)$$
$$= 3T(X_1) - 5T(X_2)$$
$$= 3[2 \ \ 1 \ \ 0]^T - 5[-1 \ \ 2 \ \ 4]^T$$
$$= [11 \ -7 \ -20]^T.$$

This is what we wanted.

Theorem 1 enables us to show that every linear transformation $\mathbb{R}^n \to \mathbb{R}^m$ is a matrix transformation. The proof requires a fact about matrix multiplication that will be used again later.

---

**LEMMA ①**

If $A$ is an $m \times n$ matrix and $AX = 0$ for all $X$ in $\mathbb{R}^n$, then $A = 0$.

**PROOF**

Write $I_n = [E_1 \quad E_2 \quad \cdots \quad E_n]$ in terms of its columns. Then $AE_i = 0$ for each $i$ by hypothesis, so

$$A = AI_n = A[E_1 \quad E_2 \quad \cdots \quad E_n] = [AE_1 \quad AE_2 \quad \cdots \quad AE_n] = [0 \quad 0 \quad \cdots \quad 0] = 0.$$

---

**THEOREM ②**

Let $T : \mathbb{R}^n \to \mathbb{R}^m$ be a linear transformation. Then $T$ is the matrix transformation induced by a uniquely determined $m \times n$ matrix $A$:

$$T(X) = AX \quad \text{for all } X \text{ in } \mathbb{R}^n.$$

Furthermore, $A$ is given in terms of its columns by

$$A = [T(E_1) \quad T(E_2) \quad \cdots \quad T(E_n)]$$

where $\{E_1, \ E_2, \ \cdots \ E_n\}$ is the standard basis of $\mathbb{R}^n$. The matrix $A$ is called the **standard matrix** of $T$.

**PROOF**

Write $X = [x_1 \quad x_2 \quad \cdots \quad x_n]^T$ in $\mathbb{R}^n$, so that $X = x_1 E_1 + x_2 E_2 + \cdots + x_n E_n$. Hence Theorem 1 and block multiplication (Theorem 3 §1.4) give

$$T(X) = x_1 T(E_1) + x_2 T(E_2) + \cdots + x_n T(E_n)$$

$$= [T(E_1) \quad T(E_2) \quad \cdots \quad T(E_n)] \begin{bmatrix} x_1 \\ x_2 \\ \vdots \\ x_n \end{bmatrix}$$

$$= AX$$

for every $X$ in $\mathbb{R}^n$. If another $m \times n$ matrix $B$ exists such that $T(X) = BX$ for all $X$ in $\mathbb{R}^n$, then $AX = T(X) = BX$ holds for all $X$. Thus $(A - B)X = 0$ for all columns $X$ in $\mathbb{R}^n$, whence $A - B = 0$ by Lemma 1.

---

In other words, Theorem 2 asserts that every linear transformation $\mathbb{R}^n \to \mathbb{R}^m$ is the matrix transformation induced by its standard matrix. Hence, linear transformations $T : \mathbb{R}^n \to \mathbb{R}^m$ and matrix transformations $\mathbb{R}^n \to \mathbb{R}^m$ are different names for the same thing. Both interpretations have virtue depending on the situation.

**Example 8**

If $T : \mathbb{R}^3 \to \mathbb{R}^2$ is linear and satisfies $T([1 \ 0 \ 1]^T) = [3 \ 2]^T$, $T([0 \ 1 \ 0]^T) = [1 \ 2]^T$, and $T([-1 \ 0 \ 1]^T) = [-2 \ 1]^T$, find $T(X)$ for all $X = [x_1 \ x_2 \ x_3]^T$ in $\mathbb{R}^3$.

**SOLUTION**

We first find the standard matrix of $T$. Consequently, let $\{E_1, \ E_2, \ E_3\}$ denote the standard basis of $\mathbb{R}^3$, so we are given $T(E_2) = [1 \ 2]^T$. If we write $P = [1 \ 0 \ 1]^T$ and $Q = [-1 \ 0 \ 1]^T$, then $E_1 = \frac{1}{2}(P - Q)$ and $E_3 = \frac{1}{2}(P + Q)$, so

$$T(E_1) = \tfrac{1}{2}(T(P) - T(Q)) = \tfrac{1}{2}[5 \ 1]^T \quad \text{and} \quad T(E_2) = \tfrac{1}{2}(T(P) + T(Q)) = \tfrac{1}{2}[1 \ 3]^T.$$

Hence the standard matrix of $T$ is $A = [T(E_1) \quad T(E_2) \quad T(E_3)] = \frac{1}{2}\begin{bmatrix} 5 & 2 & 1 \\ 1 & 4 & 3 \end{bmatrix}$ by

Theorem 2, so $T(X) = AX = \frac{1}{2}\begin{bmatrix} 5x_1 + 2x_2 + x_3 \\ x_1 + 4x_2 + 3x_3 \end{bmatrix}$.

Sometimes we can find the standard matrix of a linear transformation without using the formula in Theorem 2. Here is an example.

**Example 9**    If $U = span\{[2 \ \ 1 \ \ 2]^T, \ [1 \ \ 2 \ \ -2]^T\}$, find the standard matrix of the projection operator $T$ onto $U$ (that is $T(X) = proj_U(X)$ for all $X$ in $\mathbb{R}^3$).

**SOLUTION**    Write $F_1 = [2 \ \ 1 \ \ 2]^T$, $F_2 = [1 \ \ 2 \ \ -2]^T$, and $X = [x_1 \ \ x_2 \ \ x_3]^T$. Since $\{F_1, \ F_2\}$ is an orthogonal basis of $U$, we have

$$proj_U(X) = \frac{X \cdot F_1}{\|F_1\|^2}F_1 + \frac{X \cdot F_2}{\|F_2\|^2}F_2$$

$$= \frac{2x_1 + x_2 + 2x_3}{9}\begin{bmatrix} 2 \\ 1 \\ 2 \end{bmatrix} + \frac{x_1 + 2x_2 - 2x_3}{9}\begin{bmatrix} 1 \\ 2 \\ -2 \end{bmatrix}$$

$$= \frac{1}{9}\begin{bmatrix} 5 & 4 & 2 \\ 4 & 5 & -2 \\ 2 & -2 & 8 \end{bmatrix}\begin{bmatrix} x_1 \\ x_2 \\ x_3 \end{bmatrix}.$$

The standard matrix of $T$ is now apparent.

Let $T : \mathbb{R}^n \to \mathbb{R}^m$ and $S : \mathbb{R}^m \to \mathbb{R}^k$ be two transformations (not necessarily linear) which link together as follows: $\mathbb{R}^n \xrightarrow{T} \mathbb{R}^m \xrightarrow{S} \mathbb{R}^k$. As in Section 3.4, we define the **composite transformation** $S \circ T : \mathbb{R}^n \to \mathbb{R}^k$ by

$$(S \circ T)(X) = S[T(X)] \quad \text{for all } X \text{ in } \mathbb{R}^n.$$

Thus the action of $S \circ T$ can be described as "first $T$, then $S$". Now suppose that $S$ and $T$ are linear transformations with standard matrices $A$ and $B$ respectively. Then

$$(S \circ T)(X) = S[T(X)] = A[B(X)] = (AB)X$$

for all $X$ in $\mathbb{R}^n$. Hence $S \circ T$ is the linear transformation with standard matrix AB. We record this for reference.

**THEOREM 3**    Let $\mathbb{R}^n \xrightarrow{T} \mathbb{R}^m \xrightarrow{S} \mathbb{R}^k$ be linear transformations. Then the composite $S \circ T : \mathbb{R}^n \to \mathbb{R}^k$ is also a linear transformation. Moreover, if $S$ and $T$ have standard matrices $A$ and $B$ respectively, then $S \circ T$ has standard matrix $AB$.

As for functions in general, two linear transformations $S : \mathbb{R}^n \to \mathbb{R}^m$ and $T : \mathbb{R}^n \to \mathbb{R}^m$ are said to be **equal**, written $S = T$, if they have the same effect on every vector. In other words

$$S = T \quad \text{if and only if} \quad S(X) = T(X) \text{ for every vector } X \text{ in } \mathbb{R}^n.$$

We describe the requirement that $S(X) = T(X)$ for all $X$ by saying that $S$ and $T$ have the same **action**. The next result asserts that a linear transformation $\mathbb{R}^n \to \mathbb{R}^m$ is determined by its effect on any spanning set of $\mathbb{R}^n$. The result is an easy consequence of Theorem 1.

**LEMMA 2**    Let $S : \mathbb{R}^n \to \mathbb{R}^m$ and $T : \mathbb{R}^n \to \mathbb{R}^m$ be two linear transformations, and suppose that $\mathbb{R}^n = span\{Y_1, Y_2, \cdots, Y_k\}$. If $S(Y_i) = T(Y_i)$ for each $i$, then $S = T$.

PROOF   Given $X$ in $\mathbb{R}^n$, write $X = a_1Y_1 + a_2Y_2 + \cdots + a_kY_k$ where each $a_i$ is in $\mathbb{R}$. Then Theorem 1 gives

$$\begin{aligned} S(X) &= a_1S(Y_1) + a_2S(Y_2) + \cdots + a_kS(Y_k) \\ &= a_1T(Y_1) + a_2T(Y_2) + \cdots + a_kT(Y_k) \\ &= T(X). \end{aligned}$$

Since this holds for all $X$ in $\mathbb{R}^n$, we have $S = T$.

Lemma 2 can also be used to *construct* linear transformations. In fact, if a basis of $\mathbb{R}^n$ is given, there is a unique linear transformation $\mathbb{R}^n \to \mathbb{R}^m$ which has *any* desired effect on the basis vectors. More precisely, we have

THEOREM ④   Let $\mathcal{F} = \{F_1, F_2, \cdots, F_n\}$ be any basis of $\mathbb{R}^n$, and select any $n$ vectors $\{Z_1, Z_2, \cdots, Z_n\}$ in $\mathbb{R}^m$, possibly not distinct. Then there is a unique linear transformation $T : \mathbb{R}^n \to \mathbb{R}^m$ such that

$$T(F_i) = Z_i \quad \text{holds for each } i = 1, 2, \cdots, n.$$

In fact $T$ can be defined as follows: If $X$ is any vector in $\mathbb{R}^n$, express $X$ as a linear combination of the vectors in the basis $\mathcal{F}$, say

$$X = a_1F_1 + a_2F_2 + \cdots + a_nF_n.$$

Then $T(X)$ is given by

$$T(X) = a_1Z_1 + a_2Z_2 + \cdots + a_nZ_n. \tag{$*$}$$

PROOF   If $T : \mathbb{R}^n \to \mathbb{R}^m$ is any linear transformation with the property that $T(F_i) = Z_i$ for each $i$, it is clear from Theorem 1 that $(*)$ holds. Hence Lemma 2 shows that $T$ is uniquely determined *if it exists*. The natural temptation is to define $T$ by $(*)$. That is, if $X$ is any vector in $\mathbb{R}^n$, use the fact that $\mathcal{F}$ spans $\mathbb{R}^n$ to express $X = a_1F_1 + a_2F_2 + \cdots + a_nF_n$, and take

$$T(X) = a_1Z_1 + a_2Z_2 + \cdots + a_nZ_n.$$

However, there is one possible flaw in this definition: If $X = b_1F_1 + b_2F_2 + \cdots + b_nF_n$ is a *possibly different* linear combination for $X$, we would get

$$T(X) = b_1Z_1 + b_2Z_2 + \cdots + b_nZ_n,$$

and this might be different from $a_1Z_1 + a_2Z_2 + \cdots + a_nZ_n$. However, this is where the assumption that $\mathcal{F}$ is also independent comes in: Since

$$a_1F_1 + a_2F_2 + \cdots + a_nF_n = b_1F_1 + b_2F_2 + \cdots + b_nF_n,$$

the independence of the $F_i$ implies that $a_i = b_i$ for each $i$ (Theorem 1 §4.2). Hence the definition $(*)$ is unambiguous. It is now a routine matter to verify that $T$ is linear, so the theorem is proved.

Theorem 4 is used extensively as a way to define linear transformations by specifying their action on a basis. The definition in $(*)$ is referred to as *extending $T$ linearly* to $\mathbb{R}^n$.

Example 10   Find a linear transformation $T : \mathbb{R}^3 \to \mathbb{R}^2$ such that $T\left([1 \ \ 1 \ \ 0]^T\right) = [3 \ \ -2]^T$, $T\left([1 \ \ 0 \ \ 1]^T\right) = [2 \ \ -1]^T$, and $T\left([0 \ \ 1 \ \ 1]^T\right) = [1 \ \ 0]^T$.

SOLUTION   For convenience, write $F_1 = [1 \ \ 1 \ \ 0]^T$, $F_2 = [1 \ \ 0 \ \ 1]^T$, and $F_3 = [0 \ \ 1 \ \ 1]^T$. Then $\mathcal{F} = \{F_1, F_2, F_3\}$ is a basis of $\mathbb{R}^3$, so Theorem 4 applies. An arbitrary vector $X = [x_1 \ \ x_2 \ \ x_3]^T$ in $\mathbb{R}^3$ has the following $\mathcal{F}$-expansion:

$$X = \tfrac{x_1+x_2-x_3}{2}F_1 + \tfrac{x_1-x_2+x_3}{2}F_2 + \tfrac{-x_1+x_2+x_3}{2}F_3.$$

Hence

$$T(X) = \frac{x_1 + x_2 - x_3}{2} T(F_1) + \frac{x_1 - x_2 + x_3}{2} T(F_2) + \frac{-x_1 + x_2 + x_3}{2} T(F_3)$$

$$= \frac{x_1 + x_2 - x_3}{2} \begin{bmatrix} 3 \\ -2 \end{bmatrix} + \frac{x_1 - x_2 + x_3}{2} \begin{bmatrix} 2 \\ -1 \end{bmatrix} + \frac{-x_1 + x_2 + x_3}{2} \begin{bmatrix} 1 \\ 0 \end{bmatrix}$$

$$= \frac{1}{2} \begin{bmatrix} 4x_1 + 2x_2 \\ -3x_1 - x_2 + x_3 \end{bmatrix}.$$

Note that this shows that the standard matrix of $T$ is $\frac{1}{2} \begin{bmatrix} 4 & 2 & 0 \\ -3 & -1 & 1 \end{bmatrix}$.

Let $T : \mathbb{R}^n \to \mathbb{R}^n$ be a linear operator. A linear operator $S : \mathbb{R}^n \to \mathbb{R}^n$ is called an **inverse** of $T$ if the action of each of $S$ and $T$ "undoes" the action of the other; that is, if

$$S(T(X)) = X \quad \text{and} \quad T(S(X)) = X \text{ hold for every } X \text{ in } \mathbb{R}^n.$$

In terms of the composition operation, these conditions take the simpler form

$$S \circ T = 1_{\mathbb{R}^n} \quad \text{and} \quad T \circ S = 1_{\mathbb{R}^n}.$$

Thus if $R_\theta$ denotes rotation of $\mathbb{R}^2$ about the origin through an angle $\theta$, then $R_{-\theta}$ is an inverse of $R_\theta$ because it reverses the action.

An operator $T$ may not have an inverse (see below) but, if an inverse exists, it is uniquely determined by $T$ as the next result shows.

**LEMMA 3**    If $S$ and $S_1$ are both inverses of $T$, then $S = S_1$.

**PROOF**    We must show that $S(X) = S_1(X)$ for all $X$ in $\mathbb{R}^n$. We are assuming that $T \circ S = 1_{\mathbb{R}^n} = T \circ S_1$ so $T(S(X)) = T(S_1(X))$. Now apply $S$ to get $(S \circ T)(S(X)) = (S \circ T)(S_1(X))$. But $S \circ T = 1_{\mathbb{R}^n}$, so this becomes $S(X) = S_1(X)$, as required.

Hence, if an operator $T : \mathbb{R}^n \to \mathbb{R}^n$ has an inverse $S$, then it has exactly one which we denote as $S = T^{-1}$. In this case we say that $T$ is an **invertible** operator. If this all seems a bit like matrix algebra, the next results explains why.

**THEOREM 5**    Let $T : \mathbb{R}^n \to \mathbb{R}^n$ be an operator with standard matrix $A$. Then:
   (1) $T$ has an inverse if and only if $A$ is an invertible matrix.
   (2) In this case $A^{-1}$ is the standard matrix of $T^{-1}$.

**PROOF**    If $S$ is an inverse of $T$, let $B$ denote the standard matrix of $S$. Then $S \circ T = 1_{\mathbb{R}^n}$, and this operator has standard matrix $BA = I$ by Theorem 3. Similarly, $AB = I$, so $A$ is invertible. Conversely, if $A$ is invertible, let $S : \mathbb{R}^n \to \mathbb{R}^n$ be the operator with standard matrix $A^{-1}$. Then $(S \circ T)(X) = S(T(X)) = A^{-1}(AX) = X$ for all $X$ in $\mathbb{R}^n$, so $S \circ T = 1_{\mathbb{R}^n}$. Similarly, $T \circ S = 1_{\mathbb{R}^n}$, so $T$ is invertible with $T^{-1} = S$. This completes the proof of (1), and also proves (2).

## 4.9.2 Changing Coordinates

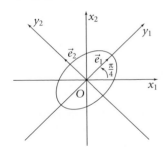

Consider the equation

$$x_1^2 + x_1x_2 + x_2^2 = 3;$$

its graph is shown in the diagram. If the plane is rotated counterclockwise through $\frac{\pi}{4}$ radians, the variables $x_1$ and $x_2$ become

$$y_1 = \tfrac{1}{\sqrt{2}}(x_1 - x_2) \text{ and } y_1 = \tfrac{1}{\sqrt{2}}(x_1 + x_2),$$

respectively. In terms of $y_1$ and $y_2$ the equation assumes a much simpler form: $y_1^2 + 3y_1^2 = 6$. The reason is that the curve is symmetric about the new $y_1$- and $y_2$-axes. Finding these axes amounts to finding the unit vectors $\vec{e}_1 = \tfrac{1}{\sqrt{2}}[1\ 1]^T$ and $\vec{e}_2 = \tfrac{1}{\sqrt{2}}[-1\ 1]^T$ along these axes,[27] that is finding a new basis $\{\vec{e}_1, \vec{e}_2\}$ of $\mathbb{R}^2$ that is compatible with the graph.

More generally, it often happens that one basis of $\mathbb{R}^n$ is more convenient than another. Changing the basis of $\mathbb{R}^n$ amounts to specifying a new "coordinate system" in $\mathbb{R}^n$, and we now investigate how the "coordinates" of a given vector change when the basis is changed. We begin by clarifying the relationship between bases and "coordinates".

Let $\mathcal{F} = \{F_1, F_2, \cdots, F_n\}$ be a basis of $\mathbb{R}^n$. Since $\mathcal{F}$ spans $\mathbb{R}^n$, every vector $X$ in $\mathbb{R}^n$ is a linear combination $X = x_1F_1 + x_2F_2 + \cdots + x_nF_n$ where the $x_i$ are real numbers. Moreover, since $\mathcal{F}$ is linearly independent, the numbers $x_i$ are uniquely determined.[28] These numbers $x_i$ are called the $\mathcal{F}$-**coordinates** of $X$, and the column

$$C_{\mathcal{F}}(X) = [x_1\ x_2\ \cdots\ x_n]^T$$

is called the $\mathcal{F}$-**coordinate vector** of $X$.

**Example 11**

If $\mathcal{E} = \{E_1, E_2, \cdots, E_n\}$ is the standard basis of $\mathbb{R}^n$, then $C_{\mathcal{E}}(X) = X$ for every $X$ in $\mathbb{R}^n$. The reason is that if $X = [x_1\ x_2\ \cdots\ x_n]^T$, then $X = x_1E_1 + x_2E_2 + \cdots + x_nE_n$.

**Example 12**

If $\mathcal{F} = \{F_1, F_2, \cdots, F_n\}$ is an orthogonal basis of $\mathbb{R}^n$, then for each $X$ in $\mathbb{R}^n$,

$$C_{\mathcal{F}}(X) = \left[ \frac{X \cdot F_1}{\|F_1\|^2}\ \ \frac{X \cdot F_2}{\|F_2\|^2}\ \ \cdots\ \ \frac{X \cdot F_n}{\|F_n\|^2} \right]^T$$

because $X = \frac{X \cdot F_1}{\|F_1\|^2}F_1 + \frac{X \cdot F_2}{\|F_2\|^2}F_2 + \cdots + \frac{X \cdot F_n}{\|F_n\|^2}F_n$ by the Expansion theorem (Theorem 6 §4.5). This formula is particularly simple if $\mathcal{F}$ is an orthonormal basis since then all the denominators $\|F_i\|^2$ equal 1.

The following simple fact about coordinate vectors will be used several times.

**LEMMA 4**

If $\mathcal{F} = \{F_1, F_2, \cdots, F_n\}$ is a basis of $\mathbb{R}^n$, let $P = [F_1\ F_2\ \cdots\ F_n]$ be the matrix with the $F_i$ as its columns. Then

$$C_{\mathcal{F}}(X) = P^{-1}X \quad \text{for all } X \text{ in } \mathbb{R}^n.$$

In particular, $C_{\mathcal{F}}$ is a linear operator $\mathbb{R}^n \to \mathbb{R}^n$ with standard matrix $P^{-1}$.

---

[27] These vectors can be *discovered* using diagonalization (see Section 4.8): $\vec{e}_1$ and $\vec{e}_2$ are the principal axes of the quadratic form $q = x_1^2 + x_1x_2 + x_2^2$.

[28] For the $x_i$ to be unique, the vectors in $\mathcal{F}$ must always be listed in the same order—we then call $\mathcal{F}$ an *ordered* basis. Thus, when working with coordinates, we assume that the bases are ordered.

**PROOF**   Let $X = x_1F_1 + x_2F_2 + \cdots + x_nF_n$, so that $C_{\mathcal{F}}(X) = [x_1 \ x_2 \ \cdots \ x_n]^T$. Block multiplication gives:

$$PC_{\mathcal{F}}(X) = [F_1 \ F_2 \ \cdots \ F_n]\begin{bmatrix} x_1 \\ x_2 \\ \vdots \\ x_n \end{bmatrix} = x_1F_1 + x_2F_2 + \cdots + x_nF_n = X.$$

The lemma follows because $P$ is invertible (its columns form a basis of $\mathbb{R}^n$).

If $\mathcal{F}$ is a basis of $\mathbb{R}^n$, the operator $C_{\mathcal{F}} : \mathbb{R}^n \to \mathbb{R}^n$ is called the **coordinate operator** corresponding to $\mathcal{F}$. Note that if the basis $\mathcal{F}$ is orthonormal, then $P$ is actually an orthogonal matrix (that is $P^{-1} = P^T$).

Now suppose that $\mathcal{F} = \{F_1, \ F_2, \ \cdots, \ F_n\}$ and $\mathcal{G} = \{G_1, \ G_2, \ \cdots, \ G_n\}$ are two bases of $\mathbb{R}^n$. We want to discover how the coordinate vectors $C_{\mathcal{F}}(X)$ and $C_{\mathcal{G}}(X)$ are related. By Lemma 4, we have

$$C_{\mathcal{F}}(X) = P^{-1}X \quad \text{and} \quad C_{\mathcal{G}}(X) = Q^{-1}X \text{ for all } X \text{ in } \mathbb{R}^n,$$

where $P = [F_1 \ F_2 \ \cdots \ F_n]$ and $Q = [G_1 \ G_2 \ \cdots \ G_n]$. Hence, for all $X$ in $\mathbb{R}^n$, we have

$$\begin{aligned} C_{\mathcal{G}}(X) = Q^{-1}X &= Q^{-1}PC_{\mathcal{F}}(X) \\ &= Q^{-1}[F_1 \ F_2 \ \cdots \ F_n]C_{\mathcal{F}}(X) \\ &= [Q^{-1}F_1 \ Q^{-1}F_2 \ \cdots \ Q^{-1}F_n]C_{\mathcal{F}}(X) \\ &= [C_{\mathcal{G}}(F_1) \ C_{\mathcal{G}}(F_2) \ \cdots \ C_{\mathcal{G}}(F_n)]C_{\mathcal{F}}(X) \end{aligned}$$

With this in mind, if $\mathcal{F}$ and $\mathcal{G}$ are any two bases of $\mathbb{R}^n$ we define the matrix

$$P_{\mathcal{G} \leftarrow \mathcal{F}} = [C_{\mathcal{G}}(F_1) \ C_{\mathcal{G}}(F_2) \ \cdots \ C_{\mathcal{G}}(F_n)].$$

The matrix $P_{\mathcal{G} \leftarrow \mathcal{F}}$ is called the **change** matrix[29] from $\mathcal{F}$ to $\mathcal{G}$. This matrix depends only on the two bases $\mathcal{F}$ and $\mathcal{G}$, and not on the vector $X$; on the other hand, the above calculation shows that

$$C_{\mathcal{G}}(X) = P_{\mathcal{G} \leftarrow \mathcal{F}}C_{\mathcal{F}}(X)$$

holds for *all* $X$ in $\mathbb{R}^n$. This is the desired relationship between the $\mathcal{F}$- and $\mathcal{G}$-coordinates of $X$, and we state it as the first part of the following theorem.

**THEOREM ⑥**   If $\mathcal{F} = \{F_1, \ F_2, \ \cdots, \ F_n\}$ and $\mathcal{G}$ are (ordered) bases of $\mathbb{R}^n$, then

$$C_{\mathcal{G}}(X) = P_{\mathcal{G} \leftarrow \mathcal{F}}C_{\mathcal{F}}(X) \quad \text{for all } X \text{ in } \mathbb{R}^n.$$

Moreover, if $\mathcal{G} = \{G_1, \ G_2, \ \cdots, \ G_n\}$, then the matrix $P_{\mathcal{G} \leftarrow \mathcal{F}}$ can be computed as follows:

$$\left[[G_1 \ G_2 \ \cdots \ G_n] \ [F_1 \ F_2 \ \cdots \ F_n]\right] \to [I_n \ P_{\mathcal{G} \leftarrow \mathcal{F}}].$$

In other words, carry the double matrix on the left to reduced row-echelon form, and $P_{\mathcal{G} \leftarrow \mathcal{F}}$ appears on the right.

**PROOF**   By the above discussion, it remains to verify the second part of the theorem. Write $P = [F_1 \ F_2 \ \cdots \ F_n]$ and $Q = [G_1 \ G_2 \ \cdots \ G_n]$ as before. Since $Q$ is invertible, we can carry $Q \to I_n = Q^{-1}Q$ by row operations, so the same series of operations carries $P \to Q^{-1}P$ by Theorem 1 §1.6. But

$$\begin{aligned} Q^{-1}P = Q^{-1}[F_1 \ F_2 \ \cdots \ F_n] &= [Q^{-1}F_1 \ Q^{-1}F_2 \ \cdots \ Q^{-1}F_n] \\ &= [C_{\mathcal{G}}(F_1) \ C_{\mathcal{G}}(F_2) \ \cdots \ C_{\mathcal{G}}(F_n)] \\ &= P_{\mathcal{G} \leftarrow \mathcal{F}} \end{aligned}$$

by Lemma 4. Hence doing the operations to the double matrix gives $[Q \ P] \to [I_n \ P_{\mathcal{G} \leftarrow \mathcal{F}}]$, as required.

---

[29] The reason for the notation $P_{\mathcal{G} \leftarrow \mathcal{F}}$ will be apparent shortly.

These change matrices are all invertible; in fact we have the following result which explains the notation $P_{\mathcal{G} \leftarrow \mathcal{F}}$.

**COROLLARY**

Let $\mathcal{F}$, $\mathcal{G}$, and $\mathcal{H}$ be three ordered bases of $\mathbb{R}^n$. Then:

**(1)** $P_{\mathcal{F} \leftarrow \mathcal{F}} = I_n$.
**(2)** $P_{\mathcal{G} \leftarrow \mathcal{F}}$ is invertible, and $(P_{\mathcal{G} \leftarrow \mathcal{F}})^{-1} = P_{\mathcal{F} \leftarrow \mathcal{G}}$.
**(3)** $P_{\mathcal{H} \leftarrow \mathcal{G}} P_{\mathcal{G} \leftarrow \mathcal{F}} = P_{\mathcal{H} \leftarrow \mathcal{F}}$.

**PROOF**

(1) is a routine computation, and (1) and (3) together imply (2) (verify). To prove (3), let $X$ be a vector in $\mathbb{R}^n$ and apply Theorem 6 three times:

$$P_{\mathcal{H} \leftarrow \mathcal{G}} P_{\mathcal{G} \leftarrow \mathcal{F}} C_{\mathcal{F}}(X) = P_{\mathcal{H} \leftarrow \mathcal{G}} C_{\mathcal{G}}(X) = C_{\mathcal{H}}(X) = P_{\mathcal{H} \leftarrow \mathcal{F}} C_{\mathcal{F}}(X).$$

Since this holds for all $X$ in $\mathbb{R}^n$, (3) follows from Lemmas 1 and 2.

**Example 13**

Find $P_{\mathcal{G} \leftarrow \mathcal{F}}$ where $\mathcal{F} = \{[2 \;\; -1]^T, \; [1 \;\; 5]^T\}$ and $\mathcal{G} = \{[1 \;\; 3]^T, \; [3 \;\; 4]^T\}$ are two bases of $\mathbb{R}^2$.

**SOLUTION**

As in the last part of Theorem 6, we compute

$$\begin{bmatrix} 1 & 3 & 2 & 1 \\ 3 & 4 & -1 & 5 \end{bmatrix} \rightarrow \begin{bmatrix} 1 & 0 & -\frac{11}{5} & \frac{11}{5} \\ 0 & 1 & \frac{7}{5} & -\frac{2}{5} \end{bmatrix}, \text{ so } P_{\mathcal{G} \leftarrow \mathcal{F}} = \frac{1}{5}\begin{bmatrix} -11 & 11 \\ 7 & -2 \end{bmatrix}.$$

The reader can verify that the columns of $P_{\mathcal{G} \leftarrow \mathcal{F}}$ are indeed $C_{\mathcal{G}}([2 \;\; -1]^T)$ and $C_{\mathcal{G}}([1 \;\; 5]^T)$.

## 4.9.3 The $\mathcal{F}$-matrix of a Linear Operator

Suppose $T$ is a rotation about the line $L$ through the origin in $\mathbb{R}^3$. Finding the standard matrix of $T$ is not convenient because the usual coordinate axes bear no relation to the line $L$. Everything is simpler if we use a new *intrinsic* set of axes, one pointing along the line $L$ and the other two orthogonal to $L$ (that is, in the plane $L^\perp$). Choosing such a set of axes amounts to choosing a basis $\mathcal{F}$ for $\mathbb{R}^3$ with one vector in $L$ and the other two vectors in $L^\perp$. The idea is to find an "$\mathcal{F}$-matrix" for the rotation that plays the same role relative to $\mathcal{F}$ that the standard matrix plays relative to the standard basis. It turns out that computing this $\mathcal{F}$-matrix is not too difficult because $\mathcal{F}$ is naturally aligned with the line $L$. Since the $\mathcal{F}$-matrix is intimately related to the basis $\mathcal{F}$, it is not surprising that the $\mathcal{F}$-coordinates $C_{\mathcal{F}}(X)$ will be used to describe each vector $X$.

To motivate the general situation, recall that if $T : \mathbb{R}^n \to \mathbb{R}^n$ is any operator, we have $T(X) = AX$ for all $X$ in $\mathbb{R}^n$ by Theorem 2 where $A$ is the standard matrix of the operator $T$. Moreover, if $\mathcal{E}$ is the standard basis of $\mathbb{R}^n$, then $C_{\mathcal{E}}(X) = X$ for every $X$ by Example 11, so the equation $T(X) = AX$ becomes

$$C_{\mathcal{E}}[T(X)] = A \, C_{\mathcal{E}}(X) \quad \text{for all } X \text{ in } \mathbb{R}^n.$$

This formula extends to other bases than $\mathcal{E}$. Hence given an operator $T : \mathbb{R}^n \to \mathbb{R}^n$ and a basis $\mathcal{F}$ of $\mathbb{R}^n$, we look for a matrix $B$ such that

$$C_{\mathcal{F}}[T(X)] = B \, C_{\mathcal{F}}(X) \quad \text{for all } X \text{ in } \mathbb{R}^n. \tag{$**$}$$

Such a matrix $B$ is unique if it exists by Lemmas 1 and 4 (verify). Moreover, $B$ describes the action of $T$ because we can easily compute $T(X)$ from the $\mathcal{F}$-coordinates $C_{\mathcal{F}}[T(X)]$—see Example 15 below for an illustration.

Furthermore, $B$ exists for *any* choice of basis $\mathcal{F} = \{F_1, F_2, \cdots, F_n\}$. Indeed, let $A$ denote the standard matrix of $T$, and write $P = [F_1 \;\; F_2 \;\; \cdots \;\; F_n]$ as in Lemma 4. Then, using Lemma 4 twice, we have

$$C_{\mathcal{F}}(T(X)) = P^{-1}(T(X)) = P^{-1}(AX) = P^{-1}A(P\,C_{\mathcal{F}}(X)) = (P^{-1}AP)\,C_{\mathcal{F}}(X).$$

Hence $B = P^{-1}AP$ works in (**). This matrix is called the $\mathcal{F}$-**matrix** of $T$ and is denoted $M_{\mathcal{F}}(T)$. We have shown that $M_{\mathcal{F}}(T)$ is the unique matrix with the property that $C_{\mathcal{F}}[T(X)] = M_{\mathcal{F}}(T)\,C_{\mathcal{F}}(X)$ for all $X$ in $\mathbb{R}^n$. This discussion proves most of the following theorem.

**THEOREM 7**

Let $T : \mathbb{R}^n \to \mathbb{R}^n$ be a linear operator, and let $\mathcal{F}$ be a basis of $\mathbb{R}^n$. Then a uniquely determined $n \times n$ matrix $M_{\mathcal{F}}(T)$ exists such that

$$C_{\mathcal{F}}[T(X)] = M_{\mathcal{F}}(T)\,C_{\mathcal{F}}(X) \quad \text{for all } X \text{ in } \mathbb{R}^n.$$

Furthermore, if $\mathcal{F} = \{F_1, F_2, \cdots, F_n\}$ and we put $P = [F_1 \;\; F_2 \;\; \cdots \;\; F_n]$, we have:

**(1)** $M_{\mathcal{F}}(T) = P^{-1}AP$ where $A$ is the standard matrix of $T$.
**(2)** $M_{\mathcal{F}}(T) = [C_{\mathcal{F}}(T(F_1)) \;\; C_{\mathcal{F}}(T(F_2)) \;\; \cdots \;\; C_{\mathcal{F}}(T(F_n))]$.
**(3)** $M_{\mathcal{F}}(T)$ is invertible if and only if $T$ is invertible.

**PROOF**

(1) is proved above. As to (2), Lemma 4 gives $P\,C_{\mathcal{F}}(T(F_j)) = T(F_j) = AF_j$ for each $j$. Hence

$$\begin{aligned}
AP &= A[F_1 \;\; F_2 \;\; \cdots \;\; F_n] \\
&= [AF_1 \;\; AF_2 \;\; \cdots \;\; AF_n] \\
&= [P\,C_{\mathcal{F}}(T(F_1)) \;\; P\,C_{\mathcal{F}}(T(F_2)) \;\; \cdots \;\; P\,C_{\mathcal{F}}(T(F_n))] \\
&= P\,[C_{\mathcal{F}}(T(F_1)) \;\; C_{\mathcal{F}}(T(F_2)) \;\; \cdots \;\; C_{\mathcal{F}}(T(F_n))]
\end{aligned}$$

Now (2) follows using (1). Finally, $T$ is invertible if and only if $A$ is invertible (by Theorem 5), and this holds if and only if $M_{\mathcal{F}}(T)$ is invertible by (1). This is (3).

**Example 14**

If $\mathcal{E}$ is the standard basis of $\mathbb{R}^n$, then $M_{\mathcal{E}}(T)$ is the standard matrix of $T$ for any linear operator $T$ on $\mathbb{R}^n$. This follows from part (1) of Theorem 7 because $P = I$ when $\mathcal{F} = \mathcal{E}$.

If we know the $\mathcal{F}$-matrix of a linear operator $T : \mathbb{R}^n \to \mathbb{R}^n$ for some basis $\mathcal{F}$ of $\mathbb{R}^n$, it is a routine matter to use Theorem 7 to find the action of $T$. Here is an example.

**Example 15**

Find the action of the operator $T : \mathbb{R}^n \to \mathbb{R}^n$ where $M_{\mathcal{F}}(T) = \begin{bmatrix} 1 & 2 & 0 \\ 2 & -1 & 3 \\ 1 & 1 & 1 \end{bmatrix}$ and $\mathcal{F} = \{[1 \;\; 0 \;\; 0]^T, \; [0 \;\; 1 \;\; 1]^T, \; [0 \;\; -1 \;\; 1]^T\}$.

**SOLUTION**

For convenience let $F_1 = [1 \;\; 0 \;\; 0]^T, F_2 = [0 \;\; 1 \;\; 1]^T$, and $F_3 = [0 \;\; -1 \;\; 1]^T$ denote the vectors in $\mathcal{F}$. By Theorem 7 we know that

$$C_{\mathcal{F}}[T(X)] = M_{\mathcal{F}}(T)\,C_{\mathcal{F}}(X)$$

for any $X = [x_1 \;\; x_2 \;\; x_3]^T$ in $\mathbb{R}^n$. Since we are given $M_{\mathcal{F}}(T)$, the hard part is to compute $C_{\mathcal{F}}(X)$ since it is necessary to express the vector $X$ as a linear combination of the vectors $F_i$ in $\mathcal{F}$. In this case $\mathcal{F}$ is actually orthogonal, so the Expansion theorem simplifies the work:

$$\begin{aligned}
X &= \frac{X \cdot F_1}{\|F_1\|^2}F_1 + \frac{X \cdot F_2}{\|F_2\|^2}F_2 + \frac{X \cdot F_3}{\|F_3\|^2}F_3 \\
&= \frac{x_1}{1}F_1 + \frac{x_2 + x_3}{2}F_2 + \frac{-x_2 + x_3}{2}F_3.
\end{aligned}$$

It follows that $C_{\mathcal{F}}(X) = \frac{1}{2}[2x_1 \quad x_2 + x_3 \quad x_3 - x_2]^T$, so

$$C_{\mathcal{F}}[T(X)] = M_{\mathcal{F}}(T)\, C_{\mathcal{F}}(X)$$

$$= \begin{bmatrix} 1 & 2 & 0 \\ 2 & -1 & 3 \\ 1 & 1 & 1 \end{bmatrix} \frac{1}{2} \begin{bmatrix} 2x_1 \\ x_2 + x_3 \\ x_3 - x_2 \end{bmatrix}$$

$$= \begin{bmatrix} x_1 + x_2 + x_3 \\ 2x_1 - 2x_2 + x_3 \\ x_1 + x_3 \end{bmatrix}.$$

Finally, this gives

$$T(X) = (x_1 + x_2 + x_3)F_1 + (2x_1 - 2x_2 + x_3)F_2 + (x_1 + x_3)F_3$$

$$= \begin{bmatrix} x_1 + x_2 + x_3 \\ x_1 - 2x_2 \\ 3x_1 - 2x_2 + 2x_3 \end{bmatrix}.$$

This is what we wanted.

The formula in part (2) of Theorem 7 is very useful because it means that we can construct the $\mathcal{F}$-matrix of an operator $T$ without knowing the standard matrix of $T$. More importantly, it means that we can often *choose* the basis $\mathcal{F}$ in such a way that the matrix $M_{\mathcal{F}}(T)$ has a relatively simple form. This is accomplished by choosing $\mathcal{F}$ so that it is naturally compatible with the action of the operator $T$. We describe this by saying that it is an **intrinsic** basis. This is a key technique in linear algebra; here are two examples.

**Example 16**    Describe the rotations of $\mathbb{R}^3$ about a line $L$ through the origin by finding an intrinsic basis $\mathcal{F}$ for $\mathbb{R}^3$ and computing $M_{\mathcal{F}}(T)$.

**SOLUTION**    Let $T$ denote the rotation about $L$ through the angle $\theta$. Choose a unit vector $\vec{f}_1$ in $L$. Since $dim(L) = 1$, we have $dim(L^{\perp}) = 2$ by Theorem 3 §4.6 so let $\{\vec{f}_2, \vec{f}_3\}$ be an orthonormal basis of $L^{\perp}$. Then $\mathcal{F} = \{\vec{f}_1, \vec{f}_2, \vec{f}_3\}$ is an orthonormal basis of $\mathbb{R}^3$, and it remains to compute $M_{\mathcal{F}}(T)$. The action of $T$ is described in the first diagram. We have

$$T(\vec{f}_1) = \vec{f}_1$$

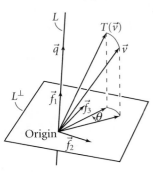

because $\vec{f}_1$ is in $L$ and $T$ is rotation about $L$. Moreover, $T(\vec{f}_2)$ and $T(\vec{f}_3)$ lie in the plane $L^{\perp}$ and so are obtained by rotating $\vec{f}_1$ and $\vec{f}_2$ in $L^{\perp}$ through the angle $\theta$. By the second diagram, the results are

$$T(\vec{f}_2) = \cos\theta\, \vec{f}_2 + \sin\theta\, \vec{f}_3$$
$$T(\vec{f}_3) = -\sin\theta\, \vec{f}_2 + \cos\theta\, \vec{f}_3.$$

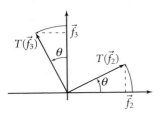

Hence Theorem 7 gives

$$M_{\mathcal{F}}[T] = \begin{bmatrix} C_{\mathcal{F}}(T(\vec{f}_1)) & C_{\mathcal{F}}(T(\vec{f}_2)) & C_{\mathcal{F}}(T(\vec{f}_3)) \end{bmatrix}$$

$$= \begin{bmatrix} 1 & 0 & 0 \\ 0 & \cos\theta & -\sin\theta \\ 0 & \sin\theta & \cos\theta \end{bmatrix}$$

This describes the operator $T$. Note that $M_{\mathcal{F}}[T]$ is actually an orthogonal matrix here. We return to this in Theorem 12 below.

**Example 17**   Describe the reflection $T$ of $\mathbb{R}^3$ in a plane $U$ through the origin by finding an intrinsic basis $\mathcal{F}$ for $\mathbb{R}^3$ and computing $M_{\mathcal{F}}(T)$.

SOLUTION

We have $dim\,U = 2$ so $dim(U^{\perp}) = 1$ by Theorem 3 §4.6. So let $\vec{f}_1$ be a unit vector in $U^{\perp}$, and choose an orthonormal basis $\{\vec{f}_2,\ \vec{f}_3\}$ of $U$. Then $\mathcal{F} = \{\vec{f}_1,\ \vec{f}_2,\ \vec{f}_3\}$ is an orthonormal basis of $\mathbb{R}^3$. If $T$ is reflection in $U$, it is clear that $T(\vec{f}_1) = -\vec{f}_1$ while $T(\vec{f}_2) = \vec{f}_2$ and $T(\vec{f}_3) = \vec{f}_3$. Hence part (2) of Theorem 7 gives

$$M_{\mathcal{F}}(T) = \begin{bmatrix} C_{\mathcal{F}}[T(\vec{f}_1)] & C_{\mathcal{F}}[T(\vec{f}_2)] & C_{\mathcal{F}}[T(\vec{f}_3)] \end{bmatrix}$$

$$= \begin{bmatrix} -1 & 0 & 0 \\ 0 & 1 & 0 \\ 0 & 0 & 1 \end{bmatrix}.$$

Thus our choice of the basis $\mathcal{F}$ has made $M_{\mathcal{F}}(T)$ into a diagonal matrix.

## 4.9.4   Similarity

Recall that two $n \times n$ matrices $A$ and $B$ are said to be **similar** if $B = P^{-1}AP$ for some invertible matrix $P$. In this case we write $A \sim B$. The similarity relation $\sim$ enjoys the following basic properties (see Section 2.3):

    **1.** $A \sim A$ for all square matrices $A$.
    **2.** If $A \sim B$, then $B \sim A$.
    **3.** If $A \sim B$ and $B \sim C$, then $A \sim C$.

Similar matrices enjoy many of the same features. For example, two similar matrices have the same determinant, trace, characteristic polynomial, and eigenvectors, and if one of them is diagonalizable, so is the other.

It turns out that there is a close connection between similarity and the matrices of a linear operator with respect to various bases. Observe that if $T : \mathbb{R}^n \to \mathbb{R}^n$ is a linear operator, part (1) of Theorem 7 shows that $M_{\mathcal{F}}(T) \sim M_{\mathcal{G}}(T)$ for any bases $\mathcal{F}$ and $\mathcal{G}$ of $\mathbb{R}^n$ (both are similar to the standard matrix of $T$). It turns out that the converse is true: If $A \sim B$, there is a linear operator $T : \mathbb{R}^n \to \mathbb{R}^n$ and bases $\mathcal{F}$ and $\mathcal{G}$ of $\mathbb{R}^n$ such that $A = M_{\mathcal{F}}(T)$ and $B = M_{\mathcal{G}}(T)$. The proof requires a preliminary fact that will be used several times.

LEMMA **5**

Let $\mathcal{F} = \{F_1,\ F_2,\ \cdots,\ F_n\}$ be any basis of $\mathbb{R}^n$, and write $P = [F_1\ \ F_2\ \cdots\ F_n]$. If $A$ is an $n \times n$ matrix, define the operator $T_A : \mathbb{R}^n \to \mathbb{R}^n$ by $T_A(X) = AX$ for all $X$ in $\mathbb{R}^n$. Then

$$M_{\mathcal{F}}(T_A) = P^{-1}AP.$$

PROOF

We use the fact that $M_{\mathcal{F}}(T_A)$ is the unique matrix $B$ such that $C_{\mathcal{F}}[T_A(X)] = BC_{\mathcal{F}}(X)$ for all $X$ in $\mathbb{R}^n$. To this end, observe that Lemma 4 asserts that $PC_{\mathcal{F}}(X) = X$ for each $X$. We use this twice, once for $X$ and once for $AX$:

$$APC_{\mathcal{F}}(X) = AX = PC_{\mathcal{F}}(AX) = PC_{\mathcal{F}}[T_A(X)].$$

Hence $C_{\mathcal{F}}[T_A(X)] = (P^{-1}AP)C_{\mathcal{F}}(X)$ for every $X$ in $\mathbb{R}^n$, so $P^{-1}AP = M_{\mathcal{F}}(T_A)$ by the uniqueness.

Now we can give our characterization of similarity in terms of operators.

**THEOREM 8** If $A$ and $B$ are $n \times n$ matrices, then $A \sim B$ if and only if there exists an operator $T : \mathbb{R}^n \to \mathbb{R}^n$ and bases $\mathcal{F}$ and $\mathcal{G}$ of $\mathbb{R}^n$ such that $A = M_{\mathcal{F}}(T)$ and $B = M_{\mathcal{G}}(T)$.

**PROOF** If such $T$, $\mathcal{F}$, and $\mathcal{G}$ exist, then $M_{\mathcal{F}}(T) \sim M_{\mathcal{G}}(T)$ because both matrices are similar to the standard matrix of $T$ by Theorem 7.

Conversely, suppose $A \sim B$, say $B = P^{-1}AP$ where $P = [F_1 \ F_2 \ \cdots \ F_n]$ is invertible. If $\mathcal{F} = \{F_1, F_2, \cdots, F_n\}$ is the basis of $\mathbb{R}^n$ consisting of the columns of $P$, then $M_{\mathcal{F}}(T_A) = P^{-1}AP = B$ by Lemma 5. On the other hand, if $\mathcal{E}$ is the standard basis of $\mathbb{R}^n$ then $M_{\mathcal{E}}(T_A) = A$ (verify). This proves the theorem.

If $A$ is a square matrix, define the **similarity class** $sim(A)$ of $A$ to be the set of all matrices similar to $A$. Thus, for example, $A$ is diagonalizable if and only if $sim(A)$ contains a diagonal matrix. In general, even for a nondiagonalizable matrix $A$, we can ask for the "nicest" matrix in $sim(A)$. Theorem 8 enters into the discussion by providing another description of $sim(A)$. It shows that, given an $n \times n$ matrix $A$, there exists an operator $T : \mathbb{R}^n \to \mathbb{R}^n$ such that $sim(A)$ is the set of all $\mathcal{F}$-matrices of $T$. More compactly,

$$sim(A) = \{M_{\mathcal{F}}(T) \mid \mathcal{F} \text{ is some basis of } \mathbb{R}^n\}.$$

Hence the task of finding the "nicest" matrix similar to $A$ becomes: Given an operator $T : \mathbb{R}^n \to \mathbb{R}^n$, find a basis such that $M_{\mathcal{F}}(T)$ is as "nice" as possible. This is illustrated in Examples 16 and 17 and, because operators have a geometric interpretation, this gives new insights into similarity. We return to it in Chapter 5.

## 4.9.5 Isometries

To illustrate all these concepts, we conclude with a look at a very important class of operators on $\mathbb{R}^n$ called isometries that arise in geometry and in other applications. Using geometrical arguments, it was established in Section 3.4 that rotations about the origin and reflections in a line through the origin are all linear operators on $\mathbb{R}^2$. Similar geometric arguments work in $\mathbb{R}^3$, but we are going to give an algebraic proof that is valid in $\mathbb{R}^n$ for any $n$. The key observation is that rotations and reflections are distance preserving in the following sense. A transformation $T : \mathbb{R}^n \to \mathbb{R}^n$ (possibly not linear) is said to be **distance preserving** if

$$\|T(X) - T(Y)\| = \|X - Y\| \quad \text{for all } X \text{ and } Y \text{ in } \mathbb{R}^n, \tag{$***$}$$

that is, the distance between $T(X)$ and $T(Y)$ is the same as the distance between $X$ and $Y$ for all vectors $X$ and $Y$ in $\mathbb{R}^n$. Distance-preserving maps need not be linear: The $Y$-translations in Example 5 are distance preserving for every choice of $Y$ (verify) but they are not linear if $Y \neq 0$. Surprisingly, the distance-preserving transformations that fix the origin are actually linear.

**THEOREM 9** Let $T : \mathbb{R}^n \to \mathbb{R}^n$ be a distance-preserving transformation with the property that $T(0) = 0$. Then $T$ is linear.

**PROOF** Let $X$ and $Y$ denote vectors in $\mathbb{R}^n$. Observe first that the equation $\|T(X) - T(Y)\|^2 = \|X - Y\|^2$ (from ($***$)) becomes

$$\|T(X)\|^2 - 2 \, T(X) \bullet T(Y) + \|T(Y)\|^2.$$

But $\|T(X)\| = \|X\|$ for every $X$ in $\mathbb{R}^n$ (take $Y = 0$ in ($***$)), so we obtain

$$T(X) \bullet T(Y) = X \bullet Y \quad \text{for all } X \text{ and } Y \text{ in } \mathbb{R}^n. \tag{$****$}$$

Now let $\{F_1, F_2, \cdots F_n\}$ be an orthogonal basis of $\mathbb{R}^n$. It follows from (****) that $\{T(F_1), T(F_2), \cdots T(F_n)\}$ is also an orthogonal basis [$T(F_i) \neq 0$ for each $i$ because $\|T(F_i)\| = \|F_i\| \neq 0$]. Hence, to prove that $T(aX) = aT(X)$, it suffices to show that $[T(aX) - aT(X)] \cdot T(F_i) = 0$ for each $i$ (see Example 3 §4.5). But (****) gives

$$[T(aX) - aT(X)] \cdot T(F_i) = T(aX) \cdot T(F_i) - aT(X) \cdot T(F_i)$$
$$= (aX) \cdot F_i - a(X \cdot F_i)$$
$$= 0,$$

as required. A similar argument shows that $T(X + Y) = T(X) + T(Y)$ for all $X$ and $Y$, so $T$ is linear.

A *linear* operator $T : \mathbb{R}^n \to \mathbb{R}^n$ which is distance preserving is called an **isometry**. This agrees with the definition in Section 3.4 for operators on $\mathbb{R}^2$. The first consequence of Theorem 9 is that, as for $\mathbb{R}^2$, rotations and reflections fixing the origin are all isometries.

**COROLLARY 1**

In $\mathbb{R}^3$ the following are all isometries:

**(1)** Rotations around a line through the origin.
**(2)** Reflections in a plane through the origin.

**PROOF**    It is clear geometrically that these transformations are all distance preserving, and they all fix the origin. Hence Theorem 9 shows that they are also linear.

Isometries are distance preserving, and it is easy to verify that translations are also distance preserving. Moreover, the composite of two distance-preserving transformations is again distance preserving (verify). The following corollary shows that *all* distance-preserving transformations arise in this way.

**COROLLARY 2**

Every distance-preserving transformation $S : \mathbb{R}^n \to \mathbb{R}^n$ is the composite of an isometry followed by a translation. More precisely, there exists a vector $Y$ in $\mathbb{R}^n$ and an isometry $T$ of $\mathbb{R}^n$ such that

$$S(X) = T(X) + Y \text{ for every } X \text{ in } \mathbb{R}^n.$$

**PROOF**    If $S : \mathbb{R}^n \to \mathbb{R}^n$ is distance preserving, write $S(0) = Y$ and define $T : \mathbb{R}^n \to \mathbb{R}^n$ by $T(X) = S(X) - Y$ for every $X$ in $\mathbb{R}^n$. Then $T(0) = 0$ because $S(0) = Y$ and, for all $X$ and $X_1$ in $\mathbb{R}^n$,

$$\|T(X) - T(X_1)\| = \|(S(X) - Y) - (S(X_1) - Y)\|$$
$$= \|S(X) - S(X_1)\|$$
$$= \|X - X_1\|$$

because $S$ is distance preserving. Hence $T$ is distance preserving, and so is an isometry by Theorem 9.

We have defined isometries as the linear operators that preserve distance. The next theorem shows that they can also be characterized as the linear operators that preserve length, or dot products, or orthonormal bases.

**THEOREM** **10** The following are equivalent for a linear operator $T : \mathbb{R}^n \to \mathbb{R}^n$:

(1) $T$ is an isometry.
(2) $\|T(X)\| = \|X\|$ for all $X$ in $\mathbb{R}^n$.
(3) $T(X) \bullet T(Y) = X \bullet Y$ for all $X$ and $Y$ in $\mathbb{R}^n$.
(4) If $\{F_1, F_2, \cdots, F_n\}$ is any orthonormal basis of $\mathbb{R}^n$, so also is $\{T(F_1), T(F_2), \cdots, T(F_n)\}$.
(5) There exists an orthonormal basis $\{F_1, F_2, \cdots, F_n\}$ of $\mathbb{R}^n$ such that $\{T(F_1), T(F_2), \cdots, T(F_n)\}$ is also an orthonormal basis.

**PROOF** (1)$\Rightarrow$(2). We have $T(0) = 0$ because $T$ is linear. If $T$ is an isometry, then $\|T(X)\| = \|T(X) - T(0)\| = \|X - 0\| = \|X\|$ for all $X$.

(2)$\Rightarrow$(3). Since $T$ is linear, (2) gives $\|T(X) - T(Y)\| = \|T(X - Y)\| = \|X - Y\|$ for all $X$ and $Y$. This means that $T$ preserves distance, so (3) follows from the proof of (****) in Theorem 9.

(3)$\Rightarrow$(4). This is clear when one remembers that $\|T(F_i)\|^2 = T(F_i) \bullet T(F_i) = F_i \bullet F_i = \|F_i\|^2 = 1$ for each $i$.

(4)$\Rightarrow$(5). This is clear since $\mathbb{R}^n$ *has* orthonormal bases.

(5)$\Rightarrow$(1). Given $X$ in $\mathbb{R}^n$, write $X = x_1 F_1 + x_2 F_2 + \cdots + x_n F_n$ where the $x_i$ are real. Then $T(X) = x_1 T(F_1) + x_2 T(F_2) + \cdots + x_n T(F_n)$, so Pythagoras' theorem and (5) give $\|T(X)\|^2 = x_1^2 + x_2^2 + \cdots + x_n^2 = \|X\|^2$. It follows that $\|T(X)\| = \|X\|$ for each $X$. Since $T$ is linear, this gives $\|T(X) - T(Y)\| = \|T(X - Y)\| = \|X - Y\|$ for all $X$ and $Y$, proving (1).

Condition (5) in Theorem 10 is a useful test that $T$ is an isometry: Simply find an orthonormal basis $\{F_1, F_2, \cdots, F_n\}$ of $\mathbb{R}^n$ (possibly the standard basis) and check whether $\{T(F_1), T(F_2), \cdots, T(F_n)\}$ is also orthonormal.

**Example 18** Let $\{F_1, F_2, F_3, F_4\}$ be an orthonormal basis of $\mathbb{R}^4$. Using Theorem 5, let $T : \mathbb{R}^4 \to \mathbb{R}^4$ be the linear operator that satisfies $T(F_1) = F_2$, $T(F_2) = -F_1$, $T(F_3) = F_4$ and $T(F_4) = -F_3$. Then this map $T$ is an isometry by Theorem 10 because $\{T(F_1), T(F_2), T(F_3), T(F_4)\} = \{F_2, -F_1, F_4, -F_3\}$ is another orthonormal basis of $\mathbb{R}^4$.

**Example 19** If $\mathcal{F} = \{F_1, F_2, \cdots, F_n\}$ is an orthonormal basis of $\mathbb{R}^n$, the corresponding coordinate operator $C_{\mathcal{F}} : \mathbb{R}^n \to \mathbb{R}^n$ is an isometry by Theorem 10 because $\{C_{\mathcal{F}}(F_1), C_{\mathcal{F}}(F_2), \cdots, C_{\mathcal{F}}(F_n),\}$ is orthonormal (it is the standard basis of $\mathbb{R}^n$).

Isometries are closely related to orthogonal matrices as the next theorem shows.

**THEOREM** **11** Let $T : \mathbb{R}^n \to \mathbb{R}^n$ be a linear operator with standard matrix $A$. The following are equivalent:

(1) $T$ is an isometry.
(2) $A$ is an orthogonal matrix.
(3) $M_{\mathcal{F}}(T)$ is an orthogonal matrix for every orthonormal basis $\mathcal{F}$ of $\mathbb{R}^n$.
(4) $M_{\mathcal{F}}(T)$ is an orthogonal matrix for some orthonormal basis $\mathcal{F}$ of $\mathbb{R}^n$.

**PROOF** $(1)\Rightarrow(2)$. Write $A = [C_1 \quad C_2 \quad \cdots \quad C_n]$ in terms of its columns. If $E_i$ is column $i$ of the identity matrix, then $C_i = AE_i$ for each $i$ and we have
$C_i \bullet C_j = (AE_i) \bullet (AE_j) = T(E_i) \bullet T(E_j) = E_i \bullet E_j$ by Theorem 10. Hence $\{C_1, \ C_2, \ \cdots, \ C_n\}$ is an orthonormal set, and (2) follows.

$(2)\Rightarrow(3)$. If $\mathcal{F} = \{F_1, \ F_2, \ \cdots, \ F_n\}$ is an orthonormal basis of $\mathbb{R}^n$, write $P = [F_1 \quad F_2 \quad \cdots \quad F_n]$—an orthogonal matrix. By Theorem 7, $M_{\mathcal{F}}(T) = P^{-1}AP$, and this is orthogonal being a product of orthogonal matrices by (2).

$(3)\Rightarrow(4)$. This is clear.

$(4)\Rightarrow(1)$. If $\mathcal{F} = \{F_1, \ F_2, \ \cdots, \ F_n\}$ is as in (4) and $P = [F_1 \quad F_2 \quad \cdots \quad F_n]$, then $M_{\mathcal{F}}(T) = P^{-1}AP$ by Theorem 7. But $P$ is an orthogonal matrix, so $A = PM_{\mathcal{F}}(T)P^{-1}$ is also an orthogonal matrix. Hence, for $X$ in $\mathbb{R}^n$,
$\|T(X)\|^2 = \|AX\|^2 = (AX)^T(AX) = X^TA^TAX = X^TX = \|X\|^2$. It follows that $\|T(X)\| = \|X\|$ for all $X$ in $\mathbb{R}^n$, so $T$ is an isometry by Theorem 10.

Since orthogonal matrices are invertible, Theorem 5 gives

**COROLLARY** Every isometry is invertible.

It was shown in Section 3.4 that every isometry of $\mathbb{R}^2$ is either a rotation about the origin or a reflection in a line through the origin. Moreover, we have seen that rotations about a line through the origin are isometries of $\mathbb{R}^3$, as are reflections in a plane through the origin. It is a remarkable fact that every isometry of $\mathbb{R}^3$ is either a rotation, a reflection, or a composite of two of these.

**THEOREM 12** Let $T : \mathbb{R}^3 \to \mathbb{R}^3$ be an isometry. Then $T$ is either a rotation about a line through the origin, a reflection in a plane through the origin, or else a composite of two of these where the line is in fact perpendicular to the plane.

**PROOF** If $A$ is the standard matrix of $T$, the characteristic polynomial of $A$ has degree 3 (as $A$ is $3 \times 3$) and so has a real root $\lambda_1$, an eigenvalue of $A$. Since $A$ is orthogonal, $\lambda_1 = 1$ or $\lambda_1 = -1$ by Theorem 3 §4.7. Let $\vec{f_1}$ be a unit eigenvector corresponding to $\lambda_1$, and let $U = \mathbb{R}\vec{f_1}$ denote the line through the origin parallel to $\vec{f_1}$. Hence $U^{\perp}$ is the plane through the origin perpendicular to $U$.

*Claim.* If $\vec{v}$ is in $U^{\perp}$, then $T(\vec{v})$ is also in $U^{\perp}$.

*Proof.* Let $\vec{u}$ be any vector in $U$, so that $A\vec{u} = \lambda_1\vec{u}$. If $\vec{v}$ is in $U^{\perp}$ then, since $A$ is an orthogonal matrix by Theorem 11,
$$\lambda_1[(A\vec{v}) \bullet \vec{u}] = (A\vec{v}) \bullet (\lambda_1\vec{u}) = (A\vec{v}) \bullet (A\vec{u}) = \vec{v}^TA^TA\vec{u} = \vec{v}^T\vec{u} = \vec{v} \bullet \vec{u} = 0$$
because $\vec{v}$ is in $U^{\perp}$ and $\vec{u}$ is in $U$. As $\lambda_1 \neq 0$, we have $(A\vec{v}) \bullet \vec{u} = \vec{0}$ for every $\vec{u}$ in $U$. Hence $T(\vec{v}) = A\vec{v}$ is in $U^{\perp}$, proving the Claim.

The Claim shows that $T$ induces a transformation
$$T_0 : U^{\perp} \to U^{\perp} \text{ given by } T_0(\vec{v}) = T(\vec{v})$$
for all $\vec{v}$ in $U^{\perp}$. This mapping $T_0$ is an isometry of the plane $U^{\perp}$ (since $T$ is an isometry) and so is either a rotation or a reflection by Theorem 7 §3.4.[30] In each case, we can choose a convenient orthonormal basis $\{\vec{f_2}, \ \vec{f_3}\}$ of $U^{\perp}$ which will show that the theorem is true.

---

[30] Strictly speaking, Theorem 7 §3.4 refers to isometries of the plane $\mathbb{R}^2$. However, as we will see in Chapter 5, the important thing is that $U^{\perp}$ and $\mathbb{R}^2$ are both *planes*, and so have the same dimension.

*Case 1.* $T_0$ is a rotation of $U^\perp$ through an angle $\theta$. In this case choose an orthonormal basis $\{\vec{f_2}, \vec{f_3}\}$ of $U^\perp$ such that $\theta$ is counter-clockwise with respect to the axes in the $\vec{f_2}$ and $\vec{f_3}$ directions. Then, as in the solution of Example 16, we have
$T(\vec{f_2}) = \cos\theta\vec{f_2} + \sin\theta\vec{f_3}$ and $T(\vec{f_3}) = -\sin\theta\vec{f_2} + \cos\theta\vec{f_3}$. Moreover $T(\vec{f_1}) = \lambda_1\vec{f_1}$.
Since $\mathcal{F} = \{\vec{f_1}, \vec{f_2}, \vec{f_3}\}$ is an orthonormal basis of $\mathbb{R}^3$, part (2) of Theorem 7 gives

$$M_\mathcal{F}(T) = \begin{bmatrix} \lambda_1 & 0 & 0 \\ 0 & \cos\theta & -\sin\theta \\ 0 & \sin\theta & \cos\theta \end{bmatrix}.$$

If $\lambda_1 = 1$, this is the rotation through $\theta$ around the line $\mathbb{R}\vec{f_1}$ by Example 16. If $\lambda_1 = -1$, then

$$M_\mathcal{F}(T) = \begin{bmatrix} -1 & 0 & 0 \\ 0 & \cos\theta & -\sin\theta \\ 0 & \sin\theta & \cos\theta \end{bmatrix} = \begin{bmatrix} 1 & 0 & 0 \\ 0 & \cos\theta & -\sin\theta \\ 0 & \sin\theta & \cos\theta \end{bmatrix} \begin{bmatrix} -1 & 0 & 0 \\ 0 & 1 & 0 \\ 0 & 0 & 1 \end{bmatrix}$$

which (using Examples 16 and 17) is reflection in the plane $U^\perp$, followed by rotation through the angle $\theta$ about the line $U$. Hence the theorem holds in this case.

*Case 2.* $T_0$ is the reflection of the plane $U^\perp$ in a line $L_0$ in $U^\perp$ through the origin. In this case choose unit vectors $\vec{g_2}$ and $\vec{g_3}$ in $U^\perp$ where $\vec{g_2}$ is parallel to $L_0$ and $\vec{g_3}$ is orthogonal to $L_0$. Hence $T(\vec{g_2}) = \vec{g_2}$ and $T(\vec{g_3}) = -\vec{g_3}$. If $\lambda_1 = 1$, we use the orthonormal basis $\mathcal{G} = \{\vec{g_3}, \vec{f_1}, \vec{g_2}\}$ of $\mathbb{R}^3$. Then Theorem 7 gives

$$M_\mathcal{G}(T) = \begin{bmatrix} -1 & 0 & 0 \\ 0 & \lambda_1 & 0 \\ 0 & 0 & 1 \end{bmatrix} = \begin{bmatrix} -1 & 0 & 0 \\ 0 & 1 & 0 \\ 0 & 0 & 1 \end{bmatrix}$$

so $T$ is reflection in the plane $span\{\vec{f_1}, \vec{g_2}\}$ (see Example 17). If $\lambda_1 = -1$, we use $\mathcal{G} = \{\vec{g_2}, \vec{g_3}, \vec{f_1}\}$. Hence

$$M_\mathcal{G}(T) = \begin{bmatrix} 1 & 0 & 0 \\ 0 & -1 & 0 \\ 0 & 0 & \lambda_1 \end{bmatrix} = \begin{bmatrix} 1 & 0 & 0 \\ 0 & -1 & 0 \\ 0 & 0 & -1 \end{bmatrix}$$

which is rotation of $\pi$ around the line $\mathbb{R}\vec{g_2}$ by Example 16.

In fact we have proved that an isometry is either a rotation about a line through the origin, a reflection in a plane through the origin, or else the composite of a reflection in a plane through the origin followed by a rotation about a line through the origin perpendicular to the plane.

---

## Exercises 4.9

1. In each case either show that the statement is true or give an example showing that it is false. Let $\{\vec{e_1}, \vec{e_2}\}$ denote the standard basis of $\mathbb{R}^2$.

   a) If $T : \mathbb{R}^n \to \mathbb{R}^m$ is linear and $T(X) = 0$ for some $X$ in $\mathbb{R}^n$, then $X = 0$.

   b) There exists a linear transformation $T : \mathbb{R}^2 \to \mathbb{R}^3$ such that $T(\vec{e_1}) = [1 \ \ 1 \ \ 0]^T$ and $T(\vec{e_2}) = [0 \ \ 1 \ \ 1]^T$.

   c) There exists a linear transformation $T : \mathbb{R}^2 \to \mathbb{R}^4$ such that $T(\vec{e_1}) = [1 \ -1 \ \ 3 \ -2]^T$ and $T(2\vec{e_1}) = [-1 \ \ 1 \ -3 \ \ 2]^T$.

   d) If $T : \mathbb{R}^2 \to \mathbb{R}^3$ is linear and $\{X, Y\}$ is orthogonal in $\mathbb{R}^2$, then $\{T(X), T(Y)\}$ is orthogonal in $\mathbb{R}^3$.

   e) If $T : \mathbb{R}^4 \to \mathbb{R}^2$ is defined by $T[x_1 \ \ x_2 \ \ x_3 \ \ x_4]^T = [x_4 \ \ x_3]^T$, then $T$ is linear.

   f) The composite of two invertible operators is invertible.

   g) If $S = 0$ is the zero operator $\mathbb{R}^n \to \mathbb{R}^n$, then $T \circ S = 0$ for all operators $T$.

2. If $Y$ and $Z$ are vectors in $\mathbb{R}^n$, a vector $X$ is said to be **between** $Y$ and $Z$ if $X = (1-f)Y + fZ$ for some number $f$ such that $0 \le f \le 1$ [see Exercise 16 §3.1]. If $T : \mathbb{R}^n \to \mathbb{R}^m$ is a linear transformation and $X$ is between $Y$ and $Z$, show that $T(X)$ is between $T(Y)$ and $T(Z)$.

3. If $a$ is a fixed real number, consider the scalar operator $T : \mathbb{R}^n \to \mathbb{R}^n$ defined by $T(X) = aX$. Show that $T$ is a linear operator, and find the standard matrix of $T$.

4. If $m < n$ define $T : \mathbb{R}^n \to \mathbb{R}^m$ by $T[x_1 \ x_2 \ \cdots \ x_n]^T = [x_1 \ x_2 \ \cdots \ x_m]^T$. Show that $T$ is a linear transformation, and find its standard matrix.

5. If $m > n$ define $T : \mathbb{R}^n \to \mathbb{R}^m$ by
$$T[x_1 \ x_2 \ \cdots \ x_n]^T = [x_1 \ x_2 \ \cdots \ x_n \ 0 \ \cdots \ 0]^T.$$
Show that $T$ is a linear transformation, and find its standard matrix.

6. a) Find $T([x_1 \ x_2 \ x_3]^T)$ if $T : \mathbb{R}^3 \to \mathbb{R}^3$ is a linear transformation satisfying $T([1 \ 1 \ 0]^T) = [3 \ -1 \ 3]^T$, $T([0 \ 1 \ 1]^T) = [2 \ 1 \ 3]^T$, and $T([0 \ 0 \ 1]^T) = [-1 \ -2 \ 0 \ ]^T$.
   b) Find $T([x_1 \ x_2 \ x_3 \ x_4]^T)$ if $T : \mathbb{R}^4 \to \mathbb{R}^2$ is a linear transformation satisfying $T([1 \ 1 \ 0 \ 0]^T) = [2 \ -1]^T$, $T([0 \ 1 \ 0 \ 0]^T) = [-2 \ 3]^T$, $T([0 \ 1 \ 0 \ 2]^T) = [1 \ 0]^T$, and $T([0 \ 0 \ 1 \ 0]^T) = [0 \ -1]^T$.

7. Let $\mathcal{F}$ denote a basis of $\mathbb{R}^3$, and let $T$ be an operator on $\mathbb{R}^3$. In each case find $T([x_1 \ x_2 \ x_3]^T)$.

   a) $\mathcal{F} = \left\{ \begin{bmatrix} 1 \\ 0 \\ 0 \end{bmatrix}, \begin{bmatrix} 1 \\ 1 \\ 0 \end{bmatrix}, \begin{bmatrix} 0 \\ 1 \\ 1 \end{bmatrix} \right\}$ and $M_{\mathcal{F}}(T) = \begin{bmatrix} 1 & 0 & -1 \\ 2 & 1 & 0 \\ -1 & 1 & 4 \end{bmatrix}$.

   b) $\mathcal{F} = \left\{ \begin{bmatrix} 1 \\ 2 \\ 2 \end{bmatrix}, \begin{bmatrix} 2 \\ 1 \\ -2 \end{bmatrix}, \begin{bmatrix} 2 \\ -2 \\ 1 \end{bmatrix} \right\}$ and $M_{\mathcal{F}}(T) = \begin{bmatrix} 3 & -5 & 1 \\ 1 & 1 & 5 \\ -2 & 2 & 7 \end{bmatrix}$.

8. If $X \neq 0$ and $Y$ are vectors in $\mathbb{R}^n$ and $\mathbb{R}^m$ respectively, show that there is a linear transformation $T : \mathbb{R}^n \to \mathbb{R}^m$ such that $T(X) = Y$. [Hint: Part (2) of Theorem 3 §4.3.]

9. A linear transformation $T : \mathbb{R}^n \to \mathbb{R}^1 = \mathbb{R}$ is called a **linear functional** on $\mathbb{R}^n$.
   a) If $X_0$ is a fixed column in $\mathbb{R}^n$, show that $T(X) = X_0 \cdot X$ defines a linear functional $T$ on $\mathbb{R}^n$.
   b) Show that every linear functional on $\mathbb{R}^n$ arises as in a) for some $X_0$ in $\mathbb{R}^n$. [Hint: Let $X_0 = [a_1, \ \cdots, \ a_n]^T$ where $a_i = T(E_i)$ and $\{E_1, \ \cdots, \ E_n\}$ is the standard basis.]

10. Suppose that $\mathbb{R}^n = span\{Y_1, Y_2, \ \cdots, \ Y_k\}$. If $T : \mathbb{R}^n \to \mathbb{R}^m$ is a linear transformation and $T(Y_i) = 0$ for each $i$, show that $T = 0$ is the zero transformation.

11. a) If $X_0$ is a fixed column in $\mathbb{R}^n$, show that $T(r) = rX_0$ defines a linear transformation $\mathbb{R} \to \mathbb{R}^n$.
    b) Show that every linear transformation $\mathbb{R} \to \mathbb{R}^n$ arises as in a) for some $X_0$ in $\mathbb{R}^n$. [Hint: Take $X_0 = T(1)$.]

12. If $T : \mathbb{R}^n \to \mathbb{R}^m$ is a linear transformation, define the **kernel** of $T$ (denoted $kerT$) by $kerT = \{X \text{ in } \mathbb{R}^n \mid T(X) = 0\}$.
    a) Show that $kerT$ is a subspace of $\mathbb{R}^n$.
    b) If $S \circ T = 1_{\mathbb{R}^n}$ for some $S : \mathbb{R}^m \to \mathbb{R}^n$, show that $kerT = 0$.

13. If $T : \mathbb{R}^n \to \mathbb{R}^m$ is a linear transformation, define the **image** of $T$ (denoted $imT$) by $imT = \{T(X) \mid X \text{ in } \mathbb{R}^n\}$.
    a) Show that $imT$ is a subspace of $\mathbb{R}^m$.
    b) If $T \circ S = 1_{\mathbb{R}^n}$ for some $S : \mathbb{R}^n \to \mathbb{R}^m$, show that $imT = \mathbb{R}^m$.

14. Let $T : \mathbb{R}^n \to \mathbb{R}^m$ be a linear transformation, and let $X_1, X_2, \ \cdots, \ X_k$ denote vectors in $\mathbb{R}^n$.
    a) If $\{T(X_1), T(X_2), \ \cdots, \ T(X_k)\}$ is independent in $\mathbb{R}^m$, show that $\{X_1, X_2, \ \cdots, \ X_k\}$ is independent in $\mathbb{R}^n$.
    b) If $\{X_1, X_2, \ \cdots, \ X_k\}$ spans $\mathbb{R}^n$, show that $\{T(X_1), T(X_2), \ \cdots, \ T(X_k)\}$ spans $imT$. (See the preceding exercise.)

15. Given bases $\mathcal{F} = \{[1 \ 1]^T, [0 \ 1]^T\}$ and $\mathcal{G} = \{[2 \ 3]^T, [1 \ 2]^T\}$ of $\mathbb{R}^2$, compute the change matrices $P_{\mathcal{G} \leftarrow \mathcal{F}}$ and $P_{\mathcal{F} \leftarrow \mathcal{G}}$. Verify that they are inverses of each other, and that $C_{\mathcal{G}}(X) = P_{\mathcal{G} \leftarrow \mathcal{F}} C_{\mathcal{F}}(X)$ holds for all $X$ in $\mathbb{R}^2$.

16. In Theorem 9, prove that $T(X + Y) = T(X) + T(Y)$ for all $X$ and $Y$ in $\mathbb{R}^n$.

17. If $\mathcal{F}$ is a basis of $\mathbb{R}^n$, find a specific formula for $C_{\mathcal{F}}^{-1} : \mathbb{R}^n \to \mathbb{R}^n$. What is the matrix of $C_{\mathcal{F}}^{-1}$?

18. a) If $T : \mathbb{R}^n \to \mathbb{R}^n$ is invertible, show that $T^{-1}$ is also invertible.
    b) If $S : \mathbb{R}^n \to \mathbb{R}^n$ and $T : \mathbb{R}^n \to \mathbb{R}^n$ are both invertible, show that $S \circ T$ is also invertible.

19. Let $\mathcal{F} = \{F_1, \ \cdots, \ F_n\}$ be a basis of $\mathbb{R}^n$, and write $P = [F_1 \ \cdots \ F_n]$. Show that:
    a) $P_{\mathcal{F} \leftarrow \mathcal{E}} = P^{-1}$;    b) $P_{\mathcal{E} \leftarrow \mathcal{F}} = P$.

20. If $T : \mathbb{R}^n \to \mathbb{R}^n$ has standard matrix $A$ and $A^2 = A$, show that $T \circ T = T$.

21. Show that the composite of two distance-preserving isometries of $\mathbb{R}^n$ is again distance preserving.

22. If $Y$ is a fixed vector in $\mathbb{R}^n$, let $Q_Y : \mathbb{R}^n \to \mathbb{R}^n$ denote translation by $Y$, that is $Q_Y(X) = X + Y$ for all $X$ in $\mathbb{R}^n$. Show that $T \circ Q_Y = Q_{T(Y)} \circ T$ holds for every linear operator $T : \mathbb{R}^n \to \mathbb{R}^n$.

23. Let $U$ denote any subspace of $\mathbb{R}^n$ and write $P(X) = proj_U(X)$ for all $X$ in $\mathbb{R}^n$. Define $T : \mathbb{R}^n \to \mathbb{R}^n$ by $T(X) = 2P(X) - X$ for all $X$ in $\mathbb{R}^n$. We call $T$ a **reflection** in the subspace $U$ by analogy with the case when $n = 3$ and $U$ is a plane in $\mathbb{R}^n$ (see Theorem 1 §3.4). Show that $T$ is an isometry of $\mathbb{R}^n$. [Hint: We have $T(X) = P(X) + [P(X) - X]$ and $X = P(X) + [X - P(X)]$. Use the Projection theorem and Pythagoras' theorem.]

24. If $T$ is the reflection of $\mathbb{R}^3$ in a plane through the origin (see the preceding exercise), show that the standard matrix of $T$ is diagonalizable.

**25.** If $A$ is an $n \times n$ matrix, define $T_A : \mathbb{R}^n \to \mathbb{R}^n$ by $T_A(X) = AX$ for all $X$ in $\mathbb{R}^n$.

  **a)** If there exists a basis $\mathcal{F} = \{F_1, F_2, \cdots, F_n\}$ of $\mathbb{R}^n$ such that $M_{\mathcal{F}}(T_A)$ is diagonal, show that $A$ is diagonalizable.

  **b)** If $A$ is diagonalizable, show that a basis $\mathcal{F} = \{F_1, F_2, \cdots, F_n\}$ of $\mathbb{R}^n$ exists such that $M_{\mathcal{F}}(T_A)$ is diagonal. [Hint: $A$ has a basis of eigenvectors.]

**26.** Let $\mathcal{F}$ be a basis of $\mathbb{R}^n$. If $A$ and $B$ are $m \times n$ matrices and $A\,C_{\mathcal{F}}(X) = B\,C_{\mathcal{F}}(X)$ for all $X$ in $\mathbb{R}^n$, show that $A = B$.

**27.** Let $\mathcal{F}$ be a basis of $\mathbb{R}^n$, and let $S : \mathbb{R}^n \to \mathbb{R}^n$ and $T : \mathbb{R}^n \to \mathbb{R}^n$ be operators.

  **a)** Show that $M_{\mathcal{F}}(S \circ T) = M_{\mathcal{F}}(S)M_{\mathcal{F}}(T)$. [Hint: For each $X$ in $\mathbb{R}^n$, right multiply by $C_{\mathcal{F}}(X)$ and use Theorem 7 and the preceding exercise.]

  **b)** If $T$ is invertible, show that $M_{\mathcal{F}}(T)$ is invertible and $[M_{\mathcal{F}}(T)]^{-1} = M_{\mathcal{F}}(T^{-1})$. [Hint: Use **a)**.]

**28.** If $T : \mathbb{R}^n \to \mathbb{R}^n$ is an operator and $a$ is a scalar, define $aT : \mathbb{R}^n \to \mathbb{R}^n$ by $(aT)(X) = aT(X)$ for all $X$ in $\mathbb{R}^n$.

  **a)** Show that $aT$ is an operator (called the **scalar multiple** of $T$ by $a$).

  **b)** If $T$ is an isometry, show that $aT$ is an isometry if and only if $a = \pm 1$.
  We say that an operator $S$ **preserves orthogonality** if $X \bullet Y = 0$ in $\mathbb{R}^n$ implies that $T(X) \cdot T(Y) = 0$.

  **c)** Show that a linear operator $S : \mathbb{R}^n \to \mathbb{R}^n$ preserves orthogonality if and only if it is a scalar multiple of an isometry. [Hint: If $S$ preserves orthogonality, show that $\|S(E)\| = \|S(F)\|$ for any unit vectors using the fact that $(E + F) \bullet (E - F) = 0$. If $\{F_1, \cdots, F_n\}$ is an orthonormal basis of $\mathbb{R}^n$, let $\|S(F_i)\| = a$ for all $i$. If $a = 0$, then $S = a1_{\mathbb{R}^n}$; if $a \neq 0$, show that $T = \frac{1}{a}S$ is an isometry.]

**29.** If $Y$ is a point in $\mathbb{R}^2$, let $S : \mathbb{R}^n \to \mathbb{R}^n$ denote rotation about $Y$ through an angle $\theta$. $S$ is clearly distance preserving. Exhibit $S$ as a rotation followed by a translation (as guaranteed by Corollary 2 of Theorem 9).

**30.** Show that the isometries of $\mathbb{R}^n$ enjoy the following properties:
  **1)** $1_{\mathbb{R}^n}$ is an isometry.
  **2)** If $T$ is an isometry, so is $T^{-1}$.
  **3)** If $S$ and $T$ are isometries, so is the composite $S \circ T$.
  By virtue of these properties, the set of all isometries is called a *group* of operators.

**31.** **a)** If $T$ is an isometry of $\mathbb{R}^n$, show that $det(M_{\mathcal{F}}(T)) = \pm 1$ for any basis $\mathcal{F}$ of $\mathbb{R}^n$.

  **b)** If $T$ is an isometry of $\mathbb{R}^3$ with standard matrix $A$, show that: $T$ is a rotation if and only if $det(M_{\mathcal{F}}(T)) = 1$ for any $\mathcal{F}$; and $T$ is a reflection if and only if $det(M_{\mathcal{F}}(T)) = -1$ for any $\mathcal{F}$. [Hint: Theorem 12 and Exercise 23.]

**32.** Let $\mathcal{F} = \{F_1, \cdots, F_n\}$ and $\mathcal{G} = \{G_1, \cdots, G_n\}$ be bases of $\mathbb{R}^n$, and write $P_{\mathcal{G} \leftarrow \mathcal{F}} = P$. If $T : \mathbb{R}^n \to \mathbb{R}^n$ is a linear operator, show that $PM_{\mathcal{F}}(T)P^{-1} = M_{\mathcal{G}}(T)$. [Hint: Show that $P M_{\mathcal{F}}(T) = M_{\mathcal{G}}(T) P$ by right multiplying by $C_{\mathcal{F}}(X)$ for all $X$ and applying Exercise 26.]

**33.** Extend Theorem 7 as follows: Let $T : \mathbb{R}^n \to \mathbb{R}^m$ be a linear transformation with standard matrix $A$. If $\mathcal{F} = \{F_1, \cdots, F_n\}$ and $\mathcal{G} = \{G_1, \cdots, G_m\}$ are bases of $\mathbb{R}^n$ and $\mathbb{R}^m$ respectively, write $P = [F_1 \cdots F_n]$ and $Q = [G_1 \cdots G_m]$.

  **a)** Show that $C_{\mathcal{G}}(T(X)) = Q^{-1}AP\,C_{\mathcal{F}}(X)$ for all $X$ in $\mathbb{R}^n$.

  **b)** Show that $Q^{-1}AP = [C_{\mathcal{G}}(T(F_1))\ \ C_{\mathcal{G}}(T(F_2))\ \cdots\ C_{\mathcal{G}}(T(F_n))]$. [Hint: Proceed as in the proof of Theorem 7 and the paragraph preceding it.]

# 4.10  COMPLEX MATRICES [31]

Ever since we introduced eigenvalues, it has been clear that complex numbers play an important role in linear algebra. Even such a simple matrix as $A = \begin{bmatrix} 0 & -1 \\ 1 & 0 \end{bmatrix}$ has characteristic polynomial $x^2 + 1$, and so has nonreal eigenvalues $i$ and $-i$. On the other hand, many of the applications of linear algebra involve only real numbers, so we have focused on $\mathbb{R}^n$ and on real matrices throughout this book. However, there is much to be gained by looking at $\mathbb{C}^n$ and at complex matrices, where $\mathbb{C}$ denotes the set of complex numbers. While a brief discussion of the complex numbers was given in Section 2.5, we give a more comprehensive discussion of complex matrices in this section, and extend several of our earlier results about real matrices. Only the basic properties of complex numbers will be needed—mainly complex arithmetic, conjugation, and absolute value.

---

[31] This section is not needed elsewhere in the book.

## 4.10.1 Complex Inner Products

Most of what we have done in Chapters 1 and 2 remains valid if the phrase "real number" is replaced by "complex number" wherever it occurs. The Gaussian algorithm works to solve systems of linear equations with complex coefficients, and the properties of matrix arithmetic and determinants remain the same. Furthermore, the proofs of theorems in the real case carry over unchanged to complex matrices. Consequently, we will not repeat those arguments again here.

The set of all $n$-tuples of complex numbers is denoted $\mathbb{C}^n$ and, as for $\mathbb{R}^n$, we denote them as columns (sometimes as rows) and refer to them as vectors. The addition and scalar multiplication of vectors in $\mathbb{C}^n$ is entirely analogous to $\mathbb{R}^n$. However, one important difference arises when the length of a complex vector is considered. For example, using the formula from $\mathbb{R}^n$ on the vector $Z = [1 \ \ i]^T$ in $\mathbb{C}^2$, we find that $Z$ has length $\sqrt{1^2 + i^2} = \sqrt{1 + (-1)} = 0$. Since it is important to retain the property that the length of a nonzero vector is positive, we see that the obvious extension of the length formula from $\mathbb{R}^n$ to $\mathbb{C}^n$ does not work. Fortunately, there is a satisfactory way to define the length of a vector in $\mathbb{C}^n$ which reduces to the usual length for vectors in $\mathbb{R}^n$. In fact, we can suitably extend the dot product.

Given $Z = [z_1 \ \ z_2 \ \cdots \ z_n]^T$ and $W = [w_1 \ \ w_2 \ \cdots \ w_n]^T$ in $\mathbb{C}^n$, define their **standard inner product** $\langle Z, \ W \rangle$ to be the complex number given by the formula

$$\langle Z, \ W \rangle = z_1 \overline{w}_1 + z_2 \overline{w}_2 + \cdots + z_n \overline{w}_n.$$

Here $\overline{w}$ denotes the conjugate of the complex number $w$. This reduces to the dot product on $\mathbb{R}^n$ since, if $Z$ and $W$ are actually in $\mathbb{R}^n$, then $\overline{w}_i = w_i$ for each $i$, so $\langle Z, \ W \rangle = Z \cdot W$. Moreover, this leads to a satisfactory notion of length in $\mathbb{C}^n$. Indeed,

$$\langle Z, \ Z \rangle = z_1 \overline{z}_1 + z_2 \overline{z}_2 + \cdots + z_n \overline{z}_n = |z_1|^2 + |z_2|^2 + \cdots + |z_n|^2$$

where $|z|$ denotes the absolute value of the complex number $z$. Hence $\langle Z, \ Z \rangle > 0$ if $Z \neq 0$ so, as in $\mathbb{R}^n$, we define the **length** $\|Z\|$ of $Z$ in $\mathbb{C}^n$ to be

$$\|Z\| = \sqrt{\langle Z, \ Z \rangle} = \sqrt{|z_1|^2 + |z_2|^2 + \cdots + |z_n|^2}.$$

Hence $\|Z\| = 0$ if and only if $Z = 0$ because $|z_1|^2 + |z_2|^2 + \cdots + |z_n|^2$ is a nonnegative real number which is zero if and only if every $z_i = 0$. This explains why the coefficients of $W$ are conjugated in the definition of $\langle Z, \ W \rangle$, and it also proves (4) of the following theorem.

**THEOREM 1**   Let $Z, Z_1, W$, and $W_1$ denote vectors in $\mathbb{C}^n$, and let $\lambda$ denote a complex number.

(1) $\langle Z + Z_1, \ W \rangle = \langle Z, \ W \rangle + \langle Z_1, \ W \rangle$   and   $\langle Z, \ W + W_1 \rangle = \langle Z, \ W \rangle + \langle Z, \ W_1 \rangle$.

(2) $\langle \lambda Z, \ W \rangle = \lambda \langle Z, \ W \rangle$   and   $\langle Z, \ \lambda W \rangle = \overline{\lambda} \langle Z, \ W \rangle$.

(3) $\langle Z, W \rangle = \overline{\langle W, Z \rangle}$.

(4) $\|Z\| \geq 0$ and $\|Z\| = 0$   if and only if   $Z = 0$.

(5) $\|\lambda Z\| = |\lambda| \ \|Z\|$.

**PROOF**   We verified (4) above, and we leave (1), (2), and (5) to the reader. To prove (3) let $Z = [z_1 \ \ z_2 \ \cdots \ z_n]^T$ and $W = [w_1 \ \ w_2 \ \cdots \ w_n]^T$. Since $\overline{z + w} = \overline{z} + \overline{w}$, $\overline{(zw)} = \overline{z} \ \overline{w}$, and $\overline{\overline{z}} = z$ hold for all complex numbers $z$ and $w$, we have

$$\overline{\langle W, Z \rangle} = \overline{w_1 \overline{z}_1 + \cdots + w_n \overline{z}_n}$$
$$= \overline{w}_1 \overline{(\overline{z}_1)} + \cdots + \overline{w}_n \overline{(\overline{z}_n)}$$
$$= \overline{w}_1 z_1 + \cdots + \overline{w}_n z_n$$
$$= \langle Z, W \rangle.$$

This proves (3).

The properties in Theorem 1 will be used frequently in what follows, usually without comment.

As for $\mathbb{R}^n$, a vector $U$ in $\mathbb{C}^n$ is called a **unit vector** if $\|U\| = 1$. With this, property (5) in Theorem 1 shows that if $Z \neq 0$ in $\mathbb{C}^n$, then $\frac{1}{\|Z\|}Z$ is a unit vector.

**Example 1**  If $Z = [i \;\; 1-i \;\; -3 \;\; 2+3i]^T$ and $W = [2-i \;\; 1 \;\; 1+i \;\; -i]^T$, compute: (1) $\|Z\|$; (2) A unit vector that is a positive scalar multiple of $Z$; (3) $\langle Z, \; W \rangle$.

SOLUTION

(1). $\|Z\|^2 = |i|^2 + |1-i|^2 + |-3|^2 + |2+3i|^2 = 1 + 2 + 9 + 13 = 25$, so $\|Z\| = 5$.

(2). Using (1), $U = \frac{1}{\|Z\|}Z = \frac{1}{5}Z$ does it.

(3). $\langle Z, \; W \rangle = (i)\overline{(2-i)} + (1-i)\overline{(1)} + (-3)\overline{(1+i)} + (2+3i)\overline{(-i)}$
$$= i(2+i) + (1-i) - 3(1-i) + (2+3i)i$$
$$= -6 + 6i.$$

## 4.10.2 Complex Matrices

Matrix multiplication, addition, transposition, and scalar multiplication (by complex scalars) are defined exactly as for real matrices, and the algebraic properties developed in Chapter 1 remain valid. However, the conjugation operation for complex numbers introduces another matrix operation. If $G = [g_{ij}]$ is a complex matrix, the **conjugate** $\overline{G}$ of $G$ is defined to be the matrix

$$\overline{G} = [\overline{g}_{ij}]$$

obtained from $G$ by conjugating every entry. It is a routine exercise to verify the following properties of matrix conjugation.

**THEOREM 2**  Let $G$ and $K$ denote complex matrices, and let $\lambda$ be a complex number.

(1) $\overline{G + K} = \overline{G} + \overline{K}$.

(2) $\overline{GK} = \overline{G}\,\overline{K}$.

(3) $\overline{(\lambda G)} = \overline{\lambda}\,\overline{G}$.

(4) $(\overline{G})^T = \overline{(G^T)}$.

(5) $\langle Z, \; W \rangle = Z^T \overline{W}$ for all columns $Z$ and $W$ in $\mathbb{C}^n$.

As for Theorem 1, these properties will be used repeatedly in what follows.

Matrix conjugation enables us to define the following notion which is essential in discussing complex analogues of the symmetric and orthogonal real matrices. If $G$ is a complex matrix, the **conjugate transpose** $G^*$ of $G$ is defined by

$$G^* = (\overline{G})^T = \overline{(G^T)}.$$

Observe that $A^* = A^T$ when $A$ is a real matrix.

**Example 2**  If $G = \begin{bmatrix} -1 & 2-i & -2i \\ 3+4i & i & 7 \end{bmatrix}$ then $G^* = \begin{bmatrix} -1 & 3-4i \\ 2+i & -i \\ 2i & 7 \end{bmatrix}$.

The following properties of the conjugate transpose follow easily from the definition of $G^*$ and the rules governing transposition. As before these properties will be used several times below.

**THEOREM 3**   Let $G$ and $K$ denote complex matrices, and let $\lambda$ be a complex number.

(1) $(G^*)^* = G$.
(2) $(G + K)^* = G^* + K^*$.
(3) $(\lambda G)^* = \overline{\lambda} G^*$.
(4) $(GK)^* = K^* G^*$.

## 4.10.3  Hermitian and Unitary Matrices

If $A$ is a real symmetric matrix, it is clear that $A^* = A$. The complex matrices that satisfy this condition turn out to be the most natural generalization of the symmetric matrices. A square complex matrix $H$ is called **Hermitian**[32] if

$$H^* = H \quad \text{equivalently} \quad H^T = \overline{H}.$$

Thus every real symmetric matrix is Hermitian. More generally, the Hermitian matrices are easy to recognize because the main diagonal entries must be real and the "reflection" of each nondiagonal entry in the main diagonal is the conjugate of that entry.

**Example 3**   $\begin{bmatrix} -1 & 2-i & -2i \\ 2+i & 7 & i \\ 2i & -i & 0 \end{bmatrix}$ is Hermitian, but $\begin{bmatrix} 0 & 3 & 1+i \\ 3 & 6 & 5i \\ 1-i & -5i & i \end{bmatrix}$ and $\begin{bmatrix} 2 & 1+i & 2-3i \\ 1-i & -4 & 5i \\ 3-2i & -5i & 3 \end{bmatrix}$ are not Hermitian.

These Hermitian matrices have an important characterization in terms of the standard inner product in $\mathbb{C}^n$.

**LEMMA 1**   An $n \times n$ complex matrix $H$ is Hermitian if and only if

$$\langle HZ, W \rangle = \langle Z, HW \rangle$$

for all columns $Z$ and $W$ in $\mathbb{C}^n$.

**PROOF**   If $H$ is Hermitian, we have $H^T = \overline{H}$. Hence, if $Z$ and $W$ are columns in $\mathbb{C}^n$, we have

$$\langle HZ, W \rangle = (HZ)^T \overline{W} = Z^T H^T \overline{W} = Z^T \overline{H}\, \overline{W} = Z^T \overline{(HW)} = \langle Z, HW \rangle.$$

Conversely, let $E_j$ denote column $j$ of $I_n$. If $H = [h_{ij}]$, some matrix arithmetic (and the hypothesis) gives

$$\overline{h}_{ij} = E_i^T \overline{H}\, E_j = E_i^T \overline{(HE_j)} = \langle E_i, HE_j \rangle$$
$$= \langle HE_i, E_j \rangle = (HE_i)^T \overline{E_j} = E_i^T H^T E_j = h_{ji}$$

for all $i$ and $j$. This shows that $\overline{H} = H^T$, as required.

Let $G$ be an $n \times n$ complex matrix. As in the real case, a complex number $\lambda$ is called an **eigenvalue** of $G$ if

$$GZ = \lambda Z \quad \text{for some nonzero column} \quad Z \neq 0 \text{ in } \mathbb{C}^n.$$

In this case $Z$ is called an **eigenvector** of $G$ corresponding to $\lambda$. The polynomial

$$c_G(x) = det(xI - G)$$

is called the **characteristic polynomial** of $G$. It has degree $n$ as in the real case, but now it has complex coefficients. The proof of Theorem 2 §2.3 goes through to show that the eigenvalues of $G$ are precisely the complex roots of $c_G(x)$. This reveals one reason why

[32] Charles Hermite (1822–1901) was a French mathematician, who proved in 1855 that the eigenvalues of these matrices are all real (see Theorem 4). Hermite is best remembered as the first to show that the number $e$ from calculus is not the root of any polynomial with coefficients that are integers.

complex linear algebra is a more natural setting for the subject than its real counterpart: The so-called Fundamental Theorem of Algebra[33] asserts that $c_G(x)$ factors completely as follows:

$$c_G(x) = (x - \lambda_1)(x - \lambda_2) \cdots (x - \lambda_n)$$

where $\lambda_1$, $\lambda_2$, $\cdots \lambda_n$ are the eigenvalues of $G$, possibly with repetitions due to multiple roots.

Theorems 3 §2.5 and 6 §4.7 assert respectively that the eigenvalues of any symmetric matrix are real, and that eigenvectors corresponding to distinct eigenvalues are orthogonal. The next result shows that this remains true for any Hermitian matrix, and so confirms that the Hermitian matrices are the correct complex analogue of the symmetric real matrices. In the complex context, two columns $Z$ and $W$ in $\mathbb{C}^n$ are said to be **orthogonal** if $\langle Z, W \rangle = 0$.

---

**THEOREM 4**

Let $H$ denote a Hermitian matrix.

(1) The eigenvalues of $H$ are real.
(2) Eigenvectors of $H$ corresponding to distinct eigenvalues are orthogonal.

PROOF

Let $\lambda$ and $\mu$ be eigenvalues of $H$ with eigenvectors $Z$ and $W$ respectively. Then $HZ = \lambda Z$ and $HW = \mu W$, so Lemma 1 and Theorem 1 give

$$\lambda \langle Z, W \rangle = \langle \lambda Z, W \rangle = \langle HZ, W \rangle = \langle Z, HW \rangle = \langle Z, \mu W \rangle = \overline{\mu} \langle Z, W \rangle. \qquad (*)$$

If $\lambda = \mu$ and $Z = W$, this reduces to $(\lambda - \overline{\lambda}) \|Z\|^2 = 0$, whence $\lambda = \overline{\lambda}$ and $\lambda$ is real, proving (1). But then $\mu = \overline{\mu}$ too, so $(*)$ becomes $(\lambda - \mu) \langle Z, W \rangle = 0$. If $\lambda \neq \mu$, this implies $\langle Z, W \rangle = 0$, proving (2).

---

The main result about symmetric matrices is the Principal Axis theorem which asserts that every symmetric real matrix $A$ can be orthogonally diagonalized: that is, there exists an orthogonal matrix $P$ such that $P^T AP$ is diagonal. Here the real matrix $P$ is orthogonal if $P^{-1} = P^T$, and the next result identifies the complex analogues of these matrices. As in the real case, a set $\{Z_1, Z_2, \cdots, Z_n\}$ of columns in $\mathbb{C}^n$ is said to be **orthogonal** if $Z_i \neq 0$ for each $i$ and $\langle Z_i, Z_j \rangle = 0$ whenever $i \neq j$. If in addition $\|Z_i\| = 1$ for each $i$, the set $\{Z_1, Z_2, \cdots, Z_n\}$ is called **orthonormal**.

---

**THEOREM 5**

The following conditions are equivalent for an $n \times n$ complex matrix $U$:

(1) $U^{-1} = U^*$.
(2) The columns of $U$ are an orthonormal set in $\mathbb{C}^n$.
(3) The rows of $U$ are an orthonormal set in $\mathbb{C}^n$.

PROOF

If $C_j$ denotes column $j$ of $U$, the $(i, j)$-entry of $U^T \overline{U}$ is $C_i^T \overline{C}_j = \langle C_i, C_j \rangle$. It follows that $U^T \overline{U} = I_n$ if and only if $\{C_1, \cdots, C_n\}$ is an orthonormal set. But $U^T \overline{U} = I_n$ if and only if $U^* U = I_n$, and $(1) \Leftrightarrow (2)$ follows. The proof that $(1) \Leftrightarrow (3)$ is similar.

---

A square complex matrix $U$ is called **unitary** if it satisfies the conditions in Theorem 5. Thus a real matrix $A$ is unitary if and only if it is orthogonal (because $A^* = A^T$). As in the real case, it is essential that the columns of $U$ be *orthonormal* in Theorem 5.

---

[33] The Fundamental Theorem of Algebra actually shows that *every* polynomial $p(x)$ with complex coefficients has a complex root. This implies that $p(x)$ is a product of linear factors (see Appendix A.3). The theorem was first proved by Gauss in 1799 at 22 years of age.

**Example 4**    $\begin{bmatrix} 1+i & 1 \\ 1-i & i \end{bmatrix}$ has orthogonal columns, but the rows are not orthogonal. However, normalizing

gives the unitary matrix $\begin{bmatrix} \frac{1}{2}(1+i) & \frac{1}{\sqrt{2}} \\ \frac{1}{2}(1-i) & \frac{i}{\sqrt{2}} \end{bmatrix} = \frac{1}{2}\begin{bmatrix} 1+i & \sqrt{2} \\ 1-i & \sqrt{2}i \end{bmatrix}$.

## 4.10.4    Unitary Diagonalization

Given a real symmetric matrix $A$, there is an orthogonal matrix $P$ such that $P^T A P$ is diagonal. This is the Principal Axis theorem, and the diagonalization algorithm in Section 4.7 provides a systematic way to find $P$. In fact, the procedure works for any Hermitian matrix, except that the matrix $P$ that we end up with is now unitary. Here is an example.

**Example 5**    Given the Hermitian matrix $H = \begin{bmatrix} 3 & 2-i \\ 2+i & 7 \end{bmatrix}$, find a unitary matrix $U$ such that $U^*HU$ is diagonal.

SOLUTION    The characteristic polynomial of $H$ is $c_H(x) = det(xI - H) = (x-2)(x-8)$, so $\lambda_1 = 2$ and $\lambda_2 = 8$ are the eigenvalues of $H$ (real as expected), with (orthogonal) eigenvectors $Z_1 = [2-i \quad -1]^T$ and $Z_2 = [1 \quad 2+i]^T$ respectively. Normalizing gives the unitary matrix $U = \left[ \frac{1}{\sqrt{6}}Z_1 \quad \frac{1}{\sqrt{6}}Z_2 \right] = \frac{1}{\sqrt{6}}\begin{bmatrix} 2-i & 1 \\ -1 & 2+i \end{bmatrix}$, and the reader can verify that

$$U^*HU = \begin{bmatrix} \lambda_1 & 0 \\ 0 & \lambda_2 \end{bmatrix} = \begin{bmatrix} 2 & 0 \\ 0 & 8 \end{bmatrix} \text{ is diagonal.}$$

In Theorem 7 we are going to show that (as in Example 5) every Hermitian matrix $H$ can be "unitarily diagonalized", that is $U^*HU$ is diagonal for some unitary matrix $U$. However, with very little extra effort, we can obtain the following important theorem which says that every square complex matrix can be "unitarily triangularized".

**THEOREM 6**    **Schur's Theorem**[34]. If $G$ is any $n \times n$ complex matrix, there exists a unitary matrix $U$ such that

$$U^*GU = T$$

where $T$ is upper triangular. Moreover, the entries on the main diagonal of $T$ are the eigenvalues $\lambda_1, \lambda_2, \cdots, \lambda_n$ of $G$, repeated according to their multiplicities.

PROOF    We use induction on $n$. If $n = 1$, $G$ is already upper triangular. If $n > 1$, assume the theorem is valid for complex matrices of size $(n-1) \times (n-1)$. Let $\lambda_1$ be an eigenvalue of $G$ with unit eigenvector $Z_1$. By (the complex analogue of) Theorem 3 §4.3, $Z_1$ is part of a basis $\{Z_1, Z_2, \cdots, Z_n\}$ of $\mathbb{C}^n$ which, by (the analogue of) the Gram-Schmidt algorithm, we may assume is orthonormal. Hence $U_1 = [Z_1 \quad Z_2 \quad \cdots \quad Z_n]$ is a unitary matrix and, as in Lemma 1 §4.7,

$$U_1^*GU_1 = \begin{bmatrix} \lambda_1 & X_1 \\ 0 & G_1 \end{bmatrix}$$

---

[34] Issai Schur (1875–1941) was a German mathematician who did fundamental work in the theory of representations of groups as complex matrices.

in block form. By induction, there is an $(n-1) \times (n-1)$ unitary matrix $V_1$ such that $V_1^*G_1V_1 = T_1$ is upper triangular. Hence $U_2 = \begin{bmatrix} 1 & 0 \\ 0 & V_1 \end{bmatrix}$ is unitary (verify), and so $U = U_1U_2$ is also unitary (using Theorem 5). Moreover

$$U^*GU = U_2^*(U_1^*GU_1)U_2$$
$$= \begin{bmatrix} 1 & 0 \\ 0 & V_1^* \end{bmatrix}\begin{bmatrix} \lambda_1 & X_1 \\ 0 & G_1 \end{bmatrix}\begin{bmatrix} 1 & 0 \\ 0 & V_1 \end{bmatrix}$$
$$= \begin{bmatrix} \lambda_1 & X_1V_1 \\ 0 & T_1 \end{bmatrix}$$

is upper triangular, as required.

Finally, if $U^*GU = T$ as in the theorem, then $T$ and $G$ are similar and so have the same eigenvalues by (the analogue of) Theorem 6 §2.3, and appear on the main diagonal of $T$ (as $T$ is triangular). This proves the last sentence of the theorem.

Schur's theorem, together with the fact that traces and determinants of similar matrices are equal, gives

**COROLLARY**  If $\lambda_1, \lambda_2, \cdots, \lambda_n$ are the eigenvalues (including multiplicities) of an $n \times n$ complex matrix $G$, then

$$det\,G = \lambda_1\lambda_2\cdots\lambda_n \quad \text{and} \quad tr\,G = \lambda_1 + \lambda_2 + \cdots + \lambda_n$$

Note that if $G$ is real, the Corollary shows that the sum and product of the (possibly nonreal) eigenvalues of $G$ are actually real. This also follows from the fact that if $\lambda$ is a complex root of any real polynomial, then $\overline{\lambda}$ is also a root.

Schur's theorem asserts that every square complex matrix can be "unitarily triangularized". This does not hold for "unitarily diagonalized". In fact, if $G = \begin{bmatrix} 1 & 1 \\ 0 & 1 \end{bmatrix}$ then $U^{-1}GU$ *cannot* be diagonal for *any* invertible complex matrix $U$ (verify). However all is well for Hermitian matrices, and Schur's theorem gives the following important extension of the Principal Axis theorem.

**THEOREM 7**  **Spectral Theorem.** If $H$ is any Hermitian matrix, there exists a unitary matrix $U$ such that $U^*HU$ is diagonal.

**PROOF**  By Schur's theorem, let $U^*HU = T$ be upper triangular where $U$ is unitary. Then $T^* = U^*H^*U^{**} = U^*HU = T$, so $T$ is also lower triangular. Hence $T$ is actually diagonal, as required.

The Principal Axis theorem asserts that a real, square matrix is symmetric if and only if it can be orthogonally diagonalized. The analogue for complex matrices would say that a matrix is Hermitian if and only if it can be unitarily diagonalized. Half of this is true by the Spectral theorem, but there exist unitarily diagonalizable matrices that are not Hermitian. Here is an example.

**Example 6**  Show that the non-Hermitian matrix $G = \begin{bmatrix} 0 & -1 \\ 1 & 0 \end{bmatrix}$ is unitarily diagonalizable.

**SOLUTION**  The characteristic polynomial is $c_G(x) = x^2 + 1$, so the eigenvalues are $\lambda_1 = i$ and $\lambda_2 = -i$ with corresponding eigenvectors $Z_1 = [-1 \ i]^T$ and $Z_2 = [i \ -1]^T$. Normalizing gives the unitary matrix $U = \begin{bmatrix} \frac{1}{\sqrt{2}}Z_1 & \frac{1}{\sqrt{2}}Z_2 \end{bmatrix}^T = \frac{1}{\sqrt{2}}\begin{bmatrix} -1 & i \\ i & -1 \end{bmatrix}$ such that $U^*GU = \begin{bmatrix} i & 0 \\ 0 & -i \end{bmatrix}$.

We conclude with a simple characterization of which square complex matrices are unitarily diagonalizable. A complex matrix $N$ is called **normal** if $NN^* = N^*N$. Hermitian and unitary matrices are easily seen to be normal, as is the matrix $G$ in Example 6. In fact we have

**THEOREM 8**    An $n \times n$ matrix is unitarily diagonalizable if and only if it is normal.

**PROOF**    If $N$ is unitarily diagonalizable, let $U^*NU = D$ be diagonal where $U$ is unitary. Since $D^*D = DD^*$ (both $D$ and $D^*$ are diagonal), we obtain $U^*(N^*N)U = U^*(NN^*)U$, and $N^*N = NN^*$ follows by cancellation. Thus $N$ is normal.

Conversely, assume that $N$ is normal, that is $N^*N = NN^*$. By Schur's theorem, let $U^*NU = T$ where $T$ is upper triangular and $U$ is unitary. We show that $T$ is diagonal by induction on $n$. If $n = 1$, the matrix $T$ is already diagonal. If $n > 1$, observe that

$$TT^* = U^*(NN^*)U = U^*(N^*N)U = T^*T.$$

If $T = [t_{ij}]$, then equating $(1, 1)$-entries in $TT^* = T^*T$ gives

$$|t_{11}|^2 + |t_{12}|^2 + \cdots + |t_{1n}|^2 = |t_{11}|^2$$

because $T$ is upper triangular. It follows that $t_{12} = \cdots = t_{1n} = 0$, so $T$ has the block form $T = \begin{bmatrix} t_{11} & 0 \\ 0 & T_1 \end{bmatrix}$. Then $TT^* = T^*T$ implies that $T_1T_1^* = T_1^*T_1$, so $T_1$ is diagonal by induction. Thus $T$ is diagonal, and the proof is complete.

We note in passing that the real normal $2 \times 2$ matrices are just the symmetric matrices together with the matrices of the form $\begin{bmatrix} a & -b \\ b & a \end{bmatrix}$.

# Exercises 4.10

1. In each case either show that the statement is true or give an example showing that it is false.
   a) Every invertible Hermitian matrix is unitary.
   b) Every unitary matrix is Hermitian.
   c) The diagonal entries of a Hermitian matrix are all real.
   d) Every Hermitian matrix has real eigenvalues.
   e) If $G$ has real eigenvalues, it is Hermitian.
   f) The product of two unitary matrices is unitary.

2. In each case compute the length of $Z$ and $W$, and determine if they are orthogonal.
   a) $Z = [-2 \quad 1 + i \quad 4i]^T$ and $W = [1 - i \quad 2i \quad 1]^T$.
   b) $Z = [2 + i \quad 3i \quad -1 - i]^T$ and $W = [2 \quad 2 - i \quad -i]^T$.
   c) $Z = [2 - i \quad -1 \quad 1 + 3i \quad 2i]^T$ and $W = [-1 \quad 2i \quad 1 + i \quad i]^T$.
   d) $Z = [1 + 2i \quad -5 \quad 2 - i \quad -2i]^T$ and $W = [2 - i \quad 1 - i \quad -3i \quad 3 - i]^T$.

3. In each case determine whether the given matrix is Hermitian, unitary or normal.
   a) $Z = \begin{bmatrix} 2 & 3 \\ -3 & 2 \end{bmatrix}$.
   b) $Z = \begin{bmatrix} 1 & -i \\ i & 2 \end{bmatrix}$.
   c) $Z = \begin{bmatrix} 1 & 1+i \\ 1+i & i \end{bmatrix}$.
   d) $Z = \frac{1}{\sqrt{2}} \begin{bmatrix} 1 & -1 \\ 1 & 1 \end{bmatrix}$.

4. In each case find a unitary matrix $U$ such that $U^*ZU$ is diagonal.
   a) $Z = \begin{bmatrix} 1 & i \\ -i & 1 \end{bmatrix}$.
   b) $Z = \begin{bmatrix} 4 & 3-i \\ 3+i & 1 \end{bmatrix}$.
   c) $Z = \begin{bmatrix} 1 & 0 & 1+i \\ 0 & 2 & 0 \\ 1-i & 0 & 0 \end{bmatrix}$.
   d) $Z = \begin{bmatrix} 1 & 0 & 0 \\ 0 & 1 & 1+i \\ 0 & 1-i & 2 \end{bmatrix}$.

5. If $A = \begin{bmatrix} 1 & 1 \\ 0 & 1 \end{bmatrix}$, show that $U^{-1}AU$ is not diagonal for any invertible complex matrix $U$. [Hint: $A$ and $U^{-1}AU$ are similar, so they have the same eigenvalues.

6. If $A = \begin{bmatrix} 0 & -1 \\ 1 & 0 \end{bmatrix}$, show that $U^{-1}AU$ is not upper triangular for any invertible *real* matrix $U$.

7. If $H$ and $H_1$ are Hermitian, show that $HH_1$ is Hermitian if and only if $HH_1 = H_1H$.

8. a) Show that every unitary matrix is normal.
   b) Show that every Hermitian matrix is normal.

**9. a)** Describe the diagonal matrices that are Hermitian.
   **b)** Describe the diagonal matrices that are unitary.

**10. a)** Show that $Z^*Z$ is Hermitian for any complex matrix $Z$ (possibly not square).
   **b)** If $ZZ^* = I$, show that $H = Z^*Z$ satisfies $H^2 = H$.

**11.** If $A$ is a real, normal $2 \times 2$ matrix, show that either $A$ is symmetric or $A = \begin{bmatrix} a & -b \\ b & a \end{bmatrix}$ for some $a$ and $b$.

**12.** Show that $\langle ZX, \ Y \rangle = \langle X, \ Z^*Y \rangle$ for all columns $X$ and $Y$ in $\mathbb{C}^n$, and all $n \times n$ complex matrices $Z$.

**13.** If $Z$ is any $n \times n$ complex matrix, show that both $Z + Z^*$ and $ZZ^*$ are Hermitian.

**14.** If $Z$ is an invertible complex matrix, show that $Z^*$ is also invertible, and $(Z^*)^{-1} = (Z^{-1})^*$.

**15.** Let $U$ denote a unitary matrix.
   **a)** Show that $U^*$ is also unitary.
   **b)** Show that $U$ is invertible and that $U^{-1}$ is also unitary.
   **c)** Show that $\|UX\| = \|X\|$ for every column $X$ in $\mathbb{C}^n$.
   **d)** Show that $|\lambda| = 1$ for every eigenvalue $\lambda$ of $U$.

**16.** If $G$ is both Hermitian and unitary, show that every eigenvalue of $G$ is 1 or $-1$.

**17.** Show that the product of two unitary matrices is again unitary.

**18.** If $N$ is normal, show that $zN$ is also normal for any complex number $z$.

**19.** If $Z$ is any complex $n \times n$ matrix, show that $U^*ZU$ is *lower* triangular for some unitary matrix $U$.

**20. a)** Show that every complex matrix $Z$ can be uniquely written as $Z = A + iB$ where $A$ and $B$ are real matrices.
   **b)** If $A$ and $B$ are real, show that $A + iB = 0$ if and only if $A = 0 = B$.
   **c)** If $Z = A + iB$ as in **a)**, show that $Z$ is Hermitian if and only if $A$ is symmetric and $B$ is skew-symmetric ($B^T = -B$).

**21.** A complex matrix $S$ is called **skew-Hermitian** if $S^* = -S$.
   **a)** Show that $Z - Z^*$ is skew-Hermitian for every complex matrix $Z$.
   **b)** If $S$ is skew-Hermitian, show that $S^2$ and $iS$ are Hermitian.
   **c)** If $S$ is skew-Hermitian, show that the eigenvalues of $S$ are pure imaginary ($i\lambda$ for real $\lambda$).
   **d)** Show that every complex matrix $Z$ can be written as $Z = H + S$ where $H$ is Hermitian and $S$ is skew-Hermitian.
   **e)** Show that the representation in **d)** is unique.

**22.** A complex matrix $Z$ is called *nilpotent* if $Z^k = 0$ for some $k \geq 1$.
   **a)** Assume $Z$ is nilpotent. Show that the only eigenvalue of $Z$ is $\lambda = 0$.
   **b)** If $Z$ is $n \times n$, conclude that $c_Z(x) = x^n$.
   **c)** If $Z \neq 0$, show that $Z$ is not diagonalizable.

**23.** If $H$ is Hermitian, show that all the coefficients of $c_H(x)$ are real numbers. [Hint: Theorem 4.]

**24. a)** Show that $det\overline{Z} = \overline{detZ}$ for every $n \times n$ complex matrix $Z$. [Hint: Use induction on $n$ and apply the cofactor expansion.]
   **b)** If $Z$ is Hermitian, show that $det(Z^*) = detZ$. [Hint: Theorem 4 and the Corollary to Schur's theorem.]

# **4.11  SINGULAR VALUE DECOMPOSITION**[35]

In this section we present an analogue of the Principal Axis theorem for non-square matrices, a result which has been described[36] as "a highlight of linear algebra". The Principal Axis theorem asserts that every symmetric matrix $A$ can be written as a product $A = PDP^T$ where $P$ is orthogonal and $D$ is diagonal. Surprisingly, this result implies that, if $A$ is *any* matrix (possibly not square), there exist orthogonal matrices $P$ and $Q$ such that $A = PDQ^T$ where $D$ is a suitable $m \times n$ diagonal matrix (entries off the main diagonal are zero). This is called a singular value decomposition of $A$ and is one of the most useful matrix factorizations in linear algebra. The decomposition is used throughout applied linear algebra, some examples being statistical pattern recognition, data analysis, artificial intelligence, and data compression.[37] Moreover, excellent algorithms exist for finding the decompostion, so it is important in numerical analysis.

---

[35] This section is not needed elsewhere in this book.

[36] Gilbert Strang, *Introduction to Linear Algebra*, Wellesley–Cambridge Press, 1993, page 325.

[37] For more on this, see *Linear Algebra with Applications* by J.T. Scheick, McGraw-Hill, 1997.

## 4.11.1 The Singular Value Decomposition

Let $A$ be an $m \times n$ matrix of rank $r$. A **singular value decomposition** (**SVD**) for $A$ is a factorization

$$A = PDQ^T$$

where $P = [P_1 \ P_2 \ \cdots \ P_m]$ and $Q = [Q_1 \ Q_2 \ \cdots \ Q_n]$ are orthogonal matrices of sizes $m \times m$ and $n \times n$ respectively, and $D = diag(d_1, d_2, \cdots)$ is an $m \times n$ diagonal matrix where $d_1 \geq d_2 \geq \cdots \geq 0$. We are going to show in this section that every matrix $A$ has an SVD, that $D$ is uniquely determined by $A$, and that the columns $P_i$ and $Q_j$ are orthonormal eigenvectors of the symmetric matrices $AA^T$ and $A^TA$, respectively. In fact, the properties of $A$ are closely related to the properties of these symmetric matrices (for example, see Theorem 5 §4.6). This is somewhat surprising because symmetric matrices are much easier to deal with than matrices in general, even square matrices. For example, the eigenvalues of any symmetric matrix are all real; in fact, the eigenvalues of $A^TA$ are real and *nonnegative* as we shall see.

If $A$ is any real $m \times n$ matrix, let $\lambda_1, \cdots, \lambda_n$ denote the (real) eigenvalues of the $n \times n$ symmetric matrix $A^TA$ and, for convenience, choose the notation so that

$$\lambda_1 \geq \lambda_2 \geq \lambda_3 \geq \cdots \geq \lambda_n.$$

By the Principal Axis theorem, let $\{Q_1, \cdots, Q_n\}$ be an orthonormal basis of $\mathbb{R}^n$ where, for each $i$, $Q_i$ is an eigenvector of $A^TA$ corresponding to $\lambda_i$. Observe first that for all $i$ and $j$

$$AQ_i \bullet AQ_j = (AQ_i)^T AQ_j = Q_i^T(A^TAQ_j) = Q_i^T(\lambda_jQ_j) = \lambda_j(Q_i \bullet Q_j) \qquad (*)$$

Since $\|Q_i\| = 1$ for each $i$, taking $j = i$ in $(*)$ shows that

$$\lambda_i = \|AQ_i\|^2 \text{ for each } i = 1, 2, \cdots, n.$$

In particular, $\lambda_i \geq 0$ for each $i$. The numbers

$$\sigma_i = \sqrt{\lambda_i} \text{ for } i = 1, 2, \cdots, n$$

are called the **singular values** of the matrix $A$. Thus

$$\sigma_1 \geq \sigma_2 \geq \cdots \geq \sigma_n \geq 0$$

and

$$\|AQ_i\|^2 = \sigma_i^2 \text{ for each } i. \qquad (**)$$

The following lemma is the first indication of the close relationship between these singular values and the structure of the matrix $A$. Recall that $colA = imA = \{AX \mid X \text{ in } \mathbb{R}^n\}$ by Theorem 5 §4.4.

---

**LEMMA 1**

Let $A$ be any $m \times n$ matrix and suppose that $rankA = r$.

(1) The matrix $A$ has exactly $r$ nonzero singular values (counting multiplicities).

(2) $\{AQ_1, AQ_2, \cdots, AQ_r\}$ is an orthogonal basis of $colA = imA$.

**PROOF**

Suppose that $s$ of the $\sigma_i$ are nonzero, so $\sigma_1 \geq \sigma_2 \geq \cdots \geq \sigma_s > 0$ and $\sigma_{s+1} = \cdots = \sigma_n = 0$. Write $\mathcal{F} = \{AQ_1, AQ_2, \cdots, AQ_s\}$. Then $\mathcal{F} \subseteq imA = colA$, and $\mathcal{F}$ is an orthogonal set by $(*)$ and $(**)$. Since $dim(colA) = r$, both (1) and (2) follow if we can show that $\mathcal{F}$ spans $imA$. But if $X$ is any vector in $\mathbb{R}^n$, write $X = r_1Q_1 + r_2Q_2 + \cdots + r_nQ_n$ where each $r_i$ is in $\mathbb{R}$. Then equation $(**)$ above gives

$$AX = r_1AQ_1 + r_2AQ_2 + \cdots + r_sAQ_s + \cdots + r_nAQ_n$$
$$= r_1AQ_1 + r_2AQ_2 + \cdots + r_sAQ_s$$

So $AX$ is in span $\mathcal{F}$. Since $imA = \{AX \mid X \text{ in } \mathbb{R}^n\}$, this shows that $\mathcal{F}$ spans $imA$, and so proves the lemma.

If $\sigma_1 \geq \cdots \geq \sigma_r$ are the nonzero singular values of the $m \times n$ matrix $A$, then $r = rankA$ by Lemma 1, and we define

$$\Sigma = \begin{bmatrix} diag(\sigma_1, \cdots, \sigma_r) & 0 \\ 0 & 0 \end{bmatrix}$$

to be the $m \times n$ block matrix with the $r \times r$ diagonal matrix $diag(\sigma_1, \cdots, \sigma_r)$ in the upper left corner and all other entries zero (some zero blocks may be absent if $r = m$ or $r = n$). In particular, $\Sigma$ is an $m \times n$ diagonal matrix.

Now observe that $(**)$ gives $\|AQ_i\| = \sigma_i \neq 0$ for each $i = 1, 2, \cdots, r$. Hence we define the unit vectors

$$P_i = \tfrac{1}{\sigma_i} AQ_i \text{ for each } i = 1, 2, \cdots, r \qquad (***)$$

in $\mathbb{R}^m$. Then Lemma 1 shows that $\{P_1, P_2, \cdots, P_r\}$ is an orthonormal set in $\mathbb{R}^m$, so extend it to an orthonormal basis $\{P_1, P_2, \cdots, P_r, P_{r+1}, \cdots, P_m\}$ of $\mathbb{R}^m$. Then

$$P = [P_1 \ P_2 \ \cdots \ P_m] \quad \text{and} \quad Q = [Q_1 \ Q_2 \ \cdots \ Q_n]$$

are orthogonal matrices, and we can now prove

**THEOREM 1**

**Singular Value Decomposition**. Let $A$ be an $m \times n$ matrix of rank $r$ with nonzero singular values $\sigma_1 \geq \cdots \geq \sigma_r$. If we define $\Sigma$, $P$, and $Q$ as above, then $P$ and $Q$ are orthogonal matrices, $\Sigma$ is a diagonal matrix, and $A$ factors as follows:

$$A = P\Sigma Q^T.$$

**PROOF** We compute

$$P\Sigma = [P_1 \ P_2 \ \cdots \ P_m] \begin{bmatrix} \sigma_1 & 0 & \cdots & 0 & 0 & \cdots & 0 \\ 0 & \sigma_2 & \cdots & 0 & 0 & \cdots & 0 \\ \vdots & \vdots & \ddots & \vdots & \vdots & & \vdots \\ 0 & 0 & \cdots & \sigma_r & 0 & \cdots & 0 \\ 0 & 0 & \cdots & 0 & 0 & \cdots & 0 \\ \vdots & \vdots & & \vdots & \vdots & \ddots & \vdots \\ 0 & 0 & \cdots & 0 & 0 & \cdots & 0 \end{bmatrix}$$

$$= [\sigma_1 P_1 \ \sigma_2 P_2 \ \cdots \ \sigma_r P_r \ 0 \ \cdots \ 0].$$

On the other hand, $Q = [Q_1 \ Q_2 \ \cdots \ Q_n]$ and we obtain

$$\begin{aligned} AQ &= A[Q_1 \ Q_2 \ \cdots \ Q_n] \\ &= [AQ_1 \ AQ_2 \ \cdots \ AQ_n] \\ &= [\sigma_1 P_1 \ \sigma_2 P_2 \ \cdots \ \sigma_r P_r \ 0 \ \cdots \ 0] \end{aligned}$$

by $(***)$ because $AQ_i = 0$ if $i > r$ by $(**)$. It follows that $P\Sigma = AQ$, whence $A = P\Sigma Q^T$ as required.

The discussion leading up to Theorem 1 provides a routine method for finding a SVD for a matrix $A$ which is useful when $A$ is not too large.

**Example 1**    Find a singular value decomposition of $A = \begin{bmatrix} 1 & 2 & 0 \\ 2 & 1 & 0 \end{bmatrix}$.

SOLUTION    $A^T A = \begin{bmatrix} 5 & 4 & 0 \\ 4 & 5 & 0 \\ 0 & 0 & 0 \end{bmatrix}$ has eigenvalues $\lambda_1 = 9$, $\lambda_2 = 1$, and $\lambda_3 = 0$ with orthonormal

eigenvectors $Q_1 = \frac{1}{\sqrt{2}}[1 \ \ 1 \ \ 0]^T$, $Q_2 = \frac{1}{\sqrt{2}}[-1 \ \ 1 \ \ 0]^T$, and $Q_3 = [0 \ \ 0 \ \ 1]^T$ respectively.
Thus the singular values are $\sigma_1 = 3$, $\sigma_2 = 1$, and $\sigma_3 = 0$. Then the formula $P_i = \frac{1}{\sigma_i} A Q_i$
gives $P_1 = \frac{1}{\sqrt{2}}[1 \ \ 1]^T$ and $P_2 = \frac{1}{\sqrt{2}}[1 \ \ -1]^T$. Hence we have

$$P = [P_1 \ \ P_2] = \frac{1}{\sqrt{2}} \begin{bmatrix} 1 & 1 \\ 1 & -1 \end{bmatrix}, \quad \Sigma = \begin{bmatrix} 3 & 0 & 0 \\ 0 & 1 & 0 \end{bmatrix}, \quad \text{and} \quad Q = \frac{1}{\sqrt{2}} \begin{bmatrix} 1 & -1 & 0 \\ 1 & 1 & 0 \\ 0 & 0 & \sqrt{2} \end{bmatrix}.$$

This gives a singular value decomposition $A = P \Sigma Q^T$, as the reader can verify.

**Remark 1.** For large matrices, the method in Example 1 is *not* the way the singular value decomposition is found in practice. One reason is that small singular values may be lost in the process due to round-off error. There are efficient, iterative algorithms that accurately produce the singular values and the matrices $P$ and $Q$.

**Remark 2.** In solving a system $AX = B$ of equations, round-off error can be a serious problem. The singular value decomposition is useful in estimating just how sensitive the system is to error. If $\sigma_1 \geq \sigma_2 \geq \cdots \geq \sigma_r$ are the nonzero singular values, the ratio

$$c = \frac{\sigma_1}{\sigma_r}$$

is called the **condition number** of the matrix $A$. It turns out that larger values of $c$ correspond to a greater danger of round-off error.

The singular value decomposition of $A = P \Sigma Q^T$ is not unique. In the above construction the columns $Q_i$ can be any orthonormal basis of eigenvectors of $A^T A$ and, even though $\{P_1, \cdots, P_r\}$ are given by (***), the only restriction on the columns $P_{r+1}, \cdots, P_m$ is that $\{P_1, \cdots, P_r, P_{r+1}, \cdots, P_m\}$ should be orthonormal. However the matrix $\Sigma$ is uniquely determined by $A$. This is part of the following theorem which also gives some constraints on the matrices $P$ and $Q$.

**THEOREM 2**    **Uniqueness of the Singular Value Decomposition.** Let $A$ be an $m \times n$ matrix of rank $r$ with nonzero singular values $\sigma_1 \geq \sigma_2 \geq \cdots \geq \sigma_r$. Suppose that $A$ can be factored as

$$A = P S Q^T$$

where $P = [P_1 \ \ P_2 \ \cdots \ P_m]$ and $Q = [Q_1 \ \ Q_2 \ \cdots \ Q_n]$ are orthogonal matrices, and $S = diag(s_1, s_2, \cdots)$ is an $m \times n$ diagonal matrix with $s_1 \geq s_2 \geq \cdots \geq 0$. Then:

(1) $S = \Sigma$.
(2) $P_i = \frac{1}{\sigma_i} A Q_i$ for $1 \leq i \leq r$, $A Q_i = 0$ for $i > r$, and $(A^T A) Q_i = \sigma_i^2 Q_i$ for $1 \leq i \leq n$.
(3) $\{P_1, P_2, \cdots, P_m\}$ and $\{Q_1, Q_2, \cdots, Q_n\}$ are (orthonormal) sets of eigenvectors of $A A^T$ and $A^T A$ respectively.

PROOF    We have $A^T A = (Q S^T P^T)(P S Q^T) = Q(S^T S)Q^T$. Hence the diagonal matrix $S^T S = diag(s_1^2, s_2^2, \cdots)$ is similar to $A^T A$ and so has the same eigenvalues $\sigma_1^2 \geq \sigma_2^2 \geq \cdots$. Since $s_1 \geq s_2 \geq \cdots \geq 0$, it follows that $s_i = \sigma_i$ for each $i$, so $S = \Sigma$, proving (1). But then $P \Sigma = P S = A Q$, so (as in the proof of Theorem 1),

$$[\sigma_1 P_1 \ \ \sigma_2 P_2 \ \cdots \ \sigma_r P_r \ \ 0 \ \cdots \ 0] = [A Q_1 \ \ A Q_2 \ \cdots \ A Q_n]$$

Comparing columns gives the first two assertions in (2).
But also $(A^TA)Q = Q(S^TS) = Q(\Sigma^T\Sigma) = Q\,diag(\sigma_1^2, \sigma_2^2, \cdots)$, and therefore it follows
(by comparing columns) that $(A^TA)Q_i = \sigma_i^2 Q_i$ for each $i$. This completes the proof
of (2) and shows that each $Q_i$ is an eigenvector of $A^TA$. Similarly $(AA^T)P = P(\Sigma\Sigma^T)$,
so the columns of $P$ are eigenvectors of $AA^T$. This proves (3).

## 4.11.2 The Fundamental Subspaces

A lot of important information about a matrix can be obtained from a singular value
decomposition. The next result, important in its own right, will be needed.

**LEMMA 2**   If $A$ is an $m \times n$ matrix, then $null(A^T) = (colA)^\perp$ in $\mathbb{R}^m$.

**PROOF**   We use the fact (Theorem 5 §4.4) that $colA = imA = \{AX \mid X \text{ in } \mathbb{R}^n\}$. If $Y$ is in $\mathbb{R}^m$,
we have

$$A^TY = 0 \Leftrightarrow X \cdot (A^TY) = 0 \quad \text{for all } X \text{ in } \mathbb{R}^n$$
$$\Leftrightarrow X^T(A^TY) = 0 \quad \text{for all } X \text{ in } \mathbb{R}^n$$
$$\Leftrightarrow (AX) \cdot Y = 0 \quad \text{for all } X \text{ in } \mathbb{R}^n.$$

It follows that $Y$ is in $null(A^T)$ if and only if $Y$ is in $(imA)^\perp$, as required.

With this we can see that the singular value decomposition of a matrix $A$ provides
orthonormal bases of four important subspaces associated with $A$, sometimes called
the **fundamental subspaces** of $A$.[38]

**THEOREM 3**   Let $A$ be an $m \times n$ matrix of rank $r$, and let $A = P\Sigma Q^T$ be a singular value decomposition
of $A$ where $P = [P_1 \quad P_2 \quad \cdots \quad P_m]$ and $Q = [Q_1 \quad Q_2 \quad \cdots \quad Q_n]$ are orthogonal matrices.
Then:

(1) $\{P_1, P_2, \cdots, P_r\}$ is an orthonormal basis of $colA = imA$.
(2) $\{P_{r+1}, P_{r+2}, \cdots, P_m\}$ is an orthonormal basis of $null(A^T)$.
(3) $\{Q_1^T, \cdots, Q_r^T\}$ is an orthonormal basis of $rowA$.
(4) $\{Q_{r+1}, \cdots, Q_n\}$ is an orthonormal basis of $nullA$.

**PROOF**   (1) By Theorem 2, $P_i = \frac{1}{\sigma_i}AQ_i = A(\frac{1}{\sigma_i}Q_i) \in imA = colA$ for each $i$. Hence (1)
follows because $\{P_1, P_2, \cdots, P_r\}$ is orthonormal and $dim(colA) = r$.

(2) By (1) and Lemma 2, we have
$$null(A^T) = (colA)^\perp = [span\{P_1, \cdots, P_r\}]^\perp = span\{P_{r+1}, \cdots, P_m\}$$
because $\{P_1, \cdots, P_r, P_{r+1}, \cdots, P_m\}$ is an orthogonal basis of $\mathbb{R}^m$.

(3) If $1 \leq i \leq r$, then, since $A^TAQ_i = \sigma_i^2 Q_i$ by Theorem 2, we have
$Q_i = A^T(\frac{1}{\sigma_i^2}AQ_i) \in im(A^T) = col(A^T)$. Now let $R_1, R_2, \cdots, R_m$ denote the rows
of $A$. Then $R_1^T, R_2^T, \cdots, R_m^T$ are the columns of $A^T$, so there exist $r_k$ in $\mathbb{R}$ such
that $Q_i = \Sigma_{k=1}^m r_k R_k^T$. Thus $Q_i^T = \Sigma_{k=1}^m r_k R_k$ is in $rowA$ for $1 \leq i \leq r$.
Since $\{Q_1^T, \cdots, Q_r^T\}$ is independent, (3) follows because $dim(rowA) = r$.

(4) We have $AQ_i = 0$ if $i > r$ by Theorem 2. Thus $\{Q_{r+1}, \cdots, Q_n\} \subseteq nullA$, and we
are done because $dim(nullA) = n - r$ by Theorem 6 §4.4.

---

[38] A more detailed discussion of these subspaces can be found in G. Strang, *Introduction to Linear Algebra*, Wellesley—
Cambridge Press, 1993.

A singular value decomposition $A = P\Sigma Q^T$ of an $m \times n$ matrix $A$ gives a way to find a type of "inverse" for $A$. Let $\sigma_1 \geq \sigma_2 \geq \cdots \geq \sigma_r$ be the singular values of $A$, and write $D = diag(\sigma_1, \sigma_2, \cdots, \sigma_r)$. Then

$$\Sigma = \begin{bmatrix} D & 0 \\ 0 & 0 \end{bmatrix}_{m \times n} \quad \text{as before, and we write } \Sigma^+ = \begin{bmatrix} D^{-1} & 0 \\ 0 & 0 \end{bmatrix}_{n \times m}$$

in block form (where the subscript gives the size of the matrix). Then define the **pseudo-inverse** (or **Moore-Penrose inverse**) $A^+$ of $A$ by

$$A^+ = Q\Sigma^+ P^T.$$

If $A$ is $n \times n$ and invertible, then $A^+ = A^{-1}$ is the usual inverse of $A$. Indeed, in this case, $r = n$ and $\Sigma$ is invertible with $\Sigma^{-1} = \Sigma^+$, so

$$A^{-1} = (P\Sigma Q^T)^{-1} = Q\Sigma^{-1}P^T = Q\Sigma^+ P^T = A^+.$$

In fact, for an arbitrary $m \times n$ matrix $A$, it is readily verified by block multiplication that

$$AA^+ = P\begin{bmatrix} I_r & 0 \\ 0 & 0 \end{bmatrix}_{m \times m} P^T \quad \text{and} \quad A^+A = Q\begin{bmatrix} I_r & 0 \\ 0 & 0 \end{bmatrix}_{n \times n} Q^T.$$

If $A$ is invertible, then $r = n$ and it follows that $AA^+ = I$ and $A^+A = I$.

## 4.11.3 The Polar Decomposition

Recall that a matrix $A$ is called positive definite if it is symmetric and $\lambda > 0$ for every eigenvalue $\lambda$ of $A$. There is an important natural generalization of this notion. The proof of Theorem 3 §4.8 goes through to show that the following conditions are equivalent for a square matrix $A$:

   **(1)**  $A$ is symmetric and $\lambda \geq 0$ for every eigenvalue $\lambda$ of $A$.
   **(2)**  $A$ is symmetric and $q(X) \geq 0$ for every $X$ in $\mathbb{R}^n$, where $q(X) = X^T A X$ is the quadratic form determined by $A$.

A square matrix $G$ is called **positive semidefinite** if it satisfies these conditions. The next example shows that these matrices occur in abundance.

**Example 2**   If $A$ is any $m \times n$ matrix, show that $A^T A$ is positive semidefinite.

**SOLUTION**   If $\lambda$ is any eigenvalue of $A^T A$, let $X$ be a corresponding eigenvector. Then $(A^T A)X = \lambda X$, so
$$\|AX\|^2 = (AX)^T AX = X^T(A^T AX) = X^T(\lambda X) = \lambda(X^T X) = \lambda \|X\|^2.$$
But $\|X\|^2 \neq 0$ because $X \neq 0$, so $\lambda = \frac{\|AX\|^2}{\|X\|^2} \geq 0$.

Every positive semidefinite matrix $G$ arises as in Example 2; in fact $G$ can be written in the form $G = C^T C$ for some square matrix $C$ with its nonzero rows orthogonal.

**THEOREM 4**   The following conditions are equivalent for an $n \times n$ matrix $G$:

   **(1)**  $G$ is positive semidefinite.
   **(2)**  $G = C^T C$ where $C$ is $n \times n$ and the nonzero rows of $C$ are orthogonal.
   **(3)**  $G = A^T A$ for some $m \times n$ matrix $A$.

PROOF  $(2) \Rightarrow (3)$ is clear, and $(3) \Rightarrow (1)$ follows from Example 2.

$(1) \Rightarrow (2)$. Given (1), let $\lambda_1, \cdots, \lambda_n$ be the (real) eigenvalues of $G$, so that $\lambda_i \geq 0$ for each $i$ by (1). Since $G$ is symmetric, the Principal Axis theorem provides an orthogonal matrix $P$ such that $P^T G P = D = diag(\lambda_1, \cdots, \lambda_n)$. Put $D_0 = diag(\sqrt{\lambda_1}, \cdots, \sqrt{\lambda_n})$ so that $D_0^2 = D$. If we take $C = D_0 P^T$, then row $i$ of $C$ is $\sqrt{\lambda_i}$ times row $i$ of $P^T$ for each $i$. Since $P^T$ is an orthogonal matrix, this shows that the nonzero rows of $C$ are orthogonal. Finally, a routine matrix calculation shows that $G = C^T C$, which proves (2).

The following theorem is one reason for the importance of these positive semidefinite matrices. The proof requires the singular value decomposition.

**THEOREM 5**  **Polar Decomposition**. If $A$ is any $n \times n$ matrix, then $A = GR$ where $G$ is positive semidefinite and $R$ is orthogonal.

PROOF  Let $A = P\Sigma Q^T$ be a singular value decomposition of $A$. By inserting $I_n = P^T P$ in this product, we obtain another factorization

$$A = P\Sigma Q^T = P\Sigma(P^T P)Q^T = (P\Sigma P^T)(PQ^T).$$

Hence $A = GR$ where $G = P\Sigma P^T$ is positive definite (because it is similar to $\Sigma$ and so has the same eigenvalues) and $R = PQ^T$ is orthogonal (because it is a product of orthogonal matrices). This is what we wanted.

The proof of Theorem 5 gives formulas for the matrices $G$ and $R$ in the polar decomposition of $A$ in terms of any singular value decomposition of $A$.

**Example 3**  Find the polar decomposition of the matrix $A = \begin{bmatrix} 1 & 1 \\ 2 & 2 \end{bmatrix}$.

SOLUTION  We begin by finding the singular value decomposition $A = P\Sigma Q^T$. The eigenvalues of $A^T A$ are $\lambda_1 = 10$ and $\lambda_2 = 0$, with orthonormal eigenvectors $Q_1 = \frac{1}{\sqrt{2}}\begin{bmatrix} 1 \\ 1 \end{bmatrix}$ and $Q_2 = \frac{1}{\sqrt{2}}\begin{bmatrix} 1 \\ -1 \end{bmatrix}$. Thus we take $Q = \frac{1}{\sqrt{2}}\begin{bmatrix} 1 & 1 \\ 1 & -1 \end{bmatrix}$. The singular values of $A$ are $\sigma_1 = \sqrt{10}$ and $\sigma_2 = 0$, so we have $P_1 = \frac{1}{\sigma_1}AQ_1 = \frac{1}{\sqrt{5}}\begin{bmatrix} 1 \\ 2 \end{bmatrix}$. Since $\sigma_2 = 0$, we can choose any $P_2$ such that $\{P_1, P_2\}$ is an orthonormal basis of $\mathbb{R}^2$, say, $P_2 = \frac{1}{\sqrt{5}}\begin{bmatrix} 2 \\ -1 \end{bmatrix}$. Hence $P = \frac{1}{\sqrt{5}}\begin{bmatrix} 1 & 2 \\ 2 & -1 \end{bmatrix}$.

Finally, $\Sigma = \begin{bmatrix} \sigma_1 & 0 \\ 0 & \sigma_2 \end{bmatrix} = \begin{bmatrix} \sqrt{10} & 0 \\ 0 & 0 \end{bmatrix}$, and the resulting singular value decomposition of $A$ is $A = P\Sigma Q^T$ as is easily verified.

Finally, as in the proof of Theorem 5, we have

$$G = P\Sigma P^T = \frac{\sqrt{10}}{5}\begin{bmatrix} 1 & 2 \\ 2 & 4 \end{bmatrix} \quad \text{and} \quad R = PQ^T = \frac{1}{\sqrt{10}}\begin{bmatrix} 3 & -1 \\ 1 & 3 \end{bmatrix}.$$

Then $G$ is positive semidefinite and $R$ is orthogonal, and the reader can verify that $A = GR$.

The name "polar decomposition" comes from the polar form for complex numbers. Recall (Section 2.5) that it is customary to identify the plane $\mathbb{R}^2$ with the set $\mathbb{C}$ of all complex numbers. This means that the complex number $z = a + ib$ is identified with the vector $\vec{z} = \begin{bmatrix} a \\ b \end{bmatrix}$ in $\mathbb{R}^2$. This being done, multiplication by $z$ is a linear transformation $T : \mathbb{R}^2 \to \mathbb{R}^2$ given by $T(\vec{v}) = \vec{z}\vec{v}$ for all $\vec{v}$ in $\mathbb{R}^2$. Here the product $\vec{z}\vec{v}$ comes from complex multiplication. More precisely, if we write $\vec{v} = \begin{bmatrix} x \\ y \end{bmatrix} = x + iy$, then

$$\vec{z}\vec{v} = (a + ib)(x + iy) = (ax - by) + i(bx + ay)$$
$$= \begin{bmatrix} ax - by \\ bx + ay \end{bmatrix} = \begin{bmatrix} a & -b \\ b & a \end{bmatrix}\begin{bmatrix} x \\ y \end{bmatrix}$$
$$= A\vec{v}$$

where $A = \begin{bmatrix} a & -b \\ b & a \end{bmatrix}$. Hence multiplication by $z = a + ib$ is matrix multiplication by $A$.

Now recall that every complex number $z = a + ib$ has the polar form $z = re^{i\theta}$ where $r = \sqrt{a^2 + b^2} \geq 0$ and $e^{i\theta} = \cos\theta + i\sin\theta$ where $\theta$ is the angle given by $\cos\theta = \frac{a}{r}$ and $\sin\theta = \frac{b}{r}$. Thus multiplication by $z = re^{i\theta}$ is first multiplication by $e^{i\theta}$ followed by multiplication by $r$, that is, multiplication by the orthogonal matrix $R = \begin{bmatrix} \cos\theta & -\sin\theta \\ \sin\theta & \cos\theta \end{bmatrix}$ followed by multiplication by the positive semidefinite matrix $G = \begin{bmatrix} r & 0 \\ 0 & r \end{bmatrix}$. Thus the polar form of $z$ becomes

$$A = GR,$$

which explains the name *polar decomposition* of $A$. Note that $G$ is actually positive definite in this case if $z \neq 0$.

## Exercises 4.11

1. Suppose that $A$ has singular values $\sigma_1 \geq \sigma_2 \geq \cdots \geq \sigma_r$. Find the singular values of $kA$ where $k$ is a real number.

2. Show that $A$ and $A^T$ have the same singular values. [Hint : Theorem 2.]

3. If $A = P\Sigma Q^T$ is a singular value decomposition of an invertible matrix $A$, find a singular value decomposition of $A^{-1}$. [If $R$ has 1's on the "antidiagonal" and zeros elsewhere, then $PDR^T$ reverses the order of the nonzero entries of the diagonal matrix $D$.]

4. If $A$ is a square matrix, show that $|detA|$ is the product of the singular values of $A$.

5. Let $A$ be an $m \times n$ matrix and let $P$ and $Q$ denote orthogonal matrices of sizes $m \times m$ and $n \times n$ respectively. Show that $PA$ and $AQ$ each have the same singular values as $A$.

6. If $A$ denotes an $m \times n$ matrix, show that $rankA = n$ if and only if all the singular values of $A$ are nonzero.

7. Let $A = P\Sigma Q^T$ be a singular value decomposition of an $m \times n$ matrix $A$. If $P_1$, $P_2$, $\cdots$, $P_m$ are the columns of $P$ and $Q_1$, $Q_2$, $\cdots$, $Q_n$ are the columns of $Q$, show that $A = \sigma_1 P_1 Q_1^T + \sigma_2 P_2 Q_2^T + \cdots + \sigma_r P_r Q_r^T$ where the $\sigma_i$ are the singular values of $A$. [Hint: Compute $A = P\Sigma Q^T$ in block form.]

8. Find a singular value decomposition of an $n \times n$ invertible matrix $A$ as follows: Apply the Principal Axis theorem to the symmetric matrix $A^T A$ to find an orthogonal matrix $Q$ such that $Q^T(A^T A)Q = diag(\lambda_1, \cdots, \lambda_n)$ where $\lambda_1 \geq \cdots \geq \lambda_n$ are the eigenvalues of $A^T A$. If $D = diag(\sqrt{\lambda_1}, \cdots, \sqrt{\lambda_n})$ and $P = AQD^{-1}$, show that $A = PDQ^T$ and that this is a singular value decomposition of $A$.

9. If $A$ is any $m \times n$ matrix show that:
   a) $AA^+$ and $A^+A$ are both symmetric matrices.
   b) $AA^+A = A$ and $A^+AA^+ = A^+$.

10. If $A$ is a positive definite matrix, show that the eigenvalues of $A$ are the singular values of $A$.

**11.** Show that the product of two positive semidefinite matrices need not be positive semidefinite.

**12.** If $G$ is positive semidefinite and orthogonal, show that $G = I$. [Hint: Theorem 3 §4.7.]

**13.** If $G$ is an invertible positive semidefinite matrix, show that $G^{-1}$ is also positive semidefinite.

**14.** If $G$ is positive semidefinite, show that $kG$ is also positive semidefinite for all real numbers $k > 0$.

**15.** If $G$ is positive semidefinite, show that $U^T G U$ is also positive semidefinite for all invertible matrices $U$.

**16.** Show that polar forms are not unique by computing $GP$ and $GQ$ where

$$G = \begin{bmatrix} 1 & 1 \\ 1 & 1 \end{bmatrix}, \quad P = \frac{1}{\sqrt{2}} \begin{bmatrix} -1 & 1 \\ 1 & 1 \end{bmatrix}, \text{ and } Q = \frac{1}{\sqrt{2}} \begin{bmatrix} 1 & 1 \\ -1 & 1 \end{bmatrix}.$$

**17.** If $A$ is any square matrix, show that it has a polar factorization of the form $A = RG$ where $R$ is orthogonal and $G$ is positive semidefinite.

**18.** Let $B$ be any symmetric $n \times n$ matrix with (real) eigenvalues $\lambda_1 \geq \cdots \geq \lambda_n$. If $X$ is any nonzero column in $\mathbb{R}^n$, define the **Rayleigh quotient** $R(X) = \frac{X^T B X}{\|X\|^2}$.

a) If $X$ is an eigenvector of $A$ corresponding to $\lambda_i$, show that $R(X) = \lambda_i$.

b) Show that $\lambda_1 \geq R(X) \geq \lambda_n$ for all $X \neq 0$ in $\mathbb{R}^n$.

c) Conclude that $\lambda_1 = max\{R(X) \mid X \neq 0 \text{ in } \mathbb{R}^n\}$ and $\lambda_n = min\{R(X) \mid X \neq 0 \text{ in } \mathbb{R}^n\}$. [Hint: For **b)**, since $B$ is symmetric let $\{Q_1, \cdots, Q_n\}$ be an orthonormal basis of $\mathbb{R}^n$ where $Q_i$ is an eigenvector corresponding to $\lambda_i$ for each $i$. Given $X$ in $\mathbb{R}^n$, write $X = r_1 Q_1 + \cdots + r_n Q_n$, so that $BX = r_1 \lambda_1 Q_1 + \cdots + r_n \lambda_n Q_n$. Use the fact that the $Q_i$ are orthonormal to compute $\|X\|^2 = X \cdot X$ and $X^T B X = X \cdot (BX)$, and exploit the assumption that $\lambda_1 \geq \lambda_i \geq \lambda_n$ for all $i$.]

**19.** Show that the largest eigenvalue $\lambda_1$ of $A^T A$ is given by

$$\lambda_1 = max \left\{ \frac{\|AX\|^2}{\|X\|^2} \mid X \neq 0 \text{ in } \mathbb{R}^n \right\}.$$

[Hint: Apply the preceding exercise with $B = A^T A$.]

# CHAPTER 5

# Vector Spaces

## 5.1 EXAMPLES AND BASIC PROPERTIES

The study of $\mathbb{R}^n$ in Chapter 4 has revealed how useful concepts like subspace, basis, and dimension can be. Surprisingly, these concepts depend on ten simple properties of $\mathbb{R}^n$ (called axioms). In other words, these ten axioms are all that is needed to define these concepts and prove their essential properties. The reason that this is important is that there are examples other than $\mathbb{R}^n$ in which which these axioms are valid, so notions like dimension can be developed in these new contexts. An abstract vector space is a "super" example in which the *only* thing we know is that the axioms hold. Nontheless, even in this full generality we can still define concepts (like subspace and dimension), and prove theorems about them (such as the Fundamental theorem). Hence, the concepts make sense and the theorems are valid in *all* the examples because each example is a specific realization of the general space. This leads to many more applications. Furthermore, clarifying the notion of an abstract vector space gives us a useful new perspective on the specific examples and points to unforeseen notions such as vector spaces of infinite dimension. This chapter is an introduction to these abstract vector spaces.

### 5.1.1 Vector Spaces

A **vector space** consists of a nonempty set $V$ of elements (called **vectors**) that can be added and multiplied by a number[1] (called a **scalar**), and for which certain properties hold (called **axioms**). We assume that any two vectors $\mathbf{v}$ and $\mathbf{w}$ in $V$ can be added and that the sum, denoted $\mathbf{v} + \mathbf{w}$, is again a vector in $V$. Similarly, we assume that any vector $\mathbf{v}$ in $V$ can be multiplied by any scalar $a$, and that the product, denoted $a\mathbf{v}$, is again in $V$. These operations are called **vector addition** and **scalar multiplication** respectively, and the following axioms are assumed to hold:

**Axioms for vector addition:**
- **A1.** If $\mathbf{v}$ and $\mathbf{w}$ are in $V$, then $\mathbf{v} + \mathbf{w}$ is in $V$.
- **A2.** $\mathbf{v} + \mathbf{w} = \mathbf{w} + \mathbf{v}$ for all $\mathbf{v}$ and $\mathbf{w}$ in $V$.
- **A3.** $\mathbf{u} + (\mathbf{v} + \mathbf{w}) = (\mathbf{u} + \mathbf{v}) + \mathbf{w}$ for all $\mathbf{u}$, $\mathbf{v}$, and $\mathbf{w}$ in $V$.
- **A4.** An element $\mathbf{0}$ in $V$ exists such that $\mathbf{v} + \mathbf{0} = \mathbf{v}$ for all $\mathbf{v}$ in $V$.
- **A5.** For each $\mathbf{v}$ in $V$ an element $-\mathbf{v}$ in $V$ exists such that and $\mathbf{v} + (-\mathbf{v}) = \mathbf{0}$.

**Axioms for scalar multiplication:**
- **S1.** If $\mathbf{v}$ is in $V$, then $a\mathbf{v}$ is in $V$ for all scalars $a$.
- **S2.** $a(\mathbf{v} + \mathbf{w}) = a\mathbf{v} + a\mathbf{w}$ for all $\mathbf{v}$ and $\mathbf{w}$ in $V$ and all scalars $a$.
- **S3.** $(a + b)\mathbf{v} = a\mathbf{v} + b\mathbf{v}$ for all $\mathbf{v}$ in $V$ and all scalars $a$ and $b$.
- **S4.** $a(b\mathbf{v}) = (ab)\mathbf{v}$ for all $\mathbf{v}$ in $V$ and all scalars $a$ and $b$.
- **S5.** $1\mathbf{v} = \mathbf{v}$ for all $\mathbf{v}$ in $V$.

---

[1] For us, the scalars will be real numbers. However, they could be complex numbers and, more generally, the scalars could be drawn from an algebraic system called a **field**. Another example is the set $\mathbb{Q}$ of all rational numbers.

We describe Axioms A1 and S1 by saying that $V$ is **closed** under vector addition and scalar multiplication, respectively. As we shall see, the property in Axiom A4 uniquely determines the vector $\mathbf{0}$, and $\mathbf{0}$ is called the **zero vector** of the vector space $V$. Similarly the vector $-\mathbf{v}$ in Axiom A5 is uniquely determined by $\mathbf{v}$, and is called the **negative** of $\mathbf{v}$. Note that combining Axiom A2 and A4 shows that $\mathbf{0} + \mathbf{v} = \mathbf{v}$ for all $\mathbf{v}$ in $V$. Similarly $(-\mathbf{v}) + \mathbf{v} = \mathbf{0}$ using Axioms A2 and A5.

Thus, to specify a vector space $V$, we must do several things. First we must say precisely what $V$ is, that is we must specify the set of vectors we are considering. Next we must define how to add two of our vectors (to get a sum in $V$), and how to multiply any vector by any scalar (to get a scalar product in $V$). Finally, we must verify that the axioms hold. Here are several examples, starting with the central concept of Chapter 4.

**Example 1**    For each $n \geq 1$, $\mathbb{R}^n$ is a vector space using the usual addition and scalar multiplication. Here the vectors are the $n$-tuples, (which we continue to write as columns). The zero vector of $\mathbb{R}^n$ is the usual zero column, and the negative of a vector is the usual negative. Note that $\mathbb{R}^1 = \mathbb{R}$ is the set of real numbers with the usual operations.

**Example 2**    If $m \geq 1$ and $n \geq 1$, the set $\mathbb{M}_{m,n}$ of all $m \times n$ matrices is a vector space using matrix addition and scalar multiplication. The set $\mathbb{M}_{m,n}$ is clearly closed under these operations, the zero vector is the zero matrix $0 = 0_{m,n}$ of size $m \times n$, and the negative of a matrix $A$ in $\mathbb{M}_{m,n}$ is $-A$, the usual negative. The other axioms are basic properties of matrix addition and scalar multiplication.

**Example 3**    The set $\mathbb{P}$ of all polynomials is a vector space. Here a **polynomial** $p(x)$ is an expression

$$p(x) = a_0 + a_1 x + a_2 x^2 + \cdots + a_n x^n$$

where $x$ is a variable called an **indeterminant**, and $a_0, a_1, a_2, \cdots, a_n$ are scalars (possibly zero), called the **coefficients** of the polynomial; $a_0$ is called the **constant coefficient** of $p(x)$. If the coefficients are all zero, the polynomial is called the **zero polynomial** and is denoted simply as 0. Two polynomials are **equal** if they are identical, that is if corresponding coefficients are all equal. In particular,

$$a_0 + a_1 x + a_2 x^2 + \cdots + a_n x^n = 0$$

means that $a_0 = a_1 = a_2 = \cdots = a_n = 0$. (This is the reason for calling $x$ an *indeterminate*.)

Two polynomials are added in the obvious way by adding corresponding coefficients, and the scalar product of a polynomial by a scalar $a$ is obtained by multiplying every coefficient by $a$. For example,

$$(2 - 5x - 3x^3) + (1 + 3x + 7x^2) = 3 - 2x + 7x^2 - 3x^3$$

and

$$4(-1 + 3x - 5x^3) = -4 + 12x - 20x^3.$$

It is a routine verification that these operations make $\mathbb{P}$ into a vector space; the zero vector is the zero polynomial, and the negative of a polynomial $p(x)$ is the polynomial $(-1)p(x)$ obtained from $p(x)$ by negating every coefficient.[2]

If $p(x)$ is a nonzero polynomial, the highest power of $x$ in $p(x)$ with nonzero coefficient is called the **degree** of $p(x)$, and is denoted as $deg\ p(x)$. The coefficient of this highest power is called the **leading coefficient** of $p(x)$. The degree of the zero polynomial is not defined. Some examples:

---

[2] More information on polynomials can be found in Appendix A.3.

$$deg(3 - x) = 1 \quad \text{with leading coefficient } -1$$
$$deg(2 + 7x^2) = 2 \quad \text{with leading coefficient } 7$$
$$deg(4) = 0 \quad \text{with leading coefficient } 4$$

The notion of degree leads to the following example.

**Example 4**    If $n \geq 0$, the set $\mathbb{P}_n = \{a_0 + a_1x + a_2x^2 + \cdots + a_nx^n \mid a_i \text{ in } \mathbb{R}\}$ consists of all polynomials of degree at most $n$, together with the zero polynomial. It is easy to see that $\mathbb{P}_n$ is a vector space using the addition and scalar multiplication of $\mathbb{P}$.

The next example is a vector space that arises frequently in analysis (the theoretical side of calculus).

**Example 5**    If $D$ is a nonempty set of real numbers, let $\mathbb{F}[D]$ denote the set of all real valued functions defined on $D$. The set $D$ is called the **domain** of the functions; usually $D$ will be either $\mathbb{R}$ or an interval

$$[a, b] = \{x \text{ in } \mathbb{R} \mid a \leq x \leq b\}.$$

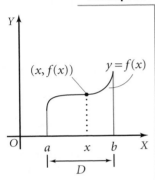

Two such functions $f$ and $g$ in $\mathbb{F}[D]$ are **equal** (written $f = g$) if they have equal values for all $x$ in $D$, that is if

$$f(x) = g(x) \quad \text{for all } x \text{ in } D,$$

and we describe this situation by saying that $f$ and $g$ have the same **action**.

Given a scalar $a$ and functions $f$ and $g$, the action of the sum $f + g$ and scalar product $af$ are defined as in calculus:

$$(f + g)(x) = f(x) + g(x) \quad \text{and} \quad (af)(x) = af(x) \quad \text{for all } x \text{ in } D.$$

Thus the value of the function $f + g$ at each $x$ is the sum of the values of $f$ and $g$ at $x$, and the value of $af$ at $x$ is $a$ times the value of $f$ at $x$. For this reason, these operations are called **pointwise** addition and scalar multiplication. They are the operations familiar from elementary algebra and calculus.

For example, if $f(x) = \sin x$ for all $x$ and $g(x) = 1 + x$ for all $x$, then the sum $f + g$ is given by

$$(f + g)(x) = f(x) + g(x)$$
$$= (1 + x) + \sin x$$

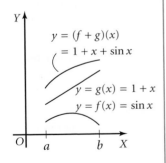

for all $x$. The graph of the equation $y = (f + g)(x)$ is shown in the diagram together with $y = f(x)$ and $y = g(x)$.

These pointwise operations make $\mathbb{F}[D]$ into a vector space. The zero vector in $\mathbb{F}[D]$ is the constant function $f_0$ that is identically zero for all values of $x$; that is, $f_0(x) = 0$ for every $x$ in $D$. The negative of a function $f$ is the function $-f$ defined by $(-f)(x) = -f(x)$ for all $x$ in $D$. The verification that axiom A4 holds in $\mathbb{F}[D]$ is as follows: If $f$ is any function in $\mathbb{F}[D]$, we compute

$$(f + f_0)(x) = f(x) + f_0(x) = f(x) + 0 = f(x) \quad \text{for all } x \text{ in } D.$$

This means that $f + f_0 = f$ so, as this is valid for every $f$ in $\mathbb{F}[D]$, it shows that axiom A4 holds in $\mathbb{F}[D]$. The reader should verify that the other axioms hold.

**Example 6**     A set $\{0\}$ with one element becomes a vector space if we define $0 + 0 = 0$ and $a0 = 0$ for all scalars $a$. The axioms are easily verified, and $\{0\}$ is called the **zero space**.

While other examples will occur later, these suffice to illustrate the diversity of the vector space concept. We now turn to the business of proving theorems in the setting of an abstract vector space $V$ (they will then hold in *any* of these examples). At the outset, all we know about $V$ is that the axioms are valid. While this does not seem to be much to go on, a surprising number of facts about $V$ can be deduced from the axioms. The idea is to build up a body of information, one theorem at a time, and to use theorems already proved to help deduce new results. Our first result is very useful.

**THEOREM 1**     **Cancellation.** If $\mathbf{w} + \mathbf{v} = \mathbf{u} + \mathbf{v}$ in a vector space $V$, then $\mathbf{w} = \mathbf{u}$.

PROOF     If $\mathbf{v}$, $\mathbf{w}$, and $\mathbf{u}$ were numbers, we would simply subtract $\mathbf{v}$ from both sides to obtain $\mathbf{w} = \mathbf{u}$. We accomplish this by adding $-\mathbf{v}$ to both sides (using Axiom A5 to guarantee that $-\mathbf{v}$ is available). The result is

$$(\mathbf{w} + \mathbf{v}) + (-\mathbf{v}) = (\mathbf{u} + \mathbf{v}) + (-\mathbf{v}).$$

Using Axiom A3, this becomes $\mathbf{w} + (\mathbf{v} + (-\mathbf{v})) = \mathbf{u} + (\mathbf{v} + (-\mathbf{v}))$, and this is $\mathbf{w} + \mathbf{0} = \mathbf{u} + \mathbf{0}$ by Axiom A5 again. Hence $\mathbf{w} = \mathbf{u}$ by Axiom A4. This is what we wanted.

Note that $\mathbf{v} + \mathbf{w} = \mathbf{v} + \mathbf{u}$ in a vector space $V$ implies that $\mathbf{w} = \mathbf{u}$ by Axiom A2 and cancellation.

In particular, Theorem 1 shows that there is only one zero vector in any vector space $V$. Indeed, if $\mathbf{0}'$ has the property in Axiom A4, then $\mathbf{v} + \mathbf{0} = \mathbf{v} = \mathbf{v} + \mathbf{0}'$ for all $\mathbf{v}$ in $V$, so $\mathbf{0} = \mathbf{0}'$ by cancellation. Moreover, a similar argument shows that there is only one negative of each vector (verify). Hence we speak of *the* zero vector in a vector space, and of *the* negative of any given vector.

The technique in the proof of Theorem 1 was to mimic the process of subtracting $\mathbf{v}$ from both sides of the equation by adding $-\mathbf{v}$ to both sides. If we do this to the equation

$$\mathbf{x} + \mathbf{v} = \mathbf{u}$$

the result is $\mathbf{x} = \mathbf{u} + (-\mathbf{v})$ as the reader can verify. Hence, by analogy with arithmetic, we define the **difference** $\mathbf{u} - \mathbf{v}$ of two vectors $\mathbf{u}$ and $\mathbf{v}$ by

$$\mathbf{u} - \mathbf{v} = \mathbf{u} + (-\mathbf{v}).$$

And, as in arithmetic, we say that $\mathbf{u} - \mathbf{v}$ is the result of having **subtracted** $\mathbf{v}$ from $\mathbf{u}$. Incidentally, we have shown that $\mathbf{x} = \mathbf{u} - \mathbf{v}$ is the unique solution to the equation $\mathbf{x} + \mathbf{v} = \mathbf{u}$.

The next theorem gives some basic facts about an arbitrary vector space $V$ that will be used extensively later.

**THEOREM 2**     Let $\mathbf{v}$ denote a vector in a vector space $V$, and let $a$ denote a scalar.

(1) $0\mathbf{v} = \mathbf{0}$.
(2) $a\mathbf{0} = \mathbf{0}$.
(3) If $a\mathbf{v} = \mathbf{0}$, then either $a = 0$ or $\mathbf{v} = \mathbf{0}$.
(4) $(-1)\mathbf{v} = -\mathbf{v}$.
(5) $(-a)\mathbf{v} = -(a\mathbf{v}) = a(-\mathbf{v})$.

PROOF     (1) By Axioms S3 and A4 we have $0\mathbf{v} + 0\mathbf{v} = (0 + 0)\mathbf{v} = 0\mathbf{v} = 0\mathbf{v} + \mathbf{0}$. Now (1) follows by cancellation.
(2) Here $a\mathbf{0} + a\mathbf{0} = a(\mathbf{0} + \mathbf{0}) = a\mathbf{0} = a\mathbf{0} + \mathbf{0}$ by Axioms S2 and A4, so again cancellation does it.

**(3)** Let $a\mathbf{v} = \mathbf{0}$. If $a = 0$, we are done. If $a \neq 0$, then multiply the equation $a\mathbf{v} = \mathbf{0}$ on both sides by $\frac{1}{a}$. Then (2) and Axioms S4 and S5 give

$$\mathbf{v} = 1\mathbf{v} = \left(\tfrac{1}{a}a\right)\mathbf{v} = \tfrac{1}{a}(a\mathbf{v}) = \tfrac{1}{a}\mathbf{0} = \mathbf{0}.$$

**(4)** Using (1) with Axioms S5, S3, A2, and A5, we have

$$(-1)\mathbf{v} + \mathbf{v} = (-1)\mathbf{v} + 1\mathbf{v} = (-1 + 1)\mathbf{v} = 0\mathbf{v} = \mathbf{0} = -\mathbf{v} + \mathbf{v}.$$

Now cancellation applies.

**(5)** For the first equation, we use (4) and Axiom S4:

$$(-a)\mathbf{v} = ((-1)a)\mathbf{v} = (-1)(a\mathbf{v}) = -(a\mathbf{v}).$$

The other equation is verified similarly, and the proof is left to the reader.

Observe that parts (1), (2), and (3) of Theorem 2 can be combined in the following statement: If $a$ is a scalar and $\mathbf{v}$ is a vector, then

$$a\mathbf{v} = \mathbf{0} \quad \text{if and only if either } a = 0 \text{ or } \mathbf{v} = \mathbf{0}.$$

This will be used several times in what follows.

Axioms S2 and S3 extend as follows. If $a, a_1, a_2, \cdots, a_n$ are scalars and $\mathbf{v}, \mathbf{v}_1, \mathbf{v}_2, \cdots, \mathbf{v}_n$ are vectors, we have[3]

$$a(\mathbf{v}_1 + \mathbf{v}_2 + \cdots + \mathbf{v}_n) = a\mathbf{v}_1 + a\mathbf{v}_2 + \cdots + a\mathbf{v}_n$$
$$(a_1 + a_2 + \cdots + a_n)\mathbf{v} = a_1\mathbf{v} + a_2\mathbf{v} + \cdots + a_n\mathbf{v}$$

These facts, together with the axioms, Theorem 2, and the definition of subtraction, enable us to manipulate and simplify sums of scalar multiples of vectors by collecting like terms, expanding, and taking out common factors. This has already been discussed for matrices and vectors in $\mathbb{R}^n$, and manipulations in an arbitrary vector space are carried out in a similar way. For example,

$$3(\mathbf{v} - 2\mathbf{w}) - 4(\mathbf{v} + 2\mathbf{u}) + 2(3\mathbf{w} + 4\mathbf{u}) = 3\mathbf{v} - 6\mathbf{w} - 4\mathbf{v} - 8\mathbf{u} + 6\mathbf{w} + 8\mathbf{u} = -\mathbf{v}.$$

Computations like this are used in every discussion of vector spaces.

## 5.1.2 Subspaces

If $V$ is a vector space, a nonempty subset $U \subseteq V$ is called a **subspace** of $V$ if $U$ is itself a vector space using the addition and scalar multiplication of $V$. Thus $\mathbb{P}_n$ is a subspace of $\mathbb{P}$ and, if we regard polynomials as real valued functions defined on $\mathbb{R}$, $\mathbb{P}$ is a subspace of $\mathbb{F}[\mathbb{R}]$. Note that if $U$ and $W$ are subspaces of some vector space $V$, and if $U \subseteq W$, then $U$ is a subspace of $W$.

Using Theorem 2, we can give a simple test for when a subset of a vector space is a subspace.

**THEOREM ③**

**Subspace Test.** A subset $U$ of a vector space $V$ is a subspace if and only if it satisfies the following three conditions:

**(1)** $\mathbf{0}$ is in $U$.

**(2)** If $\mathbf{u}$ is in $U$, then $a\mathbf{u}$ is in $U$ for all scalars $a$.
($U$ is **closed under scalar multiplication**.)

**(3)** If $\mathbf{u}$ and $\mathbf{u}_1$ are in $U$, then $\mathbf{u} + \mathbf{u}_1$ is in $U$.
($U$ is **closed under addition**.)

**PROOF**

If $U$ is a subspace, then (2) and (3) hold by Axioms S1 and A1 (applied to the vector space $U$) and, if $\mathbf{u}$ is in $U$, then Theorem 2 gives $\mathbf{0} = 0\mathbf{u}$, so $\mathbf{0}$ is in $U$ by (2), and (1) holds.[4]

---

[3] It is a consequence of Axiom A3 that we can omit parentheses when writing a sum $\mathbf{v}_1 + \mathbf{v}_2 + \cdots + \mathbf{v}_n$.

[4] This shows that condition (1) follows from (2) provided $U$ contains at least one vector. We prefer to retain (1) in Theorem 3 because checking it is usually easy and, if it fails, $U$ *cannot* be a subspace.

Conversely, suppose that (1), (2), and (3) hold for $U$. Then Axioms A1 and S1 hold by (2) and (3), and Axioms A2, A3, S2, S3, S4, and S5 all hold in $U$ because they hold in $V$. Next, Axiom A4 holds in $U$ because the zero vector of $V$ is actually in $U$ by (1), and so serves as the zero of $U$. Finally, if $\mathbf{u}$ is in $U$, then its negative $-\mathbf{u}$ in the space $V$ is actually in $U$ by (2) because $-\mathbf{u} = (-1)\mathbf{u}$. This proves Axiom A5 for $U$.

Note that the proof of Theorem 3 shows that if $U$ is a subspace of $V$, then $U$ and $V$ share the same zero vector, and the negative of a vector in $U$ is the same as its negative in $V$.

Subspaces of $\mathbb{R}^n$ were defined in Section 4.1 as subsets satisfying the conditions in Theorem 3. This gives many more examples of vector spaces.

**Example 7**    All the subspaces of $\mathbb{R}^n$ discussed in Chapter 4 are vector spaces in their own right. For example, if $A$ is any matrix, then each of $nullA$, $imA = colA$, and $rowA$ is a vector space. Another example: $U^{\perp}$ is a vector space if $U$ is any subspace of $\mathbb{R}^n$.

**Example 8**    If $V$ is any vector space, then $\{\mathbf{0}\}$ and $V$ are subspaces of $V$. The space $\{\mathbf{0}\}$ is called the **zero subspace** of $V$, and any subspace other than $\{\mathbf{0}\}$ and $V$ is called a **proper subspace** of $V$.

**Example 9**    If $\mathbf{v}$ is a vector in a vector space $V$, show that $\mathbb{R}\mathbf{v}$ is a subspace of $V$ where $\mathbb{R}\mathbf{v} = \{r\mathbf{v} \mid r \text{ in } \mathbb{R}\}$.

SOLUTION    First, $\mathbf{0} = 0\mathbf{v}$ shows that $\mathbf{0}$ is in $\mathbb{R}\mathbf{v}$. Next, given $\mathbf{u} = r\mathbf{v}$ and $\mathbf{u}_1 = r_1\mathbf{v}$ in $\mathbb{R}\mathbf{v}$, we have, for each scalar $a$,

$$a\mathbf{u} = a(r\mathbf{v}) = (ar)\mathbf{v} \quad \text{and} \quad \mathbf{u} + \mathbf{u}_1 = r\mathbf{v} + r_1\mathbf{v} = (r + r_1)\mathbf{v}.$$

Hence $a\mathbf{u}$ and $\mathbf{u} + \mathbf{u}_1$ are both in $\mathbb{R}\mathbf{v}$, so the Subspace test applies.

**Example 10**    Show that the set $U = \{A \text{ in } \mathbb{M}_{n,n} \mid A \text{ is symmetric}\}$ is a subspace of $\mathbb{M}_{n,n}$.

SOLUTION    Recall that $A$ is symmetric if $A^T = A$. Hence the zero matrix $0$ is in $U$ because $0^T = 0$. If $A$ and $B$ are in $U$, then $A^T = A$ and $B^T = B$, so $(A + B)^T = A^T + B^T = A + B$ and $(aA)^T = aA^T = aA$ for all scalars $a$. Hence $A + B$ and $aA$ are in $U$, so $U$ is a subspace by the Subspace test.

**Example 11**    If $\lambda$ is a fixed real number, let $U = \{p(x) \text{ in } \mathbb{P} \mid p(\lambda) = 0\}$ be the set of all polynomials which have $\lambda$ as a root. Show that $U$ is a subspace of $\mathbb{P}$.

SOLUTION    Clearly the zero polynomial lies in $U$. If $p(x)$ and $q(x)$ lie in $U$, then $(p + q)(\lambda) = p(\lambda) + q(\lambda) = 0 + 0 = 0$, so $p + q$ is in $U$. Similarly $ap$ is in $U$ for all scalars $a$ (verify), and the Subspace test applies.

**Example 12**[5]    Show that the set of differentiable functions is a subspace of $\mathbb{F}[\mathbb{R}]$.

SOLUTION    The zero function is differentiable (its derivative is again the zero function), and it is a theorem of calculus that sums and scalar multiples of differentiable functions are again differentiable.

---

[5] Readers not familiar with calculus may omit this example with no loss of continuity.

Let $\mathbf{v}_1$, $\mathbf{v}_2$, $\cdots$, $\mathbf{v}_n$ be vectors in a vector space $V$. As in $\mathbb{R}^n$, if $a_1$, $a_2$, $\cdots$, $a_n$ are scalars, a sum $a_1\mathbf{v}_1 + a_2\mathbf{v}_2 + \cdots + a_n\mathbf{v}_n$ is called a **linear combination** of the $\mathbf{v}_i$. The set of all such linear combinations is called the **span** of the $\mathbf{v}_i$ and is denoted

$$span\{\mathbf{v}_1,\ \mathbf{v}_2,\ \cdots,\ \mathbf{v}_n\} = \{r_1\mathbf{v}_1 + r_2\mathbf{v}_2 + \cdots + r_n\mathbf{v}_n \mid r_i \text{ in } \mathbb{R}\}.$$

In particular, $span\{\mathbf{v}\} = \{r\mathbf{v} \mid r \text{ in } \mathbb{R}\} = \mathbb{R}\mathbf{v}$ is the subspace in Example 9. These notions coincide with those in Chapter 4, and the proof of Theorem 1 §4.1 goes through verbatim to give

---

**THEOREM** ④ Let $U = span\{\mathbf{v}_1,\ \mathbf{v}_2,\ \cdots,\ \mathbf{v}_n\}$ in a vector space $V$.

    (1) $U$ is a subspace of $V$ containing each of the vectors $\mathbf{v}_1$, $\mathbf{v}_2$, $\cdots$, $\mathbf{v}_n$.

    (2) $U$ is the "smallest" subspace containing these vectors in the sense that any subspace of $V$ that contains each of $\mathbf{v}_1$, $\mathbf{v}_2$, $\cdots$, $\mathbf{v}_n$ must already contain $U$.

---

**Example 13** If $\mathbf{u}$, $\mathbf{v}$, and $\mathbf{w}$ are vectors, show that $span\{\mathbf{u},\ \mathbf{v},\ \mathbf{w}\} = span\{\mathbf{u}+\mathbf{v},\ \mathbf{v}+\mathbf{w},\ \mathbf{u}+\mathbf{w}\}$.

**SOLUTION** For convenience, write $U = span\{\mathbf{u},\ \mathbf{v},\ \mathbf{w}\}$ and $W = span\{\mathbf{u}+\mathbf{v},\ \mathbf{v}+\mathbf{w},\ \mathbf{u}+\mathbf{w}\}$. We have $W \subseteq U$ by Theorem 4 because each of $\mathbf{u}+\mathbf{v}$, $\mathbf{v}+\mathbf{w}$, and $\mathbf{u}+\mathbf{w}$ is in $U$. On the other hand,

$$\mathbf{u} = \tfrac{1}{2}(\mathbf{u}+\mathbf{v}) - \tfrac{1}{2}(\mathbf{v}+\mathbf{w}) + \tfrac{1}{2}(\mathbf{u}+\mathbf{w})$$

shows that $\mathbf{u}$ is in $W$. Similarly $\mathbf{v}$ and $\mathbf{w}$ are in $W$, so $U = span\{\mathbf{u},\ \mathbf{v},\ \mathbf{w}\} \subseteq W$, again by Theorem 4. Hence, $U = W$.

**Example 14** Show that $\mathbb{P}_3 = span\{2x + x^3,\ x^2 - x,\ 3x - 2,\ 5\}$.

**SOLUTION** For convenience, write $U = span\{2x + x^3,\ x^2 - x,\ 3x - 2,\ 5\}$. Then $U \subseteq \mathbb{P}_3$ is clear. Since $\mathbb{P}_3 = span\{1,\ x,\ x^2,\ x^3\}$, it suffices by Theorem 4 to show that each of $1$, $x$, $x^2$, and $x^3$ is in $U$.

    First $1 = \tfrac{1}{5} \cdot 5$ is in $U$ because $5$ is in $U$.

    Then $x = \tfrac{1}{3}[(3x - 2) + 2 \cdot 1]$ is in $U$ because both $3x - 2$ and $1$ are in $U$.

    Next $x^2 = (x^2 - x) + x$ is in $U$ because $x^2 - x$ and $x$ are both in $U$.

    Finally $x^3 = (2x + x^3) - 2 \cdot x$ is in $U$ because $2x + x^3$ and $x$ are in $U$.

This is what we wanted.

**Example 15** Show that $\mathbb{P}_n = span\{1,\ x,\ x^2,\ \cdots,\ x^n\}$.

**SOLUTION** We have $\mathbb{P}_n = \{a_0 + a_1 x + a_2 x^2 + \cdots + a_n x^n \mid a_i \text{ in } \mathbb{R}\}$ by Example 4.

Every $2 \times 2$ matrix $A$ in $\mathbb{M}_{2,2}$ can be written as follows:

$$A = \begin{bmatrix} a_{11} & a_{12} \\ a_{21} & a_{22} \end{bmatrix}$$

$$= a_{11}\begin{bmatrix} 1 & 0 \\ 0 & 0 \end{bmatrix} + a_{12}\begin{bmatrix} 0 & 1 \\ 0 & 0 \end{bmatrix} + a_{21}\begin{bmatrix} 0 & 0 \\ 1 & 0 \end{bmatrix} + a_{22}\begin{bmatrix} 0 & 0 \\ 0 & 1 \end{bmatrix}.$$

This reveals a useful spanning set for $\mathbb{M}_{2,2}$, and a similar decomposition works in $\mathbb{M}_{m,n}$. The $m \times n$ matrix with $(i,\ j)$-entry $1$ and the other entries $0$ is called the $(i,\ j)$-**matrix unit**, and is denoted $E_{ij}$. If $A = [a_{ij}]$ is a matrix in $\mathbb{M}_{m,n}$, it is clear that $A = \Sigma_{i,j}\, a_{ij} E_{ij}$ where the sum is taken over all $i$ and $j$ such that $1 \le i \le m$ and $1 \le j \le n$. Hence $\mathbb{M}_{m,n}$ is spanned by these matrix units. For example

$$\mathbb{M}_{2,3} = span\{E_{11},\ E_{12},\ E_{13},\ E_{21},\ E_{22},\ E_{23}\}.$$

In general, there are *mn* matrix units in $\mathbb{M}_{m,n}$, and we have

**Example 16**   $\mathbb{M}_{m,n} = span\{E_{ij} \mid 1 \le i \le m \text{ and } 1 \le j \le n\}$ and where the $E_{ij}$ are the $m \times n$ matrix units.

**Example 17**   Suppose an object is suspended by a coil spring. If the object is pulled down from its rest position and released, it will begin to oscillate. It is known that the displacement *y* about the rest position is given in terms of the time *t* by a function of the following form:
$$y = a\,\cos(\omega t) + b\,\sin(\omega t)$$
where $\omega$ is a constant depending on the spring and the mass of the object. Thus the set of all functions describing such motions is the subspace of $\mathbb{F}[\mathbb{R}]$ spanned by $\cos(\omega t)$ and $\sin(\omega t)$.

## Exercises 5.1

The problems that involve calculus concepts can be omitted by students with no calculus background.

**1.** In each case either show that the statement is true or give an example showing that it is false.
   **a)** Every vector space contains at least one vector.
   **b)** Every nonzero vector space contains an infinite number of vectors.
   **c)** $\{1 - x, 2 + x - x^2\}$ spans $\mathbb{P}_2$.
   **d)** Working in $\mathbb{F}[\mathbb{R}]$, 1 is in $span\{\cos^2 x, \sin^2 x\}$.
   **e)** Working in $\mathbb{F}[\mathbb{R}]$, $x$ is in $span\{\cos^2 x, \sin^2 x\}$.
   **f)** The set $\mathbb{Z} = \{0, \pm 1, \pm 2, \pm 3, \cdots\}$ is a vector space with the usual operations.
   **g)** The set $\mathbb{Q}$ of all rational numbers (fractions) is a vector space with the usual operations.
   **h)** Every nonzero vector space $V$ contains a proper subspace (Example 8).

**2.** Let $V$ denote the set of all $1 \times 3$ matrices, and define addition on $V$ as for $\mathbb{M}_{1,3}$. For each of the following definitions of scalar multiplication (denoted $\cdot$ in each case), decide if $V$ is a vector space.
   **a)** $a \cdot [x \ y \ z] = [ax \ ay \ z]$.
   **b)** $a \cdot [x \ y \ z] = [0 \ 0 \ 0]$.
   **c)** $a \cdot [x \ y \ z] = [ax \ ay \ 0]$.
   **d)** $a \cdot [x \ y \ z] = [-ax \ -ay \ -az]$.
   **e)** $a \cdot [x \ y \ z] = [x \ y \ z]$.
   **f)** $a \cdot [x \ y \ z] = [ay \ ax \ az]$.

**3.** In each case decide if $V$ is a vector space with the given operations. Support your answer.
   **a)** $V$ is the set of all nonnegative real numbers; usual addition and multiplication.
   **b)** $V$ is the set $\mathbb{C}$ of complex numbers; usual addition and multiplication by a real number.

**c)** $V = \{0\}$, where $0 + 0 = 0$ and $a \cdot 0 = 0$ for all real numbers $a$.
   **d)** $V = \{[x \ 1 \ z] \mid x \text{ and } z \text{ real}\}$; operations of $\mathbb{M}_{1,3}$.
   **e)** $V = \{[x \ y] \mid x \text{ and } y \text{ real}, x^2 = y^2\}$, operations of $\mathbb{M}_{1,2}$.
   **f)** $V$ is the set of all $2 \times 2$ matrices $B$; addition of $\mathbb{M}_{2,2}$ and $a \cdot B = |a| B$.
   **g)** $V$ is the set of all $2 \times 2$ matrices $B$; addition of $\mathbb{M}_{2,2}$ and $a \cdot B = aB^T$.
   **h)** $V$ is the set of functions $\mathbb{R} \to \mathbb{R}$; pointwise addition and $(a \cdot f)(x) = f(ax)$ for all $x$.

**4.** Let $V = \{[x \ y] \mid x \text{ and } y \text{ real}\}$. Show that $V$ is a vector space with the following operations:
$[x \ y] + [x_1 \ y_1] = [x + x_1 \ y + y_1 + 1]$ and
$a \cdot [x \ y] = [ax \ ay + a - 1]$.

**5.** Let $V = \{v \text{ in } \mathbb{R} \mid v > 0\}$. Show that $V$ is a vector space if the vector addition is ordinary multiplication, and scalar multiplication is defined by $a \cdot v = v^a$.

**6.** In each case determine if $U$ is a subspace of $\mathbb{P}_3$. Support your answer.
   **a)** $U = \{p(x) \mid p(1) = 1\}$
   **b)** $U = \{p(x) \mid p(x) = 0 \text{ or } \deg p(x) = 3\}$
   **c)** $U = \{x p(x) \mid p(x) \text{ in } \mathbb{P}_3\}$
   **d)** $U = \{xp(x) + (1-x)q(x) \mid p(x) \text{ and } q(x) \text{ in } \mathbb{P}_2\}$
   **e)** $U = \{xp(x) \mid p(x) \text{ in } \mathbb{P}_2\}$
   **f)** $U = \{p(x) \text{ in } \mathbb{P}_3 \mid p(x) \text{ has constant coefficient } 0\}$

**7.** In each case determine if $U$ is a subspace of $\mathbb{M}_{2,2}$. Support your answer.
   **a)** $U = \left\{ \begin{bmatrix} a & b \\ 0 & c \end{bmatrix} \mid a, b, c \text{ in } \mathbb{R} \right\}$
   **b)** $U = \left\{ \begin{bmatrix} a & b \\ c & d \end{bmatrix} \mid a + c = b + d \right\}$

**c)** $U = \{A \mid A = -A^T\}$

**d)** $U = \{A \mid AB = BA\}$, where $B$ is a fixed matrix in $\mathbb{M}_{2,2}$

**e)** $U = \{A \mid det A = 0\}$

**f)** $U = \{A \mid A^2 = A\}$

**g)** $U = \{BA_0 \mid B \text{ in } \mathbb{M}_{2,3}\}$, where $A_0$ is a fixed $3 \times 2$ matrix

**8.** In each case determine if $U$ is a subspace of $\mathbb{F}[0, 1]$. Support your answer.

**a)** $U = \{f \mid f(0) = 1\}$

**b)** $U = \{f \mid f(0) = 0\}$

**c)** $U = \{f \mid f(0) = f(1)\}$

**d)** $U = \{f \mid f(x) \geq 0 \text{ for all } x \text{ in } [0, 1]\}$

**e)** $U = \{f \mid f(x) = f(y) \text{ for all } x \text{ and } y \text{ in } [0, 1]\}$

**f)** $U = \{f \mid f(x + y) = f(x) + f(y) \text{ for all } x \text{ and } y \text{ in } [0, 1]\}$

**g)** $U = \{f \mid f'\left(\frac{1}{2}\right) = 0$, where $f'$ denotes the derivative of $f\}$

**h)** $U = \{f \mid \int_0^1 f(x)dx = 0\}$

**9.** In each case determine whether $\mathbf{v}_1$ or $\mathbf{v}_2$ lies in $span\{\mathbf{u}, \mathbf{w}\}$.

**a)** $\mathbf{u} = x^2 - 2$, $\mathbf{w} = x^2 + x$; $\mathbf{v}_1 = x^2 + 1$, $\mathbf{v}_2 = x^2 - 2x - 6$.

**b)** $\mathbf{u} = x^3 - 2x + 1$, $\mathbf{w} = x^2 + x - 2$; $\mathbf{v}_1 = x^3 + 2x^2 - 3$, $\mathbf{v}_2 = x^3$.

**c)** $\mathbf{u} = \begin{bmatrix} 1 & -1 \\ 0 & 1 \end{bmatrix}$, $\mathbf{w} = \begin{bmatrix} 0 & 1 \\ 1 & -1 \end{bmatrix}$; $\mathbf{v}_1 = \begin{bmatrix} 1 & 0 \\ 1 & 0 \end{bmatrix}$, $\mathbf{v}_2 = \begin{bmatrix} 1 & 0 \\ 0 & 1 \end{bmatrix}$.

**d)** $\mathbf{u} = \begin{bmatrix} 1 & 0 \\ 0 & 1 \end{bmatrix}$, $\mathbf{w} = \begin{bmatrix} 0 & -1 \\ -1 & 1 \end{bmatrix}$; $\mathbf{v}_1 = \begin{bmatrix} 1 & 0 \\ 0 & 0 \end{bmatrix}$, $\mathbf{v}_2 = \begin{bmatrix} 5 & 3 \\ 3 & 2 \end{bmatrix}$.

**e)** $\mathbf{u} = \sin^2 x$, $\mathbf{w} = \cos^2 x$; $\mathbf{v}_1 = \sin x$, $\mathbf{v}_2 = \cos(2x)$. (Work in $\mathbb{F}[\mathbb{R}]$.)

**f)** $\mathbf{u} = \sin^2 x$, $\mathbf{w} = \cos^2 x$; $\mathbf{v}_1 = 2$, $\mathbf{v}_2 = x$. (Work in $\mathbb{F}[\mathbb{R}]$.)

**10. a)** Does $\left\{ \begin{bmatrix} 1 & 1 \\ 0 & 0 \end{bmatrix}, \begin{bmatrix} 1 & 0 \\ 1 & 0 \end{bmatrix}, \begin{bmatrix} 0 & 0 \\ 1 & 1 \end{bmatrix}, \begin{bmatrix} 0 & 1 \\ 0 & 1 \end{bmatrix} \right\}$ span $\mathbb{M}_{2,2}$? Support your answer.

**b)** Does $\{1 + x, x + x^2, x^2 + x^3, x^3 + 1\}$ span $\mathbb{P}_3$? Support your answer.

**11.** In each case find *all* vectors $\mathbf{x}$, $\mathbf{y}$, and $\mathbf{z}$ that satisfy the given equations.

**a)**
$$\mathbf{x} - \mathbf{y} + 2\mathbf{z} = 0$$
$$2\mathbf{x} + 3\mathbf{y} - \mathbf{z} = 0$$
$$-\mathbf{x} + 6\mathbf{y} - 7\mathbf{z} = 0$$

**b)**
$$\mathbf{x} + \mathbf{y} - 2\mathbf{z} = 0$$
$$-\mathbf{x} - 3\mathbf{y} + 2\mathbf{z} = 0$$
$$-\mathbf{x} - 7\mathbf{y} + 2\mathbf{z} = 0$$

**12.** If $Y_0$ is a fixed column in $\mathbb{R}^n$, let $U = \{AY_0 \mid A \text{ in } \mathbb{M}_{m,n}\}$.

**a)** Show that $U$ is a subspace of $\mathbb{R}^m$.

**b)** Show that $U = \mathbb{R}^m$ if $Y_0 \neq 0$.

**13.** (Requires calculus) Show that the set $\mathbf{C}[a, b]$ of all continuous functions defined on the interval $[a, b]$ is a subspace of $\mathbb{F}[a, b]$.

**14. a)** Show that the zero element of a vector space is unique. That is, if $\mathbf{z} + \mathbf{v} = \mathbf{v}$ holds for all $\mathbf{v}$, show that necessarily $\mathbf{z} = 0$.

**b)** Show that the negative of a vector $\mathbf{v}$ is uniquely determined by $\mathbf{v}$. That is, if $\mathbf{v} + \mathbf{v}' = 0$, show that necessarily $\mathbf{v}' = -\mathbf{v}$.

**15.** Show that $\mathbf{x} = \mathbf{u} - \mathbf{v}$ is the *unique* solution to the equation $\mathbf{x} + \mathbf{v} = \mathbf{u}$. State all axioms used.

**16.** Prove that $a(-\mathbf{v}) = (-a)\mathbf{v}$ for all scalars $a$ and vectors $\mathbf{v}$, and so complete (5) of Theorem 2.

**17.** Show that $-0 = 0$ in every vector space $V$. Which vectors $\mathbf{v}$ satisfy $-\mathbf{v} = \mathbf{v}$?

**18.** If $\mathbf{v} \neq 0$ in a vector space $V$, show that $\mathbb{R}\mathbf{v}$ has only two subspaces, $\{0\}$ and $\mathbb{R}\mathbf{v}$.

**19.** Given $\mathbf{v}$ in a vector space $V$, show that $\mathbb{R}\mathbf{v} = span\{a\mathbf{v}\}$ for all $a \neq 0$ in $\mathbb{R}$.

**20.** Let $D$ denote a set of real numbers. If $a$ is a real number, define the **constant function** $f_a : D \rightarrow \mathbb{R}$ by $f_a(x) = a$ for all $x$ in $D$.

**a)** Show that the set $U = \{f_a \mid a \text{ in } \mathbb{R}\}$ of all constant functions is a subspace of $\mathbb{F}[D]$.

**b)** Show further that $U = span\{f_1\} = \mathbb{R}f_1$.

**c)** Show that the constant function $f_0$ is the zero vector in $\mathbb{F}[D]$ (see Example 5).

**21.** Suppose $a\mathbf{v} = b\mathbf{v}$ in a vector space.

**a)** If $a \neq b$, show that $\mathbf{v} = 0$.

**b)** If $\mathbf{v} \neq 0$, show that $a = b$.

**22.** Suppose $a\mathbf{v} = a\mathbf{w}$ in a vector space.

**a)** If $\mathbf{v} \neq \mathbf{w}$, show that $a = 0$.

**b)** If $a \neq 0$, show that $\mathbf{v} = \mathbf{w}$.

**23.** Let $U$ be a subspace of a vector space $V$.

**a)** If $a\mathbf{v}$ is in $U$ where $a \neq 0$, show that $\mathbf{v}$ is in $U$.

**b)** If $\mathbf{v}$ and $\mathbf{w} + \mathbf{v}$ are in $U$, show that $\mathbf{w}$ is in $U$.

**24. a)** Show that $\mathbb{M}_{2,2} = span \left\{ \begin{bmatrix} 1 & 0 \\ 0 & 0 \end{bmatrix}, \begin{bmatrix} 1 & 0 \\ 0 & 1 \end{bmatrix}, \begin{bmatrix} 0 & 1 \\ 1 & 0 \end{bmatrix}, \begin{bmatrix} 0 & 1 \\ 0 & 0 \end{bmatrix} \right\}$.

**b)** Show that $\mathbb{P}_3 = span\{1 + x^3, x + x^2, 1 + x + x^2, 2x - x^3\}$.

**25.** If $\mathbb{M}_{m,n} = span\{A_1, A_2, \cdots, A_k\}$, show that $\mathbb{M}_{n,m} = span\{A_1^T, A_2^T, \cdots, A_k^T\}$.

**26.** If $A_1, A_2, \cdots, A_k$ are $m \times n$ matrices and $A_1Y = A_2Y = \cdots = A_kY = 0$ for some $Y \neq 0$ in $\mathbb{R}^n$, show that $\{A_1, A_2, \cdots, A_k\}$ cannot span $\mathbb{M}_{m,n}$.

**27.** Let $X$ and $Y$ be nonempty sets of vectors in a vector space $V$. If $X \subseteq Y$, show that $span\{X\} \subseteq span\{Y\}$.

**28.** Show that no finite set of polynomials can span $\mathbb{P}$.

**29.** Let $U$ be a *nonempty* subset in a vector space $V$.
    **a)** Show that $U$ is a subspace of $V$ if and only if $\mathbf{u} + a\,\mathbf{u}_1$ is in $U$ for all $\mathbf{u}$ and $\mathbf{u}_1$ in $U$ and all $a$ in $\mathbb{R}$.
    **b)** Show that $U$ is a subspace of $V$ if and only if $U$ is closed under addition and scalar multiplication.

**30.** By calculating $(1 + 1)(\mathbf{v} + \mathbf{w})$ two ways, show that axiom A2 follows from the other axioms.

**31.** Use induction on $n$ to prove each of the following in an arbitrary vector space $V$.
    **a)** $a(\mathbf{v}_1 + \mathbf{v}_2 + \cdots + \mathbf{v}_n) = a\mathbf{v}_1 + a\mathbf{v}_2 + \cdots + a\mathbf{v}_n$ for all $a$ in $\mathbb{R}$ and $\mathbf{v}_i$ in $V$.
    **b)** $(a_1 + a_2 + \cdots + a_n)\mathbf{v} = a_1\mathbf{v} + a_2\mathbf{v} + \cdots + a_n\mathbf{v}$ for all $a_i$ in $\mathbb{R}$ and $\mathbf{v}$ in $V$.

## 5.2 INDEPENDENCE AND DIMENSION

The deepest results about vector spaces stem from the idea of dimension, and this in turn relies on the notion of linear independence. For $\mathbb{R}^n$ these concepts were discussed in Sections 4.2 and 4.3; in this section, they are extended to an arbitrary vector space $V$ in a natural way and general versions of the basic theorems are derived, usually with similar proofs.

### 5.2.1 Independence and the Fundamental Theorem

Let $\{\mathbf{v}_1, \mathbf{v}_2, \cdots, \mathbf{v}_n\}$ be a set of vectors in a vector space $V$. A linear combination $r_1\mathbf{v}_1 + r_2\mathbf{v}_2 + \cdots + r_n\mathbf{v}_n$ of these vectors is said to **vanish** if it equals $\mathbf{0}$. The **trivial** linear combination

$$0\mathbf{v}_1 + 0\mathbf{v}_2 + \cdots + 0\mathbf{v}_n$$

(with every coefficient zero) certainly vanishes. As in $\mathbb{R}^n$, the set $\{\mathbf{v}_1, \mathbf{v}_2, \cdots, \mathbf{v}_n\}$ is called **linearly independent** (or simply **independent**[6]) if the *only* linear combination that vanishes is the trivial one. Thus, to check if a set $\{\mathbf{v}_1, \mathbf{v}_2, \cdots, \mathbf{v}_n\}$ is independent, the procedure (as in $\mathbb{R}^n$) is as follows:

**Step 1.** Set a linear combination equal to zero: $r_1\mathbf{v}_1 + r_2\mathbf{v}_2 + \cdots + r_n\mathbf{v}_n = \mathbf{0}$.
**Step 2.** Show somehow that this implies that all the coefficients are zero: $r_i = 0$ for each $i$.

Of course, if some nontrivial linear combination vanishes (that is if Step 2 is impossible) then the vectors are *not* independent.

**Example 1** Show that $\{1 + x^2, 2 - x + x^2, 1 - x^2\}$ is independent in $\mathbb{P}$.

SOLUTION Suppose $r_1(1 + x^2) + r_2(2 - x + x^2) + r_3(1 - x^2) = 0$ in $\mathbb{P}$. Collecting powers of $x$ gives

$$(r_1 + 2r_2 + r_3) + (-r_2)x + (r_1 + r_2 - r_3)x^2 = 0.$$

The definition of equality in $\mathbb{P}$ means that every coefficient is zero:

$$r_1 + 2r_2 + r_3 = 0, \quad -r_2 = 0, \quad \text{and} \quad r_1 + r_2 - r_3 = 0.$$

It is a routine task to verify that the only solution of these equations is the trivial one $r_1 = r_2 = r_3 = 0$. This proves independence.

The next example gives a condition for independence in $\mathbb{P}$ which will be referred to several times below.

---

[6] The adjective "linear" is sometimes needed because there are other notions of independence in mathematics. However, none of these occur in this book.

**Example 2** Let $p_1(x)$, $p_2(x)$, $\cdots$, $p_k(x)$ be nonzero polynomials in $\mathbb{P}$ of distinct degrees. Show that $\{p_1(x), p_2(x), \cdots, p_k(x)\}$ is linearly independent.

SOLUTION There is no loss of generality in assuming that $deg\, p_1(x) < deg\, p_2(x) < \cdots < deg\, p_k(x)$. Let $deg\, p_k(x) = m_k$, and let the leading coefficient of $p_k(x)$ be $a_k$. Suppose that
$$r_1 p_1(x) + r_2 p_2(x) + \cdots + r_k p_k(x) = 0 \text{ where each } r_i \text{ is in } \mathbb{R}. \qquad (*)$$
Then $r_k a_k = 0$ because it is the coefficient of $x^{m_k}$ on the left side of $(*)$. Since $a_k \neq 0$, this means that $r_k = 0$, so $(*)$ becomes $r_1 p_1(x) + r_2 p_2(x) + \cdots + r_{k-1} p_{k-1}(x) = 0$. Now repeat the argument to conclude that $r_{k-1} = 0$. Continue to show that $r_i = 0$ for each $i$.

**Example 3** Show that $\{\sin x,\ \cos x\}$ is independent in the space $\mathbb{F}[0,\ 2\pi]$ of all real valued functions on the interval $[0,\ 2\pi]$.

SOLUTION Suppose that a linear combination vanishes in $\mathbb{F}[0,\ 2\pi]$. This means that
$$r_1 \sin(x) + r_2 \cos(x) = 0.$$
By the definition of equality in $\mathbb{F}[0,\ 2\pi]$, this must hold for *all* values of $x$ in the interval $[0,\ 2\pi]$. Taking $x = 0$ yields $r_2 = 0$ because $\sin 0 = 0$ and $\cos 0 = 1$. Similarly $r_1 = 0$ follows by taking $x = \frac{\pi}{2}$ (because $\sin\left(\frac{\pi}{2}\right) = 1$ and $\cos\left(\frac{\pi}{2}\right) = 0$).

**Example 4** If $\{\mathbf{u},\ \mathbf{v},\ \mathbf{w}\}$ is independent, show that $\{\mathbf{u} + \mathbf{v},\ \mathbf{v} + \mathbf{w},\ \mathbf{u} + \mathbf{w}\}$ is also independent.

SOLUTION Suppose a linear combination vanishes: $r_1(\mathbf{u} + \mathbf{v}) + r_2(\mathbf{v} + \mathbf{w}) + r_3(\mathbf{u} + \mathbf{w}) = \mathbf{0}$. Collecting terms in $\mathbf{u}, \mathbf{v}$, and $\mathbf{w}$, this becomes
$$(r_1 + r_3)\mathbf{u} + (r_1 + r_2)\mathbf{v} + (r_2 + r_3)\mathbf{w} = \mathbf{0}.$$
Since $\{\mathbf{u},\ \mathbf{v},\ \mathbf{w}\}$ is independent, this must be the trivial combination, that is all the coefficients must vanish:
$$r_1 + r_3 = 0, \quad r_1 + r_2 = 0, \quad r_2 + r_3 = 0.$$
Since this system of equations has only the trivial solution $r_1 = r_2 = r_3 = 0$, we have shown that $\{\mathbf{u} + \mathbf{v}, \mathbf{v} + \mathbf{w}, \mathbf{u} + \mathbf{w}\}$ is independent.

**Example 5** If $\mathbf{v} \neq \mathbf{0}$ in a vector space $V$, show that $\{\mathbf{v}\}$ is linearly independent.

SOLUTION Here a vanishing linear combination has the form $r\mathbf{v} = \mathbf{0}$. Since $\mathbf{v} \neq \mathbf{0}$, this means $r = 0$ by Theorem 2 §5.1.

**Example 6** Show that no independent set can contain the zero vector.

SOLUTION If the zero vector is in an independent set, consider the linear combination where the zero vector has coefficient one and every other coefficient is zero. This combination vanishes and is nontrivial, contrary to the independence of the set.

Remarkably, the Fundamental theorem from Chapter 4 (for $\mathbb{R}^n$) holds in every abstract vector space. In fact, both proofs of Theorem 1 §4.3 go through as written. Because of its importance we give the general version of the second proof.

**THEOREM ①**

**Fundamental Theorem.** Suppose that $V$ is a vector space and $V = span\{\mathbf{v}_1, \mathbf{v}_2, \cdots, \mathbf{v}_n\}$. If $\{\mathbf{u}_1, \mathbf{u}_2, \cdots, \mathbf{u}_m\}$ is any linearly independent subset of $V$, then $m \le n$.

**PROOF**

We assume that $m > n$ and show that this leads to a contradiction.

As $V = span\{\mathbf{v}_1, \mathbf{v}_2, \cdots, \mathbf{v}_n\}$, write $\mathbf{u}_1 = r_1\mathbf{v}_1 + r_2\mathbf{v}_2 + \cdots + r_n\mathbf{v}_n$ where the $r_i$ are in $\mathbb{R}$. Then some $r_i \ne 0$ because $\mathbf{u}_1 \ne 0$ (by Example 6). By relabelling the $\mathbf{v}_i$ if necessary, assume that $r_1 \ne 0$. Then $V = span\{\mathbf{u}_1, \mathbf{v}_2, \cdots, \mathbf{v}_n\}$ by Theorem 4 §5.1 (verify). Now use this to write $\mathbf{u}_2 = t_1\mathbf{u}_1 + r_2\mathbf{v}_2 + \cdots + r_n\mathbf{v}_n$. Then some $r_i \ne 0$ because $\{\mathbf{u}_1, \mathbf{u}_2\}$ is independent (verify), say $r_2 \ne 0$ after possibly relabelling the $\mathbf{v}_i$. Then $V = span\{\mathbf{u}_1, \mathbf{u}_2, \mathbf{v}_3, \cdots, \mathbf{v}_n\}$, again using Theorem 4 §5.1. Since we are assuming that $m > n$, this procedure continues until all the vectors $\mathbf{v}_1, \mathbf{v}_2, \cdots, \mathbf{v}_n$ are replaced by $\mathbf{u}_1, \mathbf{u}_2, \cdots, \mathbf{u}_n$ and we have $V = span\{\mathbf{u}_1, \mathbf{u}_2, \cdots, \mathbf{u}_n\}$. But then $\mathbf{u}_{n+1}$ is a linear combination of $\mathbf{u}_1, \mathbf{u}_2, \cdots, \mathbf{u}_n$, contrary to the independence of the $\mathbf{u}_i$. This contradiction completes the proof.[7]

As for $\mathbb{R}^n$, a finite set of vectors $\{\mathbf{b}_1, \mathbf{b}_2, \cdots, \mathbf{b}_n\}$ in a vector space $V$ is called a **basis** of $V$ if it is independent and $V = span\{\mathbf{b}_1, \mathbf{b}_2, \cdots, \mathbf{b}_n\}$. Then we have immediately

**THEOREM ②**

**Invariance Theorem.** If $\{\mathbf{b}_1, \mathbf{b}_2, \cdots, \mathbf{b}_n\}$ and $\{\mathbf{d}_1, \mathbf{d}_2, \cdots, \mathbf{d}_m\}$ are two bases of a vector space $V$, then $n = m$.

**PROOF**

We have $m \le n$ by the Fundamental theorem because $V = span\{\mathbf{b}_1, \mathbf{b}_2, \cdots, \mathbf{b}_n\}$ and $\{\mathbf{d}_1, \mathbf{d}_2, \cdots, \mathbf{d}_m\}$ is independent. Now interchange the roles of the $\mathbf{b}_i$ and $\mathbf{d}_j$ to get $n \le m$.

With this we can define the notion of dimension in *any* vector space that has a basis. In fact, we define the **dimension** of $V$ to be the number of vectors in any basis of $V$, and denote this number by $dim\,V$. The Invariance theorem shows that this is well-defined in the sense that we get the same dimension no matter which basis we choose. Of course this agrees with the dimension we used in Chapter 4 for subspaces of $\mathbb{R}^n$. In particular, $dim(\mathbb{R}^n) = n$ as in Section 4.3 because the standard basis contains $n$ vectors.

The zero space $\{0\}$ deserves special mention. We define its dimension to be zero:

$$dim\{0\} = 0.$$

This is equivalent to regarding $\{0\}$ as having an empty basis, and so is consistent with the fact (Example 6) that the zero vector cannot belong to any independent set. Hence the statement

$$dim\,V = n \quad \text{if and only if} \quad V \text{ has a basis of } n \text{ vectors}$$

is valid even if $n = 0$.

**Example 7**

$dim(\mathbb{P}_n) = n + 1$ because $\mathcal{B} = \{1, x, x^2, \cdots, x^n\}$ is a basis, called the **standard basis** of $\mathbb{P}_n$. Indeed, $\mathbb{P}_n = span\,\mathcal{B}$ by Example 15 §5.1, and $r_0 + r_1x + \cdots + r_nx^n = 0$ implies each $r_i = 0$ by equating corresponding coefficients (see Example 3 §5.1).

**Example 8**

$dim(\mathbb{M}_{m,n}) = mn$ because the set $\mathcal{B}$ of $mn$ matrix units in $\mathbb{M}_{m,n}$ is a basis (it spans $\mathbb{M}_{m,n}$ by Example 16 §5.1, and the reader can verify that it is independent). We call $\mathcal{B}$ the **standard basis** of $\mathbb{M}_{m,n}$.

---

[7] In addition to showing that $m \le n$, this argument proves that $m$ of the (spanning) vectors $\mathbf{v}_i$ can be replaced by the (independent) vectors $\mathbf{u}_1, \mathbf{u}_2, \cdots, \mathbf{u}_m$. In this form the result is called the **Steinitz Exchange Lemma**.

**Example 9**

The set $\mathbb{C}$ of complex numbers is a vector space using complex addition and multiplication by a real number. The set $\{1, \ i\}$ is a basis. Indeed, $\mathbb{C} = span\{1, \ i\}$ because every complex number $z$ has the form $z = a + bi$ where $a$ and $b$ are real, and $\{1, \ i\}$ is independent because the only way $a + bi = 0$ can happen is if $a = 0$ and $b = 0$ (see Section 2.5). Hence $dim(\mathbb{C}) = 2$.

**Example 10**

If $A$ is an $n \times n$ matrix, let $U = \{B \text{ in } M_{n, n} \mid AB = BA\}$.

   (1) Show that $U$ is a subspace for any choice of the matrix $A$.

   (2) If $A_0 = \begin{bmatrix} 0 & 1 \\ 0 & 0 \end{bmatrix}$ find a basis of $U$ and compute $dimU$.

**PROOF**

   (1) Clearly 0 is in $U$ because $A0 = 0 = 0A$. If $B$ and $C$ are in $U$, then $AB = BA$ and $AC = CA$, so $A(B + C) = AB + AC = BA + CA = (B + C)A$. Hence $B + C$ is in $U$. Similarly $rA$ is in $U$ for any scalar $r$, so $U$ is a subspace by the Subspace test.

   (2) If $B = \begin{bmatrix} x & y \\ z & w \end{bmatrix}$, comparing corresponding entries in the condition $A_0B = BA_0$ gives equations $z = 0$ and $w = x$, so $B = \begin{bmatrix} x & y \\ 0 & x \end{bmatrix} = xI_2 + yA_0$. It follows that $U = span\{I_2, \ A_0\}$. Moreover, $\{I_2, \ A_0\}$ is linearly independent because $rI_2 + sA_0 = \begin{bmatrix} r & s \\ 0 & r \end{bmatrix}$, and this is the zero matrix if and only if $r = s = 0$.

   Hence $\{I_2, \ A_0\}$ is a basis of $U$, so $dimU = 2$.

The next example illustrates the power of the Fundamental theorem.

**Example 11**

If $A$ is any $n \times n$ matrix, show that there is a nonzero polynomial $f(x)$ of degree at most $n^2$ such that $f(A) = 0$.

**SOLUTION**

The $n^2 + 1$ matrices $I, \ A, \ A^2, \ \cdots, \ A^{n^2}$ cannot be independent as vectors in $\mathbb{M}_{n, n}$ by the Fundamental theorem because $dim(\mathbb{M}_{n, n}) = n^2$. Hence a nontrivial linear combination vanishes; that is, there exist scalars $r_0, \ r_1, \ r_2, \ \cdots, \ r_{n^2}$, not all zero, such that $r_0I + r_1A + r_2A^2 + \cdots + r_{n^2}A^{n^2} = 0$. So take $f(x) = r_0 + r_1x + r_2x^2 + \cdots + r_{n^2}x^{n^2}$.

In fact, we can do better in Example 11. The characteristic polynomial $c_A(x)$ has degree $n$ and satisfies $c_A(A) = 0$. This is the Cayley-Hamilton theorem, and we prove it in Section 5.6. However, it is remarkable that we can deduce that $f(A) = 0$ for *some* nonzero polynomial $f(x)$ using only the Fundamental theorem.

Lest we begin thinking that *every* vector space has a basis, we have

**Example 12**

Show that the vector space $\mathbb{P}$ does not have a basis.

**SOLUTION 1**

If $\{p_1(x), \ p_2(x), \ \cdots, \ p_n(x)\}$ is a basis of $\mathbb{P}$, let $x^m$ be the highest power of $x$ appearing in any of the polynomials $p_i(x)$. Then it is clear that every $p_i(x)$ is in $\mathbb{P}_m$, and hence that $\mathbb{P} \subseteq \mathbb{P}_m$. But this is impossible because $x^{m+1}$ is in $\mathbb{P}$ but not in $\mathbb{P}_m$.

**SOLUTION 2**

If $dim\mathbb{P} = n$, then $\mathbb{P}$ is spanned by $n$ vectors, so the existence of $n + 1$ independent vectors $\{1, \ x, \ x^2, \ \cdots, \ x^n\}$ in $\mathbb{P}$ contradicts the Fundamental theorem.

We say that a vector space $V$ is **infinite dimensional** if it does not have a (finite) basis. Hence Example 12 shows that $\mathbb{P}$ is infinite dimensional. There is a way[8] to define what we mean by an infinite basis of a vector space (the set $\{1,\ x,\ x^2,\ x^3,\ \cdots,\ x^n,\ \cdots\}$ qualifies in $\mathbb{P}$) and a version of the Invariance theorem holds. But we do not pursue this here. Instead, we focus on the vector spaces which do have a finite basis. In fact, we can weaken this requirement as follows.

A vector space $V$ is called **finite dimensional** if it has a finite spanning set, that is if
$$V = span\{\mathbf{v}_1,\ \mathbf{v}_2,\ \cdots,\ \mathbf{v}_n\}$$
for some finite set $\{\mathbf{v}_1,\ \mathbf{v}_2,\ \cdots,\ \mathbf{v}_n\}$ of vectors in $V$. We are going to show that such a space $V$ *has* a basis (and so $dim V \leq n$ by the Fundamental theorem). To do this, we need two lemmas that were proved for $\mathbb{R}^n$ in Sections 4.2 and 4.3. The first of these shows how to enlarge a known independent set and the proof is identical to that of Lemma 1 §4.3.

**LEMMA  1**  **Independent Lemma**. Suppose that $\{\mathbf{v}_1,\ \mathbf{v}_2,\ \cdots,\ \mathbf{v}_n\}$ is a linearly independent set of vectors in a vector space $V$. If $\mathbf{v}$ is any vector in $V$ that is not in $span\{\mathbf{v}_1,\ \mathbf{v}_2,\ \cdots,\ \mathbf{v}_n\}$, then $\{\mathbf{v},\ \mathbf{v}_1,\ \mathbf{v}_2,\ \cdots,\ \mathbf{v}_n\}$ is also linearly independent.

As in Section 4.2, call a set of vectors **linearly dependent** (or simply **dependent**) if it is not linearly independent, that is if some nontrivial linear combination vanishes. Then the proof of Theorem 3 §4.2 goes through as written to prove

**LEMMA  2**  **Dependent Lemma**. A set $\{\mathbf{v}_1,\ \mathbf{v}_2,\ \cdots,\ \mathbf{v}_n\}$ of vectors in a vector space $V$ is linearly dependent if and only if one of the $\mathbf{v}_i$ is a linear combination of the rest.

These lemmas enable us to prove the following theorem which reveals two important ways to show that bases exist in a finite dimensional vector space. Both are used frequently.

**THEOREM  3**  Let $V \neq \{\mathbf{0}\}$ be a finite dimensional vector space and assume that $V$ is spanned by $n$ vectors.
  (1) $V$ has a basis and $dim V \leq n$.
  (2) Every linearly independent subset of $V$ is part of a basis.
  (3) Every finite spanning set of $V$ contains a basis.

**PROOF**  We first prove (2), then (1) and (3).

  (2) Let $\{\mathbf{v}_1,\ \mathbf{v}_2,\ \cdots,\ \mathbf{v}_k\}$ be independent[9] in $V$. If $span\{\mathbf{v}_1,\ \mathbf{v}_2,\ \cdots,\ \mathbf{v}_k\} = V$, we are done. Otherwise, choose $\mathbf{v}_{k+1}$ in $V$ which is not in $span\{\mathbf{v}_1,\ \mathbf{v}_2,\ \cdots,\ \mathbf{v}_k\}$. Then $\{\mathbf{v}_1,\ \mathbf{v}_2,\ \cdots,\ \mathbf{v}_k,\ \mathbf{v}_{k+1}\}$ is independent by Lemma 1, and we are done if $span\{\mathbf{v}_1,\ \mathbf{v}_2,\ \cdots,\ \mathbf{v}_k,\ \mathbf{v}_{k+1}\} = V$. If not, repeat the process to create larger and larger independent sets in $V$. Since $V$ is finite dimensional, this cannot continue indefinitely by the Fundamental theorem, so we must reach a basis of $V$ at some stage.

  (1) Since $V \neq \{\mathbf{0}\}$ let $\mathbf{v}_1$ be a nonzero vector in $V$. Then $\{\mathbf{v}_1\}$ is independent, so $V$ has a basis by (2). Finally, $dim V \leq n$ by the Fundamental theorem because $V$ is spanned by $n$ vectors.

---

[8] If $V$ is a vector space, a (possibly infinite) set $\mathcal{B}$ of vectors in $V$ is called **independent** if every finite subset of $\mathcal{B}$ is linearly independent, and $\mathcal{B}$ is said to **span** $V$ if every vector in $V$ is a linear combination of (finitely many) vectors in $\mathcal{B}$. The set $\mathcal{B}$ is called a **basis** of $V$ if it is both independent and spans $V$.

[9] Because $V$ is finite dimensional, the independence forces it to be a *finite* subset by the Fundamental theorem.

(3) Let $V = span\{\mathbf{v}_1, \mathbf{v}_2, \cdots, \mathbf{v}_n\}$. If $\{\mathbf{v}_1, \mathbf{v}_2, \cdots, \mathbf{v}_n\}$ is independent, we are done. Otherwise one of the $\mathbf{v}_i$ is in the span of the rest by Lemma 2; relabelling if necessary let $\mathbf{v}_1$ lie in $span\{\mathbf{v}_2, \cdots, \mathbf{v}_n\}$. Then $V = span\{\mathbf{v}_2, \cdots, \mathbf{v}_n\}$ by Theorem 4 §5.1, so we are done if $\{\mathbf{v}_2, \cdots, \mathbf{v}_n\}$ is independent. If not, repeat the process. If we encounter a basis at some stage, we are done. Otherwise we ultimately reach $V = span\{\mathbf{v}_n\} = \mathbb{R}\mathbf{v}_n$. But then $\{\mathbf{v}_n\}$ is the desired basis because $\mathbf{v}_n \neq \mathbf{0}$ (as $V \neq \{\mathbf{0}\}$).

Theorem 3 shows that a vector space is finite dimensional if and only if it has a finite basis. It also gives the following useful test for when a subset is a basis.

---

**THEOREM 4**    Let $V$ be a vector space with $dim V = n$. If $\mathcal{B}$ is a set of exactly $n$ vectors of $V$, then

$$\mathcal{B} \text{ is independent} \quad \text{if and only if} \quad \mathcal{B} \text{ spans } V.$$

In either case, $\mathcal{B}$ is a basis of $V$.

**PROOF**    If $\mathcal{B}$ is independent but does not span $V$, then $\mathcal{B}$ is part of a basis of more than $n$ vectors by Theorem 3, contrary to Theorem 2. Similarly, if $\mathcal{B}$ spans $V$ but is not independent, then $\mathcal{B}$ contains a basis of fewer than $n$ vectors by Theorem 3, again contrary to Theorem 2.

Theorem 4 is a very useful "labour-saving device" for a vector space $V$ of dimension $n$, since it eliminates the need to verify one or the other of independence and spanning when showing that a set $\mathcal{B}$ of $n$ vectors is a basis of $V$. Here is an example illustrating how it is used.

**Example 13**    Show that $\mathcal{B} = \{1, (x - a), (x - a)^2, \cdots, (x - a)^n\}$ is a basis of $\mathbb{P}_n$ for every scalar $a$.

**SOLUTION**    The set $\mathcal{B}$ is independent by Example 2 because it consists of polynomials of distinct degrees. Hence $\mathcal{B}$ is a basis of $\mathbb{P}_n$ by Theorem 4 because $dim(\mathbb{P}_n) = n + 1$.

It follows from Example 13 that every polynomial $p(x)$ of degree at most $n$ can be uniquely written in the form

$$p(x) = r_0 + r_1(x - a) + r_2(x - a)^2 + \cdots + r_n(x - a)^n$$

where the coefficients $r_i$ are in $\mathbb{R}$. This can be proved other ways, but this argument illustrates once more the power of these methods.

We conclude with another "labour-saving" result, this time for reducing the work in showing that a subspace of $V$ is all of $V$.

---

**THEOREM 5**    Let $U$ be a subspace of a vector space $V$, and assume $dim V = n$.

(1) $U$ has a basis and $dim U \leq n$.
(2) If $dim U = n$, then $U = V$.
(3) Every basis of $U$ is part of a basis of $V$.

**PROOF**    (1) If $U = \{\mathbf{0}\}$, then it has an empty basis and (1) holds because $dim U = 0 \leq n$. Otherwise, let $\mathbf{u}_1 \neq \mathbf{0}$ be a vector in $U$. If $span\{\mathbf{u}_1\} = U$, we are done. If not, the construction in the proof of (2) of Theorem 3 either gives a basis of $U$ or creates arbitrarily large independent subsets of $U$, and hence of $V$. Hence (1) follows because the latter possibility contradicts the Fundamental theorem.

**(2)** If $dim U = n$, then any basis $\{\mathbf{u}_1, \mathbf{u}_2, \cdots, \mathbf{u}_n\}$ of $U$ is a basis of $V$ by Theorem 4. But then $V = span\{\mathbf{u}_1, \mathbf{u}_2, \cdots, \mathbf{u}_n\} = U$, proving (2).

**(3)** This follows from Theorem 3.

**Example 14**　If $a$ is a real number, let $W = \{p(x)$ in $\mathbb{P}_n \mid p(a) = 0\}$ denote the set of all polynomials in $\mathbb{P}_n$ with $a$ as a root. Show that $W$ is a subspace of $\mathbb{P}_n$ $\{(x-a), (x-a)^2, \cdots, (x-a)^n\}$ is a basis of $W$.

**SOLUTION**　Clearly $W$ contains the zero polynomial. If $p(x)$ and $q(x)$ are both in $W$ then $(p + q)(a) = p(a) + q(a) = 0 + 0 = 0$, so $p(x) + q(x)$ is in $W$. Similarly $W$ is closed under scalar multiplication, and so is a subspace of $\mathbb{P}_n$ by the Subspace test.

Now observe that $(x-a)^k$ is in $W$ for each $k$, and that $\{(x-a), (x-a)^2, \cdots, (x-a)^n\}$ is linearly independent by Example 2.

For convenience, write
$$U = span\{(x-a), (x-a)^2, \cdots, (x-a)^n\}.$$

Then $U \subseteq W \subseteq \mathbb{P}_n$, $dim U = n$, and $dim(\mathbb{P}_n) = n + 1$. Hence $n \leq dim W \leq n + 1$ by Theorem 5; so, since $dim W$ is an integer, either $dim W = n$ or $dim W = n + 1$. If $dim W = n + 1 = dim(\mathbb{P}_n)$, then $W = \mathbb{P}_n$ by Theorem 5, a contradiction (for example, $1$ is in $\mathbb{P}_n$ but not in $W$). So $dim W = n = dim U$. But then the fact that $U$ is a subspace of $W$ gives $W = U$, again by Theorem 5. This is what we wanted.

**Example 15**　(*Requires calculus*) Let $f'$ and $f'' = (f')'$ denote the first and second derivatives of a function $f$ in $\mathbb{F}[\mathbb{R}]$. Then solutions $f$ to the differential equation
$$f'' + \omega^2 f = 0 \quad \text{where } \omega \text{ is a real constant} \qquad (**)$$
are called **simple harmonic motions** because they describe oscillations like that of a pendulum or a weight on a spring (see Example 17 §5.1). The set $H$ of all these solutions is a subspace of $\mathbb{F}[\mathbb{R}]$. In fact, if $f$ and $g$ are in $H$, the property $(f + g)'' = f'' + g''$ of differentiation gives
$$(f + g)'' + \omega(f + g) = (f'' + g'') + (\omega f + \omega g)$$
$$= (f'' + \omega f) + (g'' + \omega g)$$
$$= 0 + 0 = 0,$$
so $f + g$ is in $H$. Similarly, $rf$ is in $H$ for all $r$ in $\mathbb{R}$ because $(rf)'' = rf''$. If $f_1$ and $f_2$ are defined by $f_1(x) = \sin(\omega x)$ and $f_2(x) = \cos(\omega x)$, it is easy to see that $f_1$ and $f_2$ both lie in $H$, and that $\{f_1, f_2\}$ is independent (verify). Remarkably, it is a theorem of calculus that $H$ is two-dimensional, so $\{f_1, f_2\}$ is actually a basis of $H$ by Theorem 4. Hence every solution $f$ to $(**)$ is given by
$$f(x) = a \sin(\omega x) + b \cos(\omega x) \qquad \text{for all } x$$
where $a$ and $b$ are uniquely determined constants.

## Exercises 5.2

**1.** In each case either show that the statement is true or give an example showing that it is false.

**a)** If $\{\mathbf{u}, \mathbf{v}\}$ and $\{\mathbf{w}, \mathbf{z}\}$ are both independent, then $\{\mathbf{u}, \mathbf{v}, \mathbf{w}, \mathbf{z}\}$ is independent.

**b)** If one of $\{\mathbf{u}, \mathbf{v}\}$ and $\{\mathbf{w}, \mathbf{z}\}$ is dependent, then $\{\mathbf{u}, \mathbf{v}, \mathbf{w}, \mathbf{z}\}$ is dependent.

**c)** $\{\mathbf{u}, \mathbf{u}\}$ is never independent.

**d)** If $\{\mathbf{u}, \mathbf{v}, \mathbf{w}\}$ is independent, then $a\mathbf{u} + b\mathbf{v} + c\mathbf{w} = \mathbf{0}$ for some scalars $a$, $b$, and $c$ not all zero.

**e)** If $\{\mathbf{u}, \mathbf{v}\}$ is independent, so is $\{\mathbf{u}, \mathbf{u} + \mathbf{v}\}$.

**f)** If $\{\mathbf{u}, \mathbf{v}\}$ is independent, so is $\{\mathbf{u}, \mathbf{v}, \mathbf{u} + \mathbf{v}\}$.

**g)** If $\{\mathbf{u}, \mathbf{v}, \mathbf{w}\}$ is independent, so is $\{\mathbf{u}, \mathbf{v}\}$.

**h)** If $\{\mathbf{u}, \mathbf{v}, \mathbf{w}\}$ is dependent, so is $\{\mathbf{u}, \mathbf{v}\}$.

**i)** $\mathbb{P}_2$ has a basis of polynomials $p(x)$ such that $p(0) = 0$.

**j)** $\mathbb{P}_2$ has a basis of polynomials $p(x)$ such that $p(0) = 1$.

**k)** Every basis of $\mathbb{M}_{2,2}$ contains an invertible matrix.

**l)** No basis of $\mathbb{M}_{2,2}$ contains a matrix $A$ such that $A^2 = 0$.

**m)** No basis of $\mathbb{M}_{2,2}$ consists of matrices $A$ such that $A[1 \ 0]^T = 0$.

**2.** In each case determine if the set of vectors is independent in $V$.

**a)** $V = \mathbb{P}_2$; $\{1 + x, 2 - x^2, x^2 + 2x\}$

**b)** $V = \mathbb{P}_2$; $\{2 + x, 1 + x^2, 1 + x + x^2\}$

**c)** $V = \mathbb{M}_{2,2}$; $\left\{ \begin{bmatrix} 1 & 0 \\ 0 & 0 \end{bmatrix}, \begin{bmatrix} 0 & 1 \\ 1 & 0 \end{bmatrix}, \begin{bmatrix} 0 & 1 \\ 1 & 1 \end{bmatrix}, \begin{bmatrix} 1 & 0 \\ 0 & -1 \end{bmatrix} \right\}$

**d)** $V = \mathbb{M}_{2,2}$; $\left\{ \begin{bmatrix} 1 & 1 \\ 0 & 0 \end{bmatrix}, \begin{bmatrix} 1 & 0 \\ 1 & 0 \end{bmatrix}, \begin{bmatrix} 0 & 0 \\ 1 & 1 \end{bmatrix}, \begin{bmatrix} 0 & 1 \\ 0 & 1 \end{bmatrix} \right\}$

**e)** $V = \mathbb{F}[0, 1]$; $\left\{ \frac{1}{x^2-4}, \frac{1}{x^2+3x+2}, \frac{1}{x^2-x-2} \right\}$

**f)** $V = \mathbb{F}[1, 2]$; $\left\{ \frac{1}{x}, \frac{1}{x^2}, \frac{1}{x^3} \right\}$

**g)** $V = \mathbb{F}[0, 2\pi]$; $\{1, \cos^2 x, \sin^2 x\}$

**h)** $V = \mathbb{F}[0, 2\pi]$; $\{x, \cos^2 x, \sin^2 x\}$

**3.** In each case find a basis and calculate the dimension of the subspace $U$ of $V$.

**a)** $V = \mathbb{P}_2$; $U = span\{1 - 2x, 3x - x^2, 3 - 2x^2\}$

**b)** $V = \mathbb{P}_2$; $U = \{p(x) \mid p(1) = 0\}$

**c)** $V = \mathbb{P}_2$; $U = \{p(x) \mid p(x) = p(-x)\}$

**d)** $V = \mathbb{P}_2$; $U = span\{x - 2, x^2 + 2x, 4 - x^2\}$

**e)** $V = \mathbb{M}_{2,2}$; $U = span\left\{ \begin{bmatrix} 1 & 2 \\ -1 & 0 \end{bmatrix}, \begin{bmatrix} 2 & 0 \\ 1 & 3 \end{bmatrix}, \begin{bmatrix} 1 & -6 \\ 5 & 6 \end{bmatrix} \right\}$

**f)** $V = \mathbb{M}_{2,2}$; $U = \{A \mid AB = BA\}$ where $B = \begin{bmatrix} 1 & 1 \\ 0 & 0 \end{bmatrix}$

**g)** $V = \mathbb{M}_{2,2}$; $U = \{A \mid AB = CA\}$, $B = \begin{bmatrix} 1 & 0 \\ 0 & 0 \end{bmatrix}$, $C = \begin{bmatrix} 0 & 1 \\ 1 & 0 \end{bmatrix}$

**h)** $V = \mathbb{M}_{2,2}$; $U = \{A \mid$ the columns of $A$ have equal sums$\}$

**i)** $V = \mathbb{M}_{3,3}$; $U = \{A \mid A^T = -A\}$

**j)** $V = \mathbb{R}^4$; $U = \{X \mid X^TA = 0\}$ where $A = \begin{bmatrix} 1 & -1 & 0 & 2 \\ -2 & 3 & -1 & 4 \end{bmatrix}^T$

**k)** $V = \mathbb{R}^4$; $U = \{[a \ b \ c \ d]^T \mid a + b + c + d = 0\}$

**l)** $V = span\{\mathbf{u}, \mathbf{v}, \mathbf{w}, \mathbf{z}\}$ where $\{\mathbf{u}, \mathbf{v}, \mathbf{w}, \mathbf{z}\}$ is independent; $U = span\{\mathbf{u} + \mathbf{v}, \mathbf{v} + \mathbf{w}, \mathbf{w} + \mathbf{z}, \mathbf{z} + \mathbf{u}\}$

**4. a)** Find a basis of $\mathbb{P}_3$ consisting of polynomials whose coefficients sum to 2.

**b)** What if the coefficients sum to 0?

**5. a)** Show that any (nonempty) subset of an independent set is independent.

**b)** Show that any set of vectors that contains a dependent set is itself dependent.

**6.** Consider the space $\mathbb{C}$ of complex numbers, and let $z$ denote an element of $\mathbb{C}$.

**a)** If $z$ is not real, show that $\{z, z^2\}$ is a basis of $\mathbb{C}$.

**b)** If $z$ is not real or pure imaginary, show that $\{z, \bar{z}\}$ is a basis of $\mathbb{C}$.

**7.** Show that if $\mathcal{B} = \{p_0(x), p_1(x), p_2(x), \cdots, p_n(x)\}$ are polynomials in $\mathbb{P}_n$ of distinct degrees, then $\mathcal{B}$ is a basis of $\mathbb{P}_n$.

**8.** If $V$ is finite dimensional, show that every (possibly infinite) spanning set $\mathcal{S}$ contains a basis. [Hint: Show first that $\mathcal{S}$ contains a finite spanning set.]

**9.** Suppose that $A$ is an $n \times n$ matrix such that $A^3 = 0$ but $A^2 \neq 0$. Show that $\{I, A, A^2\}$ is independent in $\mathbb{M}_{n,n}$. Generalize.

**10.** Let $A$ and $B$ be nonzero matrices in $\mathbb{M}_{n,n}$. If $A$ is symmetric and $B$ is skew-symmetric (that is $B^T = -B$), show that $\{A, B\}$ is independent.

**11.** Let $f$ and $g$ be functions in $\mathbb{F}[0, 1]$ such that $f(0) = 0$ and $g(0) = 1$, and $f(1) = 1$ and $g(1) = 0$. Show that $\{f, g\}$ is independent.

**12.** Let $\{A_1, A_2, \cdots, A_k\}$ be a set of matrices in $\mathbb{M}_{m,n}$, and let $P$ and $Q$ be invertible matrices of sizes $m \times m$ and $n \times n$, respectively.

**a)** If $\{A_1, A_2, \cdots, A_k\}$ is independent, show that $\{PA_1Q, PA_2Q, \cdots, PA_kQ\}$ is independent.

**b)** If $\{A_1, A_2, \cdots, A_k\}$ spans $\mathbb{M}_{m,n}$, show that $\{PA_1Q, PA_2Q, \cdots, PA_kQ\}$ spans $\mathbb{M}_{m,n}$.

**13.** Let $A$ be an $m \times n$ matrix, and let $\{B_1, B_2, \cdots, B_k\}$ be independent columns in $\mathbb{R}^m$. If $X_1, X_2, \cdots, X_k$ are solutions to the equations $AX_i = B_i$ for each $i = 1, 2, \cdots, k$, show that $\{X_1, X_2, \cdots, X_k\}$ is independent in $\mathbb{R}^n$.

**14.** Let $p(x)$ and $q(x)$ be polynomials of degree at least 1. If $\{p(x), q(x)\}$ is independent, show that $\{p(x), q(x), p(x)q(x)\}$ is also independent.

**15.** Call a function $f$ in $\mathbb{F}[a, b]$ a **constant function** if there exists a constant $c$ such that $f(x) = c$ for all $x$. Show that the set of all constant functions is a subspace of $\mathbb{F}[a, b]$ of dimension 1.

**16. a)** If $\{a + bx, a_1 + b_1x\}$ is a basis of $\mathbb{P}_1$, show that $\{[a \ b]^T, [a_1 \ b_1]^T\}$ is a basis of $\mathbb{R}^2$.

**b)** If $\{[a \ b]^T, [a_1 \ b_1]^T\}$ is a basis of $\mathbb{R}^2$, show that $\{a + bx, a_1 + b_1x\}$ is a basis of $\mathbb{P}_1$.

**17.** Find the dimension of the subspace $U = span\{1, \cos^2 \theta, \cos(2\theta)\}$ of $\mathbb{F}[-\pi, \pi]$. Support your answer.

**18.** Find a basis of $\mathbb{M}_{2,2}$ consisting of matrices $A$ with the property that $A^2 = A$.

**19.** If $A$ is a $2 \times 2$ matrix and $U = \{B$ in $\mathbb{M}_{2,2} \mid AB = BA\}$, show that $dimU \geq 2$. [Hint: $I$ and $A$ are in $U$. Consider separately the case that $A = aI$ for some $a$ in $\mathbb{R}$.]

**20.** A polynomial $p(x)$ is called **even** if $p(x) = p(-x)$, and $p(x)$ is called **odd** if $p(x) = -p(-x)$. Let $E_n$ and $O_n$ denote the sets of even and odd polynomials in $\mathbb{P}_n$.
   **a)** Show $E_n$ is a subspace of $\mathbb{P}_n$, and find its dimension.
   **b)** Show $O_n$ is a subspace of $\mathbb{P}_n$, and find its dimension.

**21.** If $\mathbf{v} \neq \mathbf{0}$ in $V$, show that the only subspaces of $\mathbb{R}\mathbf{v}$ are $\{\mathbf{0}\}$ and $\mathbb{R}\mathbf{v}$.

**22.** If $dimV = 2$ and $U$ is any subspace of $V$ other than $\{\mathbf{0}\}$ and $V$, show that $U = \mathbb{R}\mathbf{v}$ for some nonzero vector $\mathbf{v}$ in $V$. What does this say about the plane $\mathbb{R}^2$?

**23.** If $U$ is a subspace of $\mathbb{P}_1$, show that $U = \{\mathbf{0}\}$, $U = \mathbb{P}_1$, $U = \mathbb{R}$, or $U = \mathbb{R}(a + x)$ for some $a$ in $\mathbb{R}$.

**24.** If $\{\mathbf{u}, \mathbf{v}, \mathbf{w}, \mathbf{z}\}$ is independent, which of the following are independent? Support your answer.
   **a)** $\{\mathbf{u} - \mathbf{v}, \mathbf{v} - \mathbf{w}, \mathbf{w} - \mathbf{u}\}$
   **b)** $\{\mathbf{u} + \mathbf{v}, \mathbf{v} + \mathbf{w}, \mathbf{w} + \mathbf{u}\}$
   **c)** $\{\mathbf{u} - \mathbf{v}, \mathbf{v} - \mathbf{w}, \mathbf{w} - \mathbf{z}, \mathbf{z} - \mathbf{u}\}$
   **d)** $\{\mathbf{u} + \mathbf{v}, \mathbf{v} + \mathbf{w}, \mathbf{w} + \mathbf{u}, \mathbf{z} + \mathbf{u}\}$

**25.** Show that $\{\mathbf{u}, \mathbf{v}\}$ is dependent if and only if one of $\mathbf{u}$ and $\mathbf{v}$ is a scalar multiple of the other.

**26.** If $\mathbf{v} \neq \mathbf{0}$ is a vector in a vector space $V$, show that there exists a basis of $V$ containing $\mathbf{v}$.

**27.** If $\{\mathbf{v}_1, \mathbf{v}_2, \mathbf{v}_3, \mathbf{v}_4\}$ is independent in $V$, show that $\{\mathbf{v}_1, \mathbf{v}_2 + a\mathbf{v}_1, \mathbf{v}_3 + c\mathbf{v}_2 + d\mathbf{v}_1, \mathbf{v}_4 + e\mathbf{v}_3 + f\mathbf{v}_2 + g\mathbf{v}_1\}$ is also independent. Generalize.

**28.** If $\{\mathbf{v}_1, \mathbf{v}_2, \cdots, \mathbf{v}_k\}$ is independent in $V$, and if $\mathbf{w}$ is not in $span\{\mathbf{v}_1, \mathbf{v}_2, \cdots, \mathbf{v}_k\}$, show that the set $\{\mathbf{w} + \mathbf{v}_1, \mathbf{w} + \mathbf{v}_2, \cdots, \mathbf{w} + \mathbf{v}_k\}$ is also independent.

**29.** Suppose that $\{\mathbf{v}_1, \mathbf{v}_2, \cdots, \mathbf{v}_k\}$ is a minimal spanning set for a vector space $V$, that is no spanning set of $V$ contains fewer than $k$ vectors. Show that $\{\mathbf{v}_1, \mathbf{v}_2, \cdots, \mathbf{v}_k\}$ is a basis of $V$.

**30.** Suppose that $\{\mathbf{v}_1, \mathbf{v}_2, \cdots, \mathbf{v}_k\}$ is a maximal independent set in a vector space $V$, that is no independent set in $V$ contains more than $k$ vectors. Show that $\{\mathbf{v}_1, \mathbf{v}_2, \cdots, \mathbf{v}_k\}$ is a basis of $V$.

**31.** Show that $\mathbb{F}[0, 1]$ is infinite dimensional. [Hint: Fundamental theorem.]

**32.** Let $U \subseteq W$ be subspaces of $V$, and suppose that $dimU = k$ and $dimW = m$ where $k < m$. If $l$ is any integer such that $k \leq l \leq m$, show that there is a subspace $X$ such that $U \subseteq X \subseteq W$ and $dimX = l$. [Hint: Theorem 3.]

**33.** Let $\mathbb{S}$ denote the set of all infinite sequences $(a_0, a_1, a_2, \cdots)$ of real numbers $a_i$. Define equality, addition, and scalar multiplication on $\mathbb{S}$ as follows:
$(a_0, a_1, a_2, \cdots) = (b_0, b_1, b_2, \cdots)$
if and only if   $a_i = b_i$   for all $i$.
$(a_0, a_1, a_2, \cdots) + (b_0, b_1, b_2, \cdots) =$
$$(a_0 + b_0, a_1 + b_1, a_2 + b_2, \cdots).$$
$r(a_0, a_1, a_2, \cdots) = (ra_0, ra_1, ra_2, \cdots)$ for all real numbers $r$.
   **a)** Show that $\mathbb{S}$ is a vector space and that $\mathbb{S}$ is infinite dimensional.
   **b)** Show that $P = \{(a_0, a_1, a_2, \cdots) \mid a_i = 0$ for all but finitely many $i\}$ is a subspace of $\mathbb{S}$. Is $dimP$ infinite? Support your answer.
   **c)** Show that the set $C = \{(a, a, a, \cdots) \mid a$ in $\mathbb{R}\}$ is a subspace of $\mathbb{S}$. What is $dimC$?
   **d)** Show that the convergent sequences are an infinite dimensional subspace of $\mathbb{S}$ (requires calculus).

**34.** Let $\mathbb{S}$ be the space of sequences in the preceding exercise, and define $U = \{(x_0, x_1, x_2, \cdots)$ in $\mathbb{S} \mid x_{k+2} = ax_k + bx_{k+1}$ for all $k \geq 0\}$ where $a$ and $b$ are fixed real numbers. Let $\mathbf{e} = (e_0, e_1, e_2, \cdots)$ be the sequence in $U$ starting with $e_0 = 1$ and $e_1 = 0$, and let $\mathbf{f} = (f_0, f_1, f_2, \cdots)$ be the sequence in $U$ starting with $f_0 = 0$ and $f_1 = 1$.
   **a)** Show that $U$ is a subspace of $S$.
   **b)** If $\mathbf{x} = (x_0, x_1, x_2, \cdots)$ is in $U$, show that $x_k = x_0 e_k + x_1 f_k$ for all $k \geq 0$.
   **c)** Show that $dimU = 2$.

**35.** If $U$ and $W$ are subspaces of $V$, define their
   **intersection** $U \cap W$ and **sum** $U + W$
by $U \cap W = \{\mathbf{v}$ in $V \mid \mathbf{v}$ is in both $U$ and $W\}$
and
   $U + W = \{\mathbf{u} + \mathbf{w} \mid \mathbf{u}$ is in $U$ and $\mathbf{w}$ is in $W\}$
   **a)** Show that $U \cap W$ and $U + W$ are subspaces of $V$.
   **b)** If $U$ and $W$ have bases $\{\mathbf{u}_1, \mathbf{u}_2\}$ and $\{\mathbf{v}_1, \mathbf{v}_2, \mathbf{v}_3\}$ respectively, and if $U \cap W = \{\mathbf{0}\}$, show that $\{\mathbf{u}_1, \mathbf{u}_2, \mathbf{v}_1, \mathbf{v}_2, \mathbf{v}_3\}$ is independent.
   **c)** If $U$ and $W$ are both finite dimensional, show that $U + W$ is also finite dimensional (even though $V$ may not be finite dimensional). [Hint: If $C$ and $D$ are (finite) bases of $U$ and $W$ respectively, show that $C \cup D = \{\mathbf{v}$ in $V \mid \mathbf{v}$ is in $C$ or $\mathbf{v}$ is in $D\}$ spans $U + W$.]
   **d)** If $V$ is finite dimensional, show that $dim(U + W) = dimU + dimW - dim(U \cap W)$. [Hint: If $\mathcal{B}$ is a basis of $U \cap W$, let (by Theorem 3) $C$ and $D$ be bases of $U$ and $W$ respectively, each containing $\mathcal{B}$. Show that $C \cup D$ is a basis of $U + W$.]

## 5.3 LINEAR TRANSFORMATIONS

In Section 3.4 we studied transformations of $\mathbb{R}^2$ such as rotations, reflections, and projections, and we found that they are all given by multiplication by certain $2 \times 2$ matrices. The key idea that connects a geometrical transformation $T : \mathbb{R}^2 \to \mathbb{R}^2$ like rotation with a matrix is that $T$ is *linear*, that is, $T$ preserves addition and scalar multiplication in the sense that

$$T(\vec{v} + \vec{w}) = T(\vec{v}) + T(\vec{w}) \quad \text{and} \quad T(r\,\vec{v}) = r\,T(\vec{v}) \tag{$*$}$$

for all vectors $\vec{v}$ and $\vec{w}$ in $\mathbb{R}^2$ and all scalars $r$. In Section 4.9 we applied this to linear transformations $T : \mathbb{R}^n \to \mathbb{R}^m$ and found that these too are given by matrix multiplication. This connection between the abstract concept of linearity and the more concrete, computational notion of matrix multiplication is very useful (one example is our analysis of the isometries of $\mathbb{R}^3$).

The reason for recalling all this here is that the conditions in ($*$) make sense much more generally, and lead to the general idea of a linear transformation $T : V \to W$ where $V$ and $W$ are arbitrary vector spaces. This is the main theme of this section (and the next). It reveals a new perspective on matrices and provides a whole new way to view similarity. It also increases the applicability of linear algebra since, once one knows what they are, linear transformations appear everywhere.

---

**5.3.1** ## Linear Transformations

If $V$ and $W$ are two vector spaces, a transformation[10] $T : V \to W$ is a rule that associates with every vector $\mathbf{v}$ in $V$ a uniquely determined vector $T(\mathbf{v})$ in $W$. A transformation $T : V \to W$ is called a **linear transformation** if it preserves addition and scalar multiplication according to the following axioms:

**T1.** $T(\mathbf{v} + \mathbf{v}_1) = T(\mathbf{v}) + T(\mathbf{v}_1)$ for all vectors $\mathbf{v}$ and $\mathbf{v}_1$ in $V$.
**T2.** $T(r\,\mathbf{v}) = r\,T(\mathbf{v})$ for all vectors $\mathbf{v}$ in $V$ and all scalars $r$.

A linear transformation $T : V \to V$ is called a **linear operator** on the vector space $V$.

The action of a linear transformation $T : V \to W$ can be visualized as in the diagram. The vector space $V$ is called the **domain** of $T$, and $W$ is the **codomain** of $T$. If $\mathbf{v}$ is a vector in $V$, the vector $T(\mathbf{v})$ is called the **image** of $\mathbf{v}$ under $T$.

**Example 1**  Every $m \times n$ matrix $A$ induces a linear transformation $T : \mathbb{R}^n \to \mathbb{R}^m$ given by

$$T(X) = AX \text{ for all } X \text{ in } \mathbb{R}^n.$$

In this case Axioms T1 and T2 are simple properties of matrix multiplication. Moreover, *every* linear transformation $\mathbb{R}^n \to \mathbb{R}^m$ is induced by some $m \times n$ matrix $A$ (Theorem 2 §4.9).

---

[10] Another word for transformation is "function", but the word "transformation" is well-established in linear algebra.

**Example 2**    Transposition is a linear transformation. That is, the map
$$T : \mathbb{M}_{m,n} \to \mathbb{M}_{n,m} \quad \text{given by} \quad T(A) = A^T \text{ for all } A \text{ in } \mathbb{M}_{m,n}$$
is a linear transformation. In this case Axioms T1 and T2 are basic properties of transposition:
$$T(A + B) = (A + B)^T = A^T + B^T = T(A) + T(B)$$
and
$$T(rA) = (rA)^T = r A^T = r T(A)$$
for all $A$ and $B$ in $\mathbb{M}_{m,n}$ and all $r$ in $\mathbb{R}$.

**Example 3**    If $a$ is a fixed real number, define a transformation $E_a : \mathbb{P}_n \to \mathbb{R}$ by $E_a[p(x)] = p(a)$ for all polynomials $p(x)$ in $\mathbb{P}_n$. Then $E_a$ is linear. Indeed, Axioms T1 and T2 are routine observations:
$$E_a[p(x) + q(x)] = p(a) + q(a) = E_a[p(x)] + E_a[q(x)]$$
and
$$E_a[r p(x)] = r p(a) = r E_a[p(x)]$$
for all $p(x)$ and $q(x)$ in $\mathbb{P}_n$ and all $r$ in $\mathbb{R}$. The transformation $E_a$ is called **evaluation** at $a$.

The next example is important in analysis.

**Example 4**    Differentiation is a linear transformation. Let $V$ denote the subspace of $\mathbb{F}[\mathbb{R}]$ consisting of all differentiable real-valued functions defined on $\mathbb{R}$ (see Example 12 §5.1). If we denote the derivative of a function $f$ by $f'$, define
$$T : V \to \mathbb{F}[\mathbb{R}] \text{ by } T(f) = f'.$$
Then $T$ is a linear transformation. In fact Axioms T1 and T2 read $(f + g)' = f' + g'$ and $(rf)' = rf'$ for all $r$ in $\mathbb{R}$, and these are basic facts about differentiation.

**Example 5**    If $V$ and $W$ are vector spaces, the following are linear transformations.

$1_V : V \to V$ given by $1_V(\mathbf{v}) = \mathbf{v}$ for all $\mathbf{v}$ in $V$.                   **(Identity operator)**
$0 : V \to W$ given by $0(\mathbf{v}) = \mathbf{0}$ for all $\mathbf{v}$ in $V$.                       **(Zero transformation)**
For $a$ in $\mathbb{R}$, $a : V \to V$ given by $a(\mathbf{v}) = a\mathbf{v}$ for all $\mathbf{v}$ in $V$.          **(Scalar operator)**

In addition to preserving sums and scalar products, linear transformations preserve the zero vector, negatives and linear combinations. These useful properties are recorded next.

**THEOREM ❶**    Let $T : V \to W$ be a linear transformation. Then:

(1) $T(\mathbf{0}) = \mathbf{0}$.[11]
(2) $T(-\mathbf{v}) = -T(\mathbf{v})$ for all $\mathbf{v}$ in $V$.
(3) $T(r_1\mathbf{v}_1 + r_2\mathbf{v}_2 + \cdots + r_k\mathbf{v}_k) = r_1T(\mathbf{v}_1) + r_2T(\mathbf{v}_2) + \cdots + r_kT(\mathbf{v}_k)$ for all $r_i$ in $\mathbb{R}$ and all $\mathbf{v}_i$ in $V$.

---

[11] Note that $\mathbf{0}$ is playing two roles here. On the left side it is the zero vector in $V$, and it is the zero vector in $W$ on the right side. This is a common abuse of notation, and is harmless once everyone knows it is going on.

PROOF

(1) $T(\mathbf{0}) = T(0\mathbf{v}) = 0\,T(\mathbf{v}) = \mathbf{0}$ for any $\mathbf{v}$ in $V$ by Axiom T2.

(2) $T(-\mathbf{v}) = T((-1)\mathbf{v}) = (-1)T(\mathbf{v}) = -T(\mathbf{v})$, again by Axiom T2.

(3) If $k = 1$, this is Axiom T2. If $k > 1$, we use induction on $k$. If it holds for $k$ vectors, then

$$T(r_1\mathbf{v}_1 + r_2\mathbf{v}_2 + \cdots + r_k\mathbf{v}_k + r_{k+1}\mathbf{v}_{k+1})$$
$$= T(r_1\mathbf{v}_1 + r_2\mathbf{v}_2 + \cdots + r_k\mathbf{v}_k) + T(r_{k+1}\mathbf{v}_{k+1})$$
$$= r_1T(\mathbf{v}_1) + r_2T(\mathbf{v}_2) + \cdots + r_kT(\mathbf{v}_k) + r_{k+1}T(\mathbf{v}_{k+1})$$

by the induction assumption and Axioms T1 and T2. Hence it holds for $k + 1$ vectors, and the induction goes through.

The next two examples illustrate the importance of the fact that linear transformations preserve linear combinations (property (3) in Theorem 1).

**Example 6**

Suppose that $T : V \rightarrow W$ is a linear transformation. If $T(\mathbf{v} + 2\mathbf{v}_1) = \mathbf{w}$ and $T(3\mathbf{v} - 5\mathbf{v}_1) = \mathbf{w}_1$, find $T(\mathbf{v})$ and $T(\mathbf{v}_1)$ in terms of $\mathbf{w}$ and $\mathbf{w}_1$.

SOLUTION

Because $T$ is linear, the given equations imply that

$$T(\mathbf{v}) + 2\,T(\mathbf{v}_1) = \mathbf{w}$$
$$3\,T(\mathbf{v}) - 5\,T(\mathbf{v}_1) = \mathbf{w}_1$$

If we subtract 3 times the first equation from the second, we get $-11T(\mathbf{v}_1) = \mathbf{w}_1 - 3\mathbf{w}$. Thus $T(\mathbf{v}_1) = \frac{3}{11}\mathbf{w} - \frac{1}{11}\mathbf{w}_1$. Substitution in the first equation gives $T(\mathbf{v}) = \frac{5}{11}\mathbf{w} + \frac{2}{11}\mathbf{w}_1$.

**Example 7**

Suppose that $T(3 - 5x) = 2$ and $T(1 - x + 2x^2) = 1 + x$ where $T : \mathbb{P}_2 \rightarrow \mathbb{P}_2$ is a linear transformation. Find $T(6 - 11x - 3x^2)$.

SOLUTION

This seems difficult until we realize that (using Theorem 1) we can find $T[p(x)]$ for any polynomial $p(x)$ that is a linear combination of $3 - 5x$ and $1 - x + 2x^2$. So we look for scalars $r$ and $s$ such that

$$6 - 11x - 3x^2 = r(3 - 5x) + s(1 - x + 2x^2).$$

Equating coefficients gives linear equations for $r$ and $s$: $6 = 3r + s$, $-11 = -5r - s$, and $-3 = 2s$. These equations have the unique solution $r = \frac{5}{2}$ and $s = -\frac{3}{2}$. Then

$$T(6 - 11x - 3x^2) = r\,T(3 - 5x) + s\,T(1 - x + 2x^2)$$
$$= \tfrac{5}{2}\,T(3 - 5x) - \tfrac{3}{2}\,T(1 - x + 2x^2)$$
$$= \tfrac{5}{2}2 - \tfrac{3}{2}(1 + x)$$
$$= \tfrac{7}{2} - \tfrac{3}{2}x.$$

This is what we wanted.

If $T : V \rightarrow W$ is a linear transformation and $T(\mathbf{v}_1)$, $T(\mathbf{v}_2)$, $\cdots$, $T(\mathbf{v}_k)$, are known, property (3) in Theorem 1 shows that $T(r_1\mathbf{v}_1 + r_2\mathbf{v}_2 + \cdots + r_k\mathbf{v}_k)$ can be computed for any choice of the scalars $r_i$. In particular, if $V = span\{\mathbf{v}_1, \mathbf{v}_2, \cdots, \mathbf{v}_k\}$ then $T(\mathbf{v})$ is determined for every $\mathbf{v}$ in $V$ by the choice of $T(\mathbf{v}_1)$, $T(\mathbf{v}_2)$, $\cdots$, $T(\mathbf{v}_k)$ in $W$. The next theorem states this differently. Recall that two transformations $S : V \rightarrow W$ and $T : V \rightarrow W$ are **equal** (written $S = T$) if $S(\mathbf{v}) = T(\mathbf{v})$ for all $\mathbf{v}$ in $V$.

**THEOREM 2**    Let $S : V \to W$ and $T : V \to W$ be two linear transformations, and assume that $V = span\{\mathbf{v}_1, \mathbf{v}_2, \cdots, \mathbf{v}_k\}$. If $S(\mathbf{v}_i) = T(\mathbf{v}_i)$ for each $i$, then $S = T$.

PROOF    Given $\mathbf{v}$ in $V$, write $\mathbf{v} = r_1\mathbf{v}_1 + r_2\mathbf{v}_2 + \cdots + r_k\mathbf{v}_k$ where $r_i$ is in $\mathbb{R}$ for each $i$. Since $S(\mathbf{v}_i) = T(\mathbf{v}_i)$ for each $i$, Theorem 1 gives

$$S(\mathbf{v}) = r_1 S(\mathbf{v}_1) + r_2 S(\mathbf{v}_2) + \cdots + r_k S(\mathbf{v}_k)$$
$$= r_1 T(\mathbf{v}_1) + r_2 T(\mathbf{v}_2) + \cdots + r_k T(\mathbf{v}_k)$$
$$= T(\mathbf{v}).$$

Since $\mathbf{v}$ was arbitrary in $V$, this shows that $S = T$.

Thus a linear transformation $T : V \to W$ is determined by its effect on a spanning set of $V$. If the spanning set is actually a basis, we can say more. The following lemma will be needed.

**LEMMA 1**    If $\{\mathbf{b}_1, \mathbf{b}_2, \cdots, \mathbf{b}_n\}$ is a basis of a vector space $V$, every vector $\mathbf{v}$ in $V$ can be uniquely represented as a linear combination

$$\mathbf{v} = r_1\mathbf{b}_1 + r_2\mathbf{b}_2 + \cdots + r_n\mathbf{b}_n \quad \text{where the } r_i \text{ are in } \mathbb{R}.$$

That is, if $\mathbf{v} = s_1\mathbf{b}_1 + s_2\mathbf{b}_2 + \cdots + s_n\mathbf{b}_n$ is another such representation, $s_i$ in $\mathbb{R}$, then $s_i = r_i$ for each $i$.

PROOF    Since $V = span\{\mathbf{b}_1, \mathbf{b}_2, \cdots, \mathbf{b}_n\}$, the vector $\mathbf{v}$ has at least one such representation. If

$$r_1\mathbf{b}_1 + r_2\mathbf{b}_2 + \cdots + r_n\mathbf{b}_n = \mathbf{v} = s_1\mathbf{b}_1 + s_2\mathbf{b}_2 + \cdots + s_n\mathbf{b}_n$$

are two such representations, then $(r_1 - s_1)\mathbf{b}_1 + (r_2 - s_2)\mathbf{b}_2 + \cdots + (r_n - s_n)\mathbf{b}_n = \mathbf{0}$. Hence $r_i - s_i = 0$ for each $i$ because $\{\mathbf{b}_1, \mathbf{b}_2, \cdots, \mathbf{b}_n\}$ is linearly independent.

**THEOREM 3**    Let $\{\mathbf{b}_1, \mathbf{b}_2, \cdots, \mathbf{b}_n\}$ be a basis of a vector space $V$, and let $\{\mathbf{w}_1, \mathbf{w}_2, \cdots, \mathbf{w}_n\}$ be arbitrary vectors in $W$ (possibly not distinct). Then there exists a unique linear transformation $T : V \to W$ such that

$$T(\mathbf{b}_i) = \mathbf{w}_i \quad \text{for each } i = 1, 2, \cdots, n.$$

Furthermore, the action of $T$ is given as follows: If

$$\mathbf{v} = r_1\mathbf{b}_1 + r_2\mathbf{b}_2 + \cdots + r_n\mathbf{b}_n$$

is in $V$ where each $r_i$ is in $\mathbb{R}$, then

$$T(\mathbf{v}) = r_1\mathbf{w}_1 + r_2\mathbf{w}_2 + \cdots + r_n\mathbf{w}_n. \tag{$**$}$$

PROOF    If $T$ exists, it is unique by Theorem 2 because $\{\mathbf{b}_1, \mathbf{b}_2, \cdots, \mathbf{b}_n\}$ spans $V$. But if we write $\mathbf{v} = r_1\mathbf{b}_1 + r_2\mathbf{b}_2 + \cdots + r_n\mathbf{b}_n$, $r_i$ in $\mathbb{R}$, then the $r_i$ are uniquely determined by $\mathbf{v}$ and so $T(\mathbf{v}) = r_1\mathbf{w}_1 + r_2\mathbf{w}_2 + \cdots + r_n\mathbf{w}_n$ is also uniquely determined by $\mathbf{v}$. Hence the formula $(**)$ defines a transformation $T : V \to W$ which clearly satisfies $T(\mathbf{b}_i) = \mathbf{w}_i$ for each $i$. The verification that this transformation $T$ is linear is left to the reader.

**Example 8**    Find a linear transformation $T : \mathbb{P}_2 \to \mathbb{P}_4$ such that

$$T(1) = x^4, \quad T(1 + x) = 1 + x^3, \quad \text{and} \quad T(1 + x^2) = 1 - x^2.$$

SOLUTION    The set $\{1, 1 + x, 1 + x^2\} \subseteq \mathbb{P}_2$ is independent because the polynomials have distinct degrees, and so it is a basis of $\mathbb{P}_2$ because $dim(\mathbb{P}_2) = 3$. By comparing coefficients, the expansion of an arbitrary polynomial in $\mathbb{P}_2$ in terms of this basis is

$$r_0 + r_1 x + r_2 x^2 = (r_0 - r_1 - r_2) \cdot 1 + r_1(1 + x) + r_2(1 + x^2).$$

Hence the required linear transformation is given by

$$T(r_0 + r_1 x + r_2 x^2) = (r_0 - r_1 - r_2) \cdot T(1) + r_1 \cdot T(1 + x) + r_2 \cdot T(1 + x^2)$$
$$= (r_0 - r_1 - r_2)x^4 + r_1(1 + x^3) + r_2(1 - x^2)$$
$$= (r_1 + r_2) - r_2 x^2 + r_1 x^3 + (r_0 - r_1 - r_2)x^4.$$

In practice, linear transformations $T : V \to W$ are usually described by simply giving the action of $T$ on some convenient basis of $V$; a complete description of $T(\mathbf{v})$ for every vector $\mathbf{v}$ in $V$ as in Example 8 is usually not provided (or needed).

## 5.3.2 Kernel and Image

Every linear transformation $T : V \to W$ gives rise to two subspaces, called the kernel and image of $T$, which encode a lot of information about $T$. The **kernel** of $T$, denoted $kerT$, is the subspace of $V$ consisting of all vectors that $T$ carries to $\mathbf{0}$; more formally,

$$kerT = \{\mathbf{v} \text{ in } V \mid T(\mathbf{v}) = \mathbf{0}\}.$$

The **image** of $T$, denoted $imT$, is the subspace of $W$ consisting of all images under $T$ of vectors in $V$:

$$imT = \{T(\mathbf{v}) \mid \mathbf{v} \text{ in } V\}.$$

We begin by verifying that $kerT$ and $imT$ are indeed subspaces of $V$ and $W$.

**LEMMA 2**

Let $T : V \to W$ be a linear transformation. Then $kerT$ is a subspace of $V$ and $imT$ is a subspace of $W$.

**PROOF**

The fact that $T(\mathbf{0}) = \mathbf{0}$ shows that both $kerT$ and $imT$ contain the zero vector. If $\mathbf{v}_1$ and $\mathbf{v}_2$ are both in $kerT$, then $T(\mathbf{v}_1) = \mathbf{0}$ and $T(\mathbf{v}_2) = \mathbf{0}$, so

$$T(\mathbf{v}_1 + \mathbf{v}_2) = T(\mathbf{v}_1) + T(\mathbf{v}_2) = \mathbf{0} + \mathbf{0} = \mathbf{0} \quad \text{and} \quad T(r\,\mathbf{v}_1) = r\,T(\mathbf{v}_1) = r\,\mathbf{0} = \mathbf{0}$$

for all $r$ in $\mathbb{R}$. Hence $\mathbf{v}_1 + \mathbf{v}_2$ and $r\,\mathbf{v}_1$ are both in $kerT$, which shows that $kerT$ is a subspace of $V$.

If $\mathbf{w}_1$ and $\mathbf{w}_2$ are both in $imT$, write $\mathbf{w}_1 = T(\mathbf{v}_1)$ and $\mathbf{w}_2 = T(\mathbf{v}_2)$ where $\mathbf{v}_1$ and $\mathbf{v}_2$ are in $V$. Then

$$\mathbf{w}_1 + \mathbf{w}_2 = T(\mathbf{v}_1) + T(\mathbf{v}_2) = T(\mathbf{v}_1 + \mathbf{v}_2) \quad \text{and} \quad r\,\mathbf{w}_1 = r\,T(\mathbf{v}_1) = T(r\,\mathbf{v}_1)$$

show that $\mathbf{w}_1 + \mathbf{w}_2$ and $r\,\mathbf{w}_1$ are both in $imT$ for all $r$ in $\mathbb{R}$. Hence $imT$ is a subspace of $W$.

The subspaces $kerT$ and $imT$, with their relationship to $T$, are depicted in the diagram.

Familiar subspaces of $\mathbb{R}^n$ and $\mathbb{M}_{n,n}$ often occur in a natural way as images or kernels of linear transformations. Here are three examples.

**Example 9**

If $A$ is an $m \times n$ matrix, let $T : \mathbb{R}^n \to \mathbb{R}^m$ be the matrix transformation given by $T(X) = AX$ for all $X$ in $\mathbb{R}^n$. The subspaces $nullA$ and $imA$ arise as follows:

$$kerT = \{X \mid AX = 0\} = nullA \quad \text{and} \quad imT = \{AX \mid X \text{ in } \mathbb{R}^n\} = imA.$$

**Example 10**  If $U$ is a subspace of $\mathbb{R}^n$, let $T : \mathbb{R}^n \to \mathbb{R}^n$ be the projection on $U$, that is define $T(X) = proj_U(X)$ for all $X$ in $\mathbb{R}^n$. Show that

$$im\,T = U \quad \text{and} \quad ker\,T = U^\perp.$$

**SOLUTION**  Note first that $T$ is a linear transformation by Example 4 §4.9. The Projection theorem (Theorem 2 §4.6) asserts that

$$T(X) \text{ is in } U \quad \text{and} \quad X - T(X) \text{ is in } U^\perp \quad \text{for all } X \text{ in } \mathbb{R}^n. \qquad (***)$$

Since $T(X)$ is in $U$ for all $X$, we have $im\,T \subseteq U$. On the other hand, if $X$ is in $U$, then $(***)$ shows that $X - T(X)$ is in $U \cap U^\perp = \{0\}$ for all $X$, so $X = T(X)$ is in $im\,T$. This proves that $im\,T = U$.

Similarly, $(***)$ shows that $X - T(X)$ is in $U^\perp$ for all $X$, so $ker\,T \subseteq U^\perp$. But if $X$ is in $U^\perp$, then $(***)$ shows that $T(X) = X - (X - T(X))$ is in $U \cap U^\perp = \{0\}$, and it follows that $X$ is in $ker\,T$. This shows that $ker\,T = U^\perp$.

Recall that a matrix $S$ is called **skew-symmetric** if $S^T = -S$.

**Example 11**  Let $T : \mathbb{M}_{n,n} \to \mathbb{M}_{n,n}$ be defined by $T(A) = A - A^T$ for all matrices $A$ in $\mathbb{M}_{n,n}$. Show that $T$ is linear,

$$ker\,T = \{A \text{ in } \mathbb{M}_{n,n} \mid A \text{ is symmetric}\},$$

and

$$im\,T = \{S \text{ in } \mathbb{M}_{n,n} \mid S \text{ is skew-symmetric}\}.$$

**SOLUTION**  The reader can verify that $T$ is linear. Clearly, $ker\,T = \{A \mid A^T = A\}$ is the set of symmetric matrices. On the other hand, $T(A) = A - A^T$ is skew-symmetric for all $A$ (verify), and every skew-symmetric matrix $S$ has the form $S = A - A^T$ where $A = \frac{1}{2}S$ (verify). This proves the second assertion.

**Example 12**  As in Example 3, if $a$ is a number, let $E_a : \mathbb{P}_n \to \mathbb{R}$ be evaluation at $a$, that is $E_a[p(x)] = p(a)$ for all polynomials $p(x)$ in $\mathbb{P}_n$. Then

$$ker(E_a) = \{p(x) \mid p(a) = 0\}$$

consists of the polynomials having $a$ as a root. On the other hand,

$$im(E_a) = \{p(a) \mid p(x) \text{ in } \mathbb{P}_n\} = \mathbb{R}$$

because every real number $b$ equals $p(a)$ for some polynomial $p(x)$ (for example, $p(x) = b$, the constant polynomial).

**Example 13**[12]  Let $T : \mathbb{P}_n \to \mathbb{P}_n$ be the differentiation operator: $T[p(x)] = p'(x)$ for all $p(x)$ in $\mathbb{P}_n$. Then $T$ is linear (see Example 4) and

$$ker\,T = \{p(x) \mid p'(x) = 0\} \text{ consists of the constant polynomials}$$

because they are the only ones whose derivative is identically zero. On the other hand,

$$im\,T = \{p'(x) \mid p(x) \text{ in } \mathbb{P}_n\} = \mathbb{P}_{n-1}.$$

Indeed, $im\,T \subseteq \mathbb{P}_{n-1}$ because $p'(x)$ is always 0 or of degree one less than $p(x)$, and $\mathbb{P}_{n-1} \subseteq im\,T$ because every polynomial in $\mathbb{P}_{n-1}$ has degree at most $n - 1$ and so is the derivative of some polynomial $p(x)$ in $\mathbb{P}_n$.

The kernel and image of a linear transformation $T$ are closely related to two other properties of $T$. If $T : V \to W$ is a linear transformation, we say that:

---

[12] Requires polynomial calculus.

$T$ is **onto** if $imT = W$.

$T$ is **one-to-one** if $T(\mathbf{v}_1) = T(\mathbf{v}_2)$ implies that $\mathbf{v}_1 = \mathbf{v}_2$.

Thus $T : V \to W$ is onto if every vector $\mathbf{w}$ in $W$ is the image of some vector in $V$, and $T$ is one-to-one if no vector $\mathbf{w}$ in $W$ is the image of two distinct vectors in $V$.

**Example 14**

The linear transformation $T : \mathbb{R}^2 \to \mathbb{R}$ given by $T([r \ \ s]^T) = r$ is onto but not one-to-one. In fact, $T$ is onto because every $r$ in $\mathbb{R}$ is the image of some vector in $\mathbb{R}^2$ (say $T([r \ \ 0]^T) = r$). But $T$ is not one-to-one. For example, $T([1 \ \ 0]^T) = T([1 \ \ 1]^T)$ but $[1 \ \ 0]^T \neq [1 \ \ 1]^T$.

**Example 15**

The linear transformation $T : \mathbb{R} \to \mathbb{R}^2$ given by $T(r) = [r \ \ 0]^T$ is one-to-one but not onto. Indeed, $T$ is one-to-one because $T(r_1) = T(r_2)$ means $[r_1 \ \ 0]^T = [r_2 \ \ 0]^T$, whence $r_1 = r_2$. But $T$ is not onto because, for example, the vector $[0 \ \ 1]^T$ in $\mathbb{R}^2$ is not the image under $T$ of any vector $r$ in $\mathbb{R}$.

The onto transformations $T : V \to W$ are the ones for which the subspace $imT$ of $W$ is as *large* as possible (namely $imT = W$). By contrast, the next result shows that the one-to-one transformations are those for which the subspace $kerT$ of $V$ is as *small* as possible in $V$ (namely $kerT = \{\mathbf{0}\}$). This is a useful test for when $T$ is one-to-one.

**THEOREM 4**

Let $T : V \to W$ be a linear transformation. Then

$$T \text{ is one-to-one} \quad \text{if and only if} \quad kerT = \{\mathbf{0}\}.$$

**PROOF**

If $T$ is one-to-one, let $\mathbf{v}$ be a vector in $kerT$, that is $T(\mathbf{v}) = \mathbf{0}$. Hence $T(\mathbf{v}) = T(\mathbf{0})$ by Theorem 1, so $\mathbf{v} = \mathbf{0}$ by the one-to-one condition. It follows that $kerT = \{\mathbf{0}\}$. Conversely, assume that $kerT = \{\mathbf{0}\}$. To show that $T$ is one-to-one, let $T(\mathbf{v}_1) = T(\mathbf{v}_2)$; we must show that $\mathbf{v}_1 = \mathbf{v}_2$. But the linearity of $T$ gives $T(\mathbf{v}_1 - \mathbf{v}_2) = T(\mathbf{v}_1) - T(\mathbf{v}_2) = \mathbf{0}$, so $\mathbf{v}_1 - \mathbf{v}_2$ is in $kerT = \{\mathbf{0}\}$. Thus $\mathbf{v}_1 = \mathbf{v}_2$, as required.

**Example 16**

If $A$ is an $m \times n$ matrix, consider the matrix transformation $T : \mathbb{R}^n \to \mathbb{R}^m$ given by $T(X) = AX$ for all $X$ in $\mathbb{R}^n$. Theorem 3 §4.4 asserts that $rankA = n$ if and only if the system $AX = 0$ has only the trivial solution $X = 0$. In our present language, this is:

$$T \text{ is one-to-one} \quad \text{if and only if} \quad rankA = n.$$

Similarly, Theorem 4 §4.4 shows that $rankA = m$ if and only if the system $AX = B$ has a solution for every column $B$ in $\mathbb{R}^m$. This gives:

$$T \text{ is one-to-one} \quad \text{if and only if} \quad rankA = m.$$

## 5.3.3 The Dimension Theorem

If $A$ is an $m \times n$ matrix, the corollary of Theorem 6 §4.4 asserts that

$$n = dim(imA) + dim(nullA).$$

This equation is important in matrix theory because $nullA$ is the set of all solutions $X$ to the system $AX = 0$ of linear equations, and $dim(imA) = dim(colA) = rankA$ (see Theorem 5 §4.4). If $T : \mathbb{R}^n \to \mathbb{R}^m$ is given by $T(X) = AX$ for all $X$ in $\mathbb{R}^n$, then Example 9 shows that $imA = imT$ and $nullA = kerT$. Consequently, the following theorem is a profound extension of the above equation.

**THEOREM 5**    **Dimension Theorem.** Let $T : V \to W$ be a linear transformation. If both $kerT$ and $imT$ are finite dimensional, then $V$ is also finite dimensional and
$$dimV = dim(imT) + dim(kerT).$$

**PROOF**    Let $\{\mathbf{b}_1, \mathbf{b}_2, \cdots, \mathbf{b}_k\}$ be a basis of $kerT$ and, since $imT = \{T(\mathbf{v}) \mid \mathbf{v} \text{ in } V\}$, let $\{T(\mathbf{d}_1), T(\mathbf{d}_2), \cdots, T(\mathbf{d}_r)\}$ be a basis of $imT$ where each $\mathbf{d}_j$ is in $V$. Then $dim(kerT) = k$ and $dim(imT) = r$, so it suffices to show that
$$\mathcal{B} = \{\mathbf{b}_1, \mathbf{b}_2, \cdots, \mathbf{b}_k, \mathbf{d}_1, \mathbf{d}_2, \cdots, \mathbf{d}_r\}$$
is a basis of $V$.

$\mathcal{B}$ *spans* $V$. If $\mathbf{v}$ is in $V$, then $T(\mathbf{v})$ is in $imT$, so write $T(\mathbf{v}) = s_1 T(\mathbf{d}_1) + s_2 T(\mathbf{d}_2) + \cdots + s_r T(\mathbf{d}_r)$ where each $s_j$ is in $\mathbb{R}$. Then $T(\mathbf{v}) = T(s_1\mathbf{d}_1 + s_2\mathbf{d}_2 + \cdots + s_r\mathbf{d}_r)$ by Theorem 1. Since $T$ is linear, this means that $\mathbf{v} - (s_1\mathbf{d}_1 + s_2\mathbf{d}_2 + \cdots + s_r\mathbf{d}_r)$ lies in $kerT$, and so is a linear combination of $\{\mathbf{b}_1, \mathbf{b}_2, \cdots, \mathbf{b}_k\}$. It follows that $\mathbf{v}$ is in $span\mathcal{B}$, as required.

$\mathcal{B}$ *is linearly independent.* Suppose that $t_i$ and $s_j$ in $\mathbb{R}$ exist such that
$$t_1\mathbf{b}_1 + t_2\mathbf{b}_2 + \cdots + t_k\mathbf{b}_k + s_1\mathbf{d}_1 + s_2\mathbf{d}_2 + \cdots + s_r\mathbf{d}_r = \mathbf{0}. \qquad (****)$$
Since each $\mathbf{b}_i$ is in $kerT$, applying $T$ gives $s_1 T(\mathbf{d}_1) + s_2 T(\mathbf{d}_2) + \cdots + s_r T(\mathbf{d}_r) = \mathbf{0}$. It follows that each $s_j = 0$, whence $(****)$ becomes $t_1\mathbf{b}_1 + t_2\mathbf{b}_2 + \cdots + t_k\mathbf{b}_k = \mathbf{0}$. Hence each $t_i = 0$, and we are done.

As noted above, the Dimension theorem provides another proof of the important fact that the dimension of the space of solutions of a system $AX = 0$ equals $n - r$ where $A$ is an $m \times n$ matrix of rank $r$. Another illustration: The result that $dimU + dimU^\perp = n$ for all subspaces $U$ of $\mathbb{R}^n$ (Theorem 3 §4.6) follows immediately from the Dimension theorem and Example 10.

The following consequence of the Dimension theorem will be referred to later.

**COROLLARY**    Let $T : V \to W$ be a linear transformation. If $dimV = dimW$ is finite, then $T$ is one-to-one if and only if $T$ is onto.

**PROOF**    The Dimension theorem gives $dim(imT) + dim(kerT) = dimV$. If $T$ is one-to-one, then $kerT = \{\mathbf{0}\}$ by Theorem 4, so $dim(kerT) = 0$. Hence $dim(imT) = dimV = dimW$, whence $imT = W$ by Theorem 5 §5.2, and $T$ is onto. Conversely, if $T$ is onto, then $imT = W$ so $dim(imT) = dimW = dimV$. It follows from the Dimension theorem that $dim(kerT) = 0$, so $kerT = \{\mathbf{0}\}$ and $T$ is one-to-one.

If $T : V \to W$ is a linear transformation where $dimV = n$ is finite, and if either of $dim(kerT)$ or $dim(imT)$ is known, then the Dimension theorem provides the other. This can be useful because one of these dimensions can be easier to compute than the other. Applications of this often depends on *defining* an appropriate linear transformation. This is illustrated in the following example which gives a different proof of an important result from Section 1.5.

**Example 17**    If $BC = I$ where $B$ and $C$ are $n \times n$ matrices, show that also $CB = I$.

**SOLUTION**    Define $T : \mathbb{M}_{n,n} \to \mathbb{M}_{n,n}$ by $T(A) = CA$ for all $A$ in $\mathbb{M}_{n,n}$. Then $T$ is linear (verify). If $A$ is in $kerT$, then $0 = T(A) = CA$, so $A = IA = BCA = B0 = 0$. This shows that $kerT = \{0\}$, so that $T$ is one-to-one. But then $T$ is onto by the Corollary to the Dimension theorem. In particular $T(D) = I$ for some matrix $D$, that is $CD = I$. Hence $D = ID = BCD = BI = B$, and so $CB = CD = I$, as asserted.

# Exercises 5.3

Throughout these exercises, $V$ always denotes a vector space.

1. True or False?
   a) $T : \mathbb{R} \to \mathbb{R}$ is a linear transformation where $T(r) = r^2$ for all $r$ in $\mathbb{R}$.
   b) $T : \mathbb{R} \to \mathbb{R}$ is a linear transformation where $T(r) = |r|$ for all $r$ in $\mathbb{R}$.
   c) $T(X) = AX$ defines a linear transformation $\mathbb{R}^6 \to \mathbb{R}^5$ for every $5 \times 6$ matrix $A$.
   d) If $T : V \to W$ is a linear transformation and $T(\mathbf{v}_1) = T(\mathbf{v}_2)$, then $\mathbf{v}_1 - \mathbf{v}_2$ is in $kerT$.
   e) If $T : V \to W$ is a linear transformation and $T(\mathbf{v}) = \mathbf{0}$ for some $\mathbf{v} \neq \mathbf{0}$, then $T = 0$ is the zero transformation.
   f) If $T : V \to W$ is linear and $kerT = V$, then $W = 0$.
   g) If $T : V \to W$ is linear and $imT = 0$, then $T = 0$ is the zero transformation.
   h) If $T : \mathbb{R}^n \to \mathbb{R}^n$ is defined by $T(X) = proj_{\mathbb{R}^n}(X)$ for all $X$ in $\mathbb{R}^n$ then $T = 1_{\mathbb{R}^n}$.
   i) A linear transformation $\mathbb{R}^n \to \mathbb{R}^m$ cannot be both one-to-one and onto.
   j) If $T : V \to W$ is one-to-one and $dimV = dimW = n$, then $imT = W$.
   k) Every linear transformation $T : \mathbb{R}^6 \to \mathbb{R}^5$ is onto.
   l) There exists a linear transformation $T$ such that $T(\mathbf{v}_1)$ is in $\mathbb{F}[0, 1]$ and $T(\mathbf{v}_2)$ is in $\mathbb{R}^3$.

2. Let $T : V \to W$ denote a linear transformation. In each case either prove the statement or give an example showing that it is false.
   a) If $dimV = 6$, $dimW = 4$ and $dim(kerT) = 2$, then $T$ is onto.
   b) If $dimV = 4$ and $dimW = 3$, then $T$ is one-to-one.
   c) If $\{\mathbf{e}_1, \mathbf{e}_2, \mathbf{e}_3, \mathbf{e}_4\}$ is a basis of $V$ and $T(\mathbf{e}_2) = \mathbf{0} = T(\mathbf{e}_4)$, then $dim(imT) \leq 2$.
   d) If $dim(kerT) \leq dimW$, then $dimW \geq \frac{1}{2}dimV$.
   e) If $T$ is one-to-one, then $dimV \leq dimW$.
   f) If $dimV \leq dimW$, then $T$ is one-to-one.
   g) If $T$ is onto, then $dimV \geq dimW$.
   h) If $dimV \geq dimW$, then $T$ is onto.

3. Show that the following are linear transformations.
   a) $T : \mathbb{C} \to \mathbb{C}$ where $T(z) = \bar{z}$, where $\bar{z}$ denotes the conjugate of the complex number $z$.
   b) $T : \mathbb{M}_{m,n} \to \mathbb{M}_{k,l}$ where $T(A) = PAQ$ where $P$ and $Q$ are fixed $k \times m$ and $n \times l$ matrices.
   c) $T : \mathbb{P}_n \to \mathbb{R}$ where $T(a_0 + a_1x + \cdots + a_nx^n) = a_n$.
   d) $T : \mathbb{R}^n \to \mathbb{R}$ where $T(X) = Y \cdot X$, where $Y$ is a fixed vector in $\mathbb{R}^n$.
   e) $T : \mathbb{P}_n \to \mathbb{P}_n$ where $T[p(x)] = p(x + 1)$ for all polynomials $p(x)$ in $\mathbb{P}_n$.
   f) $T : V \to \mathbb{R}$ where $T(r_1\mathbf{e}_1 + \cdots + r_n\mathbf{e}_n) = r_1$, where $\{\mathbf{e}_1, \cdots, \mathbf{e}_n\}$ is a fixed basis of $V$.

   g) $T : \mathbb{M}_{n,n} \to \mathbb{R}$ where $T(A) = trA$, where $trA$ denotes the trace of the matrix $A$.
   h) $T : \mathbb{M}_{m,n} \to \mathbb{R}^m$ where $T(A)$ is column 2 of $A$ and $n \geq 2$.
   i) $T : \mathbb{R} \to \mathbb{F}[D]$, where $T(a) = f_a$ is the constant function defined by $f_a(x) = a$ for all $x$ in $D$. (See Exercise 20 §5.1.)

4. Decide whether the following transformations $T : \mathbb{M}_{n,n} \to \mathbb{R}$ are linear. Support your answer.
   a) $T : \mathbb{M}_{n,n} \to \mathbb{R}$ where $T(A) = detA$
   b) $T : \mathbb{M}_{n,n} \to \mathbb{R}$ where $T(A) = rankA$

5. In each case find a linear transformation $T$ with the given properties, and compute $T(\mathbf{v})$.
   a) $T : \mathbb{P}_2 \to \mathbb{P}_4$, $T(1) = x^4$, $T(x + x^2) = 1$, $T(x - x^2) = x + x^3$; $\mathbf{v} = a + bx + cx^2$
   b) $T : \mathbb{M}_{2,2} \to \mathbb{R}^2$, $T\begin{bmatrix} 1 & 0 \\ 0 & 1 \end{bmatrix} = \begin{bmatrix} 1 \\ 2 \end{bmatrix}$, $T\begin{bmatrix} 0 & 1 \\ 0 & 0 \end{bmatrix} = \begin{bmatrix} 0 \\ 1 \end{bmatrix}$, $T\begin{bmatrix} 0 & 0 \\ 1 & 0 \end{bmatrix} = \begin{bmatrix} 1 \\ 0 \end{bmatrix}$, $T\begin{bmatrix} 0 & 0 \\ 0 & 1 \end{bmatrix} = \begin{bmatrix} 0 \\ 0 \end{bmatrix}$; $\mathbf{v} = \begin{bmatrix} a & b \\ c & d \end{bmatrix}$

6. If $T : V \to V$ is linear and $\mathbf{e}$ and $\mathbf{f}$ are in $V$, find $T(3\mathbf{e} + \mathbf{f})$ and $T(\mathbf{f})$ in terms of $\mathbf{e}$ and $\mathbf{f}$ if $T(\mathbf{e} - \mathbf{f}) = 2\mathbf{e} - \mathbf{f}$ and $T(2\mathbf{f} - \mathbf{e}) = \mathbf{e} + \mathbf{f}$.

7. Let $T : V \to V$ be a linear transformation, and let $\{\mathbf{b}_1, \mathbf{b}_2, \cdots, \mathbf{b}_n\}$ be a basis of $V$.
   a) If $T(\mathbf{b}_i) = \mathbf{b}_i$ for each $i$, show that $T = 1_V$ is the identity transformation.
   b) If $T(\mathbf{b}_i) = \mathbf{0}$ for each $i$, show that $T = 0$ is the zero transformation.

8. Let $T : V \to W$ be a linear transformation. Give a careful proof of the fact that $T(\mathbf{v} - \mathbf{v}_1) = T(\mathbf{v}) - T(\mathbf{v}_1)$ for all $\mathbf{v}$ and $\mathbf{v}_1$ in $V$. Cite every axiom or theorem used.

9. Describe all linear transformations $T : \mathbb{R} \to V$. [Hint: $r = r1$ for all $r$ in $\mathbb{R}$.]

10. A linear transformation $T : V \to \mathbb{R}$ is called a **linear functional**.
    a) Let $T : V \to \mathbb{R}$ be a linear functional where $dimV = n$. If $\mathcal{B}$ is a basis of $V$, show that there exists a vector $Y$ in $\mathbb{R}^n$ such that $T(\mathbf{v}) = Y \cdot C_\mathcal{B}(\mathbf{v})$ for all $\mathbf{v}$ in $V$. Compare with Exercise 9 §4.9.
    b) If $V$ is the set of all integrable functions in $\mathbb{F}[0, 1]$, show that $T(f) = \int_0^1 f(t)dt$ defines a linear functional $\mathbb{F}[0, 1] \to \mathbb{R}$.

11. Let $V$ be a finite dimensional vector space. Given $\mathbf{v} \neq \mathbf{0}$ in $V$ and $\mathbf{w}$ in a space $W$, show that there exists a linear transformation $T : V \to W$ such that $T(\mathbf{v}) = \mathbf{w}$. What happens if $\mathbf{v} = \mathbf{0}$?

12. Show that the following are equivalent for a linear transformation $T : V \to W$.
    (a) $kerT = V$    (b) $imT = \{\mathbf{0}\}$    (c) $T = 0$

13. Let $T : V \to W$ be a linear transformation, let
$\mathcal{B} = \{\mathbf{b}_1,\ \mathbf{b}_2,\ \cdots,\ \mathbf{b}_n\}$ be vectors in $V$, and write
$\mathcal{C} = \{T(\mathbf{b}_1),\ T(\mathbf{b}_2),\ \cdots,\ T(\mathbf{b}_n)\}$.
   a) If $\mathcal{C}$ is independent, show that $\mathcal{B}$ is independent.
   b) If $T$ is one-to-one and $\mathcal{B}$ is independent, show that $\mathcal{C}$ is independent.
   c) If $W = span\,\mathcal{C}$, show that $T$ is onto.
   d) If $V = span\,\mathcal{B}$, show that $imT = span\,\mathcal{C}$.

14. Given vectors $\mathbf{v}_1,\ \mathbf{v}_2,\ \cdots,\ \mathbf{v}_n$ in a vector space $V$, define
$T : \mathbb{R}^n \to V$ by $T[r_1\ r_2\ \cdots\ r_n]^T = r_1\mathbf{v}_1 + r_2\mathbf{v}_2 + \cdots + r_n\mathbf{v}_n$.
   a) Show that $T$ is one-to-one if and only if $\{\mathbf{v}_1,\ \mathbf{v}_2,\ \cdots,\ \mathbf{v}_n\}$ is linearly independent.
   b) Show that $T$ is onto if and only if $\{\mathbf{v}_1,\ \mathbf{v}_2,\ \cdots,\ \mathbf{v}_n\}$ spans $V$.

15. Let $U$ be a subspace of a finite dimensional vector space $V$.
   a) Show that $U = kerT$ for some linear operator $T : V \to V$. [Hint: Theorem 3.]
   b) Show that $U = imS$ for some linear operator $S : V \to V$. [Hint: Theorem 3.]

16. Let $T : \mathbb{M}_{n,n} \to \mathbb{R}$ be the trace map: $T(A) = trA$ for all $A$ in $\mathbb{M}_{n,n}$. Show that $dim(kerT) = n^2 - 1$.

17. If $A$ is an $n \times n$ matrix, let $U = \{B$ in $\mathbb{M}_{m,n} \mid BA = 0\}$ and let $W = \{BA \mid B$ in $\mathbb{M}_{m,n}\}$. Use the Dimension theorem to show that $U$ and $W$ are both subspaces of $\mathbb{M}_{m,n}$, and that $dimU + dimW = mn$.

18. Use the Dimension theorem to prove Theorem 1 §1.3: A homogeneous system of linear equations that has more variables than equations must have nontrivial solutions.

19. Let $U$ and $W$ denote the spaces of even and odd polynomials in $\mathbb{P}_n$ (see Exercise 20 §5.2). Use the Dimension theorem to show that $dimU + dimW = n + 1$. [Hint: Consider $T : \mathbb{P}_n \to \mathbb{P}_n$ where $T[p(x)] = p(x) - p(-x)$.]

20. Let $U$ and $W$ denote the spaces of symmetric and skew-symmetric $n \times n$ matrices, respectively. Show that $dimU + dimW = n^2$. [Hint: Example 11.]

21. Consider the subspace $W = \{p(x)$ in $\mathbb{P}_n \mid p(a) = 0\}$ of $\mathbb{P}_n$. Use the Dimension theorem to prove that $\{(x - a),\ (x - a)^2,\ (x - a)^3,\ \cdots,\ (x - a)^n\}$ is a basis of $W$.

[Hint: Use the evaluation map $E_a : \mathbb{P}_n \to \mathbb{R}$, Example 2 §5.2 and Theorem 4 §5.2.]

22. Use the Dimension theorem to show that every matrix $A$ in $\mathbb{M}_{n,n}$ has the form $A = B^T - 3B$ for some $B$ in $\mathbb{M}_{n,n}$.

23. Use the Dimension theorem to show that every polynomial $p(x)$ in $\mathbb{P}_{n-1}$ can be written in the form $p(x) = q(x + 1) - q(x)$ for some polynomial $q(x)$ in $\mathbb{P}_n$.

24. Show that differentiation is the only linear transformation $\mathbb{P}_n \to \mathbb{P}_{n-1}$ that satisfies $T(x^k) = kx^{k-1}$ for all $k = 0,\ 1,\ 2,\ \cdots,\ n$.

25. If $a$ and $b$ are real numbers, define $T_{a,b} : \mathbb{C} \to \mathbb{C}$ by $T_{a,b}(r + si) = ar + bsi$.
   a) Show that $T_{a,b}$ is linear and $T_{a,b}(\overline{z}) = \overline{T_{a,b}(z)}$ for all $z$ in $\mathbb{C}$. Here $\overline{z}$ denotes conjugation.
   b) If $T : \mathbb{C} \to \mathbb{C}$ is linear and satisfies $T(\overline{z}) = \overline{T(z)}$ for all $z$ in $\mathbb{C}$, show that $T = T_{a,b}$ for some $a$ and $b$. [Hint: Show that $T(1)$ is real and $T(i)$ is pure imaginary.]

26. If $a$ is in $\mathbb{R}$, consider the evaluation transformation $E_a : \mathbb{P}_n \to \mathbb{R}$ where $E_a[p(x)] = p(a)$.
   a) Show that $E_a(x^k) = [E_a(x)]^k$ for each $k = 0,\ 1,\ 2,\ \cdots,\ n$. (Note that $x^0 = 1$.)
   b) If $T : \mathbb{P}_n \to \mathbb{R}$ is linear and satisfies $T(x^k) = [T(x)]^k$ for each $k = 0,\ 1,\ 2,\ \cdots,\ n$, show that $T = E_a$ for some $a$ in $\mathbb{R}$. [Hint: If $T = E_a$ for some $a$, what is $T(x)$?]

27. If $T : \mathbb{M}_{n,n} \to \mathbb{R}$ is linear and satisfies $T(AB) = T(BA)$ for all $A$ and $B$ in $\mathbb{M}_{n,n}$, show that there exists a constant $k$ such that $T(A) = k\,tr(A)$ for all $A$ in $\mathbb{M}_{n,n}$, where $tr$ denotes the trace map (See Lemma 3 §4.7). [Hint: If $E_{ij}$ denote the matrix units in $\mathbb{M}_{n,n}$ (see Example 16 §5.1), take $k = T(E_{11})$. Use the fact that

$$E_{ij}E_{km} = \begin{cases} 0 & \text{if } j \neq k \\ E_{im} & \text{if } j = k \end{cases}$$

28. Let $V$ and $W$ denote finite dimensional vector spaces. Using Theorems 3 and 5:
   a) Show that $dimV \leq dimW$ if and only if there exists a one-to-one linear transformation $T : V \to W$.
   b) Show that $dimV \geq dimW$ if and only if there exists an onto linear transformation $T : V \to W$.

## 5.4   ISOMORPHISMS AND MATRICES

Two vector spaces can appear to be quite different but, on closer examination, be the same underlying space expressed with different symbols. In this case we say that the spaces are isomorphic. This concept is investigated in this section, and is described in two ways using linear transformations: one using the one-to-one and onto properties and the other using composition. The notion of composition leads naturally to a way of introducing the matrix of any linear transformation $T$ and using it to describe the action of $T$.

**5.4.1** **Isomorphisms**

If $V$ and $W$ are vector spaces, a linear transformation $T : V \to W$ is called an **isomorphism** if it is both one-to-one and onto. If an isomorphism $V \to W$ exists, we say that $V$ and $W$ are **isomorphic**[13] vector spaces and write

$$V \cong W.$$

**Example 1**  $V \cong V$ for every vector space $V$ because $1_V : V \to V$ is an isomorphism.

**Example 2**  $\mathbb{M}_{m,n} \cong \mathbb{M}_{n,m}$ because transposition: $\mathbb{M}_{m,n} \to \mathbb{M}_{n,m}$ is an isomorphism.

An isomorphism $T : V \to W$ induces a *pairing*

$$\mathbf{v} \longleftrightarrow T(\mathbf{v})$$

between the vectors $\mathbf{v}$ in $V$ and the vectors $T(\mathbf{v})$ in $W$. Every vector in each space is paired with exactly one vector in the other space. Moreover, this pairing preserves vector addition and scalar multiplication, and hence preserves all vector space properties. The change of notation $\mathbf{v} \longmapsto T(\mathbf{v})$ changes the vector space $V$ into $W$. Hence, roughly speaking, isomorphic spaces are the same except for notation. The following example illustrates this.

**Example 3**  We have $\mathbb{R}^2 \cong \mathbb{C}$ because, for example, the map $T : \mathbb{R}^2 \to \mathbb{C}$ given by $T([x \; y]^T) = x + iy$ is an isomorphism (verify). The pairing $T$ turns $\mathbb{R}^2$ into the complex plane $\mathbb{C}$. $\mathbb{C}$ is the same vector space as $\mathbb{R}^2$, but with a different interpretation: The ordered pair $[x \; y]^T$ is now viewed as the complex number $x + iy$. We obtain one from the other by changing symbols.

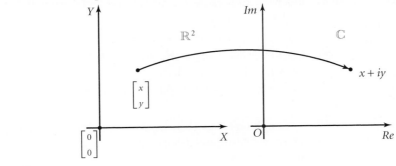

The following theorem is surprising because it shows that, up to isomorphism, there is *only one* $n$-dimensional vector space.

**THEOREM 1**  If $V$ has dimension $n$, then $V \cong \mathbb{R}^n$.

**PROOF**  Choose a basis $\mathcal{B} = \{\mathbf{b}_1, \mathbf{b}_2, \cdots, \mathbf{b}_n\}$ of $V$ and define $T : \mathbb{R}^n \to V$ by $T([r_1, r_2, \cdots, r_n]^T) = r_1\mathbf{b}_1 + r_2\mathbf{b}_2 + \cdots + r_n\mathbf{b}_n$. Then $T$ is one-to-one because $\mathcal{B}$ is independent, and $T$ is onto because $\mathcal{B}$ spans $V$ (verify). Thus $T$ is an isomorphism, so $V \cong \mathbb{R}^n$.

Because of Theorem 1, it seems at first glance that the study of an abstract $n$-dimensional vector space $V$ is a waste of time: Since $V$ is just $\mathbb{R}^n$ except for a change of symbols, simply study $\mathbb{R}^n$ and be done with it! But the meaning is in the interpretation. Viewing $\mathbb{R}^2$ as

---

[13] The word comes from two Greek words: *iso* meaning "same" and *morphos* meaning "form".

$\mathbb{C}$ casts a whole new light on the plane — for example the geometrical interpretation of complex multiplication adds new life to $\mathbb{R}^2$ and provides an important new way to study rotations in the plane, while orthogonality and projections are more naturally studied in $\mathbb{R}^2$. Similarly, one verifies that $\mathbb{R}^{n+1} \cong \mathbb{P}_n$, and interpreting vectors in $\mathbb{R}^{n+1}$ as polynomials greatly enriches the structure of $\mathbb{R}^{n+1}$ because polynomials can be multiplied and composed. Another example is $\mathbb{R}^{n^2} \cong \mathbb{M}_{n,n}$ (verify) with the attendant view of vectors in $\mathbb{R}^{n^2}$ as $n \times n$ matrices.

On the other hand, revealing the underlying vector space nature of $\mathbb{C}$, $\mathbb{P}_n$, and $\mathbb{M}_{n,n}$ allows general vector space theorems and techniques to be used to study complex numbers, polynomials, and matrices. These are powerful methods and yield new information in all three cases.

We know that every linear transformation preserves linear combinations of vectors; the isomorphisms are the ones preserving bases.

**THEOREM 2**

If $V$ and $W$ are finite dimensional vector spaces, the following are equivalent for a linear transformation $T : V \to W$:

(1) $T$ is an isomorphism.

(2) If $\{\mathbf{b}_1, \mathbf{b}_2, \cdots, \mathbf{b}_n\}$ is any basis of $V$, then $\{T(\mathbf{b}_1), T(\mathbf{b}_2), \cdots, T(\mathbf{b}_n)\}$ is a basis of $W$.

(3) There exists a basis $\{\mathbf{b}_1, \mathbf{b}_2, \cdots, \mathbf{b}_n\}$ of $V$ such that $\{T(\mathbf{b}_1), T(\mathbf{b}_2), \cdots, T(\mathbf{b}_n)\}$ is a basis of $W$.

**PROOF**

$(1) \Rightarrow (2)$. If $\{\mathbf{b}_1, \mathbf{b}_2, \cdots, \mathbf{b}_n\}$ is any basis of $V$, then $\{T(\mathbf{b}_1), T(\mathbf{b}_2), \cdots, T(\mathbf{b}_n)\}$ is independent because $T$ is one-to-one, and it spans $W$ because $T$ is onto. The verifications are left to the reader.

$(2) \Rightarrow (3)$. This follows because $V$ *has* a finite basis (being finite dimensional).

$(3) \Rightarrow (1)$. Given a basis $\{\mathbf{b}_1, \mathbf{b}_2, \cdots, \mathbf{b}_n\}$ as in (3), $T$ is one-to-one because $\{T(\mathbf{b}_1), T(\mathbf{b}_2), \cdots, T(\mathbf{b}_n)\}$ is independent and $T$ is onto because it spans $W$, so $T$ is an isomorphism. Again the details are left to the reader.

Condition (3) in Theorem 2 is a useful test for when a linear transformation is actually an isomorphism, and it can be used to *construct* isomorphisms. The next example provides an illustration.

**Example 4**

If $V$ is the space of all $2 \times 2$ symmetric matrices, find an isomorphism $T : \mathbb{P}_2 \to V$ such that $T(x) = I$.

**SOLUTION**

Write $\mathbf{d}_1 = I = \begin{bmatrix} 1 & 0 \\ 0 & 1 \end{bmatrix}$, $\mathbf{d}_2 = \begin{bmatrix} 1 & 0 \\ 0 & 0 \end{bmatrix}$ and $\mathbf{d}_3 = \begin{bmatrix} 0 & 1 \\ 1 & 0 \end{bmatrix}$. Then $\mathcal{B} = \{1, x, x^2\}$ and $\mathcal{D} = \{\mathbf{d}_1, \mathbf{d}_2, \mathbf{d}_3\}$ are bases of $\mathbb{P}_2$ and $V$ respectively. By Theorem 3 §5.3 there is a linear transformation $T : \mathbb{P}_2 \to V$ such that $T(1) = \mathbf{d}_2$, $T(x) = \mathbf{d}_1 = I$, and $T(x^2) = \mathbf{d}_3$. Since $T$ clearly takes the basis $\mathcal{B}$ to the basis $\mathcal{D}$, it is an isomorphism by Theorem 2.[14]

**THEOREM 3**

Let $V$ and $W$ denote finite dimensional vector spaces.

(1) $V \cong W$ if and only if $dim V = dim W$.

(2) If $dim V = dim W$, a linear transformation $T : V \to W$ is an isomorphism if and only if it is either one-to-one or onto.

---

[14] Of course, we can extend $T$ linearly to get the formula $T(a + bx + cx^2) = a\mathbf{d}_1 + b\mathbf{d}_2 + c\mathbf{d}_3 = \begin{bmatrix} a+b & c \\ c & a \end{bmatrix}$. However, such detail is usually unnecessary and it is customary to simply specify $T$ on a basis.

PROOF Part (2) restates the Corollary to the Dimension theorem (Theorem 5 §5.3). To prove (1), suppose $V \cong W$, let $T : V \to W$ be an isomorphism, and choose a basis $\{\mathbf{b}_1, \mathbf{b}_2, \cdots, \mathbf{b}_n\}$ of $V$. Then $\{T(\mathbf{b}_1), T(\mathbf{b}_2), \cdots, T(\mathbf{b}_n)\}$ is a basis of $W$ by Theorem 2, so $dimW = n = dimV$. Conversely, if $dimV = n = dimW$, let $\{\mathbf{b}_1, \mathbf{b}_2, \cdots, \mathbf{b}_n\}$ and $\{\mathbf{d}_1, \mathbf{d}_2, \cdots, \mathbf{d}_n\}$ be bases of $V$ and $W$ respectively. By Theorem 3 §5.3 let $T : V \to W$ be the linear transformation such that $T(\mathbf{b}_i) = \mathbf{d}_i$ for all $i$. Then $T$ is an isomorphism by Theorem 2, so $V \cong W$.

Condition (1) in Theorem 3 shows that the isomorphic relation $\cong$ has the following three properties. If $U$, $V$, and $W$ denote finite dimensional vector spaces, then:

(1) $V \cong V$.
(2) If $V \cong W$, then $W \cong V$.
(3) If $V \cong W$ and $W \cong U$, then $V \cong U$.

Because of these properties, $\cong$ is called an **equivalence relation** or, simply, an **equivalence**.[15]

---

### 5.4.2 Composition

There is another way to describe isomorphisms based on the idea of composing linear transformations. Suppose that $V \overset{T}{\to} W \overset{S}{\to} U$ are two linear transformations. As in Section 4.9 we define their **composite** $S \circ T : V \to U$ by

$$(S \circ T)(\mathbf{v}) = S[T(\mathbf{v})] \text{ for all } \mathbf{v} \text{ in } V.$$

The action of $S \circ T$ is best described by saying "first $T$, then $S$" (note the order), and is illustrated in the diagram. If $T : V \to V$, is an operator, we denote $T \circ T = T^2$, $T \circ T \circ T = T^3$, etc.

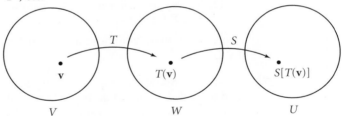

Example 5 If $A$ and $B$ are $n \times n$ matrices, define $S : \mathbb{R}^n \to \mathbb{R}^n$ and $T : \mathbb{R}^n \to \mathbb{R}^n$ by $S(X) = AX$ and $T(X) = BX$ for all $X$ in $\mathbb{R}^n$. Then $(S \circ T)(X) = (AB)X$ for all $X$ in $\mathbb{R}^n$; that is, $S \circ T$ is multiplication by the product $AB$ of the matrices for $S$ and $T$.[16] Similarly, $T \circ S$ is multiplication by the matrix $BA$. In particular, it can happen that $S \circ T \neq T \circ S$ — simply choose $A$ and $B$ such that $AB \neq BA$.

The basic properties of composition are listed in the following theorem.

---

THEOREM ④ Let $V \overset{T}{\to} W \overset{S}{\to} U \overset{R}{\to} Z$ denote linear transformations. Then:

(1) $S \circ T$ is a linear transformation.
(2) $T \circ 1_V = T = 1_W \circ T$.
(3) $(R \circ S) \circ T = R \circ (S \circ T)$.
(4) If $S$ and $T$ are both one-to-one (respectively onto), so is $S \circ T$.

---

[15] Other examples are the similarity relation $\sim$ for matrices and the equality relation $=$.

[16] We extend this in Theorem 9 below.

PROOF    Let $\mathbf{v}$ and $\mathbf{w}$ be vectors in $V$. Then the linearity of $S$ and $T$ gives

$$(S \circ T)(\mathbf{v} + \mathbf{w}) = S[T(\mathbf{v} + \mathbf{w})]$$
$$= S[T(\mathbf{v}) + T(\mathbf{w})]$$
$$= S[T(\mathbf{v})] + S[T(\mathbf{w})]$$
$$= (S \circ T)(\mathbf{v}) + (S \circ T)(\mathbf{w}).$$

Hence $S \circ T$ preserves addition; the proof that $(S \circ T)(r\mathbf{v}) = r(S \circ T)(\mathbf{v})$ for all $r$ in $\mathbb{R}$ is similar and is left to the reader. This proves (1). Now, for any vector $\mathbf{v}$ in $V$, compute

$$[(R \circ S) \circ T](\mathbf{v}) = (R \circ S)[T(\mathbf{v})]$$
$$= R\{S[T(\mathbf{v})]\}$$
$$= R\{(S \circ T)(\mathbf{v})\}$$
$$= [R \circ (S \circ T)](\mathbf{v}).$$

This proves (3). The verification of (2) is similar and is left to the reader, as is the proof of (4).

Because of property (3) in Theorem 4 we write the transformation $(R \circ S) \circ T = R \circ (S \circ T)$ simply as $R \circ S \circ T$. The proof shows that the action of $R \circ S \circ T$ is first $T$, then $S$, and finally $R$, in the order of the arrows $V \xrightarrow{T} W \xrightarrow{S} U \xrightarrow{R} Z$. We are now ready to characterize isomorphisms in terms of composition.

**THEOREM 5**    Let $V$ and $W$ be finite dimensional vector spaces. The following conditions are equivalent for a linear transformation $T : V \to W$:

(1) $T$ is an isomorphism.
(2) There exists a linear transformation $S : W \to V$ such that $S \circ T = 1_V$ and $T \circ S = 1_W$.

Moreover, if these conditions are satisfied, the transformation $S$ is uniquely determined by $T$.

PROOF    $(1) \Rightarrow (2)$. Assume that $T$ is an isomorphism. Choose a basis $\{\mathbf{b}_1, \mathbf{b}_2, \cdots, \mathbf{b}_n\}$ of $V$ and write $\mathbf{d}_i = T(\mathbf{b}_i)$ for each $i$. Then $\{\mathbf{d}_1, \mathbf{d}_2, \cdots, \mathbf{d}_n\}$ is a basis of $W$ by Theorem 2, so there exists a linear transformation $S : W \to V$ such that $S(\mathbf{d}_i) = \mathbf{b}_i$ for each $i$ (by Theorem 3 §5.3). But then $(S \circ T)(\mathbf{b}_i) = S[T(\mathbf{b}_i)] = S(\mathbf{d}_i) = \mathbf{b}_i = 1_V(\mathbf{b}_i)$ for each $i$. This means that $S \circ T = 1_V$ by Theorem 2 §5.3 because $\{\mathbf{b}_1, \mathbf{b}_2, \cdots, \mathbf{b}_n\}$ spans $V$. Similarly, $T \circ S = 1_W$ because $(T \circ S)(\mathbf{d}_i) = 1_W(\mathbf{d}_i)$ for each $i$. This proves (2).

$(2) \Rightarrow (1)$. Suppose $S$ exists as in (2). If $\mathbf{w}$ is in $W$, then $\mathbf{w} = 1_W(\mathbf{w}) = (T \circ S)(\mathbf{w}) = T[S(\mathbf{w})]$. Hence $\mathbf{w}$ is in $imT$, and it follows that $T$ is onto. On the other hand, if $T(\mathbf{v}) = \mathbf{0}$ where $\mathbf{v}$ is in $V$, then $\mathbf{v} = 1_V(\mathbf{v}) = (S \circ T)(\mathbf{v}) = S[T(\mathbf{v})] = S(\mathbf{0}) = \mathbf{0}$. This shows that $kerT = \{\mathbf{0}\}$, and so $T$ is one-to-one. Hence (1) is proved.

Finally, suppose that $S_1 : W \to V$ is another transformation satisfying the conditions in (2): $S_1 \circ T = 1_V$ and $T \circ S_1 = 1_W$. Then Theorem 4 gives $S_1 = S_1 \circ 1_W = S_1 \circ (T \circ S) = (S_1 \circ T) \circ S = 1_V \circ S = S$. This proves the uniqueness in the last sentence of the theorem.

If $T : V \to W$ is an isomorphism, the unique linear transformation $S : W \to V$ that satisfies $S \circ T = 1_V$ and $T \circ S = 1_W$ is called the **inverse** of $T$ and is denoted $T^{-1}$ (as for matrices). Hence, if it exists, the transformation $T^{-1} : W \to V$ is uniquely determined by the equations

$$T^{-1} \circ T = 1_V \quad \text{and} \quad T \circ T^{-1} = 1_W.$$

These equations mean that $T$ and $T^{-1}$ are related by the **fundamental identities**:

$$T^{-1}[T(\mathbf{v})] = \mathbf{v} \quad \text{for all} \quad \mathbf{v} \text{ in } V, \quad \text{and} \quad T[T^{-1}(\mathbf{w})] = \mathbf{w} \quad \text{for all } \mathbf{w} \text{ in } W.$$

In other words, each of $T$ and $T^{-1}$ "reverses" the action of the other. Among other things, this gives a way to find the action of $T^{-1}$ if the effect of $T$ on a basis is known. Here is an example.

**Example 6**

Let $\vec{e}_1 = [1\ 0\ 0]^T$, $\vec{e}_2 = [0\ 1\ 0]^T$, and $\vec{e}_3 = [0\ 0\ 1]^T$ denote the standard basis of $\mathbb{R}^3$. Consider the transformation $T : \mathbb{R}^3 \to \mathbb{P}_2$ given by

$$T(\vec{e}_1) = 1, \quad T(\vec{e}_2) = x - 1, \quad \text{and} \quad T(\vec{e}_3) = x^2 - 1.$$

Show that $T$ is an isomorphism and find the action of $T^{-1}$.

**SOLUTION**

Since $\{T(\vec{e}_1),\ T(\vec{e}_2),\ T(\vec{e}_3)\} = \{1,\ x - 1,\ x^2 - 1\}$ is a basis of $\mathbb{P}_2$, it follows from Theorem 2 that $T$ is an isomorphism. Hence $T^{-1} : \mathbb{P}_2 \to \mathbb{R}^3$ exists by Theorem 5. Moreover, since $T^{-1}$ reverses the action of $T$ we have

$$T^{-1}(1) = \vec{e}_1, \quad T^{-1}(x - 1) = \vec{e}_2, \quad \text{and} \quad T^{-1}(x^2 - 1) = \vec{e}_3.$$

This defines $T^{-1}$ because $\{1,\ x - 1,\ x^2 - 1\}$ is a basis of $\mathbb{P}_2$. (If a formula is needed, we extend linearly to get $T^{-1}(a + bx + cx^2) = [a + b + c\ \ b\ \ c]^T$.)

The following example illustrates the fact that it is sometimes possible to show that a linear transformation $T : V \to W$ is an isomorphism by explicitly constructing the inverse $T^{-1} : W \to V$.

**Example 7**

Consider the transpose transformation $T : \mathbb{M}_{n,n} \to \mathbb{M}_{n,n}$ where $T(A) = A^T$ for all $A$ in $\mathbb{M}_{n,n}$. Show that $T$ is an isomorphism and find $T^{-1}$.

**SOLUTION**

Rather than show that $T$ is one-to-one and onto (which is not difficult!), we utilize the fact that $(A^T)^T = A$ for every matrix $A$ in $\mathbb{M}_{n,n}$. In terms of $T$, this asserts that $(T \circ T)(A) = T[T(A)] = [A^T]^T = A = 1_{\mathbb{M}_{n,n}}(A)$ for all $A$ in $\mathbb{M}_{n,n}$. This means that $T \circ T = 1_{\mathbb{M}_{n,n}}$, and hence that $T$ is invertible and $T^{-1} = T$.

By contrast, we can sometimes show that a linear transformation is an isomorphism even though it is difficult to give an explicit formula for the inverse.

**Example 8**

Define $T : \mathbb{P}_n \to \mathbb{R}^{n+1}$ by $T[p(x)] = [p(0)\ p(1)\ p(2)\ \cdots\ p(n)]^T$ for all $p(x)$ in $\mathbb{P}_n$. Show that $T$ is an isomorphism.

**SOLUTION**

The verification that $T$ is linear is left to the reader (see Example 3 §5.3). If $T[p(x)] = 0$, then $p(k) = 0$ for each $k = 0, 1, 2, \cdots, n$, so the polynomial $p(x)$ has $n + 1$ distinct roots. But this implies that $p(x)$ is the zero polynomial because no nonzero polynomial of degree at most $n$ can have more than $n$ roots (see Appendix A.3). Hence $T$ is one-to-one. It follows by Theorem 3 that $T$ is an isomorphism because $dim(\mathbb{P}_n) = n + 1 = dim(\mathbb{R}^{n+1})$. In particular, the inverse transformation $T^{-1} : \mathbb{R}^{n+1} \to \mathbb{P}_n$ exists, but we have not given an explicit formula for it. To do so requires some ingenuity.

The following theorem contains some basic properties of inverses of linear transformations that are the analogs of the same properties for matrices (see Theorem 3 §1.5). The proof uses Theorems 4 and 5, and is left to the reader.

**THEOREM 6**     Let $V \xrightarrow{T} W \xrightarrow{S} U$ be linear transformations. Then:
> (1) $1_V : V \to V$ is an isomorphism and $(1_V)^{-1} = 1_V$.
> (2) If $T$ is an isomorphism, so is $T^{-1}$, and $(T^{-1})^{-1} = T$.
> (3) If $S$ and $T$ are both isomorphisms, so is $S \circ T$, and $(S \circ T)^{-1} = T^{-1} \circ S^{-1}$.

## 5.4.3   Coordinates

In Section 4.9 we saw that every linear transformation $S : \mathbb{R}^n \to \mathbb{R}^m$ is given by $S(X) = AX$ for some $m \times n$ matrix $A$. In much the same way we can use matrix multiplication to describe the action of *any* linear transformation $T : V \to W$. However matrices must act on $\mathbb{R}^n$, while $T$ acts on the abstract space $V$, so we must first find a way to convert vectors in $V$ or $W$ into columns. As in Section 4.9, this is done by using coordinates.

Let $\mathcal{B} = \{\mathbf{b}_1, \mathbf{b}_2, \cdots, \mathbf{b}_n\}$ be a basis of a vector space $V$. Hence every vector $\mathbf{v}$ in $V$ is represented as a linear combination $\mathbf{v} = r_1\mathbf{b}_1 + r_2\mathbf{b}_2 + \cdots + r_n\mathbf{b}_n$ where the numbers $r_i$ in $\mathbb{R}$ are uniquely[17] determined by $\mathbf{v}$. Hence, as in Section 4.9, we define the $\mathcal{B}$-**coordinate vector** $C_{\mathcal{B}}(\mathbf{v})$ of the vector $\mathbf{v}$ as follows:

$$C_{\mathcal{B}}(\mathbf{v}) = [r_1 \quad r_2 \quad \cdots \quad r_n]^T \qquad \text{when} \qquad \mathbf{v} = r_1\mathbf{b}_1 + r_2\mathbf{b}_2 + \cdots + r_n\mathbf{b}_n.$$

The numbers $r_i$ are called the $\mathcal{B}$-**coordinates** of the vector $\mathbf{v}$.

**Example 9**     This is illustrated in the diagram above where $\vec{e}_1$ and $\vec{e}_2$ are the standard basis vectors of $\mathbb{R}^2$, which give rise to the usual $X_1$- and $X_2$-axes. Thus a vector $\vec{v} = [4 \ 3]^T$ has the usual coordinates $x_1 = 4$ and $x_2 = 3$.

However, another basis $\mathcal{B} = \{\mathbf{b}_1, \mathbf{b}_2\}$ is also depicted in the diagram, giving rise to the $B_1$- and $B_2$-axes. The vector $\vec{v}$ is represented as $\vec{v} = 3\mathbf{b}_1 + 2\mathbf{b}_2$, so $C_{\mathcal{B}}(\vec{v}) = [3 \ 2]^T$.

**Example 10**     If $\mathcal{E}$ is the standard basis of $\mathbb{R}^n$ (consisting of the columns of $I_n$), then $C_{\mathcal{E}}(X) = X$ for all columns $X$ in $\mathbb{R}^n$.

**Example 11**     If $\mathcal{B} = \{1, x, x^2, \cdots, x^n\}$ in $\mathbb{P}_n$, then $C_{\mathcal{B}}(a_0 + a_1x + \cdots + a_nx^n) = [a_0 \ a_1 \ \cdots \ a_n]^T$.

**Example 12**     If $\mathcal{B} = \left\{ \begin{bmatrix} 1 & 0 \\ 0 & 0 \end{bmatrix}, \begin{bmatrix} 0 & 1 \\ 0 & 0 \end{bmatrix}, \begin{bmatrix} 0 & 0 \\ 1 & 0 \end{bmatrix}, \begin{bmatrix} 0 & 0 \\ 0 & 1 \end{bmatrix} \right\}$ in $\mathbb{M}_{2,2}$, then $C_{\mathcal{B}}\begin{bmatrix} a & b \\ c & d \end{bmatrix} = [a \ b \ c \ d]^T$.

If $\mathcal{B}$ is a basis of the $n$-dimensional vector space $V$, the formation of $\mathcal{B}$-coordinates defines a transformation

$$C_{\mathcal{B}} : V \to \mathbb{R}^n.$$

In fact, $C_{\mathcal{B}}$ is an isomorphism.

**THEOREM 7**     Let $\mathcal{B}$ be a basis of the $n$-dimensional vector space $V$. Then $C_{\mathcal{B}} : V \to \mathbb{R}^n$ is an isomorphism.

**PROOF**     Define $S : \mathbb{R}^n \to V$ by $S([r_1 \ r_2 \ \cdots \ r_n]^T) = r_1 b_1 + r_2 b_2 + \cdots + r_n b_n$ for all $[r_1 \ r_2 \ \cdots \ r_n]^T$ in $\mathbb{R}^n$, then it is routine to verify that $S \circ C_{\mathcal{B}} = 1_V$ and $C_{\mathcal{B}} \circ S = 1_{\mathbb{R}^n}$. Hence $C_{\mathcal{B}}$ is an isomorphism with inverse $S$.

---

[17] This requires that the basis $B$ is *ordered*; that is, the order in which the basis vectors are listed is taken into account. As in Section 4.9, when working with coordinates we assume (usually without saying so) that the bases are ordered.

The following useful result will be used several times below.

**LEMMA 1** Let $\mathcal{B}$ be a basis of the $n$-dimensional vector space $V$. If $A$ and $B$ are $m \times n$ matrices such that $A\, C_{\mathcal{B}}(\mathbf{v}) = B\, C_{\mathcal{B}}(\mathbf{v})$ for all $\mathbf{v}$ in $V$, then $A = B$.

**PROOF** Let $A\, C_{\mathcal{B}}(\mathbf{v}) = B\, C_{\mathcal{B}}(\mathbf{v})$ for all $\mathbf{v}$ in $V$. If we write $\mathcal{B} = \{\mathbf{b}_1,\ \mathbf{b}_2,\ \cdots,\ \mathbf{b}_n\}$, then $C_{\mathcal{B}}(\mathbf{b}_j)$ is column $j$ of the identity matrix (verify). It follows that $A\, C_{\mathcal{B}}(\mathbf{b}_j)$ is column $j$ of $A$, so the equation $A\, C_{\mathcal{B}}(\mathbf{b}_j) = B\, C_{\mathcal{B}}(\mathbf{b}_j)$ shows that column $j$ of $A$ equals column $j$ of $B$. Since this holds for each $j$, it shows that $A = B$.

## 5.4.4 The Matrix of a Linear Transformation

The results in Theorems 4 and 6 about composition of linear transformations bear a striking resemblance to the analogous properties of matrix multiplication. Furthermore, we showed in Theorem 2 §4.9 that every linear transformation $\mathbb{R}^n \to \mathbb{R}^m$ is given by multiplication by an $m \times n$ matrix. Hence it will come as no surprise that, for any linear transformation $T$, there is a way to associate a matrix to $T$ which can be used to describe the action of $T$.

So let $T : V \to W$ be a linear transformation where $dim V = n$ and $dim W = m$. We are interested in computing $T(\mathbf{v})$ for any vector $\mathbf{v}$ in $V$ using matrix multiplication. This can be accomplished as follows: Let $\mathcal{B}$ and $\mathcal{D}$ be bases of $V$ and $W$ respectively. The idea is to compute $T(\mathbf{v})$ by first finding its $\mathcal{D}$-coordinate vector $C_{\mathcal{D}}[T(\mathbf{v})]$. This is a column in $\mathbb{R}^m$, and it turns out that there exists an $m \times n$ matrix $M_{\mathcal{D}\mathcal{B}}(T)$ such that

$$C_{\mathcal{D}}[T(\mathbf{v})] = M_{\mathcal{D}\mathcal{B}}(T)\, C_{\mathcal{B}}(\mathbf{v}) \quad \text{for all } \mathbf{v} \text{ in} V.$$

The matrix $M_{\mathcal{D}\mathcal{B}}(T)$ depends only on $T$ and the bases $\mathcal{B}$ and $\mathcal{D}$, and is called the $\mathcal{D}\mathcal{B}$-**matrix** of $T$. Moreover, it is easy to compute. Indeed, if the basis $\mathcal{B}$ is listed as $\mathcal{B} = \{\mathbf{b}_1,\ \mathbf{b}_2,\ \cdots,\ \mathbf{b}_n\}$, then

$$M_{\mathcal{D}\mathcal{B}}(T) = [\,C_{\mathcal{D}}[T(\mathbf{b}_1)] \quad C_{\mathcal{D}}[T(\mathbf{b}_2)] \quad \cdots \quad C_{\mathcal{D}}[T(\mathbf{b}_n)]\,]. \tag{$*$}$$

That is, is the matrix with the coordinate vectors

$$C_{\mathcal{D}}[T(\mathbf{b}_1)],\ C_{\mathcal{D}}[T(\mathbf{b}_2)],\ \cdots,\ C_{\mathcal{D}}[T(\mathbf{b}_n)]$$

as its columns. The following theorem summarizes this discussion.

**THEOREM 8** Let $T : V \to W$ be a linear transformation where $dim V = n$ and $dim W = m$, and let $\mathcal{B}$ and $\mathcal{D}$ be bases of $V$ and $W$ respectively. Then

$$C_{\mathcal{D}}[T(\mathbf{v})] = M_{\mathcal{D}\mathcal{B}}(T)\, C_{\mathcal{B}}(\mathbf{v}) \quad \text{for all } \mathbf{v} \text{ in } V$$

where $M_{\mathcal{D}\mathcal{B}}(T) = [\,C_{\mathcal{D}}[T(\mathbf{b}_1)] \quad C_{\mathcal{D}}[T(\mathbf{b}_2)] \quad \cdots \quad C_{\mathcal{D}}[T(\mathbf{b}_n)]\,]$ is the $m \times n$ $\mathcal{D}\mathcal{B}$-matrix of $T$.

**PROOF** Let $\mathcal{B} = \{\mathbf{b}_1,\ \mathbf{b}_2,\ \cdots,\ \mathbf{b}_n\}$ and write $\mathbf{v} = r_1\mathbf{b}_1 + r_2\mathbf{b}_2 + \cdots + r_n\mathbf{b}_n$, so that $C_{\mathcal{B}}(\mathbf{v}) = [r_1\ r_2\ \cdots\ r_n]^T$. Since both $C_{\mathcal{D}}$ and $T$ are linear, we have

$$M_{\mathcal{D}\mathcal{B}}(T)\, C_{\mathcal{B}}(\mathbf{v}) = \begin{bmatrix} C_{\mathcal{D}}[T(\mathbf{b}_1)] & C_{\mathcal{D}}[T(\mathbf{b}_2)] & \cdots & C_{\mathcal{D}}[T(\mathbf{b}_n)] \end{bmatrix} \begin{bmatrix} r_1 \\ r_2 \\ \vdots \\ r_n \end{bmatrix}$$

$$= r_1 C_{\mathcal{D}}[T(\mathbf{b}_1)] + r_2 C_{\mathcal{D}}[T(\mathbf{b}_2)] + \cdots + r_n C_{\mathcal{D}}[T(\mathbf{b}_n)]$$
$$= C_{\mathcal{D}}[T(r_1\mathbf{b}_1 + r_2\mathbf{b}_2 + \cdots + r_n\mathbf{b}_n)]$$
$$= C_{\mathcal{D}}[T(\mathbf{v})]$$

using block multiplication. This is what we wanted.

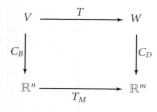

It is instructive to interpret Theorem 8 in terms of composition. Let $T : V \rightarrow W$ be linear where $dim V = n$ and $dim W = m$, and let $\mathcal{B} = \{\mathbf{b}_1, \mathbf{b}_2, \cdots, \mathbf{b}_n\}$ and $\mathcal{D}$ be bases of $V$ and $W$ respectively. For convenience, write $M = M_{\mathcal{D}\mathcal{B}}(T)$ and let $T_M : \mathbb{R}^n \rightarrow \mathbb{R}^m$ be the matrix transformation given by $T_M(X) = MX$ for all $X$ in $\mathbb{R}^n$. The diagram expresses the relationship between the transformations $T$ and $T_M$ and the two coordinate isomorphisms $C_{\mathcal{B}}$ and $C_{\mathcal{D}}$. The conclusion in Theorem 8 is that $T_M[C_{\mathcal{B}}(\mathbf{v})] = C_{\mathcal{D}}[T(\mathbf{v})]$ holds for all $\mathbf{v}$ in $V$, in other words,

$$T_M \circ C_{\mathcal{B}} = C_{\mathcal{D}} \circ T \quad \text{equivalently} \quad T = C_{\mathcal{D}}^{-1} \circ T_M \circ C_{\mathcal{B}}.$$

This is often expressed by saying that the diagram is *commutative*: That is, starting with a vector $\mathbf{v}$ in $V$, the effect on $\mathbf{v}$ of applying the transformations in the diagram is the same no matter which route is taken to $\mathbb{R}^m$.

**Example 13**     Let $\mathcal{B}$ and $\mathcal{D}$ be bases of $V$ and $W$ respectively. Then the identity operator $1_V : V \rightarrow V$ has matrix $M_{\mathcal{B}\mathcal{B}}(1_V) = I$, and the zero transformation $0 : V \rightarrow W$ has matrix $M_{\mathcal{D}\mathcal{B}}(0) = 0$.

The next theorem shows that the construction of the matrix of a linear transformation preserves composition in the sense that composing linear transformations means multiplying their matrices.

**THEOREM 9**     Let $V \xrightarrow{T} W \xrightarrow{S} U$ be linear transformations, and let $\mathcal{B}$, $\mathcal{D}$, and $\mathcal{G}$ denote bases of $V$, $W$, and $U$ respectively. Then we have the composite $S \circ T : V \rightarrow U$ and the matrices are related as follows:

$$M_{\mathcal{G}\mathcal{B}}(S \circ T) = M_{\mathcal{G}\mathcal{D}}(S) \, M_{\mathcal{D}\mathcal{B}}(T).$$

**PROOF**     Given $\mathbf{v}$ in $V$, right multiply the right side by $C_{\mathcal{B}}(\mathbf{v})$ and use Theorem 8 three times:

$$M_{\mathcal{G}\mathcal{D}}(S) \, M_{\mathcal{D}\mathcal{B}}(T) \, C_{\mathcal{B}}(\mathbf{v}) = M_{\mathcal{G}\mathcal{D}}(S) \, C_{\mathcal{D}}[T(\mathbf{v})]$$
$$= C_{\mathcal{G}}\{S[T(\mathbf{v})]\}$$
$$= C_{\mathcal{G}}\{(S \circ T)(\mathbf{v})\}$$
$$= M_{\mathcal{G}\mathcal{B}}(S \circ T)C_{\mathcal{B}}(\mathbf{v}).$$

The result now follows because we can "cancel" $C_{\mathcal{B}}(\mathbf{v})$ by Lemma 1.

We can now characterize isomorphisms in terms of their matrices.

**THEOREM 10**     The following are equivalent for a linear transformation $T : V \rightarrow W$ where $dim V = n = dim W$.

(1) $T$ is an isomorphism.
(2) $M_{\mathcal{D}\mathcal{B}}(T)$ is invertible for all bases $\mathcal{B}$ and $\mathcal{D}$ of $V$ and $W$, respectively.
(3) There exist bases $\mathcal{B}$ and $\mathcal{D}$ of $V$ and $W$ respectively such that $M_{\mathcal{D}\mathcal{B}}(T)$ is invertible.

When this is the case, $[M_{\mathcal{D}\mathcal{B}}(T)]^{-1} = M_{\mathcal{B}\mathcal{D}}(T^{-1})$.

**PROOF**     $(1) \Rightarrow (2)$. Given (1), apply Theorem 9 to $T^{-1} \circ T = 1_V$ to get $M_{\mathcal{B}\mathcal{D}}(T^{-1}) \, M_{\mathcal{D}\mathcal{B}}(T) = M_{\mathcal{B}\mathcal{B}}(1_V) = I$. Similarly $T \circ T^{-1} = 1_W$ gives the reverse equation $M_{\mathcal{D}\mathcal{B}}(T) \, M_{\mathcal{B}\mathcal{D}}(T^{-1}) = I$. This proves (2) and the last sentence.

$(2) \Rightarrow (3)$. This is clear since $V$ and $W$ have bases (they are finite dimensional).

$(3) \Rightarrow (1)$. Assume (3) and let $\mathcal{B} = \{\mathbf{b}_1, \mathbf{b}_2, \cdots, \mathbf{b}_n\}$ and $\mathcal{D}$ be bases of $V$ and $W$, respectively. Then $M_{\mathcal{DB}}(T)$ is invertible by (3), so its columns $\{C_{\mathcal{D}}[T(\mathbf{b}_1)], \ C_{\mathcal{D}}[T(\mathbf{b}_2)], \ \cdots, \ C_{\mathcal{D}}[T(\mathbf{b}_n)]\}$ form a basis of $\mathbb{R}^n$. Since $C_{\mathcal{D}}^{-1} : \mathbb{R}^n \to W$ is an isomorphism, it follows from Theorem 2 that $\{T(\mathbf{b}_1), T(\mathbf{b}_2), \cdots, T(\mathbf{b}_n)\}$ is a basis of $W$. Hence $T$ is an isomorphism, again by Theorem 2.

We conclude this section with a fact about rank: If $V$ and $W$ are finite dimensional vector spaces, the matrices of a linear transformation $T : V \to W$ all have the *same* rank. This follows from the fact (see below) that

$$rank(M_{\mathcal{BD}}(T)) = dim(imT) \qquad (**)$$

for all bases $\mathcal{B}$ and $\mathcal{D}$ of $V$ and $W$, respectively, since $imT = \{T(\mathbf{v}) \mid \mathbf{v} \text{ in } V\}$ does not depend on $\mathcal{B}$ or $\mathcal{D}$. To prove $(**)$, write $M_{\mathcal{DB}}(T) = A$ for simplicity, and recall that $rankA = dim(colA)$. Hence it suffices to show that $imT \cong colA$ and, to this end, define

$$S : imT \to colA \quad \text{by} \quad S(T(\mathbf{v})) = C_{\mathcal{D}}(T(\mathbf{v})) \text{ for all } T(\mathbf{v}) \text{ in } imT.$$

This makes sense because $C_{\mathcal{D}}(T(\mathbf{v})) = A \, C_{\mathcal{B}}(\mathbf{v})$ for all $\mathbf{v}$ by Theorem 8, and $A \, C_{\mathcal{B}}(\mathbf{v})$ is in $colA$ because $colA = \{AX \mid X \text{ in } \mathbb{R}^m\}$. Moreover $S$ is linear and one-to-one because this is true of $C_{\mathcal{B}}$, and it is onto because every $X$ in $\mathbb{R}^m$ has the form $X = C_{\mathcal{B}}(\mathbf{v})$ ($C_{\mathcal{B}}$ is onto). Hence $S$ is an isomorphism, so $(**)$ is proved.

Because of $(**)$, it is natural to define the **rank** of $T$ by

$$rankT = dim(imT).$$

Hence we have proved

**THEOREM 11**   If $T : V \to W$ is a linear transformation where $V$ and $W$ are finite dimensional, then

$$rankT = rank(M_{\mathcal{BD}}(T))$$

for all bases $\mathcal{B}$ and $\mathcal{D}$ of $V$ and $W$, respectively.

If $T : V \to W$ is a linear transformation of rank $r$, the next example exhibits a choice of bases for $V$ and $W$ that makes the matrix of $T$ as nice as possible.

**Example 14**   Let $T : V \to W$ be a linear transformation where $dimV = n$ and $dimW = m$. Show that there exist bases $\mathcal{B}$ and $\mathcal{D}$ of $V$ and $W$ such that

$$M_{\mathcal{DB}}(T) = \begin{bmatrix} I_r & 0 \\ 0 & 0 \end{bmatrix} \quad \text{where } r = rankT.$$

**SOLUTION**   By extending a basis of $kerT$, let

$$\mathcal{B} = \{\mathbf{b}_1, \mathbf{b}_2, \cdots, \mathbf{b}_r, \mathbf{b}_{r+1}, \cdots, \mathbf{b}_n\}$$

be a basis of $V$ such that $\{\mathbf{b}_{r+1}, \cdots, \mathbf{b}_n\}$ is a basis of $kerT$. One verifies that $\{T(\mathbf{b}_1), T(\mathbf{b}_2), \cdots, T(\mathbf{b}_r)\}$ is a basis of $imT$ (so $rankT = r$), and we can extend it to a basis

$$\mathcal{D} = \{T(\mathbf{b}_1), T(\mathbf{b}_2), \cdots, T(\mathbf{b}_r), \mathbf{d}_{r+1}, \cdots, \mathbf{d}_m\}$$

of $W$. Hence

$$M_{\mathcal{DB}}(T) = [C_{\mathcal{D}}(T(\mathbf{b}_1)) \quad \cdots \quad C_{\mathcal{D}}(T(\mathbf{b}_r)) \quad C_{\mathcal{D}}(T(\mathbf{b}_{r+1})) \quad \cdots \quad C_{\mathcal{D}}(T(\mathbf{b}_n))]$$

$$= \begin{bmatrix} I_r & 0 \\ 0 & 0 \end{bmatrix}$$

in block form, as required.

## Exercises 5.4

**1.** In each case either show that the statement is true or give an example showing that it is false. Assume that $T$ is linear throughout.

**a)** If $T : V \to V$ satisfies $T^2 = T$, then $T$ is an isomorphism.

**b)** If $T : V \to V$ satisfies $im\,T \subseteq ker\,T$, then $T^2 = 0$.

**c)** If $T : \mathbb{M}_{n,n} \to \mathbb{M}_{n,n}$ and $T(U) = U$ for some invertible matrix $U$, then $T$ is an isomorphism.

**d)** If $T : V \to W$ is an isomorphism, then $T^{-1}(T(\mathbf{w})) = \mathbf{w}$ for every vector $\mathbf{w}$ in $W$.

**e)** Every isomorphism $T : \mathbb{M}_{n,n} \to \mathbb{M}_{n,n}$ satisfies $T(I) = I$.

**f)** If $dim\,V = 4$ and $dim\,W = 5$, it is not possible to have an isomorphism $V \to W$.

**g)** Let $T : V \to W$ be linear and let $\mathcal{B} = \{\mathbf{b}_1,\ \mathbf{b}_2\}$ and $\mathcal{D} = \{\mathbf{d}_1,\ \mathbf{d}_2,\ \mathbf{d}_3\}$ be bases of $V$ and $W$, respectively. If $T(\mathbf{b}_2) = 5\mathbf{d}_1 + 7\mathbf{d}_2 - 2\mathbf{d}_3$, then column 2 of $M_{\mathcal{DB}}(T)$ is $[5\ 7\ {-}2]^T$.

**h)** If $T : \mathbb{M}_{2,3} \to \mathbb{M}_{3,2}$ is defined by $T(A) = A^T$ for each $A$ in $\mathbb{M}_{2,3}$, then $T^{-1} = T$.

**2.** In each case show that $T$ is an isomorphism and give a formula for the action of $T^{-1}$.

**a)** $T : \mathbb{R}^3 \to \mathbb{P}_2$ with $T([a\ b\ c]^T) = (a + b) + (b + c)x + (c + a)x^2$.

**b)** $T : \mathbb{C} \to \mathbb{C}$ with $T(z) = \bar{z}$.

**c)** $T : \mathbb{M}_{mn} \to \mathbb{M}_{mn}$ with $T(A) = UAV$, where $U$ and $V$ are fixed invertible matrices.

**d)** $T : \mathbb{P}_1 \to \mathbb{R}^2$ with $T[p(x)] = [p(0)\ p(1)]^T$.

**e)** $T : \mathbb{P}_n \to \mathbb{P}_n$ with $T[p(x)] = p(3 - x)$. [Hint: Consider $T^2$.]

**f)** $T : V \to V$ with $T(\mathbf{v}) = k\mathbf{v}$, where $k \neq 0$ is a fixed real number.

**g)** $T : \mathbb{P}_1 \to \mathbb{P}_1$ with $T(a + bx) = (b - a) - ax$. [Hint: Consider $T^3$.]

**h)** $T : \mathbb{R}^2 \to \mathbb{R}^2$ with $T([x\ y]^T) = [ky - x\ y]^T$, where $k$ is a real number.

**3.** If $T : \mathbb{R}^4 \to \mathbb{R}^4$ is defined by $T([x\ y\ z\ w]^T) = [-y\ x - y\ z\ -w]^T$, show that $T^6 = 1_{\mathbb{R}^4}$, and so determine $T^{-1}$.

**4.** If $V$ is the vector space in Exercise 5 §5.1, find an isomorphism $\mathbb{R} \to V$.

**5.** If $T : V \to V$ is linear, show that $T^2 = 1_V$ if and only if $T$ is an isomorphism and $T^{-1} = T$.

**6.** In each case find the coordinates of $\mathbf{v}$ with respect to the basis $\mathcal{B}$ of $V$.

**a)** $V = \mathbb{P}_2$, $\mathbf{v} = x^2 - 3x + 4$, $\mathcal{B} = \{x + 2,\ x^2 - 1,\ 1\}$.

**b)** $V = \mathbb{M}_{2,2}$, $\mathbf{v} = \begin{bmatrix} 2 & 1 \\ 0 & -1 \end{bmatrix}$,

$\mathcal{B} = \left\{ \begin{bmatrix} 1 & 1 \\ 0 & 0 \end{bmatrix}, \begin{bmatrix} 1 & 0 \\ 1 & 0 \end{bmatrix}, \begin{bmatrix} 1 & 0 \\ 0 & 1 \end{bmatrix}, \begin{bmatrix} 0 & 0 \\ 0 & 1 \end{bmatrix} \right\}$.

**7.** In each case find the matrix $M_{\mathcal{DB}}(T)$.

**a)** $T : \mathbb{P}_2 \to \mathbb{P}_1$ where $T(a + bx + cx^2) = b + cx$, $\mathcal{B} = \{1,\ x,\ x^2\}$ and $\mathcal{D} = \{1,\ x\}$.

**b)** $T : \mathbb{M}_{2,2} \to \mathbb{M}_{2,2}$ where $T(A) = A^T$,

$\mathcal{B} = \mathcal{D} = \left\{ \begin{bmatrix} 1 & 0 \\ 0 & 0 \end{bmatrix}, \begin{bmatrix} 0 & 1 \\ 0 & 0 \end{bmatrix}, \begin{bmatrix} 0 & 0 \\ 1 & 0 \end{bmatrix}, \begin{bmatrix} 0 & 0 \\ 0 & 1 \end{bmatrix} \right\}$.

**c)** $T : \mathbb{M}_{2,2} \to \mathbb{R}$ where $T(A) = tr\,A$,

$\mathcal{B} = \left\{ \begin{bmatrix} 1 & 0 \\ 0 & 0 \end{bmatrix}, \begin{bmatrix} 0 & 1 \\ 0 & 0 \end{bmatrix}, \begin{bmatrix} 0 & 0 \\ 1 & 0 \end{bmatrix}, \begin{bmatrix} 0 & 0 \\ 0 & 1 \end{bmatrix} \right\}$, $\mathcal{D} = \{1\}$.

**8.** In each case find the action of $T$ given the matrix $M_{\mathcal{DB}}(T)$.

**a)** $T : \mathbb{P}_2 \to \mathbb{R}^2$, $M_{\mathcal{DB}}(T) = \begin{bmatrix} 1 & -1 & 0 \\ 2 & 0 & 1 \end{bmatrix}$, $\mathcal{B} = \{1,\ x,\ x^2\}$, $\mathcal{D} = \{[1\ 1]^T,\ [1\ 0]^T\}$.

**b)** $T : \mathbb{R}^3 \to \mathbb{M}_{2,2}$, $M_{\mathcal{DB}}(T) = \begin{bmatrix} 1 & 0 & 0 \\ -1 & 2 & 1 \\ 0 & 1 & 2 \\ 1 & 0 & 0 \end{bmatrix}$, where $\mathcal{B}$ is the standard basis of $\mathbb{R}^3$ and

$\mathcal{D} = \left\{ \begin{bmatrix} 1 & 0 \\ 0 & 0 \end{bmatrix}, \begin{bmatrix} 0 & 1 \\ 0 & 0 \end{bmatrix}, \begin{bmatrix} 0 & 0 \\ 1 & 0 \end{bmatrix}, \begin{bmatrix} 0 & 0 \\ 0 & 1 \end{bmatrix} \right\}$.

**9.** Let $D : \mathbb{P}_3 \to \mathbb{P}_2$ be the differentiation map: $D[p(x)] = p'(x)$. Find $M_{\mathcal{DB}}(T)$ where $\mathcal{B} = \{1,\ x,\ x^2,\ x^3\}$ and $\mathcal{D} = \{1,\ x,\ x^2\}$, and use it to compute $D(a + bx + cx^2 + dx^3)$.

**10.** If $T : V \to \mathbb{R}^m$ and let $\mathcal{B} = \{\mathbf{b}_1,\ \mathbf{b}_2,\ \cdots,\ \mathbf{b}_n\}$ be a basis of $V$. If $\mathcal{E}$ is the standard basis of $\mathbb{R}^m$, show that $M_{\mathcal{EB}}(T) = [T(\mathbf{b}_1)\ T(\mathbf{b}_2)\ \cdots\ T(\mathbf{b}_n)]$.

**11.** Let $T : V \to W$ be an isomorphism. If $\mathcal{B} = \{\mathbf{b}_1,\ \mathbf{b}_2,\ \cdots,\ \mathbf{b}_n\}$ is a basis of $V$, Theorem 2 asserts that $\mathcal{D} = \{T(\mathbf{b}_1),\ T(\mathbf{b}_2),\ \cdots,\ T(\mathbf{b}_n)\}$ is a basis of $W$. Show that $M_{\mathcal{DB}}(T) = I_n$.

**12.** Let $U$ be any $n \times n$ invertible matrix. If $\mathcal{D} = \{\mathbf{d}_1,\ \mathbf{d}_2,\ \cdots,\ \mathbf{d}_n\}$ where $\mathbf{d}_j$ is column $j$ of $U$, show that $M_{\mathcal{ED}}(1_{\mathbb{R}^n}) = U$ where $\mathcal{E}$ is the standard basis of $\mathbb{R}^n$.

**13.** Let $V \xrightarrow{T} W \xrightarrow{S} U$ be linear transformations.

**a)** If $S$ and $T$ are both onto, prove that $S \circ T$ is onto.

**b)** If $S$ and $T$ are both one-to-one, prove that $S \circ T$ is one-to-one.

**14.** Let $V \xrightarrow{T} W \xrightarrow{S} U$ be linear transformations.

**a)** If $T$ is onto and $S \circ T = S_1 \circ T$ where $S_1 : W \to U$, show that $S = S_1$.

**b)** If $S$ is one-to-one and $S \circ T = S \circ T_1$ where $T_1 : V \to W$, show that $T = T_1$.

**15.** Let $V \xrightarrow{T} W \xrightarrow{S} U$ be linear transformations.

**a)** If $S \circ T$ is one-to-one, show that $T$ is one-to-one.

**b)** If $S \circ T$ is onto, show that $S$ is onto.

16. Let $V \xrightarrow{T} W \xrightarrow{S} U$ be linear transformations such that $S \circ T = 1_V$. If $dimV = dimW$ is finite, show that also $T \circ S = 1_W$ (so $T$ is an isomorphism and $S = T^{-1}$).

17. Let $\mathbb{S}$ be the space of all sequences $(a_0, a_1, a_2, \cdots)$ of real numbers with componentwise operations (See Exercise 33 §5.2). Define $T : \mathbb{S} \to \mathbb{S}$ and $R : \mathbb{S} \to \mathbb{S}$ by $T(a_0, a_1, a_2, \cdots) = (a_1, a_2, a_3, \cdots)$ and $R(a_0, a_1, a_2, \cdots) = (0, a_0, a_1, a_2, a_3, \cdots)$.
    a) Show that $T$ and $R$ are linear transformations.
    b) Show that $T \circ R = 1_{\mathbb{S}}$ but $R \circ T \neq 1_{\mathbb{S}}$.
    c) Show that $T$ is onto but not one-to-one, and that $R$ is one-to-one but not onto.

18. Let $T : V \to W$ be a linear transformation where $dimV$ and $dimW$ are finite. If $T$ is one-to-one, show that there exists a linear transformation $S : W \to V$ such that $S \circ T = 1_V$. [Hint: If $\mathcal{B} = \{\mathbf{b}_1, \mathbf{b}_2, \cdots, \mathbf{b}_n\}$ is a basis of $V$, show that $\{T(\mathbf{b}_1), T(\mathbf{b}_2), \cdots, T(\mathbf{b}_n)\}$ is independent.]

19. Let $T : V \to W$ be a linear transformation where $dimV$ and $dimW$ are finite. If $T$ is onto, show that there exists a linear transformation $S : W \to V$ such that $T \circ S = 1_W$. [Hint: By Theorem 3 §5.2. find a basis $\{\mathbf{b}_1, \mathbf{b}_2, \cdots, \mathbf{b}_m, \mathbf{b}_{m+1}, \cdots, \mathbf{b}_n\}$ of $V$ such that $\{\mathbf{b}_{m+1}, \cdots, \mathbf{b}_n\}$ is a basis of $kerT$. Show that $\{T(\mathbf{b}_1), T(\mathbf{b}_2), \cdots, T(\mathbf{b}_m)\}$ is independent and (as $T$ is onto) spans $W$.]

20. If $U$ is a fixed $n \times n$ matrix, define $S_U : \mathbb{M}_{n,n} \to \mathbb{M}_{n,n}$ by $S_U(A) = UA$ for all $A$ in $\mathbb{M}_{n,n}$.
    a) Show that $S_U$ is linear.
    b) If $U$ is invertible, show that $S_U$ is an isomorphism.
    c) Is every linear transformation $\mathbb{M}_{n,n} \to \mathbb{M}_{n,n}$ equal to $S_U$ for some $U$? Defend your answer.

21. If $A$ is an $m \times n$ matrix, let $T_A : \mathbb{R}^n \to \mathbb{R}^m$ be defined by $T_A(X) = AX$ for all $X$ in $\mathbb{R}^n$.
    a) Show that $T_A \circ T_B = T_{AB}$ and $T_{I_n} = 1_{\mathbb{R}^n}$.
    b) If $A$ is invertible and $n \times n$, show that $T_A$ is invertible and $(T_A)^{-1} = T_{A^{-1}}$.
    c) If $T_A = T_B$, show that $A = B$.
    d) If $A$, $B$, and $C$ are three matrices, show that $A(BC) = (AB)C$ by considering $T_A \circ (T_B \circ T_C) = (T_A \circ T_B) \circ T_C$. [Hint: a) and c).]
    e) If $\mathcal{E}$ and $\mathcal{F}$ are the standard bases of $\mathbb{R}^n$ and $\mathbb{R}^m$ respectively, show that $M_{\mathcal{FE}}(T_A) = A$. [Hint: $C_{\mathcal{F}}(Y) = Y$ for every $Y$ in $\mathbb{R}^m$.]

22. Let $T : V \to W$ and $T_1 : V \to W$ be linear transformations. If $M_{\mathcal{DB}}(T) = M_{\mathcal{DB}}(T_1)$ for some bases $\mathcal{B}$ and $\mathcal{D}$ of $V$ and $W$, show that $T = T_1$.

23. Prove Theorem 6. [Hint: Mimic the proof of Theorem 3 §1.5.]

24. Let $T : V \to W$ be a linear transformation, and suppose that $S \circ T = 1_V$ and $T \circ S = 1_W$ for some transformation $S : W \to V$ (not assumed to be linear). Show that $S$ must be linear. [Hint: To show that $S(a\mathbf{w}) = a S(\mathbf{w})$, first compute $T[S(a\mathbf{w})]$ and $T[a S(\mathbf{w})]$.]

25. Let $m$ and $n$ be fixed integers. If $A$ and $B$ are $m \times n$ matrices, define $A \overset{r}{\sim} B$ to mean that $B = UA$ for some invertible $m \times m$ matrix $U$.
    a) Show that $\overset{r}{\sim}$ is an equivalence relation on the set $\mathbb{M}_{m,n}$ of all $m \times n$ matrices. (See the discussion following Theorem 3.)
    b) Show that $A \overset{r}{\sim} B$ if and only if $A \to B$ by row operations. [Hint: Section 1.6.]

## 5.5 LINEAR OPERATORS AND SIMILARITY

While the study of linear transformations from one vector space to another is important, the central problem of linear algebra is to understand the structure of an operator from a vector space to itself. If $T : V \to V$ is any linear operator where $dimV = n$, it is possible (see Example 14 §5.4) to choose bases $\mathcal{B}$ and $\mathcal{D}$ of $V$ such that the matrix of $T$ has a very simple form indeed: $M_{\mathcal{DB}}(T) = \begin{bmatrix} I_r & 0 \\ 0 & 0 \end{bmatrix}$ where $r = rankT$. Consequently only the rank of $T$ is revealed by from examining the simplest matrices $M_{\mathcal{DB}}(T)$ where $\mathcal{B}$ and $\mathcal{D}$ can be chosen arbitrarily. But if we insist that $\mathcal{B}$ and $\mathcal{D}$ are equal, the situation changes and our central problem amounts to finding a basis $\mathcal{B}$ such that $M_{\mathcal{BB}}(T)$ is as simple as possible. The complete solution to this task is beyond the scope of this book, but we outline the main ideas in the following two sections.

## The $\mathcal{B}$-Matrix of an Operator

Let $T : V \to V$ be a linear operator where $dim\, V = n$. If $\mathcal{B}$ is a basis of $V$, we write

$$M_{\mathcal{B}}(T) = M_{\mathcal{B}\mathcal{B}}(T)$$

for simplicity, and call this the $\mathcal{B}$-**matrix** of of $T$. Hence if $\mathcal{B}$ is listed (in a fixed order) as $\mathcal{B} = \{\mathbf{b}_1,\ \mathbf{b}_2,\ \cdots,\ \mathbf{b}_n\}$ then

$$M_{\mathcal{B}}(T) = [C_{\mathcal{B}}(T(\mathbf{b}_1))\quad C_{\mathcal{B}}(T(\mathbf{b}_2))\quad \cdots\quad C_{\mathcal{B}}(T(\mathbf{b}_n))]$$

as in Section 5.4, where $C_{\mathcal{B}}(\mathbf{v})$ denotes the coordinate vector of $\mathbf{v}$ with respect to the basis $\mathcal{B}$. Note that this agrees with the notation in Theorem 7 §4.9 when $V = \mathbb{R}^n$.

**Example 1**　If $w = a + bi$ is a fixed complex number, define $T : \mathbb{C} \to \mathbb{C}$ by $T(z) = wz$ for all $z$ in $\mathbb{C}$. Then $T$ is linear (verify) and, if $\mathcal{B} = \{1,\ i\}$, we have

$$M_{\mathcal{B}}(T) = [C_{\mathcal{B}}(w)\quad C_{\mathcal{B}}(wi)] = \begin{bmatrix} a & -b \\ b & a \end{bmatrix}.$$

If $w \neq 0$, then $|w| = \sqrt{a^2 + b^2} \neq 0$ and it is instructive to write this as

$$M_{\mathcal{B}}(T) = \sqrt{a^2 + b^2}\begin{bmatrix} \frac{a}{\sqrt{a^2+b^2}} & \frac{-b}{\sqrt{a^2+b^2}} \\ \frac{b}{\sqrt{a^2+b^2}} & \frac{a}{\sqrt{a^2+b^2}} \end{bmatrix} = |w|\begin{bmatrix} \cos\theta & -\sin\theta \\ \sin\theta & \cos\theta \end{bmatrix}$$

where $\theta$ is the argument of $w$. Hence, if we identify the complex number $z = x + iy$ with the point $\begin{bmatrix} x \\ y \end{bmatrix}$, multiplication by $w$ has the effect of a rotation through the angle $\theta$ followed by expansion by $|w|$. This is depicted in the diagram.

The following theorem collects several results from Section 5.4 for reference (specialized to operators).

**THEOREM ①**　Let $T : V \to V$ be a linear operator where $dim\, V = n$. Then:

(1) $C_{\mathcal{B}}(T(\mathbf{v})) = M_{\mathcal{B}}(T)\, C_{\mathcal{B}}(\mathbf{v})$ for all $\mathbf{v}$ in $V$.

(2) If $S : V \to V$ is another operator, $M_{\mathcal{B}}(S \circ T) = M_{\mathcal{B}}(S)\, M_{\mathcal{B}}(T)$.

(3) $T$ is an isomorphism if and only if $M_{\mathcal{B}}(T)$ is invertible for some (equivalently every) basis $\mathcal{B}$ of $V$.

(4) If $T$ is an isomorphism, then $M_{\mathcal{B}}(T^{-1}) = [M_{\mathcal{B}}(T)]^{-1}$.

**Example 2**　If $A$ is an $n \times n$ matrix, define $T_A : \mathbb{R}^n \to \mathbb{R}^n$ by $T_A(X) = AX$ for all $X$ in $\mathbb{R}^n$. If $\mathcal{E}$ is the standard basis of $\mathbb{R}^n$, show that

$$M_{\mathcal{E}}(T_A) = A.$$

**SOLUTION**　We have $\mathcal{E} = \{E_1,\ E_2,\ \cdots,\ E_n\}$ where the $E_j$ are the columns of $I_n$. Since $C_{\mathcal{E}}(X) = X$ for all $X$ in $\mathbb{R}^n$, we have

$$
\begin{aligned}
M_{\mathcal{E}}(T_A) &= [C_{\mathcal{E}}(AE_1)\quad C_{\mathcal{E}}(AE_2)\quad \cdots\quad C_{\mathcal{E}}(AE_n)] \\
&= [AE_1\quad AE_2\quad \cdots\quad AE_n] \\
&= A[E_1\quad E_2\quad \cdots\quad E_n] \\
&= AI_n = A
\end{aligned}
$$

as required.

**Example 3**  Let $D : \mathbb{P}_2 \to \mathbb{P}_2$ be the differentiation operator defined by $D(p(x)) = p'(x)$ for all polynomials $p(x)$ in $\mathbb{P}_2$. If $\mathcal{B} = \{1, \ x, \ x^2\}$, we have

$$
\begin{aligned}
M_\mathcal{B}(D) &= [C_\mathcal{B}(D(1)) \ \ C_\mathcal{B}(D(x)) \ \ C_\mathcal{B}(D(x^2))] \\
&= [C_\mathcal{B}(0) \ \ C_\mathcal{B}(1) \ \ C_\mathcal{B}(2x)] \\
&= \begin{bmatrix} 0 & 1 & 0 \\ 0 & 0 & 2 \\ 0 & 0 & 0 \end{bmatrix}.
\end{aligned}
$$

Hence, if $p(x) = a + bx + cx^2$, we have

$$
C_\mathcal{B}\{D(p(x))\} = M_\mathcal{B}(D) \, C_\mathcal{B}(p(x)) = \begin{bmatrix} 0 & 1 & 0 \\ 0 & 0 & 2 \\ 0 & 0 & 0 \end{bmatrix} \begin{bmatrix} a \\ b \\ c \end{bmatrix} = \begin{bmatrix} b \\ 2c \\ 0 \end{bmatrix}.
$$

Thus Theorem 1 gives $D(p(x)) = b + (2c)x$, as it should. Note that $M_\mathcal{B}(D)$ is not invertible here, so $D$ is not invertible. Of course this can also be seen from the fact that $ker D \neq 0$ (for example, $D(1) = 0$).

**Example 4**  Let $p_0(x), \ p_1(x), \ \cdots, \ p_n(x)$ be polynomials in $\mathbb{P}_n$ with $deg(p_k(x)) = k$ for each $k$. If $T : \mathbb{P}_n \to \mathbb{P}_n$ is defined by $T(x^k) = p_k(x)$ for each $k$, find $M_\mathcal{B}(T)$ where $\mathcal{B} = \{1, \ x, \ \cdots, \ x^n\}$, and conclude that $\mathcal{D} = \{p_0(x), \ p_1(x), \ \cdots, \ p_n(x)\}$ is a basis of $\mathbb{P}_n$. (See Example 2 §5.2.)

SOLUTION  The highest power of $x$ occurring in $p_k(x)$ is $x^k$ (because $deg(p_k(x)) = k$). Hence $M_\mathcal{B}(T)$ has the form

$$
\begin{aligned}
M_\mathcal{B}(T) &= [C_\mathcal{B}(p_0(x)) \ \ C_\mathcal{B}(p_1(x)) \ \ C_\mathcal{B}(p_2(x)) \ \cdots \ C_\mathcal{B}(p_n(x))] \\
&= \begin{bmatrix} a_0 & * & * & \cdots & * \\ 0 & a_1 & * & \cdots & * \\ 0 & 0 & a_2 & \cdots & * \\ \vdots & \vdots & \vdots & & \vdots \\ 0 & 0 & 0 & \cdots & a_n \end{bmatrix}
\end{aligned}
$$

where $a_k \neq 0$ is the leading coefficient of $p_k(x)$ for each $k$. It follows that $M_\mathcal{B}(T)$ is invertible (its determinant is $a_0 a_1 a_2 \ \cdots \ a_n \neq 0$), so $T$ is an isomorphism by Theorem 1. Hence $\mathcal{D}$ is a basis of $\mathbb{P}_n$ by Theorem 2 §5.4.

The next example shows how a property of matrices can be deduced by first verifying the same property of operators, and then using Theorem 1 and Example 2. This technique is important because properties of operators are often easier to visualize.

**Example 5**  Let $T : V \to V$ be an operator on the $n$-dimensional space $V$.

(1) Show that an isomorphism $S : V \to V$ exists such that $T \circ S \circ T = T$.

(2) If $A$ is an $n \times n$ matrix, show that $AUA = A$ for some invertible matrix $U$.

SOLUTION  (1) By extending a basis of $ker T$, choose a basis $\mathcal{B} = \{\mathbf{b}_1, \ \mathbf{b}_2, \ \cdots, \ \mathbf{b}_r, \ \mathbf{b}_{r+1}, \ \cdots, \ \mathbf{b}_n\}$ of $V$ where $\{\mathbf{b}_{r+1}, \ \cdots, \ \mathbf{b}_n\}$ is a basis of $ker T$. Then $\{T(\mathbf{b}_1), \ T(\mathbf{b}_2), \ \cdots, \ T(\mathbf{b}_r)\}$ is independent (verify) so extend it to a basis $\mathcal{D} = \{T(\mathbf{b}_1), \ T(\mathbf{b}_2), \ \cdots, \ T(\mathbf{b}_r), \ \mathbf{d}_{r+1}, \ \cdots, \ \mathbf{d}_n\}$ of $V$. We use this basis $\mathcal{D}$ to define $S : V \to V$ as follows:

$$
S(T(\mathbf{b}_i)) = \mathbf{b}_i \text{ for } 1 \le i \le r \quad \text{and} \quad S(\mathbf{d}_j) = \mathbf{b}_j \text{ for } r < j \le n.
$$

Then $S$ is an isomorphism because it carries the basis $\mathcal{D}$ to the basis $\mathcal{B}$. To see that $T \circ S \circ T = T$, we verify that both $T \circ S \circ T$ and $T$ have the same effect on every vector in the basis $\mathcal{B}$. In fact:

$$(T \circ S \circ T)(\mathbf{b}_i) = T(S(T(\mathbf{b}_i))) = \begin{cases} T(\mathbf{b}_i) & \text{for } i \le r \text{ by the definition of } S, \\ T(\mathbf{b}_i) & \text{for } i > r \text{ because } T(\mathbf{b}_i) = \mathbf{0}. \end{cases}$$

(2) Given $A$, define $T = T_A : \mathbb{R}^n \to \mathbb{R}^n$ by $T_A(X) = AX$ for all $X$ in $\mathbb{R}^n$ (as in Example 2), and construct an isomorphism $S : \mathbb{R}^n \to \mathbb{R}^n$ such that $T_A \circ S \circ T_A = T_A$ as in (1). If $\mathcal{E}$ is the standard basis of $\mathbb{R}^n$, then $M_{\mathcal{E}}(T_A) = A$ by Example 2, and $U = M_{\mathcal{E}}(S)$ is invertible by Theorem 1. Finally

$$AUA = M_{\mathcal{E}}(T_A)\, M_{\mathcal{E}}(S)\, M_{\mathcal{E}}(T_A) = M_{\mathcal{E}}(T_A \circ S \circ T_A) = M_{\mathcal{E}}(T_A) = A$$

again by Theorem 1. This is what we wanted.

## 5.5.2 Change of Basis

If $T : V \to V$ is an operator, it is critical to understand how the matrix $M_{\mathcal{B}}(T)$ changes when the basis $\mathcal{B}$ changes, and it turns out that this is closely related to how the coordinates $C_{\mathcal{B}}(\mathbf{v})$ change for a vector $\mathbf{v}$. Recall (Theorem 8 §5.4) that, for all bases $\mathcal{B}$ and $\mathcal{D}$ of $V$, we have

$$C_{\mathcal{D}}[T(\mathbf{v})] = M_{\mathcal{D}\mathcal{B}}(T)C_{\mathcal{B}}(\mathbf{v}) \quad \text{for all } \mathbf{v} \text{ in } V.$$

If we take $T = 1_V$ this reads

$$C_{\mathcal{D}}(\mathbf{v}) = M_{\mathcal{D}\mathcal{B}}(1_V)C_{\mathcal{B}}(\mathbf{v}) \quad \text{for all } \mathbf{v} \text{ in } V.$$

With this in mind, define the **change matrix** $P_{\mathcal{D}\leftarrow\mathcal{B}}$ by

$$P_{\mathcal{D}\leftarrow\mathcal{B}} = M_{\mathcal{D}\mathcal{B}}(1_V)$$

for any bases $\mathcal{B}$ and $\mathcal{D}$ of $V$. This proves $(**)$ in the following theorem:

**THEOREM ②**   Let $\mathcal{B}$ and $\mathcal{D}$ denote bases of a vector space $V$. If $B = \{\mathbf{b}_1, \mathbf{b}_2, \cdots, \mathbf{b}_n\}$, the change matrix $P_{\mathcal{D}\leftarrow\mathcal{B}}$ is given in terms of its columns by

$$P_{\mathcal{D}\leftarrow\mathcal{B}} = [C_{\mathcal{D}}(\mathbf{b}_1) \quad C_{\mathcal{D}}(\mathbf{b}_2) \quad \cdots \quad C_{\mathcal{D}}(\mathbf{b}_n)], \tag{$*$}$$

and has the property that

$$C_{\mathcal{D}}(\mathbf{v}) = P_{\mathcal{D}\leftarrow\mathcal{B}}C_{\mathcal{B}}(\mathbf{v}) \quad \text{for all } \mathbf{v} \text{ in } V. \tag{$**$}$$

Moreover, if $\mathcal{F}$ is another basis of $V$, we have

(1) $P_{\mathcal{B}\leftarrow\mathcal{B}} = I_n$.
(2) $P_{\mathcal{D}\leftarrow\mathcal{B}}$ is invertible and $(P_{\mathcal{D}\leftarrow\mathcal{B}})^{-1} = P_{\mathcal{B}\leftarrow\mathcal{D}}$.
(3) $P_{\mathcal{F}\leftarrow\mathcal{D}}P_{\mathcal{D}\leftarrow\mathcal{B}} = P_{\mathcal{F}\leftarrow\mathcal{B}}$.

**PROOF**   The formula $(*)$ is immediate from the definition of $P_{\mathcal{D}\leftarrow\mathcal{B}}$ and Theorem 8 §5.4, and $(**)$ is derived above. Condition (1) is a routine verification, and (2) follows from (1) and (3). Finally, (3) follows by right multiplying $P_{\mathcal{F}\leftarrow\mathcal{D}}P_{\mathcal{D}\leftarrow\mathcal{B}}$ by $C_{\mathcal{B}}(\mathbf{v})$, applying $(**)$ three times, and using Lemma 1 §5.4.

Property (3) in Theorem 2 explains the notation $P_{\mathcal{D}\leftarrow\mathcal{B}}$.

**Example 6** If $\mathcal{B} = \{1,\ x,\ x^2\}$ and $\mathcal{D} = \{1+x,\ x+x^2,\ 1+x^2\}$ in $\mathbb{P}_2$, we have

$$P_{\mathcal{D}\leftarrow\mathcal{B}} = [C_{\mathcal{D}}(1)\ \ C_{\mathcal{D}}(x)\ \ C_{\mathcal{D}}(x^2)] = \tfrac{1}{2}\begin{bmatrix} 1 & 1 & -1 \\ -1 & 1 & 1 \\ 1 & -1 & 1 \end{bmatrix}.$$

(For example, $1 = \tfrac{1}{2}\{(1+x) - (x+x^2) + (1+x^2)\}$ gives the first column.)
Hence Theorem 1 gives

$$C_{\mathcal{D}}(a + bx + cx^2) = P_{\mathcal{D}\leftarrow\mathcal{B}}C_{\mathcal{B}}(a + bx + cx^2) = \tfrac{1}{2}\begin{bmatrix} 1 & 1 & -1 \\ -1 & 1 & 1 \\ 1 & -1 & 1 \end{bmatrix}\begin{bmatrix} a \\ b \\ c \end{bmatrix} = \tfrac{1}{2}\begin{bmatrix} a+b-c \\ -a+b+c \\ a-b+c \end{bmatrix}$$

for all $a$, $b$, and $c$ in $\mathbb{R}$. As a check, this gives the expansion of $a + bx + cx^2$ in terms of the basis $\mathcal{D}$:

$$a + bx + cx^2 = \tfrac{1}{2}(a + b - c)(1 + x) + \tfrac{1}{2}(-a + b + c)(x + x^2) + \tfrac{1}{2}(a - b + c)(1 + x^2).$$

The reader can verify that this is indeed the case.

We now show that these change matrices are also closely related to the way the matrix of the operator $T$ changes when the basis varies. More precisely, we claim that

$$P_{\mathcal{B}\leftarrow\mathcal{D}}\, M_{\mathcal{D}}(T) = M_{\mathcal{B}}(T)\, P_{\mathcal{B}\leftarrow\mathcal{D}} \qquad\qquad (\ast\ast\ast)$$

where $\mathcal{B}$ and $\mathcal{D}$ are bases. To see this, right multiply the left side by $C_{\mathcal{D}}(\mathbf{v})$ and use Theorems 1 and 2:

$$\begin{aligned} P_{\mathcal{B}\leftarrow\mathcal{D}}\, M_{\mathcal{D}}(T)\, C_{\mathcal{D}}(\mathbf{v}) &= P_{\mathcal{B}\leftarrow\mathcal{D}}\, C_{\mathcal{D}}[T(\mathbf{v})] \\ &= C_{\mathcal{B}}[T(\mathbf{v})] \\ &= M_{\mathcal{B}}(T)\, C_{\mathcal{B}}(\mathbf{v}) \\ &= M_{\mathcal{B}}(T)\, P_{\mathcal{B}\leftarrow\mathcal{D}}\, C_{\mathcal{D}}(\mathbf{v}). \end{aligned}$$

Since this holds for all vectors $\mathbf{v}$, $(\ast\ast\ast)$ follows because $C_{\mathcal{D}}(\mathbf{v})$ can be "cancelled" by Lemma 1 §5.4. Writing $P = P_{\mathcal{B}\leftarrow\mathcal{D}}$, this proves

**THEOREM 3** Let $T : V \to V$ be a linear operator where $dimV = n$. If $\mathcal{B}$ and $\mathcal{D}$ are two ordered bases of $V$, the matrices

$$M_{\mathcal{B}}(T) \quad \text{and} \quad M_{\mathcal{D}}(T) \quad \text{are similar}.$$

More precisely, if $P = P_{\mathcal{B}\leftarrow\mathcal{D}}$, then

$$M_{\mathcal{D}}(T) = P^{-1}M_{\mathcal{B}}(T)P.$$

Note that $(P_{\mathcal{B}\leftarrow\mathcal{D}})^{-1} = P_{\mathcal{D}\leftarrow\mathcal{B}}$ by Theorem 2. Hence the formula in Theorem 3 takes the easily remembered form $M_{\mathcal{D}}(T) = P_{\mathcal{D}\leftarrow\mathcal{B}}\, M_{\mathcal{B}}(T)\, P_{\mathcal{B}\leftarrow\mathcal{D}}$.

**Example 7** Define $T : \mathbb{P}_1 \to \mathbb{P}_1$ by $T(a + bx) = (a + b) - (2a + b)x$. If $\mathcal{B} = \{1,\ x\}$, then

$$M_{\mathcal{B}}(T) = [C_{\mathcal{B}}(T(1))\ \ C_{\mathcal{B}}(T(x))] = [C_{\mathcal{B}}(1 - 2x)\ \ C_{\mathcal{B}}(1 - x)] = \begin{bmatrix} 1 & 1 \\ -2 & -1 \end{bmatrix}.$$

Now let $\mathcal{D} = \{x - 1,\ x\}$ and compute

$$M_{\mathcal{D}}(T) = [C_{\mathcal{D}}(T(x - 1))\ \ C_{\mathcal{D}}(T(x))] = [C_{\mathcal{D}}(x)\ \ C_{\mathcal{D}}(1 - x)] = \begin{bmatrix} 0 & -1 \\ 1 & 0 \end{bmatrix}.$$

Thus the matrices $M_{\mathcal{B}}(T)$ and $M_{\mathcal{D}}(T)$ are similar by Theorem 3. In fact, we compute

$$P = P_{\mathcal{B}\leftarrow\mathcal{D}} = [C_{\mathcal{B}}(x - 1)\ \ C_{\mathcal{B}}(x)] = \begin{bmatrix} -1 & 0 \\ 1 & 1 \end{bmatrix},$$

so Theorem 3 gives $P^{-1}M_{\mathcal{B}}(T)P = M_{\mathcal{D}}(T)$ as the reader can verify.

Of course we knew in Theorem 8 §4.9 that *every* pair of similar $n \times n$ matrices $A$ and $B$ arises as in Theorem 3 as the matrices of some operator $T : \mathbb{R}^n \to \mathbb{R}^n$ with respect to two bases of $\mathbb{R}^n$. We restate the result here for reference.

**LEMMA 1**  If $A$ and $B$ are similar $n \times n$ matrices, say $B = P^{-1}AP$, then
$$A = M_\mathcal{E}(T_A) \quad \text{and} \quad B = M_\mathcal{F}(T_A)$$
where $T_A : \mathbb{R}^n \to \mathbb{R}^n$ is defined by $T_A(X) = AX$ for all $X$ in $\mathbb{R}^n$, $\mathcal{E}$ is the standard basis of $\mathbb{R}^n$, and $\mathcal{F} = \{F_1, F_2, \cdots, F_n\}$ consists of the columns of $P$.

**PROOF**  We have $A = M_\mathcal{E}(T_A)$ by Example 2, and $B = M_\mathcal{F}(T_A)$ by Lemma 5 §4.9.

We saw in Theorem 11 §5.4 that the rank of every matrix of a linear operator $T$ equals $dim(imT)$, and we called this number the **rank** of $T$, denoting it as $rankT$. Hence we have
$$rankT = rank(M_\mathcal{B}(T)) \quad \text{for every basis } \mathcal{B} \text{ of } V.$$
The fact that this is independent of the choice of basis can also be seen from Theorem 3 because similar matrices have the same rank (Corollary 4 of Theorem 2 §4.4). When viewed in this way, it is clear that Theorem 3 allows us to define the determinant, the trace, and the characteristic polynomial of any linear operator in the same way. The reason is that similar matrices also have the same determinant, the same trace, and the same characteristic polynomial (Theorem 6 §2.3 and Lemma 3 §4.7).

Thus, for example, Theorem 3 shows that if $T : V \to V$ is a linear operator (where $dimV$ is finite), then $det[M_\mathcal{B}(T)] = det[M_\mathcal{D}(T)]$ for all bases $\mathcal{B}$ and $\mathcal{D}$ of $V$. We call this common value the determinant of $T$; more formally, we define the **determinant** $detT$ of an operator $T : V \to V$ by
$$detT = det[M_\mathcal{B}(T)] \quad \text{for } any \text{ basis } \mathcal{B} \text{ of } V.$$
This makes sense because it is independent of the choice of the basis $\mathcal{B}$ and so depends only on the operator $T$. In the same way, similar matrices have the same trace and the same characteristic polynomial. Hence we can define the **trace** $trT$ of $T$ and the **characteristic polynomial** $c_T(x)$ of $T$ by
$$trT = tr[M_\mathcal{B}(T)] \quad \text{for any basis } \mathcal{B} \text{ of } V.$$
$$c_T(x) = c_{M_\mathcal{B}(T)}(x) \quad \text{for any basis } \mathcal{B} \text{ of } V.$$
In this way, properties of determinants, traces, and characteristic polynomials that are proved for matrices actually hold for operators. For example, we have $det(AB) = detA\,detB$ and $tr(AB) = tr(BA)$ for all square matrices $A$ and $B$. These equations translate to the fact that
$$det(S \circ T) = detS\,detT \quad \text{and} \quad tr(S \circ T) = tr(T \circ S)$$
hold for any operators $S$ and $T$ on the same space $V$. The verifications of these properties (and others) are left as exercises for the reader.

### 5.5.3  Diagonalization

Possibly the most important application of the concept of similarity is its use when discussing diagonalization. Recall that an $n \times n$ matrix $A$ is said to be *diagonalizable* if it is similar to a diagonal matrix $D$. By Theorem 3 this happens if there exists an operator $T : V \to V$ and bases $\mathcal{B}$ and $\mathcal{D}$ of $V$ such that $A = M_\mathcal{B}(T)$ and $D = M_\mathcal{D}(T)$. This, in turn, focuses our attention on how to find a basis $\mathcal{D}$ of $V$ for which $M_\mathcal{D}(T)$ is diagonal.

To this end, list the basis $\mathcal{D}$ as $\mathcal{D} = \{\mathbf{d}_1, \mathbf{d}_2, \cdots, \mathbf{d}_n\}$ and assume that $M_{\mathcal{D}}(T) = diag(\lambda_1, \lambda_2, \cdots, \lambda_n)$ where the $\lambda_i$ are real numbers. Then

$$[C_{\mathcal{D}}(T(\mathbf{d}_1)) \quad C_{\mathcal{D}}(T(\mathbf{d}_2)) \quad \cdots \quad C_{\mathcal{D}}(T(\mathbf{d}_n))] = M_{\mathcal{D}}(T) = diag(\lambda_1, \lambda_2, \cdots, \lambda_n).$$

Comparing columns shows that $C_{\mathcal{D}}(T(\mathbf{d}_i)) = [0 \ \cdots \ \lambda_i \ \cdots \ 0]^T$, and hence that

$$T(\mathbf{d}_i) = \lambda_i \mathbf{d}_i \quad \text{for each } i = 1, 2, \cdots, n.$$

This prompts the following definitions: As for matrices, we say that a number $\lambda$ is an **eigenvalue** of the operator $T$ if

$$T(\mathbf{v}) = \lambda \mathbf{v} \quad \text{for some vector } \mathbf{v} \neq \mathbf{0},$$

and in this case we say that $\mathbf{v}$ is an **eigenvector** of $T$ corresponding to $\lambda$. Finally, the subspace

$$E_\lambda(T) = \{\mathbf{v} \text{ in } V \mid T(\mathbf{v}) = \lambda \mathbf{v}\}$$

of $V$ is called the **eigenspace** of $T$ corresponding to $\lambda$. All this terminology is consistent with our earlier usage for matrices because, as the reader can verify, the eigenvalues and eigenvectors of an $n \times n$ matrix $A$ are the same as the eigenvalues and eigenvectors of the operator $T_A : \mathbb{R}^n \to \mathbb{R}^n$, where $T_A(X) = AX$ for all $X$ in $\mathbb{R}^n$.

We can now characterize when the $\mathcal{D}$-matrix of an operator is diagonal.

**THEOREM 4**

The following conditions are equivalent for an operator $T : V \to V$ where $dim V = n$:

(1) $M_{\mathcal{D}}(T)$ is diagonal for some basis $\mathcal{D}$.

(2) There exists a basis of $V$ consisting of eigenvectors of $T$.

Moreover, if $\mathcal{D} = \{\mathbf{d}_1, \mathbf{d}_2, \cdots, \mathbf{d}_n\}$ is a basis of eigenvectors and $T(\mathbf{d}_i) = \lambda_i \mathbf{d}_i$ for each $i$, then

$$M_{\mathcal{D}}(T) = diag(\lambda_1, \lambda_2, \cdots, \lambda_n).$$

**PROOF**

The preceding discussion proves that $(1) \Rightarrow (2)$. Conversely, if $\mathcal{D}$ and the $\lambda_i$ are as in the theorem, we compute $M_{\mathcal{D}}(T)$ in terms of its columns:

$$M_{\mathcal{D}}(T) = [C_{\mathcal{D}}(T(\mathbf{d}_1)) \quad C_{\mathcal{D}}(T(\mathbf{d}_2)) \quad \cdots \quad C_{\mathcal{D}}(T(\mathbf{d}_n))]$$
$$= [C_{\mathcal{D}}(\lambda_1 \mathbf{d}_1) \quad C_{\mathcal{D}}(\lambda_2 \mathbf{d}_2) \quad \cdots \quad C_{\mathcal{D}}(\lambda_n \mathbf{d}_n)]$$
$$= diag(\lambda_1, \lambda_2, \cdots, \lambda_n).$$

This proves that $(2) \Rightarrow (1)$, and also proves the last sentence in the theorem.

A linear operator $T : V \to V$ is called **diagonalizable** if it satisfies the conditions in Theorem 4.

**Example 8**

Let $R_\theta : \mathbb{R}^2 \to \mathbb{R}^2$ denote rotation about the origin through the angle $\theta \neq 0$, (a linear operator by Theorem 3 §3.4). Show geometrically and algebraically that $R_\theta$ is not diagonalizable.

**SOLUTION**

If $\vec{v} \neq \vec{0}$ is a vector in $\mathbb{R}^2$, then $R_\theta(\vec{v})$ is the result of rotating $\vec{v}$ through the angle $\theta \neq 0$, and so cannot equal $\lambda \vec{v}$ for any real number $\lambda$ (see the diagram). Hence $R_\theta$ has no eigenvectors, and so is not diagonalizable.

For an algebraic argument, recall that the matrix of $R_\theta$ is $\begin{bmatrix} \cos\theta & -\sin\theta \\ \sin\theta & \cos\theta \end{bmatrix}$ with respect to the standard basis (Theorem 3 §3.4), so the characteristic polynomial is $c_{R_\theta}(x) = x^2 - 2\cos\theta\, x + 1$ with roots $\cos\theta + i\sin\theta$. This means that $R_\theta$ has no real eigenvalues, and so is not diagonalizable.

Before characterizing the diagonalizable operators, we need the next result which reveals the close connection between the eigenvalues, eigenvectors, and eigenspaces of $T$ and those of each of its matrices.

**LEMMA 2**

Let $T : V \rightarrow V$ be a linear operator where $dim V = n$, let $\mathcal{B}$ be a basis of $V$, and let $\lambda$ be a real number. Then:

(1) $C_{\mathcal{B}} : E_{\lambda}(T) \rightarrow E_{\lambda}(M_{\mathcal{B}}(T))$ is an isomorphism.
(2) The eigenvalues $\lambda$ of $T$ are precisely the eigenvalues of the matrix $M_{\mathcal{B}}(T)$.
    Moreover, if $\mathbf{v}$ is in $V$ and we write $\mathbf{v} = C_{\mathcal{B}}(X)$ where $X$ is in $\mathbb{R}^n$, then
    $\mathbf{v}$ is in $E_{\lambda}(T)$    if and only if    $X$ is in $E_{\lambda}(M_{\mathcal{B}}(T))$.

**PROOF**

For convenience, write $A = M_{\mathcal{B}}(T)$, so that $C_{\mathcal{B}}(T(\mathbf{v})) = AC_{\mathcal{B}}(\mathbf{v})$ for all $\mathbf{v}$ in $V$. If $T(\mathbf{v}) = \lambda\mathbf{v}$ where $\mathbf{v} \neq \mathbf{0}$, then the linearity of $C_{\mathcal{B}}$ gives

$$AC_{\mathcal{B}}(\mathbf{v}) = C_{\mathcal{B}}(T(\mathbf{v})) = C_{\mathcal{B}}(\lambda\mathbf{v}) = \lambda C_{\mathcal{B}}(\mathbf{v}),$$

so $\lambda$ is an eigenvalue of $A$ and $C_{\mathcal{B}}(\mathbf{v})$ is a corresponding eigenvector. This shows that every eigenvalue $\lambda$ of $T$ is also an eigenvalue of $A$, and also that $C_{\mathcal{B}} : E_{\lambda}(T) \rightarrow E_{\lambda}(A)$ is a linear transformation (one-to-one). On the other hand, if $AX = \lambda X$ where $X \neq 0$, write $X = C_{\mathcal{B}}(\mathbf{v})$ where $\mathbf{v} \neq \mathbf{0}$ is in $V$. Then

$$C_{\mathcal{B}}(T(\mathbf{v})) = AC_{\mathcal{B}}(\mathbf{v}) = AX = \lambda X = \lambda C_{\mathcal{B}}(\mathbf{v}) = C_{\mathcal{B}}(\lambda\mathbf{v}).$$

Hence $T(\mathbf{v}) = \lambda\mathbf{v}$ because $C_{\mathcal{B}}$ is one-to-one. This shows that every eigenvalue $\lambda$ of $A$ is also an eigenvalue of $T$, and that $C_{\mathcal{B}} : E_{\lambda}(T) \rightarrow E_{\lambda}(A)$ is onto. This proves all our assertions.

**Example 9**

Let $\mathcal{B} = \{\mathbf{v}, \ \mathbf{w}\}$ be a basis of $V$, and define $T : V \rightarrow V$ by $T(\mathbf{v}) = 2\mathbf{v}$ and $T(\mathbf{w}) = -3\mathbf{v} + 5\mathbf{w}$. Find the matrix $M_{\mathcal{B}}(T)$ and use it to find the eigenvectors of $T$.

**SOLUTION**

We have $M_{\mathcal{B}}(T) = [C_{\mathcal{B}}(T(\mathbf{v})) \ \ C_{\mathcal{B}}(T(\mathbf{w}))] = \begin{bmatrix} 2 & -3 \\ 0 & 5 \end{bmatrix}$. Hence the eigenvalues of $M_{\mathcal{B}}(T)$ are $\lambda_1 = 2$ and $\lambda_2 = 5$, with corresponding eigenvectors $X_1 = \begin{bmatrix} 1 \\ 0 \end{bmatrix}$ and $X_2 = \begin{bmatrix} -1 \\ 1 \end{bmatrix}$. Since $X_1 = C_{\mathcal{B}}(\mathbf{v})$ and $X_2 = C_{\mathcal{B}}(\mathbf{w} - \mathbf{v})$, it follows by Lemma 2 that $\mathbf{v}_1 = \mathbf{v}$ is an eigenvector of $T$ corresponding to $\lambda_1$ (as is clear from the definition of $T$) and $\mathbf{v}_2 = \mathbf{w} - \mathbf{v}$ is an eigenvector of $T$ corresponding to $\lambda_2$. Of course, we could have discovered these eigenvectors $\mathbf{v}_1$ and $\mathbf{v}_2$ for $T$ directly, and then used them to find $X_1$ and $X_2$ using Lemma 2.

Lemma 2 enables us to characterize when an operator is diagonalizable in terms of its matrices. The result is not surprising.

**THEOREM 5**

If $T : V \rightarrow V$ is a linear operator where $dim V = n$, the following are equivalent:

(1) $T$ is diagonalizable.
(2) $M_{\mathcal{B}}(T)$ is a diagonalizable matrix for every basis $\mathcal{B}$ of $V$.
(3) $M_{\mathcal{B}}(T)$ is a diagonalizable matrix for some basis $\mathcal{B}$ of $V$.

Moreover, we can pass back and forth between the eigenvector bases of $V$ for $T$ and those of $\mathbb{R}^n$ for $M_{\mathcal{B}}(T)$ using the coordinate isomorphism $C_{\mathcal{B}} : V \rightarrow \mathbb{R}^n$.

**PROOF**

$(1) \Rightarrow (2)$. If $T$ is diagonalizable, there exists a basis $\mathcal{D}$ of $V$ such that $M_{\mathcal{D}}(T)$ is diagonal. Since $M_{\mathcal{B}}(T)$ is similar to $M_{\mathcal{D}}(T)$ by Theorem 3, it follows that $A$ is diagonalizable.

$(2) \Rightarrow (3)$. This is clear.

$(3) \Rightarrow (1)$. Given (3), $M_\mathcal{B}(T)$ is a diagonalizable matrix, so let $\{X_1, X_2, \cdots, X_n\}$ be a basis of $\mathbb{R}^n$ consisting of eigenvectors of $M_\mathcal{B}(T)$. If $X_j = C_\mathcal{B}(\mathbf{b}_j)$ for each $j$, then $\mathcal{D} = \{\mathbf{b}_1, \mathbf{b}_2, \cdots, \mathbf{b}_n\}$ is a basis of $V$ (because $C_\mathcal{B}^{-1}$ is an isomorphism and $\mathbf{b}_j = C_\mathcal{B}^{-1}(X_j)$ for each $j$). Since $\mathcal{D}$ consists of eigenvectors of $T$ by Lemma 2, this proves (1).

Finally, the last sentence follows from Lemma 2.

**Example 10**   Let $T : \mathbb{M}_{2,2} \to \mathbb{M}_{2,2}$ be defined by $T(A) = A^T$ for each matrix $A$ in $\mathbb{M}_{2,2}$. Determine if $T$ is diagonalizable and, if so, find a basis of $\mathbb{M}_{2,2}$ consisting of eigenvectors of $T$.

**SOLUTION**   Consider the standard basis $\mathcal{B} = \{E_{11}, E_{12}, E_{21}, E_{22}\}$ of $\mathbb{M}_{2,2}$ where $E_{ij}$ is the matrix unit with 1 in the $(i, j)$-position and zeros elsewhere. Then

$$T(E_{11}) = E_{11}, \ T(E_{12}) = E_{21}, \ T(E_{21}) = E_{12}, \text{ and } T(E_{22}) = E_{22}$$

as is easily verified. Hence

$$M_\mathcal{B}(T) = \begin{bmatrix} 1 & 0 & 0 & 0 \\ 0 & 0 & 1 & 0 \\ 0 & 1 & 0 & 0 \\ 0 & 0 & 0 & 1 \end{bmatrix}$$

so the characteristic polynomial of $T$ is $c_T(x) = c_{M_\mathcal{B}(T)}(x) = (x-1)^3(x+1)$. Thus $M_\mathcal{B}(T)$ (and hence $T$) has eigenvalues $\lambda_1 = 1$ of multiplicity 3 and $\lambda_2 = -1$ of multiplicity 1. Moreover, corresponding eigenvectors of $M_\mathcal{B}(T)$ are

$$X_1 = \begin{bmatrix} 1 \\ 0 \\ 0 \\ 0 \end{bmatrix}, \ Y_1 = \begin{bmatrix} 0 \\ 1 \\ 1 \\ 0 \end{bmatrix}, \text{ and } Z_1 = \begin{bmatrix} 0 \\ 0 \\ 0 \\ 1 \end{bmatrix} \text{ for } \lambda_1, \text{ and } X_2 = \begin{bmatrix} 0 \\ -1 \\ 1 \\ 0 \end{bmatrix} \text{ for } \lambda_2.$$

Hence $M_\mathcal{B}(T)$ is diagonalizable by Theorem 10 §4.7, so $T$ is diagonalizable by Theorem 5. In addition, we have

$$X_1 = C_\mathcal{B}\begin{bmatrix} 1 & 0 \\ 0 & 0 \end{bmatrix}, \ Y_1 = C_\mathcal{B}\begin{bmatrix} 0 & 1 \\ 1 & 0 \end{bmatrix}, \ Z_1 = C_\mathcal{B}\begin{bmatrix} 0 & 0 \\ 0 & 1 \end{bmatrix}, \ X_2 = C_\mathcal{B}\begin{bmatrix} 0 & -1 \\ 1 & 0 \end{bmatrix}.$$

Hence Lemma 2 shows that $\mathcal{D} = \left\{ \begin{bmatrix} 1 & 0 \\ 0 & 0 \end{bmatrix}, \begin{bmatrix} 0 & 1 \\ 1 & 0 \end{bmatrix}, \begin{bmatrix} 0 & 0 \\ 0 & 1 \end{bmatrix}, \begin{bmatrix} 0 & -1 \\ 1 & 0 \end{bmatrix} \right\}$ is a basis of eigenvectors of $T$ (showing again that $T$ is diagonalizable).

If $T : \mathbb{M}_{2,2} \to \mathbb{M}_{2,2}$ is the transposition operator in Example 10, observe further that the eigenspaces of $T$ are

$$E_{\lambda_1}(T) = \{A \mid A^T = A\} \quad \text{and} \quad E_{\lambda_2}(T) = \{A \mid A^T = -A\},$$

and so consist of all symmetric and all skew-symmetric matrices, respectively. Furthermore, if we separate out the vectors in $\mathcal{D}$ corresponding to different eigenvalues, we obtain bases for the eigenspaces:

$$E_{\lambda_1}(T) = span\left\{ \begin{bmatrix} 1 & 0 \\ 0 & 0 \end{bmatrix}, \begin{bmatrix} 0 & 1 \\ 1 & 0 \end{bmatrix}, \begin{bmatrix} 0 & 0 \\ 0 & 1 \end{bmatrix} \right\}$$

$$= \left\{ \begin{bmatrix} a & b \\ b & c \end{bmatrix} \mid a, b, \text{ and } c \text{ in } \mathbb{R} \right\},$$

$$E_{\lambda_2}(T) = span\left\{ \begin{bmatrix} 0 & -1 \\ 1 & 0 \end{bmatrix} \right\}$$

$$= \left\{ \begin{bmatrix} 0 & -d \\ d & 0 \end{bmatrix} \mid d \text{ in } \mathbb{R} \right\}.$$

This phenomenon holds more generally, and depends only upon the fact that $T^2 = 1_{\mathbb{M}_{2,2}}$. To see this, we need the techniques developed in the next section (this result is Exercise 26 §5.6).

## Exercises 5.5

**1.** In each case either show that the statement is true or give an example showing that it is false. Throughout, $A$ and $B$ denote $n \times n$ matrices and $S$ and $T$ are operators $V \to V$.

**a)** If $det A = det B$, then $A$ and $B$ are similar.

**b)** If $A$ is similar to $B$, then $kA$ is similar to $kB$ for any scalar $k$.

**c)** If $T = R^{-1} \circ S \circ R$ for some isomorphism $R : V \to V$, then $S$ and $T$ have the same eigenvalues.

**d)** If $A$ is diagonalizable and $A = M_{\mathcal{B}}(T)$ for some basis $\mathcal{B}$, then $V$ has a basis of $T$-eigenvectors.

**e)** If $\mathbf{v}$ and $\mathbf{w}$ are distinct eigenvectors of $T$, then $\{\mathbf{v}, \mathbf{w}\}$ is independent.

**f)** If $A = M_{\mathcal{B}}(T)$ and $B = M_{\mathcal{B}}(S)$ for some basis $\mathcal{B}$, then $A$ and $B$ are similar.

**g)** If $\mathbf{v}$ is an eigenvector of $T$, the same is true of $k\mathbf{v}$ for all scalars $k \neq 0$.

**h)** If $\mathbf{v}$ is an eigenvector of $T$ and $\mathbf{w}$ is in $ker T$, then $\mathbf{v} + \mathbf{w}$ is an eigenvector of $T$.

**2.** Let $T : \mathbb{P}_2 \to \mathbb{P}_2$ be given by
$T(a + bx + cx^2) = (2a - b) + (a + b + c)x + (c - a)x^2$.

**a)** Find $M_{\mathcal{B}}(T)$ where $\mathcal{B} = \{1, x, x^2\}$.

**b)** Is $T$ invertible? Defend your answer.

**3.** Let $T : \mathbb{M}_{2,2} \to \mathbb{M}_{2,2}$ be given by

$T\begin{bmatrix} a & b \\ c & d \end{bmatrix} = \begin{bmatrix} a+b+c & 2c-a \\ b+d & c-d \end{bmatrix}$.

**a)** Find $M_{\mathcal{B}}(T)$ where $\mathcal{B} = \{E_{11}, E_{12}, E_{21}, E_{22}\}$ is the standard basis of matrix units.

**b)** Is $T$ invertible? Defend your answer.

**4.** If $T : \mathbb{P}_1 \to \mathbb{P}_1$ has matrix $M_{\mathcal{B}}(T) = \begin{bmatrix} 1 & -2 \\ -2 & 3 \end{bmatrix}$ where $\mathcal{B} = \{1, x\}$, find $T^{-1}(2x - 3)$.

**5.** If $T : \mathbb{M}_{2,2} \to \mathbb{M}_{2,2}$ has matrix

$M_{\mathcal{B}}(T) = \begin{bmatrix} 1 & 0 & 1 & 1 \\ 3 & 1 & -1 & 0 \\ 0 & 0 & -1 & 0 \\ 2 & 0 & 0 & 1 \end{bmatrix}$ where $\mathcal{B} = \{E_{11}, E_{12}, E_{21}, E_{22}\}$

is the standard basis of matrix units, find $T^{-1}\begin{bmatrix} 2 & 3 \\ -1 & 1 \end{bmatrix}$.

**6.** If $B$ is a fixed $2 \times 2$ matrix, define $T : \mathbb{M}_{2,2} \to \mathbb{M}_{2,2}$ by $T(A) = BA$ for all $A$ in $\mathbb{M}_{2,2}$. If $B = \begin{bmatrix} a & b \\ c & d \end{bmatrix}$,

find a basis $\mathcal{B}$ of $\mathbb{M}_{2,2}$ such that $M_{\mathcal{B}}(T) = \begin{bmatrix} aI_2 & bI_2 \\ cI_2 & dI_2 \end{bmatrix}$.

**7. a)** Find the eigenvalues and eigenvectors of the matrix $A = \begin{bmatrix} 1 & 2 \\ 3 & 2 \end{bmatrix}$.

**b)** If $T : \mathbb{R}^2 \to \mathbb{R}^2$ is defined by $T(X) = AX$ for all $X$ in $\mathbb{R}^2$, use **a)** to find a basis of eigenvectors of $T$.

**c)** If $T : \mathbb{P}_1 \to \mathbb{P}_1$ is defined by $T(a + bx) = (a + 2b) + (3a + 2b)x$, use **a)** to find a basis of eigenvectors of $T$.

**8.** Let $D : \mathbb{P}_3 \to \mathbb{P}_3$ be the differentiation operator given by $D(p(x)) = p'(x)$.

**a)** Find the matrix of $D$ with respect to the basis $\mathcal{B} = \{1, x, x^2, x^3\}$.

**b)** Is $D$ an isomorphism? Justify your answer.

**c)** Show that $D^4 = 0$ is the zero operator.

**9.** Let $\mathcal{B} = \{\mathbf{b}_1, \mathbf{b}_2, \mathbf{b}_3\}$ be a basis of $V$, and let $\mathbf{d}_1 = 2\mathbf{b}_1 - \mathbf{b}_2$, $\mathbf{d}_2 = \mathbf{b}_2 - \mathbf{b}_3$, and $\mathbf{d}_3 = \mathbf{b}_1 + \mathbf{b}_2 + \mathbf{b}_3$.

**a)** Show that $\mathcal{D} = \{\mathbf{d}_1, \mathbf{d}_2, \mathbf{d}_3\}$ is a basis of $V$.

**b)** Find the change matrix $P_{\mathcal{B} \leftarrow \mathcal{D}}$.

**10.** Let $\mathcal{B} = \{\mathbf{b}_1, \mathbf{b}_2, \mathbf{b}_3\}$ be a basis of $V$. If $T : V \to V$ is an operator satisfying $T(\mathbf{b}_1) = \mathbf{b}_1 + \mathbf{b}_2$, $T(\mathbf{b}_2) = \mathbf{b}_2 + \mathbf{b}_3$, and $T(\mathbf{b}_3) = \mathbf{b}_1$, find:

**a)** $M_{\mathcal{B}}(T)$.    **b)** $T^{-1}(\mathbf{b}_2)$.

**11.** Let $\mathcal{B} = \{\mathbf{b}_1, \mathbf{b}_2, \mathbf{b}_3\}$ be a basis of $V$. If $T : V \to V$ is an operator with matrix $M_{\mathcal{B}}(T) = \begin{bmatrix} 1 & 0 & 3 \\ -2 & 1 & 4 \\ 0 & -1 & 1 \end{bmatrix}$,

find $T(2\mathbf{b}_2 - 5\mathbf{b}_3)$.

**12.** Let $\mathcal{B}$ be a basis of $V$ where $dim V = n$. If $A$ is $n \times n$, show that $A = M_{\mathcal{B}}(T)$ for some operator $T : V \to V$.

**13.** Describe the operator $T : V \to V$ if it has the property that $V = E_\lambda(T)$ for some real number $\lambda$.

**14.** Let $\mathcal{B} = \{\mathbf{v}, \mathbf{w}\}$ be a basis of $V$, and let $T : V \to V$ be defined by $T(\mathbf{v}) = \mathbf{v}$ and $T(\mathbf{w}) = \mathbf{v} - \mathbf{w}$.

**a)** Find a basis of eigenvectors of $T$.

**b)** Find the matrix $M_{\mathcal{B}}(T)$ and diagonalize it.

**15.** Let $T : \mathbb{P}_n \to \mathbb{P}_n$ be defined by $T(p(x)) = p(x) + x p'(x)$ where $p'$ indicates the derivative. Show that $T$ is an isomorphism by considering $M_{\mathcal{B}}(T)$ where $\mathcal{B} = \{1, x, x^2, \cdots, x^n\}$.

**16.** If $k$ is any number, define $T : \mathbb{M}_{2,2} \to \mathbb{M}_{2,2}$ by $T(A) = A + kA^T$. Show that $T$ is an isomorphism if $k \neq \pm1$ by considering $M_{\mathcal{B}}(T)$ where

$\mathcal{B} = \left\{ \begin{bmatrix} 1 & 0 \\ 0 & 0 \end{bmatrix}, \begin{bmatrix} 0 & 0 \\ 0 & 1 \end{bmatrix}, \begin{bmatrix} 0 & 1 \\ 1 & 0 \end{bmatrix}, \begin{bmatrix} 0 & -1 \\ 1 & 0 \end{bmatrix} \right\}$.

**17.** Let $Q : \mathbb{R}^2 \to \mathbb{R}^2$ denote reflection in some line $L$ through the origin.

**a)** Show geometrically that any direction vector $\vec{d}$ for $L$ is an eigenvector of $Q$.

**b)** Use a geometrical argument to find eigenvectors independent of $\vec{d}$.

**c)** Show that $Q$ is diagonalizable.

**18.** Let $S$ and $T$ be linear operators $V \to V$ where $dim V$ is finite. Show that $det(S \circ T) = det S \, det T$.

**19.** Let $S$ and $T$ be linear operators $V \to V$ where $dim V$ is finite. Show that $tr(S \circ T) = tr(T \circ S)$. [Hint: Lemma 3 §4.7.]

**20.** Let $S$ and $T$ be two operators on a vector space $V$ of dimension $n$.

**a)** If $T = R \circ S$ for some isomorphism $R : V \to V$, show that $kerS = kerT$.

**b)** If $kerS = kerT$ show that $T = R \circ S$ for some isomorphism $R : V \to V$. [Hint: Let $\mathcal{B} = \{\mathbf{b}_1, \cdots, \mathbf{b}_r, \mathbf{b}_{r+1}, \cdots, \mathbf{b}_n\}$ be a basis of $V$ such that $\{\mathbf{b}_{r+1}, \cdots, \mathbf{b}_n\}$ is a basis of $kerT$. Then show that there exist bases of $V$ of the form $\mathcal{D} = \{S(\mathbf{b}_1), \cdots, S(\mathbf{b}_r), \mathbf{d}_{r+1}, \cdots, \mathbf{d}_n\}$ and $\mathcal{F} = \{T(\mathbf{b}_1), \cdots, T(\mathbf{b}_r), \mathbf{f}_{r+1}, \cdots, \mathbf{f}_n\}$. Define $R$ on the basis $\mathcal{D}$ by taking $R[S(\mathbf{b}_i)] = T(\mathbf{b}_i)$ for $1 \le i \le r$ and $R(\mathbf{d}_j) = \mathbf{f}_j$ for $r < j \le n$.]

**c)** If $A$ and $B$ are $n \times n$ matrices, show that $nullA = nullB$ if and only if $B = CA$ for some invertible matrix $C$. [Hint: See Example 2 §5.3.]

**21.** Show that $M_{\mathcal{D}'\mathcal{B}'}(T)P_{\mathcal{B}' \leftarrow \mathcal{B}} = P_{\mathcal{D}' \leftarrow \mathcal{D}}M_{\mathcal{D}\mathcal{B}}(T)$ where $T : V \to W$ is a linear transformation, $\mathcal{B}$ and $\mathcal{B}'$ are bases of $V$, and $\mathcal{D}$ and $\mathcal{D}'$ and are bases of W. [Hint: Right multiply by $C_{\mathcal{B}}(\mathbf{v})$ for arbitrary $\mathbf{v}$ in $V$, and use Lemma 1 and Theorems 1 and 2.]

## 5.6 INVARIANT SUBSPACES

As noted earlier, the central problem in linear algebra is to find a way to discover the "simplest" matrix for a linear operator $T : V \to V$. Often the way to do this is to regard $T$ as an operator on subspaces of $V$ of smaller dimension, and then find a way to build up the matrix of $T$ on $V$ from smaller matrices. The notion of a $T$-invariant subspace of $V$ is a basic tool in this endeavour.

### 5.6.1 Invariant Subspaces

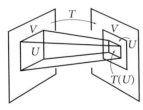

Suppose $T : V \to V$ is a linear operator. A subspace $U$ of $V$ is called $T$-**invariant** if
$$T(\mathbf{u}) \text{ is in } U \text{ for every vector u in } U.$$
If we write $T(U) = \{T(\mathbf{u}) \mid \mathbf{u} \text{ in } U\}$, this condition can be compactly expressed as
$$T(U) \subseteq U.$$
This is shown in the diagram.

**Example 1** $\{\mathbf{0}\}$ and $V$ are $T$-invariant subspaces of $V$ for every operator $T : V \to V$.

**Example 2** If $\lambda$ is an eigenvalue of the operator $T : V \to V$, show that the eigenspace $E_\lambda(T)$ is a $T$-invariant subspace of $V$.

**SOLUTION** If $\mathbf{u}$ is in $E_\lambda(T)$, then $T(\mathbf{u}) = \lambda\mathbf{u}$ is again in $E_\lambda(T)$ because $E_\lambda(T)$ is a subspace.

**Example 3** Let $V$ be a 2-dimensional vector space with basis $\{\mathbf{v}, \mathbf{w}\}$, and define $T : V \to V$ by taking $T(\mathbf{v}) = -\mathbf{w}$ and $T(\mathbf{w}) = \mathbf{v}$. Show that $V$ has no $T$-invariant subspace other than $\{\mathbf{0}\}$ and $V$.

**SOLUTION** Suppose that a $T$-invariant subspace $U$ exists other than $\{\mathbf{0}\}$ or $V$; we show that this leads to a contradiction. Since $dimV = 2$, we have $dimU = 1$, so $U = \mathbb{R}\mathbf{u}$ for some vector $\mathbf{u} \neq \mathbf{0}$ in $V$. Since $T(\mathbf{u})$ is in $U$, we have $T(\mathbf{u}) = \lambda\mathbf{u}$ for some real number $\lambda$. On the other hand, $\mathbf{u}$ is in $V = span\{\mathbf{v}, \mathbf{w}\}$, so we can write $\mathbf{u} = a\mathbf{v} + b\mathbf{w}$ where $a$ and $b$ are in $\mathbb{R}$. Hence the equation $\lambda\mathbf{u} = T(\mathbf{u})$ becomes
$$\lambda a\mathbf{v} + \lambda b\mathbf{w} = \lambda\mathbf{u} = T(\mathbf{u}) = T(a\mathbf{v} + b\mathbf{w}) = aT(\mathbf{v}) + bT(\mathbf{w}) = -a\mathbf{w} + b\mathbf{v}.$$
Since $\{\mathbf{v}, \mathbf{w}\}$ is independent, it follows that $\lambda a = b$ and $\lambda b = -a$. But then $\lambda^2 a = \lambda(\lambda a) = \lambda b = -a$, so that $(\lambda^2 + 1)a = 0$. Hence $a = 0$ because $\lambda$ is real. Similarly $\lambda^2 b = -b$, so $b = 0$, and we obtain $\mathbf{u} = a\mathbf{v} + b\mathbf{w} = \mathbf{0}$, a contradiction.

We now give a useful test for when a subspace $U$ is $T$-invariant: Simply check that $T$ carries every vector in some (usually finite) spanning set for $U$ back into $U$.

**LEMMA 1**   If $T : V \to V$ is a linear operator and $U = span\{\mathbf{u}_1, \mathbf{u}_2, \cdots, \mathbf{u}_m\}$ is a subspace of $V$, then $U$ is $T$-invariant if and only if $T(\mathbf{u}_i)$ is in $U$ for each $i = 1, 2, \cdots, m$.

**PROOF**   If $U$ is $T$-invariant, then certainly $T(\mathbf{u}_i)$ is in $U$ for each $i$. Conversely, assume that each $T(\mathbf{u}_i)$ is in $U$, and let $\mathbf{u}$ be any vector in $U$. Then $\mathbf{u} = r_1\mathbf{u}_1 + r_2\mathbf{u}_2 + \cdots + r_m\mathbf{u}_m$ for some $r_i$ in $\mathbb{R}$, so $T(\mathbf{u}) = r_1T(\mathbf{u}_1) + r_2T(\mathbf{u}_2) + \cdots + r_mT(\mathbf{u}_m)$ is also in $U$ because $U$ is a subspace. Thus $U$ is $T$-invariant.

**Example 4**   Write $B = \begin{bmatrix} 1 & -1 \\ 1 & 0 \end{bmatrix}$, define $T : \mathbb{M}_{2,2} \to \mathbb{M}_{2,2}$ by $T(A) = BA$ for all $A$ in $\mathbb{M}_{2,2}$, and consider the subspace $U = span \{I, B\}$ of $\mathbb{M}_{2,2}$. Show that $U$ is a $T$-invariant subspace of $\mathbb{M}_{2,2}$ which is not an eigenspace of $T$.

**SOLUTION**   We have $T(I) = B$ is in $U$, and $T(B) = B^2$, so it is enough (by Lemma 1) to show that $B^2$ is in $U$. This follows from the fact that $B^2 = B - I$ (verify). [This can be seen as follows: The characteristic polynomial of $B$ is $c_B(x) = x^2 - x + 1$, and the Cayley-Hamilton theorem (Theorem 7 below) asserts that $c_B(B) = 0$.] To see that $U$ is not an eigenspace of $T$, simply observe that $T(B) = B^2 = B - I \neq \lambda B$ for any number $\lambda$ (otherwise $(1 - \lambda)B = I$, a contradiction).

We now come to the principal reason for the importance of $T$-invariant subspaces. If $T : V \to V$ is an operator and $U$ is a $T$-invariant subspace of $V$, the fact that $T(\mathbf{u})$ is in $U$ for each vector $\mathbf{u}$ in $U$ means that

$$T : U \to U \text{ is an operator } on \text{ the vector space } U,$$

called the **restriction** of $T$ to $U$. This is very important for two reasons: First, the dimension of $U$ will in general be smaller than the dimension of $V$ so it will be easier to find a "simple" matrix for $T : U \to U$ (possibly by induction on the dimension). Second, if a basis of $U$ is extended to a basis of $V$, the corresponding matrix of $T : V \to V$ takes a simplified, block upper triangular form. This is the content of the following theorem.

**THEOREM 1**   Let $T : V \to V$ be a linear operator where $dim V = n$, and let $U$ be a $T$-invariant subspace of $V$. If $\mathcal{B}_0 = \{\mathbf{u}_1, \mathbf{u}_2, \cdots, \mathbf{u}_m\}$ is any basis of $U$, extend it to a basis

$$\mathcal{B} = \{\mathbf{u}_1, \mathbf{u}_2, \cdots, \mathbf{u}_m, \mathbf{v}_{m+1}, \cdots, \mathbf{v}_n\}$$

of $V$ in any way at all. Then

$$M_\mathcal{B}(T) = \begin{bmatrix} M_{\mathcal{B}_0}(T) & Y \\ 0 & A \end{bmatrix}$$

in block triangular form, where $M_{\mathcal{B}_0}(T)$ is the $\mathcal{B}_0$-matrix of the restriction of $T$ to $U$.

**PROOF**   First write the $n \times n$ matrix $M_\mathcal{B}(T)$ in terms of its columns:

$$M_\mathcal{B}(T) = [C_\mathcal{B}(T(\mathbf{u}_1)) \quad C_\mathcal{B}(T(\mathbf{u}_2)) \quad \cdots \quad C_\mathcal{B}(T(\mathbf{u}_m)) \quad C_\mathcal{B}(T(\mathbf{v}_{m+1})) \quad \cdots \quad C_\mathcal{B}(T(\mathbf{v}_n))].$$

Since $\mathbf{u}_i$ is in $U$ for each $i = 1, 2, \cdots, m$, the fact that $U$ is $T$-invariant means $T(\mathbf{u}_i)$ is in $U$. Hence the $\mathcal{B}$-expansion of $T(\mathbf{u}_i)$ has the form

$$T(\mathbf{u}_i) = r_1\mathbf{u}_1 + r_2\mathbf{u}_2 + \cdots + r_m\mathbf{u}_m + 0\mathbf{v}_{m+1} \cdots + 0\mathbf{v}_n$$

where $r_1, r_2, \cdots, r_m$ are real numbers. It follows that

$$C_\mathcal{B}(T(\mathbf{u}_i)) = [r_1 \ r_2 \ \cdots \ r_m \ 0 \ 0 \ \cdots \ 0]^T \quad \text{and} \quad C_{\mathcal{B}_0}(T(\mathbf{u}_i)) = [r_1 \ r_2 \ \cdots \ r_m]^T.$$

This means that, for each $i$, the column $C_\mathcal{B}(T(\mathbf{u}_i))$ in $M_\mathcal{B}(T)$ consists of the column $C_{\mathcal{B}_0}(T(\mathbf{u}_i))$ on top and zeros below. Since our hypotheses give us no information about the remaining columns $C_\mathcal{B}(T(\mathbf{v}_k))$ in $M_\mathcal{B}(T)$, this shows that $M_\mathcal{B}(T)$ has the triangular form given in the Theorem.

**Example 5**

In Example 4 we considered the operator $T : \mathbb{M}_{2,2} \to \mathbb{M}_{2,2}$ defined by $T(A) = BA$ for all $A$ in $\mathbb{M}_{2,2}$, where $B = \begin{bmatrix} 1 & -1 \\ 1 & 0 \end{bmatrix}$, and we showed that $U = span\{I, B\}$ is a $T$-invariant subspace of $\mathbb{M}_{2,2}$. Complete $\mathcal{B}_o = \{I, B\}$ to a basis $\mathcal{B}$ of $\mathbb{M}_{2,2}$, and so illustrate Theorem 1.

**SOLUTION**

$\mathcal{B}_o$ is independent because $B \neq kI$ for any real number $k$. One possible completion (verify) of $\mathcal{B}_o$ to a basis of $\mathbb{M}_{2,2}$ is $\mathcal{B} = \{I, B, E_{11}, E_{12}\}$ where $E_{11} = \begin{bmatrix} 1 & 0 \\ 0 & 0 \end{bmatrix}$ and $E_{12} = \begin{bmatrix} 0 & 1 \\ 0 & 0 \end{bmatrix}$. Since $B^2 = B - I$ (see Example 4), we see that

$$
\begin{aligned}
T(I) &= BI = B & &\text{so} & C_\mathcal{B}(T(I)) &= [0\ 1\ 0\ 0]^T, \\
T(B) &= B^2 = B - I & &\text{so} & C_\mathcal{B}(T(B)) &= [-1\ 1\ 0\ 0]^T, \\
T(E_{11}) &= BE_{11} = B + E_{12} & &\text{so} & C_\mathcal{B}(T(E_{11})) &= [0\ 1\ 0\ 1]^T, \\
T(E_{12}) &= BE_{12} = I - E_{11} + E_{12} & &\text{so} & C_\mathcal{B}(T(E_{12})) &= [1\ 0\ -1\ 1]^T.
\end{aligned}
$$

Hence the $\mathcal{B}$-matrix of $T$ is
$$
\begin{aligned}
M_\mathcal{B}(T) &= [C_\mathcal{B}(T(I))\ \ C_\mathcal{B}(T(B))\ \ C_\mathcal{B}(T(E_{11}))\ \ C_\mathcal{B}(T(E_{12}))] \\
&= \begin{bmatrix} 0 & -1 & 0 & 1 \\ 1 & 1 & 1 & 0 \\ 0 & 0 & 0 & -1 \\ 0 & 0 & 1 & 1 \end{bmatrix}.
\end{aligned}
$$

This has the block upper triangular form expected from Theorem 1.

Theorem 1 has an important consequence that will be used later.

**THEOREM 2**

Let $T : V \to V$ be an operator where $dim V = n$, let $U$ be a $T$-invariant subspace of $V$, and let $q(x)$ denote the characteristic polynomial of the restriction of $T$ to $U$. Then

$$c_T(x) = p(x)\, q(x) \quad \text{for some polynomial } p(x).$$

**PROOF**

Let $dim U = m$. Choose a basis $\mathcal{B}_0$ of $U$, so that $q(x) = det[xI_m - M_{\mathcal{B}_0}(T)]$. Extend $\mathcal{B}_0$ to a basis $\mathcal{B}$ of $V$. Then $M_\mathcal{B}(T)$ has the block form $M_\mathcal{B}(T) = \begin{bmatrix} M_{\mathcal{B}_0}(T) & Y \\ 0 & A \end{bmatrix}$ by Theorem 1 for some $(n-m) \times (n-m)$ matrix $A$. Now a routine matrix computation (using Theorem 3 §2.2) gives

$$
\begin{aligned}
c_T(x) &= c_{M_\mathcal{B}(T)}(x) \\
&= det[xI_n - M_\mathcal{B}(T)] \\
&= det\begin{bmatrix} xI_m - M_{\mathcal{B}_0}(T) & -Y \\ 0 & xI_{n-m} - A \end{bmatrix} \\
&= det[xI_m - M_{\mathcal{B}_0}(T)]\ det[xI_{n-m} - A] \\
&= q(x)\, c_A(x).
\end{aligned}
$$

Theorem 2 follows with $p(x) = c_A(x)$.

If $T : V \to V$ is a linear operator, the triangular matrix

$$M_{\mathcal{B}}(T) = \begin{bmatrix} M_{\mathcal{B}_0}(T) & Y \\ 0 & A \end{bmatrix}$$

in Theorem 1 results from choosing the basis $\mathcal{B}$ so that the first few vectors are a basis $\mathcal{B}_0$ of some $T$-invariant subspace of $V$. If we can choose the remaining vectors in $\mathcal{B}$ so that they also span a $T$-invariant subspace, we can make $Y = 0$ and so get a block diagonal matrix for $T$. Describing how and when this can be done requires the idea of a direct sum decomposition of $V$, and we digress slightly to discuss this concept.

## 5.6.2  Direct Sums

Let $V$ be any vector space. If $U$ and $W$ are subspaces of $V$, define their **sum** $U + W$ and **intersection** $U \cap W$ as follows:

$$U + W = \{\mathbf{u} + \mathbf{w} \mid \mathbf{u} \text{ in } U \text{ and } \mathbf{w} \text{ in } W\}$$

and

$$U \cap W = \{\mathbf{v} \text{ in } V \mid \mathbf{v} \text{ is in both } U \text{ and } W\}.$$

Thus $U + W$ consists of all sums of a vector in $U$ and a vector in $W$, while $U \cap W$ consists of all vectors in both $U$ and $W$. The routine proof of the following lemma is left to the reader.

**LEMMA 2**

If $U$ and $W$ are subspaces of $V$, then $U + W$ and $U \cap W$ are both subspaces of $V$.

The most important pairs of subspaces $U$ and $W$ of a vector space $V$ turn out to be those for which

$$U + W = V \quad \text{and} \quad U \cap W = \{\mathbf{0}\}.$$

When this is the case we say that $V$ is the **direct sum** of $U$ and $W$ and we indicate this by writing

$$V = U \oplus W.$$

In this case we call each of $U$ and $W$ a **complement** of the other in $V$. A subspace $U$ can have more than one complement.

**Example 6**

If $U = \mathbb{R}[1 \ \ 0]$ in $V = \mathbb{M}_{1,2}$, show that each of $W_1 = \mathbb{R}[0 \ \ 1]$, $W_2 = \mathbb{R}[1 \ \ 1]$, and $W_3 = \mathbb{R}[1 \ \ -1]$ are complements of $U$ in $\mathbb{M}_{1, 2}$.

**SOLUTION**

We show that $V = U \oplus W_2$, and leave the other two cases to the reader. To see that $V = U + W_2$, observe that

$$[x \ \ y] = (x - y)[1 \ \ 0] + y[1 \ \ 1] \text{ is in } U + W_2 \text{ for all } x \text{ and } y.$$

Now suppose that $[x \ \ y]$ is in both $U = \mathbb{R}[1 \ \ 0]$ and $W_2 = \mathbb{R}[1 \ \ 1]$. Then $[x \ \ y] = r[1 \ \ 0] = s[1 \ \ 1]$ for some $r$ and $s$ in $\mathbb{R}$, so comparing second entries gives $0 = s$. Thus $[x \ \ y] = s[1 \ \ 1] = 0$, and we have proved that $U \cap W_2 = 0$. Hence $V = U \oplus W_2$.

Direct sum decompositions occur frequently as the following examples show.

**Example 7**

Consider the subspaces $U = \{A \text{ in } \mathbb{M}_{n,n} \mid A^T = A\}$ and $W = \{A \text{ in } \mathbb{M}_{n,n} \mid A^T = -A\}$ of $\mathbb{M}_{n,n}$ consisting of all symmetric and all skew-symmetric matrices, respectively. Show that $\mathbb{M}_{n,n} = U \oplus W$.

SOLUTION | Here the fact that $U \cap W = \{0\}$ is the observation that only the zero matrix is both symmetric and skew-symmetric (verify). To see that $U + W = \mathbb{M}_{n,n}$, let $A$ be any matrix in $\mathbb{M}_{n,n}$ and consider the decomposition $A = \frac{1}{2}(A + A^T) + \frac{1}{2}(A - A^T)$. It is a routine exercise to verify that $\frac{1}{2}(A + A^T)$ is symmetric (and so is in $U$) and $\frac{1}{2}(A - A^T)$ is skew-symmetric (and so is in $W$). Hence $\mathbb{M}_{n,n} = U + W$.

**Example 8** | Consider the space $\mathbb{F}[a, b]$ of all real valued functions defined on the interval $[a, b]$. A function $f$ in $\mathbb{F}[a, b]$ is called **even** if $f(-x) = f(x)$ for all $x$ in $[a, b]$, and $f$ is called **odd** if $f(-x) = -f(x)$ for all $x$ in $[a, b]$. If $U$ consists of all even functions and $W$ consists of all odd functions, show that $\mathbb{F}[a, b] = U \oplus W$.

SOLUTION | We leave to the reader the verification that $U$ and $W$ are both subspaces of $\mathbb{F}[a, b]$. If $f$ is in $U \cap W$, then $f$ is both even and odd, so $f(-x) = f(x)$ and $f(-x) = -f(x)$ both hold for all $x$ in $[a, b]$. It follows that $f(x) = 0$ for all $x$; in other words, $f = f_0$ is the zero vector in $\mathbb{F}[a, b]$ (see Example 5 §5.1). Hence $U \cap W = \{f_0\}$ is the zero subspace of $\mathbb{F}[a, b]$.

Next, if $f$ is any function in $\mathbb{F}[a, b]$, define $g : [a, b] \to \mathbb{R}$ and $h : [a, b] \to \mathbb{R}$ by

$$g(x) = \tfrac{1}{2}\{f(x) + f(-x)\} \quad \text{and} \quad h(x) = \tfrac{1}{2}\{f(x) - f(-x)\}$$

for all $x$ in $[a, b]$. Then $g$ and $h$ both lie in $\mathbb{F}[a, b]$, and the reader can verify that $g$ is even and $h$ is odd. Moreover $g(x) + h(x) = f(x)$ holds for all $x$ in $[a, b]$, which means that $f = g + h$. Hence $\mathbb{F}[a, b] = U + W$.

**Example 9** | If $U$ denotes any subspace of $\mathbb{R}^n$, show that

$$\mathbb{R}^n = U \oplus U^\perp$$

where $U^\perp = \{X \text{ in } \mathbb{R}^n \mid X \bullet Y = 0 \text{ for all } Y \text{ in } U\}$ is the orthogonal complement of $U$.

SOLUTION | If $X$ is in both $U$ and $U^\perp$, then $0 = X \bullet X = \|X\|^2$, so $X = 0$. Hence $U \cap U^\perp = \{0\}$. To see that $\mathbb{R}^n = U + U^\perp$, let $X$ be any vector in $\mathbb{R}^n$ and write $P = proj_U(X)$. Then $X = P + (X - P)$, and the Projection theorem (Theorem 2 §4.6) asserts that $P$ is in $U$ and $(X - P)$ is in $U^\perp$. It follows that $\mathbb{R}^n = U + U^\perp$.

Clearly, $V = U + W$ if and only if every vector $\mathbf{v}$ in $V$ has the form $\mathbf{v} = \mathbf{u} + \mathbf{w}$ for some $\mathbf{u}$ in $U$ and $\mathbf{w}$ in $W$. We are interested in the situation where this representation is *unique*, that is where condition (1) in the following lemma holds.

LEMMA ❸ | The following conditions are equivalent for subspaces $U$ and $W$ of a vector space $V$:

(1) If $\mathbf{u} + \mathbf{w} = \mathbf{u}_1 + \mathbf{w}_1$ where $\mathbf{u}, \mathbf{u}_1$ in $U$, and $\mathbf{w}, \mathbf{w}_1$ in $W$, then $\mathbf{u} = \mathbf{u}_1$ and $\mathbf{w} = \mathbf{w}_1$.
(2) If $\mathbf{u} + \mathbf{w} = \mathbf{0}$ where $\mathbf{u}$ is in $U$ and $\mathbf{w}$ is in $W$, then $\mathbf{u} = \mathbf{0}$ and $\mathbf{w} = \mathbf{0}$.
(3) $U \cap W = \{\mathbf{0}\}$.

PROOF | $(1) \Rightarrow (2)$. If $\mathbf{u} + \mathbf{w} = \mathbf{0}$, (2) follows by taking $\mathbf{u}_1 = \mathbf{0}$ and $\mathbf{w}_1 = \mathbf{0}$ in (1).

$(2) \Rightarrow (3)$. If $\mathbf{v}$ is in $U \cap W$, then $\mathbf{v} + (-\mathbf{v}) = \mathbf{0}$ so $\mathbf{v} = \mathbf{0}$ by (2) because $\mathbf{v}$ is in $U$ and $-\mathbf{v}$ is in $W$.

$(3) \Rightarrow (1)$. Suppose that $\mathbf{u} + \mathbf{w} = \mathbf{u}_1 + \mathbf{w}_1$ as in (1). Then $\mathbf{u} - \mathbf{u}_1 = \mathbf{w}_1 - \mathbf{w}$, and this vector is in both $U$ (left side) and $W$ (right side). Hence $\mathbf{u} - \mathbf{u}_1 = \mathbf{0}$ and $\mathbf{w}_1 - \mathbf{w} = \mathbf{0}$ by (3), proving (1).

We can now give some important characterizations of direct sum decompositions.

**THEOREM 3**   If $U$ and $W$ are subspaces of $V$ where $dim\,V$ is finite, the following are equivalent:

    (1) $V = U \oplus W$.

    (2) Every vector $\mathbf{v}$ in $V$ has a unique representation in the form $\mathbf{v} = \mathbf{u} + \mathbf{w}$ with $\mathbf{u}$ in $U$ and $\mathbf{w}$ in $W$.

    (3) If $\mathcal{B}_1 = \{\mathbf{u}_1, \mathbf{u}_2, \cdots, \mathbf{u}_m\}$ and $\mathcal{B}_2 = \{\mathbf{w}_1, \mathbf{w}_2, \cdots, \mathbf{w}_k\}$ are any bases of $U$ and $W$ respectively, then $\mathcal{B} = \{\mathbf{u}_1, \mathbf{u}_2, \cdots, \mathbf{u}_m, \mathbf{w}_1, \mathbf{w}_2, \cdots, \mathbf{w}_k\}$ is a basis of $V$.

    (4) There exist bases $\mathcal{B}_1 = \{\mathbf{u}_1, \mathbf{u}_2, \cdots, \mathbf{u}_m\}$ and $\mathcal{B}_2 = \{\mathbf{w}_1, \mathbf{w}_2, \cdots, \mathbf{w}_k\}$ of $U$ and $W$ respectively such that $\mathcal{B} = \{\mathbf{u}_1, \mathbf{u}_2, \cdots, \mathbf{u}_m, \mathbf{w}_1, \mathbf{w}_2, \cdots, \mathbf{w}_k\}$ is a basis of $V$.

**PROOF**   $(1) \Rightarrow (2)$. Since $V = U + W$, every vector $\mathbf{v}$ in $V$ has the form $\mathbf{v} = \mathbf{u} + \mathbf{w}$ for some vectors $\mathbf{u}$ in $U$ and $\mathbf{w}$ in $W$. This representation is unique by Lemma 3 because $U \cap W = \{\mathbf{0}\}$.

$(2) \Rightarrow (3)$. If $\mathbf{v}$ is in $V$, use (2) to write $\mathbf{v} = \mathbf{u} + \mathbf{w}$ where $\mathbf{u}$ is in $U$ and $\mathbf{w}$ is in $W$. Since $\mathbf{u}$ and $\mathbf{w}$ are in $span\{\mathcal{B}_1\}$ and $span\{\mathcal{B}_2\}$ respectively, it follows that $\mathbf{v} = \mathbf{u} + \mathbf{w}$ is in $span\{\mathcal{B}\}$. Hence $V = span\{\mathcal{B}\}$.

To see that $\mathcal{B}$ is independent, suppose numbers $s_i$ and $t_j$ exist such that

$$s_1\mathbf{u}_1 + s_2\mathbf{u}_2 + \cdots + s_m\mathbf{u}_m + t_1\mathbf{w}_1 + t_2\mathbf{w}_2 + \cdots + t_k\mathbf{w}_k = \mathbf{0}.$$

If we write $\mathbf{u} = s_1\mathbf{u}_1 + s_2\mathbf{u}_2 + \cdots + s_m\mathbf{u}_m$ and $\mathbf{w} = t_1\mathbf{w}_1 + t_2\mathbf{w}_2 + \cdots + t_k\mathbf{w}_k$, we have $\mathbf{u} + \mathbf{w} = \mathbf{0}$ with $\mathbf{u}$ in $U$ and $\mathbf{w}$ in $W$. The uniqueness in (2) implies that $\mathbf{u} = \mathbf{0}$ and $\mathbf{w} = \mathbf{0}$, and then the independence of $\mathcal{B}_1$ and $\mathcal{B}_2$ forces $s_i = 0$ for each $i$ and $t_j = 0$ for each $j$. Hence $\mathcal{B}$ is independent.

$(3) \Rightarrow (4)$. This is is clear since $U$ and $W$ have bases.

$(4) \Rightarrow (1)$. Given (4), every vector $\mathbf{v}$ in $V$ is in $span\{\mathcal{B}\}$, and so there exist $s_i$ and $t_j$ in $\mathbb{R}$ such that

$$\mathbf{v} = s_1\mathbf{u}_1 + s_2\mathbf{u}_2 + \cdots + s_m\mathbf{u}_m + t_1\mathbf{w}_1 + t_2\mathbf{w}_2 + \cdots + t_k\mathbf{w}_k.$$

Hence $\mathbf{v} = \mathbf{u} + \mathbf{w}$ where $\mathbf{u} = s_1\mathbf{u}_1 + s_2\mathbf{u}_2 + \cdots + s_m\mathbf{u}_m$ is in $U$ and $\mathbf{w} = t_1\mathbf{w}_1 + t_2\mathbf{w}_2 + \cdots + t_k\mathbf{w}_k$ is in $W$. It follows that $V = U + W$. If $\mathbf{v}$ is in $U \cap W$, (4) gives $\mathbf{v} = s_1\mathbf{u}_1 + s_2\mathbf{u}_2 + \cdots + s_m\mathbf{u}_m$ and $\mathbf{v} = t_1\mathbf{w}_1 + t_2\mathbf{w}_2 + \cdots + t_k\mathbf{w}_k$ with $s_i$ and $t_j$ in $\mathbb{R}$. Then the independence of $\mathcal{B}$ implies that $s_i = 0 = t_j$ for all $i$ and $j$, so $\mathbf{v} = \mathbf{0}$. This proves that $U \cap W = \{\mathbf{0}\}$, and so proves (1).

If $\mathcal{B} = \{\mathbf{u}_1, \mathbf{u}_2, \cdots, \mathbf{u}_m, \mathbf{w}_1, \mathbf{w}_2, \cdots, \mathbf{w}_k\}$ is any basis of a vector space $V$, then condition (4) of Theorem 3 shows that $V = U \oplus W$ where $U = span\{\mathbf{u}_1, \mathbf{u}_2, \cdots, \mathbf{u}_m\}$ and $W = span\{\mathbf{w}_1, \mathbf{w}_2, \cdots, \mathbf{w}_k\}$. Moreover, every direct sum decomposition of $V$ arises in this way by partitioning some basis $\mathcal{B}$ of $V$ into two parts.

The following useful result is an immediate consequence of Condition (4) in Theorem 3.

**THEOREM 4**   If $V$ is a finite dimensional vector space and $V = U \oplus W$ where $U$ and $W$ are subspaces of $V$, then $dim\,V = dim\,U + dim\,W$.

## 5.6.3   Reducible Operators

Returning to $T$-invariant subspaces, we can now prove the promised refinement of Theorem 1 wherein we outline how to find a basis such that the matrix of the operator $T$ is block diagonal.

**THEOREM 5**  Let $T : V \to V$ be a linear operator where $dim V = n$, let
$$V = U \oplus W \text{ where both } U \text{ and } W \text{ are } T\text{-invariant,}$$

and let $\mathcal{B}_1 = \{\mathbf{u}_1, \mathbf{u}_2, \cdots, \mathbf{u}_m\}$ and $\mathcal{B}_2 = \{\mathbf{w}_1, \mathbf{w}_2, \cdots, \mathbf{w}_k\}$ be bases of $U$ and $W$ respectively. Then $\mathcal{B} = \{\mathbf{u}_1, \mathbf{u}_2, \cdots, \mathbf{u}_m, \mathbf{w}_1, \mathbf{w}_2, \cdots, \mathbf{w}_k\}$ is a basis of $V$ and $M_{\mathcal{B}}(T)$ has the block diagonal form

$$M_{\mathcal{B}}(T) = \begin{bmatrix} M_{\mathcal{B}_1}(T) & 0 \\ 0 & M_{\mathcal{B}_2}(T) \end{bmatrix}$$

where $M_{\mathcal{B}_1}(T)$ and $M_{\mathcal{B}_2}(T)$ are the matrices of the restrictions of $T$ to $U$ and $W$ respectively.

**PROOF**  Theorem 4 shows that $\mathcal{B}$ is a basis of $V$, and Theorem 1 shows that the columns in $M_{\mathcal{B}}(T)$ corresponding to $\mathcal{B}_1$ have the required form because $U$ is $T$-invariant. Since $W$ is also $T$-invariant, a similar argument works for the remaining columns of $M_{\mathcal{B}}(T)$.

An operator $T : V \to V$ is called **reducible** if a direct sum decomposition $V = U \oplus W$ can be found such that both $U$ and $W$ are $T$-invariant.

**Example 10**  In Example 4 we considered the operator $T : \mathbb{M}_{2,2} \to \mathbb{M}_{2,2}$ defined by $T(A) = BA$ for all $A$ in $\mathbb{M}_{2,2}$, where $B = \begin{bmatrix} 1 & -1 \\ 1 & 0 \end{bmatrix}$. Show that $T$ is reducible.

**SOLUTION**  In Example 4 we showed that $U = span\{I, B\}$ is a $T$-invariant subspace of $\mathbb{M}_{2,2}$ with basis $\mathcal{B}_1 = \{I, B\}$. In Example 5 we completed $\mathcal{B}_1$ to a basis $\{I, B, E_{11}, E_{12}\}$ and found that $\mathbb{M}_{2,2} = U \oplus W_1$ where $W_1 = span\{E_{11}, E_{12}\}$. The problem is that $W_1$ is *not* $T$-invariant.

However a different completion of the basis $\mathcal{B}_1$ *does* produce a $T$-invariant complement of $U$. If we write $\mathcal{B}_2 = \{E_{11}, E_{21}\}$, we find that $W = span\{\mathcal{B}_2\}$ is $T$-invariant by Lemma 1 because $T(E_{11}) = E_{11} + E_{21}$ and $T(E_{21}) = -E_{11}$. Moreover, $\mathcal{B} = \{I, B, E_{11}, E_{21}\}$ is a basis of $V$ (verify) so $V = U \oplus W$. Hence $T$ is reducible. In fact,

$$M_{\mathcal{B}}(T) = \begin{bmatrix} M_{\mathcal{B}_1}(T) & 0 \\ 0 & M_{\mathcal{B}_2}(T) \end{bmatrix}$$

where $M_{\mathcal{B}_1}(T) = \begin{bmatrix} 0 & -1 \\ 1 & 1 \end{bmatrix}$ because $T(I) = B$ and $T(B) = B - I$, and $M_{\mathcal{B}_2}(T) = \begin{bmatrix} 1 & -1 \\ 1 & 0 \end{bmatrix}$ because $T(E_{11}) = E_{11} + E_{21}$ and $T(E_{21}) = -E_{11}$.

The operator $T$ in Example 10 has a natural $T$-invariant subspace $U$, but the first complement $W_1$ of $U$ we considered was not $T$-invariant. However we succeeded in finding a $T$-invariant complement $W$ of $U$, so $T$ is in fact reducible. Sometimes, however, an operator can fail to be reducible even though it has an invariant subspace; the problem is that *no* invariant complement exists. Here is an example.

**Example 11**  Let $\{\mathbf{u}, \mathbf{v}\}$ be a basis of $V$ and define $T : V \to V$ by $T(\mathbf{u}) = \mathbf{u}$ and $T(\mathbf{v}) = \mathbf{u} + \mathbf{v}$. Show that $U = \mathbb{R}\mathbf{u}$ is $T$-invariant but has no $T$-invariant complement in $V$.

**SOLUTION**  $U$ is $T$-invariant by Lemma 1 because $T(\mathbf{u}) = \mathbf{u}$. Suppose that $V = U \oplus W$ where $W$ is also $T$-invariant. Then $dim W = 1$ by Theorem 4 (because $dim V = 2$), say $W = \mathbb{R}\mathbf{w}$. Write $\mathbf{w} = a\mathbf{u} + b\mathbf{v}$ where $a$ and $b$ are in $\mathbb{R}$. Then $b \neq 0$ because otherwise $\mathbf{w} = a\mathbf{u}$ is in $U \cap W = \{\mathbf{0}\}$. Since $W$ is $T$-invariant, there is a real number $\lambda$ such that $T(\mathbf{w}) = \lambda \mathbf{w}$.

Since $\mathbf{w} = a\mathbf{u} + b\mathbf{v}$, the fact that $T$ is linear gives

$$\lambda a\mathbf{u} + \lambda b\mathbf{v} = \lambda\mathbf{w} = T(\mathbf{w}) = aT(\mathbf{u}) + bT(\mathbf{v}) = a\mathbf{u} + b(\mathbf{u} + \mathbf{v}) = (a + b)\mathbf{u} + b\mathbf{v}.$$

Since $\{\mathbf{u}, \mathbf{v}\}$ is independent, this gives $\lambda a = a + b$ and $\lambda b = b$. Hence $\lambda = 1$ (because $b \neq 0$) so that $a + b = \lambda a = a$. This gives $b = 0$, a contradiction.

Sometimes, a direct sum decomposition can be found into two eigenspaces of an operator. This not only shows that the operator is reducible, but diagonalizable as well. The next example is an illustration.

**Example 12**     An operator $T : V \to V$ is called an **idempotent** if $T^2 = T$.

(1) If $T$ is an idempotent, show that $T$ is diagonalizable by showing that $V$ is the direct sum of its eigenspaces.

(2) If $A = A^2$ is any idempotent matrix, show that $A$ is similar to $\begin{bmatrix} 0 & 0 \\ 0 & I_m \end{bmatrix}$ for some integer $m$ (so $A$ is diagonalizable).

**SOLUTION**     (1) If $\lambda$ is an eigenvalue of $T$, we first show that either $\lambda = 0$ or $\lambda = 1$. Indeed, if $T(\mathbf{v}) = \lambda\mathbf{v}$ where $\mathbf{v} \neq \mathbf{0}$, then

$$\lambda\mathbf{v} = T(\mathbf{v}) = T^2(\mathbf{v}) = T(T(\mathbf{v})) = T(\lambda\mathbf{v}) = \lambda T(\mathbf{v}) = \lambda^2\mathbf{v}.$$

Hence $\lambda = \lambda^2$ because $\mathbf{v} \neq \mathbf{0}$, so $\lambda = 0$ or $\lambda = 1$. Hence there are two eigenspaces:

$$E_0(T) = \{\mathbf{u} \text{ in } V \mid T(\mathbf{u}) = \mathbf{0}\} \quad \text{and} \quad E_1(T) = \{\mathbf{w} \text{ in } V \mid T(\mathbf{w}) = \mathbf{w}\}.$$

We claim that $V = E_0(T) \oplus E_1(T)$. It is clear that $E_0(T) \cap E_1(T) = \{\mathbf{0}\}$. Moreover $V = E_0(T) + E_1(T)$ because any vector $\mathbf{v}$ in $V$ can be written in the form $\mathbf{v} = (\mathbf{v} - T(\mathbf{v})) + T(\mathbf{v})$ and (since $T^2 = T$) it is easy to verify that $\mathbf{v} - T(\mathbf{v})$ is in $E_0(T)$ and that $T(\mathbf{v})$ is in $E_1(T)$. Hence

$$V = E_0(T) \oplus E_1(T)$$

and both $E_0(T)$ and $E_1(T)$ are $T$-invariant (by Example 2). Moreover, if $\mathcal{B}_0$ and $\mathcal{B}_1$ are bases of $E_0(T)$ and $E_1(T)$ respectively, the matrices of the restriction of $T$ to $E_0(T)$ and $E_1(T)$ are $M_{\mathcal{B}_0}(T) = 0$ and $M_{\mathcal{B}_1}(T) = I_m$ where $m = dim(E_1(T))$. Thus Theorem 5 provides a basis $\mathcal{B}$ of $V$ such that

$$M_{\mathcal{B}}(T) = \begin{bmatrix} M_{\mathcal{B}_0}(T) & 0 \\ 0 & M_{\mathcal{B}_1}(T) \end{bmatrix} = \begin{bmatrix} 0 & 0 \\ 0 & I_m \end{bmatrix}.$$

In particular, $T$ is diagonalizable.

(2) If $A$ is $n \times n$ and $A^2 = A$, define $T_A : \mathbb{R}^n \to \mathbb{R}^n$ by $T_A(X) = AX$ for all $X$ in $\mathbb{R}^n$. Then $T_A^2 = T_A$ (verify) so, by part (1), there exists a basis $\mathcal{B}$ of $\mathbb{R}^n$ such that

$$M_{\mathcal{B}}(T_A) = \begin{bmatrix} 0 & 0 \\ 0 & I_m \end{bmatrix} \text{ where } m = dim(E_1(A)).$$

Since $A = M_{\mathcal{E}}(T_A)$ where $\mathcal{E}$ is the standard basis of $\mathbb{R}^n$, we have $A \sim M_{\mathcal{B}}(T_A)$ by Theorem 3 §5.5, as required.

A subspace $U$ of a vector space $V$ is called a **proper subspace** if $U \neq \{\mathbf{0}\}$ and $U \neq V$. Hence a one-dimensional subspace has no proper subspaces. Our interest here is in how this relates to direct sum decompositions. The vector space $V$ always has the trivial direct decomposition

$$V = \{\mathbf{0}\} \oplus V.$$

We call all other direct decompositions $V = U \oplus W$ **proper**. Thus $V = U \oplus W$ is proper if $U$ is a proper subspace of $V$ (equivalently, if $W$ is a proper subspace of $V$).

If $T : V \rightarrow V$ is an operator, the vector space $V$ is called $T$-**irreducible** if no proper direct decomposition $V = U \oplus W$ can be found where both $U$ and $W$ are $T$-invariant. This certainly happens if $V$ has *no* $T$-invariant subspaces except $\{\mathbf{0}\}$ and $V$, and such a space is called $T$-**simple**.[18] If $dim V = 1$, then $V$ is $T$-simple for every operator $T : V \rightarrow V$ (verify), and Example 3 describes a 2-dimensional, $T$-simple space. However, even though every $T$-simple space is $T$-irreducible, it is possible to have a $T$-irreducible space that is not $T$-simple.

**Example 13**    As in Example 11, let $\mathcal{B} = \{\mathbf{u}, \mathbf{v}\}$ be a basis of $V$ and define $T : V \rightarrow V$ by $T(\mathbf{u}) = \mathbf{u}$ and $T(\mathbf{v}) = \mathbf{u} + \mathbf{v}$. Show that $V$ is a $T$-irreducible space that is not $T$-simple.

PROOF    The subspace $\mathbb{R}\mathbf{u}$ is $T$-invariant by Lemma 1, so $V$ is not $T$-simple because $dim V = 2$. In Example 11 we showed that $\mathbb{R}\mathbf{u}$ has no $T$-complement in $V$; we now show that $V$ is in fact irreducible. Suppose on the contrary that $T$ is reducible, say $V = Z \oplus W$, where $Z$ and $W$ are proper, $T$-invariant subspaces of $V$. Since $dim V = 2$, the fact that $Z$ and $W$ are proper means that $dim Z = 1$ and $dim W = 1$, say $Z = \mathbb{R}\mathbf{z}$ and $W = \mathbb{R}\mathbf{w}$ where $\mathbf{z} \neq \mathbf{0}$ and $\mathbf{w} \neq \mathbf{0}$. Moreover, since $Z$ and $W$ are $T$-invariant, it follows that $\mathbf{z}$ and $\mathbf{w}$ are both eigenvectors of $T$ (verify). But $\{\mathbf{z}, \mathbf{w}\}$ is independent because $\mathbb{R}\mathbf{z} \cap \mathbb{R}\mathbf{w} = \{\mathbf{0}\}$, and so is a basis of eigenvectors of $T$. This means that $T$ is diagonalizable, and hence that $M_\mathcal{B}(T) = \begin{bmatrix} 1 & 1 \\ 0 & 1 \end{bmatrix}$ is a diagonalizable matrix, a contradiction by Example 7 §2.3.

So $T$ is irreducible.

## 5.6.4   The Cayley-Hamilton Theorem

We conclude this section by using invariant subspaces to prove a classical theorem of linear algebra, the Cayley-Hamilton theorem, which asserts that every matrix satisfies its characteristic polynomial. The following example provides an illustration.

**Example 14**    If $A = \begin{bmatrix} 1 & -1 \\ 2 & 3 \end{bmatrix}$, then the characteristic polynomial is $c_A(x) = det(xI - A) = x^2 - 4x + 5$. If we substitute $A$ for $x$ in $c_A(x)$ the result is $c_A(A) = A^2 - 4A + 5I$.[19] The Cayley-Hamilton theorem asserts that in fact $c_A(A) = 0$; indeed

$$c_A(A) = A^2 - 4A + 5I = \begin{bmatrix} -1 & -4 \\ 8 & 7 \end{bmatrix} - \begin{bmatrix} 4 & -4 \\ 8 & 12 \end{bmatrix} + \begin{bmatrix} 5 & 0 \\ 0 & 5 \end{bmatrix} = \begin{bmatrix} 0 & 0 \\ 0 & 0 \end{bmatrix},$$

as expected.

Our approach to the Cayley-Hamilton theorem is to prove it first for operators by showing that every operator $T : V \rightarrow V$ satisfies its characteristic polynomial. This clearly entails defining what is meant by $p(T)$ where $p(x)$ is any polynomial. This, in turn, means that we must define sums and scalar multiples of operators.

Suppose that $S : V \rightarrow V$ and $T : V \rightarrow V$ are operators and that $a$ is a scalar. We define new operators $S + T : V \rightarrow V$ and $aT : V \rightarrow V$ as follows:

$$(S + T)(\mathbf{v}) = S(\mathbf{v}) + T(\mathbf{v}) \quad \text{for all } \mathbf{v} \text{ in } V$$

and

$$(aT)(\mathbf{v}) = a\,T(\mathbf{v}) \quad \text{for all } \mathbf{v} \text{ in } V.$$

---

[18] $T$-simple spaces are sometimes call **completely** $T$-**irreducible**, especially in older books, but we prefer the term $T$-simple.

[19] Note that we are thinking of 5 in $c_A(x)$ as $5x^0$, and then taking $5A^0 = 5I$ in $c_A(A)$.

It is a routine matter to verify that $S + T$ and $aT$ are both linear; they are called the **sum** of $S$ and $T$, and the **scalar product** of $T$ by $a$, respectively. If we recall that the power $T^k$ means the composition of $T$ with itself $k$ times, we can define $p(T)$ for every polynomial $p(x)$. Indeed, if

$$p(x) = a_0 + a_1 x + a_2 x^2 + a_3 x^3 + \cdots + a_m x^m \text{ where each } a_i \text{ is in } \mathbb{R},$$

we define the operator $p(T) : V \to V$ by

$$p(T) = a_0 + a_1 T + a_2 T^2 + a_3 T^3 + \cdots + a_m T^m.$$

Thus, for example, if $p(x) = 3 - 5x + 7x^3$, the action of $p(T)$ on a vector $\mathbf{v}$ is

$$p(T)(\mathbf{v}) = (3 - 5T + 7T^3)(\mathbf{v}) = 3\mathbf{v} - 5T(\mathbf{v}) + 7T^3(\mathbf{v}) \text{ for all } \mathbf{v} \text{ in } V.$$

Of course, $T^3(\mathbf{v})$ means $T(T(T(\mathbf{v})))$ for each vector $\mathbf{v}$.

Now suppose that $\mathbf{v}$ is any nonzero vector in $V$ where $dim V$ is finite, and consider the subsets

$$\{\mathbf{v}\}, \ \{\mathbf{v}, \ T(\mathbf{v})\}, \ \{\mathbf{v}, \ T(\mathbf{v}), \ T^2(\mathbf{v})\}, \ \cdots.$$

The first of these sets is independent because $\mathbf{v} \neq \mathbf{0}$. But they cannot all be independent because $dim V$ is finite, so one of them is dependent. Let $m \geq 1$ be the smallest integer such that $\{\mathbf{v}, \ T(\mathbf{v}), \ T^2(\mathbf{v}), \ \cdots, \ T^m(\mathbf{v})\}$ is dependent. Then

$$\mathcal{B}_0 = \{\mathbf{v}, \ T(\mathbf{v}), \ T^2(\mathbf{v}), \ \cdots, \ T^{m-1}(\mathbf{v})\}$$

is independent. (If $m = 1$, this is taken to mean that $\mathcal{B}_0 = \{\mathbf{v}\}$.)[20] Having found $\mathcal{B}_0$, write

$$U = span\mathcal{B}_0 = span\{\mathbf{v}, \ T(\mathbf{v}), \ T^2(\mathbf{v}), \ \cdots, \ T^{m-1}(\mathbf{v})\} \tag{$*$}$$

so that $dim(U) = m$ and $\mathcal{B}_0$ is a basis of $U$.

Since $\{\mathbf{v}, \ T(\mathbf{v}), \ T^2(\mathbf{v}), \ \cdots, \ T^{m-1}(\mathbf{v}), \ T^m(\mathbf{v})\}$ is dependent, Lemma 2 §5.2 implies that $T^m(\mathbf{v})$ is in $U$. This has two consequences: First we can write

$$T^m(\mathbf{v}) = -r_0\mathbf{v} - r_1 T(\mathbf{v}) - r_2 T^2(\mathbf{v}) - \cdots - r_{m-1}T^{m-1}(\mathbf{v}) \tag{$**$}$$

where each $r_i$ is in $\mathbb{R}$. (The reason for the minus signs will be apparent shortly.) The second consequence of the fact that $T^m(\mathbf{v})$ is in $U$ is that it shows (using Lemma 1) that $U$ is a $T$-invariant subspace of $V$, called the $T$-**cyclic subspace** generated by $\mathbf{v}$. Hence $T : U \to U$ is an operator on $U$, and $(*)$ and $(**)$ show that the $\mathcal{B}_0$-matrix of $T$ is

$$M_{\mathcal{B}_0}(T) = \begin{bmatrix} 0 & 0 & \cdots & 0 & -r_0 \\ 1 & 0 & \cdots & 0 & -r_1 \\ 0 & 1 & \cdots & 0 & -r_2 \\ \vdots & \vdots & \vdots & \vdots \\ 0 & 0 & \cdots & 1 & -r_{m-1} \end{bmatrix}. \tag{$***$}$$

If we let $q(x)$ denote the characteristic polynomial of the restriction of $T$ to $U$, we have

$$q(x) = c_{M_{\mathcal{B}_0}(T)}(x) = det[xI - M_{\mathcal{B}_0}(T)] \tag{$****$}$$
$$= r_0 + r_1 x + r_2 x^2 + \cdots + r_{m-1} x^{m-1} + x^m$$

as the reader can verify.[21] For this reason, the matrix in $(***)$ is called the **companion matrix** of the polynomial $q(x) = r_0 + r_1 x + r_2 x^2 + \cdots + r_{m-1} x^{m-1} + x^m$.

We can now prove the operator version of the Cayley-Hamilton theorem.

**THEOREM 6**　If $T : V \to V$ is a linear operator where $dim V = n$, then $T$ satisfies its characteristic polynomial, that is

$$c_T(T) = 0.$$

---

[20] Note that $\{\mathbf{v}, \ T(\mathbf{v})\}$ is dependent if and only if $\mathbf{v}$ is an eigenvector of $T$ (verify).

[21] This is the reason for the negative signs in $(**)$.

PROOF We are to prove that $c_T(T)(\mathbf{v}) = \mathbf{0}$ for every vector $\mathbf{v}$ in This is clear if $\mathbf{v} = \mathbf{0}$ because $c_T(T)$ is linear. If $\mathbf{v} \neq \mathbf{0}$, we use it to construct $U$, $\mathcal{B}_0$, and $q(x)$ as above. If we complete $\mathcal{B}_0$ to a basis of $V$ in any way at all, then Theorem 2 asserts that

$$c_T(x) = p(x)\,q(x) \quad \text{for some polynomial } p(x).$$

Hence $c_T(T)(\mathbf{v}) = p(T)\,[q(T)(\mathbf{v})]$ so it suffices to show that $q(T)(\mathbf{v}) = \mathbf{0}$. But (****) and (**) give

$$q(T)(\mathbf{v}) = r_0\mathbf{v} + r_1 T(\mathbf{v}) + r_2 T^2(\mathbf{v}) + \cdots + r_{m-1}T^{m-1}(\mathbf{v}) + T^m(\mathbf{v}) = \mathbf{0}.$$

This completes the proof.

The Cayley-Hamilton theorem is the matrix version of Theorem 6. We need one more observation about matrices of operators: If $S : V \to V$ and $T : V \to V$ are operators on a vector space $V$ and $\mathcal{B}$ is any basis of $V$, then

$$M_{\mathcal{B}}(S + T) = M_{\mathcal{B}}(S) + M_{\mathcal{B}}(T)$$
$$M_{\mathcal{B}}(aT) = a\,M_{\mathcal{B}}(T) \quad \text{for any scalar } a$$
$$M_{\mathcal{B}}(S \circ T) = M_{\mathcal{B}}(S)M_{\mathcal{B}}(T)$$

The last of these is part of Theorem 1 §5.5, and the routine verifications of the first two are left as an exercise.

THEOREM 7 **Cayley-Hamilton Theorem.**[22] Every square matrix $A$ satisfies its characteristic polynomial, that is

$$c_A(A) = 0.$$

PROOF Define $T_A : \mathbb{R}^n \to \mathbb{R}^n$ by $T_A(X) = AX$ for all $X$ in $\mathbb{R}^n$. Then $c_{T_A}(T_A) = 0$ by Theorem 6. If $\mathcal{E}$ is the standard basis of $\mathbb{R}^n$, then

$$M_{\mathcal{E}}(T_A) = A \quad \text{and} \quad c_{T_A}(x) = c_{M_{\mathcal{E}}(T_A)}(x) = c_A(x).$$

Now write $c_A(x) = r_0 + r_1 x + r_2 x^2 + \cdots + r_n x^n$, and compute

$$
\begin{aligned}
c_A(A) &= r_0 I + r_1 A + r_2 A^2 + \cdots + r_n A^n \\
&= r_0 M_{\mathcal{E}}(1_{\mathbb{R}^n}) + r_1 M_{\mathcal{E}}(T_A) + r_2 M_{\mathcal{E}}(T_A)^2 + \cdots + r_n M_{\mathcal{E}}(T_A)^n \\
&= M_{\mathcal{E}}(r_0 1_{\mathbb{R}^n} + r_1 T_A + r_2 T_A^2 + \cdots + r_n T_A^n) \\
&= M_{\mathcal{E}}(c_{T_A}(T_A)) \\
&= M_{\mathcal{E}}(0) \\
&= 0.
\end{aligned}
$$

This is what we wanted.

## Exercises 5.6

Throughout these exercises let $V$ denote a vector space and let $U$ and $W$ denote subspaces of $V$.

1. In each case either show that the statement is true or give an example showing that it is false. Throughout, let $T : V \to V$ denote an operator.
   a) $\ker T$ is $T$-invariant.
   b) $\operatorname{im} T$ is $T$-invariant.
   c) It can happen that $V$ has no $T$-invariant subspace.

   d) $U + W$ contains both $U$ and $W$, and is contained in every such subspace.
   e) $U \cap W$ is contained in both $U$ and $W$, and contains every such subspace.
   f) If $T$ is an isomorphism, the only $T$-invariant subspaces are $V$ and $\{0\}$.
   g) If $\mathbf{v}$ is an eigenvector of $T$, then $\mathbb{R}\mathbf{v}$ is $T$-invariant.
   h) If $\mathbb{R}\mathbf{v}$ is $T$-invariant where $\mathbf{v} \neq \mathbf{0}$, then $\mathbf{v}$ is an eigenvector of $T$.

---

[22] This theorem is named after the English mathematician Arthur Cayley (1821–1895), one of the pioneers in matrix theory, and William Rowan Hamilton (1805–1865), an Irish mathematician famous for his work on physical dynamics. The theorem was established for a special class of matrices by Hamilton in 1853, and Cayley announced the general version five years later.

**2.** Let $V$ have basis $\mathcal{B} = \{\mathbf{b}_1, \mathbf{b}_2, \mathbf{b}_3, \mathbf{b}_4\}$, and define $T : V \to V$ by $T(\mathbf{b}_1) = \mathbf{b}_1$, $T(\mathbf{b}_2) = \mathbf{b}_2 + 2\mathbf{b}_4$, $T(\mathbf{b}_3) = \mathbf{b}_1$, $T(\mathbf{b}_4) = \mathbf{0}$. Show that $T^2 = T$, find a basis $\mathcal{B}_o$ of $E_1(T)$, extend it to a basis of $V$, and illustrate Theorem 1.

**3.** Let $T : \mathbb{P}_2 \to \mathbb{P}_2$ be defined by
$T(a + bx + cx^2) = (a - 2b - c) - (a + 3b + c)x - (a + 4b)x^2$.
   **a)** Show that $U = span\{1, \ x + x^2\}$ is $T$-invariant.
   **b)** Use **a)** to find a basis $\mathcal{B}$ of such that $M(T)$ is block upper triangular.
   **c)** Use **b)** to compute $c_T(x)$.

**4. a)** Show that every subspace of $E_\lambda(T)$ is $T$-invariant.
   **b)** Describe the action of the restriction of $T$ to $E_\lambda(T)$.

**5.** If $T : V \to V$ is an operator and $U$ is a subspace of $V$, show that $T(U)$ is a subspace of $V$, $T$-invariant if $U$ is $T$-invariant.

**6.** In each case show that $V = U \oplus W$.
   **a)** $V = \mathbb{M}_{2,2}$, $U = \left\{ \begin{bmatrix} a & b \\ a & b \end{bmatrix} \mid a, b \text{ in } \mathbb{R} \right\}$, and
   $W = \left\{ \begin{bmatrix} c & c \\ d & -d \end{bmatrix} \mid c, d \text{ in } \mathbb{R} \right\}$.
   **b)** $V = \mathbb{P}_3$, $U = \{a + b(x - x^2) \mid a, \ b \text{ in } \mathbb{R}\}$ and $W = \{c(1 + x) + dx^3 \mid c, d \text{ in } \mathbb{R}\}$.
   **c)** $V = \mathbb{R}^4$, $U = \{[a \ -b \ b \ a]^T \mid a, b \text{ in } \mathbb{R}\}$ and $W = \{[c \ d \ c \ d]^T \mid c, d \text{ in } \mathbb{R}\}$.

**7.** Define $T : \mathbb{P}_3 \to \mathbb{P}_3$ by $T(a + bx + cx^2 + dx^3)$
$= (2a + b) + (-5a - 2b)x + (c + 2d)x^2 - (c + d)x^3$
for all $a, b, c$, and $d$. Show that $\mathbb{P}_3$ has two 2-dimensional $T$-invariant subspaces $U$ and $W$ such that $\mathbb{P}_3 = U \oplus W$, but that $T$ has no real eigenvalue. [Hint: First find $M_\mathcal{B}(T)$ where $\mathcal{B} = \{1, \ x, \ x^2, \ x^3\}$.]

**8.** Let $T : V \to V$ be an operator.
   **a)** Show that every subspace of $V$ is $T$-invariant if and only if every nonzero vector $\mathbf{v}$ in $V$ is an eigenvector of $T$.
   **b)** If **a)** holds, show that there is a constant $\lambda$ such that $T(\mathbf{v}) = \lambda \mathbf{v}$ for every vector $\mathbf{v}$ in $V$.

**9.** Let $V$ be a finite dimensional vector space. Show that $V$ and $\{\mathbf{0}\}$ are the only subspaces of $V$ that are $T$-invariant for every operator $T : V \to V$. [Hint: Theorem 3 §5.3.]

**10.** Let $V = U \oplus W$ and $V = U \oplus W_1$.
   **a)** If $dimV$ is finite, show that $dimW = dim(W_1)$.
   **b)** If $W_1 \subseteq W$, show that $W_1 = W$.
   **c)** If $V = U_1 \oplus W_1$ where $U_1 \subseteq U$ and $W_1 \subseteq W$, show that $U_1 = U$ and $W_1 = W$.

**11.** Show that $U \cap W = \{\mathbf{0}\}$ if and only if $\{\mathbf{u}, \ \mathbf{w}\}$ is independent for all nonzero vectors $\mathbf{u}$ in $U$ and $\mathbf{w}$ in $W$.

**12.** If $U \cap W = \{\mathbf{0}\}$ and $dimU + dimW = dimV$, show that $V = U \oplus W$.

**13.** Suppose that $V = U \oplus W$ where both $U$ and $W$ are $T$-invariant. If the restrictions of $T$ to $U$ and $W$ are both diagonalizable, show that $T$ is diagonalizable.

**14.** Let $T : V \to V$ be a linear operator where $dimV = n$.
   **a)** Show that $E_\lambda(T) \cap E_\mu(T) = \{\mathbf{0}\}$ if $\lambda \neq \mu$.
   **b)** If $\lambda \neq \mu$ are the only eigenvalues of an operator $T$, and if $dim(E_\lambda(T)) + dim(E_\mu(T)) = n$, show that $T$ is diagonalizable.
   **c)** If $\lambda \neq \mu$ are the only eigenvalues of $T$ and $dimV = 2$, show that $T$ is diagonalizable.
   **d)** Give an example where $T$ has exactly two distinct eigenvalues but is not diagonalizable.

**15.** Let $T : V \to V$ be an operator. If $V$ is $T$-simple and $T \neq 0$, show that $T$ is an isomorphism. (**Schur's Lemma.**) [Hint: Consider $kerT$ and $imT$.]

**16.** Let $S$ and $T$ be linear operators $V \to V$, and assume that $S \circ T = T \circ S$.
   **a)** If $\lambda$ is an eigenvalue of $S$, show that $E_\lambda(S)$ is $T$-invariant.
   **b)** Show that $imS$ and $kerS$ are both $T$-invariant.
   **c)** If $U$ is a $T$-invariant subspace of $V$, show that $S(U)$ is also $T$-invariant.

**17.** Let $T : V \to V$ be an operator, and let $U$ be the $T$-cyclic subspace generated by $\mathbf{v} \neq \mathbf{0}$ in $V$ (see the discussion preceding Theorem 6). If $W$ is any $T$-invariant subspace that contains $\mathbf{v}$, show that $U \subseteq W$. [Thus $U$ is the "smallest" $T$-invariant subspace containing $\mathbf{v}$.]

**18.** Determine if the following are true or false. Justify your answer.
   **a)** If $dimV = 3$, and $V = U \oplus W$ where both $U$ and $W$ are $T$-invariant, then $T$ is diagonalizable.
   **b)** If $dimV = 2$, and $V = U \oplus W$ where both $U$ and $W$ are $T$-invariant, then $T$ is diagonalizable.

**19.** If $T : V \to V$ is an operator and $dimV = 1$, show that $V$ is $T$-simple.

**20.** Let $T : V \to V$ be an operator. If $U = \mathbb{R}\mathbf{u}$ is a 1-dimensional subspace of $V$, show that $U$ is $T$-invariant if and only if $\mathbf{u}$ is an eigenvector of $T$.

**21. a)** If $dimV > 1$ and $V$ is $T$-simple, show that $T$ has no real eigenvalue.
   **b)** Let $A = \begin{bmatrix} B & 0 \\ 0 & C \end{bmatrix}$ where $B = \begin{bmatrix} 2 & 1 \\ -5 & -2 \end{bmatrix}$ and $C = \begin{bmatrix} -1 & 1 \\ -2 & 1 \end{bmatrix}$.
   Show the converse of **a)** is false as follows: Consider the operator $T_A : \mathbb{R}^4 \to \mathbb{R}^4$ such that $T_A(X) = AX$ for all $X$ in $\mathbb{R}^4$. Show that $T_A$ has no real eigenvalue, but that $\mathbb{R}^4 = U \oplus W$ where $U$ and $W$ are both 2-dimensional and $T_A$-invariant.

**22.** Let $T : V \to V$ be an operator where $dimV = 2$. Show that $V$ is $T$-reducible if and only if $T$ is diagonalizable.

23. Let $T : V \rightarrow V$ be an operator with real eigenvalue $\lambda$. If $\mathcal{B}_0$ is any basis of $E_\lambda(T)$, show that $M_{\mathcal{B}_0}(T) = \lambda I$.

24. If $T : V \rightarrow V$ is an operator and $U$ and $W$ are $T$-invariant subspaces of $V$, show that $U \cap W$ and $U + W$ are also $T$-invariant.

25. a) If $V = U \oplus W$, define $T : V \rightarrow V$ as follows: Given $\mathbf{v}$ in $V$, write it as $\mathbf{v} = \mathbf{u} + \mathbf{w}$ with $\mathbf{u}$ in $U$ and $\mathbf{w}$ in $W$, and take $T(\mathbf{v}) = \mathbf{u}$. Show that $T$ is an operator, $T^2 = T$, $imT = U$, and $kerT = W$.

   b) Show that every operator $T : V \rightarrow V$ that satisfies $T^2 = T$ arises as in **a)** for some direct decomposition $V = U \oplus W$. [Hint : Try $U = imT$ and $W = kerT$.]

26. a) An operator $T : V \rightarrow V$ is called an **involution** if $T^2 = 1_V$. (Examples: conjugation on $\mathbb{C}$; transposition on $\mathbb{M}_{n,n}$; $T(p(x)) = p(1 - x)$ on $\mathbb{P}_n$.) If $T$ is an involution, show that $T$ is diagonalizable by showing that $V$ is the direct sum of the eigenspaces of $T$. [Hint: Example 12.]

   b) If $A^2 = I$ where $A$ is a square matrix, show that $A$ is similar to a block matrix of the form $\begin{bmatrix} I_m & 0 \\ 0 & -I_k \end{bmatrix}$. [Hint: Example 12.]

27. Let $S : V \rightarrow V$ and $T : V \rightarrow V$ be operators, let $\mathcal{B}$ be a basis of $V$, and let $a$ be a scalar.

   a) Show that $S + T$ and $aT$ are operators.

   b) Show that $M_{\mathcal{B}}(S + T) = M_{\mathcal{B}}(S) + M_{\mathcal{B}}(T)$ and $M_{\mathcal{B}}(aT) = aM_{\mathcal{B}}(T)$.

## 5.7   GENERAL INNER PRODUCTS

The importance of the dot product in $\mathbb{R}^n$ would be difficult to exaggerate. It leads to the idea of an orthogonal basis and hence to a number of results of both theoretical and practical interest, including the Projection theorem, approximation and least squares, the Principal Axis theorem, and the Principal Value Decomposition. In this section we generalize the dot product to the idea of an inner product on an arbitrary vector space, and so see that the techniques we developed for $\mathbb{R}^n$ in Chapter 4 using the dot product will generalize to a much wider context.

### 5.7.1   Inner Products

An **inner product** $\langle \ , \ \rangle$ on a vector space $V$ is a function that assigns a real number $\langle \mathbf{v}, \ \mathbf{w} \rangle$ to every pair of vectors $\mathbf{v}$ and $\mathbf{w}$ in $V$ in such a way that the following axioms are satisfied for all vectors $\mathbf{u}$, $\mathbf{v}$, and $\mathbf{w}$ in $V$ and all scalars $r$:

> **P1.** $\langle \mathbf{v}, \ \mathbf{w} \rangle = \langle \mathbf{w}, \ \mathbf{v} \rangle$.
> **P2.** $\langle \mathbf{u}, \ \mathbf{v} + \mathbf{w} \rangle = \langle \mathbf{u}, \ \mathbf{v} \rangle + \langle \mathbf{u}, \ \mathbf{w} \rangle$.
> **P3.** $\langle r\mathbf{v}, \ \mathbf{w} \rangle = r \langle \mathbf{v}, \ \mathbf{w} \rangle$.
> **P4.** $\langle \mathbf{v}, \ \mathbf{v} \rangle > 0$ whenever $\mathbf{v} \neq \mathbf{0}$.

A vector space endowed with an inner product is called an **inner product space**. Note that $\langle \mathbf{v}, \ r\mathbf{w} \rangle = r \langle \mathbf{v}, \ \mathbf{w} \rangle$ is an immediate consequence of Axioms P1 and P3, and that $\langle \mathbf{v} + \mathbf{w}, \ \mathbf{u} \rangle = \langle \mathbf{v}, \ \mathbf{u} \rangle + \langle \mathbf{w}, \ \mathbf{u} \rangle$ follows from Axioms P1 and P2.

Before giving examples, we clarify Axiom P4. Axiom P2 gives $\langle \mathbf{v}, \ \mathbf{0} \rangle = \langle \mathbf{v}, \ \mathbf{0} + \mathbf{0} \rangle = \langle \mathbf{v}, \ \mathbf{0} \rangle + \langle \mathbf{v}, \ \mathbf{0} \rangle$, so

$$\langle \mathbf{v}, \ \mathbf{0} \rangle = 0 \text{ for every vector } \mathbf{v} \text{ in } V.$$

With Axiom P1, this proves the first part of the following theorem and gives some useful refinements of Axiom P4 (parts (2) and (3) in the theorem).

**THEOREM 1**   If $\mathbf{v}$ is any vector in an inner product space, then:

(1) $\langle \mathbf{0}, \ \mathbf{v} \rangle = 0 = \langle \mathbf{v}, \ \mathbf{0} \rangle$.

(2) $\langle \mathbf{v}, \ \mathbf{v} \rangle \geq 0$ for all vectors $\mathbf{v}$.

(3) $\langle \mathbf{v}, \ \mathbf{v} \rangle = 0$ if and only if $\mathbf{v} = \mathbf{0}$.

**Example 1**    The dot product $\langle X, Y \rangle = X \cdot Y$ is an inner product on $\mathbb{R}^n$ (see Theorem 1 §4.5).

**Example 2**    Define $\langle \, , \, \rangle$ on $\mathbb{M}_{n,n}$ by $\langle A, B \rangle = tr(AB^T)$, where $trA$ denotes the trace of the square matrix $A$. Then Axioms P1, P2, and P3 follow from the fact that $tr(\ )$ is a linear transformation which satisfies $tr(A^T) = trA$ for all $A$. If $R_1, \cdots, R_n$ are the rows of $A$, then the $(i, j)$-entry of $AA^T$ is $R_i \cdot R_j$ (verify), so

$$\langle A, A \rangle = tr(AA^T) = R_1 \cdot R_1 + R_2 \cdot R_2 + \cdots + R_n \cdot R_n$$
$$= \|R_1\|^2 + \|R_2\|^2 + \cdots + \|R_n\|^2.$$

Hence if $A \neq 0$, then some $R_i \neq 0$, so $\|R_i\|^2 > 0$ and $\langle A, A \rangle > 0$. This is Axiom P4. We note in passing that $\langle A, A \rangle$ is the sum of the squares of all $n^2$ entries of the $n \times n$ matrix $A$.

**Example 3**    Any $n + 1$ distinct real numbers $a_0, a_1, a_2, \cdots, a_n$ determine an inner product $\langle \, , \, \rangle$ on $\mathbb{P}_n$ as follows: Define

$$\langle p(x), q(x) \rangle = p(a_0)q(a_0) + \cdots + p(a_n)q(a_n)$$

for all polynomials $p(x)$ and $q(x)$ in $\mathbb{P}_n$. Axioms P1, P2, and P3 are routine verifications. If $p(x) \neq 0$, then $p(a_i) \neq 0$ for some $i$ because nonzero polynomials of degree at most $n$ can have at most $n$ distinct roots (see Appendix A.3). But then $\langle p(x), p(x) \rangle = p(a_0)^2 + p(a_1)^2 + \cdots + p(a_n)^2 > 0$, proving Axiom P4.

The next example involves calculus.

**Example 4**    The set $\mathbf{C}[a, b]$ of all continuous real valued functions on the interval $[a, b]$ is a subspace of $\mathbb{F}[a, b]$. Define $\langle \, , \, \rangle$ on $\mathbf{C}[a, b]$ by

$$\langle f, g \rangle = \int_a^b f(x)g(x)dx.$$

Again, all the axioms except P4 are routine properties of integrals. But if $\langle f, f \rangle = 0$, then $\int_a^b f(x)^2 dx = 0$. Since $f(x)^2$ is continuous and $f(x)^2 \geq 0$ for all $x$ in $[a, b]$, it is a theorem of calculus (see the diagram) that $f(x)^2 = 0$ for all $x$. Hence $f = 0$ is the zero function, proving that Axiom P4 holds.

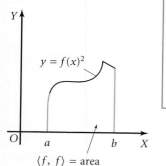

$\langle f, f \rangle$ = area

Because of axioms P2 and P3, every inner product is linear in the first variable, that is

$$\langle r_1\mathbf{v}_1 + r_2\mathbf{v}_2 + \cdots + r_m\mathbf{v}_m, \mathbf{w} \rangle = r_1 \langle \mathbf{v}_1, \mathbf{w} \rangle + r_2 \langle \mathbf{v}_2, \mathbf{w} \rangle + \cdots + r_m \langle \mathbf{v}_m, \mathbf{w} \rangle$$

for all vectors $\mathbf{v}_i$ and $\mathbf{w}$. Of course, linearity in the second variable now follows from Axiom P1. Hence computations with an inner product are carried out much as for the dot product in $\mathbb{R}^n$.

In fact, the linearity enables us to easily describe *all* inner products $\langle \, , \, \rangle$ on $\mathbb{R}^n$. Let $\mathcal{E} = \{E_1, E_2, \cdots, E_n\}$ denote the standard basis of $\mathbb{R}^n$, and consider the $n \times n$ matrix $P = [p_{ij}]$ where

$$p_{ij} = \langle E_i, E_j \rangle \quad \text{for all } i \text{ and } j.$$

If $X = [x_1 \, x_2 \, \cdots \, x_n]^T$ and $Y = [y_1 \, y_2 \, \cdots \, y_n]^T$ are any two vectors in $\mathbb{R}^n$, then $X = \Sigma_i x_i E_i$ and $Y = \Sigma_j y_j E_j$, so the linearity of $\langle \, , \, \rangle$ gives $\langle X, Y \rangle$ as a matrix product:

$$\langle X, Y \rangle = \langle \Sigma_i x_i E_i, \Sigma_j y_j E_j \rangle = \Sigma_{i,j} x_i y_j \langle E_i, E_j \rangle = \Sigma_{i,j} x_i p_{ij} y_j = X^T P Y.$$

In particular $X^T P X = \langle X, X \rangle > 0$ whenever $X \neq 0$ by P4. Since $P$ is symmetric (by Axiom P1), this shows that $P$ is a positive definite matrix (see Section 4.8). On the other hand, if $P$ is any positive definite matrix and we define $\langle X, Y \rangle = X^T P Y$ for all $X$ and $Y$ in $\mathbb{R}^n$, it is a routine matter to verify that $\langle \, , \, \rangle$ is an inner product on $\mathbb{R}^n$. Hence we have

**THEOREM 2**

Every positive definite matrix $P$ gives rise to an inner product $\langle\ ,\ \rangle$ on $\mathbb{R}^n$ given by
$$\langle X,\ Y\rangle = X^T PY \quad \text{for all } X \text{ and } Y \text{ in } \mathbb{R}^n.$$
Moreover, every inner product on $\mathbb{R}^n$ arises in this way from some positive definite matrix $P$.

Note that the dot product on $\mathbb{R}^n$ corresponds to $P = I$ in Theorem 2.

Theorem 2 can be stated also in the language of quadratic forms in Section 4.8: If $\langle\ ,\ \rangle$ is an inner product on $\mathbb{R}^n$, then $q(X) = \langle X,\ X\rangle$ is a positive definite quadratic form in $n$ variables, and every such quadratic form arises in this way from an inner product.

**Example 5**

Define $\langle\ ,\ \rangle$ on $\mathbb{R}^2$ by
$$\big\langle [x_1,\ x_2]^T,\ [y_1,\ y_2]^T\big\rangle = x_1y_1 - 2x_1y_2 - 2x_2y_1 + 5x_2y_2.$$

Hence $P = \begin{bmatrix} 1 & -2 \\ -2 & 5 \end{bmatrix}$ in Theorem 2. Since $P$ is a positive definite matrix by Theorem 4 §4.8, it follows that $\langle\ ,\ \rangle$ is an inner product. Note that Axioms P1, P2, and P3 hold for *any* symmetric matrix $P$; Axiom P4 also can be seen directly in this case because
$$\big\langle [x_1,\ x_2]^T,\ [x_1,\ x_2]^T\big\rangle = x_1^2 - 4x_1x_2 + 5x_2^2 = (x_1 - 2x_2)^2 + x_2^2.$$

## 5.7.2 Norms and Orthogonality

We showed in Theorem 1 that $\langle \mathbf{v},\ \mathbf{v}\rangle \geq 0$ holds for all vectors $\mathbf{v}$ in any inner product space $V$. Hence, as in $\mathbb{R}^n$, we can define the **norm**[23] of $\mathbf{v}$ by
$$\|\mathbf{v}\| = \sqrt{\langle \mathbf{v},\ \mathbf{v}\rangle}.$$
A vector of norm 1 is called a **unit vector**.

**Example 6**

Compute $\|p(x)\|$ if $p(x) = 2 - x - x^2$ where $\mathbb{P}_2$ has the norm in Example 3 determined by the numbers $-1$, $1$, and $2$.

SOLUTION

$\|p(x)\|^2 = \big\langle p(x),\ p(x)\big\rangle = p(-1)^2 + p(1)^2 + p(2)^2 = 2^2 + 0^2 + (-4)^2 = 20,$
so $\|p(x)\| = 2\sqrt{5}$.

The proof of Theorem 2 §4.5 and its Corollary goes through as written to prove the following fundamental inequalities about norms.

**THEOREM 3**

Let $V$ be an inner product space.
(1) **Cauchy-Schwarz Inequality.** $|\langle \mathbf{v},\ \mathbf{w}\rangle| \leq \|\mathbf{v}\|\ \|\mathbf{w}\|$ for all $\mathbf{v}$ and $\mathbf{w}$ in $V$.
(2) **Triangle Inequality.** $\|\mathbf{v} + \mathbf{w}\| \leq \|\mathbf{v}\| + \|\mathbf{w}\|$ for all $\mathbf{v}$ and $\mathbf{w}$ in $V$.

Again motivated by the situation in $\mathbb{R}^n$, we say that two vectors $\mathbf{v}$ and $\mathbf{w}$ are **orthogonal** if
$$\langle \mathbf{v},\ \mathbf{w}\rangle = 0$$

---

[23] The term *length* is used for $\mathbb{R}^n$, reflecting the geometric interpretation of $\mathbb{R}^2$ and $\mathbb{R}^3$, but *norm* is used more commonly for spaces of polynomials or functions.

and we say that a set $\{\mathbf{e}_0, \mathbf{e}_1, \cdots, \mathbf{e}_m\}$ of vectors is an **orthogonal set**[24] if
$$\mathbf{e}_i \neq \mathbf{0} \text{ for each } i, \text{ and } \langle \mathbf{e}_i, \mathbf{e}_j \rangle = 0 \text{ for all } i \neq j.$$
An orthogonal set consisting of unit vectors is called an **orthonormal** set. If $\{\mathbf{e}_0, \mathbf{e}_1, \cdots, \mathbf{e}_m\}$ is an orthogonal set of vectors, we can **normalize** it to obtain the orthonormal set $\left\{ \frac{1}{\|\mathbf{e}_0\|}\mathbf{e}_0, \frac{1}{\|\mathbf{e}_1\|}\mathbf{e}_1, \cdots, \frac{1}{\|\mathbf{e}_m\|}\mathbf{e}_m \right\}$.

**Example 7**    Show that $\{\sin x, \cos x\}$ is orthogonal in the space $\mathbf{C}[-\pi, \pi]$ of all continuous real valued functions on the interval $[-\pi, \pi]$.

SOLUTION    We have $\langle \sin x, \cos x \rangle = \int_{-\pi}^{\pi} \sin x \, \cos x \, dx = -\frac{1}{4}[\cos(2x)]_{-\pi}^{\pi} = 0$.

The following theorem includes Theorems 4 and 5 in Section 4.5 in the case of the dot product in $\mathbb{R}^n$, and the same proofs go through.

**THEOREM 4**    Let $\{\mathbf{e}_0, \mathbf{e}_1, \cdots, \mathbf{e}_m\}$ be an orthogonal set in an inner product space $V$.
(1) $\{\mathbf{e}_0, \mathbf{e}_1, \cdots, \mathbf{e}_m\}$ is independent.
(2) **Pythagoras' Theorem.** $\|\mathbf{e}_0 + \mathbf{e}_1 + \cdots + \mathbf{e}_m\|^2 = \|\mathbf{e}_0\|^2 + \|\mathbf{e}_1\|^2 + \cdots + \|\mathbf{e}_m\|^2$.

Part (1) in Theorem 4 suggests the concept of an orthogonal basis for an arbitrary inner product space $V$, and the general version of the Expansion theorem (Theorem 6 in Section 4.5) remains valid.

**THEOREM 5**    **Expansion Theorem.** Let $\{\mathbf{e}_1, \mathbf{e}_2, \cdots, \mathbf{e}_n\}$ be an orthogonal basis of an inner product space $V$. Then
$$\mathbf{v} = \frac{\langle \mathbf{v}, \mathbf{e}_0 \rangle}{\|\mathbf{e}_0\|^2}\mathbf{e}_0 + \frac{\langle \mathbf{v}, \mathbf{e}_1 \rangle}{\|\mathbf{e}_1\|^2}\mathbf{e}_1 + \cdots + \frac{\langle \mathbf{v}, \mathbf{e}_n \rangle}{\|\mathbf{e}_n\|^2}\mathbf{e}_n \quad \text{for every vector } \mathbf{v} \text{ in } V.$$

We are going to make use of this general version of the Expansion theorem in several situations. The first application requires an important orthogonal basis of $\mathbb{P}_n$.

**Example 8**    **Lagrange polynomials.**[25] Consider the inner product on $\mathbb{P}_n$ determined (see Example 3) by $n + 1$ distinct real numbers, $a_0, a_1, a_2, \cdots, a_n$:
$$\langle p(x), q(x) \rangle = p(a_0)q(a_0) + p(a_1)q(a_1) + \cdots + p(a_n)q(a_n).$$
Define the **Lagrange polynomials** $\delta_0(x), \delta_1(x), \delta_2(x), \cdots, \delta_n(x)$ by
$$\delta_k(x) = \frac{\Pi_{i \neq k}(x - a_i)}{\Pi_{i \neq k}(a_k - a_i)}$$
where $\Pi_{i \neq k}(x - a_i)$ means the product of all the terms $(x - a_i)$ with $(x - a_k)$ omitted. Note that the denominator of $\delta_k(x)$ is nonzero because the $a_i$ are distinct. Clearly each $\delta_k(x)$ has degree $n$, and we have
$$\delta_k(a_k) = 1 \quad \text{and} \quad \delta_k(a_i) = 0 \quad \text{if } i \neq k. \tag{$*$}$$

---

[24] We are labelling from 0 here in anticipation of the main application to infinite orthogonal sets, where this is customary.

[25] Joseph Louis Lagrange (1736–1813) was one of the great mathematicians. He made important contributions to many parts of mathematics. He is also remembered for his skill at exposition, and his *Méchanique Analytique* was described by Hamilton as a "scientific poem".

If $p(x)$ is any polynomial in $\mathbb{P}_n$, this shows that

$$\langle p(x),\ \delta_k(x)\rangle = p(a_k) \quad \text{for each } k.$$

In particular, if we take $p(x) = \delta_i(x)$ it follows that

$$\mathcal{D} = \{\delta_0(x),\ \delta_1(x),\ \delta_2(x),\ \cdots,\ \delta_n(x)\}$$

is an orthonormal basis of $\mathbb{P}_n$. Hence the Expansion theorem gives the following simple formula for the expansion of any polynomial $p(x)$ in terms of the basis $\mathcal{D}$:

$$p(x) = p(a_0)\ \delta_0(x) + p(a_1)\ \delta_1(x) + \cdots + p(a_n)\ \delta_n(x).$$

More generally, let $f$ be any function in $\mathbb{F}[a,\ b]$, where $[a,\ b]$ is an interval containing all the $a_i$. Then (∗) shows that the polynomial

$$p_f(x) = f(a_0)\ \delta_0(x) + f(a_1)\ \delta_1(x) + \cdots + f(a_n)\ \delta_n(x) \tag{∗∗}$$

agrees with $f$ on the $n+1$ numbers $a_i$; that is,

$$f(a_i) = p_f(a_i) \quad \text{holds for each } i = 1,\ 2,\ \cdots,\ n.$$

The polynomial $p_f(x)$ is called the **Lagrange interpolation polynomial** for $f$, and is the unique polynomial in $\mathbb{P}_n$ that coincides with $f$ at each $a_i$.

The Lagrange polynomials have many applications, one of which we explore in the following example.

**Example 9** **Numerical integration.** Suppose that we want to approximate $\int_a^b f(x)dx$. The idea is to choose $n+1$ distinct numbers $a_0,\ a_1,\ a_2,\ \cdots,\ a_n$ in the interval $[a,\ b]$, let $\delta_0(x),\ \delta_1(x),\ \delta_2(x),\ \cdots,\ \delta_n(x)$ be the corresponding Lagrange polynomials, and approximate $f$ by the Lagrange interpolation polynomial $p_f$ given in (∗∗). Then we obtain the approximation (called a **Gaussian quadrature formula**)

$$\int_a^b f(x)dx \approx \int_a^b p_f(x)dx = f(a_0)\ d_0 + f(a_1)\ d_1 + \cdots + f(a_n)\ d_n$$

where $\approx$ means "approximately equal", and $d_i = \int_a^b \delta_i(x)dx$ for each $i$. This approximation is exact by Example 8 if $f$ is a polynomial of degree at most $n$ and, with careful choice of the $a_i$, is an important tool in numerical analysis.

The techniques for finding orthogonal sets in an arbitrary inner product space $V$ are entirely analogous to those in $\mathbb{R}^n$. The following lemma shows how to enlarge any orthogonal set; the proof of Lemma 1 §4.5 goes through as written.

**LEMMA 1** **Orthogonal Lemma.** Let $\{\mathbf{e}_0,\ \mathbf{e}_1,\ \cdots,\ \mathbf{e}_m\}$ be an orthogonal set in an inner product space $V$. Given $\mathbf{v}$ in $V$, write

$$\mathbf{e}_{m+1} = \mathbf{v} - \left( \frac{\langle \mathbf{v},\ \mathbf{e}_0\rangle}{\|\mathbf{e}_0\|^2}\mathbf{e}_0 + \frac{\langle \mathbf{v},\ \mathbf{e}_1\rangle}{\|\mathbf{e}_1\|^2}\mathbf{e}_1 + \cdots + \frac{\langle \mathbf{v},\ \mathbf{e}_m\rangle}{\|\mathbf{e}_m\|^2}\mathbf{e}_m \right).$$

Then $\langle \mathbf{e}_{m+1},\ \mathbf{e}_i\rangle = 0$ for each $i = 1,\ 2,\ \cdots,\ m$, and if $\mathbf{v}$ is not in $span\{\mathbf{e}_0,\ \mathbf{e}_1,\ \cdots,\ \mathbf{e}_m\}$ then $\{\mathbf{e}_0,\ \mathbf{e}_1,\ \cdots,\ \mathbf{e}_m,\ \mathbf{e}_{m+1}\}$ is an orthogonal set.

As in $\mathbb{R}^n$, the Orthogonal lemma leads to the general version of the Gram-Schmidt algorithm for obtaining an orthogonal basis of a subspace from any convenient basis. Again the proof of Theorem 8 §4.5 goes through.

**THEOREM 6**    **Gram-Schmidt Algorithm.** If $\{\mathbf{b}_0, \mathbf{b}_1, \cdots, \mathbf{b}_m\}$ is any basis of a subspace $U$ of an inner product space $V$, construct $\mathbf{e}_0, \mathbf{e}_1, \cdots, \mathbf{e}_m$, in $U$ successively as follows:

$$\mathbf{e}_0 = \mathbf{b}_0$$

$$\mathbf{e}_1 = \mathbf{b}_1 - \frac{\langle \mathbf{b}_1, \mathbf{e}_0 \rangle}{\|\mathbf{e}_0\|^2} \mathbf{e}_0$$

$$\mathbf{e}_2 = \mathbf{b}_2 - \frac{\langle \mathbf{b}_2, \mathbf{e}_0 \rangle}{\|\mathbf{e}_0\|^2} \mathbf{e}_0 - \frac{\langle \mathbf{b}_2, \mathbf{e}_1 \rangle}{\|\mathbf{e}_1\|^2} \mathbf{e}_1$$

$$\vdots$$

$$\mathbf{e}_k = \mathbf{b}_k - \frac{\langle \mathbf{b}_k, \mathbf{e}_0 \rangle}{\|\mathbf{e}_0\|^2} \mathbf{e}_0 - \frac{\langle \mathbf{b}_k, \mathbf{e}_1 \rangle}{\|\mathbf{e}_1\|^2} \mathbf{e}_1 - \cdots - \frac{\langle \mathbf{b}_k, \mathbf{e}_{k-1} \rangle}{|\mathbf{e}_{k-1}\|^2} \mathbf{e}_{k-1}$$

$$\vdots$$

for each $k = 1, 2, 3, \cdots, m$. Then:

(1) $\{\mathbf{e}_0, \mathbf{e}_1, \cdots, \mathbf{e}_m\}$ is an orthogonal basis of $U$.
(2) $span\{\mathbf{e}_0, \mathbf{e}_1, \cdots, \mathbf{e}_k\} = span\{\mathbf{b}_0, \mathbf{b}_1, \cdots, \mathbf{b}_k\}$ for each $k = 0, 1, \cdots, m$.

**Example 10**    If $V = \mathbb{P}_3$ with the inner product $\langle p, q \rangle = \int_{-1}^{1} p(x)q(x)dx$, apply the Gram-Schmidt algorithm to the basis $\{1, x, x^2, x^3\}$ of $\mathbb{P}_3$ to obtain the orthogonal basis $\left\{1, x, \frac{1}{3}(3x^2 - 1), \frac{1}{5}(5x^3 - 3x)\right\}$.

**SOLUTION**    By the Gram-Schmidt algorithm, we obtain polynomials $q_0, q_1, \cdots$ one after the other.[26] To begin with, we have $q_0(x) = 1$. The algorithm continues:

$$q_1(x) = x - \frac{\langle x, q_0 \rangle}{\|q_0\|^2} q_0 = x - \frac{0}{2} q_1 = x,$$

$$q_2(x) = x^2 - \frac{\langle x^2, q_0 \rangle}{\|q_0\|^2} q_0 - \frac{\langle x^2, q_1 \rangle}{\|q_1\|^2} q_1$$

$$= x^2 - \frac{2/3}{2} q_0 - 0 q_1 = x^2 - \frac{1}{3} = \frac{1}{3}(3x^2 - 1),$$

$$q_3(x) = x^3 - \frac{\langle x^3, q_0 \rangle}{\|q_0\|^2} q_0 - \frac{\langle x^3, q_1 \rangle}{\|q_1\|^2} q_1 - \frac{\langle x^3, q_2 \rangle}{\|q_2\|^2} q_2$$

$$= x^3 - 0 q_0 - \frac{2/5}{2/3} q_1 - 0 q_2$$

$$= x^3 - \frac{3}{5} x = \frac{1}{5}(5x^3 - 3x).$$

The polynomials in Example 10 all have leading coefficient 1. In applications (for example differential equations) it is customary to take scalar multiples $p(x)$ of these polynomials such that $p(1) = 1$. The result is the orthogonal basis

$$\left\{1, x, \frac{1}{2}(3x^2 - 1), \frac{1}{2}(5x^3 - 3x)\right\} \quad \text{of } \mathbb{P}_3.$$

These are the first four **Legendre polynomials**, and we will return to them below.

Let $U$ be a subspace of an arbitrary inner product space $V$. As in $\mathbb{R}^n$, define the **orthogonal complement** $U^{\perp}$ of $U$ in $V$ by

$$U^{\perp} = \{\mathbf{v} \text{ in } V \mid \langle \mathbf{u}, \mathbf{v} \rangle = \mathbf{0} \text{ for all } \mathbf{u} \text{ in } U\}.$$

This is a subspace of $V$ even if $V$ is not finite dimensional, and we have

**THEOREM 7**    If $U$ is a finite dimensional subspace of an arbitrary inner product space $V$, then $V = U \oplus U^{\perp}$.

---

[26] For simplicity of notation, we occasionally write a polynomial $q(x)$ simply as $q$. This amounts to thinking of $q$ as a function in $\mathbb{F}[\mathbb{R}]$.

**PROOF**  If $\mathbf{v}$ is in $U \cap U^\perp$, then $\|\mathbf{v}\|^2 = \langle \mathbf{v}, \mathbf{v} \rangle = 0$, so $\mathbf{v} = \mathbf{0}$. Hence $U \cap U^\perp = \{\mathbf{0}\}$. Now choose an orthogonal basis $\{\mathbf{e}_0, \mathbf{e}_1, \cdots, \mathbf{e}_m\}$ of $U$. If $\mathbf{v}$ is any vector in $V$, the Orthogonal lemma shows that $\mathbf{v} - (r_0\mathbf{e}_0 + r_1\mathbf{e}_1 + \cdots + r_m\mathbf{e}_m)$ is in $U^\perp$ for certain real numbers $r_i$. It follows that $V = U + U^\perp$.

If $U$ is a subspace of $\mathbb{R}^n$, we showed that $U^{\perp\perp} = U$ (Theorem 3 §4.6 ). However, lest the reader get the idea that *all* properties of orthogonal complements in $\mathbb{R}^n$ extend to the general case, we now give an example where $U^{\perp\perp} \neq U$.

**Example 11**  Consider $\mathbb{P}$ with the inner product $\langle p, q \rangle = \int_0^1 p(x)q(x)dx$, and let $U$ denote the subspace of all polynomials with zero constant term. Show that $U^\perp = \{0\}$, so $U^{\perp\perp} = \mathbb{P} \neq U$.

**SOLUTION**  Observe first that $U = \{x\, q(x) \mid q(x) \text{ in } \mathbb{P}\}$. Suppose that $p(x)$ is in $U^\perp$. Then $\langle p(x), x\, q(x) \rangle = 0$ for all $q(x)$ in $\mathbb{P}$, so

$$\langle x\, p(x), q(x) \rangle = \int_0^1 x\, p(x)q(x)dx = \int_0^1 p(x)(x\, q(x))dx = \langle p(x), x\, q(x) \rangle = 0$$

for all $q(x)$ in $\mathbb{P}$. Taking $q(x) = x\, p(x)$ gives $\|x\, p(x)\|^2 = 0$, whence $x\, p(x) = 0$. This implies that $p(x) = 0$, and hence that $U^\perp = \{0\}$, as asserted. Note also that $U \oplus U^\perp \neq \mathbb{P}$.

It is no coincidence that the subspace $U$ is infinite dimensional in Example 11; we show below (Theorem 9) that $U^{\perp\perp} = U$ holds for all finite dimensional subspaces of any inner product space.

## 5.7.3  Projections and Approximation

Let $\mathbf{v}$ be a vector in an inner product space $V$, and let $U$ be a finite dimensional subspace of $V$ with orthogonal basis $\{\mathbf{e}_0, \mathbf{e}_1, \cdots, \mathbf{e}_m\}$. As in Section 4.6, define the **projection** of $\mathbf{v}$ on $U$ by

$$proj_U(\mathbf{v}) = \frac{\langle \mathbf{v}, \mathbf{e}_0 \rangle}{\|\mathbf{e}_0\|^2}\mathbf{e}_0 + \frac{\langle \mathbf{v}, \mathbf{e}_1 \rangle}{\|\mathbf{e}_1\|^2}\mathbf{e}_1 + \cdots + \frac{\langle \mathbf{v}, \mathbf{e}_m \rangle}{\|\mathbf{e}_m\|^2}\mathbf{e}_m. \qquad (*\!*\!*)$$

As the notation implies, $proj_U(\mathbf{v})$ depends on $U$ and $\mathbf{v}$ only, and not on the choice of orthogonal basis $\{\mathbf{e}_0, \mathbf{e}_1, \cdots, \mathbf{e}_m\}$ of $U$ (the proof of Lemma 1 §4.6 goes through). The basic properties of projections are given in the following result which extends the Projection theorem for $\mathbb{R}^n$ (Theorem 2 §4.6) to the case where $V$ is an arbitrary inner product which need not be finite dimensional.

**THEOREM 8**  **Projection Theorem**. Let $U$ be a finite dimensional subspace of an arbitrary inner product space $V$. Then:

(1) $proj_U(\mathbf{v})$ is in $U$ and $\mathbf{v} - proj_U(\mathbf{v})$ is in $U^\perp$ for every vector $\mathbf{v}$ in $V$.
(2) $proj_U : V \to V$ is a linear transformation with image $U$ and kernel $U^\perp$.

**PROOF**  (1) That $proj_U(\mathbf{v})$ is in $U$ for every vector $\mathbf{v}$ follows from $(*\!*\!*)$ because each $\mathbf{e}_i$ is in $U$, and $\mathbf{v} - proj_U(\mathbf{v})$ is in $U^\perp$ because it is orthogonal to each $\mathbf{e}_i$ by the Orthogonal lemma.
(2) The linearity of $proj_U : V \to V$ follows from Axioms P2 and P3 (verify). We have $im(proj_U) \subseteq U$ by (1); and $U \subseteq im(proj_U)$ because $\mathbf{u} = proj_U(\mathbf{u})$ by the Expansion theorem for every $\mathbf{u}$ in $U$. Hence $U = im(proj_U)$. Turning to the kernel, if $\mathbf{v}$ is in $ker(proj_U)$, then $\mathbf{v} = \mathbf{v} - proj_U(\mathbf{v})$ is in $U^\perp$ by the Projection theorem. Thus $ker(proj_U) \subseteq U^\perp$. On the other hand, if $\mathbf{v}$ is in $U^\perp$, then $(*\!*\!*)$ shows that $proj_U(\mathbf{v}) = \mathbf{0}$, so $U^\perp \subseteq ker(proj_U)$. Hence $U^\perp = ker(proj_U)$.

The projection theorem has an important consequence which we record in the following theorem.

**THEOREM 9**    If $U$ is a finite dimensional subspace in an arbitrary inner product space $V$, then $V = U \oplus U^{\perp}$ and $U^{\perp\perp} = U$.

**PROOF**    The fact that $V = U \oplus U^{\perp}$ restates the projection theorem, and the inclusion $U \subseteq U^{\perp\perp}$ always holds (verify). If $\mathbf{v}$ is in $U^{\perp\perp}$, write $\mathbf{p} = proj_U(\mathbf{v})$. Then $\mathbf{p}$ is in $U$ by the Projection theorem so $\mathbf{p}$ is in $U^{\perp\perp}$. Hence $\mathbf{v} - \mathbf{p}$ is in $U^{\perp\perp}$. But $\mathbf{v} - \mathbf{p}$ is in $U^{\perp}$ by the Projection theorem, and it follows that $\mathbf{v} = \mathbf{p}$. Thus $\mathbf{v}$ is in $U$, proving that $U^{\perp\perp} \subseteq U$. This is what we wanted.

Here is an example outlining how the Projection theorem can be applied to time series analysis in statistics.

**Example 12**    **Statistical Trend Analysis.** Let $f$ be a function of time $t$, say some measure of stock market value or the degree of global warming, and assume that $f$ is known only at times $t_0, t_1, \cdots, t_m$. One constantly arising question is whether the data values $f(t_0), f(t_1), \cdots, f(t_m)$ exhibit a "linear trend", in which case we might use the least squares line $f(t) = a_0 + a_1 t$ to estimate other values of $f(t)$. If there is a "quadratic trend" we would fit a quadratic $f(t) = a_0 + a_1 t + a_2 t^2$, and so on for a "cubic trend", etc. The difficulty in deciding which trend is appropriate is that, when a polynomial is fitted to the data, the coefficients may not be "statistically independent".

To remedy this, one procedure is to use the inner product on $\mathbb{P}_m$ (from Example 3) given by

$$\langle p, \ q \rangle = p(t_0)q(t_0) + p(t_1)q(t_1) + \cdots + p(t_m)q(t_m),$$

and apply the Gram-Schmidt algorithm to the basis $\{1, \ x, \ x^2, \ \cdots, \ x^m\}$ of $\mathbb{P}_m$ to obtain an orthogonal basis $\{q_0, \ q_1, \ \cdots, \ q_m\}$. Let $p_f$ be the Lagrange interpolation polynomial for $f$ in Example 8 (that is, the unique polynomial in $\mathbb{P}_m$ such that $p_f(t_i) = f(t_i)$ for each $i$). Assuming (as is usually the case) that we are only interested in trends of degree no more than 4, we compute the projection of $p_f$ onto $\mathbb{P}_4$:

$$proj_{\mathbb{P}_4}(p_f) = b_0 q_0 + b_1 q_1 + b_2 q_2 + b_3 q_3 + b_4 q_4.$$

The coefficients $b_i = \frac{\langle p_f, \ q_i \rangle}{\|q_i\|^2}$ are called the **trend coefficients** for $f$. They are easily computed, statistically independent (for most data), and $b_1$ measures the linear trend, $b_2$ measures the quadratic trend, etc. More information on this can be found in books on linear models in statistics.[27]

We now turn to an extension of the Approximation theorem for $\mathbb{R}^n$ (Section 4.6) which is valid for inner product spaces that may not be finite dimensional.

**THEOREM 10**    **Approximation Theorem.** Let $U$ be a finite dimensional subspace of an arbitrary inner product space $V$. If $\mathbf{v}$ is any vector in $V$, then $proj_U(\mathbf{v})$ is the vector in $U$ that is closest to $\mathbf{v}$ in the sense that

$$\|\mathbf{v} - proj_U(\mathbf{v})\| < \|\mathbf{v} - \mathbf{u}\| \quad \text{for all } \mathbf{u} \text{ in } U, \ \mathbf{u} \neq proj_U(\mathbf{v}).$$

**PROOF**    For convenience write $\mathbf{p} = proj_U(\mathbf{v})$. Given $\mathbf{u}$ in $U$, we can write $\mathbf{v} - \mathbf{u} = (\mathbf{v} - \mathbf{p}) + (\mathbf{p} - \mathbf{u})$ where $\mathbf{v} - \mathbf{p}$ is in $U^{\perp}$ and $\mathbf{p} - \mathbf{u}$ is in $U$. Hence Pythagoras' theorem gives $\|\mathbf{v} - \mathbf{u}\|^2 = \|\mathbf{v} - \mathbf{p}\|^2 + \|\mathbf{p} - \mathbf{u}\|^2$. This does it because $\|\mathbf{p} - \mathbf{u}\| > 0$ if $\mathbf{u} \neq \mathbf{p}$.

---

[27] For example, I. Guttman, *Linear Models, An Introduction*, Wiley, 1982.

In Chapter 4 we used the $\mathbb{R}^n$-version of the Approximation theorem to find all best approximations to a solution of an inconsistent system of linear equations and to find least squares approximating polynomials for finite sets of data pairs. Both these applications required the finite dimensionality of the underlying vector space $\mathbb{R}^n$. However, much can be said when the underlying space is infinite dimensional.

An infinite set $\{\mathbf{e}_0,\ \mathbf{e}_1,\ \mathbf{e}_2,\ \cdots,\ \mathbf{e}_n,\ \cdots\}$ in an inner product space $V$ is called **orthogonal** if

$$\mathbf{e}_i \neq 0 \text{ for each } i \quad \text{and} \quad \langle \mathbf{e}_i,\ \mathbf{e}_j \rangle = 0 \text{ whenever } i \neq j.$$

Such orthogonal sets can be found by an extension of the Gram-Schmidt algorithm.

We need the following idea. An infinite set $\{\mathbf{b}_0,\ \mathbf{b}_1,\ \mathbf{b}_2,\ \cdots,\ \mathbf{b}_m,\ \cdots\}$ of vectors in an inner product space is called (linearly) **independent** if $\{\mathbf{b}_0,\ \mathbf{b}_1,\ \mathbf{b}_2,\ \cdots,\ \mathbf{b}_m\}$ is independent for each $m$. One example is $\{1,\ x,\ x^2,\ \cdots,\ x^n,\ \cdots\}$ in $\mathbb{P}$.

**Example 13**

Suppose that $\{\mathbf{b}_0,\ \mathbf{b}_1,\ \mathbf{b}_2,\ \cdots,\ \mathbf{b}_m,\ \cdots\}$ is an infinite independent set of vectors in an inner product space. Apply the Gram-Schmidt process to the vectors $\mathbf{b}_i$ as follows: Take

$$\mathbf{e}_0 = \mathbf{b}_0$$

and, having determined $\mathbf{e}_0,\ \mathbf{e}_1,\ \mathbf{e}_2,\ \cdots,\ \mathbf{e}_{m-1}$ for some $m \geq 1$, take

$$\mathbf{e}_m = \mathbf{b}_m - \frac{\langle \mathbf{b}_m,\ \mathbf{e}_0 \rangle}{\|\mathbf{e}_0\|^2}\mathbf{e}_0 - \frac{\langle \mathbf{b}_m,\ \mathbf{e}_1 \rangle}{\|\mathbf{e}_1\|^2}\mathbf{e}_1 - \cdots - \frac{\langle \mathbf{b}_{m-1},\ \mathbf{e}_{m-1} \rangle}{\|\mathbf{e}_{m-1}\|^2}\mathbf{e}_{m-1}.$$

This produces an infinite orthogonal set $\{\mathbf{e}_0,\ \mathbf{e}_1,\ \mathbf{e}_2,\ \cdots,\ \mathbf{e}_m,\ \cdots\}$ by Theorem 6. In fact we have $span\{\mathbf{e}_0,\ \mathbf{e}_1,\ \mathbf{e}_2,\ \cdots,\ \mathbf{e}_m\} = span\{\mathbf{b}_0,\ \mathbf{b}_1,\ \mathbf{b}_2,\ \cdots,\ \mathbf{b}_m\}$ for each $m$.

Now assume that an infinite orthogonal set

$$\mathcal{E} = \{\mathbf{e}_0,\ \mathbf{e}_1,\ \mathbf{e}_2,\ \cdots,\ \mathbf{e}_n,\ \cdots\}$$

is given in an inner product space $V$. We use it to define subspaces $U_m$ as follows:

$$U_m = span\{\mathbf{e}_0,\ \mathbf{e}_1,\ \mathbf{e}_2,\ \cdots,\ \mathbf{e}_m\} \quad \text{for each } m \geq 1.$$

Clearly $U_0 \subseteq U_1 \subseteq U_2 \subseteq \cdots \subseteq U_m \subseteq \cdots$, and $dim(U_m) = m + 1$ for each $m$ because $\{\mathbf{e}_0,\ \mathbf{e}_1,\ \mathbf{e}_2,\ \cdots,\ \mathbf{e}_m\}$ is an (orthogonal) basis of $U_m$. Given $\mathbf{v}$ in $V$ and $m \geq 1$, define

$$\mathbf{v}_m = proj_{U_m}(\mathbf{v}) = \frac{\langle \mathbf{v},\ \mathbf{e}_0 \rangle}{\|\mathbf{e}_0\|^2}\mathbf{e}_0 + \frac{\langle \mathbf{v},\ \mathbf{e}_1 \rangle}{\|\mathbf{e}_1\|^2}\mathbf{e}_1 + \cdots + \frac{\langle \mathbf{v},\ \mathbf{e}_m \rangle}{\|\mathbf{e}_m\|^2}\mathbf{e}_m. \qquad (****)$$

The vector $\mathbf{v}_m$ is in $U_m$ for each $m$, and is called the $m^{\text{th}}$ $\mathcal{E}$-**approximation** of $\mathbf{v}$. The reason for the name will be apparent soon.

Recall that Theorem 10 with $U = U_{m+1}$ gives $\|\mathbf{v} - \mathbf{v}_{m+1}\| \leq \|\mathbf{v} - \mathbf{u}\|$ for each vector $\mathbf{u}$ in $U_{m+1}$. But $\mathbf{v}_m$ is in $U_{m+1}$ because $U_m \subseteq U_{m+1}$, so we have

$$\|\mathbf{v} - \mathbf{v}_{m+1}\| \leq \|\mathbf{v} - \mathbf{v}_m\| \quad \text{for each } m = 0,\ 1,\ 2,\ \cdots.$$

Using this we obtain

$$\|\mathbf{v} - \mathbf{v}_0\| \geq \|\mathbf{v} - \mathbf{v}_1\| \geq \|\mathbf{v} - \mathbf{v}_2\| \geq \cdots \geq \|\mathbf{v} - \mathbf{v}_m\| \geq \cdots.$$

Thus the vectors $\mathbf{v}_0,\ \mathbf{v}_1,\ \mathbf{v}_2,\ \cdots$ are better and better approximations to $\mathbf{v}$ (this is the reason for calling these $\mathbf{v}_i$ the $\mathcal{E}$-approximations of $\mathbf{v}$). Furthermore, each vector $\mathbf{v}_m$ is easily computable from $\mathbf{v}$ and the $\mathbf{e}_i$ because of $(****)$.

Hence the following is a general method in approximation theory: Given a vector $\mathbf{v}$ in an inner product space $V$, find an orthogonal set $\{\mathbf{e}_0,\ \mathbf{e}_1,\ \mathbf{e}_2,\ \cdots,\ \mathbf{e}_n,\ \cdots\}$ in $V$ such that the vectors $\mathbf{v}_m$ approach $\mathbf{v}$, that is

$$\lim_{m \to \infty} \|\mathbf{v} - \mathbf{v}_m\| = 0.$$

Then $\mathbf{v}_m$ is a suitable approximation to $\mathbf{v}$ if $m$ is large enough.[28]

---

[28] We say that a sequence of numbers $a_1,\ a_2,\ a_3,\ \cdots$ approaches a number $a$, and write $\lim_{n \to \infty} a_n = a$, if the numbers $a_i$ become and remain very close to $a$ as $n$ increases. More precisely: We ask that for any small number $\varepsilon > 0$ there should exist an integer $N$ such that $|a_n - a| < \varepsilon$ for all $n \geq N$. This notion is fundamental in calculus.

Of course, this may not be the case for some particular infinite orthogonal set of vectors $\mathbf{e}_i$. But there are several situations where it does happen, many in spaces of functions, and the results are important tools for a variety of applications in science and engineering. A complete discussion of this topic is not possible here, and we content ourselves with two examples, one where the $\mathbf{e}_i$ are trigonometric functions and the other where the $\mathbf{e}_i$ are orthogonal polynomials.

### 5.7.4  Fourier Approximation

Consider the space $\mathbf{C}[-\pi, \; \pi]$ of all continuous functions on the interval $[-\pi, \; \pi]$, endowed with the inner product

$$\langle f, \; g \rangle = \int_{-\pi}^{\pi} f(x)g(x)dx.$$

Then standard integration techniques (using trigonometric identities) show that

$$\mathcal{F} = \{1, \; \cos x, \; \sin x, \; \cos(2x), \; \sin(2x), \; \cdots, \; \cos(mx), \; \sin(mx), \; \cdots\}$$

is an orthogonal set in $\mathbf{C}[-\pi, \; \pi]$, and that the norms of these functions are given by

$$\|1\|^2 = 2\pi \quad \text{and} \quad \|\cos(mx)\|^2 = \|\sin(mx)\|^2 = \pi \text{ for each } m \geq 1.$$

Slightly modifying our earlier notation, we write

$$\mathcal{F}_m = span\{1, \; \cos x, \; \sin x, \; \cos(2x), \; \sin(2x), \; \cdots, \; \cos(mx), \; \sin(mx)\}.$$

If $f$ is any function in $\mathbf{C}[-\pi, \; \pi]$, the $m$th $\mathcal{F}$-approximation of $f$ (called the $m$th **Fourier approximation**[29]) is the function $f_m$ given (as in Equation (****) above) by

$$f_m(x) = a_0 + a_1 \cos x + b_1 \sin x + a_2 \cos(2x) + b_2 \sin(2x) + \cdots + a_m \cos(mx) + b_m \sin(mx)$$

for all $x$ in $[-\pi, \; \pi]$. Here the $a_i$ and $b_i$ are called the **Fourier coefficients** of $f$, and are given by

$$a_0 = \frac{\langle f, 1 \rangle}{\|1\|^2} = \frac{1}{2\pi} \int_{-\pi}^{\pi} f(x)dx,$$

$$a_k = \frac{\langle f, \cos(kx) \rangle}{\|\cos(kx)\|^2} = \frac{1}{\pi} \int_{-\pi}^{\pi} f(x)\cos(kx)dx \qquad \text{for } k = 1, \; 2, \; 3, \; \cdots,$$

$$b_k = \frac{\langle f, \sin(kx) \rangle}{\|\sin(kx)\|^2} = \frac{1}{\pi} \int_{-\pi}^{\pi} f(x)\sin(kx)dx \qquad \text{for } k = 1, \; 2, \; 3, \; \cdots,$$

Calculating these coefficients can be tedious (integration by parts is often useful), but one simplification is worth noting. Recall (see Example 8 §5.6) that

$$\int_{-\pi}^{\pi} f(x)dx = 0 \text{ if } f \text{ is an odd function}$$

and

$$\int_{-\pi}^{\pi} f(x)dx = 2 \int_0^{\pi} f(x)dx \text{ if } f \text{ is even.}$$

This is useful in calculating Fourier coefficients because $\cos(kx)$ is even and $\sin(kx)$ is odd for every $k \geq 1$, and because the product of an even function and an odd function is odd while the product of two odd (or two even) functions is even. In particular, we have

$$a_k = 0 \text{ for all } k \text{ if } f \text{ is odd} \quad \text{and} \quad b_k = 0 \text{ for all } k \text{ if } f \text{ is even.}$$

This simplifies the work in the following example.

---

[29] Named for J.B.J. Fourier (1768–1830), a French mathematician, who used these techniques in 1822 to investigate heat conduction in solids.

**Example 14**  Let $f$ be the absolute value function defined by $f(x) = |x|$ for all $x$ in the interval $[-\pi, \ \pi]$. Then $f$ is even, so $b_k = 0$ for all $k$, and we have

$$a_0 = \tfrac{\pi}{2} \quad \text{and} \quad a_k = \begin{cases} 0 & \text{if } k \geq 1 \text{ is even} \\ -\frac{4}{\pi k^2} & \text{if } k \geq 1 \text{ is odd} \end{cases}.$$

Hence the fifth Fourier approximation of $f$ is

$$f_5(x) = \tfrac{\pi}{2} - \tfrac{4}{\pi}\left\{\cos x + \tfrac{1}{9}\cos(3x) + \tfrac{1}{25}\cos(5x)\right\}.$$

The graphs of $f$ and $f_5$ are plotted in the diagram.

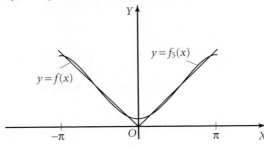

The reason for the importance of Fourier approximations is the following fundamental theorem.[30]

**THEOREM 11**  Let $f$ be any function in $\mathbf{C}[-\pi, \ \pi]$ such that $f(-\pi) = f(\pi)$. Then

$$\|f - f_m\| \quad \text{approaches } 0.$$

In fact, we have more:

$$f_m(x) \quad \text{approaches } f(x) \text{ for all } x \text{ in the interval } [-\pi, \ \pi].$$

This is an important result in the theory of Fourier approximation, and shows that $f$ has a representation as an infinite series

$$f(x) = a_0 + a_1 \cos x + b_1 \sin x + a_2 \cos(2x) + b_2 \sin(2x) + \cdots$$

called the **Fourier series** of $f$.

**Example 15**  Expand $f(x) = x^2$ on the interval $[-\pi, \ \pi]$ in a Fourier series, and so obtain a series expansion of $\frac{\pi^2}{6}$.

**SOLUTION**  Here $f$ is an even function, so all the Fourier sine coefficients $b_k$ are zero. As to the cosine coefficients:

$$a_0 = \tfrac{1}{2\pi}\int_{-\pi}^{\pi} x^2 dx = \tfrac{\pi^2}{3} \quad \text{and} \quad a_k = \tfrac{1}{\pi}\int_{-\pi}^{\pi} x^2 \cos(kx)dx = \tfrac{4}{k^2}(-1)^k \text{ for } k \geq 1,$$

where the second integral requires integration by parts twice, and uses the fact that $\cos(k\pi) = (-1)^k$ for every integer $k$. Hence the Fourier series for $x^2$ is

$$x^2 = \tfrac{\pi^2}{3} - 4\left[\cos x - \tfrac{1}{2^2}\cos(2x) + \tfrac{1}{3^2}\cos(3x) - \tfrac{1}{4^2}\cos(4x) + \cdots\right]$$

for $-\pi \leq x \leq \pi$. In particular, taking $x = \pi$ gives an infinite series for $\frac{\pi^2}{6}$:

$$\tfrac{\pi^2}{6} = 1 + \tfrac{1}{2^2} + \tfrac{1}{3^2} + \tfrac{1}{4^2} + \cdots.$$

Many other such formulas can be proved using Theorem 11.

---

[30] See, for example, R.V. Churchill and J.W. Brown, *Fourier Series and Boundary Value Problems*, 4th ed., McGraw-Hill, 1987.

Finally, we observe that the functions $\cos(kx)$ and $\sin(kx)$ are easily generated as an electric voltage (if $x$ is time), so the Fourier approximations to $f$ can be obtained in electronic form. Hence Fourier series play a fundamental role in electrical engineering. In fact, they are an effective tool for solving many of the partial differential equations that arise throughout Science and Engineering.

## 5.7.5 Legendre Polynomials

Consider the inner product space $\mathbb{P}$ with the inner product

$$\langle p,\ q \rangle = \int_{-1}^{1} p(x)q(x)dx.$$

If we apply the Gram-Schmidt algorithm to the infinite independent set $\{1,\ x,\ x^2,\ \cdots x^m,\ \cdots\}$ in $\mathbb{P}$, we obtain an infinite orthogonal set $\{q_0,\ q_1,\ q_2,\ \cdots,\ q_m,\ \cdots\}$ in $\mathbb{P}$ given recursively as follows:

$$q_0 = 1,$$
$$q_m = x^m - \frac{\langle x^m,\ q_0 \rangle}{\|q_0\|^2}q_0 - \frac{\langle x^m,\ q_1 \rangle}{\|q_1\|^2}q_1 - \cdots - \frac{\langle x^m,\ q_{m-1} \rangle}{\|q_{m-1}\|^2}q_{m-1} \quad \text{for } m \geq 1.$$

We investigated these polynomials in Example 10, and calculated the first few: $q_0 = 1, q_1 = x,\ q_2 = \frac{1}{3}(3x^2 - 1)$, and $q_3 = \frac{1}{5}(5x^3 - 3x)$. It is easy to see (by induction) that $q_m$ has degree $m$ for each $m$ and has leading coefficient 1. In particular, $\{q_0,\ q_1,\ q_2,\ \cdots,\ q_m\}$ is an orthogonal basis of $\mathbb{P}_m$ for each $m$ because $dim(\mathbb{P}_m) = m + 1$.

Surprisingly, there is a length 2 recurrence for the $q_m$ that is much easier to use than the Gram-Schmidt formula. To derive it, consider the polynomial $xq_m$. It has degree $m + 1$ so it lies in $\mathbb{P}_{m+1}$, and hence the Expansion theorem gives

$$xq_m = r_0q_0 + r_1q_1 + r_2q_2 + \cdots + r_{m+1}q_{m+1} \quad \text{where } r_i = \frac{\langle xq_m,\ q_i \rangle}{\|q_i\|^2} \text{ for each } i.$$

But $\langle xq_m,\ q_i \rangle = \langle q_m,\ xq_i \rangle$ (verify), and $\langle q_m,\ xq_i \rangle = 0$ for $i \leq m - 2$ because then $xq_i$ is in $\mathbb{P}_{m-1}$. It follows that $r_i = 0$ for all $i \leq m - 2$, so

$$xq_m = r_{m-1}q_{m-1} + r_mq_m + r_{m+1}q_{m+1}.$$

Moreover, since $deg\ q_k = k$ for each $k$, comparing the coefficients of $x^{m+1}$ on each side of this equation gives $r_{m+1} = 1$. Hence we obtain the following length 2 recurrence for the polynomials $q_i$ :

$$q_{m+1} = \left\{ x - \frac{\langle xq_m,\ q_m \rangle}{\|q_m\|^2} \right\} q_m - \frac{\langle xq_m,\ q_{m-1} \rangle}{\|q_{m-1}\|^2}q_{m-1} \quad \text{for } m \geq 1.$$

Since we know that $q_0 = 1$ and $q_1 = x$, this formula gives every $q_m$ one after the other. In particular, we find that $q_4 = \frac{1}{35}(35x^4 - 30x^2 + 3)$ and $q_5 = \frac{1}{63}(63x^5 - 70x^3 + 15x)$.

However, there is an even simpler recurrence, but it is for the scalar multiples $p_m$ of the $q_m$ which have the property that

$$p_m(1) = 1 \quad \text{for all } m.$$

These are the **Legendre polynomials**[31] $p_0(x),\ p_1(x),\ p_2(x),\ \cdots p_m(x),\ \cdots$ and they are defined by

$$p_m(x) = \frac{1}{q_m(1)}q_m(x) \quad \text{for each } m = 1,\ 2,\ \cdots.$$

---

[31] A.M. Legendre (1752–1833) was a French mathematician known for his work in number theory and differential equations.

Using the above expressions for the $q_i$, the first six Legendre polynomials are

$$p_0 = 1$$

$$p_1 = x$$

$$p_2 = \tfrac{1}{2}(3x^2 - 1)$$

$$p_3 = \tfrac{1}{2}(5x^3 - 3x)$$

$$p_4 = \tfrac{1}{8}(35x^4 - 30x^2 + 3)$$

$$p_5 = \tfrac{1}{8}(63x^5 - 70x^3 + 15x).$$

The Legendre polynomials form an orthogonal set (because they are scalar multiples of the $q_m$) and find use in a wide variety of applications.

Moreover, an argument much like the one above for the $q_m$ shows that the polynomials $p_m$ also satisfy a length 2 recurrence. However, it is much simpler than the recurrence for the $q_m$ and we record the result in the following theorem.

**THEOREM 12**

The Legendre polynomials satisfy the recurrence

$$p_{m+1} = \tfrac{2m+1}{m+1} x p_m - \tfrac{m}{m+1} p_{m-1} \quad \text{for all } m \geq 1.$$

This gives an easy way to derive the polynomials $p_m$, beginning with $p_0 = 1$ and $p_1 = x$.

We omit the proof of Theorem 12 here. It is usually derived from the fact that the Legendre polynomials first arose as solutions to differential equations. Indeed, $y = p_m(x)$ is a solution to **Legendre's differential equation**

$$(1 - x^2)\tfrac{d^2 y}{dx^2} - 2x\tfrac{dy}{dx} + m(m+1)y = 0,$$

and many treatments of these polynomials begin with this fact. Another observation is that these polynomials are the coefficients in the power series expansion of the function $(1 - 2xz + z^2)^{-\frac{1}{2}}$ regarded as a function of $z$, that is

$$(1 - 2xz + z^2)^{-\frac{1}{2}} = p_0(x) + p_1(x)z + p_2(x)z^2 + \cdots + p_m(x)z^m + \cdots$$

when $|z| \leq 1$. Because of this the function $(1 - 2xz + z^2)^{-\frac{1}{2}}$ is called a **generating function** for the Legendre polynomials $p_m(x)$, and it leads quickly to many of their properties. One elegant example is **Rodrigue's formula**:

$$p_m(x) = \tfrac{1}{2^m m!} \tfrac{d^m}{dx^m}(x^2 - 1)^m \quad \text{for each } m \geq 1.$$

All these facts, and many others, can be found in books on special functions[32] where other families of orthogonal polynomials are discussed.

## Exercises 5.7

Throughout these exercises, $V$ denotes an inner product space with inner product $\langle \ , \ \rangle$.

**1.** In each case determine which of Axioms P1–P4 fail.

**a)** $V = \mathbb{R}^2$, $\langle [x_1 \ x_2]^T, [y_1 \ y_2]^T \rangle = x_1 x_2 y_1 y_2$.

**b)** $V = \mathbb{R}^3$, $\langle [x_1 \ x_2 \ x_3]^T, [y_1 \ y_2 \ y_3]^T \rangle = x_1 y_1 - x_2 y_2 + x_3 y_3$.

**c)** $V = \mathbb{C}$, $\langle z, w \rangle = re(z\overline{w})$ where $re(z)$ denotes the real part of the complex number $z$.

**d)** $V = \mathbb{P}_3$, $\langle p(x), q(x) \rangle = p(1)q(1)$.

**e)** $V = M_{2,2}$, $\langle A, B \rangle = det(AB^T)$.

**f)** $V = \mathbb{F}[0, 1]$, $\langle f, g \rangle = f(0)g(1) + f(1)g(0)$.

**2.** Deduce that $\langle \mathbf{v}, \mathbf{0} \rangle = 0$ for all $\mathbf{v}$ in $V$ using only Axioms P1 and P3.

**3.** Show that $\mathbf{v}$ and $\mathbf{w}$ are orthogonal in $V$ if and only if $\|\mathbf{v} + \mathbf{w}\| = \|\mathbf{v} - \mathbf{w}\|$.

---

[32] See, for example, N.N. Lebedev, *Special Functions and their Applications*, Dover, 1972.

4. **a)** Show that $\langle \mathbf{v}, \mathbf{w} \rangle = \frac{1}{4}\{\|\mathbf{v} + \mathbf{w}\|^2 - \|\mathbf{v} - \mathbf{w}\|^2\}$ for all $\mathbf{v}$ and $\mathbf{w}$ in $V$.

   **b)** If two inner products on $V$ have the same norm functions, show that they are equal.

5. If $V = span\{\mathbf{v}_1, \mathbf{v}_2, \cdots, \mathbf{v}_n\}$ and $\langle \mathbf{v}, \mathbf{v}_i \rangle = \langle \mathbf{w}, \mathbf{v}_i \rangle$ for each $i$, show that $\mathbf{v} = \mathbf{w}$.

6. Let $\mathbb{D}_n$ denote the space of all functions $f : \{1, 2, \cdots, n\} \to \mathbb{R}$ with pointwise addition and scalar multiplication (see Example 5 §5.1). For each $k = 1, 2, \cdots, n$, define $\delta_k$ in $\mathbb{D}_n$ by $\delta_k(i) = \begin{cases} 1 & \text{if } i = k \\ 0 & \text{if } i \neq k \end{cases}$.

   **a)** If $\langle f, g \rangle = f(1)g(1) + f(2)g(2) + \cdots + f(n)g(n)$, show that $\langle \; , \; \rangle$ is an inner product on $\mathbb{D}_n$.

   **b)** Show that $\{\delta_1, \delta_2, \cdots, \delta_n\}$ is an orthonormal basis of $\mathbb{D}_n$.

   **c)** Show that $f = f(1)\,\delta_1 + f(2)\,\delta_2 + \cdots + f(n)\,\delta_n$ for every $f$ in $\mathbb{D}_n$.

7. Show that every inner product on $\mathbb{R}^n$ has the form $\langle X, Y \rangle = (UX) \cdot (UY)$ for some invertible upper triangular matrix $U$ with positive diagonal entries. [Hint: Theorem 4 §4.8.]

8. If $V$ is any finite dimensional vector space and $\mathcal{B}$ is any basis of $V$, show that $\langle \; , \; \rangle$ is an inner product on $V$ if we define $\langle \mathbf{v}, \mathbf{w} \rangle = C_{\mathcal{B}}(\mathbf{v}) \cdot C_{\mathcal{B}}(\mathbf{w})$ for all $\mathbf{v}$ and $\mathbf{w}$ in $V$.

9. If $P$ is any $n \times n$ matrix, define $\langle \; , \; \rangle$ on $\mathbb{R}^n$ by $\langle X, Y \rangle = X^T P Y$ for all $X$ and $Y$ in $\mathbb{R}^n$.

   **a)** Show that $\langle \; , \; \rangle$ satisfies Axioms P2 and P3, and that Axiom P1 holds if $P$ is symmetric.

   **b)** Show that if Axiom P1 holds, then $P$ is necessarily symmetric. [Hint: Show that $X^T(P - P^T)Y = 0$ for all $X$ and $Y$ in $\mathbb{R}^n$.]

10. In each case use the Gram-Schmidt process to convert the basis $\{1, x, x^2\}$ of $\mathbb{P}_2$ into an orthogonal basis of $\mathbb{P}_2$.

    **a)** $\langle p, q \rangle = p(0)q(0) + p(1)q(1) + p(2)q(2)$.

    **b)** $\langle p, q \rangle = \int_0^1 p(x)q(x)dx$.

11. Using the inner product $\langle A, B \rangle = tr(AB^T)$ on $M_{2,2}$, find the matrix in $U = span\{I, J\}$ closest to $A = \begin{bmatrix} 1 & -1 \\ 2 & 3 \end{bmatrix}$, where $I = \begin{bmatrix} 1 & 0 \\ 0 & 1 \end{bmatrix}$ and $J = \begin{bmatrix} 1 & 1 \\ 1 & 1 \end{bmatrix}$.

12. If $\mathcal{B} = \{\mathbf{b}_1, \mathbf{b}_2, \cdots, \mathbf{b}_n\}$ is any orthonormal basis of $V$, show that $\langle \mathbf{v}, \mathbf{w} \rangle = C_{\mathcal{B}}(\mathbf{v}) \cdot C_{\mathcal{B}}(\mathbf{w})$ for all $\mathbf{v}$ and $\mathbf{w}$ in $V$.

13. If $\mathcal{B} = \{\mathbf{b}_1, \mathbf{b}_2, \cdots, \mathbf{b}_n\}$ is any basis of $V$, show that there exists a positive definite matrix $P$ such that $\langle \mathbf{v}, \mathbf{w} \rangle = C_{\mathcal{B}}(\mathbf{v})^T P C_{\mathcal{B}}(\mathbf{w})$ for all $\mathbf{v}$ and $\mathbf{w}$ in $V$. [Hint: Theorem 2 and Exercise 12.]

14. If $\{\mathbf{b}_1, \mathbf{b}_2, \cdots, \mathbf{b}_n\}$ is an orthonormal basis of $V$ and $\mathbf{v} = r_1\mathbf{b}_1 + r_2\mathbf{b}_2 + \cdots + r_n\mathbf{b}_n$, show that $\|\mathbf{v}\|^2 = r_1^2 + r_2^2 + \cdots + r_n^2$. (**Parseval's formula.**)

15. If $\{\mathbf{b}_1, \mathbf{b}_2, \cdots, \mathbf{b}_n\}$ is an orthogonal basis of $V$, show that $\langle \mathbf{v}, \mathbf{w} \rangle = \frac{\langle \mathbf{v}, \mathbf{b}_1 \rangle \langle \mathbf{w}, \mathbf{b}_1 \rangle}{\|\mathbf{b}_1\|^2} + \cdots + \frac{\langle \mathbf{v}, \mathbf{b}_n \rangle \langle \mathbf{w}, \mathbf{b}_n \rangle}{\|\mathbf{b}_n\|^2}$ for all $\mathbf{v}$ and $\mathbf{w}$ in $V$. [Hint: Theorem 5 and Exercise 12.]

16. Show that $(U + W)^\perp = U^\perp \cap W^\perp$ for all subspaces $U$ and $W$ of $V$.

17. **a)** Show that $U^\perp + W^\perp \subseteq (U \cap W)^\perp$ for all subspaces $U$ and $W$ of $V$.

    **b)** If $dimV = n$, show that $dim(U^\perp) = n - dimU$ for every subspace $U$ of $V$. [Hint: Theorem 4 §5.6.]

    **c)** If $dimV = n$, show that $(U \cap W)^\perp = U^\perp + W^\perp$. [Hint: Use **b)**, Exercise 16, and Exercise 25 §4.3 (twice) to show that $dim(U^\perp + W^\perp) = dim((U \cap W)^\perp)$. Then apply **a)**.]

18. Let $U_1 \subseteq U_2$ be finite dimensional subspaces of an inner product space $V$. If $\mathbf{v}$ is any vector in $V$, show that $\|\mathbf{v} - proj_{U_2}(\mathbf{v})\| \leq \|\mathbf{v} - proj_{U_1}(\mathbf{v})\|$.

19. Let $W_1 \subseteq W_2 \subseteq \cdots$ be subspaces of an inner product space $V$ such that $W_m \neq W_{m+1}$ for each $m$. Choose $\mathbf{b}_1$ in $W_1$ and, for $m \geq 2$, choose $\mathbf{b}_m$ in $W_m$ but not in $W_{m-1}$. Show that $\{\mathbf{b}_1, \mathbf{b}_2, \mathbf{b}_3, \cdots\}$ is an infinite independent set. [Hint: Lemma 1 §5.2.]

20. Consider the space $\mathbf{C}[a, b]$ with the inner product $\langle f, g \rangle = \int_a^b f(x)g(x)dx$. Show that $\langle fh, g \rangle = \langle f, hg \rangle$ for all $f, g$, and $h$ in $\mathbf{C}[a, b]$.

21. **a)** Show that $\frac{\pi^2}{8} = 1 + \frac{1}{3^2} + \frac{1}{5^2} + \cdots$. [Hint: Example 14 and Theorem 11.]

    **b)** Show that $\frac{\pi^2}{12} = 1 - \frac{1}{2^2} + \frac{1}{3^2} - \frac{1}{4^2} + \frac{1}{5^2} - \cdots$. [Hint: Example 15 and Theorem 11.]

22. If $V$ is a vector space, a function $\| \; \| : V \to \mathbb{R}$ is called a **norm** if:

    (1) $\|\mathbf{v}\| \geq 0$ for all in $V$
    (2) $\|\mathbf{v}\| = 0$ if and only if $\mathbf{v} = \mathbf{0}$
    (3) $\|r\mathbf{v}\| = |r|\,\|\mathbf{v}\|$ for all $r$ in $\mathbb{R}$ and all $\mathbf{v}$ in $V$ and
    (4) $\|\mathbf{v} + \mathbf{w}\| \leq \|\mathbf{v}\| + \|\mathbf{w}\|$ for all $\mathbf{v}$ and $\mathbf{w}$ in $V$.

    **a)** Show that $\|\mathbf{v}\| = \sqrt{\langle \mathbf{v}, \mathbf{v} \rangle}$ is a norm in this sense for any inner product $\langle \; , \; \rangle$ on $V$.

    **b)** Show that $\|[x \; y]^T\| = |x| + |y|$ is a norm on $\mathbb{R}^2$ which does not arise from any inner product. [Hint: Theorem 2.]

23. Show that $\{1, \cos x, \cos(2x), \cos(3x), \cdots\}$ is an orthogonal set in $C[0, \pi]$ using the inner product $\langle f, g \rangle = \int_0^\pi f(x)g(x)dx$. [Hint: $\cos x \cos y = \frac{1}{2}[\cos(x + y) + \cos(x - y)]$.]

24. Compute the Legendre polynomials $p_6$ and $p_7$.

25. Show that the Legendre polynomial $p_m$ is an even function if $m$ is even, and an odd function if $m$ is odd. [Hint: Use Theorem 12 to prove "$p_{2n}$ is even and $p_{2n+1}$ is odd" by induction on $n$.]

26. Show that $p_m(-1) = (-1)^m$ for each $m = 0,\ 1,\ 2,\ \cdots$ where $p_m$ is the Legendre polynomial.

27. Let $a_m$ denote the leading coefficient of the Legendre polynomial $p_m$.
    a) Show that $a_{m+1} = \frac{2m+1}{m+1} a_m$ for each $m \geq 1$. [Hint: Theorem 12.]
    b) Deduce that $a_m = \frac{(2m-1)(2m-3)\ \cdots\ (3)(1)}{m!}$ for each $m \geq 1$.

# Appendix

## A.1 BASIC TRIGONOMETRY

Figure A.1

Figure A.2

Figure A.3

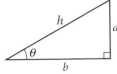

Figure A.4

The traditional degree measure for angles is not convenient for mathematics or science. A much better measure, called radian measure, is given geometrically as follows. Suppose a standard Cartesian coordinate system is given in the plane. The circle with radius 1, centred at the origin, is called the **unit circle**. An angle $\theta$ is said to be in **standard position** if it is measured counterclockwise from the positive $X$-axis as in Figure A.1. The **radian measure** of $\theta$ is the length of the arc on the unit circle determined by the angle, as in Figure A.2. Since the circumference of the unit circle is $2\pi$, we have the basic relationship between degree and radian measures:

$$2\pi = 360°$$

Other important equivalences follow:

$$180° = \pi, \quad 90° = \tfrac{\pi}{2}, \quad 60° = \tfrac{\pi}{3}, \quad 45° = \tfrac{\pi}{4}, \quad 30° = \tfrac{\pi}{6}$$

Note that negative angles are measured *clockwise* from the positive $X$-axis, and their radian measure is the negative of the length of the corresponding arc on the unit circle. For example, $-\tfrac{\pi}{3}$ radians is the angle $60°$ measured clockwise from the positive $X$-axis (shown in Figure A.3). Unless otherwise mentioned, all angles in this book are given in radians.[1]

The word "trigonometry" comes from the Greek word *trigonon* for triangle, and so means "measurement of triangles". The trigonometric functions *cosine* and *sine* were invented for this purpose. For **acute** angles (between 0 and $\tfrac{\pi}{2}$) these functions are based upon the fundamental **similarity property** of triangles:[2]

> *If two triangles are similar (have the same angles) the ratios of the lengths of corresponding sides are equal.*

We need this fact for right triangles only; that is, triangles in which one angle is a **right angle** $\tfrac{\pi}{2}$.

Now suppose that $\theta$ is an acute angle in a right triangle as in Figure A.4. The side opposite the right angle is called the **hypotenuse**, and the other two sides are opposite and adjacent to the angle $\theta$. The **sine** and **cosine** functions $\sin \theta$ and $\cos \theta$ are defined as ratios of the lengths of these sides:

$$\sin \theta = \tfrac{a}{h} = \tfrac{\text{opposite}}{\text{hypotenuse}} \qquad \cos \theta = \tfrac{b}{h} = \tfrac{\text{adjacent}}{\text{hypotenuse}}$$

These ratios do not depend on the size of the right triangle (by the similarity property) and so depend only on the angle $\theta$. This fact makes them very useful for practical applications of geometry such as surveying.

---

[1] A good way to visualize radian measure is to think of the $\theta$-axis as wrapped around the unit circle with the origin at the point $P(1, 0)$. The positive $\theta$-axis is wrapped counterclockwise, and the negative $\theta$-axis is wrapped clockwise.

[2] Pythagoras' theorem also follows from the similarity property (see Theorem 4 in §3.1).

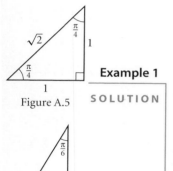

Figure A.5

Figure A.6

The numerical values of $\sin \theta$ and $\cos \theta$ can be obtained from any scientific calculator, but these values are only approximations (albeit to several decimal places). The following example gives two special triangles for which the exact ratios are known.

**Example 1**  Find $\sin \theta$ and $\cos \theta$ for $\theta = \frac{\pi}{4} = 45°$, $\theta = \frac{\pi}{6} = 30°$, and $\theta = \frac{\pi}{3} = 60°$.

**SOLUTION**  The triangle in Figure A.5 is half the square of side 1, so both acute angles equal $\frac{\pi}{4}$. The hypotenuse has length $\sqrt{2}$ by Pythagoras' theorem, so $\cos \frac{\pi}{4} = \frac{1}{\sqrt{2}}$ and $\sin \frac{\pi}{4} = \frac{1}{\sqrt{2}}$.

The triangle in Figure A.6 is half the equilateral triangle of side 2, and so the acute angles are $\frac{\pi}{6}$ and $\frac{\pi}{3}$. The hypotenuse has length 2 and the short side has length 1, so the third side has length $\sqrt{3}$ by Pythagoras' theorem. Hence $\cos \frac{\pi}{3} = \frac{1}{2} = \sin \frac{\pi}{6}$ and $\sin \frac{\pi}{3} = \frac{\sqrt{3}}{2} = \cos \frac{\pi}{6}$.

While the trigonometric functions of an acute angle are useful for many practical purposes, it soon becomes apparent that it is necessary to extend these functions to arbitrary angles. That is, we must make clear what we mean by $\sin \theta$ and $\cos \theta$ where $\theta$ is a possible nonacute angle.

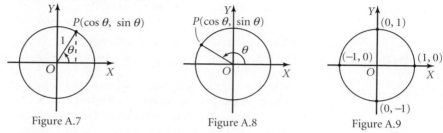

Figure A.7            Figure A.8            Figure A.9

Consider an acute angle $\theta$ in standard position as in Figure A.7. The angle $\theta$ determines a unique point $P$ on the unit circle, and $P$ has coordinates $P(\cos \theta, \sin \theta)$ because the unit circle has radius 1. This gives a simple way to define $\cos \theta$ and $\sin \theta$ for any angle $\theta$:

> Given an angle $\theta$ in standard position, $\cos \theta$ and $\sin \theta$ are defined to be the
> $X$- and $Y$-coordinates of the point $P$ on the unit circle determined by $\theta$. Thus

$$P = P(\cos \theta, \sin \theta).$$

This is illustrated in Figure A.8, and makes it easy to find the cosine and sine of basic angles.

**Example 2**  Find the cosine and sine of the following angles: $\theta = 0$, $\theta = \frac{\pi}{2}$, $\theta = \pi$, $\theta = \frac{3\pi}{2}$, and $\theta = 2\pi$.

**SOLUTION**  The coordinates of the points on the unit circle corresponding to these angles are plotted in Figure A.9. Hence

$$\cos 0 = 1 \quad \cos \frac{\pi}{2} = 0 \quad \cos \pi = -1 \quad \cos \frac{3\pi}{2} = 0 \quad \cos 2\pi = 1$$
$$\sin 0 = 0 \quad \sin \frac{\pi}{2} = 1 \quad \sin \pi = 0 \quad \sin \frac{3\pi}{2} = -1 \quad \sin 2\pi = 0$$

Notice that the angles $\theta = 0$ and $\theta = 2\pi$ have the same cosine and sine because they determine the same point on the unit circle.

Figure A.10

Figure A.11

Figure A.12

**Example 3** Find $\cos \theta$ and $\sin \theta$ for the angles $\theta = \frac{2\pi}{3}$ and $\theta = -\frac{\pi}{4}$.

**SOLUTION** The angle $\theta = \frac{2\pi}{3}$ is given in standard position in Figure A.10. The triangle $AOP$ has angles $\frac{\pi}{3}$ and $\frac{\pi}{6}$ at vertices $O$ and $P$ respectively. Since the hypotenuse has length 1 (it is in the unit circle), the sides $AO$ and $AP$ have lengths $\frac{1}{2}$ and $\frac{\sqrt{3}}{2}$ respectively. Hence $P = P(-\frac{1}{2}, \frac{\sqrt{3}}{2})$, so $\cos \frac{2\pi}{3} = -\frac{1}{2}$ and $\sin \frac{2\pi}{3} = \frac{\sqrt{3}}{2}$.

The angle $-\frac{\pi}{4}$ is shown in standard position in Figure A.11. Now the triangle $AOP$ has angles $\frac{\pi}{4}$ at both vertices $O$ and $P$ so, as the hypotenuse has length 1, sides $AO$ and $PA$ both have length $\frac{1}{\sqrt{2}}$. Thus $P = P(\frac{1}{\sqrt{2}}, -\frac{1}{\sqrt{2}})$, and so $\cos(-\frac{\pi}{4}) = \frac{1}{\sqrt{2}}$ and $\sin(-\frac{\pi}{4}) = -\frac{1}{\sqrt{2}}$.

The observation in Example 2 that the angles $\theta = 0$ and $\theta = 2\pi$ have the same cosine and sine is a special case of a more general fact: If $\theta$ is any angle, then adding any multiple of $2\pi$ to $\theta$ (even a negative multiple) does not change the point on the unit circle because $2\pi$ corresponds to a full rotation. This is an important property of the cosine and sine functions.

**THEOREM** **1** If $\theta$ is any angle, then

$$\cos(\theta + 2k\pi) = \cos \theta \qquad \text{for all } k = 0, \pm 1, \pm 2, \cdots.$$
$$\sin(\theta + 2k\pi) = \sin \theta$$

Because of Theorem 1, the cosine and sine functions are said to be **periodic functions** with **period** $2\pi$.

The geometrical relationship between $\cos \theta$ and $\sin \theta$ leads to many identities for these functions. For example, the points $P(\cos \theta, \sin \theta)$ and $P(\cos(-\theta), \sin(-\theta))$ corresponding to $\theta$ and $-\theta$ are reflections of each other in the $X$-axis (see Figure A.12). This gives the following important formulas:

$$\cos(-\theta) = \cos \theta \qquad \text{for all angles } \theta.$$
$$\sin(-\theta) = -\sin \theta$$

Moreover the equation of the unit circle is $x^2 + y^2 = 1$, so the fact that the point $P(\cos \theta, \sin \theta)$ lies on the unit circle for any angle $\theta$ gives the following basic relationship between these functions:

**THEOREM** **2** $\cos^2 \theta + \sin^2 \theta = 1$ for all angles $\theta$.

This is the fundamental identity of trigonometry.

## A.2  INDUCTION

Suppose you are given the following sequence of equations:

$$1 = 1$$
$$1 + 3 = 4$$
$$1 + 3 + 5 = 9$$
$$1 + 3 + 5 + 7 = 16$$
$$1 + 3 + 5 + 7 + 9 = 25$$
$$\vdots$$

There is a clear pattern: If $k = 1,\ 2,\ 3,\ 4,$ or 5, the sum of the first $k$ odd integers equals $k^2$. It is intriguing to speculate whether this is true for *every* integer $k \geq 1$. At first glance this seems to be a difficult question since the answer entails checking infinitely many cases $k = 1,\ 2,\ 3,\ \cdots$. But a clever idea makes it easy to prove that it is indeed true for every $k$.

Since the sum of the first $k$ odd integers can be written $1 + 3 + 5 + \cdots + (2k - 1)$, the conjecture takes the form

$$1 + 3 + 5 + \cdots + (2k - 1) = k^2 \quad \text{for every integer } k \geq 1.$$

For convenience, think of the assertion that $1 + 3 + 5 + \cdots + (2k - 1) = k^2$ as a statement $p_k$ about the integer $k$. Then we have an infinite sequence of statements

$$p_1,\ p_2,\ p_3,\ \cdots,\ p_k,\ \cdots,$$

and we want to verify that $p_k$ is true for every $k \geq 1$. The idea is as follows: First show that $p_1$ is true, then show that if $p_1$ is true, then $p_2$ must be true, then show that the truth of $p_2$ forces the truth of $p_3$, then that $p_3$ implies $p_4$, and so on indefinitely. In general we show that if $p_k$ is true, then $p_{k+1}$ is also true. If we can do this, then $p_k$ must be true for every $k \geq 1$.

In the above example, $p_k$ is the assertion that $1 + 3 + 5 + \cdots + (2k - 1) = k^2$. If this holds for some $k$, we want to show that $p_{k+1}$ must also be true. In this case, the statement $p_{k+1}$ reads

$$1 + 3 + 5 + \cdots + (2k - 1) + (2k + 1) = (k + 1)^2$$

by replacing $k$ by $k + 1$ in $p_k$. If we assume that $p_k$ is true, the left side becomes

$$1 + 3 + 5 + \cdots + (2k - 1) + (2k + 1) = k^2 + (2k + 1) = (k + 1)^2,$$

so $p_{k+1}$ is true, as required. Hence $p_k$ is indeed true for each $k$.

There is a compact way to formulate this method. If $p$ and $q$ are statements, the assertion "*If p is true, then q is true*" is expressed by saying that $p$ **implies** $q$, and is written as

$$p \Rightarrow q.$$

Hence the above method can be given as follows:

**THEOREM 1**   **Principle of Mathematical Induction.** Suppose a sequence of statements $p_1,\ p_2,\ p_3,\ \cdots,\ p_k,\ \cdots$ is given such that:

1. $p_1$ is true.
2. $p_k \Rightarrow p_{k+1}$ for every $k \geq 1$.

Then $p_k$ is true for every $k \geq 1$.

This is usually taken as one of the axioms for the positive integers, and is one of the most useful techniques in all of mathematics.[3] The following examples give a small indication of its range of application.

**Example 1**  Show that $1 + 2 + 3 + \cdots + k = \frac{1}{2}k(k + 1)$ for all $k \geq 1$.

SOLUTION   Let $p_k$ be the statement that $1 + 2 + 3 + \cdots + k = \frac{1}{2}k(k + 1)$. Then $p_1$ is true–it reads $1 = \frac{1}{2}1(2)$. To prove that $p_k \Rightarrow p_{k+1}$, we assume that $p_k$ is true for some value of $k \geq 1$, and verify that $p_{k+1}$ is also true; that is, $1 + 2 + 3 + \cdots + (k + 1) = \frac{1}{2}(k + 1)(k + 2)$. But the left side is

$$1 + 2 + 3 + \cdots + k + (k + 1) = \frac{1}{2}k(k + 1) + (k + 1) = \frac{1}{2}(k + 1)(k + 2)$$

using $p_k$. Hence $p_k \Rightarrow p_{k+1}$, as required.

**Example 2**  If $x \neq 1$ is a number, show that
$$1 + x + x^2 + \cdots + x^k = \frac{x^{k+1}-1}{x-1} \quad \text{for all } k \geq 1.$$

SOLUTION   Let $p_k$ denote the statement $1 + x + x^2 + \cdots + x^k = \frac{x^{k+1}-1}{x-1}$. If $k = 1$, then $p_1$ reads $1 + x = \frac{x^2-1}{x-1}$ which is true because $x^2 - 1 = (x + 1)(x - 1)$. If we assume that $p_k$ is true for some $k \geq 1$, then we must prove $p_{k+1} : 1 + x + x^2 + \cdots + x^{k+1} = \frac{x^{k+2}-1}{x-1}$. But the left side of $p_{k+1}$ reads

$$1 + x + x^2 + \cdots + x^k + x^{k+1} = \frac{x^{k+1}-1}{x-1} + x^{k+1} = \frac{(x^{k+1}-1)+x^{k+1}(x-1)}{x-1}$$
$$= \frac{x^{k+2}-1}{x-1}$$

using $p_k$. This is $p_{k+1}$, so we have proved that $p_k \Rightarrow p_{k+1}$.

When using the principle of induction we verify that $p_1$ is true and that $p_k \Rightarrow p_{k+1}$ for each $k \geq 1$, and so conclude that $p_k$ is true for all $k \geq 1$. There is nothing special about 1 here; we can start the induction at any fixed integer $m$. This extended principle is as follows: Suppose that

1. $p_m$ is true.
2. $p_k \Rightarrow p_{k+1}$ for each $k \geq m$.

Then $p_k$ is true for all $k \geq m$.

Here is an example where this extended principle is needed. Recall that the number $k!$ (called "$k$-factorial") denotes the product of all the numbers from $k$ to 1:
$$k! = k(k - 1)(k - 2)(k - 3) \cdots 3 \cdot 2 \cdot 1.$$
Hence $2! = 2$, $3! = 6$, $4! = 24$, etc.

**Example 3**  Show that $2^k < k!$ for all $k \geq 4$.

SOLUTION   If we let $p_k$ denote the statement $2^k < k!$ for each $k \geq 1$, then $p_1$, $p_2$, and $p_3$ are all false as is easily verified. But $p_4$ is true because $2^4 = 16 < 24 = 4!$. If we assume that $p_k$ is true for some $k \geq 4$, then $2^k < k!$ and we compute:
$$2^{k+1} = 2 \cdot 2^k < 2 \cdot k! < (k + 1)k! = (k + 1)!$$
because $2 < k + 1$. This completes the induction.

---

[3] A graphic description of the principle of induction can be given as follows: View the statements $p_k$ as a series of dominos lined up in such a way that if $p_k$ falls over, it knocks $p_{k+1}$ over. Then, if we push $p_1$ over, every domino will fall.

## A.3   POLYNOMIALS

Expressions like $3 - 5x$ and $1 + 3x - 2x^2$ are examples of polynomials. In general, a **polynomial** is an expression of the form

$$f(x) = a_0 + a_1x + a_2x^2 + \cdots + a_nx^n$$

where the $a_i$ are numbers, called the **coefficients** of the polynomial, and $x$ is a variable called an **indeterminate**. The number $a_0$ is called the **constant** term of the polynomial. The polynomial with every coefficient zero is called the **zero polynomial**, and is denoted simply as 0.

If $f(x) \neq 0$, the coefficient of the highest power of $x$ appearing in $f(x)$ is called the **leading term** of $f(x)$, and the highest power itself is called the **degree** of the polynomial and is denoted $deg(f(x))$. Hence

$$-1 + 5x + 3x^2 \quad \text{has constant term } -1, \text{ leading term } 3, \text{ and degree } 2,$$
$$7 \quad \text{has constant term } 7, \text{ leading term } 7, \text{ and degree } 0,$$
$$6x - 3x^3 + x^4 - x^5 \quad \text{has constant term } 0, \text{ leading term } -1, \text{ and degree } 5.$$

We do not define the degree of the zero polynomial.

Two polynomials $f(x)$ and $g(x)$ are called **equal** if every coefficient of $f(x)$ is the same as the corresponding coefficient of $g(x)$. More precisely, if

$$f(x) = a_0 + a_1x + a_2x^2 + \cdots \quad \text{and} \quad g(x) = b_0 + b_1x + b_2x^2 + \cdots$$

are polynomials, then

$$f(x) = g(x) \quad \text{if and only if} \quad a_0 = b_0, a_1 = b_1, a_2 = b_2, \cdots.$$

In particular, this means that

$$f(x) = 0 \text{ is the zero polynomial if and only if } a_0 = 0, a_1 = 0, a_2 = 0, \cdots.$$

This is the reason for calling $x$ an indeterminate.

Let $f(x)$ and $g(x)$ denote nonzero polynomials of degrees $n$ and $m$ respectively, say

$$f(x) = a_0 + a_1x + a_2x^2 + \cdots + a_nx^n \quad \text{and} \quad g(x) = b_0 + b_1x + b_2x^2 + \cdots + b_mx^m$$

where $a_n \neq 0$ and $b_m \neq 0$. If these expressions are multiplied, the result is

$$f(x)g(x) = a_0b_0 + (a_0b_1 + a_1b_0)x + (a_0b_2 + a_1b_1 + a_2b_0)x^2 + \cdots + a_nb_mx^{n+m}.$$

Since $a_n$ and $b_m$ are nonzero numbers, their product $a_nb_m \neq 0$ and we have

**THEOREM 1**

If $f(x)$ and $g(x)$ are nonzero polynomials of degrees $n$ and $m$ respectively, their product $f(x)g(x)$ is also nonzero and

$$deg[f(x)g(x)] = n + m.$$

**Example 1**   $(2 - x + 3x^2)(3 + x^2 - 5x^3) = 6 - 3x + 11x^2 - 11x^3 + 8x^4 - 15x^5.$

If $f(x)$ is any polynomial, the next theorem shows that $f(x) - f(a)$ is a multiple of the polynomial $x - a$. In fact we have the

**THEOREM 2**

**Remainder Theorem.** If $f(x)$ is a polynomial of degree $n$ and $a$ is any number, then there exists a polynomial $q(x)$ such that

$$f(x) = (x - a)q(x) + f(a)$$

where either $q(x) = 0$ or $deg(q(x)) \leq n - 1$.

PROOF  Write $f(x) = a_0 + a_1x + a_2x^2 + \cdots + a_nx^n$ where the $a_i$ are numbers, so that $f(a) = a_0 + a_1a + a_2a^2 + \cdots + a_na^n$. If these expressions are subtracted, the constant terms cancel and we obtain

$$f(x) - f(a) = a_1(x - a) + a_2(x^2 - a^2) + \cdots + a_n(x^n - a^n).$$

Hence it suffices to show that, for each $k \geq 1$, $x^k - a^k = (x - a)p(x)$ for some polynomial $p(x)$ of degree $k - 1$. This is clear if $k = 1$. If it holds for some value $k$, the fact that

$$x^{k+1} - a^{k+1} = (x - a)x^k + a(x^k - a^k)$$

shows that it holds for $k + 1$. Hence the proof is complete by induction.

There is a systematic procedure for finding the polynomial $q(x)$ in the Remainder theorem. It is illustrated below for $f(x) = x^3 - 3x^2 + x - 1$ and $a = 2$. The polynomial $q(x)$ is generated on the top line one term at a time as follows: First $x^2$ is chosen because $x^2(x - 2)$ has the same $x^3$-term as $f(x)$, and this is subtracted from $f(x)$ to leave a "remainder" of $-x^2 + x - 1$. Next, the second term on top is $-x$ because $-x(x - 2)$ has the same $x^2$-term, and this is subtracted to leave $-x - 1$. Finally, the third term on top is $-1$, and the process ends with a "remainder" of $-3$.

$$
\require{enclose}
\begin{array}{r}
x^2 - \phantom{x}x - 1 \phantom{x} \\
x - 2 \enclose{longdiv}{x^3 - 3x^2 + \phantom{x}x - 1} \\
\underline{x^3 - 2x^2} \phantom{+ x - 1xxx} \\
-x^2 + \phantom{x}x - 1 \\
\underline{-x^2 + 2x} \phantom{- 1x} \\
-x - 1 \\
\underline{-x + 2} \\
-3
\end{array}
$$

Hence $x^3 - 3x^2 + x - 1 = (x - 2)(x^2 - x - 1) + (-3)$. The final remainder is $-3 = f(2)$ as is easily verified. This procedure is called the **division algorithm**.[4]

A real number $a$ is called a **root** of the polynomial $f(x)$ if $f(a) = 0$. Hence for example, 1 is a root of $f(x) = 2 - x + 3x^2 - 4x^3$, but $-1$ is not a root because $f(-1) = 10 \neq 0$. If $f(x)$ is a multiple of $x - a$, we say that $x - a$ is a **factor** of $f(x)$. Hence the Remainder theorem shows immediately that if $a$ is a root of $f(x)$, then $x - a$ is a factor of $f(x)$. But the converse is also true: If $x - a$ is a factor of $f(x)$, say $f(x) = (x - a)q(x)$, then $f(a) = (a - a)q(a) = 0$. This proves the

THEOREM ③  **Factor Theorem.** If $f(x)$ is a polynomial and $a$ is a number, then $x - a$ is a factor of $f(x)$ if and only if $a$ is a root of $f(x)$.

Example 2  If $f(x) = x^3 - 2x^2 - 6x + 4$, then $f(-2) = 0$, so $x - (-2) = x + 2$ is a factor of $f(x)$. In fact, the division algorithm gives $f(x) = (x + 2)(x^2 - 4x + 2)$.

Consider the polynomial $f(x) = x^3 - 3x + 2$. Then 1 is clearly a root of $f(x)$, and the division algorithm gives $f(x) = (x - 1)(x^2 + x - 2)$. But 1 is also a root of $x^2 + x - 2$; in fact, $x^2 + x - 2 = (x - 1)(x + 2)$. Hence

$$f(x) = (x - 1)^2(x + 2)$$

and we say that the root 1 has **multiplicity** 2.

---

[4] This procedure can be used to divide $f(x)$ by *any* nonzero polynomial $d(x)$ in place of $x - a$; the remainder then is a polynomial that is either zero or of degree less than the degree of $d(x)$.

Note that every nonzero constant polynomial $f(x) = b \neq 0$ has *no* roots. However, there do exist nonconstant polynomials with no roots. For example, if $g(x) = x^2 + 1$, then $g(a) = a^2 + 1 \geq 1$ for every real number $a$, so $a$ is not a root. However, the *complex* number $i$ is a root of $g(x)$; we return to this below.

Now suppose that $f(x)$ is any nonzero polynomial. We claim that it can be factored in the following form:

$$f(x) = (x - a_1)(x - a_2) \cdots (x - a_m)g(x)$$

where $a_1$, $a_2$, $\cdots$, $a_m$ are the roots of $f(x)$ and $g(x)$ has no root (where the $a_i$ may have repetitions, and may not appear at all if $f(x)$ has no real root).

The above calculation shows that the polynomial $f(x) = x^3 - 3x + 2$ has roots 1 and $-2$, with 1 of multiplicity two. Counting the root $-2$ once, we say that $f(x)$ has three roots counting multiplicities. The next theorem shows that no polynomial can have more roots than its degree even if multiplicities are counted.

**THEOREM 4**    If $f(x)$ is a nonzero polynomial of degree $n$, then $f(x)$ has at most $n$ roots counting multiplicities.

PROOF    If $n = 0$, then $f(x)$ is a constant and has no roots. So the theorem is true if $n = 0$. (It also holds for $n = 1$ because, if $f(x) = a + bx$ where $b \neq 0$, then the only root is $-\frac{a}{b}$.) In general, suppose inductively that the theorem holds for some value of $n \geq 0$, and let $f(x)$ have degree $n + 1$. We must show that $f(x)$ has at most $n + 1$ roots counting multiplicities. This is certainly true if $f(x)$ has no root. On the other hand, if $a$ is a root of $f(x)$, the Factor theorem shows that $f(x) = (x - a)q(x)$ for some polynomial $q(x)$, and $q(x)$ has degree $n$ by Theorem 1. By induction, $q(x)$ has at most $n$ roots. But if $b$ is any root of $f(x)$, then

$$(b - a)q(b) = f(b) = 0$$

so either $b = a$ or $b$ is a root of $q(x)$. It follows that $f(x)$ has at most $n$ roots. This completes the induction and so proves Theorem 4.

As we have seen, a polynomial may have *no* root, for example $f(x) = x^2 + 1$. Of course $f(x)$ has complex roots $i$ and $-i$, where $i$ is the complex number such that $i^2 = -1$. But Theorem 4 even holds for complex roots: the number of complex roots (counting multiplicities) cannot exceed the degree of the polynomial. Moreover, the Fundamental Theorem of Algebra asserts that the only nonzero polynomials with no complex root are the constant polynomials.

# Selected Answers and Solutions

CHAPTER **1**

## Linear Equations and Matrices

## 1.1 MATRICES

**1. b)**
$$\begin{bmatrix} 70 & 70 \\ 137 & 175 \\ 129 & 176 \\ 78 & 111 \end{bmatrix}$$
**d)**
$$\begin{bmatrix} 268 & 259 \\ 455 & 552 \\ 439 & 574 \\ 275 & 384 \end{bmatrix}$$

**2. b)** $A$ and $B$ must the same size.

**4. b)**
$$\begin{bmatrix} 624 & 5933 & 8754 \\ -2059 & 4115 & 3715 \\ 9864 & 8264 & 1172 \end{bmatrix}$$

**6. b)** False. $A = \begin{bmatrix} 1 & 0 \\ 2 & 1 \end{bmatrix}$, $B = \begin{bmatrix} -1 & 0 \\ -2 & -1 \end{bmatrix}$.

   **d)** True. If $A = -A$, then $2A = A + A = A + (-A) = 0$, so $A = \frac{1}{2}(2A) = \frac{1}{2}(0) = 0$.

**7. b)** $a = \frac{11}{9}$, $b = \frac{4}{3}$.

**8. b)** $A = \begin{bmatrix} \frac{7}{3} & \frac{-10}{3} & \frac{1}{3} \\ 2 & \frac{5}{3} & \frac{11}{3} \end{bmatrix}$

**9. b)** $-4A$

**10. b)** Write $A = diag(a_1, \cdots, a_n)$ where $a_1, \cdots, a_n$ are the entries along the main diagonal. Then $cA = diag(ca_1, \cdots, ca_n)$.

**11. b)** Write $A = [a_{ij}]$. Then $(cd)A = [(cd)a_{ij}] = [c(da_{ij})] = c[da_{ij}] = c(dA)$.

## 1.2 LINEAR EQUATIONS

**1. b)** If $X = [-s - t \quad -1 + s - 2t \quad s \quad t]^T$, we have $x_1 = -s - t$, $x_2 = -1 + s - 2t$, $x_3 = s$ and $x_4 = t$. Then $x_1 - 2x_2 + 3x_3 - 3x_4 = (-s - t) - 2(-1 + s - 2t) + 3s - 3t = 2$, so $X$ is a solution to the first equation. The other equations can be verified similarly.

**3. b)**
$$\begin{bmatrix} 1 & 0 & 0 & 0 & 0 & \frac{1}{4} & \frac{1}{4} \\ 0 & 1 & -3 & 0 & 0 & -\frac{1}{2} & -\frac{1}{2} \\ 0 & 0 & 0 & 1 & 0 & \frac{1}{2} & -\frac{1}{2} \\ 0 & 0 & 0 & 0 & 1 & \frac{1}{4} & \frac{1}{4} \end{bmatrix}$$

**4. b)** $x_1 = 3$, $x_2 = \frac{2}{3}$.
   **d)** $x_1 = 5 - t$, $x_2 = \frac{5}{2}t - \frac{9}{2}$, $x_3 = t$.
   **f)** The system has no solution.
   **h)** $x_1 = 9 + 2s + 13t$, $x_2 = s$, $x_3 = -2$, $x_4 = -2 - 4t$, $x_5 = t$.

**6.** Let $r$ be the rank of the augmented matrix and let $n$ be the number of variables.
   **b)** $r = 2$, $n = 2$. The system has 1 solution.
   **d)** $r = 2$, $n = 3$. The system has infinitely many solutions with 1 parameter.
   **f)** $r = 3$, $n = 3$. The system has no solutions.
   **h)** $r = 3$, $n = 5$. The system has infinitely many solutions with 2 parameters.

**8.** Unique solution if $a \neq 0$.

**9. b)** If $c \neq 3a + b$, no solutions. If $c = 3a + b$, the solutions are $x_1 = \frac{4}{3}t - \frac{a+2b}{3}$, $x_2 = \frac{5}{3}t - \frac{2a+b}{3}$, $x_3 = t$.

**d)** If $a = 2$ and $b \neq -1$, no solutions. If $a = 2$ and $b = -1$, the solutions are $x_1 = -\frac{b}{2} - \frac{1}{2}$, $x_2 = t$. If $a \neq 2$, the unique solution is $x_1 = \frac{b+1}{2-a}$, $x_2 = \frac{-2-ab}{2-a}$.

**10.** $\operatorname{rank} A = \operatorname{rank} A^T = 2$. A proof that this always works will come in Section 4.4.

**12.** $f_1 = t + 70$, $f_2 = t + 10$, $f_3 = t + 60$, and $f_4 = t$. $f_1$ is the largest traffic flow.

**14.** Assume that $A$ is $m \times n$ and let $\operatorname{rank} A = r$.

**b)** False. $A = \begin{bmatrix} 1 & 0 & 1 \\ 0 & 1 & 1 \\ 0 & 0 & 0 \end{bmatrix}$    **d)** False. $A = \begin{bmatrix} 1 & 0 & 1 \\ 0 & 1 & 1 \end{bmatrix}$

**f)** False. $A = R = \begin{bmatrix} 1 & 0 & 1 & 1 & 1 \\ 0 & 1 & 1 & 1 & 1 \\ 0 & 0 & 0 & 0 & 1 \end{bmatrix}$

**h)** True. The reduced row-echelon form of $A$ contains all the leading ones in the same form as $C$.

**j)** False. $A = \begin{bmatrix} 1 & 0 & 1 & 1 \\ 0 & 1 & 1 & 1 \\ 0 & 0 & 0 & 1 \end{bmatrix}$

## 1.3   HOMOGENEOUS SYSTEMS

**1.** Let $X_1$, $X_2$, $\cdots$ denote the basic solutions, and let $X$ denote the general solution.

**b)** No basic solutions; $X = 0$.

**d)** $X_1 = \begin{bmatrix} 2 \\ 1 \\ 0 \\ 0 \end{bmatrix}$, $X_2 = \begin{bmatrix} -3 \\ 0 \\ -5 \\ 1 \end{bmatrix}$; $X = sX_1 + tX_2$.

**f)** $X_1 = \begin{bmatrix} 9 \\ 24 \\ 1 \\ 0 \\ 0 \\ 0 \end{bmatrix}$, $X_2 = \begin{bmatrix} -5 \\ -10 \\ 0 \\ 1 \\ 0 \\ 0 \end{bmatrix}$, $X_3 = \begin{bmatrix} -2 \\ -7 \\ 0 \\ 0 \\ 1 \\ 0 \end{bmatrix}$, $X_4 = \begin{bmatrix} 11 \\ 29 \\ 0 \\ 0 \\ 0 \\ 1 \end{bmatrix}$;

$X = rX_1 + sX_2 + tX_3 + uX_4$.

**h)** $X_1 = \begin{bmatrix} -\frac{13}{4} \\ \frac{23}{4} \\ \frac{11}{4} \\ 1 \\ 0 \end{bmatrix}$, $X_2 = \begin{bmatrix} -2 \\ 6 \\ 3 \\ 0 \\ 1 \end{bmatrix}$; $X = sX_1 + tX_2$.

**2. b)** Let $w, x, y, z$ be the number of octane, oxygen, carbon dioxide, and water molecules present, respectively. Then $X = \begin{bmatrix} \frac{1}{9}s & \frac{25}{18}s & \frac{8}{9}s & s \end{bmatrix}^T$.

**4. b)** $n = 6$ and $r \leq 4$, so $n - r \geq 2$. There may be 2, 3, 4, 5, or 6 parameters.

**d)** There are $n - r = 6 - 4 = 2$ parameters.

**6.** If the system is consistent and the rank of the augmented matrix is the same as the number of variables, then there is a unique solution.

**7. b)** False.   $\begin{aligned} x + 2y &= 0 \\ -x - 2y &= 0 \end{aligned}$    **d)** False. $A = \begin{bmatrix} 1 & 0 & 2 & 0 \\ 0 & 1 & 1 & 0 \end{bmatrix}$.

**f)** False. $A = \begin{bmatrix} 1 & 0 & 0 \\ 0 & 1 & 0 \\ 0 & 0 & 0 \end{bmatrix}$.

**8.** Since the system is homogeneous, we don't have to worry about having no solution. If any system has a solution, $\operatorname{rank}(A) = \operatorname{rank}(C)$ (see Exercise 14(g), §1.2).

## 1.4   MATRIX MULTIPLICATION

**1. b)** $\begin{bmatrix} 34 \\ 0 \end{bmatrix}$    **d)** $\begin{bmatrix} -20 \end{bmatrix}$    **f)** $\begin{bmatrix} 4 & 2 & 22 \\ -2 & 11 & -31 \\ 1 & -13 & 28 \end{bmatrix}$

**2. b)** Not defined.    **d)** Not defined.

**f)** Not defined.    **g)** $\begin{bmatrix} -5 & 18 \\ -12 & 19 \end{bmatrix}$

**4.** If $AB$ is $n \times n$, then $A$ is $n \times m$ for some $m$, and $B$ is $p \times n$ for some $p$. But $AB$ is defined only if $m = p$, so $BA$ is $m \times m$.

**6.** Let $A$ be $n \times m$. Then $C$ is $m \times p$ for some $p$, and $AC$ is $n \times p$. Now $B$ is $p \times q$ for some $p$, and $CB$ is $m \times q$. But $AC = CB$, so $n = m$ and $p = q$, so $A$ is $n \times n$ and $B$ is $p \times p$.

**7. b)** $-3AB + 2AC + A^2B$.

**8. b)** If $AB \neq BA$, then $(A + B)^2 = A^2 + AB + BA + B^2 \neq A^2 + AB + AB + B^2 = A^2 + 2AB + B^2$.

**10. b)** True. If $A$ is $n \times m$ and $AB$ is square, then $B$ must be $m \times n$. Hence $BA$ is defined too.

**d)** True. If $A$ is $n \times m$ and $A^2$ can be formed, then $A$ must also be $m \times n$, so $n = m$.

**f)** False. $\begin{bmatrix} 1 & 0 \\ 3 & 0 \end{bmatrix}\begin{bmatrix} 1 \\ 1 \end{bmatrix} = \begin{bmatrix} 1 \\ 3 \end{bmatrix}$.

**h)** False. If $A = \begin{bmatrix} 1 & 0 \\ 1 & 0 \end{bmatrix}$ and $B = \begin{bmatrix} 1 & 1 \\ 0 & 0 \end{bmatrix}$, then $(AB)^2 = \begin{bmatrix} 2 & 2 \\ 2 & 2 \end{bmatrix}$, but $A^2B^2 = \begin{bmatrix} 1 & 1 \\ 1 & 1 \end{bmatrix}$.

**j)** False. If $A = \begin{bmatrix} 0 & 0 \\ 0 & 1 \end{bmatrix}$ and $J = \begin{bmatrix} 1 & 1 \\ 0 & 1 \end{bmatrix}$, then $AJ = A$, but $J \neq I$.

**l)** False. $A = \begin{bmatrix} 1 & 0 \\ 0 & 1 \\ 0 & 0 \end{bmatrix}$.

**12. b)** This is the converse of Example 8.

**13. b)** $\begin{bmatrix} -2 \\ 2 \\ 0 \end{bmatrix} + t \begin{bmatrix} 1 \\ -3 \\ 1 \end{bmatrix}$    **d)** $\begin{bmatrix} 3 \\ -9 \\ -2 \\ 0 \end{bmatrix} + t \begin{bmatrix} -1 \\ 4 \\ 1 \\ 1 \end{bmatrix}$

**14.** Let $A = \begin{bmatrix} 1 & 1 & 0 & 1 \\ 1 & 1 & 0 & 1 \\ 0 & 0 & 1 & 1 \\ 0 & 0 & 1 & 1 \end{bmatrix}$ and $B = \begin{bmatrix} 1 \\ 1 \\ 0 \\ 1 \end{bmatrix}$. Then $AX = B$ has no

solutions, and $AX = 0$ has a 2-parameter family of solutions.

**16.** Since $A$ is $4 \times 6$ and $rankA = 4$, the reduced row-echelon form of $A$ has a leading one in every row. Hence every system $AX = B$ has a solution.

**18. b)** If $R_i$ and $C_i$ denote the $i^{\text{th}}$ row and column of $A$, respectively, then $AD = [a_1C_1 \ a_2C_2 \ a_3C_3]$, and

$$DA = \begin{bmatrix} a_1R_1 \\ a_2R_2 \\ a_3R_3 \end{bmatrix}.$$

**20.** For each odd $n = 1, 3, 5, \cdots$, $A^n = A$. For each even

$n = 2, 4, 6, \cdots$, $A^n = \begin{bmatrix} 1 & 0 & 0 \\ 1 & 0 & 2 \\ 0 & 0 & 1 \end{bmatrix}$.

**21. b)** $A^2 - 5A + 4I = \begin{bmatrix} -29 & 30 \\ -45 & 46 \end{bmatrix} - \begin{bmatrix} -25 & 30 \\ -45 & 50 \end{bmatrix} + \begin{bmatrix} 4 & 0 \\ 0 & 4 \end{bmatrix} = 0.$

**22. b)** If $A = \begin{bmatrix} a & b \\ b & c \end{bmatrix}$, then the $(1, 1)$- and $(2, 2)$-entries of $A^2$ are $a^2 + b^2$ and $b^2 + c^2$, respectively. Hence if $A^2 = 0$, then $a = b = c = 0$, so $A = 0$.

**24. b)** For each even $n = 2, 4, 6, \cdots$, $A^n = \left[ \begin{array}{cc|cc} 1 & -n & 0 & 0 \\ 0 & 1 & 0 & 0 \\ \hline 0 & 0 & 1 & 0 \\ 0 & 0 & 0 & 1 \end{array} \right]$.

For each odd $n = 1, 3, 5, \cdots$, $A^n = \left[ \begin{array}{cc|cc} 1 & -n & 2 & -1 \\ 0 & 1 & 0 & 0 \\ \hline 0 & 0 & -1 & 1 \\ 0 & 0 & 0 & 1 \end{array} \right]$.

**25. b)** $\begin{bmatrix} I & 0 \\ 0 & I \end{bmatrix} = I$    **d)** $[0] = 0$

**26.** For each odd $n = 1, 3, 5, \cdots$, $A^n = \begin{bmatrix} 0 & B^{\frac{n-1}{2}} \\ B^{\frac{n+1}{2}} & 0 \end{bmatrix}$.

For each even $n = 2, 4, 6, \cdots$, $A^n = \begin{bmatrix} B^{\frac{n}{2}} & 0 \\ 0 & B^{\frac{n}{2}} \end{bmatrix}$.

**27. b)** $A = \begin{bmatrix} 1 & 1 & 1 & 1 \\ 0 & 0 & 1 & 1 \\ 0 & 0 & 0 & 1 \\ 1 & 0 & 0 & 0 \end{bmatrix}$, $A^4 = \begin{bmatrix} 10 & 5 & 7 & 10 \\ 3 & 2 & 3 & 4 \\ 2 & 1 & 2 & 3 \\ 5 & 2 & 3 & 5 \end{bmatrix}$, $\cdots$,

$A^{10} = \begin{bmatrix} 765 & 370 & 549 & 815 \\ 266 & 129 & 191 & 283 \\ 179 & 87 & 129 & 191 \\ 370 & 179 & 266 & 395 \end{bmatrix}$.

**28.** Write $I_n = [C_1 \ C_2 \ \cdots \ C_n]$ where $C_j$ is column $j$ of $I_n$. Then $B = BI_n = B[C_1 \ C_2 \ \cdots \ C_n] = [BC_1 \ BC_2 \ \cdots \ BC_n] = [C_1 \ C_2 \ \cdots \ C_n] = I_n$.

**30.** The $(1, 1)$ and $(2, 2)$-entries of $AA^T$ are $a^2 + b^2 + c^2$ and $(a')^2 + (b')^2 + (c')^2$, so if $AA^T = 0$, $a = b = c = a' = b' = c' = 0$. Thus $A = 0$.

**32.** If $x = \frac{c}{a}$, $y = a$, $z = b$, and $w = \frac{da - bc}{a}$,

then $\begin{bmatrix} 1 & 0 \\ x & 1 \end{bmatrix} \begin{bmatrix} y & z \\ 0 & w \end{bmatrix} = \begin{bmatrix} a & b \\ c & d \end{bmatrix}.$

**34.**

| Matrix | $A \rightarrow A^T \rightarrow KA^T \rightarrow (KA^T)K$ |
| --- | --- |
| $(i, j)$-entry | $a_{ij} \quad a_{ji} \quad a_{n-j \ i} \quad a_{n-j \ n-i}$ |

Now note that $a_{n-j \ n-i}$ is the entry across the "anti-diagonal" from $a_{ij}$.

## 1.5  MATRIX INVERSES

**2. b)** $\begin{bmatrix} 2 & -5 \\ -1 & 3 \end{bmatrix}$    **d)** $\begin{bmatrix} 0 & -1 & 0 \\ 1 & 1 & -1 \\ 2 & 2 & -3 \end{bmatrix}$

**f)** $\begin{bmatrix} \frac{-1}{2} & \frac{-3}{2} & \frac{-1}{2} \\ 1 & 4 & 2 \\ \frac{1}{2} & \frac{-1}{2} & \frac{-1}{2} \end{bmatrix}$    **h)** $\begin{bmatrix} \frac{1}{3} & \frac{-4}{3} & \frac{-1}{3} & 0 \\ \frac{2}{3} & \frac{-8}{3} & \frac{-2}{3} & -1 \\ \frac{1}{3} & \frac{-1}{3} & \frac{-1}{3} & 0 \\ 1 & -6 & -2 & 0 \end{bmatrix}$

**4.** $A^{-1} = \begin{bmatrix} -6 & -4 & 11 \\ -3 & -2 & 5 \\ -1 & -1 & 1 \end{bmatrix}$, and $(A^{-1})^{-1} = A$.

**5. b)** $A = \frac{1}{25} \begin{bmatrix} 4 & 1 \\ 3 & 2 \end{bmatrix}$    **d)** $A = \frac{1}{34} \begin{bmatrix} 23 & -15 \\ -10 & 8 \end{bmatrix}$

**f)** $A = \frac{21}{40} \begin{bmatrix} 3 & -1 \\ 2 & 1 \end{bmatrix}$

**6.** Left multiply the equation $AX = B$ by $A^{-1}$ to get $A^{-1}B = X$.

**7. b)** $X = \begin{bmatrix} -5 \\ \frac{9}{2} \end{bmatrix}$    **d)** $X = \begin{bmatrix} -4 \\ \frac{7}{5} \\ \frac{11}{5} \end{bmatrix}$

**8.** If $A$ is invertible, then the system $AX = B$ has the unique solution $X = A^{-1}B$, clearly dependent on $B$.

**9. b)** Here you must right multiply the given equation by $A^{-1}$. The result is $X = \begin{bmatrix} 2 & 1 & 4 \\ 14 & 25 & 13 \end{bmatrix}$.

**11. b)** Here $C$ is playing a role similar to $A^{-1}$, even though it may not be square. Suppose that $AX = B = AX'$. Then $CAX = CB = CAX'$, so $X = CB = X'$, and the solution is uniquely determined.

**13. b)** True. If $A = 0$ is $n \times n$, then $CA = 0$ for every $n \times n$ matrix $C$, so there is no matrix $C$ such that $CA = I$.

    **d)** False. $\begin{bmatrix} 1 & 0 \\ 0 & 0 \end{bmatrix}\begin{bmatrix} 1 & 0 \\ 0 & 0 \end{bmatrix} = \begin{bmatrix} 1 & 0 \\ 0 & 0 \end{bmatrix}$, but $\begin{bmatrix} 1 & 0 \\ 0 & 0 \end{bmatrix}$ is not invertible.

    **f)** False. $\begin{bmatrix} 1 & 0 \\ 0 & 0 \end{bmatrix}\begin{bmatrix} 0 & 0 \\ 1 & 1 \end{bmatrix} = \begin{bmatrix} 0 & 0 \\ 0 & 0 \end{bmatrix}$.

    **h)** False. $AX = 0$ always has the trivial solution.

    **j)** True. $A(A(A^2)^{-1}) = (AA)(A^2)^{-1} = I$ and $((A^2)^{-1}A)A = (A^2)^{-1}AA = I$. Thus $A^{-1} = (A^2)^{-1}A$.

**14.** $U$ is invertible. Now if $AU = 0$, then $A = A(UU^{-1}) = (AU)U^{-1} = 0U^{-1} = 0$.

**16. b)** $(I - A)(I + A + A^2) = I + A + A^2 - A - A^2 - A^3 = I$, and $(I + A + A^2)(I - A) = I - A + A - A^2 + A^2 - A^3 = I$.

    **c)** If $A^n = 0$, then $(I - A)$ is invertible and $(I - A)^{-1} = 1 + A + A^2 + \cdots + A^{n-1}$.

**18. b)** $A = P(P^{-1}AP)P^{-1}$, a product of invertible matrices, so $A$ is invertible.

**19. b)** Given $P(A + X)Q = B$, left multiply by $P^{-1}$ and right multiply by $Q^{-1}$ to get $A + X = P^{-1}BQ^{-1}$. Hence $X = P^{-1}BQ^{-1} - A$.

**20. b)** $B$ commutes with $A^{-1}$ by **a)**; then $B^{-1}$ commutes with $A^{-1}$, again by **a)**.

**21. b)** If $A$ has a column of zeros, then $BA$ has a column of zeros for any compatible matrix $B$. Hence there is no matrix $C$ for which $CA = I$.

**22.** $\begin{bmatrix} \sin\theta & -\cos\theta \\ \cos\theta & \sin\theta \end{bmatrix}$

**24.** Left multiply by $A^{-1}$.

**26.** If $B^{-1}$ exists, then $A = (AB)B^{-1} = 0B^{-1} = 0$, contrary to assumption. Similarly, in the case that $A^{-1}$ exists.

**28. b)** $A(BC(ABC)^{-1}) = (ABC)(ABC)^{-1} = I$, so $A$ is invertible by Corollary 1 of Theorem 5. Similarly, $C$ is invertible. Finally, $B = A^{-1}(ABC)C^{-1}$, a product of invertible matrices. Hence $B$ is invertible.

**30.** $(cA)(\frac{1}{c}A^{-1}) = \frac{c}{c}AA^{-1} = 1I = I$, and $(\frac{1}{c}A^{-1})(cA) = \frac{c}{c}A^{-1}A = 1I = I$.

## 1.6 ELEMENTARY MATRICES

**1. b)** The row operation is: Add 5 (row 1) to row 2.
$$E^{-1} = \begin{bmatrix} 1 & 0 & 0 \\ -5 & 1 & 0 \\ 0 & 0 & 1 \end{bmatrix}$$

    **d)** The row operation is: Subtract 3 (row 3) from row 1.
$$E^{-1} = \begin{bmatrix} 1 & 0 & 3 \\ 0 & 1 & 0 \\ 0 & 0 & 1 \end{bmatrix}$$

**2. b)** $E = \begin{bmatrix} 1 & 1 \\ 0 & 1 \end{bmatrix} \quad F = E^{-1} = \begin{bmatrix} 1 & -1 \\ 0 & 1 \end{bmatrix}$.

    **d)** $E = \begin{bmatrix} 1 & 0 & 0 \\ 0 & 1 & 0 \\ 2 & 0 & 1 \end{bmatrix} \quad F = E^{-1} = \begin{bmatrix} 1 & 0 & 0 \\ 0 & 1 & 0 \\ -2 & 0 & 1 \end{bmatrix}$.

**3. b)** $E_1 = \begin{bmatrix} 0 & 1 \\ 1 & 0 \end{bmatrix} \quad E_2 = \begin{bmatrix} 1 & 0 \\ -2 & 1 \end{bmatrix}$.

**4.** Elementary matrices are used to perform elementary row operations by matrix multiplication.

**5. b)** Proceed as in Example 4.
$$\begin{bmatrix} 2 & -1 \\ 1 & -1 \end{bmatrix} = \begin{bmatrix} 1 & 1 \\ 0 & 1 \end{bmatrix}\begin{bmatrix} 1 & 0 \\ 1 & 1 \end{bmatrix}\begin{bmatrix} 1 & 0 \\ 0 & -1 \end{bmatrix}.$$

**6. b)** $U = \frac{1}{8}\begin{bmatrix} 3 & -1 \\ 1 & -3 \end{bmatrix}$     **d)** $U = \frac{1}{16}\begin{bmatrix} 0 & 1 & 5 \\ 0 & 3 & -1 \\ 16 & 2 & -6 \end{bmatrix}$

Proceed as in Example 3.

**8. b)** Find row operations carrying $A$ to $B$, and use Theorem 1.
$$U = \begin{bmatrix} -2 & 1 \\ 1 & 0 \end{bmatrix}$$

**10.** $AF$ is the result of doing a *column* operation to $A$. Use the fact that $B = AF$ if and only if $B^T = F^TA^T$, so we get $F^T$ from a *row* operation on $A^T$. The result is $F = \begin{bmatrix} 1 & 0 \\ -2 & 1 \end{bmatrix}$.

**12. b)** False. $E_1 = \begin{bmatrix} 1 & -1 \\ 0 & 1 \end{bmatrix}$, $E_2 = \begin{bmatrix} 1 & 0 \\ 4 & 1 \end{bmatrix}$.

    **d)** False. $A = \begin{bmatrix} 2 & -1 \\ 6 & 3 \end{bmatrix}$, $B = \begin{bmatrix} 4 & -2 \\ 2 & 1 \end{bmatrix}$.

**13. b)** Proceed as in Example 5. $\text{rank}A = 2$,
$$U = \begin{bmatrix} 0 & 0 & 1 \\ 0 & \frac{1}{2} & \frac{-3}{2} \\ 1 & \frac{-1}{2} & \frac{1}{2} \end{bmatrix}, \text{ and } V = \begin{bmatrix} 0 & 0 & 1 & 0 \\ 0 & 0 & 0 & 1 \\ 4 & 3 & -4 & -3 \\ -1 & -1 & 1 & 1 \end{bmatrix}$$

**14.** If $U$ is the product of the corresponding elementary matrices, then $P = UA$ and $Q = UI = U$, so $P = QA$.

**16.** Since every invertible matrix can be written as a product of elementary matrices, $A$ can be carried to $B$ by the corresponding row operations.

**18.** We can carry $A \to B$ by one column operation by first carrying $A^T \to B^T$ by one row operation and then transposing. But then $B^T = EA^T$ where $E$ is elementary, so $B = AE^T$. If we write $F = E^T$, then $F$ is elementary (Exercise 11) and $B = AF$. Now, if $A \to B$ by a series of column operations, then $A \to AF_1 \to AF_1F_2 \to \cdots \to AF_1F_2 \cdots F_m = B$ where the $F_i$ are elementary. Write $V = F_1F_2 \cdots F_m$, so $V$ is invertible because the $F_i$ are invertible. If we do the same sequence of column operations to $I$, then $I \to IV = V$, so we have $[A \ I] \to [B \ V]$.

## 1.7 LU-FACTORIZATION

**1. b)** $LY = B$ gives $Y = \frac{1}{3} \begin{bmatrix} 6 \\ 11 \\ -9 \\ -65 \end{bmatrix}$, but then $UX = Y$ has no solution.

**2. b)** $\begin{bmatrix} -3 & 0 & 0 \\ 2 & 7 & 0 \\ -1 & -10 & 0 \end{bmatrix} \begin{bmatrix} 1 & -2 & -3 \\ 0 & 0 & 1 \\ 0 & 0 & 0 \end{bmatrix}$

**d)** $\begin{bmatrix} 3 & 0 & 0 \\ -2 & 5 & 0 \\ -1 & 2 & \frac{18}{5} \end{bmatrix} \begin{bmatrix} 1 & -3 & 2 & 0 & -3 \\ 0 & 0 & 1 & \frac{1}{5} & -1 \\ 0 & 0 & 0 & 1 & \frac{5}{9} \end{bmatrix}$

**f)** $\begin{bmatrix} 1 & 0 & 0 & 0 \\ 2 & 1 & 0 & 0 \\ 0 & -1 & -4 & 0 \\ -1 & 0 & 4 & 13 \end{bmatrix} \begin{bmatrix} 1 & -3 & 2 & 0 \\ 0 & 1 & -7 & 1 \\ 0 & 0 & 1 & -2 \\ 0 & 0 & 0 & 1 \end{bmatrix}$

**h)** $\begin{bmatrix} -1 & 0 & 0 & 0 \\ 1 & 1 & 0 & 0 \\ 1 & -1 & 0 & 0 \\ 0 & -2 & 0 & 0 \end{bmatrix} \begin{bmatrix} 1 & 3 & -1 & 0 & 1 \\ 0 & 1 & 2 & 1 & 0 \\ 0 & 0 & 0 & 0 & 0 \\ 0 & 0 & 0 & 0 & 0 \end{bmatrix}$

**4. b)** $P = \begin{bmatrix} 0 & 0 & 0 & 1 \\ 0 & 0 & 1 & 0 \\ 0 & 1 & 0 & 0 \\ 1 & 0 & 0 & 0 \end{bmatrix}$  $PA = \begin{bmatrix} 1 & 0 & 0 & 0 \\ 0 & -1 & 0 & 0 \\ 0 & 0 & 2 & 0 \\ 0 & 0 & 0 & 3 \end{bmatrix} \begin{bmatrix} 1 & 3 & -2 & 3 \\ 0 & 1 & 0 & -5 \\ 0 & 0 & 1 & -2 \\ 0 & 0 & 0 & 1 \end{bmatrix}$

**5. b)** False. $A = \begin{bmatrix} 2 & 0 & 0 \\ 0 & 0 & 1 \\ 0 & 3 & 0 \end{bmatrix}$

**d)** True. $(LBL^T)^T = L^{TT}B^TL^T = LBL^T$ because $B^T = B$.

**6.** Suppose $LU = \begin{bmatrix} 0 & 1 \\ 1 & 0 \end{bmatrix}$, with $L = \begin{bmatrix} a & 0 & \cdots \\ x & b & \cdots \end{bmatrix}$, and $U = \begin{bmatrix} p & y \\ 0 & q \\ \vdots & \vdots \end{bmatrix}$ where $L$ is $2 \times n$ and $U$ is $n \times 2$. Then comparing entries in $LU = \begin{bmatrix} 0 & 1 \\ 1 & 0 \end{bmatrix}$ gives $ap = 0$, $xp = 1$, and $ay = 1$, a contradiction.

**8.** $A = \begin{bmatrix} R_1 \\ R_2 \end{bmatrix} \to \begin{bmatrix} R_1 + R_2 \\ R_2 \end{bmatrix} \to \begin{bmatrix} R_1 + R_2 \\ -R_1 \end{bmatrix} \to \begin{bmatrix} R_2 \\ -R_1 \end{bmatrix} \to \begin{bmatrix} R_2 \\ R_1 \end{bmatrix}$

**10. b)** If $P = E_1E_2 \cdots E_n$ where each $E_1$ is elementary corresponding to a row interchange, then $E_i^{-1} = E_i$ for each $i$, so $P^{-1} = E_n \cdots E_2E_1$ is a permutation matrix.

**12. b)** Let $A = LU = L_1U_1$ be two such factorizations of the invertible matrix $A$. Then $L_1^{-1}L = U_1U^{-1}$ is lower triangular (left side) and upper triangular (right side) and so is diagonal. But $L_1^{-1}L$ must be the identity because $L_1$ and $L$ are unit lower triangular.

## 1.8 APPLICATION TO MARKOV CHAINS

**2. b)** $\frac{1}{3} \begin{bmatrix} 2 \\ 1 \end{bmatrix}$, $S_3 = \frac{1}{8} \begin{bmatrix} 5 \\ 3 \end{bmatrix}$

**d)** $\frac{1}{3} \begin{bmatrix} 1 \\ 1 \\ 1 \end{bmatrix}$, $S_3 = \begin{bmatrix} 0.350 \\ 0.312 \\ 0.338 \end{bmatrix}$

**f)** $\frac{1}{20} \begin{bmatrix} 5 \\ 7 \\ 8 \end{bmatrix}$, $S_3 = \begin{bmatrix} 0.244 \\ 0.306 \\ 0.450 \end{bmatrix}$

**3. b)** not regular          **d)** regular

**4. b)** $S = \begin{bmatrix} 0.4 \\ 0.3 \\ 0.3 \end{bmatrix}$

**6. b)** $\frac{2}{9}$

**8. b)** 50% middle, 25% upper, 25% lower

**10.** $\frac{1}{2}$, for any $p \neq 0$

**12. b)** False. $P = \begin{bmatrix} \frac{1}{2} & 0 & \frac{1}{3} \\ \frac{1}{4} & 1 & \frac{1}{3} \\ \frac{1}{4} & 0 & \frac{1}{3} \end{bmatrix}$

**d)** False. $P = \begin{bmatrix} \frac{1}{2} & 0 & \frac{1}{3} \\ \frac{1}{4} & 1 & \frac{1}{3} \\ \frac{1}{4} & 0 & \frac{1}{3} \end{bmatrix}$ is invertible but not regular.

**14.** Here $P^n$ has zero entries for every $n$, so the chain is not regular. You can get from any state to any other because State 1→ State 2→ State 3→State 1 are all possible.

## 1.9 FINITE FIELDS AND CRYPTOGRAPHY

**1. b)**

| + | 0 | 1 | 2 | 3 | 4 | 5 | 6 |
|---|---|---|---|---|---|---|---|
| 0 | 0 | 1 | 2 | 3 | 4 | 5 | 6 |
| 1 | 1 | 2 | 3 | 4 | 5 | 6 | 0 |
| 2 | 2 | 3 | 4 | 5 | 6 | 0 | 1 |
| 3 | 3 | 4 | 5 | 6 | 0 | 1 | 2 |
| 4 | 4 | 5 | 6 | 0 | 1 | 2 | 3 |
| 5 | 5 | 6 | 0 | 1 | 2 | 3 | 4 |
| 6 | 6 | 0 | 1 | 2 | 3 | 4 | 5 |

| × | 0 | 1 | 2 | 3 | 4 | 5 | 6 |
|---|---|---|---|---|---|---|---|
| 0 | 0 | 0 | 0 | 0 | 0 | 0 | 0 |
| 1 | 0 | 1 | 2 | 3 | 4 | 5 | 6 |
| 2 | 0 | 2 | 4 | 6 | 1 | 3 | 5 |
| 3 | 0 | 3 | 6 | 2 | 5 | 1 | 4 |
| 4 | 0 | 4 | 1 | 5 | 2 | 6 | 3 |
| 5 | 0 | 5 | 3 | 1 | 6 | 4 | 2 |
| 6 | 0 | 6 | 5 | 4 | 3 | 2 | 1 |

**2. b)** $37 = 2 \cdot 17 + 3$

$\quad\quad 17 = 5 \cdot 3 + 2$

$\quad\quad 3 = 1 \cdot 2 + 1$

$\quad\quad 1 = 3 - 1 \cdot 2 = 3 - 1 \cdot (17 - 5 \cdot 3)$

$\quad\quad\quad = 6 \cdot 3 - 1 \cdot 17 = 6(37 - 2 \cdot 17) - 1 \cdot 17$

$\quad\quad\quad = 6 \cdot 37 - 13 \cdot 17.$

$\quad\quad$ So $\frac{1}{17} = -13 = 24$ in $\mathbb{F}_{32}$.

**d)** As in (b), $\frac{1}{8} = 11$ in $\mathbb{F}_{29}$.

**3. b)** $det A = 24 - 28 = -4 = 7$, and $\frac{1}{7} = 8$ because $7 \cdot 8 = 1$.

$\quad\quad$ So $A^{-1} = \frac{1}{7}\begin{bmatrix} 8 & -4 \\ -7 & 3 \end{bmatrix} = 8\begin{bmatrix} 8 & 7 \\ 4 & 3 \end{bmatrix} = \begin{bmatrix} 9 & 1 \\ 10 & 2 \end{bmatrix}$.

**d)** $det A = 8 - 3 = 5 = 0$, so $A$ has no inverse.

**4.** $[4 \ \ 23] \ [20 \ \ 25] \ [23 \ \ 27] \ [15 \ \ 2] \ [14 \ \ 8] \ [8 \ \ 3]$.

$\quad A^{-1} = \begin{bmatrix} 2 & 16 \\ 9 & 10 \end{bmatrix}$.

# CHAPTER 2

## Determinants and Eigenvalues

## 2.1 COFACTOR EXPANSION

**1. b)** $-5$        **d)** $13$

**2.** $(-1)^{n+n} = 1$

**3. b)** $0$    **d)** $-1$    **f)** $119$

  **h)** $2abc$    **j)** $-56$    **l)** $abcd$

**4.** See §2.1.1.

**6. b)** $87$

**7. b)** $det(AB) = det A \cdot det B$ for all $A$ and $B$—this is the Product theorem in §2.2.

**8. b)** False. $A = \begin{bmatrix} 1 & 2 \\ 0 & 4 \end{bmatrix}$

  **d)** True. By Theorem 3, $det(5A) = 5^2 \, det A$ because $A$ is $2 \times 2$.

  **f)** False. $A = \begin{bmatrix} 0 & 1 \\ 1 & 0 \end{bmatrix}$

**10.** $det(A) = 126$.

  **b)** $42$    **d)** $126$    **f)** $126$

**12. b)** $513$      **d)** $-756$

**13. b)** $x = \pm\sqrt{\frac{3}{2}}$      **d)** $x = \pm y$

**14.** $1$

**16.** If $A$ is $n \times n$, $det(-A) = (-1)^n \, det A = det A$ if $n$ is even; if $n$ is odd, then $det(-A) = -det A$ so this happens only if $det A = 0$.

**18. b)** $det(P + Q)$ may not equal $det P + det Q$.

**20.** Subtract row 1 from rows 2 and 3 to get $det\begin{bmatrix} 1 & x & x^2 \\ 1 & y & y^2 \\ 1 & z & z^2 \end{bmatrix} =$

$det\begin{bmatrix} y-x & y^2-x^2 \\ z-x & z^2-x^2 \end{bmatrix}$. Since $(y - x)$ is a common factor in row 1, and $(z - x)$ is a common factor in row 2, we get

$det\begin{bmatrix} 1 & x & x^2 \\ 1 & y & y^2 \\ 1 & z & z^2 \end{bmatrix} = (y - x)(z - x) \, det\begin{bmatrix} 1 & y+x \\ 1 & z+x \end{bmatrix}$

$\quad\quad\quad\quad\quad = (y - x)(z - x)(z - y)$.

## 2.2 DETERMINANTS AND INVERSES

**2.** If $A$ is invertible, then $A^{-1}A = I$ so the product theorem gives $det(A^{-1}) \, det(A) = det(A^{-1}A) = det(I) = 1$. Hence $det(A) \neq 0$, so we can divide by it to obtain $det(A^{-1}) = 1/det(A)$.

**4. b)** If $det(A^3) = det(I)$, then $(det A)^3 = 1$, so $det A = 1$.

  **d)** If $det A = det(-A^T)$, then $det A = (-1)^n \, det A$, so $[1 - (-1)^n] \, det A = 0$. If $n$ is even, then $det A$ can have any value, while $det A = 0$ if $n$ is odd.

  **f)** If $det(A^3) = det A$, then $det A[(det A)^2 - 1] = 0$, so $det A = 0$ or $det A \pm 1$.

**6.** We use the fact that $det(A^T) = det A$ and $det(A^{-1}) = \frac{1}{det A}$.

  **b)** $det[B^T A^{-1} B^{-1} C A^2 (C^{-1})^T]$

$\quad\quad = det B \frac{1}{det A} \frac{1}{det B} \, det C \, (det A)^2 \frac{1}{det C} = det A = -2.$

**8. b)** $adj(A) = \begin{bmatrix} -8 & 2 & 6 \\ -2 & 2 & 4 \\ 10 & 10 & 0 \end{bmatrix}$, and $det A = 20$. Hence

$\quad\quad A \, adj(A) = 20 \, I$, so $A^{-1} = \frac{1}{20}\begin{bmatrix} -8 & 2 & 6 \\ -2 & 2 & 4 \\ 10 & 10 & 0 \end{bmatrix}$.

**d)** $adj(A) = \begin{bmatrix} 711 & 146 & 327 & -333 \\ 262 & -300 & 528 & -208 \\ 715 & 210 & -145 & -79 \\ -704 & -34 & -30 & 216 \end{bmatrix}$, and $detA = 2246$.

Hence $A\,adj(A) = 2246\,I$, so

$$A^{-1} = \frac{1}{2246} \begin{bmatrix} 711 & 146 & 327 & -333 \\ 262 & -300 & 528 & -208 \\ 715 & 210 & -145 & -79 \\ -704 & -34 & -30 & 216 \end{bmatrix}.$$

**10. b)** False. $adj(A)$ *always* exists, but $A$ may not be invertible.

**d)** False. $det(3A) = 3^n\,detA$ if $A$ is $n \times n$, and this equals $3\,detA$ only if $n = 1$ or $detA = 0$.

**f)** True. Using the adjoint formula, $(detA)I = A(adjA) = AA^{-1} = I$, and so $detA = 1$.

**h)** False. Let $A = \begin{bmatrix} 1 & 1 & 1 \\ 1 & 1 & 1 \\ 1 & 1 & 1 \end{bmatrix}$. Then for all $i$ and $j$,

$$C_{ij}(A) = (-1)^{i+j} det \begin{bmatrix} 1 & 1 \\ 1 & 1 \end{bmatrix} = (-1)^{i+j} \cdot 0 = 0. \text{ Thus}$$

$adjA = 0$, but clearly $A \neq 0$.

**11. b)** $\frac{1}{det(A)} C_{32} = \frac{1}{216}(-1)^5 det \begin{bmatrix} -2 & 6 \\ 7 & 3 \end{bmatrix} = \frac{1}{216}(-1)(-6-42) = \frac{2}{9}$.

**12. b)** $det(B^{-1}A^{-1}BA) = \frac{1}{detB}\frac{1}{detA}\,detB\,detA = 1$.

**14. b)** Expanding along the first row, $det(A) = 1 - c^2$. Thus $A$ is not invertible when $c = \pm 1$. If $c \neq \pm 1$, the adjoint formula gives $A^{-1} = \frac{1}{1-c^2} \begin{bmatrix} 1-c & c-c^2 & 0 \\ -1-c & 1-c^2 & 1+c \\ -2c & c-c^2 & 1+c \end{bmatrix}$.

**15. b)** $y = \frac{-4}{263}$.

**16. b)** Apply Theorem 3 twice: $det \begin{bmatrix} A & 0 & 0 \\ X & B & 0 \\ Z & Y & C \end{bmatrix} = det \begin{bmatrix} A & 0 \\ X & B \end{bmatrix} \cdot detC = (detA\ detB)\ detC = -6$.

**17. b)** $det \begin{bmatrix} 0 & -1 & 2 & 0 \\ 0 & 3 & 1 & 0 \\ -1 & 2 & -7 & 5 \\ 0 & -3 & 6 & 0 \end{bmatrix} = -det \begin{bmatrix} 2 & -1 & 0 & 0 \\ 1 & 3 & 0 & 0 \\ -7 & 2 & -1 & 5 \\ 6 & -3 & 0 & 0 \end{bmatrix} = $

$-det \begin{bmatrix} 2 & -1 \\ 1 & 3 \end{bmatrix} det \begin{bmatrix} -1 & 5 \\ 0 & 0 \end{bmatrix} = -(7)(0) = 0$.

**18.** $det(kA) = det[(kI_n)A] = det(kI_n)\ detA = k^n\ detA$.

**20.** Apply the product theorem to the adjoint formula: $A(adjA) = (detA)I$, and use Theorem 3 §2.1.

**22.** We have $adjA = (detA)\,A^{-1} = 2A^{-1}$, so $det[A^{-1} + adjA] = det[3A^{-1}] = \frac{3^n}{detA} = \frac{3^n}{2}$.

## 2.3 DIAGONALIZATION AND EIGENVALUES

**2. b)** $c_A(x) = x^2 - x - 6$; Eigenvalues: $-2$, $3$; Corresponding Eigenvectors: $[1\ \ 4]^T$, $[-1\ \ 1]^T$; $P = \begin{bmatrix} 1 & -1 \\ 4 & 1 \end{bmatrix}$ and $D = \begin{bmatrix} -2 & 0 \\ 0 & 3 \end{bmatrix}$.

**d)** $c_A(x) = x^3 - 6x^2 + 11x - 6$; Eigenvalues: 1, 2, 3; Corresponding Eigenvectors: $[1\ \ 1\ \ 1]^T$, $[2\ \ 1\ \ 1]^T$, $[1\ \ 1\ \ 2]^T$; $P = \begin{bmatrix} 1 & 2 & 1 \\ 1 & 1 & 1 \\ 1 & 1 & 2 \end{bmatrix}$ and $D = \begin{bmatrix} 1 & 0 & 0 \\ 0 & 2 & 0 \\ 0 & 0 & 3 \end{bmatrix}$.

**f)** $c_A(x) = x^3 - 7x^2 + 15x - 9$; Eigenvalues: 1, 3, 3; Corresponding Eigenvectors: $[2\ \ 1\ \ 1]^T$, $[4\ \ 0\ \ 1]^T$, $[0\ \ 4\ \ 1]^T$; $P = \begin{bmatrix} -2 & 4 & 0 \\ 1 & 0 & 4 \\ 1 & 1 & 1 \end{bmatrix}$ and $D = \begin{bmatrix} 1 & 0 & 0 \\ 0 & 3 & 0 \\ 0 & 0 & 3 \end{bmatrix}$.

**4.** The diagonal entries of $D$ are the eigenvalues of $A$, and column $i$ of $P$ is an eigenvector corresponding to the diagonal entry $d_{ii}$ of $D$.

**6. b)** Not diagonalizable. There are two eigenvalues: 3 (multiplicity 1), and $-1$ (multiplicity 2). However the only basic solution to $(-1I - A)X = 0$ is $[2\ \ -1\ \ 4]^T$ so, by Theorem 5, $A$ is not diagonalizable.

**7. b)** False. Consider Exercise 2(f).

**d)** True. If $A = PDP^{-1}$, then $A^T = (PDP^{-1})^T = (P^{-1})^TD^TP^T = (P^{-1})^TDP^T$, since $D^T = D$. Now let $Q = (P^{-1})^T = (P^T)^{-1}$, then $A^T = QDQ^{-1}$ and so $A^T$ is diagonalizable.

**f)** False. Consider $A = \begin{bmatrix} 1 & 1 \\ 0 & 2 \end{bmatrix}$.

**h)** True. If $\lambda = 0$ is an eigenvalue, then $AX = 0X = 0$ for some $X \neq 0$. Hence $A$ is not invertible.

**8.** If $AX = \lambda X$, then $A_1X = (A - \alpha I)X = AX - \alpha X = \lambda X - \alpha X = (\lambda - \alpha)X$, so the same eigenvector works.

**10.** $A(X + Y) = AX + AY = \lambda X + \lambda Y = \lambda(X + Y)$; Similarly $A(aX) = \lambda(aX)$.

**12. b)** Let $AX = \lambda X$ where $X \neq 0$. Suppose that $A^{-1}Y = \mu Y$ for some $Y \neq 0$. Then $Y = \mu(AY)$ so $AY = \frac{1}{\mu}Y$ ($\mu \neq 0$ because $A^{-1}$ is invertible). Hence $\frac{1}{\mu} = \lambda$ is an eigenvalue of $A$, so $\mu = \frac{1}{\lambda}$, as required.

**14. b)** Suppose that $P^{-1}AP = D$ is diagonal for some invertible $P$. Transposing gives $P^TA^T(P^T)^{-1} = D^T = D$. Since $P^T$ is invertible, this shows that $A^T$ is diagonalizable.

**16. b)** Let $P^{-1}AP = D = diag(\lambda_1, \lambda_2, \cdots, \lambda_n)$, where the $\lambda_i$ are the eigenvalues of $A$. Since $\lambda_i = \pm 1$ for each $i$, it follows that $D^2 = I$ (see Exercise 13). Since $A = PDP^{-1}$, this gives $A^2 = (PDP^{-1})^2 = PD^2P^{-1} = PIP^{-1} = I$. Hence $A^{-1} = A$.

**18. b)** Given the $n \times n$ matrix $A$, we can write $A = IAI = I^{-1}AI$ where $I$ is the $n \times n$ identity matrix. Thus $A \sim A$.

**d)** Suppose that $A \sim B$ and $B \sim C$. Then, for invertible matrices $P$ and $Q$, $A = P^{-1}BP$ and $B = Q^{-1}CQ$. Eliminating $B$ gives $A = P^{-1}(Q^{-1}CQ)P = (QP)^{-1}C(QP)$. Hence $A \sim C$ because $QP$ is invertible.

**f)** If $A \sim B$, then $A = P^{-1}BP$ for some invertible matrix $P$. Then $A^{-1} = (P^{-1}BP)^{-1} = P^{-1}B^{-1}P$, and so $A^{-1} \sim B^{-1}$.

**h)** If $A \sim B$, then $A = P^{-1}BP$ for some invertible matrix $P$. Then $A^T = (P^{-1}BP)^T = P^TB^T(P^{-1})^T = P^TB^T(P^T)^{-1} = Q^{-1}B^TQ$ for $Q = (P^T)^{-1}$. Hence $A^T \sim B^T$.

**19. b)** $A = -\frac{1}{3}\begin{bmatrix} 2 & 1 \\ 2 & 1 \end{bmatrix}$.

**20. b)** Use the quadratic formula to solve the characteristic polynomial in **a)** for $x$:

$$
\begin{aligned}
x &= \tfrac{1}{2}\left[(a+d) \pm \sqrt{(a+d)^2 - 4(ad-bc)}\right] \\
&= \tfrac{1}{2}\left[(a+d) \pm \sqrt{a^2 + 2ad + d^2 - 4ad + 4bc}\right] \\
&= \tfrac{1}{2}\left[(a+d) \pm \sqrt{a^2 - 2ad + d^2 + 4bc}\right] \\
&= \tfrac{1}{2}\left[(a+d) \pm \sqrt{(a-d)^2 + 4bc}\right].
\end{aligned}
$$

## 2.4 LINEAR DYNAMICAL SYSTEMS

**2. b)** The eigenvalues are $\lambda_1 = \frac{1}{60}[9 + \sqrt{2481}] = 0.980$ and $\lambda_2 = \frac{1}{60}[9 - \sqrt{2481}] = -0.680$; an eigenvector corresponding to $\lambda_i$ is $\begin{bmatrix} \lambda_i \\ 2 \end{bmatrix}$. Hence $P = \begin{bmatrix} 0.980 & -0.680 \\ 2 & 2 \end{bmatrix}$.

**d)** $V_5 = \begin{bmatrix} 58.288 \\ 141.21 \end{bmatrix}$.

**f)** $V_k \approx (70.439)(0.980)^k\begin{bmatrix} 0.981 \\ 2 \end{bmatrix}$.

**h)** They differ by $(45.439)(-0.680)^{10}\begin{bmatrix} -0.680 \\ 2 \end{bmatrix} = \begin{bmatrix} -0.65317 \\ 1.9211 \end{bmatrix}$.

**j)** No, in this model it becomes extinct for every initial population.

**4.** As above, if $\lambda_1$ is the dominant eigenvalue, the conditions for stability, extinction, and becoming large are that $\lambda_1 = 1$, $\lambda_1 < 1$, or $\lambda_1 > 1$ respectively.

**b)** We have $\lambda_1 = \frac{1}{2}(\alpha + r)$ where $r = \sqrt{\alpha^2 + 4\beta m}$, so $\lambda_1 < 1$ means $r < 2 - \alpha$. Squaring gives $\alpha^2 + 4\beta m < 4 - 4\alpha + \alpha^2$, which has solution $\beta < \frac{1-\alpha}{m}$.

**6.** If $V_0 = \begin{bmatrix} a \\ b \end{bmatrix}$, then $V_1 = \begin{bmatrix} b \\ -a \end{bmatrix}$, $V_2 = \begin{bmatrix} -a \\ -b \end{bmatrix}$, $V_3 = \begin{bmatrix} -b \\ a \end{bmatrix}$, $V_4 = \begin{bmatrix} a \\ b \end{bmatrix}$, so $V_4 = V_0$ and the process starts over. Hence the process repeats endlessly.

## 2.5 COMPLEX EIGENVALUES

**1. b)** $(\sqrt{13} - 18) + 13i$.    **d)** $-46 - 9i$.

**f)** $\frac{13}{25} - \frac{16}{25}i$.    **h)** $\frac{37}{5} + \frac{16}{5}i$.

**j)** $\left|\frac{3+4i}{|3+4i|}\right| = \frac{1}{|3+4i|}|3 + 4i| = 1$ because $|3 + 4i|$ is real and positive.

**2. b)** Expanding the left-hand side gives $-x^2 + 4xi = 0$. Equating the real and imaginary parts gives $-x^2 = 0$ and $4x = 0$, the solution being $x = 0$. [Note that $x = 4i$ is a nonreal solution.]

**3. b)** $z = \frac{11}{5} + \frac{3}{5}i$.

**d)** If $z = a + bi$, $a$ and $b$ real, then $\bar{z} = a - bi$. Comparing real and imaginary parts in the equation gives two equations in $a$ and $b$. The solution is $a = b = 1$, so $z = 1 + i$.

**4.** $z = \pm i$.

**6. b)** $zw = (aa' - bb') + (ab' + a'b)i$, so $\overline{(zw)} = (aa' - bb') - (ab' + a'b)i = (a - bi)(a' - b'i) = \bar{z}\,\bar{w}$.

**7. b)** False. For example, $1^2 + i^2 = 0$.

**d)** False. Consider $z = \cos\theta + i\sin\theta$ for any angle $\theta$. Then $|z| = \sqrt{\sin^2\theta + \cos^2\theta} = \sqrt{1} = 1$.

**f)** True. If $z = a + bi$, equating real parts in $\bar{z} = -z$ gives $a = -a$, so $a = 0$. Hence $z = bi$.

**8. b)** $x^2 + 1 = x^2 - i^2 = (x - i)(x + i)$, so the roots are $x = \pm i$.

**d)** By the quadratic formula: $x = \frac{1}{6}\left[-5 \pm \sqrt{-95}\right] = \frac{1}{6}\left[-5 \pm i\sqrt{95}\right]$.

**f)** By the quadratic formula: $x = \frac{1}{-6}\left[-5 \pm \sqrt{145}\right] = \frac{1}{6}\left[5 \pm \sqrt{145}\right]$.

**9. b)** Here $\lambda + \bar{\lambda} = 0$ and $\lambda\bar{\lambda} = 9$, so the quadratic is $x^2 + 9$.

**d)** Here $\lambda + \bar{\lambda} = 4$ and $\lambda\bar{\lambda} = 29$, so the quadratic is $x^2 - 4x + 29$.

**10. b)** Carry the augmented matrix to reduced row-echelon form:

$$
\begin{bmatrix} 1 & 1+i & -i & 2 \\ 1+i & 0 & 1-i & 3+i \\ -1 & 1-i & i & i-1 \end{bmatrix} \rightarrow \begin{bmatrix} 1 & 0 & -i & 2-i \\ 0 & 1 & 0 & \frac{1}{2}+\frac{1}{2}i \\ 0 & 0 & 0 & 0 \end{bmatrix},
$$

which gives $x = (2 - i) + is$, $y = \frac{1}{2} + \frac{1}{2}i$ and $z = s$, where $s$ is a complex parameter.

**11. b)** $A^{-1} = \frac{1}{2i}\begin{bmatrix} 3+i & -1+i \\ -2-i & 1 \end{bmatrix}$.

**d)** $A^{-1} = \frac{1}{-13+7i}\begin{bmatrix} 1+22i & -7+i & 3-7i \\ -2-27i & 7+3i & -4+6i \\ 2+9i & -4 & 1-i \end{bmatrix}$.

**12. b)** $P = \begin{bmatrix} i & -i \\ 1 & 1 \end{bmatrix}$, $D = \begin{bmatrix} 2 & 0 \\ 0 & 0 \end{bmatrix}$.

**d)** $P = \begin{bmatrix} 2 & 2 & 1 \\ 1 & 1 & 1 \\ 2-i & 2+i & 0 \end{bmatrix}$, $D = \begin{bmatrix} -i & 0 & 0 \\ 0 & i & 0 \\ 0 & 0 & 1 \end{bmatrix}$.

**f)** $P = \begin{bmatrix} -4.07 & 0.83 & -0.43 \\ 0.98 & 1.85 & -0.04 \\ 1 & 1 & 1 \end{bmatrix}$, $D = \begin{bmatrix} -102.26 & 0 & 0 \\ 0 & 37.02 & 0 \\ 0 & 0 & 30.76 \end{bmatrix}$.

**h)** $P = \begin{bmatrix} -0.19+0.20i & 1.39-0.53i & -0.67-2.32i \\ 0.90-0.05i & -2.03+0.01i & 0.99+0.54i \\ 1 & 1 & 1 \end{bmatrix}$,

$D = \begin{bmatrix} 7.01+8.72i & 0 & 0 \\ 0 & -3.74+0.98i & 0 \\ 0 & 0 & 1.73-0.29i \end{bmatrix}$.

**14.** Suppose $Z = [z_{ij}]$ is $m \times n$ and $W = [w_{ij}]$ is $n \times p$. The $(i, j)$-entry of $ZW$ is $z_{i1}w_{1j} + z_{i2}w_{2j} + \cdots + z_{in}w_{nj}$ so, using C1 and C2, the $(i, j)$-entry of $\overline{ZW}$ is $\overline{z_{i1}w_{1j} + z_{i2}w_{2j} + \cdots + z_{in}w_{nj}} = \overline{z}_{i1}\overline{w}_{1j} + \overline{z}_{i2}\overline{w}_{2j} + \cdots + \overline{z}_{in}\overline{w}_{nj}$. This is the $(i, j)$-entry of $\overline{Z}\,\overline{W}$ because $\overline{Z} = [\overline{z}_{ij}]$ and $\overline{W} = [\overline{w}_{ij}]$. As this is true for all $i$ and $j$, it follows that $\overline{ZW} = \overline{Z}\,\overline{W}$.

**16.** Write $f(x) = a_0 + a_1x + a_2x^2 + \cdots + a_nx^n$ where the coefficients $a_i$ are real. We have $f(\lambda) = 0$ so $0 = \overline{f(\lambda)} = \overline{a_0 + a_1\lambda + a_2\lambda^2 + \cdots + a_n\lambda^n} = \overline{a}_0 + \overline{a}_1\overline{\lambda} + \overline{a}_2\overline{\lambda}^2 + \cdots + \overline{a}_n\overline{\lambda}^n$. But $\overline{a}_i = a_i$ for each $i$ because $a_i$ is real, so this gives $0 = a_0 + a_1\overline{\lambda} + a_2\overline{\lambda}^2 + \cdots + a_n\overline{\lambda}^n = f(\overline{\lambda})$.

**18.** The condition that $|z| = 1$ means that the distance of $z$ from the origin 0 equals 1, that is $z$ lies on the unit circle.

**20. b)** $z = 2e^{3\pi i/2}$     **d)** $z = 2\sqrt{3}e^{7\pi i/6}$     **f)** $z = 5\sqrt{2}e^{\pi i/4}$

**21. b)** $z = \frac{1}{2} - \frac{\sqrt{3}}{2}i$     **d)** $z = 1 - i$     **f)** $z = \sqrt{3} - 3i$

**22. b)** $(\sqrt{3} + i)^{-4} = (2e^{i\pi/6})^{-4} = \frac{1}{16}e^{-2i\pi/3}$
$$= \frac{1}{16}\left(-\frac{1}{2} - \frac{\sqrt{3}}{2}i\right) = -\frac{1}{32} - \frac{\sqrt{3}}{32}i.$$

**d)** $(1 - i)^{10} = (\sqrt{2}e^{7\pi i/4})^{10} = 32e^{35\pi/2} = -32i$.

**f)** $(\sqrt{3} - 1)^9(2 - 2i)^5 = (2e^{-i\pi/6})^9(2\sqrt{2}e^{-\pi i/4})^5$
$$= (2^9 e^{-3\pi i/2})(2^7\sqrt{2}e^{-5\pi i/4})$$
$$= 2^{16}\sqrt{2}e^{-3\pi i/4}$$
$$= 2^{16}\sqrt{2}\left(-\frac{1}{\sqrt{2}} - \frac{1}{\sqrt{2}}i\right)$$
$$= -2^{16}(1 + i).$$

**23. b)** $z = e^{2\pi ki/3}$ for $k = 0, 1, 2$.

**24. b)** $z = \sqrt{2}e^{i\left(\frac{\pi}{6} + \frac{k\pi}{2}\right)}$ for any integer $k$.

**d)** $z = 2e^{i\left(\frac{\pi}{6} + \frac{k\pi}{3}\right)}$ for any integer $k$.

**26.** Use Euler's formula and the fact that $\cos(-\theta) = \cos\theta$ and $\sin(-\theta) = -\sin\theta$ for any angle $\theta$.

## 2.6   LINEAR RECURRENCES

**2. b)** Here $V_{k+1} = AV_k$ where $A = \begin{bmatrix} 0 & 1 \\ 2 & -1 \end{bmatrix}$. Diagonalizing $A$ gives $P = \begin{bmatrix} 1 & 1 \\ 1 & -2 \end{bmatrix}$ and $D = \begin{bmatrix} 1 & 0 \\ 0 & -2 \end{bmatrix}$, so $\begin{bmatrix} b_1 \\ b_2 \end{bmatrix} = P^{-1}V_0 = \begin{bmatrix} \frac{4}{3} \\ -\frac{1}{3} \end{bmatrix}$. Hence $\begin{bmatrix} x_k \\ x_{k+1} \end{bmatrix} = \frac{4}{3}1^k\begin{bmatrix} 1 \\ 1 \end{bmatrix} - \frac{1}{3}(-2)^k\begin{bmatrix} 1 \\ -2 \end{bmatrix}$ for each $k$. Comparing top entries gives $x_k = \frac{4}{3} - \frac{1}{3}(-2)^k = -\frac{(-2)^k}{3}\left[1 - 4\left(\frac{1}{-2}\right)^k\right] \approx -\frac{(-2)^k}{3}$ for large $k$.

**d)** Here $V_{k+1} = AV_k$ where $A = \begin{bmatrix} 0 & 1 \\ 6 & -1 \end{bmatrix}$. Diagonalizing $A$ gives $P = \begin{bmatrix} 1 & 1 \\ 2 & -3 \end{bmatrix}$ and $D = \begin{bmatrix} 2 & 0 \\ 0 & -3 \end{bmatrix}$, so $\begin{bmatrix} b_1 \\ b_2 \end{bmatrix} = P^{-1}V_0 = \begin{bmatrix} \frac{4}{5} \\ \frac{1}{5} \end{bmatrix}$. Hence $\begin{bmatrix} x_k \\ x_{k+1} \end{bmatrix} = \frac{4}{5}2^k\begin{bmatrix} 1 \\ 2 \end{bmatrix} + \frac{1}{5}(-3)^k\begin{bmatrix} 1 \\ -3 \end{bmatrix}$, and so $x_k = \frac{4}{5}2^k + \frac{1}{5}(-3)^k = \frac{(-3)^k}{5}\left[1 + 4\left(\frac{2}{-3}\right)^k\right] \approx \frac{(-3)^k}{5}$ for large $k$.

**3. b)** Let $V_k = \begin{bmatrix} x_k \\ x_{k+1} \\ x_{k+2} \end{bmatrix}$. Then $A = \begin{bmatrix} 0 & 1 & 0 \\ 0 & 0 & 1 \\ 2 & 1 & -2 \end{bmatrix}$, and diagonalization gives $P = \begin{bmatrix} 1 & 1 & 1 \\ -1 & -2 & 1 \\ 1 & 4 & 1 \end{bmatrix}$ and $D = \begin{bmatrix} -1 & 0 & 0 \\ 0 & -2 & 0 \\ 0 & 0 & 1 \end{bmatrix}$. Then $\begin{bmatrix} b_1 \\ b_2 \\ b_3 \end{bmatrix} = P^{-1}V_0 = \begin{bmatrix} \frac{1}{2} \\ 0 \\ \frac{1}{2} \end{bmatrix}$, giving the general formula $V_k = \frac{1}{2}(-1)^k\begin{bmatrix} 1 \\ -1 \\ 1 \end{bmatrix} + (0)(-2)^k\begin{bmatrix} 1 \\ -2 \\ 4 \end{bmatrix} + \frac{1}{2}1^k\begin{bmatrix} 1 \\ 1 \\ 1 \end{bmatrix}$. Thus first entries give $x_k = \frac{1}{2}(-1)^k + \frac{1}{2}1^k = \frac{1}{2}[(-1)^k + 1]$. Note that the sequence $x_k$ here is 0, 1, 0, 1, 0, 1, $\cdots$ which does not converge to any fixed value for large $k$.

**4. b)** Note that, while complex numbers are used to analyze the problem, the results are ordinary integers! Let $V_k = \begin{bmatrix} x_k \\ x_{k+1} \\ x_{k+2} \end{bmatrix}$. Then $A = \begin{bmatrix} 0 & 1 & 0 \\ 0 & 0 & 1 \\ -2 & -1 & 2 \end{bmatrix}$, and diagonalization gives

$P = \begin{bmatrix} 2.06 & 0.09+0.33i & 0.09-0.33i \\ -1.43 & 0.46+0.35i & 0.46-0.35i \\ 1 & 1 & 1 \end{bmatrix}$ and

$D = \begin{bmatrix} -0.6956 & 0 & 0 \\ 0 & 1.3405-1.0289i & 0 \\ 0 & 0 & 1.3405+1.0289i \end{bmatrix}$.

Then, with 2 decimal approximation, we get

$\begin{bmatrix} b_1 \\ b_2 \\ b_3 \end{bmatrix} = P^{-1}V_0 = \begin{bmatrix} 0.26652 & -0.25129 & 0.09 \\ -0.131-0.7196i & 0.12565-0.75008i & 0.4542+0.4098i \\ -0.13326+0.7196i & 0.12565+0.75008i & 0.4542-0.4098i \end{bmatrix} \times$

$\begin{bmatrix} 1 \\ 1 \\ 1 \end{bmatrix} = \begin{bmatrix} 0.1068 \\ 0.4466-1.0599i \\ 0.4466+1.0599i \end{bmatrix}$ giving the general formula

$$V_k \approx (0.1068)(-0.6956)^k \begin{bmatrix} 2.06 \\ -1.43 \\ 1 \end{bmatrix}$$

$$+ (0.4466 - 1.0599i)(1.3405 - 1.0289i)^k \begin{bmatrix} 0.09 + 0.33i \\ 0.46 + 0.35i \\ 1 \end{bmatrix}$$

$$+ (0.4466 + 1.0599i)(1.3405 + 1.0289i)^k \begin{bmatrix} 0.09 - 0.33i \\ 0.46 - 0.35i \\ 1 \end{bmatrix}.$$

Thus,

$$x_k = (0.1068)(-0.6956)^k(2.06) +$$
$$(0.4466 - 1.0599i)(1.3405 - 1.0289i)^k(0.09 + 0.33i) +$$
$$(0.4466 + 1.0599i)(1.3405 + 1.0289i)^k(0.09 - 0.33i).$$

**6.** Let $x_k$ denote the number of ways to form words of $k$ letters. A word of $k + 2$ letters must end in either $a$ or $b$. The number of words that end in $b$ is $x_{k+1}$—just add a $b$ to a $(k + 1)$-letter word. But the number ending in $a$ is $x_k$ since the second-last letter must be a $b$ (no adjacent $a$'s) so we simply add $ba$ to any $k$-letter word. This gives the recurrence $x_{k+2} = x_{k+1} + x_k$ which is the same as in Example 3, but with different initial conditions: $x_0 = 1$ (since the "empty" word is the only one formed with no letters) and $x_1 = 2$. The eigenvalues, eigenvectors, and diagonalization remain the same, and so

$$V_k = b_1 \lambda_1^k \begin{bmatrix} 1 \\ \lambda_1 \end{bmatrix} + b_2 \lambda_2^k \begin{bmatrix} 1 \\ \lambda_2 \end{bmatrix}$$

where $\lambda_1 = \frac{1}{2}(1 + \sqrt{5})$ and $\lambda_2 = \frac{1}{2}(1 - \sqrt{5})$. Comparing top entries gives $x_k = b_1 \lambda_1^k + b_2 \lambda_2^k$. By Theorem 1 §2.4, the constants $b_1$ and $b_2$ come from $[b_1 \ b_2]^T = P^{-1}V_0$. However, we vary the method and use the initial conditions to determine the values of $b_1$ and $b_2$ directly. More precisely, $x_0 = 1$ means $1 = b_1 + b_2$ while $x_1 = 2$ means $2 = b_1\lambda_1 + b_2\lambda_2$. These equations have unique solution $b_1 = \frac{\sqrt{5}+3}{2\sqrt{5}}$ and $b_2 = \frac{\sqrt{5}-3}{2\sqrt{5}}$. It follows that

$$x_k = \frac{1}{2\sqrt{5}}\left[(3 + \sqrt{5})\left(\frac{1+\sqrt{5}}{2}\right)^k + (-3 + \sqrt{5})\left(\frac{1-\sqrt{5}}{2}\right)^k\right] \text{ for}$$

each $k \geq 0$.

**8.** In a stack of $k + 2$ chips, if the last chip is gold, then (to avoid having two gold chips together) the second last chip must be either red or blue. This can happen in $2x_k$ ways. But there are $x_{k+1}$ ways that the last chip is red (or blue) so there are $2x_{k+1}$ ways these possibilities can occur. Hence $x_{k+2} = 2x_k + 2x_{k+1}$. The matrix is $A = \begin{bmatrix} 0 & 1 \\ 2 & 2 \end{bmatrix}$ with eigenvalues $\lambda_1 = 1 + \sqrt{3}$ and $\lambda_2 = 1 - \sqrt{3}$ and corresponding eigenvectors $X_1 = \begin{bmatrix} 1 \\ \lambda_1 \end{bmatrix}$ and $X_2 = \begin{bmatrix} 1 \\ \lambda_2 \end{bmatrix}$. Given the initial conditions $x_0 = 1$ and $x_1 = 3$, we get $\begin{bmatrix} b_1 \\ b_2 \end{bmatrix} = P^{-1}V_0 = \frac{1}{2\sqrt{3}}\begin{bmatrix} 2+\sqrt{3} \\ -2+\sqrt{3} \end{bmatrix}$. Since Theorem 1 §2.4 gives $V_k = b_1\lambda_1^k \begin{bmatrix} 1 \\ \lambda_1 \end{bmatrix} + b_2\lambda_2^k \begin{bmatrix} 1 \\ \lambda_2 \end{bmatrix}$, comparing top entries gives

$$x_k = b_1\lambda_1^k + b_2\lambda_2^k = \frac{1}{2\sqrt{3}}[(2 + \sqrt{3})(1 + \sqrt{3})^k + (-2 + \sqrt{3})(1 - \sqrt{3})^k].$$

**10.** Let $y_k$ be the yield for year $k$. Then the yield for year $k + 2$ is $y_{k+2} = \frac{y_k + y_{k+1}}{2} = \frac{1}{2}y_k + \frac{1}{2}y_{k+1}$. The eigenvalues are $\lambda_1 = 1$ and $\lambda_2 = -\frac{1}{2}$, with corresponding eigenvectors $X_1 = \begin{bmatrix} 1 \\ 1 \end{bmatrix}$ and $X_2 = \begin{bmatrix} -2 \\ 1 \end{bmatrix}$. Given that $k = 0$ for the year 1990, we have the initial conditions $y_0 = 10$ and $y_1 = 12$. Thus $\begin{bmatrix} b_1 \\ b_2 \end{bmatrix} = P^{-1}V_0 = \frac{1}{3}\begin{bmatrix} 34 \\ 2 \end{bmatrix}$.

Since $V_k = \frac{34}{3}(1)^k \begin{bmatrix} 1 \\ 1 \end{bmatrix} + \frac{2}{3}\left(-\frac{1}{2}\right)^k \begin{bmatrix} -2 \\ 1 \end{bmatrix}$ comparing top entries gives $x_k = \frac{34}{3}(1)^k + \frac{2}{3}(-2)\left(-\frac{1}{2}\right)^k = \frac{34}{3} - \frac{4}{3}\left(-\frac{1}{2}\right)^k$.

For large $k$, $x_k \approx \frac{34}{3}$ so the long-term yield is $11\frac{1}{3}$ million tons of wheat.

**12. b)** Here $A = \begin{bmatrix} 0 & 1 \\ a & b \end{bmatrix}$, so $c_A(x) = x(x - b) - a = x^2 - bx - a$. The quadratic formula gives $\lambda_1 = \frac{1}{2}[b + \sqrt{b^2 + 4a}]$ and $\lambda_2 = \frac{1}{2}[b - \sqrt{b^2 + 4a}]$, and $X_i = \begin{bmatrix} 1 \\ \lambda_i \end{bmatrix}$ is a $\lambda_i$-eigenvector because $\lambda_i^2 = a + b\lambda_i$ $(c_A(\lambda_i) = 0)$. Hence $P = \begin{bmatrix} 1 & 1 \\ \lambda_1 & \lambda_2 \end{bmatrix}$, so $\begin{bmatrix} b_1 \\ b_2 \end{bmatrix} = P^{-1}V_0 = \frac{-1}{\sqrt{b^2+4a}}\begin{bmatrix} -1 \\ 1 \end{bmatrix}$.

Thus Theorem 1 §2.4 gives

$$V_k = \frac{1}{\sqrt{b^2+4a}}\lambda_1^k X_1 - \frac{1}{\sqrt{b^2+4a}}\lambda_2^k X_2$$
$$= \frac{1}{\sqrt{b^2+4a}}\lambda_1^k \begin{bmatrix} 1 \\ \lambda_1 \end{bmatrix} - \frac{1}{\sqrt{b^2+4a}}\lambda_2^k \begin{bmatrix} 1 \\ \lambda_2 \end{bmatrix}.$$

Comparing top entries gives

$$x_k = \frac{1}{\sqrt{b^2+4a}}\lambda_1^k - \frac{1}{\sqrt{b^2+4a}}\lambda_2^k$$
$$= \frac{1}{\sqrt{b^2+4a}}[\lambda_1^k - \lambda_2^k].$$

**14.** For the recurrence $x_{k+1} = rx_k + c$, given $x_0$

$$x_1 = rx_0 + c$$
$$x_2 = rx_1 + c = r(rx_0 + c) + c$$
$$= r^2x_0 + rc + c$$
$$x_3 = rx_2 + c = r(r^2x_0 + rc + c) + c$$
$$= r^3x_0 + r^2c + rc + c$$
$$x_4 = rx_3 + c = r(r^3x_0 + r^2c + rc + c) + c$$
$$= r^4x_0 + r^3c + r^2c + rc + c$$
$$\vdots \quad \vdots \qquad \vdots$$
$$x_k = r^kx_0 + r^{k-1}c + r^{k-2}c + \cdots + rc + c$$
$$= r^kx_0 + c\sum_{i=0}^{k-1}r^i.$$

Clearly, if $r = 1$, then $x_k = x_0 + kc$. If $r \neq 1$, then $\sum_{i=0}^{k-1}r^i = \frac{r^k-1}{r-1}$, so $x_k = r^kx_0 + c\frac{r^k-1}{r-1}$ for every $k$.

Hence $x_k = \begin{cases} x_0 + kc & \text{if } r = 1 \\ r^kx_0 + c\frac{r^k-1}{r-1} & \text{if } r \neq 1 \end{cases}$.

**16. b)** If $r_k$ is any solution of (*), then $r_{k+2} = ar_{k+1} + br_k + c(k)$. Define $q_k = r_k - p_k$ for each $k$. Then it suffices to show that $q_k$ is a solution of (**).
But $q_{k+2} = r_{k+2} - p_{k+2}$
$$= (ar_{k+1} + br_k + c(k)) - (ap_{k+1} + bp_k + c(k))$$
$$= aq_k + bq_k$$
which is what we wanted.

## 2.7 POLYNOMIAL INTERPOLATION

**1. b)** $p(x) = 5 - 4x + 2x^2$, and so $p(1.5) = 3.5$.
**d)** $p(x) = 1 - \frac{5}{3}x + \frac{1}{2}x^2 + \frac{7}{6}x^3$, and so $p\left(-\frac{1}{2}\right) = \frac{29}{16}$.

**3. b)** The angle of $0°$ at the horizontal means that $p'(35) = 0$. This fourth condition requires that we use an interpolating polynomial of degree 3, that is $p(x) = a + bx + cx^2 + dx^3$. So then $p'(35) = 0$ becomes a fourth equation $b + 2c(35) + 3d(35)^2 = 0$. This with $p(0) = 40$, $p(30) = 7.5$, and $p(35) = 5$ gives the following system of equations for $a$, $b$, $c$ and $d$:
$$\begin{bmatrix} 1 & 0 & 0 & 0 \\ 1 & 30 & 30^2 & 30^3 \\ 1 & 35 & 35^2 & 35^3 \\ 0 & 1 & 2 \cdot 35 & 3 \cdot 35^2 \end{bmatrix} \begin{bmatrix} a \\ b \\ c \\ d \end{bmatrix} = \begin{bmatrix} 40 \\ 7.5 \\ 5 \\ 0 \end{bmatrix}$$
The unique solution, to 2 decimal places, is $p(x) = 40 + 0.92x - 0.14x^2 + 0.0024x^3$.

**d)** Insisting that the second derivatives match implies that $p''(25) = 0$ since the second derivative of a straight line is zero. This is the fourth condition on the system, requiring us to use a third degree polynomial to interpolate the data: $p(x) = ax^3 + bx^2 + cx + d$. The equations are
$$\begin{bmatrix} 1 & 25 & 25^2 & 25^3 \\ 1 & 35 & 35^2 & 35^3 \\ 0 & 1 & 2 \cdot 25 & 3 \cdot 25^2 \\ 0 & 0 & 2 & 6 \cdot 25 \end{bmatrix} \begin{bmatrix} a \\ b \\ c \\ d \end{bmatrix} = \begin{bmatrix} 20 \\ 5 \\ -\frac{4}{5} \\ 0 \end{bmatrix}$$
The unique solution is
$p(x) = \frac{1195}{8} - \frac{557}{40}x + \frac{21}{40}x^2 - \frac{7}{1000}x^3$.

**4.** If two of the $x_i$ are equal, then two rows of the Vandermonde matrix are equal, so that the matrix is not invertible (it has determinant zero).

**5. b)** The 8 conditions result in the following 8 equations:
$s_1(1000) = 1.5$
$\Rightarrow 10^9 a_1 + 10^6 b_1 + 10^3 c_1 + d_1 = 1.5$

$s_1(2000) = 2.1$
$\Rightarrow 2000^3 a_1 + 2000^2 b_1 + 2000c_1 + d_1 = 2.1$
$s_2(2000) = 2.1$
$\Rightarrow 2000^3 a_2 + 2000^2 b_2 + 2000c_2 + d_2 = 2.1$
$s_2(3000) = 3.4$
$\Rightarrow 3000^3 a_2 + 3000^2 b_2 + 3000c_2 + d_2 = 3.4$
$s'_1(2000) = s'_2(2000)$
$\Rightarrow 3(2000)^2 a_1 + 4000b_1 + c_1 - 3(2000)^2 a_2 - 4000b_2 - c_2 = 0$
$s''_1(2000) = s''_2(2000)$
$\Rightarrow 12000a_1 + 2b_1 - 12000a_2 - 2b_2 = 0$
$s''_1(1000) = 0$
$\Rightarrow 6000a_1 + 2b_1 = 0$
$s''_2(3000) = 0$
$\Rightarrow 18000a_2 + 2b_2 = 0$
The unique solution to all of these equations gives the two polynomials:
$s_1(x) = (1.75E{-}10)x^3 - (5.25E{-}7)x^2 + (9.5E{-}4)x + 0.9$
$s_2(x) = (-1.75E{-}10)x^3 + (1.575E{-}6)x^2 - (3.25E{-}3)x + 3.7$.

## 2.8 SYSTEMS OF DIFFERENTIAL EQUATIONS

**1. b)** The general solution is $\mathbf{f}(x) = c_1 \begin{bmatrix} 1 \\ 1 \end{bmatrix} e^{4x} + c_2 \begin{bmatrix} 5 \\ -1 \end{bmatrix} e^{-2x}$.
If $f_1(0) = 1$ and $f_2(0) = -1$, then
$\mathbf{f}(x) = -\frac{2}{3} \begin{bmatrix} 1 \\ 1 \end{bmatrix} e^{4x} + \frac{1}{3} \begin{bmatrix} 5 \\ -1 \end{bmatrix} e^{-2x}$.

**d)** The general solution is
$\mathbf{f}(x) = c_1 \begin{bmatrix} 1 \\ 0 \\ 1 \end{bmatrix} e^{4x} + c_2 \begin{bmatrix} -8 \\ 10 \\ 7 \end{bmatrix} e^{-x} + c_3 \begin{bmatrix} 1 \\ -2 \\ 1 \end{bmatrix} e^{2x}$.
The conditions $f_1(0) = f_2(0) = f_3(0) = 1$ imply that
$\mathbf{f}(x) = \frac{3}{2} \begin{bmatrix} 1 \\ 0 \\ 1 \end{bmatrix} e^{4x} - \frac{1}{2} \begin{bmatrix} 1 \\ -2 \\ 1 \end{bmatrix} e^{2x}$.

**2.** If $f$ is a solution consider $f_1 = f$, $f_2 = f' - af$, and $f_3 = f'' - af' - bf$. Show that $\begin{cases} f'_1 = f_2 + af_1 \\ f'_2 = f_3 + bf_1 \\ f'_3 = cf_1 \end{cases}$.
Conversely, given any solution of these equations, show that $f = f_1$ is a solution of the given equation.

CHAPTER **3**

# Vector Geometry

## 3.1 GEOMETRIC VECTORS

**2. b)** $\vec{p} + \vec{q} = \begin{bmatrix} 2 \\ 4 \end{bmatrix}$ and $\|\vec{p} + \vec{q}\| = 2\sqrt{5}$;

$\vec{p} - \vec{q} = \begin{bmatrix} 4 \\ -8 \end{bmatrix}$ and $\|\vec{p} - \vec{q}\| = 4\sqrt{5}$;

$3\vec{p} = \begin{bmatrix} 9 \\ -6 \end{bmatrix}$ and $\|3\vec{p}\| = 3\|\vec{p}\| = 3\sqrt{13}$;

$-2\vec{q} = \begin{bmatrix} 2 \\ -12 \end{bmatrix}$ and $\|-2\vec{q}\| = |-2|\,\|\vec{q}\| = 2\sqrt{37}$.

$3\vec{p} - 2\vec{q} = \begin{bmatrix} 11 \\ -18 \end{bmatrix}$ and

$\|3\vec{p} - 2\vec{q}\| = \sqrt{11^2 + (-18)^2} = \sqrt{445}$.

**3. b)** $\overrightarrow{PQ} = [4 \ -3 \ 5]^T$ and $\|\overrightarrow{PQ}\| = 5\sqrt{2}$.
**d)** $\overrightarrow{PQ} = [2 \ 0 \ -2]^T$ and $\|\overrightarrow{PQ}\| = 2\sqrt{2}$.

**4. b)** Position vector of $P$ is $\vec{p} = \vec{q} + \overrightarrow{QP} = \vec{q} - \overrightarrow{PQ} =$
$[0 \ -1 \ 3]^T - [3 \ 0 \ -2]^T = [-3 \ -1 \ 5]^T$.

**5. b)** $[39 \ -8 \ 5]^T$        **d)** $\frac{1}{2}\sqrt{89}$

**6. b)** $\left\| \frac{1}{\|\vec{v}\|}\vec{v} \right\| = \left| \frac{1}{\|\vec{v}\|} \right| \|\vec{v}\| = 1$; $\frac{1}{\|\vec{v}\|}\vec{v}$ is in the same direction as $\vec{v}$ because $\frac{1}{\|\vec{v}\|} > 0$.

**7. b)** $-10\vec{u} - \vec{v} - 17\vec{w}$.

**8. b)** $\frac{1}{3}[-2 \ 3 \ -5]^T$        **d)** $-\frac{1}{2}[1 \ 6 \ 10]^T$

**9. b)** $[2 \ -1 \ 3]^T = 1[2 \ -1 \ 3]^T + 0[0 \ 2 \ -3]^T + 0[6 \ 1 \ 3]^T$
**d)** No solution.

**10. b)** Parallel: $\vec{v} = -2\vec{u}$        **d)** Not parallel.

**12. b)** The condition $\overrightarrow{PQ} = a\vec{d}$ gives $5 = \|\overrightarrow{PQ}\| = |a|\,\|\vec{d}\| = |a|\sqrt{13}$. Hence $a = \pm\frac{5}{\sqrt{13}}$ so, since the position
vector of $Q$ is $\vec{q} = \vec{p} + \overrightarrow{PQ} = \vec{p} + a\vec{d}$, we get
$Q(1, \ 2 - \frac{15}{\sqrt{13}}, \ -5 + \frac{10}{\sqrt{13}})$ and $Q(1, \ 2 + \frac{15}{\sqrt{13}}, \ -5 - \frac{10}{\sqrt{13}})$.

**13. b)** $S(7, -5, -2)$

**14. b)** $M\left(-1, \frac{7}{2}, 0\right)$

**15. b)** $\vec{b} = \frac{1}{4}\vec{p} + \frac{3}{4}\vec{q}$, so $B = B\left(\frac{1}{4}, -2, \frac{9}{2}\right)$

**16.** The condition means that $\overrightarrow{PA} = r\overrightarrow{PQ}$. In terms of position vectors, this is $\vec{a} - \vec{p} = r(\vec{q} - \vec{p})$. Hence $\vec{a} = (1 - r)\vec{p} + r\vec{q}$.

**17. b)** True. The condition $\|\vec{v} - \vec{w}\| = \vec{0}$ implies $\vec{v} - \vec{w} = \vec{0}$, so $\vec{v} = \vec{w}$.
**d)** False. For example, $\|[2 \ 2 \ 1]^T\| = \|[2 \ 1 \ -2]^T\|$.
**f)** False. If $\vec{w} = t\vec{v}$, the direction is the same only if $t > 0$.
**h)** False. Always, $\|(-5)\vec{v}\| = |-5|\,\|\vec{v}\| = 5\|\vec{v}\|$, and this cannot equal $-5\|\vec{v}\|$ if $\vec{v} \neq \vec{0}$.
**j)** False. For example, $\vec{v} = [1 \ 1]^T$ and $\vec{w} = [1 \ -1]^T$.

**18. b)** Using **a)**: $\overrightarrow{EF} = \overrightarrow{EB} + \overrightarrow{BF} = \frac{1}{2}\overrightarrow{AB} + \frac{1}{2}\overrightarrow{BC} = \frac{1}{2}\overrightarrow{AC}$. Hence the segment $EF$ is parallel to $AC$ and half as long.

**20.** First $\overrightarrow{AE} = \overrightarrow{BD}$ by the symmetry of the hexagon. Hence $\overrightarrow{AB} + \overrightarrow{AE} = \overrightarrow{AB} + \overrightarrow{BD} = \overrightarrow{AD}$. Similarly, $\overrightarrow{AC} + \overrightarrow{AF} = \overrightarrow{AD}$.

**22. b)** If $P = P(x, y, z)$, then the distance from $P$ to the origin $O$ is $\|\vec{p}\|$. If $\|\vec{p}\| = r$, then $P$ is at a distance $r$ from the origin, so it lies on $\mathcal{S}$ (by the definition of $\mathcal{S}$).

## 3.2 DOT PRODUCT AND PROJECTIONS

**1. b)** 4        **d)** 26        **f)** 0        **h)** −43

**2. b)** $\vec{v} \cdot \vec{v} = 9$, $\|\vec{v}\| = 3$        **d)** $\vec{v} \cdot \vec{v} = 3$, $\|\vec{v}\| = \sqrt{3}$

**3. b)** $\vec{v} \cdot \vec{w} = 4$, acute        **d)** $\vec{v} \cdot \vec{w} = 26$, acute (zero)
**f)** $\vec{v} \cdot \vec{w} = 0$, right angle        **h)** $\vec{v} \cdot \vec{w} = -43$, obtuse

**4. b)** $\cos\theta = \frac{\vec{v} \cdot \vec{w}}{\|\vec{v}\|\|\vec{w}\|} = \frac{0}{\sqrt{2}\sqrt{2}} = 0$, $\theta = \frac{\pi}{2}$
**d)** $\cos\theta = \frac{\vec{v} \cdot \vec{w}}{\|\vec{v}\|\|\vec{w}\|} = \frac{22}{\sqrt{29}\sqrt{74}} = 0.4749$, $\theta = 1.076 = 61.65°$
**f)** $\cos\theta = \frac{\vec{v} \cdot \vec{w}}{\|\vec{v}\|\|\vec{w}\|} = \frac{-1}{\sqrt{2}\sqrt{2}} = -\frac{1}{2}$, $\theta = \frac{2\pi}{3}$

**6. b)** We have $\|\overrightarrow{AB}\| = \sqrt{29}$, $\|\overrightarrow{AC}\| = \sqrt{86}$, and $\|\overrightarrow{BC}\| = \sqrt{57}$, so $\|\overrightarrow{AB}\|^2 + \|\overrightarrow{BC}\|^2 = 29 + 57 = 86 = \|\overrightarrow{AC}\|^2$, confirming Pythagoras' theorem.

**8. b)** $\overrightarrow{AB} \cdot \overrightarrow{CD} = -12$, neither
**d)** $\overrightarrow{AB} \cdot \overrightarrow{CD} = 0$, perpendicular **f)** $\overrightarrow{CD} = -2\overrightarrow{AB}$, parallel

**9. b)** $proj_{\vec{w}}(\vec{v}) = \frac{\vec{v} \cdot \vec{w}}{\|\vec{w}\|^2}\vec{w} = -\frac{1}{5}\vec{w}$.    **d)** $proj_{\vec{w}}(\vec{v}) = \frac{\vec{v} \cdot \vec{w}}{\|\vec{w}\|^2}\vec{w} = \frac{38}{57}\vec{w}$.
**f)** $proj_{\vec{w}}(\vec{v}) = \frac{\vec{v} \cdot \vec{w}}{\|\vec{w}\|^2}\vec{w} = \frac{3}{7}\vec{w}$.

**10. b)** $\vec{v}_1 = proj_{\vec{w}}(\vec{v}) = \frac{1}{7}[-2 \ 1 \ 4]^T$,
$\vec{v}_2 = \vec{v} - \vec{v}_1 = \frac{1}{7}[23 \ 6 \ 10]^T$.
**d)** $\vec{v}_1 = proj_{\vec{w}}(\vec{v}) = -\frac{1}{3}[1 \ 4 \ -1]^T$,
$\vec{v}_2 = \vec{v} - \vec{v}_1 = \frac{2}{3}[5 \ -1 \ 1]^T$.

**12.** Using position vectors, $\vec{m} - \vec{a} = \overrightarrow{AM} = [2 \;\; 0 \;\; -1]^T$. Hence $\vec{m} = \vec{a} + [2 \;\; 0 \;\; -1]^T = [4 \;\; -1 \;\; 2]^T$. But then $\vec{b} = \vec{m} + \overrightarrow{MB} = \vec{m} + \frac{1}{2}\overrightarrow{CB} = \vec{m} - \frac{1}{2}\overrightarrow{BC} = [2 \;\; -\frac{3}{2} \;\; 3]^T$. Now we have $\vec{c} = \vec{b} + \overrightarrow{BC} = [6 \;\; -\frac{1}{2} \;\; 1]^T$.

**14.** In the diagram, write $\vec{v} = \overrightarrow{CO}$ and $\vec{w} = \overrightarrow{OB}$ so that $\|\vec{v}\| = \|\vec{w}\|$ is the radius of the circle. Then $\overrightarrow{CB} = \overrightarrow{CO} + \overrightarrow{OB} = \vec{v} + \vec{w}$ and $\overrightarrow{CA} = \overrightarrow{CO} + \overrightarrow{OA} = \vec{v} - \overrightarrow{AO} = \vec{v} - \overrightarrow{OB} = \vec{v} - \vec{w}$. Hence
$$\overrightarrow{CA} \cdot \overrightarrow{CB} = (\vec{v} - \vec{w}) \cdot (\vec{v} + \vec{w}) = \|\vec{v}\|^2 - \|\vec{w}\|^2 = 0,$$
which shows that the angle at $C$ is a right angle. This is what we wanted.

**16. b)** True. If $\vec{v} \cdot \vec{w} = 0$, then $(2\vec{v}) \cdot (-3\vec{w}) = -6(\vec{v} \cdot \vec{w}) = -6(0) = 0$.

**d)** True. If $\vec{v} \cdot \vec{w} > 0$, then $(-\vec{v}) \cdot \vec{w} = -(\vec{v} \cdot \vec{w}) < 0$.

**f)** True. If $0 = proj_{\vec{d}}(\vec{v}) = \frac{\vec{v} \cdot \vec{d}}{\|\vec{d}\|^2}\vec{d}$, then $\vec{v} \cdot \vec{d} = 0$, so $\vec{v}$ is orthogonal to $\vec{d}$.

**18.** $\vec{v} \cdot \vec{w} = 0$, so $\|\vec{v} + \vec{w}\|^2 = (\vec{v} + \vec{w}) \cdot (\vec{v} + \vec{w}) = \|\vec{v}\|^2 + 2(\vec{v} \cdot \vec{w}) + \|\vec{w}\|^2 = \|\vec{v}\|^2 + \|\vec{w}\|^2$.

**20.** We have $\|\vec{v} + \vec{w}\|^2 = \|\vec{v}\|^2 + 2(\vec{v} \cdot \vec{w}) + \|\vec{w}\|^2$ by Exercise 18. Replacing $\vec{w}$ by $-\vec{w}$ in this equation gives $\|\vec{v} - \vec{w}\|^2 = \|\vec{v}\|^2 - 2(\vec{v} \cdot \vec{w}) + \|\vec{w}\|^2$. Adding these equations gives $\|\vec{v} + \vec{w}\|^2 + \|\vec{v} - \vec{w}\|^2 = 2(\|\vec{v}\|^2 + \|\vec{w}\|^2)$.

**22. b)** $proj_{\vec{d}}(a\vec{v}) = \frac{(a\vec{v}) \cdot \vec{d}}{\|\vec{d}\|^2}\vec{d} = \frac{a(\vec{v} \cdot \vec{d})}{\|\vec{d}\|^2}\vec{d} = a\frac{\vec{v} \cdot \vec{d}}{\|\vec{d}\|^2}\vec{d} = a\,proj_{\vec{d}}(\vec{v})$.

**24. b)** By **a)**, $\vec{d}_1 = [1 \;\; m_1]^T$ and $\vec{d}_2 = [1 \;\; m_2]^T$ point along the lines. Then the lines are perpendicular if and only if $\vec{d}_1$ and $\vec{d}_2$ are orthogonal vectors; if and only if $\vec{d}_1 \cdot \vec{d}_2 = 0$; if and only if $1 + m_1 m_2 = 0$.

**26. b)** Let $\theta$ be the angle between $\vec{v}$ and $\vec{w}$. If there is equality in **a)**, $\cos\theta = 1$ or $\cos\theta = -1$. Since $0 \le \theta \le \pi$, this means $\theta = 0$ or $\theta = \pi$. Hence $\vec{v}$ and $\vec{w}$ have the same or opposite direction; in other words, $\vec{v}$ and $\vec{w}$ are parallel.

## 3.3 LINES AND PLANES

**2. b)** $x = -3 + 3t$
$y = -2t$
$z = 7$

**f)** $x = 1 - 7t$
$y = 2 + 3t$
$z = -5t$

**j)** $x = 1$
$y = 2 + 4t$
$z = -9 - 3t$

**d)** $x = 9 - 11t$
$y = -3 + 9t$
$z = 1 - 5t$

**h)** $x = 2 - 5t$
$y = -5 + 6t$
$z = 7 + t$

**l)** $x = -7 - t$
$y = 2 + 2t$
$z = 7 + 3t$

**4. b)** The line through $A$ and $B$ has equations $\begin{cases} x = 1 - t \\ y = 2 + 3t \\ z = -1 + 2t \end{cases}$;
this line contains $C(2, -1, -3)$.

**6.** $C(3, 1, 0)$ and $C\left(\frac{5}{3}, \frac{-1}{3}, \frac{4}{3}\right)$

**7. b)** $P(-2, 5, 4)$

**d)** Use $s$ for the parameter in the second line. These lines do not intersect.

**8.** Proceed as in Example 9.

**b)** $Q\left(-\frac{3}{13}, \frac{15}{13}, \frac{52}{13}\right)$, Distance: $\frac{1}{13}\sqrt{1846}$

**d)** $Q\left(\frac{53}{13}, \frac{21}{13}, -\frac{39}{13}\right)$, Distance: $\frac{6}{13}\sqrt{65}$

**10. b)** $x - 2y + 3z = 17$    **d)** $2x + y - 7z = -27$

**f)** $5x - y + 5z = \frac{29}{2}$

**11. b)** $\vec{v} \times \vec{w} = [-11 \;\; -5 \;\; 8]^T$  **d)** $\vec{v} \times \vec{w} = [-5 \;\; 7 \;\; -2]^T$

**12. b)** $x = 2 - 3t$
$y = -1 + 2t$
$z = 3 + 21t$

**d)** $x = \frac{3}{7} + 2t$
$y = -\frac{1}{7} + t$
$z = t$

**13. b)** $x - 2y - 7z = 1$   **d)** $4x - y = 5$

**f)** $13x + 3y + 4z = 43$   **h)** $3x + y = -2$

**j)** $8x - 3y - 6z = 17$

**14. b)** The plane through $A$, $B$, and $C$ has equation $2x - y + 3z = 1$. $D(-1, 3, 2)$ is *on* this plane.

**d)** The plane through $A$, $B$, and $C$ has equation $2x - 4y - 3z = 8$. $D(5, 1, 1)$ is *not* on this plane.

**15. b)** Distance: $\frac{8}{13}\sqrt{13}$.

**16.** In each case proceed as in Example 16.

**b)** Distance: $\frac{13}{42}\sqrt{42}$.    **d)** Distance: $\frac{3}{5}\sqrt{5}$.

**18.** $z = k$, $k$ any constant.

**19. b)** No. $\vec{d} = [0 \;\; 2 \;\; 5]^T$ is not orthogonal to the normal $\vec{n} = [1 \;\; -1 \;\; 1]^T$ of the plane.

**20. b)** No. The point $P_0(3, 7, 5)$ on the line is not on the plane.

**21. b)** $P\left(5, -\frac{39}{17}, -\frac{1}{17}\right)$

**22.** The line through $A$ and $B$ has equations $x = 1 + 3t$, $y = -3 + 8t$, and $z = 2 - 2t$, and lies in every plane through $A$ and $B$. Since $C = C(1 + 3t, -3 + 8t, 2 - 2t)$ where $t = 2$, it is on the line, and so is on every such plane.

**23. b)** Distance: $\frac{9}{26}\sqrt{26}$; $Q = Q\left(\frac{51}{26}, -\frac{17}{26}, -\frac{10}{26}\right)$

**d)** Distance: $\frac{5}{2}\sqrt{14}$; $Q = Q\left(-\frac{19}{2}, -\frac{5}{2}, 4\right)$

**24. b)** This system has solution $[3 - 4t \quad 2 - t \quad t]^T$ which is the line $[3 \quad 2 \quad 0]^T + t[-4 \quad -1 \quad 1]^T$. Writing $P_0(3, 2, 0)$ and $\vec{d} = [-4 \quad -1 \quad 1]^T$, we have $\vec{v} = \overrightarrow{P_0P} = [-4 \quad 0 \quad 0]^T$ and we compute $\vec{v}_1 = proj_{\vec{d}}(\vec{v}) = \frac{8}{9}[-4 \quad -1 \quad 1]^T$. Hence $\vec{v} - \vec{v}_1 = \frac{4}{9}[-1 \quad 2 \quad -2]^T$. Hence the shortest distance is $\|\vec{v} - \vec{v}_1\| = 4$.

**d)** Here the solutions are $[1 + 3s - 5t \quad s \quad t]^T$ which is a plane $x = 1 + 3y - 5z$, or $x - 3y + 5z = 1$. Choose any point in this plane, say $P_0(1, 0, 0)$, write $\vec{v} = \overrightarrow{P_0P} = [-2 \quad 4 \quad 0]^T$. If $\vec{n}$ is the normal to the plane, compute $\vec{v}_1 = proj_{\vec{n}}(\vec{v}) = \frac{-2}{5}[1 \quad -3 \quad 5]^T$. Then the point $Q$ on the plane closest to $P$ has position vector $\vec{q} = \vec{p} - \vec{v}_1 = \frac{1}{5}[-3 \quad 14 \quad 10]^T$. So $Q\left(\frac{-3}{5}, \frac{14}{5}, 2\right)$ is the solution closest to $P$.

**25. b)** False. They could meet in a line (if they are like the pages of a book).

**d)** False. The plane $x - y + 2z = 3$ also has equation $2x - 2y + 4z = 6$.

**f)** False. If the line is parallel to the normal of a plane, then it is *perpendicular* to the plane.

**h)** True. If a line is perpendicular to the normal of a plane, it is parallel to that plane.

**26.** Hint: The plane has equation $\vec{n} \cdot \vec{v} = k$ where $\vec{n}$ is a normal. If $\vec{a}$ and $\vec{b}$ are the position vectors of $A$ and $B$, show that the line through $A$ and $B$ has equation $\vec{p} = \vec{a} + t\vec{d}$ where $\vec{d} = \vec{b} - \vec{a}$. Then show that $\vec{n} \cdot \vec{p} = k$.

**28. b)** Hint: Let $\vec{n} = [a \quad b \quad c]^T \neq \vec{0}$ denote a normal to the plane, and write $\vec{p} = [x \quad y \quad z]^T$. Let $\vec{p}_0$ be the position vector of an arbitrary point $P_0$ in the plane, so that $\vec{n} \cdot \vec{p}_0 = k$. As in Example 12, the magnitude of $proj_{\vec{n}}(\vec{a} - \vec{p}_0)$ is the required shortest distance.

## 3.4   MATRIX TRANSFORMATIONS OF $\mathbb{R}^2$

**1. b)** $A = \begin{bmatrix} 3 & -1 \\ 1 & 2 \end{bmatrix}$      **d)** $A = \begin{bmatrix} -1 & 1 \\ 1 & 1 \end{bmatrix}$

**2.** Proceed as in Example 4.

**b)** $A = \frac{1}{7}\begin{bmatrix} 5 & 1 \\ -1 & -10 \end{bmatrix}$, $T\begin{bmatrix} x \\ y \end{bmatrix} = \frac{1}{7}\begin{bmatrix} 5x+y \\ -x-10y \end{bmatrix}$.

**d)** $A = \begin{bmatrix} 3 & 2 \\ -3 & -3 \end{bmatrix}$, $T\begin{bmatrix} x \\ y \end{bmatrix} = \begin{bmatrix} 3x+2y \\ -3x-3y \end{bmatrix}$.

**3. b)** $A = \frac{1}{5}\begin{bmatrix} -3 & 4 \\ 4 & 3 \end{bmatrix}$      **d)** $A = \begin{bmatrix} 0 & 1 \\ -1 & 0 \end{bmatrix}$

**f)** $A = \frac{1}{2}\begin{bmatrix} 1 & -1 \\ -1 & 1 \end{bmatrix}$

**4. b)** Axiom T2 fails. For example, $T\left(2\begin{bmatrix} 1 \\ 1 \end{bmatrix}\right) = \begin{bmatrix} 0 \\ 4 \end{bmatrix}$, while $2T\left(\begin{bmatrix} 1 \\ 1 \end{bmatrix}\right) = \begin{bmatrix} 0 \\ 2 \end{bmatrix}$.

**5. b)** $T\begin{bmatrix} x \\ y \end{bmatrix} = x\begin{bmatrix} p \\ q \end{bmatrix} + y\begin{bmatrix} r \\ s \end{bmatrix} = \begin{bmatrix} px+ry \\ qx+sy \end{bmatrix} = \begin{bmatrix} p & r \\ q & s \end{bmatrix}\begin{bmatrix} x \\ y \end{bmatrix}$ for all $\begin{bmatrix} x \\ y \end{bmatrix}$ in $\mathbb{R}^2$. So $A = \begin{bmatrix} p & r \\ q & s \end{bmatrix}$.

**6.** In each case we obtain the matrix $A$ of $T$ from either Theorem 1 or Theorem 3.

**b)** $A = \frac{1}{5}\begin{bmatrix} -4 & 3 \\ 3 & 4 \end{bmatrix}$ so $T\begin{bmatrix} 1 \\ 1 \end{bmatrix} = \frac{1}{5}\begin{bmatrix} -1 \\ 7 \end{bmatrix}$, and $T\begin{bmatrix} 2 \\ -1 \end{bmatrix} = \frac{1}{5}\begin{bmatrix} -11 \\ 2 \end{bmatrix}$.

**d)** $A = \frac{\sqrt{2}}{2}\begin{bmatrix} 1 & 1 \\ -1 & 1 \end{bmatrix}$ so $T\begin{bmatrix} 1 \\ 1 \end{bmatrix} = \begin{bmatrix} \sqrt{2} \\ 0 \end{bmatrix}$, and $T\begin{bmatrix} 2 \\ -1 \end{bmatrix} = \frac{\sqrt{2}}{2}\begin{bmatrix} 1 \\ -3 \end{bmatrix}$.

**f)** $A = \frac{1}{5}\begin{bmatrix} 4 & -2 \\ -2 & 1 \end{bmatrix}$ so $T\begin{bmatrix} 1 \\ 1 \end{bmatrix} = \frac{1}{5}\begin{bmatrix} 2 \\ -1 \end{bmatrix}$, and $T\begin{bmatrix} 2 \\ -1 \end{bmatrix} = \begin{bmatrix} 2 \\ -1 \end{bmatrix}$.

**7.** In each case, read off the matrix $A$ of $T$, and use Theorems 1 and 3.

**b)** $A = \frac{1}{2}\begin{bmatrix} 1 & -1 \\ -1 & 1 \end{bmatrix}$, so $T$ is projection on the line $y = -x$.

**d)** $A = \frac{1}{5}\begin{bmatrix} -3 & 4 \\ 4 & 3 \end{bmatrix}$, so $T$ is reflection in the line $y = 2x$.

**f)** $A = \frac{1}{2}\begin{bmatrix} 1 & -\sqrt{3} \\ \sqrt{3} & 1 \end{bmatrix} = \begin{bmatrix} \frac{1}{2} & -\frac{\sqrt{3}}{2} \\ \frac{\sqrt{3}}{2} & \frac{1}{2} \end{bmatrix}$, so $T$ is a rotation of $\theta = \frac{\pi}{3}$.

**8. b)** $T(\vec{v}) = kR(\vec{v}) = k(A\vec{v}) = (kA)\vec{v}$ holds for all $\vec{v}$ in $\mathbb{R}^2$, so $T$ has matrix $kA$.

**10. b)** $P_L(\vec{v} + \vec{w}) = \frac{(\vec{v}+\vec{w}) \cdot \vec{d}}{\|\vec{d}\|^2}\vec{d} = \left(\frac{\vec{v} \cdot \vec{d} + \vec{w} \cdot \vec{d}}{\|\vec{d}\|^2}\right)\vec{d} = $
$\frac{\vec{v} \cdot \vec{d}}{\|\vec{d}\|^2}\vec{d} + \frac{\vec{w} \cdot \vec{d}}{\|\vec{d}\|^2}\vec{d} = P_L(\vec{v}) + P_L(\vec{w})$ for all $\vec{v}$ and $\vec{w}$ in $\mathbb{R}^2$. Hence Axiom T1 holds for $P_L$. Similarly $P_L(a\vec{v}) = aP_L(\vec{v})$, so T2 also holds for $P_L$.

**12. b)** Let $\vec{v} = \begin{bmatrix} x \\ y \end{bmatrix}$ be the position vector of a point on the line. As in the proof of Theorem 1, $Q_L(\vec{v}) = 2P_L(\vec{v}) - \vec{v}$. Using **a)** this gives (with some matrix arithmetic)
$$Q_L(\vec{v}) = \frac{2}{a^2+b^2}\begin{bmatrix} a^2 & ab \\ ab & b^2 \end{bmatrix}\begin{bmatrix} x \\ y \end{bmatrix} - \begin{bmatrix} x \\ y \end{bmatrix}$$
$$= \frac{1}{a^2+b^2}\begin{bmatrix} a^2-b^2 & 2ab \\ 2ab & b^2-a^2 \end{bmatrix}\begin{bmatrix} x \\ y \end{bmatrix}.$$

**13.** In each case $A^{-1}$ is the matrix of the transformation $T^{-1}$ that "reverses" the action of $T$.

**b)** Here $T\begin{bmatrix} x \\ y \end{bmatrix} = \begin{bmatrix} x+5y \\ y \end{bmatrix}$ so $T^{-1}\begin{bmatrix} x \\ y \end{bmatrix} = \begin{bmatrix} x-5y \\ y \end{bmatrix}$, whence $A^{-1} = \begin{bmatrix} 1 & -5 \\ 0 & 1 \end{bmatrix}$.

**d)** Now $A = \frac{1}{\sqrt{2}}\begin{bmatrix} 1 & -1 \\ 1 & 1 \end{bmatrix}$ so $T$ is rotation through $\frac{\pi}{4}$. It follows that $T^{-1}$ is rotation through $-\frac{\pi}{4}$, with matrix $A^{-1} = \frac{1}{\sqrt{2}}\begin{bmatrix} 1 & 1 \\ -1 & 1 \end{bmatrix}$.

**f)** Here $A = \frac{1}{1+m^2}\begin{bmatrix} 1-m^2 & 2m \\ 2m & m^2-1 \end{bmatrix}$ so $T$ is reflection in the line $y = mx$. Hence $T^{-1} = T$ and it follows that $A^{-1} = A$.

**14.** In each case we have two transformations $\mathbb{R}^2 \xrightarrow{T} \mathbb{R}^2 \xrightarrow{S} \mathbb{R}^2$ and we identify the composite $S \circ T$ by first finding its matrix and then using Theorems 1 and 3.

  **b)** $T$ is rotation through $\pi$ with matrix $A = \begin{bmatrix} -1 & 0 \\ 0 & -1 \end{bmatrix}$,

    $S$ is reflection in the $X$-axis with matrix $B = \begin{bmatrix} 1 & 0 \\ 0 & -1 \end{bmatrix}$, so

    $S \circ T$ has matrix $BA = \begin{bmatrix} -1 & 0 \\ 0 & 1 \end{bmatrix}$. Hence $S \circ T$ is reflection in the $Y$-axis.

  **d)** $T$ is reflection in the $X$-axis with matrix $A = \begin{bmatrix} 1 & 0 \\ 0 & -1 \end{bmatrix}$,

    $S$ is rotation through $\frac{\pi}{2}$ with matrix $B = \begin{bmatrix} 0 & -1 \\ 1 & 0 \end{bmatrix}$, so

    $S \circ T$ has matrix $BA = \begin{bmatrix} 0 & 1 \\ 1 & 0 \end{bmatrix}$. Hence $S \circ T$ is reflection in the line $y = x$.

**16.** In each case we are given two transformations $\mathbb{R}^2 \xrightarrow{T} \mathbb{R}^2 \xrightarrow{S} \mathbb{R}^2$, we find $S \circ T$ by finding the matrices $A$ and $B$ of $S$ and $T$ respectively, and using Theorems 1 and 3.

  **b)** $A = \begin{bmatrix} 0 & 1 \\ 1 & 0 \end{bmatrix}$, $B = \begin{bmatrix} 0 & -1 \\ -1 & 0 \end{bmatrix}$, so $BA = \begin{bmatrix} -1 & 0 \\ 0 & -1 \end{bmatrix}$ and $(S \circ T)(\vec{v}) = -\vec{v}$ for all $\vec{v}$ in $\mathbb{R}^2$.

  **d)** $A = \begin{bmatrix} 1 & 0 \\ 0 & 2 \end{bmatrix}$, $B = \begin{bmatrix} 0 & -1 \\ 1 & 0 \end{bmatrix}$, $BA = \begin{bmatrix} 0 & -2 \\ 1 & 0 \end{bmatrix}$, so $(S \circ T)\begin{bmatrix} x \\ y \end{bmatrix} = \begin{bmatrix} -2y \\ x \end{bmatrix}$.

**18.** See the solution of Exercise 20.

**20.** Let $T$ be reflection in a line $L$ through the origin.

  **a)** If $X$ is any point in the plane, write $Y = T(X)$. Then $Y$ and $X$ are reflections of each other in the line $L$ and, in particular, $X = T(Y)$. Eliminating $Y$ gives $T(T(X)) = X$, that is $(T \circ T)(X) = X$. Now think of this as $(T \circ T)(X) = 1_{\mathbb{R}^2}(X)$ so, as this holds for every $X$ in $\mathbb{R}^2$, we have $T \circ T = 1_{\mathbb{R}^2}$. Hence $T = T^{-1}$ by Theorem 6, as required.

  **b)** Assume first that $L$ has a slope $m$ (that is, $L$ is not vertical). Then $T = Q_m$ has matrix $A = \frac{1}{1+m^2}\begin{bmatrix} 1 - m^2 & 2m \\ 2m & m^2 - 1 \end{bmatrix}$ and we compute

$$A^2 = \left(\tfrac{1}{1+m^2}\right)^2 \begin{bmatrix} 1 - m^2 & 2m \\ 2m & m^2 - 1 \end{bmatrix}^2$$
$$= \left(\tfrac{1}{1+m^2}\right)^2 \begin{bmatrix} (1+m^2)^2 & 0 \\ 0 & (1+m^2)^2 \end{bmatrix} = \begin{bmatrix} 1 & 0 \\ 0 & 1 \end{bmatrix} = I.$$

Hence $A^{-1} = A$, so $T^{-1} = T$. Finally, if $L$ is the $Y$-axis, then $A = \begin{bmatrix} -1 & 0 \\ 0 & 1 \end{bmatrix}$ and it is easy to see that $A^2 = I$ in this case too.

**22.** Let $S : \mathbb{R}^2 \to \mathbb{R}^2$ denote reflection in the $X$-axis. If $T : \mathbb{R}^2 \to \mathbb{R}^2$ is any reflection, write $T \circ S = R$. Then $R$ is a rotation by the preceding exercise and, composing on the right by $S^{-1}$, we see that $T = R \circ S^{-1}$. But $S^{-1} = S$ by Exercise 20, so this gives $T = R \circ S$. Hence $T$ is $S$ followed by $R$, as required.

**24.** **b)** Suppose that $A^{-1} = A^T$, so that $A^T A = I$. Then given any $\vec{v}$ in $\mathbb{R}^2$ we have $\|A\vec{v}\|^2 = (A\vec{v})^T(A\vec{v}) = \vec{v}^T A^T A\vec{v} = \vec{v}^T I \vec{v} = \vec{v}^T \vec{v} = \|\vec{v}\|^2$. It follows that $\|A\vec{v}\| = \|\vec{v}\|$ for every $\vec{v}$ in $\mathbb{R}^2$. Replacing $\vec{v}$ by $\vec{v} - \vec{w}$ gives $\|A\vec{v} - A\vec{w}\| = \|\vec{v} - \vec{w}\|$, that is $\|T(\vec{v}) - T(\vec{w})\| = \|\vec{v} - \vec{w}\|$. This shows that the matrix transformation $T$ induced by $A$ is an isometry.

**25.** **b)** $\begin{bmatrix} 0 & -1 & 1 \\ 1 & 0 & 1 \\ 0 & 0 & 1 \end{bmatrix}$.   **d)** $\frac{1}{2}\begin{bmatrix} 1 & -\sqrt{3} & 1 - 2\sqrt{3} \\ \sqrt{3} & 1 & 2 + \sqrt{3} \\ 0 & 0 & 2 \end{bmatrix}$.

**26.** **b)** $\frac{1}{1+m^2}\begin{bmatrix} 1 & m & -mb \\ m & m^2 & b \\ 0 & 0 & 1+m^2 \end{bmatrix}$.

## 3.5   THE CROSS PRODUCT: OPTIONAL

**1.** **b)** $\frac{7}{2}\sqrt{2}$.

**3.** **b)** 7.

**4.** **b)** True. If $\vec{v} \times \vec{w} = \vec{w} \times \vec{v}$ then, since $\vec{w} \times \vec{v} = -(\vec{v} \times \vec{w})$, we obtain $2(\vec{v} \times \vec{w}) = \vec{0}$, whence $\vec{v} \times \vec{w} = \vec{0}$. By Theorem 5, $\vec{v}$ and $\vec{w}$ are parallel.

  **d)** True by Theorem 1 because
$$det[\vec{u} \ \ \vec{v} \ \ \vec{w}]^T$$
$$= -det[\vec{u} \ \ \vec{w} \ \ \vec{v}]^T$$
$$= det[\vec{w} \ \ \vec{u} \ \ \vec{v}]^T.$$

**6.** $\vec{u} \times \vec{v} = \vec{u} \times (-\vec{u} - \vec{w})$
$$= -[\vec{u} \times (\vec{u} + \vec{w})]$$
$$= -[(\vec{u} \times \vec{u}) + (\vec{u} \times \vec{w})]$$
$$= -(\vec{u} \times \vec{w}) = \vec{w} \times \vec{u}.$$
Similarly, $\vec{u} \times \vec{v} = \vec{v} \times \vec{w}$.

**8.** $A$, $B$, and $C$ are colinear if and only if the parallelogram determined by $\vec{AB}$ and $\vec{AC}$ has zero area; if and only if $\|\vec{AB} \times \vec{AC}\| = 0$; if and only if $\vec{AB} \times \vec{AC} = \vec{0}$.

**10.** Theorem 1 gives $|\vec{u} \cdot (\vec{v} \times \vec{w})| = |det[\vec{u} \ \ \vec{v} \ \ \vec{w}]|$. Rearranging columns in a determinant merely changes the sign of the determinant.

**12.** If $h$ is the shortest distance, the parallelogram determined by $\vec{P_0P}$ and $\vec{d}$ has area $\|\vec{P_0P} \times \vec{d}\|$ by Theorem 5, so $\|\vec{P_0P} \times \vec{d}\| = \|\vec{d}\| \cdot h$. The desired formula follows.

# CHAPTER 4

# The Vector Space $\mathbb{R}^n$

## 4.1 SUBSPACES AND SPANNING

**2. b)** $\mathbb{R}^6$.          **d)** $\mathbb{R}^1 = \mathbb{R}$.

**3. b)** Not equal.          **d)** Equal.

**4. b)** $nullA = \mathbb{R}^4$, and $imA = \{0\}$.

  **d)** $nullA = span\left\{ \begin{bmatrix} -1 \\ 0 \\ 0 \\ 5 \end{bmatrix} \right\}$, and

    $imA = span\left\{ \begin{bmatrix} 0 \\ 1 \\ 0 \\ 0 \end{bmatrix}, \begin{bmatrix} -7 \\ 4 \\ 1 \\ 0 \end{bmatrix}, \begin{bmatrix} 1 \\ 0 \\ 0 \\ 1 \end{bmatrix} \right\}$.

**5. b)** Subspace. $U = span\{[0 \ 1 \ 0]^T\}$.
  **d)** Not a subspace. $\vec{0}$ is not in $U$.
  **f)** Subspace. $U = span\{[1 \ 1 \ 0]^T, [0 \ 0 \ 1]^T\}$.

**6. b)** $X$ is not in $span\{Y, \ Z\} : [1 \ 2 \ 15 \ 11]^T = s[2 \ -1 \ 0 \ 2]^T + t[1 \ -1 \ -3 \ 1]^T$ has no solution.
  **d)** $X = 3Y + 4Z$.

**7. b)** The vectors span $\mathbb{R}^4$. One way to see this is to observe that the matrix $A$ with these vectors as columns is invertible (the determinant is $-42$), so each $E_i = AX$ for some column in $\mathbb{R}^n$. Hence $E_i$ is in the span of these vectors by Example 12.

**8. b)** False. Let $X$ be any vector in $\mathbb{R}^n$ that is not in $U$. If $r = 0$, then $rX = 0X = 0 \in U$.
  **d)** True. Use axiom S3 with $r = -1$.
  **f)** False. 0 is not in this set.

**10.** Clearly $aX \in span\{X\}$ so $span\{aX\} \subseteq span\{X\}$ by Theorem 1. Since $a \neq 0$, $X = a^{-1}(aX) \in span\{aX\}$, so $span\{X\} \subseteq span\{aX\}$, again by Theorem 1. Hence $span\{X\} = span\{aX\}$.

**12.** No. For example, $[0 \ 0 \ 0]^T$ is not in $\mathbb{R}^2$.

**14.** Let $Y$ be in $U$, say $Y = a_1X_1 + a_2X_2 + \cdots + a_kX_k, a_i$ in $\mathbb{R}$. If $AX_i = 0$ for each $i$, then
$$AY = a_1AX_1 + a_2AX_2 + \cdots + a_kAX_k$$
$$= a_10 + a_20 + \cdots + a_k0 = 0.$$

**15. b)** Always $im(AV) \subseteq imA$ because any vector in $im(AV)$ has the form $(AV)X = A(VX)$ and so is in $imA$. Conversely, let $Y \in imA$, say $Y = AX$ where $X$ is in $\mathbb{R}^n$. Since $V$ is invertible, we can write $Y = AX = A(VV^{-1})X = AV(V^{-1}X)$. Hence $Y$ is in $im(AV)$ because $V^{-1}X$ is in $\mathbb{R}^n$. It follows that $imA \subseteq im(AV)$. Thus we have $im(AV) = imA$.

**16. b)** By axiom S2, $-X = (-1)X \in U$. Then we can write $Y = (X + Y) + (-X)$ which is in $U$ by axiom S3.

**18.** S1. The zero matrix 0 is in $U$ because $A0 = 0 = B0$.
  S2. If $X, Y$ are in $U$, then $A(X + Y) = AX + AY = BX + BY = B(X + Y)$; thus $X + Y$ is in $U$.
  S3. If $X$ is in $U$ and $r$ is in $\mathbb{R}$, then $A(rX) = r(AX) = r(BX) = B(rX)$; thus $rX$ is in $U$.

**20.** The fact that the zero vector is in $U$ forces $B = 0$.

**22.** We have $span\{V\} = \{aV \mid a \text{ in } \mathbb{R}\}$. If $U \neq \{0\}$ is a subspace, say $0 \neq aV \in U$, then $V = a^{-1}(aV) \in U$. Hence $span\{V\} \subseteq U$, and it follows that $U = span\{V\}$.

**24. b)** S2. Given $X, Y$ in $U + W$, write $X = X_1 + X_2$ and $Y = Y_1 + Y_2$, where $X_1, Y_1$ are in $U$ and $X_2, Y_2$ are in $W$. Then $X + Y = (X_1 + Y_1) + (X_2 + Y_2)$ is in $U + W$ because $X_1 + Y_1$ is in $U$ and $X_2 + Y_2$ is in $W$.

## 4.2 LINEAR INDEPENDENCE

**1. b)** If $r[1 \ 1 \ 1]^T + s[2 \ 1 \ 0]^T = [0 \ 0 \ 0]^T$, then $r + 2s = 0, r + s = 0$, and $r = 0$. This implies that $r = s = 0$, so the vectors are independent by the Independence test.

  **d)** If $r[1 \ 0 \ 1 \ 0]^T + s[0 \ 1 \ 0 \ 1]^T + t[3 \ -1 \ 2 \ -1]^T = [0 \ 0 \ 0 \ 0]^T$, equating coefficients gives the equations $r + 3t = 0, s - t = 0, r + 2t = 0$, and $s - t = 0$. This implies that $r = s = t = 0$, and so the vectors are independent by the Independence test.

**2. b)** Dependent. $[2 \ -6]^T + 2[-1 \ 3]^T = 0$.
  **d)** Independent.          **f)** Independent.

**3. b)** Invertible. The columns are independent.
  **d)** Not invertible. If $C_1, C_2, C_3, C_4$, and $C_5$ denote the columns of $A$, then $C_1 - 2C_2 - C_3 + C_4 + 0C_5 = 0$.

**4. b)** Independent. $det \begin{bmatrix} 1 & 1 & 0 \\ 1 & -1 & 0 \\ -1 & 1 & 1 \end{bmatrix} = -2 \neq 0.$

**d)** Independent. $det \begin{bmatrix} 1 & 4 & 0 & 1 \\ 0 & 4 & 1 & 3 \\ -2 & -3 & 0 & 3 \\ 5 & 2 & -3 & 1 \end{bmatrix} = -55 \neq 0.$

**5. b)** Does not span $\mathbb{R}^3$. $det \begin{bmatrix} 0 & 8 & -9 \\ -6 & -3 & 7 \\ -6 & 5 & -2 \end{bmatrix} = 0.$

**d)** Does not span $\mathbb{R}^4$. $det \begin{bmatrix} 2 & 3 & 4 & -5 \\ 1 & 5 & -4 & 0 \\ 7 & 4 & -3 & 6 \\ -2 & 5 & -3 & -4 \end{bmatrix} = 0.$

**6.** The sets where the given vectors would form a square matrix.

**7. b)** True. If $rX + sY = 0$, then $rX + sY + 0Z = 0$, whence $r = s = 0$ by the independence of $\{X, Y, Z\}$.

**d)** False. One of the $X_i$ could be a linear combination of the other vectors.

**f)** False. We could take $a = b = c = 0$ for *any* set $\{X, Y, Z\}$ of vectors.

**h)** True. This is the negation of the definition of independence.

**9. b)** Independent. If $a(X + Y) + b(Y + Z) + c(Z + X) = 0$, then $(a + c)X + (a + b)Y + (b + c)Z = 0$. Since $\{X, Y, Z, W\}$ is independent, $\{X, Y, Z\}$ is also independent which forces $a + c = 0$, $a + b = 0$, and $b + c = 0$. The only solution is $a = b = c = 0$.

**d)** Dependent. $(X + Y) - (Y + Z) + (Z + W) - (W + X) = 0$.

**10.** Use Theorems 2 and 3 on the columns of $A$.

**12.** Set a linear combination equal to zero and obtain a set of equations in the coefficients that has only the trivial solution.

**14.** Suppose $rY + r_1X_1 + r_2X_2 + \cdots + r_kX_k = 0$. If $r \neq 0$, show that $Y$ is in $span\{X_1, X_2, \cdots, X_k\}$, contrary to assumption. So $r = 0$ and the independence of $\{X_1, X_2, \cdots, X_k\}$ can be applied.

**16. b)** Since $\mathcal{X}$ is dependent, some nontrivial linear combination $r_1X_1 + r_2X_2 + \cdots + r_mX_m = 0$, where each $X_i$ is in $\mathcal{X}$. But each $X_i$ is also in $\mathcal{Y}$ (because $\mathcal{X} \subseteq \mathcal{Y}$) so this is a nontrivial linear combination of vectors in $\mathcal{Y}$ that vanishes. Hence $\mathcal{Y}$ is dependent too.

## 4.3 DIMENSION

**1. b)** 3      **d)** 3

**2. b)** Independent (and so span $\mathbb{R}^3$ by Theorem 4) because

$det \begin{bmatrix} -1 & 1 & 1 \\ 1 & -1 & 1 \\ 1 & 1 & -1 \end{bmatrix} = 4 \neq 0.$

**d)** Independent (and so span $\mathbb{R}^4$ by Theorem 4) because

$det \begin{bmatrix} 1 & 1 & 0 & 0 \\ 1 & 0 & 1 & 1 \\ 0 & 0 & 1 & 0 \\ 0 & 1 & 0 & 1 \end{bmatrix} = -2 \neq 0.$

**3. b)** The first four vectors are independent by Theorem 2 §4.2 because $det \begin{bmatrix} 2 & 6 & 3 & 1 \\ 3 & 1 & 6 & 0 \\ 5 & -4 & 3 & 0 \\ -4 & -7 & 2 & 0 \end{bmatrix} = 289 \neq 0$, and so these vectors form a basis of $\mathbb{R}^4$ by Theorem 4. The given set has too many vectors to be a basis for $\mathbb{R}^4$ by the Corollary to Theorem 1.

**d)** These vectors are independent by Theorem 2 §4.2, since $det \begin{bmatrix} 1 & 5 & 3 & 2 \\ -6 & 1 & 9 & 0 \\ 3 & 5 & 0 & 3 \\ -7 & 9 & 2 & 0 \end{bmatrix} = 441$, and so form a basis of $\mathbb{R}^4$ by Theorem 4.

**4. b)** False. Both sets are independent, so the subspaces they span are of dimension 3 and 2, respectively, and so cannot be equal.

**d)** False. Both sets are independent, so the subspaces they span have dimensions 2 and 4, respectively, and so cannot be equal.

**5. b)** $\left\{ \begin{bmatrix} 1 \\ 0 \\ 0 \end{bmatrix}, \begin{bmatrix} 0 \\ 1 \\ 0 \end{bmatrix}, \begin{bmatrix} 0 \\ 0 \\ 1 \end{bmatrix} \right\}$      **d)** $\{1\}$

**6.** In each case we find a basis inside the given spanning set (as guaranteed by Theorem 3).

**b)** $\{[2 \ 1 \ 0 \ -1]^T, [-1 \ 1 \ 1 \ 2]^T\}$; dimension is 2. In fact, $[2 \ 7 \ 4 \ 5]^T = 3[2 \ 1 \ 0 \ -1]^T + 4[-1 \ 1 \ 1 \ 2]^T.$

**d)** $\{[-2 \ 0 \ 3 \ 1]^T, [1 \ 2 \ -1 \ 0]^T, [1 \ 2 \ 2 \ 1]^T\}$; dimension is 3. In fact, $[-2 \ 8 \ 5 \ 3]^T = 3[-2 \ 0 \ 3 \ 1]^T + 4[1 \ 2 \ -1 \ 0]^T.$

**7. b)** $\{[1 \ 1 \ 0 \ 1]^T, [1 \ -1 \ 1 \ 0]^T\}$. Dimension is 2.

**d)** $\{[1 \ 0 \ 1 \ 0]^T, [-1 \ 1 \ 0 \ 1]^T, [0 \ 1 \ 0 \ 1]^T\}$. Dimension is 3.

**8. b)** $\{[-2 \ 4 \ 1 \ 0 \ 0]^T, [7 \ -11 \ 0 \ 2 \ 1]^T\}$. $dim(nullA) = 2.$

**9. b)** The condition can be written as $A^TX = 0$. The solution is $[-2s - 2t \ \ -2s + t \ \ s \ \ t]^T$, so the corresponding basis of $U$ is $\{[-2 \ -2 \ 1 \ 0]^T, [-2 \ 1 \ 0 \ 1]^T\}$ and $dimU = 2.$

**d)** Rearrange the condition $X^TA = (BX)^T$ to give $(A^T - B)X = 0$. As in part **b)**, a basis of $U$ is $\{[1 \ 1 \ 1 \ 0 \ ]^T, [1 \ -4 \ 0 \ 2]^T\}$ and $dimU = 2.$

**10. b)** False. Take the standard basis of $\mathbb{R}^3$ as a counterexample.

**d)** False. $\left\{ \begin{bmatrix} 1 \\ 0 \\ 0 \\ 0 \\ 1 \end{bmatrix}, \begin{bmatrix} 0 \\ 1 \\ 0 \\ 0 \\ 1 \end{bmatrix}, \begin{bmatrix} 0 \\ 0 \\ 1 \\ 0 \\ 1 \end{bmatrix}, \begin{bmatrix} 0 \\ 0 \\ 0 \\ 1 \\ 1 \end{bmatrix}, \begin{bmatrix} 0 \\ 0 \\ 0 \\ -1 \\ 1 \end{bmatrix} \right\}$ is a basis of $\mathbb{R}^5$.

**f)** False. Take $Y_i = -X_i$ for $1 \le i \le 4$. Then $\{X_1 + Y_1, X_2 + Y_2, X_3 + Y_3, X_4 + Y_4\} = \{0, 0, 0, 0\} = \{0\}$ which is not a basis of $\mathbb{R}^4$.

**12.** Let $X$ be a vector in $\mathbb{R}^n$ such that $X \ne 0$. Then the set $\{X\}$ is linearly independent, and so by Theorem 3 can be enlarged to a basis of $\mathbb{R}^n$.

**14.** The condition is that $\{X, Y, Z, W\}$ must span $\mathbb{R}^3$. In this case the set $\{X, Y, Z, W\}$ contains a basis of $\mathbb{R}^3$ by Theorem 3. Conversely, if $\{X, Y, Z, W\}$ contains a basis of $\mathbb{R}^3$, then the three vectors of that basis will span $\mathbb{R}^3$. Hence $\{X, Y, Z, W\}$ spans $\mathbb{R}^3$ by Theorem 1 §4.1.

**16. b)** If $t_1(X + W) + t_2(Y + W) + t_3(Z + W) + t_4W = 0$, collecting coefficients gives $t_1 = t_2 = t_3 = 0$ and $t_1 + t_2 + t_3 + t_4 = 0$. Hence $t_1 = t_2 = t_3 = t_4 = 0$.

**17. b)** Assume that $\{aX + bY, cX + dY\}$ is independent. To show that $A$ is invertible, let $AX = 0$; we must show that $X = 0$. If we write $X = \begin{bmatrix} r \\ s \end{bmatrix}$, the condition $AX = 0$ implies that $ra + sc = 0$ and $rb + sd = 0$, and hence that $r(aX + bY) + s(cX + dY) = (ra + sc)X + (rb + sd)Y = 0$. This yields $r = s = 0$ by the independence of $\{X, Y\}$, as required.

**18. b)** $\{E_1, E_2, \cdots, E_n\}$ is not contained in $span\{X_1, \cdots, X_k\}$ since $dim[span\{X_1, \cdots, X_k\}] \le k < n$. So some $E_{l_{k+1}}$ is not in $span\{X_1, \cdots, X_k\}$, whence $\{X_1, \cdots, X_k, E_{l_{k+1}}\}$ is independent by the Independent lemma. If $\{X_1, \cdots, X_k, E_{l_{k+1}}\}$ spans $\mathbb{R}^n$, we are done. Otherwise, $\{E_1, E_2, \cdots, E_n\}$ is not contained in $span\{X_1, \cdots, X_k, E_{l_{k+1}}\}$ so some $E_{l_{k+2}}$ is not in $span\{X_1, \cdots, X_k, E_{l_{k+1}}\}$. Hence $\{X_1, \cdots, X_k, E_{l_{k+1}}, E_{l_{k+2}}\}$ is independent. Continuing, we eventually reach a basis of $\mathbb{R}^n$.

**20.** Since $dimU \le dimW$ and $dimW = 1$, either $dimU = 0$ or $dimU = 1$. If $dimU = 0$, then $U = \{0\}$; if $dimU = 1$, then $U = W$ by part (2) of Theorem 4.

**22.** $U \cap W$ is a subspace of $U$, so $dim(U \cap W) \le dimU = 2$. Thus $dim(U \cap W) = 0, 1,$ or $2$. If $dim(U \cap W) = 2$, then $U \cap W = U$ by Theorem 4; in particular $U \subseteq W$.

**23. b)** Let $\{Y_1, \cdots, Y_k\}$ be a basis of $im(UA)$, say $Y_i = (UA)X_i$ for some $X_i$ in $\mathbb{R}^n$. It follows that $U^{-1}Y_i = AX_i$ is in $imA$, and it suffices to show that $\mathcal{D} = \{U^{-1}Y_1, \cdots, U^{-1}Y_k\}$ is a basis of $imA$. If $t_1(U^{-1}Y_1) + \cdots + t_k(U^{-1}Y_k) = 0$, then $t_1Y_1 + \cdots + t_kY_k = 0$, so each $t_i = 0$ by the independence of the $Y_i$, proving that $\mathcal{D}$ is independent. Finally, given $Y$ in $imA$, write $Y = AX$ where $X$ is in $\mathbb{R}^n$, so $UY = UAX$ is in $im(UA)$. Hence write $UY = s_1Y_1 + \cdots + s_kY_k, s_i$ in

$\mathbb{R}$, so that $Y = s_1U^{-1}Y_1 + \cdots + s_kU^{-1}Y_k$. This shows that $\mathcal{D}$ spans $imA$.

**24.** We show that $span\{X_1, X_2, \cdots, X_n\} = U$. If not, choose $Y$ in $U$ which is not in $span\{X_1, X_2, \cdots, X_n\}$. Then by Lemma 1, $\{Y, X_1, X_2, \cdots, X_n\}$ is linearly independent, contrary to the maximality of $\{X_1, X_2, \cdots, X_n\}$. So $span\{X_1, X_2, \cdots, X_n\} = U$ after all.

**26. b)** Let $AX_i = \lambda_iX_i$ for each $i = 1, 2, \cdots, n + m$. It follows that $BY_i = \lambda_iY_i$ and $CZ_i = \lambda_iZ_i$ for each $i$. Moreover, if $Y$ and $Z$ are arbitrary columns in $\mathbb{R}^n$ and $\mathbb{R}^m$ respectively, then $\begin{bmatrix} Y \\ Z \end{bmatrix}$ is a linear combination of the $X_i = \begin{bmatrix} Y_i \\ Z_i \end{bmatrix}$. It follows that $Y$ is the same linear combination of the $Y_i$, and hence that $\mathbb{R}^n = span\{Y_1, \cdots, Y_{n+m}\}$. Thus $\{Y_1, \cdots, Y_{n+m}\}$ contains a basis of $\mathbb{R}^n$ consisting of eigenvectors of $B$ (by Theorem 3); that is, $B$ is diagonalizable by Theorem 3 §2.3. Similarly, $C$ is diagonalizable.

## 4.4 RANK

**1. b)** $rowA = span\{[-5 \ \ -4], [8 \ \ 0]\}$.
$colA = span\{[-5 \ \ 8 \ \ -1]^T, [-4 \ \ 0 \ \ -3]^T\}$.

**d)** $rowA = span\{[9 \ \ -5 \ \ 1 \ \ -1], [8 \ \ -5 \ \ -3 \ \ 1],$
$[0 \ \ 4 \ \ 7 \ \ 6], [0 \ \ 2 \ \ -7 \ \ 7]\}$.
$colA = span\{[9 \ \ 8 \ \ 0 \ \ 0]^T, [-5 \ \ -5 \ \ 4 \ \ 2]^T,$
$[1 \ \ -3 \ \ 7 \ \ -7]^T, [-1 \ \ 1 \ \ 6 \ \ 7]^T\}$.

**2.** (c) and (b) share the same column space since the columns of matrix (b) can be obtained by adding columns 1 and 2 of (a) to column 3, and adding column 1 to column 4 and then subtracting column 2 from column 4. However, the two matrices do not share the same row space. For matrix (c), $rowA = span\{[1 \ \ 0 \ \ 1 \ \ 1], [0 \ \ 1 \ \ 1 \ \ -1]\}$, whereas for matrix (b), $rowA = span\{[1 \ \ 0 \ \ 0 \ \ 0], [0 \ \ 1 \ \ 0 \ \ 0]\}$.

**3. b)** $rankA = dim(rowA) = 2$. Indeed: $[4 \ \ -13 \ \ 5 \ \ 6]^T = -2[1 \ \ 2 \ \ -1 \ \ 3]^T + 3[2 \ \ -3 \ \ 1 \ \ 4]^T$.

**4. b)** $\{[1 \ \ 2 \ \ -1 \ \ 0 \ \ 4]^T, [-3 \ \ 1 \ \ 0 \ \ 2 \ \ 5]^T, [-7 \ \ 7 \ \ -2 \ \ 6 \ \ 13]^T\}$.

$$\begin{bmatrix} 1 & -3 & -7 & -2 \\ 2 & 1 & 7 & 3 \\ -1 & 0 & -2 & -1 \\ 0 & 2 & 6 & 2 \\ 4 & 5 & 13 & 9 \end{bmatrix} \rightarrow \begin{bmatrix} 1 & 0 & 0 & 1 \\ 0 & 1 & 0 & 1 \\ 0 & 0 & 1 & 0 \\ 0 & 0 & 0 & 0 \\ 0 & 0 & 0 & 0 \end{bmatrix}.$$

**5. b)** $rowA : \{[1 \ \ 2 \ \ 0 \ \ 5 \ \ 7 \ \ 1], [2 \ \ 1 \ \ 1 \ \ -1 \ \ 0 \ \ 3],$
$[0 \ \ 0 \ \ 0 \ \ 0 \ \ 0 \ \ 1]\}; colA : \{[1 \ \ 2 \ \ -5 \ \ 1]^T,$
$[2 \ \ 1 \ \ 2 \ \ -1]^T, [1 \ \ 3 \ \ 9 \ \ 2]^T\}; rankA = 3;$

$$A \rightarrow \begin{bmatrix} 1 & 2 & 0 & 5 & 7 & 1 \\ 0 & 1 & -\frac{1}{3} & \frac{11}{3} & \frac{14}{3} & 0 \\ 0 & 0 & 0 & 0 & 0 & 1 \\ 0 & 0 & 0 & 0 & 0 & 0 \end{bmatrix}$$

**6. b)** $nullA = span\{[6 \ 0 \ -4 \ 1 \ 0]^T, [5 \ 0 \ -3 \ 0 \ 1]^T\};$

$$A \rightarrow \begin{bmatrix} 1 & 0 & 0 & -6 & -5 \\ 0 & 1 & 0 & 0 & 0 \\ 0 & 0 & 1 & 4 & 3 \\ 0 & 0 & 0 & 0 & 0 \end{bmatrix} \text{ so } rankA = 3;$$

$dim(nullA) = n - rankA = 5 - 3 = 2.$

**7. b)** False. Consider $A = \begin{bmatrix} 1 & 0 & 0 \\ 0 & 1 & 0 \end{bmatrix}$.

Here, $rowA = span\{[1 \ 0 \ 0]^T, [0 \ 1 \ 0]^T\} \neq \mathbb{R}^n$.

**d)** False. $\begin{bmatrix} 1 & 0 & 0 \\ 0 & 0 & 0 \end{bmatrix}$ has rank 1.

**f)** False. Let $A = \begin{bmatrix} 1 & 0 \\ 0 & 1 \end{bmatrix}$ and $B = \begin{bmatrix} -1 & 0 \\ 0 & -1 \end{bmatrix}$. Then

$rankA = 2 = rankB$ but $rank(A + B) = rank\begin{bmatrix} 0 & 0 \\ 0 & 0 \end{bmatrix} = 0.$

**8.** The rows could be independent, so $rankA = dim(rowA) = 3$, But $A$ can have at most 3 independent columns ($rankA \leq 3$ — the number of rows). So the 4 columns of $A$ cannot be independent.

**10.** No, the rank cannot be equal to both the number of rows and the number of columns.

**12.** If $A$ is $m \times n$ and both the rows and columns are independent, then $m = rankA = n$, a contradiction.

**14. b)** It follows from **a)** that $col(AB) \subseteq colA$. Taking dimensions gives $rank(AB) \leq rankA$.

**16. b)** Write $A = [C_1 \ C_2 \ \cdots \ C_n]$ where $C_j$ is column $j$ of $A$. We have $colA = span\{C_1, C_2, \cdots, C_n\}$ and $col([A \ B]) = span\{C_1, C_2, \cdots, C_n, B\}$. By **a)**, the system $AX = B$ has a solution if and only if $B$ is in $colA$; that is, if and only if $colA = col([A \ B])$. But $colA \subseteq col([A \ B])$, so $colA = col([A \ B])$ if and only if $dim[colA] = dim\{col([A \ B])\}$; that is, if and only if $rankA = rank[A \ B]$.

**18. b)** If $AB = I_m$, then $rankA \geq rank(AB) = m$ using Exercise 14. But $rankA \leq m$ because $A$ is $m \times n$.

**20. b)** If $rankA = m$, then $AA^T$ is invertible by Theorem 4. The rest is routine.

# 4.5  ORTHOGONALITY

**1. b)** As in Example 2: $\|3X - 4Y\| = \sqrt{76} = 2\sqrt{19}.$
**d)** $(3X + 4Y) \cdot (X - 5Y) = 1.$

**2. b)** Verify that the dot product of each pair of vectors is zero.

**3. b)** $\{\frac{1}{\sqrt{14}}[-3 \ 1 \ 2]^T, \frac{1}{\sqrt{3}}[1 \ 1 \ 1]^T, \frac{1}{\sqrt{42}}[1 \ -5 \ 4]^T\}.$

**5. b)** Setting the dot products of the three vectors with $[a \ b \ c \ d]^T$ equal to zero produces a homogeneous system of linear equations.
Solution: $[a \ b \ c \ d]^T = [-5s \ 4s \ 2s \ 5s]^T.$

**6. b)** $[a \ b \ c]^T = \frac{-4a+b+5c}{42}[-4 \ 1 \ 5]^T + \frac{-a+b-c}{3}[-1 \ 1 \ -1]^T + \frac{2a+3b+c}{14}[2 \ 3 \ 1]^T.$

**7. b)** $F_1 = [2 \ 1]^T, F_2 = [-1 \ 2]^T.$
**d)** $F_1 = [0 \ 1 \ 1]^T, F_2 = [1 \ 0 \ 0]^T, F_3 = [0 \ -1 \ 1]^T.$

**8. b)** $Q = \frac{\sqrt{5}}{5}\begin{bmatrix} 2 & 1 \\ 1 & -2 \end{bmatrix}, R = \frac{\sqrt{5}}{5}\begin{bmatrix} 5 & 3 \\ 0 & 1 \end{bmatrix}.$

**d)** $Q = \frac{\sqrt{3}}{3}\begin{bmatrix} 1 & 1 & 0 \\ -1 & 0 & 1 \\ 0 & 1 & 1 \\ 1 & -1 & 1 \end{bmatrix}, R = \frac{\sqrt{3}}{3}\begin{bmatrix} 3 & 0 & -1 \\ 0 & 3 & 1 \\ 0 & 0 & 2 \end{bmatrix}.$

**9. b)** False. Consider $\{X, Y\} = \{[1 \ 0]^T, [0 \ 1]^T\}$. Then $\{X, X + Y\} = \{[1 \ 0]^T, [1 \ 1]^T\}$, so $X \cdot (X + Y) = [1 \ 0]^T \cdot [1 \ 1]^T = 1 + 0 = 1 \neq 0.$ Thus $\{X, X + Y\}$ is not orthogonal.

**d)** True. In order for $\{X_1, X_2, Y_1, Y_2, Y_3\}$ to be orthogonal, $X_1 \cdot X_2 = 0$, $Y_i \cdot Y_j = 0$, and $X_i \cdot Y_j = 0$ for all $i$ and $j$. This is the case here since $\{X_1, X_2\}$ and $\{Y_1, Y_2, Y_3\}$ are orthogonal, and since we are assuming that $X_i \cdot Y_j = 0$ for all $i$ and $j$.

**f)** False. $E = [1 \ 0]^T$ and $X = [2 \ 1]^T$.

**h)** True. If $Q$ has columns $C_i$, then the $(i, j)$-entry of $Q^T Q$ is $C_i \cdot C_j$.

**10.** Each $a_i X_i \neq 0$ as $a_i \neq 0$ and $X_i \neq 0$; and $(a_i X_i) \cdot (a_j X_j) = a_i a_j (X_i \cdot X_j) = a_i a_j(0) = 0$ if $i \neq j$.

**12. b)** We have $X + Y + Z = [2 \ 1 \ 1]^T$, so $\|X + Y + Z\|^2 = 6$, while $\|X\|^2 = \|Y\|^2 = \|Z\|^2 = 2$. However, $X \cdot Z = 1 + 1 + 0 = 2$ so $X$ and $Z$ are not orthogonal.

**14.** Show that $\|AX\|^2 = \lambda\|X\|^2.$

**16.** If $X = a_1 F_1 + \cdots + a_m F_m,$
then $\|X\|^2 = a_1(X \cdot F_1) + \cdots + a_m(X \cdot F_m) = 0.$
So $\|X\| = 0$, $X = 0.$

**18. b)** If $\{F_1, \cdots, F_{m-1}, H\}$ is another orthonormal basis of $U$, expand $H$ in terms of the orthonormal basis $\{F_1, \cdots, F_{m-1}, G\}$ using the Expansion theorem, and conclude that $H = aG$ where $a = H \cdot G$ is a real number. Show that $a = \pm 1$ because $H$ and $G$ are unit vectors.

**20.** Let $A$ be the matrix with the $X_i^T$ as its rows, and let $B = [Y_1 \ Y_2 \ \cdots \ Y_n]$ be the matrix with the $Y_j$ as its columns. Show that $AB = I_n$ and so $B$ is invertible by Theorem 5 §1.5. Hence $B$ has independent columns by Theorem 2 §4.2.

**22.** By Theorem 1 §1.6, it is enough to show that $A \to B$ by row operations, where $B$ is row-echelon with orthogonal rows. We know that $A \to R$ where $R$ is row-echelon. Apply the Gram-Schmidt process to the nonzero rows of $R$, starting at the bottom. The algorithm replaces each of these rows by itself minus a linear combination of the rows below it, and so preserves the row-echelon character of the matrix. Since this is done using row operations, and since the resulting rows are orthogonal, we can get to a matrix $B$ of the desired form.

## 4.6  PROJECTIONS AND APPROXIMATION

**1. b)** $X = \frac{1}{21}\begin{bmatrix} 23 \\ 10 \\ 64 \end{bmatrix} + \frac{1}{21}\begin{bmatrix} -2 \\ 11 \\ -1 \end{bmatrix}$. **d)** $X = \frac{1}{3}\begin{bmatrix} 6 \\ -5 \\ 8 \\ -2 \end{bmatrix} + \frac{4}{3}\begin{bmatrix} 0 \\ 2 \\ 1 \\ -1 \end{bmatrix}$.

**f)** $X = \frac{1}{12}\begin{bmatrix} 5a - 5b + c - 3d \\ -5a + 5b - c + 3d \\ a - b + 11c + 3d \\ -3a + 3b + 3c + 3d \end{bmatrix} + \frac{1}{12}\begin{bmatrix} 7a + 5b - c + 3d \\ 5a + 7b + c - 3d \\ -a + b + c - 3d \\ 3a - 3b - 3c + 9d \end{bmatrix}$.

**2. b)** First note that $\begin{bmatrix} 1 \\ 3 \\ -1 \\ -2 \end{bmatrix} = -\begin{bmatrix} 1 \\ -1 \\ 0 \\ 2 \end{bmatrix} + \begin{bmatrix} 2 \\ 2 \\ -1 \\ 0 \end{bmatrix}$ and

$\begin{bmatrix} 7 \\ 1 \\ -2 \\ 6 \end{bmatrix} = 3\begin{bmatrix} 1 \\ -1 \\ 0 \\ 2 \end{bmatrix} + 2\begin{bmatrix} 2 \\ 2 \\ -1 \\ 0 \end{bmatrix}$, so the new vectors are indeed

in $U$, and so form a basis (as $\dim U = 2$). Using this basis:

$$proj_U(X) = \frac{4}{15}\begin{bmatrix} 1 \\ 3 \\ -1 \\ 2 \end{bmatrix} + \frac{8}{90}\begin{bmatrix} 7 \\ 1 \\ -2 \\ 6 \end{bmatrix} = \frac{4}{9}\begin{bmatrix} 2 \\ 2 \\ -1 \\ 0 \end{bmatrix}.$$

**3. b)** By Gram-Schmidt, use the orthogonal basis $\{[1 \;\; -1 \;\; 0]^T, [1 \;\; 1 \;\; -2]^T\}$. Then the closest vector to $X$ is

$proj_U(X) = \frac{1}{2}\begin{bmatrix} 1 \\ -1 \\ 0 \end{bmatrix} + \frac{-3}{6}\begin{bmatrix} 1 \\ 1 \\ -2 \end{bmatrix} = \begin{bmatrix} 0 \\ -1 \\ 1 \end{bmatrix}$.

**d)** By Gram-Schmidt, use the orthogonal basis $\{[1 \;\; -1 \;\; 1 \;\; 0]^T, [1 \;\; 1 \;\; 0 \;\; 0]^T, [0 \;\; 0 \;\; 0 \;\; 1]^T\}$. Then the closest vector to $X = [3 \;\; 0 \;\; 5 \;\; -1]^T$ is

$proj_U(X) = \frac{8}{3}\begin{bmatrix} 1 \\ -1 \\ 1 \\ 0 \end{bmatrix} + \frac{3}{2}\begin{bmatrix} 1 \\ 1 \\ 0 \\ 0 \end{bmatrix} + \frac{-1}{1}\begin{bmatrix} 0 \\ 0 \\ 0 \\ 1 \end{bmatrix} = \frac{1}{6}\begin{bmatrix} 25 \\ -7 \\ 16 \\ -6 \end{bmatrix}$.

**4. b)** $A^T A = \begin{bmatrix} 6 & 6 \\ 6 & 27 \end{bmatrix}$, $A^T B = \begin{bmatrix} 7 \\ 14 \end{bmatrix}$; $(A^T A)Z = A^T B$ has

solution $Z = \begin{bmatrix} \frac{5}{6} \\ \frac{1}{3} \end{bmatrix}$.

**d)** $A^T A = \begin{bmatrix} 23 & 16 & -13 \\ 16 & 19 & -20 \\ -13 & -20 & 30 \end{bmatrix}$, $A^T B = \begin{bmatrix} 6 \\ 6 \\ -14 \end{bmatrix}$; $(A^T A)Z = A^T B$

has solution $Z = \begin{bmatrix} \frac{722}{1339} \\ -\frac{1722}{1339} \\ -\frac{1460}{1339} \end{bmatrix}$.

**5. b)** $M^T M = \begin{bmatrix} 4 & 3 & 11 \\ 3 & 11 & 27 \\ 11 & 27 & 83 \end{bmatrix}$, $M^T Y = \begin{bmatrix} 2 \\ -2 \\ -8 \end{bmatrix}$, so $Z = \begin{bmatrix} \frac{74}{55} \\ \frac{137}{220} \\ -\frac{21}{44} \end{bmatrix}$.

Hence $y = \frac{74}{55} + \frac{137}{220}x - \frac{21}{44}x^2$.

**6. b)** $M^T M = \begin{bmatrix} 5 & 6 & 30 & 84 \\ 6 & 30 & 84 & 354 \\ 30 & 84 & 354 & 1236 \\ 84 & 354 & 1236 & 4890 \end{bmatrix}$, $M^T Y = \begin{bmatrix} 5 \\ 3 \\ 9 \\ 39 \end{bmatrix}$,

$Z = \begin{bmatrix} \frac{429}{133} \\ -\frac{743}{798} \\ -\frac{219}{266} \\ \frac{13}{57} \end{bmatrix}$, $y = \frac{429}{133} - \frac{743}{798}x - \frac{219}{266}x^2 + \frac{13}{57}x^3$.

**8. b)** $U^\perp = span\{[5 \;\; 3 \;\; 1 \;\; 0], [-5 \;\; -2 \;\; 0 \;\; 1]\}$.

**9.** Throughout this exercise, write $proj_U(X) = P$.
**b)** True. If $P$ is in $U^\perp$, then $P$ is in $U \cap U^\perp = \{0\}$, so $P = 0$. Hence $X = X - P$ is in $U^\perp$ by the Projection theorem.
**d)** True. If $P = 0$, then $X = X - P$ is in $U^\perp$ by the Projection theorem.
**f)** False. $0 \subseteq \mathbb{R}^n$ but $0^\perp = \mathbb{R}^n \nsubseteq 0 = (\mathbb{R}^n)^\perp$.

**10.** $P = proj_U(X)$ is in $U$ for any choice of $X$, so if $X = P$, then $X$ is in $U$. Conversely, if $X$ is in $U$, then $P - X$ is in $U$ so, since $P - X$ is always in $U^\perp$, it follows that $P - X = 0$. Hence $P = X$, as required.

**12.** If $X$ is in $W^\perp$, then $X \cdot Y = 0$ for every $Y$ in $W$. Since $U \subseteq W$, this means that $X \cdot Y = 0$ for every $Y$ in $U$, that is $X$ is in $U^\perp$.

**14. b)** If $U = \mathbb{R}^n$, then any vector in $U^\perp$ is orthogonal to every vector in $U = \mathbb{R}^n$, in particular to itself. So $U^\perp = \{0\}$. Conversely, if $U^\perp = \{0\}$ and $X$ is in $\mathbb{R}^n$, let $P = proj_U(X)$. Then the Projection theorem shows that $X - P$ is in $U^\perp = \{0\}$, so $X = P$. Hence $X$ is in $U$.

**16. b)** Let $\{F_1, \cdots, F_m\}$ be an orthogonal basis of $U$, so that $proj_U(X) = \sum_{i=1}^m \frac{X \cdot F_i}{\|F_i\|^2} F_i$ for every $X$ in $\mathbb{R}^n$. Then

$proj_U(aX) = \sum_{i=1}^m \frac{(aX) \cdot F_i}{\|F_i\|^2} F_i = \sum_{i=1}^m a(\frac{X \cdot F_i}{\|F_i\|^2})F_i = a\sum_{i=1}^m \frac{X \cdot F_i}{\|F_i\|^2}F_i$
$= a \, proj_U(X)$.

**18.** Let $\{F_1, \cdots, F_k\}$ be an orthogonal basis of $U$, and extend it to an orthogonal basis $\{F_1, \cdots, F_k, F_{k+1}, \cdots, F_n\}$ of $\mathbb{R}^n$. By the preceding exercise we have $U^\perp = span\{F_{k+1}, \cdots, F_n\}$.

Thinking of the $F_i$ as columns, consider the $n \times n$ matrix $A$ with $F_{k+1}^T, \cdots, F_n^T$ as its first $n - k$ rows, and the rest of the rows zero. Then $AX = 0$ if and only if $F_i^T X = 0$ for $k + 1 \leq i \leq n$, if and only if $F_i \cdot X = 0$ for $k + 1 \leq i \leq n$, if and only if $X$ is in $(U^\perp)^\perp = U$. Hence $U = \{X \mid AX = 0\}$, as required.

**20. b)** $(I - E)^2 = I - 2E + E^2 = I - 2E + E = I - E$, and $(I - E)^T = I^T - E^T = I - E$.

# 4.7  ORTHOGONAL DIAGONALIZATION

**1. b)** False. Consider the matrix $\begin{bmatrix} \frac{1}{2} & -1 \\ \frac{1}{2} & 1 \end{bmatrix}$.

**d)** False. Consider the matrix $\frac{1}{\sqrt{2}}\begin{bmatrix} 1 & 1 \\ -1 & 1 \end{bmatrix}$.

**f)** True. $(P^T A P)^T = P^T A^T P^{TT} = P^T A^T P = P^T A P$ because $A^T = A$.

**2. b)** $\frac{1}{5}\begin{bmatrix} 3 & 4 \\ -4 & 3 \end{bmatrix}$  **d)** $\frac{1}{\sqrt{6}}\begin{bmatrix} \sqrt{2} & 0 & 2 \\ -\sqrt{2} & \sqrt{3} & 1 \\ \sqrt{2} & \sqrt{3} & -1 \end{bmatrix}$

**f)** $\frac{1}{7}\begin{bmatrix} 2 & 3 & -6 \\ 6 & 2 & 3 \\ -3 & 6 & 2 \end{bmatrix}$

**4.** The expansion is $Y = (Y \cdot X_1)X_1 + (Y \cdot X_2)X_2 + \cdots + (Y \cdot X_n)X_n$, so $Y = (Y \cdot X_1)X_1$ because $Y$ is orthogonal to $X_2, \cdots, X_n$. Since $\|Y\| = 1 = \|X_1\|$, this implies that $|Y \cdot X_1| = 1$, so $Y \cdot X_1 = \pm 1$. Hence $Y = \pm X_1$.

**5. b)** $P = \frac{1}{\sqrt{2}}\begin{bmatrix} -1 & 1 & 0 \\ 0 & 0 & \sqrt{2} \\ 1 & 1 & 0 \end{bmatrix}$; $P^T A P = diag(-4, 10, 5)$.

**d)** $P = \frac{1}{\sqrt{182}}\begin{bmatrix} 2\sqrt{14} & 3\sqrt{13} & 3 \\ 0 & -\sqrt{13} & 13 \\ -3\sqrt{14} & 2\sqrt{13} & 2 \end{bmatrix}$; $P^T A P = diag(1, 0, 14)$.

**f)** $P = \frac{1}{2}\begin{bmatrix} 1 & -1 & 0 & \sqrt{2} \\ -1 & 1 & 0 & \sqrt{2} \\ -1 & -1 & \sqrt{2} & 0 \\ 1 & 1 & \sqrt{2} & 0 \end{bmatrix}$; $P^T A P = diag(0, -4, 8, 8)$.

**6.** If $cP$ is orthogonal then $(cP)^{-1} = (cP)^T$, so $I = (cP)(cP)^T = c^2 PP^T = c^2 I$. Hence $c^2 = 1$, and $c = \pm 1$. Conversely, $-P$ is orthogonal because $(-P)^T(-P) = P^T P = I$.

**8. b)** Each row of $P$ contains exactly one 1 because no two columns of $I_n$ have a 1 in the same row. It follows that the rows of $P$ are the rows of $I_n$ in some order.

**10.** The $(i, j)$-entry of $A^T A$ equals $C_i^T C_j = C_i \cdot C_j$.

**12.** (ii) $\Rightarrow$ (iii). If $A^2 = I$, let $\lambda$ be an eigenvalue of $A$, say $AX = \lambda X$ for some column $X \neq 0$. Then $X = IX = A^2 X = A(\lambda X) = \lambda(AX) = \lambda(\lambda X) = \lambda^2 X$. Since $X \neq 0$ this means that $\lambda^2 = 1$, so $\lambda = \pm 1$.

**14.** If $\lambda \geq 0$ for every eigenvalue $\lambda$ of $A$, let $P^T A P = D = diag(\lambda_1, \lambda_2, \cdots, \lambda_n)$ where the $\lambda_i$ are the eigenvalues of $A$. Since the $\lambda_i$ are nonnegative by

hypothesis, write $D_1 = diag(\sqrt{\lambda_1}, \sqrt{\lambda_2}, \cdots, \sqrt{\lambda_n})$, so that $D_1^2 = D$. Now use the Hint.

**16. b)** If $A \overset{\circ}{\sim} B$, say $B = P^T A P$ where $P$ is orthogonal, then $B^T = P^T A^T P$; that is, $A^T \overset{\circ}{\sim} B^T$.

**d)** If $A \overset{\circ}{\sim} B$, say $B = P^T A P$ where $P$ is orthogonal, then $A$ and $B$ are similar (because $P^T = P^{-1}$) and so have the same eigenvalues by Theorem 6 §2.3.

**18. b)** If $P$ is orthogonal and symmetric, then $P^{-1} = P^T = P$. If $E = \frac{1}{2}(I - P)$, then $E^2 = \frac{1}{4}(I - 2P + P^2) = \frac{1}{4}(I - 2P + I) = E$, and $E^T = \frac{1}{2}(I^T - P^T) = \frac{1}{2}(I - P) = E$.

**20. b)** If $A^m = 0$, let $AX = \lambda X$, $X \neq 0$. Then $A^2 X = A(\lambda X) = \lambda(AX) = \lambda^2 X$. Continue to get $A^k X = \lambda^k X$ for every $k \geq 1$. Then $\lambda^m X = A^m X = 0$, so $\lambda^m = 0$, $\lambda = 0$. Hence, by Theorem 7, let $P^T A P = U$ be upper triangular where $P$ is orthogonal. Then $U^m = P^T A^m P = 0$. Because $U$ is triangular, it follows that every element $a$ on the main diagonal of $U$ also satisfies $a^m = 0$, so $a = 0$. Hence $U$ has zeros on the main diagonal.

**22.** Apply Theorem 7 to $A^T$, then transpose.

# 4.8  QUADRATIC FORMS

**1. b)** False. $\begin{bmatrix} 1 & 1 \\ 0 & 2 \end{bmatrix}$.

**d)** True. See the discussion following Theorem 3.

**f)** False. $\begin{bmatrix} 1 & 2 \\ 0 & 2 \end{bmatrix}$.

**2. b)** $\begin{bmatrix} 2 & 3 & 3 \\ 3 & 1 & -1 \\ 3 & -1 & 5 \end{bmatrix}$.

**3. b)** $y_1 = \frac{1}{\sqrt{5}}(x_1 - 2x_2)$, $y_2 = \frac{1}{\sqrt{5}}(2x_1 + x_2)$, $q = -2y_1^2 + 3y_2^2$.

**d)** $y_1 = \frac{1}{\sqrt{4+2\sqrt{2}}}[x_1 + (1 + \sqrt{2})x_2]$, $y_2 = \frac{1}{\sqrt{4-2\sqrt{2}}}[x_1 + (1 - \sqrt{2})x_2]$, $q = (1 + 2\sqrt{2})y_1^2 + (1 - 2\sqrt{2})y_2^2$.

**f)** $y_1 = \frac{1}{3}(2x_1 + x_2 + 2x_3)$, $y_2 = \frac{1}{\sqrt{5}}(x_1 - 2x_2)$, $y_3 = \frac{1}{\sqrt{5}}(-2x_2 + x_3)$, $q = 9y_2^2 + 9y_3^2$.

**4. b)** Ellipse. $y_1 = \frac{1}{\sqrt{5}}(-2x_1 + x_2)$, $y_1 = \frac{1}{\sqrt{5}}(x_1 + 2x_2)$.

**5. b)** Maximum $\frac{1}{2}(-1 + 5\sqrt{2})$, Minimum $-\frac{1}{2}(1 + 5\sqrt{2})$.

**d)** Maximum $-1 + 2\sqrt{3}$, Minimum $-1 - 2\sqrt{3}$.

**6. b)** $U = \frac{\sqrt{3}}{3}\begin{bmatrix} 3 & -2 \\ 0 & \sqrt{2} \end{bmatrix}$.  **d)** $U = \frac{\sqrt{5}}{5}\begin{bmatrix} 5 & 3 & -2 \\ 0 & 1 & 1 \\ 0 & 0 & \sqrt{15} \end{bmatrix}$.

**8. b)** $A = \begin{bmatrix} 3 & 2 \\ 2 & 1 \end{bmatrix}$ is not positive definite because $det A = -1$ is not positive (Theorem 4). However, $A^2 = \begin{bmatrix} 13 & 8 \\ 8 & 5 \end{bmatrix}$ is positive definite by Theorem 4 because $det A = 1 > 0$.

**d)** If $k \geq 3$ is odd and $A^k$ is positive definite, we show that $A^{k-2}$ is also positive definite. If $X \neq 0$ in $\mathbb{R}^n$, let $Y = A^{-1}X$ so that $Y \neq 0$ and $X = AY$. As before, the fact that $A$ is symmetric gives
$X^T A^{k-2} X = (AY)^T A^{k-2}(AY) = Y^T A^k Y > 0$ because $A^k$ is positive definite and $Y \neq 0$. Hence $A^{k-2}$ is positive definite.

**10.** Let $A$ and $B$ be positive definite $n \times n$ matrices. If $X \neq 0$ in $\mathbb{R}^n$, then $X^T(A + B)X = X^T AX + X^T BX > 0$ because both $X^T AX > 0$ and $X^T BX > 0$. Hence $A + B$ is positive definite.

**12. b)** If $(A^T A)X = \lambda X$ where $X \neq 0$, then
$\|AX\|^2 = (AX)^T(AX) = X^T(A^T AX) = X^T(\lambda X) = \lambda \|X\|^2$.
Since $\|X\| \neq 0$, this gives $\lambda = \frac{\|AX\|^2}{\|X\|^2} \geq 0$.

**14.** Follow the Hint, and take $B = PD_0 P^T$.

**16. b)** Suppose $A$ is positive definite with Cholesky factorization $A = U^T U$. Let $D$ be the diagonal matrix with the same (positive) diagonal entries as $U$, and write $U_1 = D^{-1}U$. Since the diagonal entries of $D^{-1}$ are the reciprocals of those of $U$, it follows that $U_1$ is unit upper triangular. But then $A = U^T U = (DU_1)^T(DU_1) = U_1^T D^2 U_1$ is a factorization of the desired type.

Conversely, suppose $A = U^T DU$ is such a factorization. If $X \neq 0$ in $\mathbb{R}^n$, then $X^T AX = (UX)^T D(UX) > 0$ because $UX \neq 0$ (as $U$ is invertible) and $D$ is positive definite (by Theorem 4). Hence $A$ is positive definite. [Note that this shows that $U^T PU$ is positive definite whenever $P$ is positive definite and $U$ is invertible.]

**18. b)** We have $X = UY$, so
$q = X^T AX = (UY)^T A(UY) = Y^T(U^T AU)Y = Y^T BY$
where $B = U^T AU$.

## 4.9 LINEAR TRANSFORMATIONS

**1. b)** True. By Theorem 4.
**d)** False. Consider the transformation in **b)** with $X = \vec{e}_1$ and $Y = \vec{e}_2$.
**f)** True. If $S$ and $T$ have invertible standard matrices $A$ and $B$, respectively, then $S \circ T$ has matrix $AB$, which is invertible.

**2.** If $X = (1 - f)Y + fZ$ where $0 \leq f \leq 1$, then $T(X) = (1 - f)T(Y) + fT(Z)$.

**4.** The standard matrix is $A = [I_m \ \ 0]_{m \times n}$ in block form.

**6. b)** Standard matrix $\begin{bmatrix} 4 & -2 & 0 & \frac{3}{2} \\ -4 & 3 & -1 & -\frac{3}{2} \end{bmatrix}$.

**7. b)** $T\begin{bmatrix} x_1 \\ x_2 \\ x_3 \end{bmatrix} = \frac{1}{9}\begin{bmatrix} 53x_1 - 47x_2 + 25x_3 \\ -29x_1 + 23x_2 + 41x_3 \\ -20x_1 - 4x_2 + 23x_3 \end{bmatrix}$

**8.** Since $X \neq 0$, the set $\{X\}$ is independent and so is part of a basis $\{X, X_2, \cdots, X_n\}$ of $\mathbb{R}^n$ by Theorem 3 §4.3. Now use Theorem 4.

**10.** Write $X = r_1 Y_1 + \cdots + r_k Y_k$ where each $r_i$ is in $\mathbb{R}$. Hence $T(X) = r_1 T(Y_1) + \cdots + r_k T(Y_k) = r_1 0 + \cdots + r_k 0 = 0$ for every $X$ in $\mathbb{R}^n$.

**12. b)** If $X$ is in $\ker T$, then
$X = 1_{\mathbb{R}^n}(X) = (S \circ T)(X) = S(T(X)) = S(0) = 0$.

**14. b)** Let $Y = T(X)$, $X$ in $\mathbb{R}^n$.
By hypothesis, $X = r_1 X_1 + \cdots + r_k X_k$ for $r_i$ in $\mathbb{R}$, so $Y = T(X) = r_1 T(X_1) + \cdots + r_k T(X_k)$.
Hence $im T \subseteq span\{T(X_1), \cdots, T(X_k)\}$; the other inclusion follows from Theorem 1 §4.1.

**16.** Using the notation of Theorem 9, show that
$[T(X + Y) - T(X) - T(Y)] \cdot T(F_i) = 0$ for each $i$.
Since $\{T(F_1), T(F_2), \cdots, T(F_n)\}$ is a basis of $\mathbb{R}^n$, we have $T(X + Y) = T(X) + T(Y)$.

**18. b)** If $A$ and $B$ are the (invertible) matrices of $S$ and $T$, respectively, then the matrix of $S \circ T$ is $AB$ which is invertible. Use Theorem 5.

**20.** $(T \circ T)(X) = T(T(X)) = A(AX) = A^2 X = AX = T(X)$ for all $X$ in $\mathbb{R}^n$. So $T \circ T = T$.

**22.** $(T \circ Q_Y)(X) = T[Q_Y(X)] = T[X + Y] = T(X) + T(Y) = Q_{T(Y)}[T(X)] = (Q_{T(Y)} \circ T)(X)$ for all $X$.

**24.** Choose a unit vector $\vec{n}$ normal to the plane, and choose two orthogonal unit vectors $\vec{u}$ and $\vec{v}$ in the plane. Then $\mathcal{B} = \{\vec{n}, \vec{u}, \vec{v}\}$ is a basis of $\mathbb{R}^3$ (it is orthonormal),
$M_{\mathcal{B}}(T) = \begin{bmatrix} -1 & 0 & 0 \\ 0 & 1 & 0 \\ 0 & 0 & 1 \end{bmatrix}$ because $T(\vec{n}) = -\vec{n}, T(\vec{u}) = \vec{u}$,
$T(\vec{v}) = \vec{v}$, and $M_{\mathcal{B}}(T)$ is similar to the standard matrix of $T$.

**26.** By Lemma 4, $C_{\mathcal{F}}(X) = P^{-1}X$, so $AC_{\mathcal{F}}(X) = BC_{\mathcal{F}}(X)$ becomes $AP^{-1}X = BP^{-1}X$ for all $X$. Hence $AP^{-1} = BP^{-1}$ by Lemma 1, so $A = B$.

**28. b)** $\|(aT)(X)\|^2 = a^2 \|T(X)\|^2$ for all $X$ in $\mathbb{R}^n$, so $\|T(X)\|^2 = \|X\|^2$ for all $X$ if and only if $a^2 = 1$; that is, if and only if $a = \pm 1$.

**30.** 2) Given $X$ in $\mathbb{R}^n$, write $X = T(Y)$ where $Y$ is in $\mathbb{R}^n$. Then $T^{-1}(X) = T^{-1}[T(Y)] = Y$ by the preceding exercise. But then, $\|T^{-1}(X)\| = \|Y\| = \|T(Y)\| = \|X\|$, where $\|T(Y)\| = \|Y\|$ because $T$ is an isometry. This shows that $T^{-1}$ is an isometry.

**32.** Theorem 7 states that $C_{\mathcal{F}}[T(X)] = M_{\mathcal{F}}(T)C_{\mathcal{F}}(X)$ for all $X$ in $\mathbb{R}^n$, and Theorem 6 states that $P_{\mathcal{G} \leftarrow \mathcal{F}} C_{\mathcal{F}}(X) = C_{\mathcal{G}}(X)$ for all $X$. By the Hint, we are to show that
$P_{\mathcal{G} \leftarrow \mathcal{F}} M_{\mathcal{F}}(T) = M_{\mathcal{G}}(T)P_{\mathcal{G} \leftarrow \mathcal{F}}$. If $X$ is arbitrary in $\mathbb{R}^n$, multiply the left side by $C_{\mathcal{F}}(X)$ to get:
$$P_{\mathcal{G} \leftarrow \mathcal{F}} M_{\mathcal{F}}(T)C_{\mathcal{F}}(X) = P_{\mathcal{G} \leftarrow \mathcal{F}} C_{\mathcal{F}}[T(X)] = C_{\mathcal{G}}[T(X)]$$
$$= M_{\mathcal{G}}(T)C_{\mathcal{G}}(X) = M_{\mathcal{G}}(T)P_{\mathcal{G} \leftarrow \mathcal{F}} C_{\mathcal{F}}(X).$$
Hence $[P_{\mathcal{G} \leftarrow \mathcal{F}} M_{\mathcal{F}}(T)]C_{\mathcal{F}}(X) = [M_{\mathcal{G}}(T)P_{\mathcal{G} \leftarrow \mathcal{F}}]C_{\mathcal{F}}(X)$ for all $X$ in $\mathbb{R}^n$. Since we may "cancel" $C_{\mathcal{F}}(X)$ from both sides (by Exercise 26), the result follows.

## 4.10   COMPLEX MATRICES

**1. b)** False. $\frac{1}{\sqrt{2}}\begin{bmatrix} 1 & -1 \\ 1 & 1 \end{bmatrix}$ is unitary but not Hermitian.

   **d)** True. This is part of Theorem 4.

   **f)** True. If $U$ and $V$ are unitary, then
$(UV)^* = V^*U^* = V^{-1}U^{-1} = (UV)^{-1}$.

**2. b)** $\|Z\| = \sqrt{5 + 9 + 2} = 4$, $\|W\| = \sqrt{4 + 5 + 1} = \sqrt{10}$,
$\langle Z, W \rangle = 2 + 7i$, so not orthogonal.

   **d)** $\|Z\| = \sqrt{5 + 25 + 5 + 4} = \sqrt{39}$,
$\|W\| = \sqrt{5 + 2 + 9 + 10} = \sqrt{26}$,
$\langle Z, W \rangle = 0$, so orthogonal.

**3. b)** Hermitian (and normal), not unitary.

   **d)** Unitary (and normal), not Hermitian.

**4. b)** $U = \frac{1}{\sqrt{14}}\begin{bmatrix} -2 & 3-i \\ 3+i & 2 \end{bmatrix}$, $U^*ZU = \begin{bmatrix} -1 & 0 \\ 0 & 6 \end{bmatrix}$

   **d)** $U = \frac{1}{\sqrt{3}}\begin{bmatrix} \sqrt{3} & 0 & 0 \\ 0 & 1+i & 1 \\ 0 & -1 & 1-i \end{bmatrix}$, $U^*ZU = \begin{bmatrix} 1 & 0 & 0 \\ 0 & 0 & 0 \\ 0 & 0 & 3 \end{bmatrix}$

**6.** Write $U = \begin{bmatrix} a & b \\ c & d \end{bmatrix}$ and suppose that $U^{-1}AU = T = \begin{bmatrix} p & z \\ 0 & q \end{bmatrix}$.
We have $AU = UT$ and comparing entries gives $-c = ap$
and $a = cp$. Hence $a(1 + p^2) = 0 = c(1 + p^2)$, so $a = 0 = c$
if $p$ is real; contradiction. But $p$ is indeed real because both
$a$ and $c$ are real and they are not both zero ($U$ is invertible).

**8. b)** If $H$ is Hermitian, then $H^* = H$, so it is clear that
$HH^* = H^*H$. Hence $H$ is normal.

**9. b)** $U = diag(u_1, \cdots, u_n)$ is unitary if and only if $|u_i| = 1$
for each $i$.

**10. b)** If $ZZ^* = I$ and $H = Z^*Z$, then
$H^2 = Z^*ZZ^*Z = Z^*IZ = Z^*Z = H$.

**12.** Observe that $\langle X, Y \rangle = X^T \overline{Y}$ for all columns $X$ and $Y$
in $\mathbb{C}^n$. We have $\langle X, Z^*Y \rangle = X^T \overline{(Z^*Y)} = X^T \overline{Z^*}\, \overline{Y}$.
But $\overline{Z^*} = \overline{\left[\overline{Z^T}\right]} = Z^T$, so we get
$\langle X, Z^*Y \rangle = X^T Z^T \overline{Y} = (ZX)^T \overline{Y} = \langle ZX, Y \rangle$.

**14.** Using Theorem 3: $Z^*(Z^{-1})^* = (Z^{-1}Z)^* = I^* = I$.

**16.** If $G$ is both Hermitian and unitary, then $G^{-1} = G^* = G$,
so $G^2 = I$. Hence if $GX = \lambda X$ for some $X \neq 0$, then
$X = G^2X = G(\lambda X) = \lambda^2 X$. It follows that $\lambda^2 = 1$, so
$\lambda = \pm 1$.

**18.** Since $NN^* = N^*N$, Theorem 3 gives
$(zN)(zN)^* = z\bar{z}NN^* = \bar{z}zN^*N = \bar{z}N^*(zN) = (zN)^*(zN)$.

**20. b)** If $A = [a_{jk}]$ and $B = [b_{jk}]$, then $A + iB = 0$ becomes
$[a_{jk} + ib_{jk}] = 0$, whence $a_{jk} + ib_{jk} = 0$ for all $j$ and $k$.
This forces $a_{jk} = 0 = b_{jk}$ for all $j$ and $k$, and hence
$A = 0 = B$.

**22. b)** Let $Z$ be nilpotent, say $Z^k = 0$ for some $k \geq 1$. The
polynomial $c_Z(x)$ has only the eigenvalues of $Z$ as its
roots, and so its only root is 0. Hence $c_Z(x) = x^n$.

**24. b)** First $detZ = \lambda_1\lambda_2 \cdots \lambda_n$ by the Corollary to Schur's
theorem, and so $detZ$ is real by Theorem 4 (because
$Z$ is Hermitian). Hence $det(Z^*) = det[(\overline{Z})^T] =$
$det(\overline{Z}) = detZ$.

## 4.11   SINGULAR VALUE DECOMPOSITION

**2.** If $\sigma_1 \geq \cdots \geq \sigma_r$ are the nonzero singular values of $A$, write
$D = diag(\sigma_1, \cdots, \sigma_r)$. If $A$ is $m \times n$, let $A = P\begin{bmatrix} D & 0 \\ 0 & 0 \end{bmatrix}_{m \times n} Q^T$
be a singular value decomposition of $A$. Then
$A^T = Q\begin{bmatrix} D^T & 0 \\ 0 & 0 \end{bmatrix}_{n \times m} P^T = Q\begin{bmatrix} D & 0 \\ 0 & 0 \end{bmatrix}_{n \times m} P^T$. It follows by
Theorem 2 that the entries of $D$ are the singular values of $A^T$.

**4.** If $A = P\Sigma Q^T$ is a singular value decomposition of $A$ then,
since the determinant of any orthogonal matrix is $\pm 1$, we
have $detA = (\pm 1) det\Sigma$. Hence $|detA| = det\Sigma$ is the prod-
uct of the singular values of $A$.

**6.** By Theorem 3 §4.4, $rankA = n$ if and only if $A^TA$ is
invertible. But $A^TA$ is invertible if and only if all its
eigenvalues are nonzero, and this holds if and only if the
singular values of $A$ are all nonzero.

**8.** Let $\lambda_1, \cdots, \lambda_n$ be the eigenvalues of the symmetric
matrix $A^TA$, and observe that each $\lambda_i > 0$ by Example 2
and the fact that $A$ is invertible. Since $A^TA$ is symmetric,
let $Q^T(A^TA)Q = diag(\lambda_1, \cdots, \lambda_n)$ where $Q$ is orthogonal
and (by changing the order of the columns of $Q$)
$\lambda_1 \geq \cdots \geq \lambda_n$. Hence $D = diag(\sqrt{\lambda_1}, \cdots, \sqrt{\lambda_n})$ has
positive entries and, writing $P = AQD^{-1}$, we have
$PDQ^T = (AQ)Q^T = A$. One verifies that $P^TP = I$.

**10.** If $A$ is symmetric with (real) eigenvalues $\lambda_1, \cdots, \lambda_n$,
we show more generally that the singular values of $A$
are $|\lambda_1|, \cdots, |\lambda_n|$. If $AX = \lambda_iX$ where $X \neq 0$, then
$(A^TA)X = A^2X = A(\lambda_iX) = \lambda_i(AX) = \lambda_i^2X$. This shows
that $|\lambda_i| = \sqrt{\lambda_i^2}$ is a singular value of $A$. Conversely,
observe that $x^2I - A^2 = -(xI - A)(-xI - A)$, so that
$c_{A^2}(x^2) = (-1)^n c_A(x)c_A(-x)$. Now suppose that $\sigma$ is a
singular value of $A$, that is $\sigma$ is an eigenvalue of $A^TA = A^2$.
Hence $0 = c_{A^2}(\sigma^2) = (-1)^n c_A(\sigma)c_A(-\sigma)$, so either
$c_A(\sigma) = 0$ or $c_A(-\sigma) = 0$. In other words, either $\sigma = \lambda_i$
or $-\sigma = \lambda_i$ for some $i$; that is, $\sigma = |\lambda_i|$ for some $i$.

**12.** Let $\lambda$ be an eigenvalue of $G$. Then $|\lambda| = 1$ because $G$ is
orthogonal (by Theorem 3 §4.7), so $\lambda = 1$ because $\lambda \geq 0$
($G$ is positive definite). Since $G$ is diagonalizable (being
symmetric), let $P^{-1}GP = D$ where $D$ is diagonal. But
$D = I$ because the entries of $D$ are eigenvalues of $G$, so
$G = PDP^{-1} = I$.

**14.** If $\lambda$ is an eigenvalue of $kG$, then $(kG)X = \lambda X$ for some $X \neq 0$. Hence $GX = \frac{\lambda}{k}X$, so $\frac{\lambda}{k}$ is an eigenvalue of $G$. Hence $\frac{\lambda}{k} \geq 0$ because $G$ is positive semidefinite, so $\lambda \geq 0$ because $k > 0$. Since $kG$ is symmetric (because $G$ is symmetric), we are done.

**16.** $GP = \begin{bmatrix} 0 & \sqrt{2} \\ 0 & \sqrt{2} \end{bmatrix} = GQ$, $P$ and $Q$ are orthogonal, and $G$ is positive semidefinite (eigenvalues are 0 and 2).

**18. b)** Let $\{Q_1, \cdots, Q_n\}$ be an orthonormal basis of $\mathbb{R}^n$. Given $X = r_1Q_1 + \cdots + r_nQ_n$, $r_i$ in $\mathbb{R}$, then $\|X\|^2 = \Sigma_i r_i^2$. Since $BX = \Sigma_i r_i BQ_i = \Sigma_i r_i \lambda_i Q_i$, a similar computation gives $X^TBX = \Sigma_i r_i^2 \lambda_i$. Hence the fact that $\lambda_i \geq \lambda_n$ for each $i$ gives $X^TBX \geq \lambda_n \|X\|^2$. It follows that $R(X) \geq \lambda_n$. Similarly, $R(X) \leq \lambda_1$.

# CHAPTER 5

## Vector Spaces

## 5.1 EXAMPLES AND BASIC PROPERTIES

**1. b)** True. Every multiple of any nonzero vector.
**d)** True. $1 = \sin^2 x + \cos^2 x$ for all real $x$.
**f)** False. $\frac{1}{2} \cdot 3$ not in this set.
**h)** False. $\mathbb{R}$ has no proper subspaces.

**2. b)** Not a vector space. Axiom S5 fails (and no others).
**d)** Not a vector space. Axioms S4 and S5 fail (and no others).
**f)** Not a vector space. Axioms S4 and S5 fail (and no others).

**3. b)** This is a vector space. The axioms are rules of complex arithmetic.
**d)** Not a vector space. Axioms A1, A4, A5, and S1 all fail.
**f)** Not a vector space. Axiom S3 fails (and no others).
**h)** Not a vector space. Axioms S2 and S3 fail (and no others).

**4.** The zero vector is $[0 \ \ -1]$, the negative of $[x \ \ y]$ is $[-x \ \ -2 - y]$.

**6. b)** Not a subspace.   **d)** Subspace.
**f)** Subspace.

**7. b)** Subspace.   **d)** Subspace.
**f)** Not a subspace.

**8. b)** Subspace.   **d)** Not a subspace.
**f)** Subspace.   **h)** Subspace.

**9.** Write $U = span\{\mathbf{u}, \mathbf{w}\}$.
**b)** $\mathbf{v}_1$ is in $U$; $\mathbf{v}_2$ is not in $U$.
**d)** $\mathbf{v}_1$ is not in $U$; $\mathbf{v}_2$ is in $U$.
**f)** $\mathbf{v}_1$ is in $U$; $\mathbf{v}_2$ is not in $U$.

**10. b)** No. 1 is not in the span of these vectors.

**11. b)** $\mathbf{x} = 2\mathbf{t}$, $\mathbf{y} = 0$, $\mathbf{z} = \mathbf{t}$, $\mathbf{t}$ arbitrary.

**12. b)** If $Y_0 = [y_1 \ \cdots \ y_n]$ and $y_k \neq 0$, and if $\{E_1, \cdots, E_n\}$ is the standard basis of $\mathbb{R}^m$, then $E_j = AY_0$ where $A$ has $(i, j)$-entry $y_k^{-1}$ and all other entries zero. So $E_j \in U$, whence $\mathbb{R}^m \subseteq U$.

**14. b)** If $\mathbf{v} + \mathbf{v}' = \mathbf{0}$ then $-\mathbf{v} + (\mathbf{v} + \mathbf{v}') = -\mathbf{v} + \mathbf{0}$, so $(-\mathbf{v} + \mathbf{v}) + \mathbf{v}' = -\mathbf{v}$. Thus $\mathbf{0} + \mathbf{v}' = -\mathbf{v}$, that is $\mathbf{v}' = -\mathbf{v}$.

**16.** $a(-\mathbf{v}) = a[(-1)\mathbf{v}] = [a(-1)]\mathbf{v} = (-a)\mathbf{v}$.

**18.** Let $U \neq 0$ be a subspace of $V$. If $\mathbf{0} \neq \mathbf{u} \in U$, then $\mathbf{u} = a\mathbf{v}$, $a \in \mathbb{R}$, so $\mathbf{v} = a^{-1}\mathbf{u} \in \mathbb{R}\mathbf{u}$. Hence $\mathbb{R}\mathbf{v} \subseteq U$, so $\mathbb{R}\mathbf{v} = U$.

**20. b)** $f_a = af_1$ is in $\mathbb{R}f_1$ for all $a$, so $U \subseteq \mathbb{R}f_1$. Clearly $\mathbb{R}f_1 \subseteq U$.

**22. b)** If $a \neq 0$, then $\mathbf{v} - \mathbf{w} = \mathbf{0}$, so $\mathbf{v} = \mathbf{w}$.

**24. b)** Write $U = span\{1 + x^3, x + x^2, 1 + x + x^2, 2x - x^3\}$. Then $1 = (1 + x + x^2) - (x + x^2)$ is in $U$; so $x^3 = (1 + x^3) - 1$ is in $U$, whence $x = \frac{1}{2}\{(2x - x^3) + x^3\}$ is in $U$, and finally $x^2 = (x + x^2) - x$ is in $U$. Hence $\mathbb{P}_3 = span\{1, x, x^2, x^3\} \subseteq U$. Clearly $U \subseteq \mathbb{P}_3$.

**26.** Since $Y \neq 0$, let $AY \neq 0$ with $A \in \mathbb{M}_{m,n}$ (if entry $j$ of $Y$ is nonzero, let the $(1, j)$-entry of $A$ be 1 and let the other entries be 0). Now suppose that $\mathbb{M}_{m,n} = span\{A_1, \cdots, A_k\}$, so that $A = r_1A_1 + \cdots + r_kA_k$, $r_i \in \mathbb{R}$. Then $AY = r_1A_1Y + \cdots + r_kA_kY = 0$ by hypothesis, a contradiction. Hence $\mathbb{M}_{m,n} \neq span\{A_1, \cdots, A_k\}$.

**28.** If $\mathcal{F}$ is a finite set of polynomials, let $n$ be the maximum of the degrees of the polynomials in $\mathcal{F}$. Then each of these polynomials is in $\mathbb{P}_n$, so $span\{\mathcal{F}\} \subseteq \mathbb{P}_n$. Hence $span\{\mathcal{F}\}$ cannot equal $\mathbb{P}$.

**30.** Compute $(1 + 1)(\mathbf{v} + \mathbf{w})$ twice, first using Axiom S2 first, and then using Axiom S3 first.

## 5.2   INDEPENDENCE AND DIMENSION

**1. b)** True.      **d)** False.      **f)** False.
   **h)** False.      **j)** True.       **l)** False.

**2. b)** Independent.      **d)** Not independent.
   **f)** Independent.      **h)** Independent.

**3. b)** Dimension 2.   **d)** Dimension 3.   **f)** Dimension 2.
   **h)** Dimension 3.   **j)** Dimension 2.   **l)** Dimension 3.

**4. b)** If $\mathbb{P}_n = span\{p_1(x), \cdots, p_k(x)\}$ where the coefficients of each $p_i(x)$ sum to zero, then $p_i(1) = 0$ for all $i$. But then $p(1) = 0$ for every polynomial $p(x)$ in $span\{p_1(x), \cdots, p_k(x)\} = \mathbb{P}_n$, a contradiction.

**6. b)** Let $z = a + bi$ where $a \neq 0$ and $b \neq 0$ by hypothesis. If $rz + s\bar{z} = 0$, $r$, $s$ in $\mathbb{R}$, then equating real and imaginary parts gives $(r + s)a = 0$ and $(r - s)b = 0$. Thus $r + s = 0 = r - s$, which implies that $r = s = 0$. Thus $\{z, \bar{z}\}$ is independent, and so is a basis of $\mathbb{C}$ because $dim(\mathbb{C}) = 2$.

**8.** As $V$ is finite dimensional, let $V = span\{v_1, \cdots, v_m\}$. Then each $v_i$ is a linear combination of a finite set $S_i$ of vectors in $S$. Hence $S_1 \cup S_2 \cup \cdots \cup S_m$ is a finite spanning set of $V$, and so contains a basis of $V$ by Theorem 3.

**10.** If $rA + sB = 0$, $r$, $s$ in $\mathbb{R}$, transpose to get $rA - sB = 0$. Adding these equations gives $2rA = 0$, whence $r = 0$ because $A \neq 0$. Then $sB = 0$ from the first equation, so $s = 0$ because $B \neq 0$.

**12. b)** Given $C$ in $\mathbb{M}_{m,n}$, the matrix $P^{-1}CQ^{-1}$ is also in $\mathbb{M}_{m,n}$ so write $P^{-1}CQ^{-1} = s_1A_1 + \cdots + s_kA_k$. Then $C = s_1PA_1Q + \cdots + s_kPA_kQ$, $s_i \in \mathbb{R}$, and it follows that $span\{PA_1Q, \cdots, PA_kQ\} = \mathbb{M}_{m,n}$.

**14.** Let $p(x)$ and $q(x)$ have, respectively, degrees $n$ and $m$, and leading coefficients $a_n \neq 0$ and $b_m \neq 0$. If $rp(x) + sq(x) + tp(x)q(x) = 0$, $r$, $s$, $t$ in $\mathbb{R}$, the highest power of $x$ on the left is $x^{m+n}$, and so its coefficient $ta_nb_m = 0$. Hence $t = 0$, so $rp(x) + sq(x) = 0$. Now $r = s = 0$ by hypothesis.

**16. b)** If $r(a + bx) + s(a_1 + b_1x) = 0$, then $ra + sa_1 = 0 = rb + sb_1$, so $r[a \ b]^T + s[a_1 \ b_1]^T = 0$. Hence $r = s = 0$ because $\{[a \ b]^T, [a_1 \ b_1]^T\}$ is independent. This proves that $\{a + bx, a_1 + b_1x\}$ is independent, and so is a basis of $\mathbb{P}_1$, by Theorem 4 because $dim(\mathbb{P}_1) = 2$.

**18.** $\left\{\begin{bmatrix} 1 & 0 \\ 0 & 0 \end{bmatrix}, \begin{bmatrix} 0 & 0 \\ 0 & 1 \end{bmatrix}, \begin{bmatrix} 1 & 1 \\ 0 & 0 \end{bmatrix}, \begin{bmatrix} 1 & 0 \\ 1 & 0 \end{bmatrix}\right\}$.

**20. b)** Show that $O_n = span\{x, x^3, x^5, \cdots\}$ by comparing coefficients in $p(x) = -p(-x)$. If $n = 2k$ is even, then $O_n = span\{x, x^3, x^5, \cdots, x^{2k-1}\}$ so $dim(O_n) = k = \frac{n}{2}$. If $n = 2k+1$ is odd, then $O_n = span\{x, x^3, x^5, \cdots, x^{2k+1}\}$, so $dim(O_n) = k + 1 = \frac{n+1}{2}$.

**22.** If $U \neq 0$ and $U \neq V$, then $dimU = 1$. So $U = \mathbb{R}v$ where $\{v\}$ is any basis of $U$. If $V = \mathbb{R}^2$, this says that the proper subspaces $U$ are lines $U = \mathbb{R}v$ (with direction vector $v$).

**24. b)** Yes. If $r(u + v) + s(v + w) + t(w + u) = 0$, then $0 = r + t = r + s = s + t$, so $r = s = t = 0$.
   **d)** Yes. If $r(u + v) + s(v + w) + t(w + u) + x(z + u) = 0$, then $r + t + x = 0$, $r + s = 0$, $s + t = 0$ and $x = 0$, whence $r = s = t = x = 0$.

**26.** $\{v\}$ is independent because $v \neq 0$, so Theorem 3 applies.

**28.** If $r_1(w + v_1) + r_2(w + v_2) + \cdots + r_k(w + v_k) = 0$, $(r_1 + r_2 + \cdots + r_k)w + r_1v_1 + r_2v_2 + \cdots + r_kv_k = 0$. Hence each $r_i = 0$ by Lemma 1.

**30.** Let $U = span\{v_1, v_2, \cdots, v_k\}$. If $U \neq V$, choose $v \in V$ such that $v \notin U$. Then $\{v, v_1, v_2, \cdots, v_k\}$ is independent by Lemma 1, contrary to the maximality of the independent set $\{v_1, v_2, \cdots, v_k\}$.

**32.** Let $\{v_1, v_2, \cdots, v_k\}$ be a basis of $U$, and extend it to a basis $\{v_1, v_2, \cdots, v_k, v_{k+1}, \cdots, v_m\}$ of $W$ by Theorem 3. Take $X = span\{v_1, v_2, \cdots, v_k, v_{k+1}, \cdots, v_l\}$.

**34. b)** We prove that $x_k = x_0e_k + x_1f_k$ for all $k \geq 0$ by induction on $k$. It holds for $k = 0$ and for $k = 1$ by the choice of $e_0$, $e_1$, $f_0$, and $f_1$. If it holds for $k$ and $k + 1$, then one verifies that $x_{k+2} = ax_k + bx_{k+1} = x_0e_{k+1} + x_1f_{k+1}$, so the result holds for $k + 1$, and the induction goes through.

## 5.3   LINEAR TRANSFORMATIONS

**1. b)** False.      **d)** True.      **f)** False.
   **h)** True.       **j)** True.      **l)** False.

**2. b)** False.      **d)** True.
   **f)** False.      **h)** False.

**3. b)** $T(A + B) = P(A + B)Q = PAQ + PBQ = T(A) + T(B)$, and $T(rA) = P(rA)Q = rPAQ = rT(A)$.
   **d)** $T(X + X_1) = Y \bullet (X + X_1) = Y \bullet X + Y \bullet X_1 = T(X) + T(X_1)$, and $T(rX) = Y \bullet (rX) = r(Y \bullet X) = rT(X)$.
   **f)** Adding two linear combinations adds the coefficients of $e_1$; and multiplying a linear combination by $r$ multiplies the coefficient of $e_1$ by $r$.
   **h)** Adding two matrices adds their second columns; and multiplying a matrix by $r$ multiplies its second column by $r$.

**4. b)** Not linear. $rank(A + B)$ need not equal $rankA + rankB$. For example $A = \begin{bmatrix} 1 & 0 \\ 0 & 1 \end{bmatrix}$ and $B = \begin{bmatrix} 0 & 1 \\ 1 & 0 \end{bmatrix}$. Note that $rank(rA) = r \cdot rankA$ always holds.

**5. b)** $T(v) = \begin{bmatrix} a+c \\ 2a+b \end{bmatrix}$.

**6.** $T(3e + f) = 18e - 3f$ and $T(f) = 3e$.

**8.** $T(\mathbf{v} - \mathbf{v}_1) = T[\mathbf{v} + (-1)\mathbf{v}_1]$     Definition of subtraction

$\qquad = T(\mathbf{v}) + T[(-1)\mathbf{v}_1]$     Axiom T1

$\qquad = T(\mathbf{v}) + (-1)T(\mathbf{v}_1)$     Axiom T2

$\qquad = T(\mathbf{v}) - T(\mathbf{v}_1).$     Definition of subtraction

**10. b)** $T$ is clearly a transformation $\mathbb{F}[0, 1] \to \mathbb{R}$; it is linear by properties of the integral.

**12.** (b)$\Rightarrow$(c). Given any vector $\mathbf{v}$ in $V$, $T(\mathbf{v})$ is in $imT = \{\mathbf{0}\}$, so $T(\mathbf{v}) = \mathbf{0}$. Thus $T = 0$.

**14. b)** $imT = \{T(X) \mid X \text{ in } \mathbb{R}^n\} = \{r_1\mathbf{v}_1 + \cdots + r_n\mathbf{v}_n \mid r_i \text{ in } \mathbb{R}\}$
$= span\{\mathbf{v}_1, \cdots, \mathbf{v}_n\}$. So $T$ is onto if and only if $imT = V$, if and only if $\{\mathbf{v}_1, \cdots, \mathbf{v}_n\}$ spans $V$.

**15. b)** Let $\{\mathbf{b}_1, \cdots, \mathbf{b}_k\}$ be a basis of $U$, and extend it to a basis $\{\mathbf{b}_1, \cdots, \mathbf{b}_k, \mathbf{c}_1, \cdots, \mathbf{c}_m\}$ of $V$. Define $S : V \to V$ by $S(\mathbf{b}_i) = \mathbf{b}_i$ for each $i$ and $S(\mathbf{c}_j) = \mathbf{0}$ for all $j$. If $\mathbf{v} = r_1\mathbf{b}_1 + \cdots + r_k\mathbf{b}_k + s_1\mathbf{c}_1 + \cdots + s_m\mathbf{c}_m$, then $S(\mathbf{v}) = r_1\mathbf{b}_1 + \cdots + r_k\mathbf{b}_k$ is in $U$. Thus $imS \subseteq U$. But each $\mathbf{b}_i = S(\mathbf{b}_i)$ is in $imS$, so $U \subseteq imS$ because $U = span\{\mathbf{b}_1, \cdots, \mathbf{b}_k\}$.

**16.** $T$ is onto because, given $r$ in $\mathbb{R}$, $r = tr(A)$ where $A$ has $(1, 1)$-entry $r$ and all other entries 0. So the Dimension theorem gives $dim(\mathbb{M}_{n, n}) = dim(kerT) + dim(imT)$, that is $n^2 = dim(kerT) + 1$.

**18.** If the system is $AX = 0$ where $A$ is $m \times n$, define $T : \mathbb{R}^n \to \mathbb{R}^m$ by $T(X) = AX$. Then $T$ is linear and $kerT$ is the space of all solutions. The Dimension theorem gives $n = dim(\mathbb{R}^n) = dim(kerT) + dim(imT) \leq dim(kerT) + m$ because $imT \subseteq \mathbb{R}^m$. Hence $dim(kerT) \geq n - m > 0$ by hypothesis, so $kerT \neq 0$. This means that there are non-trivial solutions to $AX = 0$.

**20.** Define $T : \mathbb{M}_{n, n} \to \mathbb{M}_{n, n}$ by $T(A) = A - A^T$. By Example 11, $kerT = U$ and $imT = W$. Thus the Dimension theorem gives $n^2 = dim(\mathbb{M}_{n, n}) = dimU + dimW$.

**22.** Define $T : \mathbb{M}_{n, n} \to \mathbb{M}_{n, n}$ by $T(B) = B^T - 3B$. Then $T$ is linear (verify) and we claim that $T$ is one-to-one. For if $T(B) = 0$ then $B^T = 3B$, whence $B = (B^T)^T = (3B)^T = 3B^T = 9B$, so $B = 0$. Hence $T$ is onto (Corollary to the Dimension theorem) so every matrix $A$ in $\mathbb{M}_{n, n}$ has the form $A = T(B) = B^T - 3B$ for some $B \in \mathbb{M}_{n, n}$.

**24.** If $D : \mathbb{P}_n \to \mathbb{P}_{n-1}$ denotes differentiation, then $T(x^k) = kx^{k-1} = D(x^k)$ for each $k = 0, 1, 2, \cdots n$. [If $k = 0$, this is $T(1) = 0 = D(1)$.] Hence $T = D$ by Theorem 2.

**26. b)** Suppose that $T : \mathbb{P}_n \to \mathbb{R}$ is linear and satisfies $T(x^k) = a^k$ for each $k \geq 0$. If we define $a = T(x)$, then $T(x^k) = a^k = E_a(x^k)$ for all $k \geq 1$. Moreover, $T(1) = T(x^0) = a^0 = 1 = E_a(1)$. Hence $T$ and $E_a$ agree on the basis $\{1, x, x^2, \cdots, x^n\}$, and so $T = E_a$ by Theorem 2.

**28. b)** If $T : V \to W$ is onto, then $dimV = dim(kerT) + dim(imT) = dim(kerT) + dim(W) \geq dimW$. Conversely, if $m = dimW \leq dimV = n$, let $\{\mathbf{w}_1, \cdots, \mathbf{w}_m\}$ be a basis of $W$, and let $\{\mathbf{v}_1, \cdots, \mathbf{v}_m, \mathbf{v}_{m+1}, \cdots, \mathbf{v}_n\}$ be a basis of $V$. Define $T : V \to W$ by $T(\mathbf{v}_i) = \mathbf{w}_i$ for $1 \leq i \leq m$ and $T(\mathbf{v}_j) = \mathbf{0}$ for $m < j \leq n$. Then $T$ is onto (if $\mathbf{w}$ is in $W$, then $\mathbf{w} = r_1\mathbf{w}_1 + \cdots + r_m\mathbf{w}_m = r_1T(\mathbf{v}_1) + \cdots + r_mT(\mathbf{v}_m) = T(r_1\mathbf{v}_1 + \cdots + r_m\mathbf{v}_m) \in imT$). Since $\mathbf{w}$ was arbitrary in $W$, it follows that $imT = W$.

## 5.4   ISOMORPHISMS AND MATRICES

**1. b)** True.    **d)** False.    **f)** True.    **h)** False.

**2. b)** $T^{-1}(z) = \bar{z}$      **d)** $T^{-1}\{[r \ \ s]^T\} = r + (s - r)x$
**f)** $T^{-1}(\mathbf{v}) = \frac{1}{k}\mathbf{v}$
**h)** $T^{-1}[x \ \ y]^T = [ky - x \ \ y]^T$. Here $T^2 = 1_{\mathbb{R}^2}$ so $T^{-1} = T$.

**4.** $T(x) = 2^x$.

**6. b)** $C_{\mathcal{B}}(\mathbf{v}) = [1 \ \ 0 \ \ 1 \ \ -2]^T$.

**7. b)**
$$\begin{bmatrix} 1 & 0 & 0 & 0 \\ 0 & 0 & 1 & 0 \\ 0 & 1 & 0 & 0 \\ 0 & 0 & 0 & 1 \end{bmatrix}$$

**8. b)** $T\begin{bmatrix} x \\ y \\ z \end{bmatrix} = \begin{bmatrix} x & -x+2y+z \\ y+2z & x \end{bmatrix}$

**10.** $M_{\mathcal{E}\mathcal{B}}(T) = [C_{\mathcal{E}}[T(\mathbf{b}_1)] \ \ C_{\mathcal{E}}[T(\mathbf{b}_2)] \ \ \cdots \ \ C_{\mathcal{E}}[T(\mathbf{b}_n)]]$
$\qquad = [T(\mathbf{b}_1) \ \ T(\mathbf{b}_2) \ \ \cdots \ \ T(\mathbf{b}_n)].$

**12.** $M_{\mathcal{E}\mathcal{D}}(1_{\mathbb{R}^n}) = [C_{\mathcal{E}}(\mathbf{b}_1) \ \ C_{\mathcal{E}}(\mathbf{b}_2) \ \ \cdots \ \ C_{\mathcal{E}}(\mathbf{b}_n)]$
$\qquad = [\mathbf{b}_1 \ \ \mathbf{b}_2 \ \ \cdots \ \ \mathbf{b}_n] = U.$

**14. b)** Given $\mathbf{v}$ in $V$, we have $(S \circ T)(\mathbf{v}) = (S \circ T_1)(\mathbf{v})$, that is $S[T(\mathbf{v})] = S[T_1(\mathbf{v})]$. Since $S$ is one-to-one, this gives $T(\mathbf{v}) = T_1(\mathbf{v})$, proving that $T = T_1$.

**16.** If $S \circ T = 1_V$, then $S \circ T$ is one-to-one and onto, so (Exercise 15) $T$ is one-to-one and $S$ is onto. But then $S$ and $T$ are both isomorphisms by Theorem 3. Thus $S \circ T = 1_V$ gives $S \circ T \circ T^{-1} = 1_V \circ T^{-1}$; so $S \circ 1_W = T^{-1}$, that is $S = T^{-1}$. But then $T \circ S = T \circ T^{-1} = 1_W$.

**18.** If $T : V \to W$ is one-to-one and $\mathcal{B} = \{\mathbf{b}_1, \cdots, \mathbf{b}_n\}$ is a basis of $V$, Then $\{T(\mathbf{b}_1), \cdots, T(\mathbf{b}_n)\}$ is independent (verify), so extend it to a basis $\mathcal{D} = \{T(\mathbf{b}_1), \cdots, T(\mathbf{b}_n), \mathbf{d}_1, \cdots, \mathbf{d}_m\}$ of $W$. Now define $S : W \to V$ by $S[T(\mathbf{b}_i)] = \mathbf{b}_i$ for $1 \leq i \leq n$ and $S(\mathbf{d}_j) = \mathbf{0}$ for $1 \leq j \leq m$. Show that $S \circ T$ and $1_V$ agree on the basis $\mathcal{B}$, so $S \circ T = 1_V$.

**20. b)** If $U^{-1}$ exists, then $(S_U)^{-1} = S_{U^{-1}}$.

**22.** Equate columns in $M_{DB}(T) = M_{DB}(T_1)$ to get $C_D[T(\mathbf{b}_j)] = C_D[T_1(\mathbf{b}_j)]$ for each $j$. Use the fact that $C_D$ is one-to-one.

**24.** Use the Hint.

**25. b)** If $A \to B$ by row operations, then $B = UA$ where $U = E_k \cdots E_2 E_1$. Conversely, if $B = UA$ where $U$ is invertible, then $U \to I_m$ so $U = E_k \cdots E_2 E_1$ by Theorem 1 §1.6. But then $B = E_k \cdots E_2 E_1 A$ so $B \to A$.

## 5.5 LINEAR OPERATORS AND SIMILARITY

**1. b)** True.  **d)** True.  **f)** False.  **h)** False.

**2. b)** Invertible because $M_B(T)$ is invertible (determinant 4).

**4.** $T^{-1}(2x - 3) = 5 + 4x$.

**6.** $B = \{E_{11}, E_{12}, E_{21}, E_{22}\}$.

**8. b)** No because $M_B(D)$ is not invertible. Alternatively, $D(1) = 0$ so $1 \in kerD$.

**10. b)** $T^{-1}(\mathbf{b}_2) = \mathbf{b}_1 - \mathbf{b}_3$.

**12.** Define $T$ so that $T(\mathbf{b}_j)$ has column $j$ of $A$ as its $B$-coordinate vector.

**14.** $M_B(T) = \begin{bmatrix} 1 & 1 \\ 0 & -1 \end{bmatrix}$.

**16.** $M_B(T) = \begin{bmatrix} k+1 & 0 & 0 & 0 \\ 0 & k+1 & 0 & 0 \\ 0 & 0 & k+1 & 0 \\ 0 & 0 & 0 & 1-k \end{bmatrix}$.

**18.** If $B$ is any basis of $V$, then $detT = det[M_B(T)]$ for any operator $T : V \to V$. Hence Theorem 1 gives $det(S \circ T) = det[M_B(S \circ T)] = det[M_B(S) M_B(T)] = det[M_B(S)] det[M_B(T)] = detS \, detT$.

**20. b)** Assume that $kerS = kerT$, and let $B = \{\mathbf{b}_1, \cdots, \mathbf{b}_m, \mathbf{b}_{m+1}, \cdots, \mathbf{b}_n\}$ be a basis of $V$ such that $\{\mathbf{b}_{m+1}, \cdots, \mathbf{b}_n\}$ is a basis of $kerS = kerT$. Then $\{S(\mathbf{b}_1), \cdots, S(\mathbf{b}_m)\}$ is independent (verify), so we enlarge it to a basis $D = \{S(\mathbf{b}_1), \cdots, S(\mathbf{b}_m), \mathbf{d}_{m+1}, \cdots, \mathbf{d}_n\}$ of $V$. Similarly construct a basis $F = \{T(\mathbf{b}_1), \cdots, T(\mathbf{b}_m), \mathbf{f}_{m+1}, \cdots, \mathbf{f}_n\}$ of $V$, and define $R$ on the basis $D$ by setting $R[S(\mathbf{b}_i)] = T(\mathbf{b}_i)$ for $1 \le i \le m$, and $R(\mathbf{d}_j) = \mathbf{f}_j$ for $m < j \le n$. Then $R$ is an isomorphism because it carries the basis $D$ to the basis $F$, and $R \circ S = T$ because they agree on the basis $B$ (verify).

## 5.6 INVARIANT SUBSPACES

**1. b)** True. If $\mathbf{w}$ is in $imT$, say $\mathbf{w} = T(\mathbf{v})$, then $T(\mathbf{w}) = T[T(\mathbf{v})]$ is also in $imT$.

**d)** True. If $\mathbf{u} \in U$, then $\mathbf{u} = \mathbf{u} + \mathbf{0}$ is in $U + W$, so $U \subseteq U + W$. Similarly $W \subseteq U + W$. On the other hand, if $U \subseteq Y$ and $W \subseteq Y$ where $Y$ is a subspace, then $\mathbf{u} + \mathbf{w}$ is in $Y$ for all $\mathbf{u} \in U$ and $\mathbf{w} \in W$, that is $U + W \subseteq Y$.

**f)** False. If $T : \mathbb{M}_{2,2} \to \mathbb{M}_{2,2}$ is defined by $T(A) = A^T$, then $T$ is an isomorphism (in fact $T^{-1} = T$) but $U = \{A \mid A^T = A\}$ is a $T$-invariant subspace for which $U \ne 0$ and $U \ne \mathbb{M}_{2,2}$.

**h)** True. If $\mathbb{R}\mathbf{v}$ is $T$-invariant, then $T(\mathbf{v})$ is in $\mathbb{R}\mathbf{v}$, so $T(\mathbf{v}) = \lambda\mathbf{v}$ for some $\lambda$ in $\mathbb{R}$.

**2.** Since $B$ is a basis of $V$, show that $T^2 = T$ by showing that $T^2(\mathbf{b}_i) = T(\mathbf{b}_i)$ for every $\mathbf{b}_i$ in $B$. We have $E_1(T) = span\{\mathbf{b}_1, \mathbf{b}_2 + 2\mathbf{b}_4\}$; extend $\{\mathbf{b}_1, \mathbf{b}_2 + 2\mathbf{b}_4\}$ to a basis $C$ of $V$, say $C = \{\mathbf{b}_1, \mathbf{b}_2 + 2\mathbf{b}_4, \mathbf{b}_3, \mathbf{b}_4\}$ (verify that $C$ is independent, hence a basis because $dimV = 4$). Then

$$M_C(T) = \begin{bmatrix} 1 & 0 & 1 & 0 \\ 0 & 1 & 0 & 0 \\ 0 & 0 & 0 & 0 \\ 0 & 0 & 0 & 0 \end{bmatrix}.$$

**4. b)** The restriction of $T$ to the $T$-invariant subspace $E_\lambda(T)$ is the scalar operator $\lambda$.

**6. b)** $\{1, x - x^2\}$ is a basis of $U$ and $\{1 + x, x^3\}$ is a basis of $W$, and one verifies that $\{1, x - x^2, 1 + x, x^3\}$ is a basis of $\mathbb{P}_3$ — verifying independence is easiest. So Theorem 3 applies.

**8. b)** Let $\{\mathbf{v}_1, \mathbf{v}_2, \cdots, \mathbf{v}_n\}$ be a basis of $V$ and, by hypothesis, let $T(\mathbf{v}_1) = \lambda\mathbf{v}_1$ and, given $i > 1$, let $T(\mathbf{v}_i) = \mu\mathbf{v}_i$. Then $T(\mathbf{v}_1 + \mathbf{v}_i) = \gamma(\mathbf{v}_1 + \mathbf{v}_i)$ for some $\gamma \in \mathbb{R}$, so $\lambda\mathbf{v}_1 + \mu\mathbf{v}_i = T(\mathbf{v}_1) + T(\mathbf{v}_i) = T(\mathbf{v}_1 + \mathbf{v}_i) = \gamma(\mathbf{v}_1 + \mathbf{v}_i) = \gamma\mathbf{v}_1 + \gamma\mathbf{v}_i$. The fact that $\{\mathbf{v}_1, \mathbf{v}_i\}$ is independent shows that $\lambda = \gamma$ and $\mu = \gamma$, whence $\lambda = \mu$. But then $T(\mathbf{v}_i) = \lambda\mathbf{v}_i$ holds for every $i$. Now show that $T(\mathbf{v}) = \lambda\mathbf{v}$ holds for every vector $\mathbf{v}$ in $V$.

**10. b)** We are assuming that $W_1 \subseteq W$. If $w$ is in $W$ then, since $V = U + W_1$, write $w = u + w_1$, $u$ in $U$ and $w_1$ in $W_1$. Then $w - w_1 = u$ is in $U$ (right side) and in $W$ (left side as $W_1 \subseteq W$), so $w - w_1$ is in $U \cap W = (\mathbf{0})$. Hence $w = w_1$ is in $W_1$, and it follows that $W \subseteq W_1$. Since $W_1 \subseteq W$, this shows that $W = W_1$.

**12.** Consider $U + W = U \oplus W$, a subspace of $V$. Then Theorem 4 gives $dim(U \oplus W) = dimU + dimW = dimV$ by hypothesis, so $U \oplus W = V$.

**14. b)** By **a)**, $E_\lambda(T) \cap E_\mu(T) = \{0\}$. If $dim\{E_\lambda(T)\} + dim\{E_\mu(T)\} = n$, then $V = E_\lambda(T) \oplus E_\mu(T)$ by Exercise 12. If $C = \{\mathbf{c}_1, \mathbf{c}_2, \cdots, \mathbf{c}_k\}$ and $D = \{\mathbf{d}_{k+1}, \mathbf{d}_{k+2}, \cdots, \mathbf{d}_n\}$ are bases of $E_\lambda(T)$ and $E_\mu(T)$ respectively, then $M_C(T) = \lambda I_k$ and $M_D(T) = \mu I_{n-k}$. Moreover, $B = \{\mathbf{c}_1, \mathbf{c}_2, \cdots, \mathbf{c}_k, \mathbf{d}_{k+1}, \mathbf{d}_{k+2}, \cdots, \mathbf{d}_n\}$ is a basis of $V$ and $M_B(T) = \begin{bmatrix} M_C(T) & 0 \\ 0 & M_D(T) \end{bmatrix} = \begin{bmatrix} \lambda I_k & 0 \\ 0 & \mu I_{n-k} \end{bmatrix}$ is diagonal.

**d)** Write $B = \begin{bmatrix} 1 & 1 & 0 \\ 0 & 1 & 0 \\ 0 & 0 & 1 \end{bmatrix}$, $U = \left\{ \begin{bmatrix} r \\ s \\ 0 \end{bmatrix} \mid r,\ s \text{ in } \mathbb{R} \right\}$ and

$W = \left\{ \begin{bmatrix} 0 \\ 0 \\ t \end{bmatrix} \mid t \text{ in } \mathbb{R} \right\}$. Define $T : \mathbb{R}^3 \to \mathbb{R}^3$

by $T(X) = BX$. Then $U$ and $W$ are $T$-invariant, $U \oplus W = \mathbb{R}^3$, but $T$ is not diagonalizable because its standard matrix $B$ is not diagonalizable.

**16. b)** If $\mathbf{v}$ is in $imS$, say $\mathbf{v} = S(\mathbf{w})$, then
$T(\mathbf{v}) = T[S(\mathbf{w})] = (T \circ S)(\mathbf{w}) = (S \circ T)(\mathbf{w}) = S[T(\mathbf{w})]$
is also in $imS$. Hence $imS$ is $T$-invariant. If $\mathbf{v}$ is in $kerS$, then
$S[T(\mathbf{v})] = (S \circ T)(\mathbf{v}) = (T \circ S)(\mathbf{v}) = T[S(\mathbf{v})] = T(\mathbf{0}) = \mathbf{0}$.
Hence $T(\mathbf{v})$ is in $kerS$, proving that $kerS$ is $T$-invariant.

**18. b)** True. We have $dimU = 1 = dimW$, and the matrix of $T$ restricted to either of them is diagonal (it is $1 \times 1$). So $T$ is diagonalizable by Theorem 5.

**20.** If $U = \mathbb{R}\mathbf{u}$ is $T$-invariant, $\mathbf{u} \neq \mathbf{0}$, then $T(\mathbf{u})$ is in $U$, that is $T(\mathbf{u}) = \lambda\mathbf{u}$ for some $\lambda$ in $\mathbb{R}$. Conversely, if $T(\mathbf{u}) = \lambda\mathbf{u}$, then $T(r\mathbf{u}) = rT(\mathbf{u}) = r\lambda\mathbf{u}$ is in $\mathbb{R}\mathbf{u}$; that is, $\mathbb{R}\mathbf{u} = U$ is $T$-invariant.

**22.** Let $T : V \to V$ be an operator where $dimV = 2$. If $V$ is $T$-reducible, let $V = U \oplus W$ where $U$ and $W$ are both $T$-invariant. Then $dimU = 1 = dimW$, say $U = \mathbb{R}\mathbf{u}$ and $W = \mathbb{R}\mathbf{w}$. Then $\mathcal{B} = \{\mathbf{u},\ \mathbf{w}\}$ is a basis of eigenvectors of $V$, so $T$ is diagonalizable by Theorem 4 §5.5.

Conversely, if $T$ is diagonalizable, let $\mathcal{B} = \{\mathbf{u},\ \mathbf{w}\}$ be a basis of eigenvectors. Then both $\mathbb{R}\mathbf{u}$ and $\mathbb{R}\mathbf{w}$ are $T$-invariant, and $V = \mathbb{R}\mathbf{u} \oplus \mathbb{R}\mathbf{w}$, so $V$ is $T$-reducible.

**24.** If $\mathbf{v}$ is in $U \cap W$, then $T(\mathbf{v})$ is in $U$ because $U$ is $T$-invariant and $\mathbf{v}$ is in $U$. Similarly $T(\mathbf{v})$ is in $W$, so $T(\mathbf{v})$ is in $U \cap W$. This shows that $U \cap W$ is $T$-invariant. Similarly, $U + W$ is $T$-invariant.

**26. b)** If $A$ is an $n \times n$ matrix such that $A^2 = I$, define $T : \mathbb{R}^n \to \mathbb{R}^n$ by $T(X) = AX$. Show that $T^2 = 1_{\mathbb{R}^n}$ so, by **a)**, there is a basis $\mathcal{B}$ of $\mathbb{R}^n$ such that $M_{\mathcal{B}}(T) = \begin{bmatrix} I_m & 0 \\ 0 & -I_k \end{bmatrix}$. Then $A = M_{\mathcal{E}}(T)$ is the standard matrix of $T$, and so $A \sim M_{\mathcal{B}}(T)$ by Theorem 2 §5.3.

## 5.7   GENERAL INNER PRODUCTS

**1.**

|    | b)    | d)    | f)    |
|----|-------|-------|-------|
| P1 | True  | True  | True  |
| P2 | True  | True  | True  |
| P3 | True  | True  | True  |
| P4 | False | False | False |

**2.** $\langle \mathbf{v},\ \mathbf{0} \rangle = \langle \mathbf{0},\ \mathbf{v} \rangle = \langle 0\mathbf{v},\ \mathbf{v} \rangle = 0\langle \mathbf{v},\ \mathbf{v} \rangle = 0$.

**4. b)** If $\langle \_,\ \_ \rangle$ and $\langle \_,\ \_ \rangle_1$ are two inner products, the norm functions are $\|\mathbf{v}\| = \sqrt{\langle \mathbf{v},\ \mathbf{v} \rangle}$ and $\|\mathbf{v}\|_1 = \sqrt{\langle \mathbf{v},\ \mathbf{v} \rangle_1}$. Thus, $\|\mathbf{v}\| = \|\mathbf{v}\|_1$ for every vector $\mathbf{v}$ in $V$ if and only if $\langle \mathbf{v},\ \mathbf{v} \rangle = \langle \mathbf{v},\ \mathbf{v} \rangle_1$ for all $\mathbf{v}$.

**6. b)** $\langle \delta_k,\ \delta_m \rangle = \sum_{i=1}^{n} \delta_k(i)\delta_m(i)$. If $k \neq m$, then $\delta_k(i)\delta_m(i) = 0$ for every $i$ because one of the factors is zero. If $k = m$, then $\langle \delta_k,\ \delta_k \rangle = \sum_{i=1}^{n} \delta_k(i)^2 = 1$ because exactly one term equals 1, and the rest equal 0.

**8.** Define $\langle \mathbf{v},\ \mathbf{w} \rangle = C_{\mathcal{B}}(\mathbf{v}) \cdot C_{\mathcal{B}}(\mathbf{w})$ for all $\mathbf{v}$ and $\mathbf{w}$ in $V$. Then $\langle \mathbf{v},\ \mathbf{w} \rangle = \langle \mathbf{w},\ \mathbf{v} \rangle$ because $X \cdot Y = Y \cdot X$ for all $X$ and $Y$ in $\mathbb{R}^n$. Since $C$ is linear, we obtain
$\langle r\mathbf{v},\ \mathbf{w} \rangle = C_{\mathcal{B}}(r\mathbf{v}) \cdot C_{\mathcal{B}}(\mathbf{w}) = r[C_{\mathcal{B}}(\mathbf{v}) \cdot C_{\mathcal{B}}(\mathbf{w})] = r\langle \mathbf{v},\ \mathbf{w} \rangle$,
and $\langle \mathbf{u},\ \mathbf{v} + \mathbf{w} \rangle = C_{\mathcal{B}}(\mathbf{u}) \cdot C_{\mathcal{B}}(\mathbf{v} + \mathbf{w})$
$$= C_{\mathcal{B}}(\mathbf{u}) \cdot C_{\mathcal{B}}(\mathbf{v}) + C_{\mathcal{B}}(\mathbf{u}) \cdot C_{\mathcal{B}}(\mathbf{w})$$
$$= \langle \mathbf{u},\ \mathbf{v} \rangle + \langle \mathbf{u},\ \mathbf{w} \rangle.$$
Finally, $\langle \mathbf{v},\ \mathbf{v} \rangle = C_{\mathcal{B}}(\mathbf{v}) \cdot C_{\mathcal{B}}(\mathbf{v}) = \|C_{\mathcal{B}}(\mathbf{v})\|^2 > 0$ if $\mathbf{v} \neq \mathbf{0}$ because $C_{\mathcal{B}}(\mathbf{v}) \neq \mathbf{0}$ ($C_{\mathcal{B}}$ is one-to-one).

**10. b)** Write $b_0 = b_0(x) = 1$, $b_1 = b_1(x) = x$, and $b_2 = b_2(x) = x^2$, and construct the orthogonal basis $e_0 = e_0(x)$, $e_1 = e_1(x)$, and $e_2 = e_2(x)$ using the Gram-Schmidt algorithm. First $e_0 = b_0 = 1$; then
$e_1 = b_1 - \frac{\langle b_1,\ e_0 \rangle}{\|e_0\|^2} e_0 = x - \frac{1/2}{1} 1 = x - \frac{1}{2}$;
$e_2 = b_2 - \frac{\langle b_2,\ e_0 \rangle}{\|e_0\|^2} e_0 - \frac{\langle b_2,\ e_1 \rangle}{\|e_1\|^2} e_1$
$\quad = x^2 - \frac{1/3}{1} 1 - \frac{1/12}{1/12}\left( x - \frac{1}{2} \right)$
$\quad = x^2 - x + \frac{1}{6}$.

**12.** Write $\mathbf{v} = \sum_{i=1}^{n} r_i\mathbf{b}_i$ and $\mathbf{w} = \sum_{j=1}^{n} s_j\mathbf{b}_j$, so that $C_{\mathcal{B}}(\mathbf{v}) = [r_1 \ r_2 \ \cdots \ r_n]^T$ and $C_{\mathcal{B}}(\mathbf{w}) = [s_1 \ s_2 \ \cdots \ s_n]^T$. Hence $\langle \mathbf{v},\ \mathbf{w} \rangle = \left\langle \sum_{i=1}^{n} r_i\mathbf{b}_i,\ \sum_{j=1}^{n} s_j\mathbf{b}_j \right\rangle = \sum_{i,j} r_i s_j \langle \mathbf{b}_i,\ \mathbf{b}_j \rangle$. But $\langle \mathbf{b}_i,\ \mathbf{b}_j \rangle = 0$ or 1 according as $i \neq j$ or $i = j$, so $\langle \mathbf{v},\ \mathbf{w} \rangle = \sum_{i=1}^{n} r_i s_i = C_{\mathcal{B}}(\mathbf{v}) \cdot C_{\mathcal{B}}(\mathbf{w})$.

**14.** Using the notation in the exercise,
$\langle \mathbf{v},\ \mathbf{v} \rangle = \left\langle \sum_{i=1}^{n} r_i\mathbf{b}_i, \sum_{j=1}^{n} r_j\mathbf{b}_j \right\rangle = \sum_{i,j} r_i r_j \langle \mathbf{b}_i,\ \mathbf{b}_j \rangle = \sum_{i=1}^{n} r_i^2$
because $\langle \mathbf{b}_i,\ \mathbf{b}_j \rangle = 0$ or 1 according as $i \neq j$ or $i = j$.

**16.** If $\mathbf{v}$ is in $(U + W)^{\perp}$, then $\langle \mathbf{v},\ \mathbf{z} \rangle = 0$ for all $\mathbf{z}$ in $U + W$. Since $U \subseteq U + W$, this shows that $\mathbf{v}$ is in $U^{\perp}$; similarly $\mathbf{v}$ is in $W^{\perp}$, so $\mathbf{v}$ is in $U^{\perp} \cap W^{\perp}$. Thus $(U + W)^{\perp} \subseteq U^{\perp} \cap W^{\perp}$. On the other hand, if $\mathbf{v}$ is in $U^{\perp} \cap W^{\perp}$, then $\mathbf{v}$ is in $U^{\perp}$ so $\langle \mathbf{v},\ \mathbf{u} \rangle = 0$ for all $\mathbf{u}$ in $U$. Similarly $\langle \mathbf{v},\ \mathbf{w} \rangle = 0$ for all $\mathbf{w}$ in $W$, so $\langle \mathbf{v},\ \mathbf{u} + \mathbf{w} \rangle = \langle \mathbf{v},\ \mathbf{u} \rangle + \langle \mathbf{v},\ \mathbf{w} \rangle = 0 + 0 = 0$ for all $\mathbf{u} + \mathbf{w}$ in $U + W$. Thus $\mathbf{v}$ is in $(U + W)^{\perp}$, which proves that $U^{\perp} \cap W^{\perp} \subseteq (U + W)^{\perp}$. This completes the proof.

**18.** By Theorem 10, $\|\mathbf{v} - proj_{U_2}(\mathbf{v})\| \leq \|\mathbf{v} - \mathbf{u}\|$ for all $\mathbf{u}$ in $U_2$. If we take $\mathbf{u} = proj_{U_1}(\mathbf{v})$, then $\mathbf{u}$ is in $U_1$ and $U_1 \subseteq U_2$, so $\mathbf{u}$ is in $U_2$. The result follows.

**19.** We must show that $\{\mathbf{b}_1, \cdots, \mathbf{b}_n\}$ is independent for each $n \geq 1$, and we do it by induction on $n$. If $n = 1$, it is because $\mathbf{b}_1 \neq \mathbf{0}$. If $\{\mathbf{b}_1, \cdots, \mathbf{b}_n\}$ is independent, observe that $\{\mathbf{b}_1, \cdots, \mathbf{b}_n\} \subseteq W_n$ because $\mathbf{b}_i \in W_i \subseteq W_n$ for each $i \leq n$. Since $\mathbf{b}_{n+1} \notin W_n$, it follows that $\{\mathbf{b}_1, \cdots, \mathbf{b}_n, \mathbf{b}_{n+1}\}$ is independent by the Independent lemma (Lemma 1 §5.2). This completes the induction.

**20.** $\langle fh, g \rangle = \int_a^b (fh)(x)\, g(x)\, dx$

$= \int_a^b [f(x)h(x)]g(x)\, dx$

$= \int_a^b f(x)[h(x)g(x)]\, dx$

$= \int_a^b f(x)[(hg)(x)]\, dx$

$= \langle f, hg \rangle.$

**22. a)** $\|\mathbf{v}\| = \sqrt{\langle \mathbf{v}, \mathbf{v} \rangle} \geq 0$ for all $\mathbf{v}$ because $\langle \mathbf{v}, \mathbf{v} \rangle \geq 0$ by P4 and Theorem 1, and $\sqrt{\phantom{x}}$ denotes the positive square root. This is (1). Next (2) is (3) of Theorem 1, and (4) is the triangle inequality (Theorem 3). Finally, $\|r\mathbf{v}\| = \sqrt{\langle r\mathbf{v}, r\mathbf{v} \rangle} = \sqrt{r^2 \langle \mathbf{v}, \mathbf{v} \rangle} = \sqrt{r^2}\sqrt{\langle \mathbf{v}, \mathbf{v} \rangle} = |r|\, \|\mathbf{v}\|.$ This is (3).

**b)** Here, (1) is clear because $|x| \geq 0$ for all $x$; (2) is because $|x| = 0$ if and only if $x = 0$; (3) is because $|rx| = |r|\, |x|$ for all $r$ and $x$; and (4) is because $|x + y| \leq |x| + |y|$ for all real numbers $x$ and $y$. Hence this is a norm on $\mathbb{R}^2$.

Suppose that $\|\_\|$ arises from an inner product $\langle \_, \_ \rangle$ on $\mathbb{R}^2$. Then Theorem 2 shows that there is a positive definite matrix $P = \begin{bmatrix} a & b \\ b & c \end{bmatrix}$ such that

$\langle X, Y \rangle = X^T P Y$ for all $X, Y$ in $\mathbb{R}^2$. Writing $X = [x \ \ y]^T$, this means that

$(|x| + |y|)^2 = [x \ \ y]\, P \begin{bmatrix} x \\ y \end{bmatrix} = ax^2 + 2bxy + cy^2$ for all $x$ and $y$ in $\mathbb{R}$.

Taking $[x \ \ y]^T = [1 \ \ 0]^T$, $[0 \ \ 1]^T$, and $[1 \ \ 1]^T$ gives respectively, $a = 1$, $c = 1$, and $b = 1$. But then $(|x| + |y|)^2 = x^2 + 2xy + y^2 = (x + y)^2$ for all $x$ and $y$, a contradiction (take $x = 1$ and $y = -1$).

**24.** Taking $m = 5$ in Theorem 12 gives

$p_6 = \frac{11}{6}p_5 - \frac{5}{6}p_4$

$= \frac{11}{6}x \cdot \frac{1}{8}(63x^5 - 70x^3 + 15x) - \frac{5}{6} \cdot \frac{1}{8}(35x^4 - 30x^2 + 3)$

$= \frac{1}{48}(693x^6 - 945x^4 + 315x^2 - 15)$

$= \frac{1}{16}(231x^6 - 315x^4 + 105x^2 - 5).$

Then taking $m = 6$ gives

$p_7 = \frac{1}{16}(429x^7 - 693x^5 + 315x^3 - 35x).$

**26.** We prove that $p_m(-1) = (-1)^m$ by induction on $m \geq 0$. It holds for $m = 0$ because $p_0 = 1 = (-1)^0$ and $p_1(-1) = -1 = (-1)^1$. If it holds for $m - 1$ and $m$, Theorem 12 (with $x = -1$) gives

$p_{m+1}(-1) = \frac{1}{m+1}\{(2m + 1)(-1)p_m(-1) - mp_{m-1}(-1)\}$

$= \frac{1}{m+1}\{(2m + 1)(-1)^{m+1} - m(-1)^{m-1}\}$

$= \frac{(-1)^{m+1}}{m+1}\{(2m + 1) - m\} = (-1)^{m+1}$

where we used the fact that $(-1)^{m-1} = (-1)^{m+1}$ for every integer $m$.

# Index